Geology in Coal Resource Utilization

Sponsored by:
Energy Minerals Division
American Association of Petroleum Geologists

Edited by
Douglas C. Peters

TechBooks

ISBN 1-878907-22-0

TechBooks
4012 Williamsburg Court
Fairfax, VA 22032
(703) 352-0001

Preface

The 37 papers in this book, both invited ones and those initiated on the basis of calls for papers in late 1988 and early 1989, were compiled with an overriding theme in mind: to provide the coal industry with a comprehensive source of information on how geology and geologic concepts can be applied to the many facets of coal resource location, extraction, and utilization. This book, therefore, is designed to provide industry managers with an idea of the applications, and limitations, of coal geology and related fields through general review-type papers as well as more specific case studies which may provide solutions to their specific problems and concerns. The book also is designed to address the needs of the coal geologist, especially the mine geologist and geological technician, by providing case histories from which specific techniques can be derived and a broad coverage of all subfields of applied coal geology so that more specialized individuals can maintain contact and knowledge of other aspects of the industry.

The original impetus for the compilation of this collection was the strong backing of coal geology as a primary focus of the Energy Minerals Division (EMD) of the American Association of Petroleum Geologists by the membership of EMD, as noted by Platt in the Introduction for this book. However, more than that, I have perceived from many discussions with members of the coal geoscience community that there is an underappreciation within the coal producing and utilizing industry for how coal geology can be applied to coal industry problems and concerns. Moreover, there appears to be a sense in the industry that coal geoscience research is predominantly theoretical and not of practical use to the industry.

Both of these apparent industry viewpoints are in error! As I hope that this collection points out, there are myriad practical ways that coal geoscience applies to the daily, mine-site, plant-site, and field work concerns and problems of the coal industry. In addition, the theory of today will be the practical knowledge of tomorrow.

The chapters have been arranged to address the major coal geology subfields of Exploration and Reserve Definition, Reserve Estimation, Coalbed Methane, Underground Coal Gasification, Mining, Coal Quality Concerns, and Environmental Impacts, with papers distributed on the basis of their primary emphasis. There most certainly are holes in terms of subjects not covered (e.g., a paper entirely on sulfur content/variations in coal) or covered incompletely. However, even if the collection contained 50 papers, there probably still would be incomplete coverage of all possible worthwhile subjects. We (I and the authors) have attempted to be as complete and explicit as possible in covering those subjects that are included.

To help guide you through the collection, I have included prefaces at the beginning of each chapter. They are intended as a brief lead-in to the subject of the chapter and an acknowledgement of the papers' connections to the subject and contributions to the chapter. In addition, a brief "cross-reference" section has been included in each preface to help you find papers of interest in other chapters. The subfields of coal geology are intimately intertwined, and investigations in one area may impact problems in another area. Some subfields tend to blur at their edges, such as with reserve definition and reserve estimation.

So, without further palavering, I commend you to the reading of this collection, the compilation of which I have found rewarding both as an editor and as an interested geologist/engineer. Before closing, I would like to thank the U.S. Bureau of Mines for permission to act as the editor of this collection and for their support in terms of allowing me the time and computer use to apply to the effort. I, the officials of the EMD, and especially the authors hope you will find the articles both intellectually stimulating and ultimately of practical use!

Douglas C. Peters
Denver, Colorado
May 30, 1991

Table Of Contents

Preface v

Units Used In This Book x

List of Reviewers xi

List of Contributors xii

Introduction: Fitting Coal Geoscience Into The Big Picture xiii
by Jeremy Platt

CHAPTER 1: EXPLORATION AND RESERVE DEFINITION

Preface 1

Overview Paper: Commentary On The Evolution Of The Exploration And Evaluation Process 3
by Stephen V. Tivy and John M. Mercier

Surface Geophysics For Coal Exploration In The Alberta Plains: Recent Research 13
by James D. Henderson, Anthony N. Sartorelli, Mark M. Fenton, John G. Pawlowicz and Georgia L. Hoffman

Formation Of Clinker And Its Effects On Locating And Limiting Coal Resources 29
by Joseph C. Sarnecki

Glacially Deformed Bedrock: Implications For Coal Exploration, Mining, And Highwall Stability 37
by Mark M. Fenton and John G. Pawlowicz

CHAPTER 2: RESERVE ESTIMATION

Preface 53

Overview Paper: Coal Resources and Reserve Estimates in the United States in 1991 — A Status Report 55
by James E. Palmer

Techniques For Estimating The Coal Reserve Base Of A Region Using A Geographic Information System: A Case Study From Illinois *59*
by Colin G. Treworgy

Use Of Geostatistics In Planning An Optimum Drilling Program *69*
by S.N. Ghose

Restrictions To Mining: Their Effect On Available Coal Resources *81*
by Roy S. Sites, Elizabeth V.M. Campbell, Karen K. Hostettler, Nancy K. Gardner and Susan J. Tewalt

CHAPTER 3: COALBED METHANE

Preface *95*

Overview Paper: Coalbed Methane: Natural Gas Resources From Coal Seams *97*
by D. Keith Murray

Structural Properties Of Coal That Control Coalbed Methane Production *105*
by T. Ertekin, W. Sung and H.I. Bilgesu

The Effects Of Overburden Stress On Coalbed Methane Production *125*
by Daniel J. Soeder

Coalbed Methane Production Potential In Complex Geologic Settings *137*
by Terry E. Mitchell and Steven P. Pappajohn

CHAPTER 4: UNDERGROUND COAL GASIFICATION

Preface *153*

Overview Paper: Underground Coal Gasification *155*
by Robert L. Oliver and G.F. "Pete" Dana

The Role Of Hydrogeology In Underground Coal Gasification With An Example From The Rocky Mountain 1 (RM1) Test, Carbon County, Wyoming *169*
by F.W. Beaver, G.H. Groenewold, C.R. Schmit, D.J. Daly and R.L. Oliver

CHAPTER 5: MINING

Preface *187*

Overview Paper: Application Of Depositional Models To Mining Problems *189*
by Gerald A. Weisenfluh and John C. Ferm

Geologic Diagnosis For Reducing Coal Mine Roof Failure *203*
by Gary P. Sames and Noel N. Moebs

Cleat And Joint System Evaluation In Coal Seams And Zones Of The Blackhawk Formation And Ferron Sandstone Member Of The Mancos Shale Formation For Mine Planning And Site Characterization In The Book Cliffs, Wasatch Plateau, And Emery Coal Fields, Utah 225
by Brigitte P. Hucka

Roof Falls And Hazard Prediction In Eastern Kentucky Coal Mines 245
by Stephen F. Greb

Geological Modeling Techniques For Evaluation Of Productivity-related Longwall Mining Roof Conditions: A Case Study 263
by J. Scott Nelson, Ronald H. Mullennex, and Marshall S. Miller

Effects Of Structural Faults On Ground Control In Selected Coal Mines In Southwest Virginia 287
by Gregory M. Molinda and David K. Ingram

Stress In Underground Coal Mines: A Practical Review Of Origin, Recognition, And Control 313
by Bruce A. Collins

Coping With Tectonic Stress In The Illinois Basin Coal Field 321
by W. John Nelson and Robert A. Bauer

The Effects Of Geology On Coal Mine-induced Subsidence In The United States 335
by William F. Kane and Larry R. Powell

CHAPTER 6: COAL QUALITY CONCERNS

Preface 357

Overview Paper: Coal Quality And Geology 359
by Maurice Deul

Coal Quality Of The Mary Lee Coal Group And Its Relationship To Coal Resource Development, Coalbed Methane, And Geologic History In The Warrior Coal Field, Alabama 369
by Richard B. Winston

Geological Controls And Constraints On The Concentration Of Elements In Western Canadian Coals 389
by F. Goodarzi and E. van der Flier-Keller

Structural Controls On The Petrology And Geochemistry Of The Pond Creek Coal Seam, Pike And Martin Counties, Eastern Kentucky 413
by James C. Hower, James D. Pollock and Thomas B. Griswold

Lithotype Control Of Trace Element Concentrations In Some Southeastern West Virginia Coals 427
by Jeffrey W. Martin and Victor V. Cavaroc

CHAPTER 7: ENVIRONMENTAL IMPACTS

Preface 451

Overview Paper: Coal Utilization, The Environment, And The Role Of Geologists 453
by L.A.F. Dutcher and R.R. Dutcher

Environmental Impact Of Surface Coal Mining And Reclamation In The Plains Region Of Alberta, Canada 457
by S.R. Moran, M.R. Trudell, T.M. Macyk, D.B. Cheel and A.E. Howard

Coal As A Host And As An Indicator Of Mineral Resources 471
Robert B. Finkelman and Robert D. Brown, Jr.

Analysis Of Aromatic Hydrocarbons In Overburden From Coal Mines: Assessment Of The Environmental Impact 483
by Wilhelm Püttmann, Kai Steffens, and Wolfgang Goßel

Drainage From Coal Mines: Chemistry And Environmental Problems 499
by Thomas Wildeman

Review Of Processes Operating In Constructed Wetlands To Attenuate Acid Mine Drainage 513
by Ronald W. Klusman and Steven D. Machemer

The Role Of Geology In Managing Coal Conversion Byproducts In The Northern Great Plains 541
by Frank W. Beaver, Gerald H. Groenewold and Robert D. Koob

Geological Controls On Coal Seams Burning "In Place" 559
by Fariborz Goodarzi and Thomas Gentzis

Author Index

Units Used In This Book

The below units and abbreviations are those used in this book. Most units are given in the text as both metric and English equivalents, except for where the generally accepted form of the unit is either normally just metric or just English (e.g., "bbl" for barrels of oil). Units for which only magnitudes are given are not converted from the authors' original units. Units in Tables and Figures also are not converted from the authors' original units.

acre = acre
angstrom = Å
atmosphere = atm
barrel = bbl
billion cubic feet = bcf
British thermal unit = Btu
centimeter = cm
degrees Centigrade = °C
electron volts = eV
foot = ft
U.S. gallon = gal.
gram = g
hectare = ha
inch = in.
kilogram = kg
kilopascals = kPa
liter = L
megapascals = MPa
meter = m
metric ton = tonne
micrometer = μm
micromho = μmho
microsecond = μsec
mile = mi.
milliamps = ma
millidarcy = md
milligram = mg
milliliter = ml
millimeter = mm
millisecond = msec
minute = min.
molar = M
ounce = oz
parts per billion = ppb
parts per million = ppm
pound = lb
pounds per square inch = psi
second = sec
short ton = ton
standard cubic foot = scf
standard unit (acidity) = s.u.
thousand cubic feet per day = mcfd
trillion cubic feet = tcf
year = yr

List Of Reviewers

The colleagues listed below were instrumental in the completion of this collection of papers by acting as technical reviewers for the papers. Their help and investment of their time are appreciated very much by the editor and by the entire Energy Minerals Division-AAPG Executive Council and membership.

Ronald H. Affolter
U.S. Geological Survey

Walter B. Ayers Jr.
Bureau of Economic Geology

Donna L. Boreck
U.S. Bureau of Mines

Charles Byrer
U.S. Department of Energy

Robert F. Chaiken
U.S. Bureau of Mines

Donald A. Coates
U.S. Geological Survey

John Crelling
Southern Illinois University

Patricia Erickson
U.S. Bureau of Mines

Rodger C. Fry
Utah Power & Light Co.

Harold Ganow
U.S. Geological Survey

Joseph R. Hatch
U.S. Geological Survey

Ricky T. Hildebrand
U.S. Geological Survey

David M. Hyman
U.S. Bureau of Mines

Ann G. Kim
U.S. Bureau of Mines

Edwin R. Landis
U.S. Geological Survey

Joseph W. Martin
U.S. Department of Energy

Eric Matheson
U.S. Bureau of Mines

Frederick D. McCosh
F.D. McCosh Resource Consultants

Timothy Rohrbacher
U.S. Bureau of Mines

Dale D. Teeters
U.S. Bureau of Mines

Carol M. Tremain
Colorado Geological Survey

Charles E. Wier
Charles E. Wier and Associates

List Of Contributors

The below companies and organizations have generously donated funds toward the publication of this book. The editor of this book and the officers and members of the Energy Minerals Division and the American Association of Petroleum Geologists gratefully acknowledge their support for this endeavor.

AMAX Coal Industries

ARCO Coal Company

Cyprus Coal Company

Eastern Section—AAPG

Grand Junction Geological Society

Raven Ridge Resources

Rocky Mountain Section—AAPG Foundation

INTRODUCTION: FITTING COAL GEOSCIENCE INTO THE BIG PICTURE

Jeremy Platt
Integrated Energy Systems Division
Electric Power Research Institute
P.O. Box 10412
Palo Alto, CA 94303

The Energy Minerals Division (EMD) of the American Association of Petroleum Geologists surveyed its members in 1988 to identify their priorities among many, diverse energy topics. Over 400 individuals responded. Many felt very strongly about their own areas of interest (spanning remote sensing, nuclear minerals, geothermal energy, oil shale, oil sands, and coal). Yet they reached a surprising level of unanimity that coal should be EMD's top priority — 65% ranking coal at or near the top of their list. The first preparations for this volume were begun later that year at the instigation of Doug Peters, who has shown vision and fortitude to carry this project to completion. To his credit, he has created a volume of unusual scope and timeliness which should further the skills and appreciation of all interested readers and professionals.

My purpose in these introductory comments is to outline a few strands of the changing economic and environmental context of the United States coal industry. These forces, while not the specific focus of the volume, nevertheless drive much of what takes place in the coal industry.

ENERGY FLOWS

The role of coal in the U.S. economy and energy infrastructure is surprising to many who learn about it for the first time. The value to consumers of the coal produced in the U.S. is about $30,000,000,000 per year. Including mining and transportation, coal is big business and a pivotal part of the U.S. energy infrastructure. Coal production has grown progressively (although not steadily) since 1975, reaching 730,000,000 tonnes (800,000,000 tons) per year for the first time in 1980 and 910,000,000 tonnes (1,000,000,000 tons) per year in 1990. Its growth over this period has been fueled primarily by growing demand for electricity.

Electric utility markets are the mainstay of the coal industry. About 85% of the coal consumed in the U.S., or approximately 705,000,000 tonnes (775,000,000 tons), is used for making electricity. The importance of coal within the electric utility industry is somewhat smaller. About 55% of electricity generation is derived from coal. Yet this amount is far and away greater than the next largest sources: nuclear = 20%; hydro = 10%; and other fossil fuels = 16% (gas = 10%; oil = 6%).

Perhaps even more surprising is coal's importance in the nation's energy budget. This is expressed best in quadrillion Btus (10^{15} Btu or "quads"). Drawing on 1989 statistics, the U.S. currently consumes over 80 quads of energy in all forms for all purposes per year. Coal provides 23% (19 quads), natural gas provides 24% (19.5 quads), and oil provides 42% (34 quads).

Coal consumption is projected to continue to grow, with electricity markets continuing to act as the driving force. In the Department of Energy's (DOE) National Energy Strategy (NES) proposed in February, 1991, coal use was projected to grow modestly over the next 10 yrs. Coal used in electricity would grow from 16 quads in 1990 to about 19 quads in 2000. Then, depending on the success of various energy conservation measures and other programs, its use could grow by another 5-10 quads over the following 10 years. The higher number would be twice the rate of growth experienced during the 1980s. In either case, it appears that total coal production (allowing for steelmaking, exports, and other uses) would reach about 1,270,000,000-1,550,000,000 tonnes (1,400,000,000-1,700,000,000 tons) by 2010.

The NES ventures beyond 2010 where the pace of further growth in coal use can be seen to depend critically on revitalization of the nuclear industry and, to a lesser extent, other programs such as conservation and renewable sources. Again, depending on the success of these efforts, coal use for electricity alone over the 2010-2030 period is projected to grow by another 227,000,000-910,000,000 tonnes (250,000,000-1,000,000,000 tons).

These projections make the simple point that, not only is coal big business, it is getting bigger. Yet this simple fact is not a bellwether for geologists. Coal is also a highly competitive business. Prices have declined in real terms for about 12 yrs. The industry has been forced to restructure operations, adopt new technologies, improve procedures and heighten labor productivity. Under pressure of persistent excess capacity — which has been exacerbated by steps to enhance competitiveness — the challenge has been to control costs at existing properties or, through acquisition or development, to open new mines which can be operated at a profit. This has been an exacting environment for geologists, who often have had to move closer to mine operations and support. Exploration activities have declined.

1990 CLEAN AIR ACT AMENDMENTS

Two events in 1990 mark the tension in the nation's love-hate relationship with coal. As mentioned, the 910,000,000-tonne (1,000,000,000-ton) mark was reached, a number that underscores the degree to which the nation has come to rely on coal. The second event was enactment of the 1990 Clean Air Act Amendments (CAAA), which will have far reaching effects on coal production and utilization.

Faced with stringent requirements to reduce SO_2 emissions at existing boilers, the nation's utilities must take effective and often costly steps to retrofit technology controls, such as flue-gas desulfurization (called "scrubbers"), and to switch to cleaner, lower-sulfur coals. Reductions are imposed in two steps: one by 1995 and the other by 2000. When all is said and done, SO_2 emissions will be reduced to about one-half of 1980 levels. No further growth in emissions will be allowed thereafter — the nation will be operating under a "cap", requiring any growth or new sources of emissions to be offset by reductions elsewhere. Companies have been given freedom to choose how they can best achieve the required reductions, yet most would agree the job ahead is enormous, perhaps costing in the neighborhood of $5,000,000,000/yr.

About one-third of the SO_2 reductions to be achieved likely will come from switching to lower-sulfur coals. This option also includes blending and cleaning coals. Primary beneficiaries, it appears, include the Powder River Basin, where estimates of increased demand on top of normal business trends are as high as 55,000,000 tonnes/yr (60,000,000 tons/yr), and Central Appalachia, where a small group of counties may receive about 27,000,000 tonnes (30,000,000 tons) or so of increased demand. Different estimators, of course, differ in their views of how the CAAA will affect coal markets and regional production. However, the basic message is that turbulent times and uncertain price movements lie ahead. Wildcards include the direction of changes in mining productivity as stronger demands materialize for lower-sulfur coals, success in burning subbituminous coal in boilers designed for bituminous coal, and questions of how railroad companies will price their services because more coal will be traveling greater distances.

About two-thirds of the SO_2 reductions to be achieved likely will come from technology controls. If so, these scrubbers, plus a selection of novel approaches that still are under development, will help to preserve the markets for high-sulfur coal. Some analyses suggest that production from Midwest high-sulfur coal-producing regions actually may hold quite close to current levels, forgoing growth but, perhaps, not experiencing dramatic decline either, whereas other high-sulfur regions may face real, but limited, cutbacks. A wildcard is predicting how individual utilities will act — regardless of the evidence and economic analyses that have gone before.

Other changes to be wrought by the CAAA are even more uncertain. Studies are underway to determine whether controls should be applied on electric utility emissions of air toxics, such as mercury, lead, chlorine, and other elements and compounds. If so, control costs could escalate in multiples over the costs of reducing SO_2 emissions. From a geoscience perspective, such measures have obvious implications for data on trace elements in coal and other geochemical studies. Reductions in nitrogen oxides also have been imposed. Much of the nitrogen supplied during combustion comes from the atmosphere; however, nitrogen also comes from the coal itself. This too could spawn a new series of coal quality studies.

An important new economic variable also has been introduced into coal companies' and electric utilities' planning by the CAAA. This is the emerging market in emissions allowances. The CAAA will be implemented through distributing a yearly budget of "emissions allowances" to the affected utilities. Altogether, the allowances issued will not exceed the national emissions cap (8,140,000 tonnes (8,950,000 tons) in 2000). Each allowance represents 0.9 tonne (1 ton) of SO_2 emissions. If companies find it possible to reduce emissions by more than the amounts required, they generate credits which they can bank for their own future use or sell to others who find purchasing allowances to be cheaper than achieving 100% of their requirements at their own powerplants.

The value of emissions allowances is hard to predict. The higher the premium for low-sulfur coal, the higher the allowance price and vice versa. Many other things also will come into play, including trends in technology costs for reducing emissions, how rapidly needs may grow to obtain offsets, and short-term factors, such as extreme weather, which could place unanticipated burdens on generation using fossil fuels. One thing is certain, though, and that is that emissions allowances will become an added — perhaps routine — consideration in all sorts of business decisions and transactions. The coal industry will take an active interest in the value of emissions allowances because allowances will, in a sense, "compete" with low-sulfur coal, provide a measure of the new economic costs of burning higher-sulfur coals, and perhaps open new market opportunities for those who can acquire them at advantageous costs.

PREPARING FOR CHANGE

With so much recent attention given to acid rain legislation, it is prudent to remind ourselves of the different energy/economic concerns which have affected coal in the recent past. Acid rain is one in a succession of different issues that have affected the coal industry and, likewise, the activities of coal geoscientists. A rough chronology follows.

The CAAA of 1970 marked the awakening of coal quality concerns. For the first time, data on sulfur content became a national priority, predating concerns over acidic deposition by almost 10 yrs. The motivation was to improve ambient air quality via standards imposed on the States. Rules limited emissions from new powerplants, first (in 1971) requiring that emissions not exceed 0.54 kg (1.2 lb) per million Btu, a level that permitted very low-sulfur coal to be used without further controls, and later (in 1978) requiring that the "best available" technology controls be installed, regardless of sulfur content of the coal.

From 1975 to 1980, national concerns were dominated by energy security and supply issues. Various models and databases were developed, oil prices skyrocketed, and coal prices did too, at least at the beginning of this period. The multinational energy companies got into coal. The Powder River Basin was opened up. Coal mining productivity plummeted. The ability of the coal industry to gear up to meet ever-rising demand expectations became a question. Applying the power of hindsight, we see now that coal price projections made at the time were more than twice as high as actually occurred in 1990. Likewise, coal production was estimated to reach 1,360,000,000 tonnes (1,500,000,000 tons) by 1990, off by 454,000,000 tonnes (500,000,000 tons).

1980 to 1985 was a period of adjustment to the prior high expectations. High coal demands did not materialize and coal prices began to fall, with spot prices increasingly diverging below contract prices. High fuel costs and escalation provisions in coal contracts became a top-level management concern throughout the utility industry. Previous decisions were scrutinized closely and methods developed to better understand and manage risks, all the way to optimizing coal selection for powerplant performance. As might be expected, this adjustment phase included a great deal of litigation. Mining productivity rose steadily. An important event was deregulation of the railroads pursuant to the Staggers Act. Great strides also were made in coal geoscience, particularly through the development and application of depositional environment concepts to understanding the distribution and quality of coal. A growing body of work came to light on how coal geoscience could provide valuable insurance in major coal development and acquisition decisions and, as part of an integrated approach to mining, support more efficient operations.

The 1985-1990 period has been dominated by coal quality concerns. The acid rain debate finally came to conclusion with passage of the 1990 CAAA. The period was characterized by a progressive ratcheting of legislative proposals to lower and lower sulfur emissions standards. These proposals, all differing in detail, first called for SO_2 reductions in the neighborhood of 4,500,000-7,300,000 tonnes (5,000,000-8,000,000 tons). In later years, they called for reductions of 9,100,000, 11,000,000, and even 13,000,000 tonnes (10,000,000, 12,000,000, and even 14,000,000 tons). The CAAA, because of its special capping provisions, is one of the more stringent measures to have been proposed. Coal prices continued their decline, and as noted earlier, coal mining productivity continued to improve, almost to an astonishing extent. Company survival and jobs became the persistent concerns of management and labor. Finally, last — but by no means least — environmental attention has shifted to the problematic, multinational issues of global warming and control of greenhouse gases.

Actions on carbon dioxide, if enacted, will certainly affect coal, other fossil fuels, the entire energy system, and the economy. Like a snake swallowing an elephant, it is not at all clear what the outcome may be. What other issues or concerns does the future hold in store for coal? Already in the pipeline (or digestive tract), as noted, are air toxics and nitrogen oxide emissions. Yet another important new area of concern, still understood only dimly, is coal supply.

A growing number of coal professionals and energy analysts have noted that several of the top-quality coals in the East will become virtually exhausted in 25-40 yrs. Recent evaluations illustrate how various technical, economic, geologic, and institutional considerations whittle away at estimates of coal resources in-the-ground, yielding a much more sober picture of minable and marketable coal supplies. In this case, it appears history may repeat itself, with a rekindling of concern over energy security and supply issues. What remains unclear is whether the findings that have been obtained in selected areas or for the lowest-sulfur coals extrapolate to other districts and coal qualities. It is premature to push the panic button, but it now is becoming clear that multidisciplinary reassessments of coal supplies may be warranted. Such studies will go beyond the traditional domain of geoscientists into engineering, economics, and other facets of the problem. Nonetheless, many geoscientists may be well qualified to take on the additional challenges.

Chapter 1

Exploration And Reserve Definition

PREFACE

Before we can make use of coal resources, we have to know where they are, their extent, and the basic conditions of the deposit (e.g., quality, geologic structure, etc.) which could impact their utilization. This Exploration and Reserve Definition chapter is intended as a sort of gateway to the rest of the book. Truly, it is difficult where geology is concerned to define where exploration ends and other aspects pick up as all geological aspects are concerned with understanding the coal deposit in some way.

A considerable amount of literature exists on methods of coal exploration. Indeed, it seems sometimes that there is little new in this area under the sun. To give perspective on where we have been and where we are going in the exploration segment of the industry, the overview paper by Tivy and Mercier reviews progress in the application of geology to the finding and delimiting of coal resources, and considerations for the future in this field.

An area of exploration techniques which has received more attention in recent years is that of geophysical exploration and definition of coal resources and conditions. The expense of detailed geophysical surveys tends to put off smaller coal operators. However, the application of such techniques can provide very useful information on the size and condition of deposits, as pointed out by Henderson et al.

The paper by Sarnecki is included with a view to limits on useable coal deposits caused by natural burning of seams and by their remains (clinker). Clinker and burning coal can be guides to subsurface deposits, but more importantly for the eventual reserve estimation and mining processes, they also dissect and/or destroy coal resources.

An example of the carry-through of geology and geologic knowledge from exploration concerns through the mining process is provided by Fenton and Pawlowicz. The specific case discussed is of the impact of glaciation on coal deposits and resources. However, the paper provides a good example of how geologic knowledge and site characterization can provide useful information from deposit identification on through exploitation.

Cross-Reference

As noted above, the application of geology to coal exploration slides easily into areas such as reserve estimation and mining. Especially in the field of reserves, it is difficult to define where reserve definition ends and reserve estimation starts, a distinction that often is made for administrative rather than technical reasons. Therefore, the papers by Treworgy and by Sites et al. will be of interest for coal explorationists as well.

In the Mining chapter, the paper by Weisenfluh and Ferm provides information to the explorationist on coal deposition and its impact on deposit distribution. Likewise, although they discuss specific cases, the papers by Mitchell and Pappajohn (in the Coalbed Methane chapter) and Molinda and Ingram (in the Mining chapter) provide insight on the impact of geologic structure on coal deposits.

The paper by Finkelman and Brown, in the Environmental Impacts section, brings up the interesting possibility of exploitation of coal seams for their non-coal values (coal as a byproduct?). Fariborz and Gentzis, in the same chapter, provide a discussion of natural coal combustion that is complementary to Sarnecki's paper.

COMMENTARY ON THE EVOLUTION OF THE EXPLORATION AND EVALUATION PROCESS

Stephen V. Tivy and John M. Mercier
Cyprus Coal Company
P.O. Box 3299
Englewood, CO 80155

INTRODUCTION

Technical personnel in the coal industry have witnessed a significant restructuring and refocusing of individual and corporate activities in recent years. This new focus has been brought about by fluctuating economic and political conditions. In particular, activities associated with coal exploration, management, and reserve definition have been profoundly affected. Whereas the positive and growing coal industry of the late 1970s and the early 1980s spawned innovative and generally well-funded research and operationally supportive programs, the late 1980s have seen the evolution of a strongly pragmatic industrial course and a declining emphasis on geologic research and academic involvement.

In spite of the constrained nature of the industry, individuals in the field of coal geology today find the profession is weathering the recessional cycle in good shape. Although the ranks of participants have been thinned, the spirit of geologic intuition, adaptability, and creativity is strong and ready to meet the challenges of the future. It is evident that the role of the coal geologist in exploration, reserve analysis, operations support, research, and management is vital and appropriate.

This review re-examines industrial cycles, the basic goals of reserve evaluation, and the importance of new tools, attitudes, and management methods in today's practice of coal exploration and geology.

THE GOAL OF COAL EXPLORATION

Successful coal exploration has as its goal the definition of a quantifiable resource both in terms of physical parameters and potential marketability. Technically, the realm of coal exploration ends at the point when resource evaluation becomes reserve estimation. As will be discussed below, geologists may find that this point is not sharply defined or static when the definitions of resource and reserve are examined.

The term "coal resource" usually reflects the presence of coal without regard to eventual recovery. U.S. Geological Survey Circular 891 (Wood et al., 1983), the current archetypal reference with regard to resource classification, lists "coal resource" as "naturally occurring concentrations or deposits of coal in the Earth's crust in such forms and amounts that economic extraction is currently or potentially feasible." As this includes "potentially feasible" deposits, it is essentially all-inclusive.

As a matter of practice, coal professionals rarely devote a significant amount of time and energy to the determination of a resource without the application of some currently acceptable categories. For example, the breakdown of bituminous coal resources into ranges of coal seam thickness, such as 0-71 cm (0-28 in.), 71-107 cm (28-42 in.), 107-213 cm (42-84 in.), etc., and further divisions by depth of burial has survived many reviews. Here the use of measurement in centimeters or inches is strongly regional or traditional in nature, whereas the classification of overburden is complexly related to parameters such as economics, technical mineability, experience, and marketability.

The classification of a resource into proven, probable, and inferred does not necessarily indicate that the resource has crossed the line to become a reserve (such as a "measured subeconomic resource"). A resource does not become a part of the reserve base until site specific economic analyses are applied to the resource. Today's reserve may well become tomorrow's resource simply by changes in the marketplace. Therefore, coal geologists may be called in to re-evaluate a resource at any time. The work of establishing definable and verifiable reports of resources, which may be recovered if the right economic and engineering criteria are met, is the goal of coal exploration.

COAL INDUSTRY CYCLES

Historically, the coal industry has displayed a very short and pronounced cycle, which has had a profound effect upon employment, production, productivity, and exploration goals (Paull, 1983). More recently, this industry has been driven by demand created by an affluent post-war society and a rising "baby boom" population. Cyclic recovery in the industry often has been accompanied by adoption of new technology. The graph presented as Figure 1 reflects the sensitivity of the coal industry to the shifting demands of manufacturing, heating, and energy consumptions (Energy Information Administration, 1987).

This cycle has notable social impact, and its period often is reflected more severely in human terms than in statistical measures of gross production. Figure 1 also shows that the total employment levels of the 1920s (Energy Information Administration, 1979) never recovered from technological advances and the increase in size of operations. The average price of coal (Figure 2) essentially remained flat from 1920 through World War II, followed by a slight increase through 1968. An outcome of the early part of this period was the excellent and extensive exploration of U.S. coal fields.

F.O.B. coal prices from 1920 through 1965, as shown in Figure 2, remained relatively constant, perhaps losing ground slightly in terms of inflation (National Coal Association, 1979). In fact, the coal market was considered a non-growth industry at this time. However, during the same period of time, production experienced significant fluctuations over short periods (Figure 1). Bursts or peaks of production were followed closely by sharp production declines.

The increases in production from 1962 through 1973, as shown in Figure 2, were obtained by improved productivity through technology and management, not by exploration and market expansion. This period saw relaxed exploration throughout the U.S.

During the early to mid-1970s, the price of coal increased by four to five times the pre-1970 price and fueled the "coal boom" with which most of us currently in the industry were closely associated. This boom rapidly doubled employment in most coal fields, but witnessed a decrease in per-man productivity (Figure 2). The jump in price was in response to the world petroleum situation and particularly the OPEC oil embargo.

During the early 1970s, coal geologists opened up the Western U.S. coal fields, expanded the reserve base in the Southeastern U.S., developed interest in many foreign projects, and saw the application of new tools and concepts in the field. The exploration of this period was generally an extension of the exploration of the

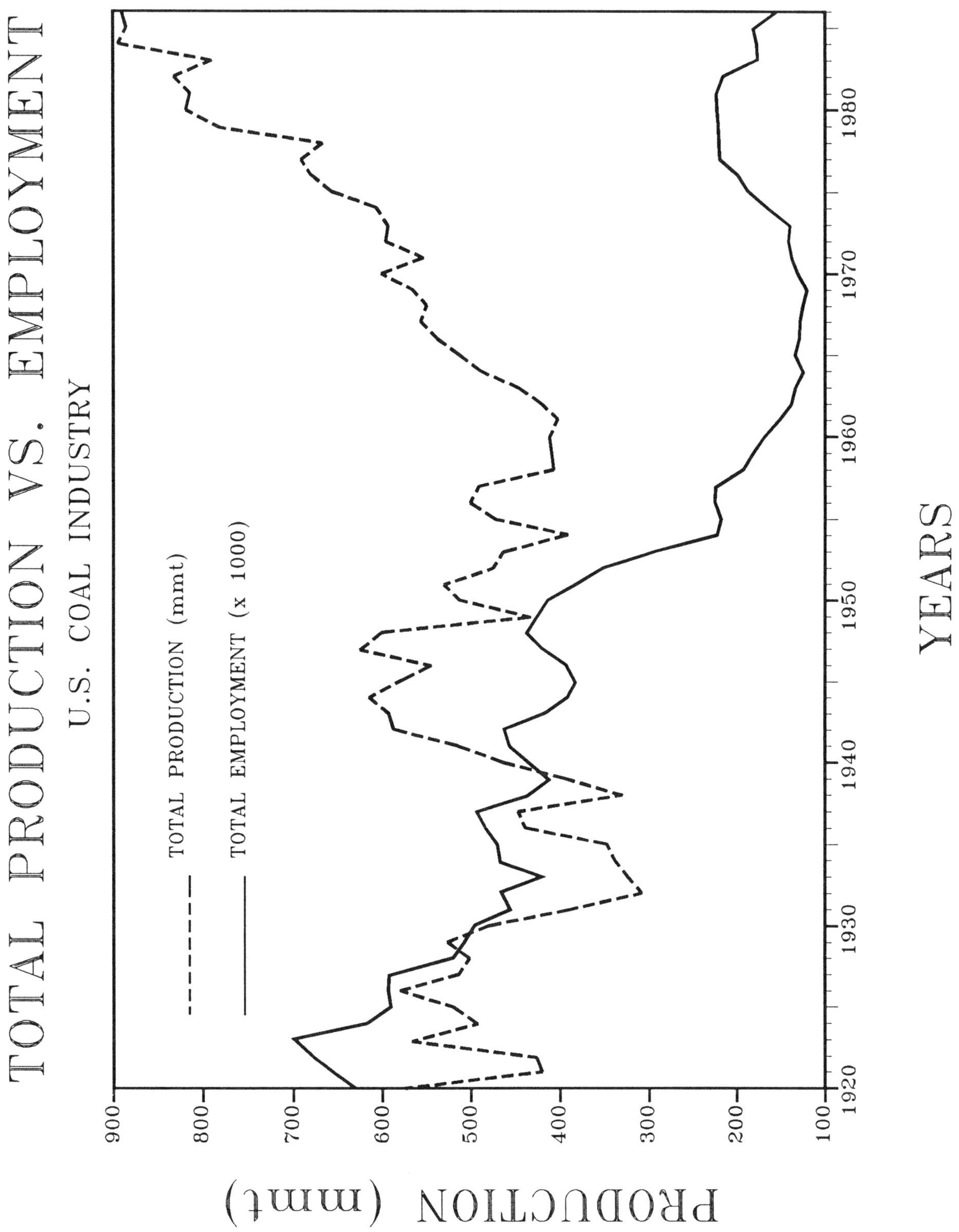

Figure 1. *Graphs of total production and total employment in the U.S. coal industry, 1920-1986.*

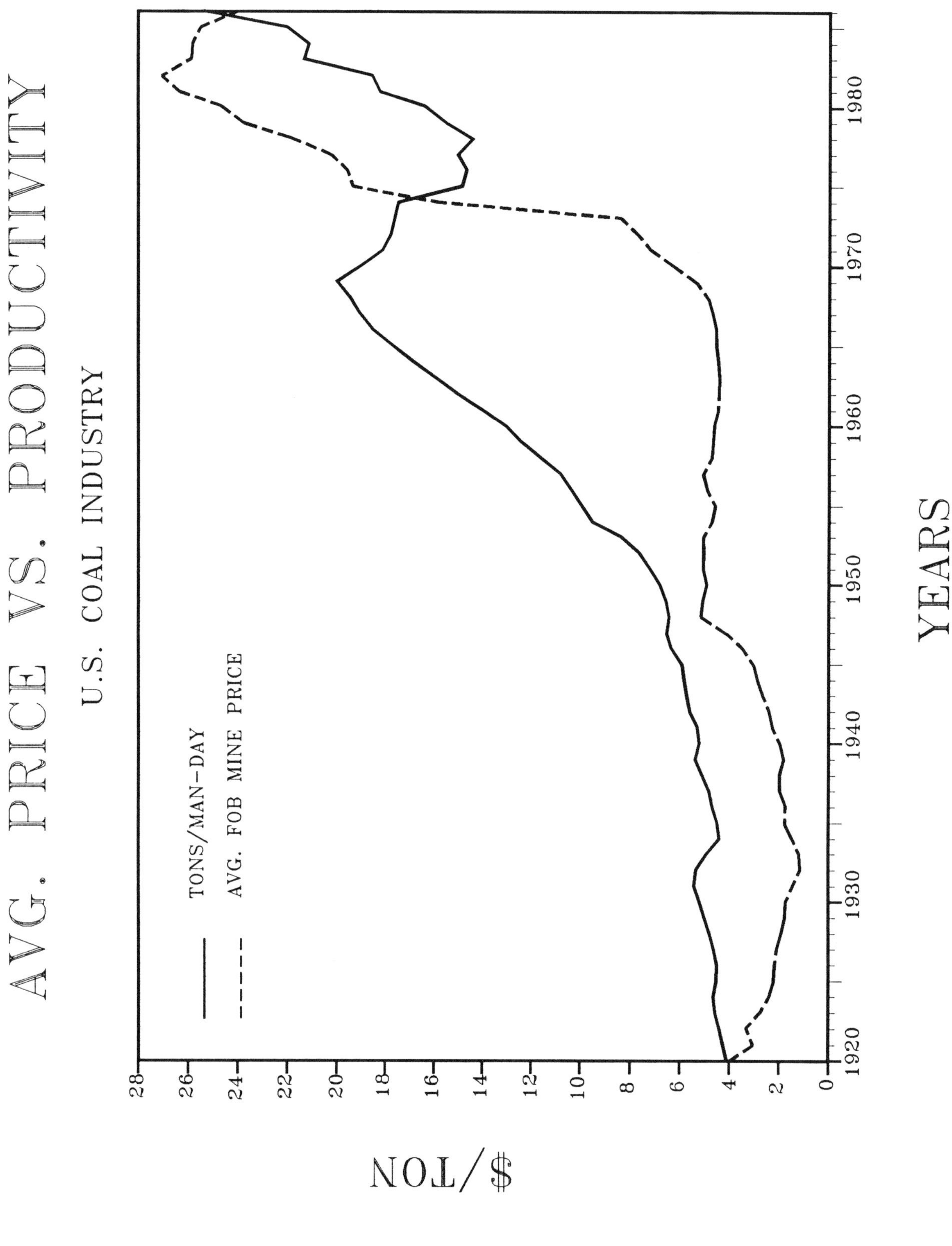

Figure 2. *Graphs of average F.O.B. price per ton of coal and productivity in tons per man per day, 1920-1986.*

1920s, rather than "grass-roots" projects. The coal boom leveled off in the late 1970s, but geologists were kept busy as a result of environmental legislation and marketing constraints. For example, Clean Air legislation focused interest on quality aspects such as low sulfur while the identification of resources for mine-mouth utilization in areas of suitable airshed brought new concerns to the field such as remoteness, combustion performance, and "integral vistas" (Energy Information Association, 1989).

Environmental sensitivities precluded active consideration of vast areas such as Utah's Kaiparowits Plateau and much of the Powder River Basin. Transportation limited consideration of remote areas such as the Henry Mountains of Utah and restricted development in New Mexico and Wyoming to those who could deal successfully with the railroads. New and/or more stringent reclamation legislation changed the face of mining in the Appalachian coal fields as well as other fields across the country. In this environment, coal geologists became as sensitive to political and economic criteria as to the presence of coal.

The decade of the 1980s has seen yet another perturbation of the cycle. Increased capital that was available because of the presence of "big oil" in the coal fields has given way to management practices to extend the coal boom in terms of production through the engendering of a sharp decrease in employment and the maximization of short-term profit. Once again, the role of the geologist has changed. This time, the edict has been to glean the "high-productivity" reserves from the total resource. This swing may indicate a near-sightedness industrialized nations may one day regret: the "high grading" of resources to assuage the market.

EVOLUTION OF EXPLORATION

The earliest booms brought about some of the finest exploration work ever documented. The pioneers responsible for the publication of government folios, monographs, and various state-sponsored bulletins provided a basis that has been used for many decades. The exploration work of this era indicated that the federal government controlled much desirable coal, particularly in the Western U.S. Interior.

The most recent boom, occurring in the early to mid-1970s, saw the evolution of exploration from raw field work to broad-based reworking of the pioneer data. Companies across the country had teams of geologists performing basin analyses, from the Pacific Northwest to the Black Warrior Basin of Alabama and from Alaska to New Mexico. International attention was concentrated on Australia, Colombia, South Africa, and Indonesia. Domestic efforts depended upon the pioneer work for their form and targeting. Additional research was done through the offices of state mining agencies to identify areas of remaining reserves. Oil and gas well logs were examined for signs of coal in the subsurface. Firms were created to track lease activity and status in target areas. Anyone who could finance a drill rig opened a drilling company as exploration moved from the office to the field. Many saw this as the zenith of coal exploration in the United States.

On the federal level, programs were developed that allowed companies to apply for federal leases and to conduct exploration at their own cost. If favorable results were received, the companies were given preferential rights to convert the application to a lease. These were known as PRLAs (Preference Right Lease Applications). Millions of dollars were spent in hope of consummating a PRLA, but at the height of activity, the government recognized the inherent problems associated with a non-competitive leasing process involving federal lands and froze the process. By the time the federal government finally came out with a revised program, much of the interest in western deposits was waning. Under the new program the government identified target tracts, drilled them, and put them up for lease. One of the criteria for targeting was "expressions of interest" from coal companies. Again, due to the tremendous time lag between the initial expressions and final determination, often most of the initial interest was gone. The State of Utah's North Horn Mountain Lease sale illustrates what happens when part of the exploration process is allowed to slip between the cracks. In this case, a basic market drive (coal quality) was ignored, and no one came to the lease sale!

Early booms ended with a basic change in the needs of the market which could be measured in kilowatts, tons of steel produced, or homes heated. The most recent bust came in response to the re-evaluation of inflated growth curves, a weakened economy, and to a market that appeared to have been partially contrived on Wall Street.

For the present, the coal industry needs to determine in which direction it is evolving so as to better plan and develop the energy resources of the future (McMahan and Knutson, 1989). Efforts must be made to make the energy resources available on a proactive scale, not a reactive one. This likely will never happen until Congress works to develop a realistic national energy policy.

THE EXPLORATION PROCESS OF TODAY

We can see from the above that the basic roles and goals of the coal geologists of today remain unchanged from those of the pioneers. The job of the coal geologist is to identify a resource for which there may exist the potential for development. The methods, however, to define and reach these goals may differ depending upon geographic location.

Exploration in remote areas of the world may look a lot like the efforts of the mid-1970s in North America. Basin analysis, literature research, and rough field reconnaissance are the tools of the trade.

In North America today, exploration takes on a different tenor. Many companies use acquisition rather than exploration as the preferred tool for expansion. Companies involved in active coal mining include an explorationist/coal geologist on a team which identifies targets for acquisition. The evaluation process of the team incorporates a matrix that addresses strategic significance, marketability, transportation, and coal availability, as well as favorable geology and mineability.

Strategic Significance

It is important to identify the strategic need of a company. Is it the acquisition of low-sulfur coals, metallurgical coals, presence in an unserved transportation corridor, or a favorable labor market that will best serve a company's short- or long-range plans? The identification of these needs may culminate in a table of desired products or result in a circle on a map. Perhaps the financial picture requires a resource re-evaluation of the company's active reserves. Whatever the target or combination of company needs, the geologist must be able to identify these concerns to focus his energies effectively.

Firms or agencies involved in non-traditional resource recovery or by- products may be concerned with very different cutoff parameters. Location may be less significant than regional characteristics. Vitrinite reflectance for the coal-gas geologist may be more important than proximity to rail transport, which is important for the traditional geologist. The end result, in any case, is a location on which to concentrate the efforts of the coal geologist.

Marketability

What makes a coal marketable? Coal quality alone seldom makes a resource marketable. There are markets for high-sulfur as well as low-sulfur coals, for slagging and non-slagging coals, markets for coals with "exotic" mineral components, coals of every range of volatility, moisture, and ash. The trick for the coal geologist is to find the right coal, in the right place, that can be mined at the right cost. Aside from strategic-niche consideration (the conscious decision of a company to serve a particular market), the determination of marketability may be front-end or back-end loaded. In the former sense, the company may undertake a complete market analysis and develop several product mixes to broaden the influence of the company's resource base. In the latter case, an existing client or market may indicate the potential for expanded business opportunities. The expansion may require the company to find or develop a product that meets specifications within the sphere of influence of the market. In this case, the market analysis yields a general location or locations and a general set of quality specifications for use by the explorationist.

Transportation

The role of transportation in modern exploration and mine siting cannot be overstated. Entire foreign coal operations have developed as a result of transportation advantages. Transportation can play just as important a role in non-traditional resource utilization as in conventional mine development. The location of gas pipeline networks would play a significant role in the siting of in situ gasification targets as well as projects intended to extract methane from coal. The presence of railways, navigable waterways, and/or high-quality roadways are vital components to a potential mining operation which would depend on the competitive and efficient transportation of mine product and the timely delivery of essential equipment.

Coal Availability

Details of coal availability may be the final question to be asked. It is here that the conceptual skills of the coal geologist come to bear. The determination of a resource of a suitable size, quality, and condition generally is attained in a three-step sequence. First comes the determination of regional character, second is development of a model, and third is the risk analysis or application of realistic expectations to the property.

Homework on Regional Character

Published exploration work, if available, still always is the first stop for the coal geologist. The work of the pioneers is invaluable, as are such follow-up works originating from federal, state, and academic sources. The more information that is available without costly field work, the better for the company and the project. Mining history, if applicable, is a basic step in analysis of regional character. Some of the most detailed exploration data for an "old" field can be found on existing mine maps. Data displayed on mine maps can show trends caused by either structural or depositional discontinuities, as well as seam thickness and general mineability. Within the U.S., the task of locating these maps often is simplified by the existence of state mine agencies from which detailed records may be obtained.

In a truly green-field enterprise, field reconnaissance and photogeology may be the primary tasks. It is difficult to imagine anywhere in the U.S. where this is true, but involvement in Colombia, South America, and parts of Australia have shown this to be a practical first step. Field geology, detailed mapping, and re-thinking old fields and resources have brought new life back to a number of areas. Structural control is very often an overriding consideration. We all know of the millions of tons of coal that would be available, for example, if only we could determine a practical way to mine coal that is steeply dipping or highly dissected by faulting. The role of structure in evaluation also may be quite subtle. Small folds may provide surface access to a seam that everywhere else is unattainable due to depth of cover. Always, the coal seam structure map and a good regional understanding are tools of preference in this endeavor. Interpolation and extrapolation of coal elevation data should be tempered by knowledge of regional dip and depositional character. Additional sources of information regarding structural discontinuities should be reviewed and combined with basic, reliable elevation data. Aerial photography or other remote sensing imagery may be useful in detailing structure. Governmental publications such as Geologic Quadrangle Maps are always beneficial.

The understanding of coal seam continuity, thickness, and depositional character is of paramount importance. It is in this task that the interpretive skill of the coal geologist can be brought to the forefront. Interpretation is the key word, for nowhere is the science of geology more closely akin to an art than in the interpretation of often meager exploration information (Levey, 1985). Once a geologist has become involved in a developmental role, it frequently is difficult to move "backward" to dealing with exploratory data, but it is in this realm where the geologist is most free to ply his trade.

At this point, the exploration task may enter a repetitious stage of data acquisition, analysis, and reporting. The goal is to tighten the interpretation and further limit the area of examination to high-quality prospects. Secondary steps put primary emphasis on structural and stratigraphic interpretation, for all other considerations should have been addressed. At this stage, data integrity becomes an overriding concern. The data generated during exploration provide a framework for each step of later exploration and development.

Significantly, first impressions mean a lot to management, so it is important that these impressions do nothing to unfairly characterize a prospect. Do your homework! The identification of all aspects of a potential target, both positive and negative, is significant and deserving of consideration. Areas of deficiency in the geologic or geographic database have to be recognized and resolved or carefully weighed.

Development Of A Model

The iterative or repetitive process also is necessary in the activity of quantitative analysis. The quantitative model in its crudest form is nothing more than an isopach and structure map of the seam or seams being considered. In a more elaborate system, the maps may include non-coal isopachs and/or structures. There may be a computer involved, in which case the isopachs and structures could be reduced to grids or block models for volumetric calculations. Some field data to augment the basic information usually is required before much advanced modeling can be considered, if for no other reason than to "ground truth" early hypotheses and models.

Risk Analysis

Coal geology in the modern sense is much more closely tied to the development of a mine plan than was the work of the earlier investigators. After completion of the quantitative model, the coal geologist must be willing to assess his confidence in his new creation. The time-honored "influence distance from data points" method still is widely used, although geologists now see wider application of geostatistical methods for these purposes. Here, the application of mining criteria, cost, and marketability take the process beyond resource exploration and into reserve estimation or development.

Within the coal industry today, it would be highly unusual for a coal geologist to limit himself to pure exploration. The efforts of the U.S. Geological Survey, the U.S. Bureau of Mines, and State Surveys also reflect this change in the role of exploration (Boreck, 1988). These agencies emphasize refining estimates and developing tools and models for use by the industry rather than true exploration.

Risk analysis involves the application of economic and marketing "filters" to the results of the exploratory process. While determining whether or not to proceed with consideration of a property, companies often use resource estimates as the first and coarsest filter. The question, "Are the reserves sufficient to provide a basis for production as anticipated?", requires a simple yes or no answer. The apparent simplicity of the answer belies the importance of the question. If, and only if, the question is answered in the affirmative will the company continue consideration of the property. Continued consideration means attention to questions of cost or anticipated cost, markets, legal and labor considerations, and geographic/environmental factors.

CONCLUSIONS

The field of coal geology is not what it used to be. However, very little in the world of the early 1990s is as it used to be in the early parts of the century. The role of the geologist has changed to keep pace with the industry, and this role is as vital today as in 1900. The individual geologist who struggles to interface with government regulators, operational engineers, and company managers is an essential component in today's coal industry. Geologic work still progresses today through the utilization of traditional field activities and elbow grease, innovative applications of new and old technology, critical reviews of existing concepts and programs, changes in data handling systems, and the identification of new products and markets, to name but a few procedures. Advanced tools, concepts, and attitudes are available to assist in this work. Geologists need to recognize that they occupy important positions as stewards of the coal resources of their industries or countries. It is the geologist's responsibility to understand these resources, to document exploration efforts, and to make every effort to insure complete, concise data results and communication from those efforts.

REFERENCES

Boreck, D.L., 1988, Increasing the recovery of thick and closely spaced coal seams in the Western United States--Some geologic and longwall considerations: Mining Engineering, v. 40, no. 3, p. 168-173.

Energy Information Administration, 1987, Coal Data: A reference: Washington, Report #DOE/EIA-0064(85), 84 p.

Energy Information Administration, 1989, Coal Data: A reference: Washington, Report #DOE/EIA-0064(87), p. 27-28.

Levey, R.A., 1985, Depositional model for understanding geometry of Cretaceous coals: Major coal seams, Rock Springs Formation, Green River Basin, Wyoming: American Association of Petroleum Geologists Bulletin, v. 69, no. 9, p. 1359-1380.

McMahan, R.L., and K. Knutson, 1989, A look back at the 80s and a glimpse of the 90s: Coal, v. 94, no. 12, p. 33-34.

National Coal Association, 1979, Coal facts; 1978-1979: Washington, 95 p.

Paull, M.K., 1984, Annual outlook for U. S. coal, 1984, with projections to 1995: Washington, Energy Information Administration, Report #DOE/EIA-0333(84), 38 p.

Wood, G.H., Jr., T.M. Kehn, M.D. Carter, and W.C. Culbertson, 1983, Coal resource classification system of the U.S. Geological Survey: U.S. Geological Survey Circular 891, 65 p.

SURFACE GEOPHYSICS FOR COAL EXPLORATION IN THE ALBERTA PLAINS: RECENT RESEARCH

James D. Henderson and Anthony N. Sartorelli
Geo-Physi-Con Co., Ltd.
5800 2nd St. S.W., Ste. 205
Calgary, Alberta T2H 0H2
Canada

Mark M. Fenton and John G. Pawlowicz
Terrain Sciences Department
Alberta Research Council
Box 8330, Station F
Edmonton, Alberta T6H 5X2
Canada

Georgia L. Hoffman
Coal Mining Research Co.
One Oil Patch Drive
P.O. Bag No. 1400
Devon, Alberta T0C 1E0
Canada

ABSTRACT

This paper summarizes a 4-yr long project to evaluate the applicability of electrical and seismic techniques to coal exploration and mining in the Alberta Plains. Investigation targets were in the 10-70 m (33-230 ft) depth range and included buried channels, coal subcrops, deformed bedrock, and continuity of strata. Test sites were chosen at four different coal mines and 27 tests were run. Conclusions are that reflection seismic, refraction seismic, and direct current sounding and profiling are useful for locating the shallow targets and that surface geophysical techniques are most effective when integrated with drilling programs. Recommendations are that 1) surface geophysics should be added to the principal tools for plains coal exploration and development, 2) reflection seismic and direct current sounding should be considered for coal exploration (for mine planning, method selection depends on the specific target or problem), and 3) selection of geophysical techniques should consider operating environment, geologic framework, and a firm statement of information needed, including depth of the horizon of interest and resolution desired.

INTRODUCTION

The primary objective of this study was to develop and demonstrate the application, for target depths of 10-70 m (33-230 ft), of cost-effective surface geophysical techniques to plains coal exploration and surface mine planning. This paper summarizes the results of the study which was conducted over a 4-year period in the Alberta portion of the Great Plains (Geo-Physi-Con, 1985a and b, 1986a and b, and 1987; Green et al., 1988; Moell et al., 1984 and 1985a; Pawlowicz et al., 1986a, b, c and d; and Rogan et al., 1985).

Approach

The geological targets to be identified by the selected geophysical methods included buried channels, coal subcrops, deformed bedrock, and continuity of strata. A series of test sites were chosen at four different surface coal mines (Figure 1). At least two methods were tested at each site for a total of 27 tests. Electrical methods, reflection seismic, and refraction seismic were selected for testing.

Figure 1. *Map of Alberta showing location of study sites.*

At each test site, there were enough pre-existing data to confirm the existence of one or more of the chosen target types. The geophysical surveys then were run, and the interpreted data were used to select the location of new geological core holes. Each hole was logged using standard slim-line downhole geophysical techniques. Cross-hole seismic tests were run at some sites to determine the vertical variations in the shear and compression wave velocities, which were used in refining the geological and geophysical models.

This paper presents the results of the application of one or more geophysical methods to each target type. The examples presented here were chosen, from the results of all the tests, to best illustrate the use of a particular geophysical method for the detection of a specific target or targets.

GEOPHYSICAL METHODS

The successful application of surface geophysical measurements relies on the existence of suitable contrasts between the physical properties of target strata and those of surrounding materials. Coal is distinct from other rock types in that it exhibits very low density and very high resistivity. Although the sonic velocity in coal from this area is not substantially different from that of the surrounding rock, the impedance contrast (product of density and velocity) is largely due to the low density of coal. These resistivity and impedance properties make coal present an excellent target for surface-based electrical and reflection seismic techniques. A more comprehensive discussion of the topics in this section is included in Green et al. (1988), and a general description of the physics behind each of the methods can be found in Redpath (1973) and Telford et al. (1978).

Seismic Methods

Three types of seismic methods were applied during this study: reflection, refraction, and cross-hole. Each method relies on the propagation of seismic energy along a different wave path.

The reflection method records two-way travel time of seismic waves reflected from subsurface strata boundaries. Coal seams provide an excellent reflecting surface because of their characteristic acoustic impedance. However, use of the method is hampered at very shallow depths (less than 10 m (33 ft)) because of interference from energy arriving via other travel paths. High-frequency content of the seismic signal is critical for shallow compression-wave reflection seismic studies; higher frequency provides better resolution (Sartorelli et al., 1986).

The refraction seismic method provides compression and shear wave measurements of the thickness of unconsolidated sediment overlying bedrock and the mechanical properties of the sediment and bedrock. Successful use of the method requires that the seismic velocity in bedrock be faster than the overlying sediment velocity. During the field investigation for this project, both compression and shear wave data were recorded.

Cross-hole seismic studies utilize the propagation of energy waves along a direct path, where the source and geophone are placed at the same depth in adjacent boreholes spaced 3 m (10 ft) apart (Geo-Physi-Con, 1981). Compression and shear wave data are acquired from cross-hole seismic tests to confirm the thickness of velocity layers determined from refraction seismic measurements.

Electrical Methods

The electrical resistivity of subsurface materials can be measured from the surface using a wide variety of techniques. Direct current sounding and direct current profiling were found to be the most effective methods during this project (Green et al., 1988). In both cases, the measurements are sensitive to the large resistivity contrast between coal and adjacent rocks.

Direct current sounding measurements usually are obtained using a Schlumberger electrode configuration (Figure 2), where current-injection electrodes and potential-measurement electrodes are symmetrically placed with respect to the array center. Depth of exploration is controlled by the distance between the current electrodes. The method is useful for determining the presence of coal and the thickness of overlying strata. Because of the very high resistivity of coal, seam thickness and information about strata underlying the uppermost seam often cannot be determined easily.

The direct current profiling method was performed using a pole-dipole array, which is similar to the Schlumberger array except that one of the current electrodes is located at a very large distance from the other current electrode and the potential electrodes. The measurements yield very local estimates of earth resistivity and can be used to determine the lateral position of coal subcrop.

TARGET RECOGNITION

The major targets were buried channels, coal subcrop, deformed bedrock, and continuity of strata. The examples presented below are from two sites at the Genesee Mine and one site each from the Highvale

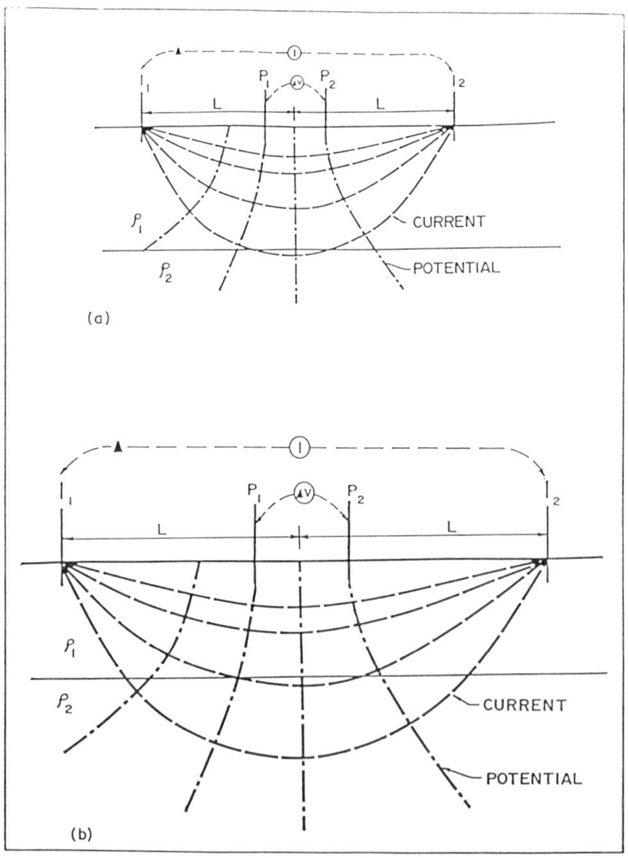

Figure 2. *Schematic of the current and potential electrode configuration of the Schlumberger array used for the direct current sounding method.*

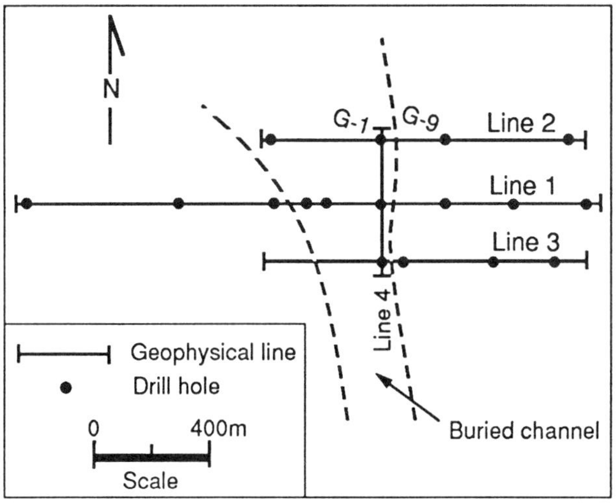

Figure 3. Location map of buried channel site (Site 1) at the Genesee Mine showing geophysical survey line and drill hole coverage.

Mine and Paintearth Mine (Figure 1). Results of additional work carried out at the Whitewood Mine can be found in Green et al. (1988).

Buried Channel

This target type consisted of channels eroded into the bedrock surface and buried by poorly consolidated and unconsolidated sediment of Quaternary age. The selected target was a north-south trending buried channel at Site 1 in the Genesee Mine area. The channel was cut into the bedrock and later completely filled with glacial sediment. The local terrain is essentially flat with no geomorphic manifestation of the channel. Direct current sounding, seismic reflection, and seismic refraction data were collected from four survey lines (Figure 3). The interpretations from these data were used to locate the drill holes.

The interpreted geophysical and geological data from Line 2 between stations 400 and 900 are shown in Figures 4 and 5. The test hole data show that the channel is located west of station 600 and that the channel is greater than 500 m (1,600 ft) wide; the western edge was not intersected. Depth to bedrock is about 25 m (80 ft) within the channel and 10 m (33 ft) outside the channel. The channel fill consists primarily of clayey till containing large blocks of displaced bedrock material. This fill is overlain by clay till and a thin mantle of lacustrine clay and silt.

The bedrock is composed of sandstone, siltstone, mudstone, and coal typical of the Ardley coal-bearing sequence. The only evidence of deformation in the bedrock is a high degree of fracturing in the upper few meters, likely due to glacial loading. The coal seams are flat-lying and the High Seam and Upper Main Seam subcrop at the channel edge between stations 550 and 650.

The direct current soundings were used to construct a geoelectric section (Figure 4). These soundings traced the High Seam to the channel edge at about station 600. The sediment filling the channel is characterized by relatively low resistivity in comparison to the surrounding bedrock. The absence of shallow, high-resistivity zones (coal) west of station 600 indicates that the channel extends across the entire western portion of the survey line.

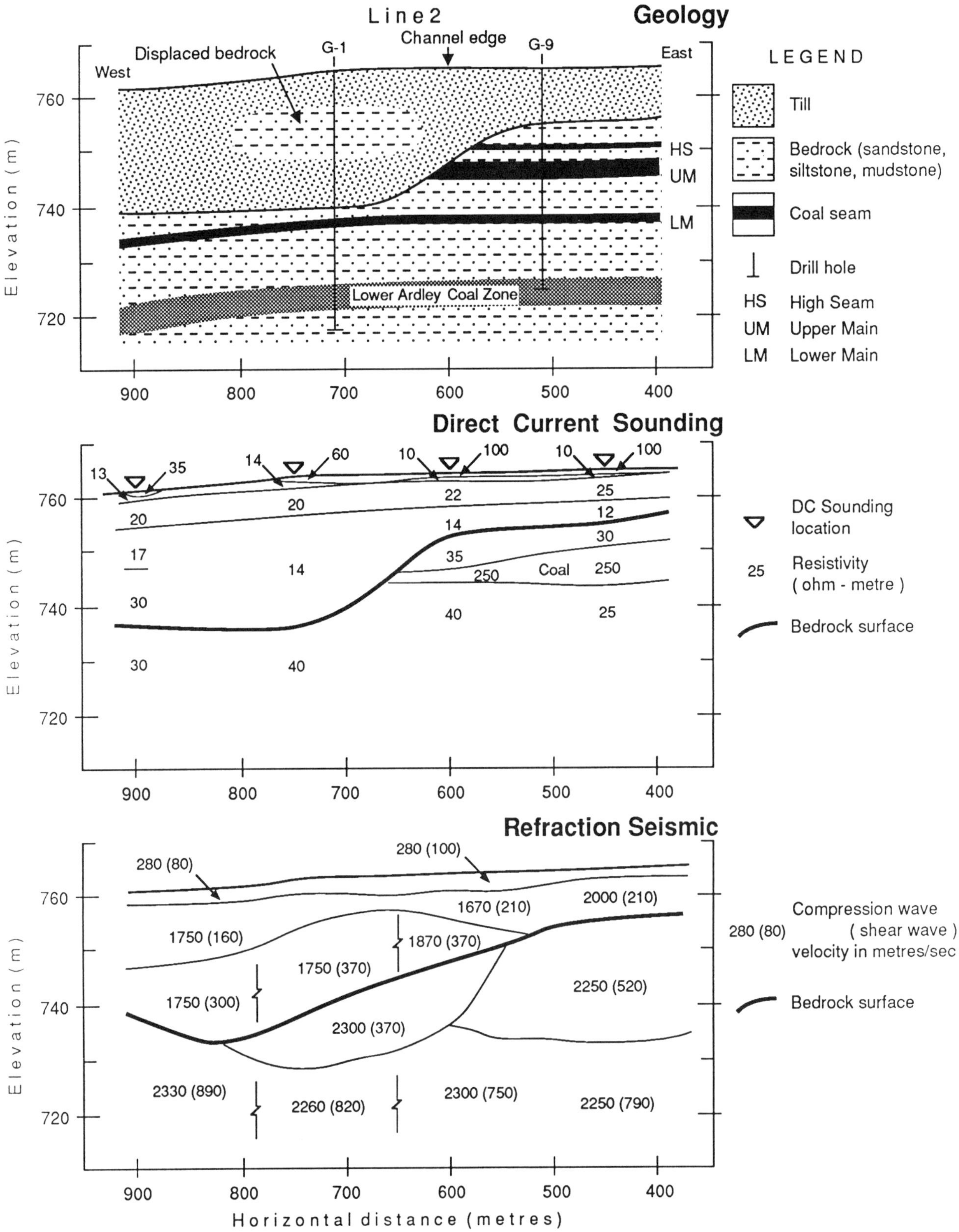

Figure 4. Geology, direct current sounding, and refraction seismic interpreted sections of buried channel along Line 2, Site 1, Genesee Mine.

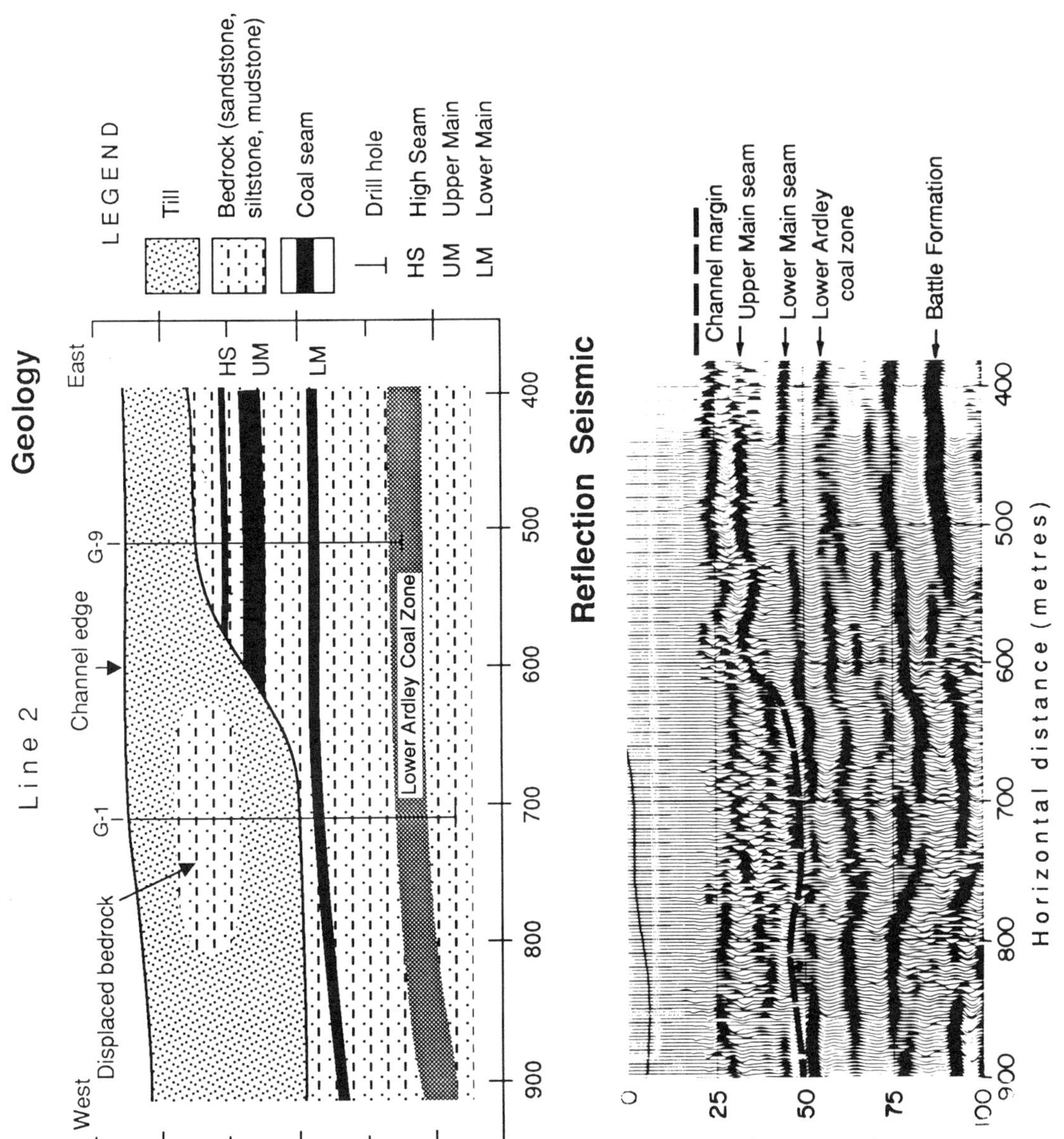

Figure 5. *Reflection seismic section of buried channel along Line 2, Site 1, Genesee Mine.*

The refraction seismic section in Figure 4 shows the interpretation of shear and compression wave measurements. The approximate location of the bedrock surface is marked by the upper boundary of material exhibiting a compression wave velocity of greater than 2,000 m/sec (6,500 ft/sec). Bedrock materials show a range of shear wave velocities. The low shear-wave velocity observed below the base of the channel indicates that this bedrock is less competent. This is corroborated by the abundant fractures observed in the core obtained from below the base of the channel. The increase in shear-wave velocity within the channel-fill sediment west of station 550, as compared to that of the overlying unconsolidated sediments, is indicative of the presence of displaced bedrock.

The darkened events on the reflection seismic section in Figure 5 correspond to the upper surfaces of low impedance zones which are characteristic of the coal seams. Each trace represents data from up to six different source-to-geophone separations. Traces correspond to 2.5-m (8.2-ft) intervals along the survey lines. The channel edge is associated with the disappearance of the reflector that represents the subcrop of the Upper Main Seam at about station 600. The reflector representing the Lower Main Seam extends beneath the channel. All the test holes drilled at the site (Figure 3) confirmed the location of the channel, channel margin, coal subcrop, displaced bedrock, and the thickening of the fill within the channel.

Subcrop

The selected target was situated at a different site (Site 2) near the southeast edge of the Genesee Mine area beneath a thin overburden cover. The topography is flat with a slight eastward slope. Geophysical data were collected along four lines oriented east-west (Figure 6). The direct current profiling method was used

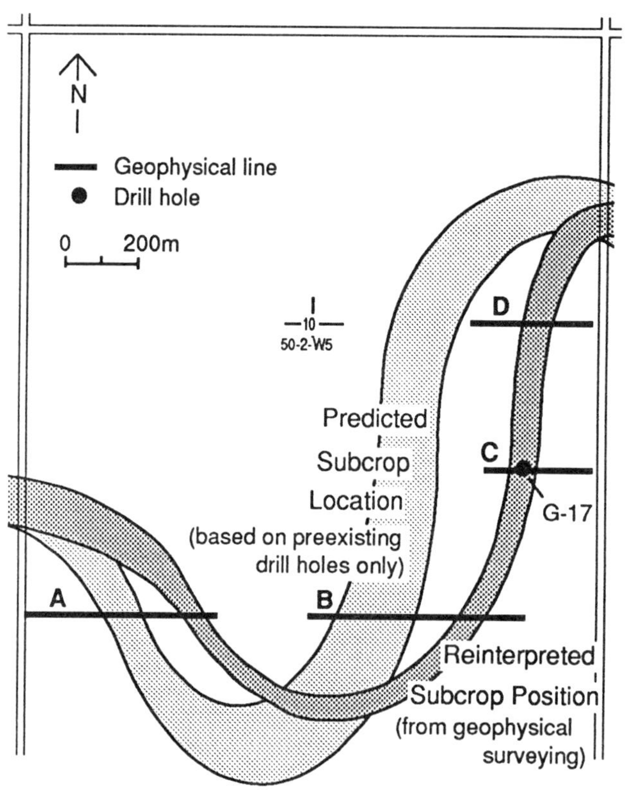

Figure 6. Map showing direct current profiling line coverage and revised subcrop position at Site 2, Genesee Mine.

to locate the subcrop of the High Seam and Upper Main Seam (Figure 7). Basically, the method detects the presence of electrically resistive coal amidst less resistive surrounding material, with the edge of the coal identified from the lateral variation in the data.

The resistivity data from Line C (Figure 7) are highest west of station 90 for all array geometries, indicating that the area contains both the High and Upper Main seams. Between stations 90 and 150, data are grouped at an intermediate resistivity level, suggesting the presence of only the Upper Main Seam. The marked decrease in resistivity east of station 150 indicates that the subcrop of the Upper Main Seam has been passed. On the basis of these data, a hole was drilled at station 110 and found the High Seam to be absent as predicted.

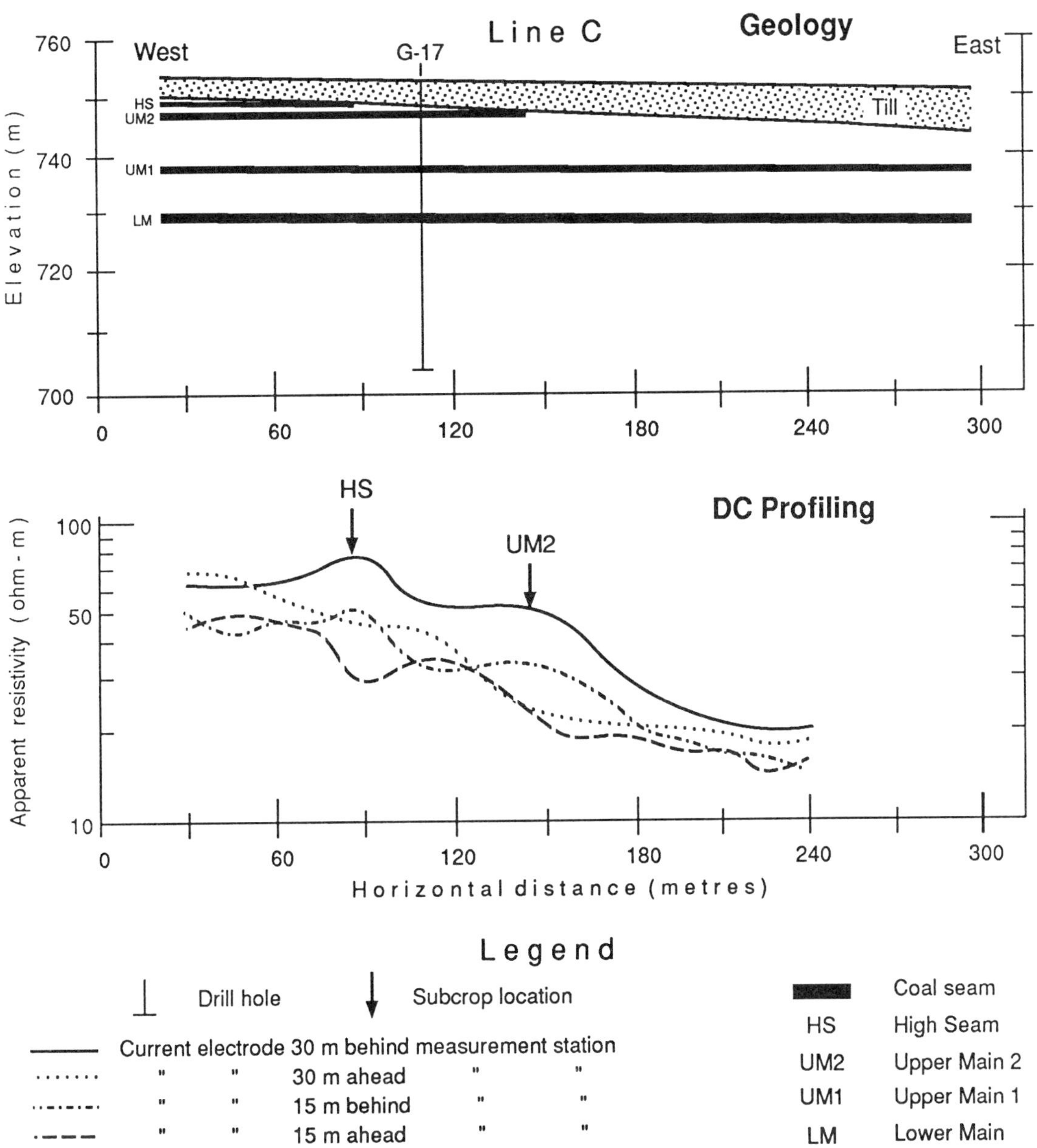

Figure 7. *Geology and direct current profiling results showing coal seam subcrop along Line C, Site 2, Genesee Mine.*

Prior to this study, the subcrop position was interpreted solely from preexisting drill hole data. The integrated surface geophysical and drilling approach greatly increased the reliability of the prediction of subcrop location. Figure 6 shows the subcrop location of the two coal seams prior to and after the direct current profiling survey. The revised subcrop, based on the geophysical results, increased the anticipated coal reserves in this section of land.

Deformed Bedrock

Continental glaciers have deformed the coal-bearing strata in many areas of North America from the Arctic Islands to Illinois and Pennsylvania (Fenton et al., 1986). The presence of this deformed bedrock has serious ramifications for coal exploration and mining (Fenton and Pawlowicz, 1989). At the Highvale Mine (Figure 1), a portion of the area was known to include deformed bedrock and, therefore, was chosen to test the ability of seismic refraction techniques to recognize this material.

The overburden above the uppermost coal seam was divided into four units (Figure 8; Moell et al., 1985b). Surficial deposits, mostly of till, are thickest on top of the upland and at the base. Deformed bedrock underlies the surficial sediment in all but the extreme north end of the section and was divided into two units: strongly deformed and moderately deformed. The base of translational deformation is identified by a shear zone at a depth of about 26 m (85 ft) near station 300. This shear zone appears to begin at the base of the hill near station 100. The coal sequence and remaining bedrock underlying the basal shear were believed to be undeformed.

The refraction seismic section in Figure 8 shows the interpretation of the compression and shear wave data gathered at the site. The compression wave data indicate the presence of low-velocity unsaturated sediments overlying high-velocity saturated bedrock at shallow depth. The shear-wave velocity data define an anomalous incompetent bedrock zone, characterized by intermediate shear wave velocity, forming the upper 30 m (100 ft) of bedrock and thinning towards the base of the slope. Core and downhole geophysical log data show that this material is deformed. The refraction data indicate that the zone of incompetent bedrock extends below the basal shear zone identified from drilling. This suggests that the effects of deformation are present below this basal shear zone. This supports the observations of Moell et al. (1985b) that fractures due to glacial loading exist below the basal shear zone of other thrust masses in the Highvale Mine.

The compression and shear wave velocity structure was confirmed by cross-hole seismic measurements made at drill site 84-504 (Figure 8). The base of deformation corresponds to the increase in shear velocity and decrease in velocity ratio at about 28 m (92 ft). The high velocity ratio above 9 m (30 ft) is produced by unconsolidated (low shear-wave velocity) and saturated (high compression-wave velocity) sediment. Note that the compression-wave velocity profile, unlike the shear-wave profile, is affected by the water table at about a depth of 4 m (13 ft). Piezometric data collected from observation wells in the area indicate depth to groundwater is less than 3 m (10 ft) on the highland and greater than 5 m (16 ft) on the lowland.

Stratigraphic Continuity And/Or Disruption

The selected target consists of a mass of glacially deformed sediment beneath flat terrain at the Paintearth Mine (Figure 1). The major geologic features consist of a mantle of till overlying deformed bedrock (including coal) which is underlain by undeformed and stratigraphically continuous coal seams (Figure 9). The principal evidence of deformation in this area is the absence of the uppermost seam (Seam 4) near station 400 and the folded, shattered, and overthickened condition of the coal seam between stations 450 and 650.

The reflection seismic section in Figure 9 shows data which were obtained using a 12-channel binary-gain seismograph. These results represent an early (1984) attempt to obtain reflection seismic data in support of

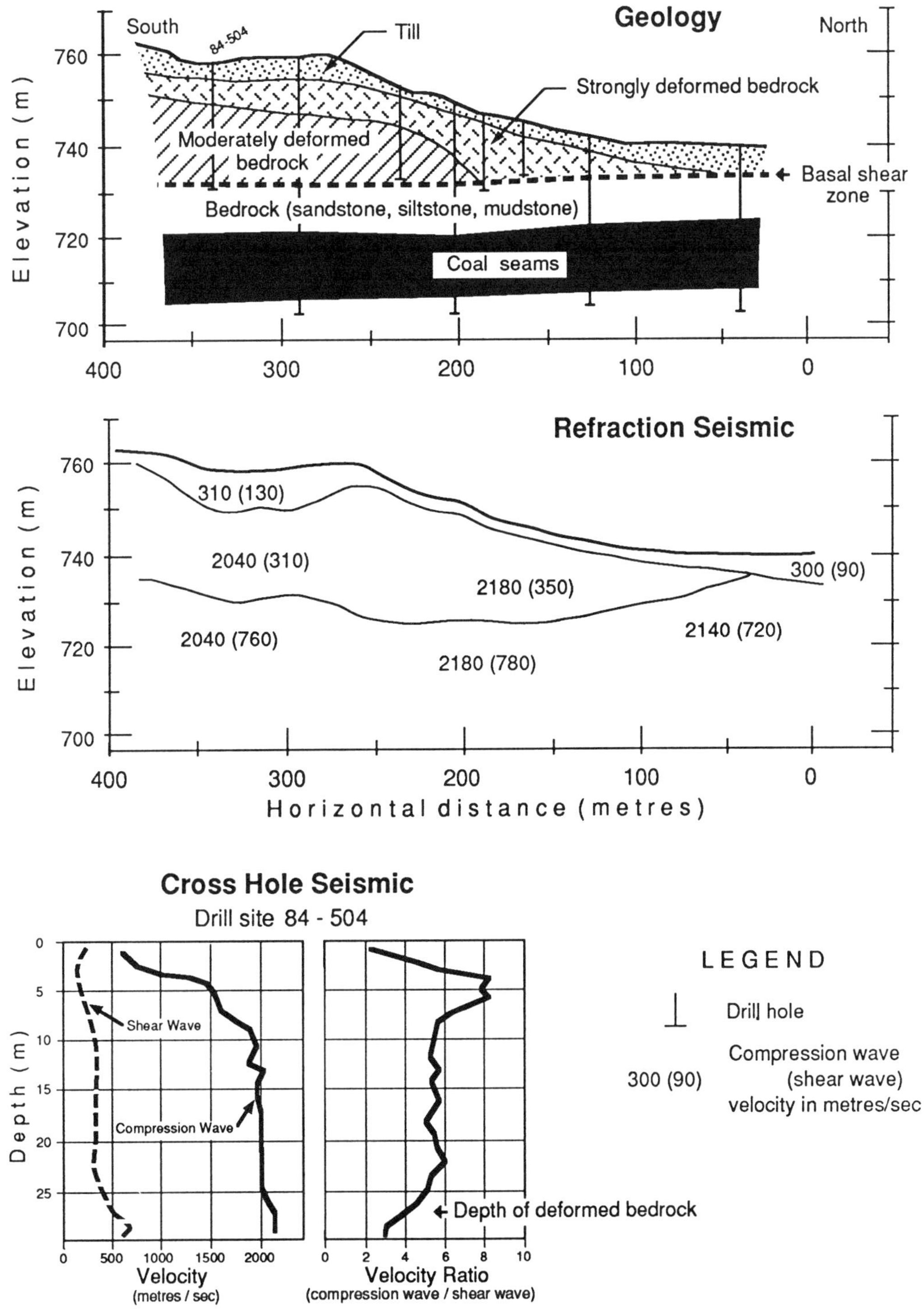

Figure 8. *Comparison of refraction seismic interpretation with geology derived from drilling data for detection of glacially deformed bedrock from the Highvale Mine. Cross-hole seismic results show confirmation of velocity layers.*

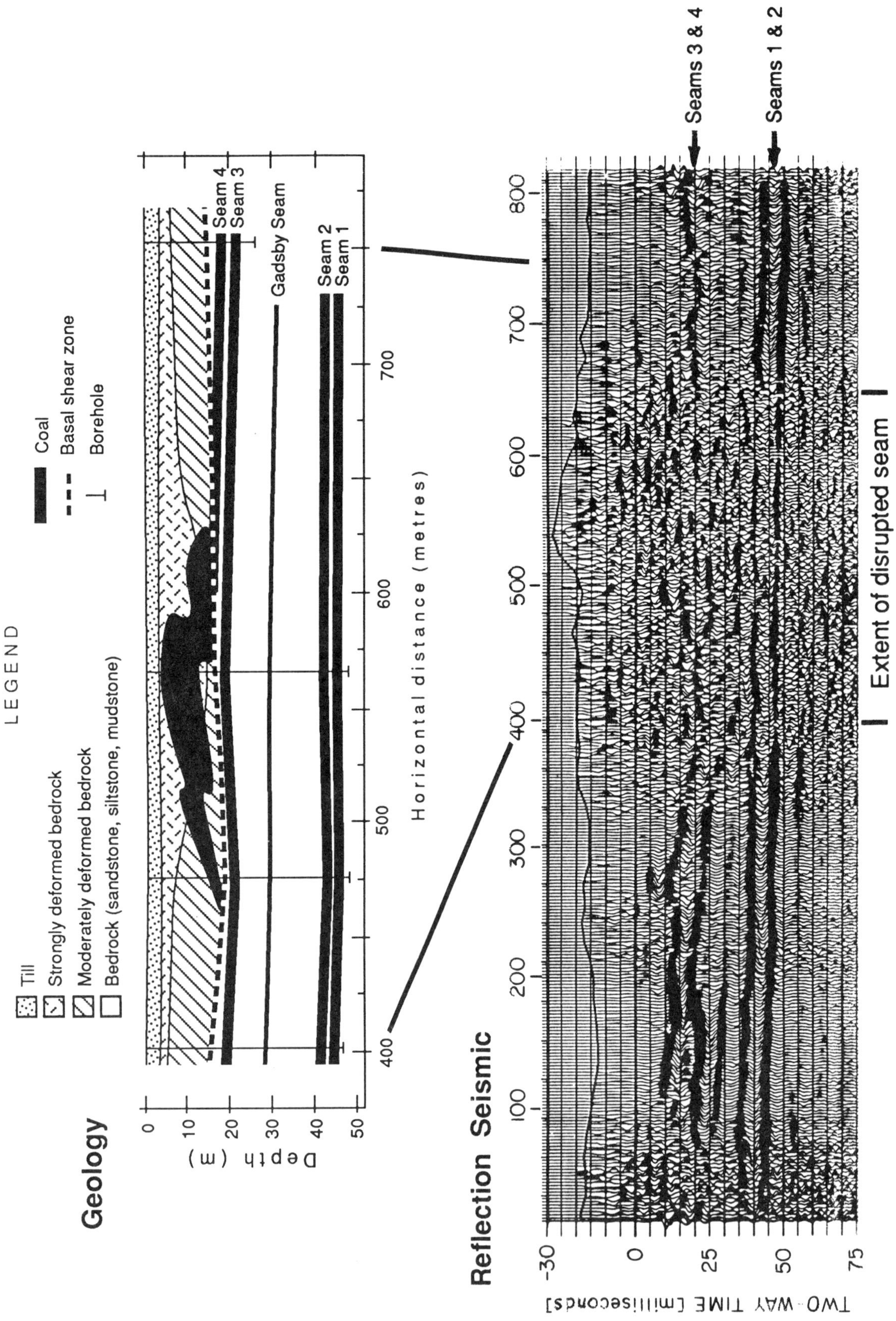

Figure 9. *Use of reflection seismic data to show discontinuous nature of coal seams at the Paintearth Mine site.*

shallow coal studies. Nonetheless, the data clearly illustrate the continuity of Seams 1 and 2 (at about 45 msec) across the site. The disruption of the reflectors between stations 450 to 650 is attributed to the deformation of Seam 4. The data traces represent up to four different source-to-geophone separations. Traces occur at a 2.5-m (8.2-ft) interval along the survey line. Darkened areas of the traces correspond to zones of decreased acoustic impedance and show the upper and lower surfaces of two coal seams.

The use of electrical methods to define stratigraphic continuity is illustrated at Site 1 from the Genesee Mine by the direct current sounding section in Figure 4. Several resistive layers were detected, indicating the presence of different geologic strata and demonstrating their continuity. The High Seam, for example, is shown to extend only to about station 600. The electrical data also give a good representation of the thickness and depth of subsurface layers.

APPLICATION OF GEOPHYSICAL TECHNOLOGY

Exploration Strategy

During the very early stages of exploration, regional geologic studies, airphoto analysis, and drilling over a coarse grid are carried out to gather preliminary data on the overall structure and stratigraphy of a coal prospect. The next stage of exploration requires more detailed geologic and airphoto studies, and an increase in drill hole density. At this stage surface geophysical technology can be valuable because it provides continuous subsurface information between the original exploratory holes, thus indicating both the uniform and anomalous features over the target area. Infill drilling then can be planned in an efficient and cost-effective manner. The two principal geophysical methods to consider at this stage are reflection seismic and direct current sounding and profiling. Both can be used to monitor the continuity of coal and surrounding strata.

The subsequent stages of exploration mainly are concerned with mine development. Targets such as channel location, subcrop location, presence of deformed bedrock masses, and other local geologic features which may affect the economics of coal extraction can be identified at this stage using geophysical technology.

The economics of incorporating geophysical technology into the exploration strategy have been discussed by Henderson and Nikols (1988). It is reasonable to expect that, as the depth to economically mineable coal increases, the cost-benefit of geophysical exploration also will increase.

Technological Improvements And Decreasing Costs

The cost of performing geophysical exploration was decreased over the 3-yr term of the research program by the development of more efficient data acquisition procedures. These procedures decrease the amount of field effort, which is a major component of cost. Recording instrumentation also was improved. Larger data-channel capacity and increased dynamic range for the reflection seismograph produced higher quality data at a lower cost per kilometer during the latter stages of the project.

Marked improvements also were made in the data processing capabilities applied to shallow reflection seismic data. Further improvements yet can be achieved. An important point to remember, however, is that the quality of data collected in the field has a great influence on the results of the processing.

Limitations To The Use Of Geophysical Methods

Geophysical technology cannot be considered as being available "off the shelf". Successful application of the appropriate techniques requires an understanding of the limiting factors which can be grouped into three general categories: 1) site-specific conditions and geology, 2) limitations of the methods themselves, and 3) tradeoffs between cost, depth of penetration, and required resolution.

In terms of site conditions, a first consideration is time of year. The presence of seasonal frost can adversely affect the use of direct current techniques and reflection seismic. Highly irregular topography and the presence of swampy ground (which precludes shear wave propagation) are other important site-specific constraining factors. The complexity of the geology, such as steeply dipping strata, also can pose some limitations. In areas where there are no significant contrasts in physical properties of the various lithostratigraphic units of interest, geophysical exploration cannot be used to advantage.

Each geophysical method is characterized by certain geometrical constraints. For instance, in direct current soundings, depth of penetration is gained at the cost of lateral resolution and requirements for large power sources. A similar restraint exists for refraction seismic because the limited data-channel capacity requires larger spacing between geophones to attain larger depths of exploration. Also, the thickness resolution of the refraction seismic is limited in a thin-layer sequence in which deeper strata are characterized by lower velocity than overlying strata. Cross-hole seismic measurements are used to reduce the possibilities of serious depth conversion errors in these situations. The main limitation for the reflection seismic method is data acquisition at very shallow depth. Ground roll, airwaves, refracted, direct, and reflected waves all occur virtually simultaneously within depths of less than 10 m (33 ft).

The limitations described above are not unresolvable. Often data acquisition geometries can be tailored to optimize data recovery from very shallow or very deep zones of interest, but the increased field effort required generally will affect the cost of the survey. For example, there are cases of reflection seismic sections being developed for the upper 5 m (16 ft) of the subsurface, gained by placing geophones at spacings of less than 1 m (3 ft).

The successful application of surface geophysics must take into account these limitations and requires a firm statement of specific mapping objectives, including the depth range of interest and the lateral and vertical resolution desired. Theoretical modelling of anticipated target responses can be performed to judge the feasible use of appropriate methods prior to any field studies. In new or geologically complex areas, a field test often can be designed to demonstrate the compliance of geophysical measurements with the specific objectives. This strategy will ensure that exploration budgets can be allocated wisely and enable optimal data recovery from both geologic and geophysical studies.

Integration Of Geological And Geophysical Exploration

The roles of geological and geophysical techniques in coal exploration are closely interrelated. Proper planning of a geophysical exploration program in terms of applicable methods, data acquisition geometries, and appropriate target strata depends on the availability of a reasonable geologic model for the area in question. Geophysical techniques provide a comparatively high number of data stations per kilometer (50 for direct current profiling and 400 for seismic reflection, for example) that results in a relatively clear definition of both the anomalous and uniform areas. The availability of continuous geophysical information can aid in planning the proper location and density of geologic exploration drill holes to satisfy specific information needs, therefore yielding the most economical and efficient use of drilling funds.

Specific geologic information, such as that available from downhole geophysical and lithologic logs, often can be readily incorporated into the interpretation of surface geophysical data. Perhaps the best case of this is the use of downhole log data to support the interpretation of high resolution reflection seismic data. Sonic and bulk density logs can be used to produce a detailed log of the acoustic impedance structure in the subsurface. These data are used to create synthetic seismograms of the reflection seismic response of the

subsurface by convolution with theoretical seismic wavelets that match the dominant frequency content of the seismic data (Sartorelli et al., 1986). By this process major features imaged in the reflection seismic data can be assigned definite geologic significance. Conversely, a well-interpreted reflection seismic section basically makes available a sonic log at every trace location along the seismic survey lines.

CONCLUSIONS

1) The primary conclusion is that surface geophysical techniques, particularly seismic and direct current sounding and profiling, have proven to be successful for locating subcrop, buried channels, deformed bedrock, and for determining stratigraphic continuity in the 10-70-m (33-230-ft) depth range.

   This project also demonstrated the cost-effectiveness of these techniques (Green et al., 1988; Henderson and Nikols, 1988). The cost effectiveness increases with target depth because drilling costs increase substantially more with depth than geophysical costs.

2) Surface geophysical techniques are most effective when integrated with a drilling program. Some pre-existing subsurface information, particularly downhole geophysical logs, is useful in selecting the most appropriate geophysical techniques. After the surface geophysical survey has been completed, i) the interpreted data can be used as guides for locating future test holes, and ii) the data from these test holes can be used to refine the interpretation of the surface geophysical data.

3) The costs of performing geophysical exploration, for the above shallow (less than 70 m (230 ft)) targets decreased during this research project due to experience gained in data acquisition and data processing. Continued development likely will decrease costs even more.

RECOMMENDATIONS

1) Surface geophysics should be adopted as one of the principal tools for plains coal exploration and development.

2) Reflection seismic and direct current sounding methods should be considered first for the intermediate coal exploration stage. At the mine planning stage, methods selection will depend more on the specific target type or problem to be solved.

3) Selection of the appropriate geophysical techniques should take into consideration i) the operating environment, ii) the geologic framework, and iii) a firm statement of specific information needed, including the depth of interest and the lateral and vertical resolution desired.

ACKNOWLEDGMENTS

This was a cooperative venture involving members of the coal industry and government. Half of the funds were provided by an industry joint-venture group consisting of Alberta Research Council, Esso Resources Canada Ltd., Fording Coal Ltd., Luscar Ltd., and Saskatchewan Power Corp., under the leadership of TransAlta Utilities Corp., and half came from the Alberta/Canada Energy Resources Research Fund, administered by the Alberta Office of Coal Research and Technology. The research was carried out by a team assembled from the Coal Mining Research Company (which served as the operations manager),

Geo-Physi-Con Co. Ltd., (geophysics), and the Terrain Sciences Department of the Alberta Research Council (geology). Thanks also to the two EMD-AAPG reviewers; their comments were appreciated.

REFERENCES

Fenton, M.M., and J.G. Pawlowicz, 1989, Glacially thrust bedrock: Styles and ramifications for coal mining, in W. Langenberg, comp., Advances in Western Canadian coal geoscience--Forum proceedings: Alberta Research Council Information Series No. 103, p. 359.

Fenton, M.M., M.R. Trudell, J.G. Pawlowicz, C.E. Jones, S.R Moran, and D.J. Nikols, 1986, Glaciotectonic deformation and geotechnical stability in open pit coal mining, in R.K. Singhal, ed., Geotechnical stability in surface mining: Rotterdam, A.A. Balkema, p. 225-234.

Geo-Physi-Con Co., Ltd., 1981, Seismic cross-hole shear wave measurements: Calgary, Internal Technical Note, 10 p.

Geo-Physi-Con Co., Ltd., 1985a, Surface geophysical project 1984, combined electrical and seismic surface geophysical test surveys, Highvale Mine, Pit 02, and Paintearth Mine: Calgary, Unpublished Report #C84-25, Prepared for Coal Mining Research Co., February, 50 p.

Geo-Physi-Con Co., Ltd., 1985b, Surface geophysical project 1984, shallow reflection seismic studies, Highvale Mine and Paintearth Mine: Calgary, Unpublished Report #C84-28, Prepared for Coal Mining Research Co., February, 50 p.

Geo-Physi-Con Co., Ltd., 1986a, Surface geophysical project 1985, combined refraction seismic and electrical surface geophysical study, Genesee Site: Calgary, Unpublished Report #C85-24, Prepared for Coal Mining Research Co., March, 31 p.

Geo-Physi-Con Co., Ltd., 1986b, Surface geophysical project 1985, shallow reflection seismic study, Genesee Site: Calgary, Unpublished Report #C85-24, Prepared for Coal Mining Research Co., March, 25 p.

Geo-Physi-Con Co., Ltd., 1987, Surface geophysical joint venture (1986), Genesee Mine: Calgary, Unpublished Report #C86-17, Prepared for Alberta Research Council, January, 40 p.

Green, D.H., J.G. Pawlowicz, M.M. Fenton, C. Sterenberg, T. Sartorelli, J. Henderson, and M. Pesowski, 1988, Surface geophysical coal research project, 1984-86, Final Report: Alberta Office of Coal Research and Technology, Report AOCRT# 2835-SP-84/3, and Devon, Alberta, Coal Mining Research Co., Report #MI-8605, 148 p.

Henderson, J.D., and D.J. Nikols, 1988, Cost effectiveness of surface geophysics: Alberta Research Council, Contribution Series 1868, 20 p.

Moell, C.E., J.G. Pawlowicz, and M.M. Fenton, 1984, Surface geophysics for coal exploration, preliminary geology report: Alberta Research Council, Terrain Sciences Department, Unpublished Report, Prepared for Coal Mining Research Co., 17 p.

Moell, C.E., J.G. Pawlowicz, M.M. Fenton, and S.R. Moran, 1985a, Surface geophysics geology report, February 1985, Highvale Mine Pit 02 and Paintearth Mine: Alberta Research Council, Terrain Sciences Department, Unpublished Report, Prepared for Coal Mining Research Co., 29 p.

Moell, C.E., J.G. Pawlowicz, M.R. Trudell, M.M. Fenton, C.W. Langenberg, G.J. Sterenberg, and C.E. Jones, 1985b, Highwall stability project, Highvale Mine, report of 1984 activities: Terrain Sciences Department, Alberta Research Council, Unpublished Report Prepared for TransAlta Utilities Corporation, 256 p.

Pawlowicz, J.G., M.M. Fenton, and C.E. Jones, 1986b, Surface geophysics project, preliminary geology report March, 1986, Paintearth Mine area, Draft 1: Alberta Research Council, Terrain Sciences Department, Unpublished Report, Prepared for Coal Mining Research Co., 11 p.

Pawlowicz, J.G., M.M. Fenton, C.E. Jones, D.E. Jardine, and C.E. Moell, 1986c, Surface geophysics project, final geology report 1985-86 program, Genesee Mine and Paintearth Mine areas: Alberta Research Council, Terrain Sciences Department, Unpublished Report, Prepared for Coal Mining Research Co., April, 79 p.

Pawlowicz, J.G., M.M. Fenton, and C.E. Sterenberg, 1986d, Surface geophysics project, geology report 1986-87 program, Genesee Mine Area: Alberta Research Council, Terrain Sciences Department, Unpublished Report, Prepared for Coal Mining Research Co., December, 37 p.

Pawlowicz, J.G., D.E. Jardine, M.M. Fenton, C.E. Jones, and C.E. Moell, 1986a, Surface geophysics project, preliminary geology report, February 1986, Genesee Mine area, Draft 1: Alberta Research Council, Terrain Sciences Department, Unpublished Report, Prepared for Coal Mining Research Co., 41 p.

Redpath, B.B., 1973, Seismic refraction exploration for engineering site investigations: Springfield, Virginia, National Technical Information Service, 51 p.

Rogan, M.N., C.E. Moell, J.G. Pawlowicz, and M.M. Fenton, 1985, Surface geophysics study geology report, Paintearth Mine: Luscar Ltd. and Alberta Research Council, Unpublished Report, Prepared for Coal Mining Research Co., 23 p.

Sartorelli, A.N., J.D. Henderson and D.C. Lawton, 1986, Development of high resolution reflection seismic techniques for shallow depth in Alberta: Calgary, Geo-Physi-Con Co., Ltd., Unpublished Presentation at the 1986 National Convention of the Canadian Society of Exploration Geophysicists, Calgary, Alberta, 19 p.

Telford, W.M., L.P. Geldart, R.E. Sheriff, and D.A. Keys, 1978, Applied geophysics: Cambridge, Cambridge University Press, 860 p.

FORMATION OF CLINKER AND ITS EFFECTS ON LOCATING AND LIMITING COAL RESOURCES

Joseph C. Sarnecki
420 Hackberry Lane
P. O. Box 0
Geological Survey of Alabama
Tuscaloosa, AL 35486-9780

ABSTRACT

Coal burns occur in nature primarily due to spontaneous combustion and the result is baked overburden known as clinker. Understanding occurrences of natural coal burns and formation of clinker is important for coal resource exploration and development. Clinker is an indication of potential commercial coal reserves, and can be located easily due to its difference from the unaltered coal section, especially color, geochemical properties, and aquifer properties. The high porosity and variable material strength of clinker create impacts for aspects of mine development such as foundation planning, slope stability, and water handling.

INTRODUCTION

Behind every clinker deposit is, or was, a thick coal. By its nature, clinker, the resulting rock sequence of natural coal burns, can be used to indicate minable coal. Therefore, understanding more about the formation of clinker from natural coal burns not only gives the coal exploration geologist information on locating new commercial coal but it also gives the mining engineer information crucial for developing the coal resource efficiently. Clinker's effect on surface-mine development includes initial pit, water resource, and final reclamation planning.

FORMATION

Terms like "scoria", "scorio", "shale", and "baked and fused bedrock", have been applied to the lithologic result of natural coal burns, but the term "clinker" is in common geologic use and comes closest to the true meaning of this rock type (Coates and Naeser, 1984; Sarnecki, 1981). Clinker may be thought of as a rock sequence because the overburden from which it originates is a mix of different rock types, and heat from the burning coal creates several new rock types.

Field studies in western North America have shown that natural coal burns usually occur in beds thicker than 1.5 m (5 ft) (Mapel et al., 1975, p. 71). Natural coal burns have been attributed to lightning, prairie fires, and man-made fires, but the most significant cause is spontaneous combustion.

Spontaneous combustion has been observed in different ranks of coal in widespread areas. In western surface coal mines, volatile, low-rank subbituminous coal slakes or breaks down rapidly because of moisture loss in arid conditions and typically is susceptible to natural burning. The speed of spontaneous combustion is related to gas desorption or release, which is natural to the coalification process, and to self-heating tendency (Kuchta et al., 1980). The self-heating temperature in coal is the minimum temperature that starts producing a sustained exothermic reaction (Smith, 1989).

Coal which is most susceptible to self heating is characterized by high intrinsic moisture and oxygen content, as found in low-rank coal such as subbituminous coal and lignite. The heat evolved in coal capable of self heating is up to 2.5 times greater in moist air than in dry air. This moisture-attendant reactivity leads to a catalyst-like condition known as the "heat of wetting" which can be greater than the heat of oxidation (Kuchta et al., 1980, p. 4). Free moisture in coal can be responsible for this heat-of-wetting reaction. Many major western coal beds are aquifers, and the water level within the beds fluctuates seasonally, thereby increasing the heat-of-wetting potential. When water levels drop in these unconfined coal aquifers, oxidation increases and the self-heating of coal accelerates until smoldering or flame combustion occurs. The minimum self-heating temperature above which a coal mass can produce exothermic reaction under adiabatic conditions varies with rank. Self-heating temperatures for lignite and subbituminous coal are as low as 30°C, whereas they are higher for bituminous coal (as low as 60°C) (Kuchta et al., 1980, p. 28).

In coal resource development (mining), oxidation goes on continually. As such, spontaneous combustion may occur whenever the rate of release of energy as heat exceeds the rate of dissipation of the heat released and when the temperature of the coal increases to the level at which active combustion occurs (Duzy and Land, 1985, p. 139). During combustion, oxygen is depleted to produce carbon-oxygen complexes, and carbon monoxide and carbon dioxide are liberated along with methane as gases of complete and incomplete combustion (Figure 1a). Other complexes such as carbon disulfide and carbon oxysulfide have been detected from some active coal burns (Dunrud and Osterwald, 1980, p. 31), and sulfur compounds may crystalize from the hot gases at the edge of burn vents.

When sufficient natural combustion occurs, the consumed coal leaves a white ash (Figure 1a), which rarely is found in a clinker section because it is soft, not resistant to erosion, and typically buried beneath the baked overburden. The rocks underlying the ash generally are unaltered. With steady retreat of coal consumed during natural combustion, a fire chamber or void is produced. Continued burning within this void produces blast furnace-like circulation due to heat differences.

Sufficient heat changes the overlying bedrock in a zone of alteration as shown in Figure 1. The clinker section rock types are variable. Fine-grained rocks such as siltstone and shale become altered to a dense, glassy brick-like rock called buchite (Fullerton, 1977) or, perhaps more rightly, porcellainite (Coates, written communication, 1990) particularly where these rocks are adjacent to the fire chamber. Coarse-grained rocks like sandstone and rocks distant from the fire chamber are bake hardened and change color in these oxidizing conditions.

Color of the clinker ranges from light yellow to deep red and depends on the oxidation state of the iron (Cosca et al., 1989). The more welded or glassy portions may be black. Pyrite and siderite found in shale and associated strata near coal were deposited in reducing environments and contain iron in the reduced ferrous (Fe^{2+}) state. Iron is changed during combustion to the oxidized ferric (Fe^{3+}) state, indicated by the reddish colors.

During natural combustion of coal, the altered overburden eventually collapses into the resulting void (Figure 1b). In some places, the collapse smothers the fire chamber and combustion ceases until erosion exposes new coal outcrops. In other places, large collapses set up a joint or vent system at the surface, and fresh air

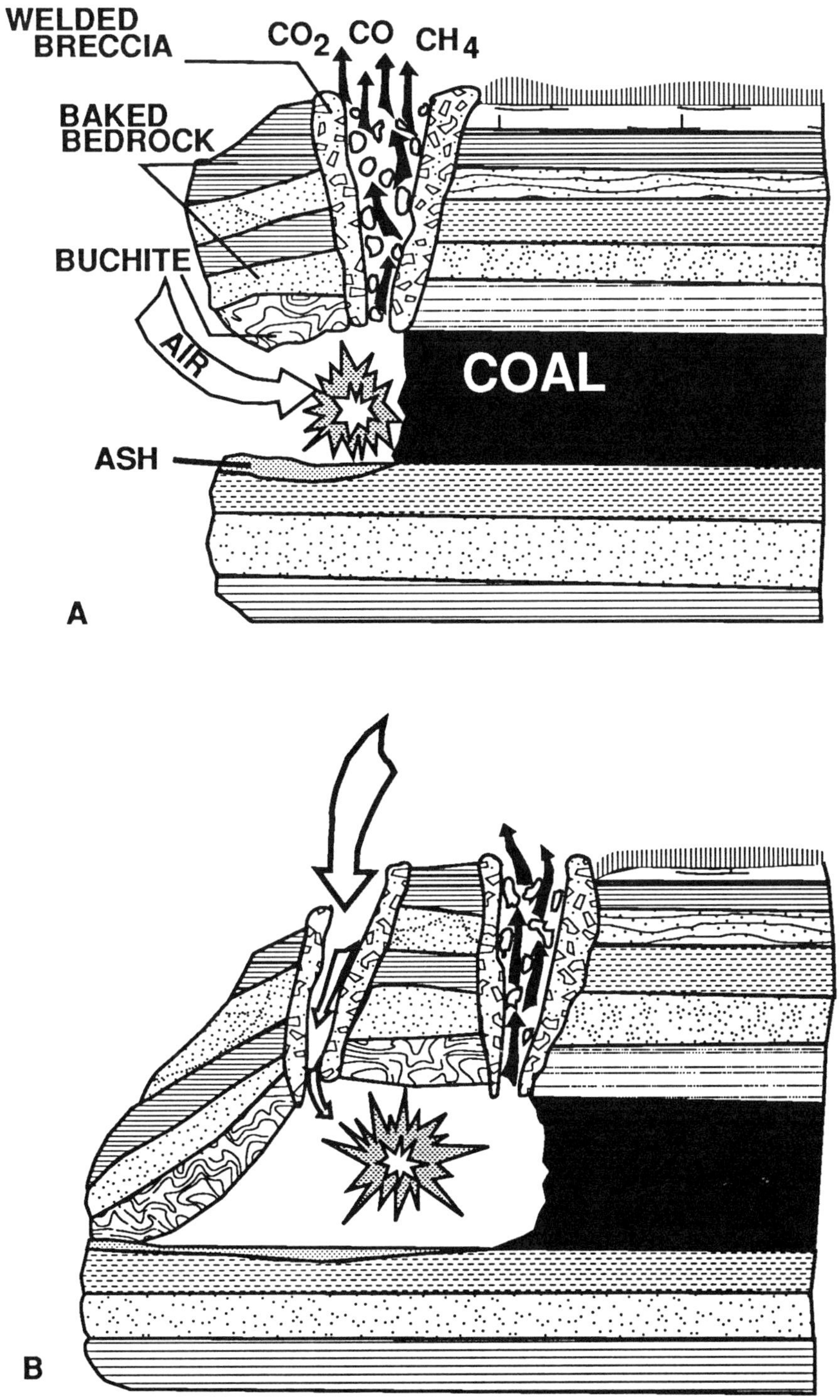

Figure 1. *Clinker formation sequence: A) Initial clinker formation due to natural combustion at the coal outcrop showing resultant rock types, and B) post-collapse burn continuation set up by circulation of air to fire chamber through vents. Fresh-air flow is shown by white arrows, and exhaust-gas flow is shown by black arrows in exhaust vents. Zone of alteration occurs to the left of, and includes, the exhaust vent.*

is brought in through these vents. A series of vents may set up a circulation system with air intake at some openings and exhaust at others. The cycle of natural coal burn propagation and smothering has been occurring on and off in the Powder River Basin over the last million years (Coates and Naeser, 1981, 1984). This cycle is seen in the creation of a clinker basin where erosion removes part of a coal seam behind a cooled clinker and re-exposes it to natural combustion (Coates, 1980, p. 80; Heffern, 1980, p. 40).

Combustion may take place in the fire chamber immediately adjacent to the coal, or gases released from the coal may ignite within the vent system. Combustion in the exhaust vents creates enough heat to melt the surrounding overburden and other material which subsequently breaks up and falls into the vent. Molten material in these vents cements any broken particles to form welded breccia (Figure 1). Combustion along these vents and joints can weld long columns of rock into black pipes or chimney structures which are even more erosion resistant than the rest of the bake-hardened overburden.

USE OF CLINKER FOR COAL DEPOSIT LOCATION

The discussion on clinker formation indicated several key attributes of clinker of prime interest to the coal explorationist for the location of large tracts of coal with remote sensing tools such as aerial photographs and magnetometers. These key attributes are marked color change of the clinker relative to the unburned section, erosion resistance of the clinker, and its oxidized chemical state change.

Because the results of natural coal burns are thick, reddish to black rock masses, these clinker features are easy to identify in aerial color photographs, especially in the treeless plains of western North America. Plotting these color changes on topographic maps also gives data on regional dip and the clinker zone's areal extent.

Because the clinker is bake hardened, it is tough, brittle, fractured, and often very porous. This highly porous and fractured nature of clinker gives it hydrologic transmissivities many times greater than the associated, unaltered coal overburden (Davis, 1976, p. 204). Therefore, even where vegetated, clinker's water-bearing capacity may show profound cover differences over adjacent non-burned tracts. For example, burns of the Anderson coal seam in the western Powder River Basin form thick clinker units which, because of their water-bearing capacity, support dense stands of pines adjacent to otherwise sparsely vegetated tracts.

Detailed field mapping of clinker areas verify the areal extent of the burn and establish the structural base of the former coal seam. For example, clinker outcrops are often well preserved in the Powder River Basin because they form erosion-resistant buttes or ridges with steep updip slopes overlying the softer, unaltered coal floor rock. Measurement of this type of outcrop section can disclose clinker thickness, lithologic variabilities and strengths, and aquifer behavior. Reconnaissance field mapping also can be useful in establishing hydrologic baseline data such as spring locations, recharge and discharge points, and surface aquifer water-level changes because these features are very evident in porous, surface material.

Where the coal overburden is baked to create clinker, it becomes significantly more magnetic (Miller, 1979, p. 27). This magnetic property aids initially in locating clinker/coal through aerial magnetic surveys, but is even more important when using surface magnetometer studies. Location of the back of the clinker/edge of coal burn and general thickness trends in the clinker can be determined by analyzing surface magnetometer profiles, especially if drill-hole data are incorporated within the magnetometer sections. In seismic studies, Miller (1979, p. 5) also found that compressional wave velocities were noticeably less for clinker than for unaltered, overburden. Thus, seismic profiles may serve as additional tools to delimit clinker.

EFFECT OF CLINKER ON COAL QUALITY

Some of the changes that occur in coal quality in response to nearby combustion are as follows: 1) loss of volatile matter and carbon content, 2) increase in moisture, especially with weathering, and 3) decrease in calorific value in the altered coal. Even where burned, however, an underlying, unburned portion of the coal bed may still be commercially valuable. The Kerr McGee Coal Company has removed clinker in the Powder River Basin, Wyoming, to develop the unburned part of the Wyodak coal in its Jacob's Ranch Mine (Coates and Naeser, 1984).

The effect of nearby combustion on the unburned coal varies with weathering and coal aquifer behavior. In time, weathering makes coal quality changes more profound, especially along the clinker outcrop where the coal subcrop is exposed to oxidation due to frequent water-level changes. In the Acme area, north of Sheridan, Wyoming, coal burns occur seasonally in the old underground mine workings. Spontaneous combustion occurs during the low water-level periods and combustion ceases as the water level rises during the wet periods. Exposure to oxidation can be reduced by maintaining high water levels in the coal aquifer. Therefore, if water levels are maintained in the coal until it is mined, clinker formation potential is reduced and the original quality of the coal is preserved.

Because water is stored readily in the clinker, usually under unconfined aquifer conditions, some exploration effort should be devoted to water level monitoring in clinker wells. Where weathering potential due to variable water levels is recognized in either the clinker or the coal aquifers, the subcropping coal should be cored more densely, and sampled and analyzed more than the deeper part of the coal deposit which probably exists under confined aquifer conditions. The usefulness of closely spaced coring may be extended near the coal/clinker transition when coupled with surface magnetometer profiles located adjacent to geophysically logged core holes. Magnetometer surveys are cheaper than detailed coring, so coal quality data from cored areas may be extended to uncored areas with magnetic profile data to help the mine planner target future development zones.

EFFECT OF CLINKER ON COAL RESOURCE DEVELOPMENT

To address clinker impact on surface mine development, mine engineers should become aware of clinker's variable material strengths, water storage in clinker, and associated slope stability problems. Material strength variabilities should be understood for any facilities foundation planning. Clinker does not provide good foundations for plant structures, shops, or even paved roadways because its strength is so variable. Clinker sections even may swell and heave differentially during the winter from variable freezing and thawing. The cored and sampled clinker should be analyzed by standard engineering tests and these analyses compared to the geophysical log data and magnetic and seismic surveys to extend test results from cored areas to untested areas where the cheaper surveys have been run. Material strength data also can help in pit highwall and spoil-pile slope stability planning for determining planes of weakness. Strength tests also provide information on the usefulness of the clinker as road-base material in haul road construction.

Pumping tests of selected, aquifer water-monitoring wells are necessary to determine optimal dewatering rates to prevent pit flooding and ensure stable equipment movement. These tests also can determine communication between aquifers and the movement of water during various mine operations and can be used to plan aquifer reconstruction during reclamation. Highwall slopes also are more stable where excess water is removed in advance of pit opening. The water removed from the clinker aquifer has applications for local water use such as dust control.

USE OF CLINKER IN SURFACE MINE RECLAMATION

Aquifer restoration is one of the main focuses of the federal reclamation plan and includes restoring near-surface, alluvial aquifers as well as the major coal-measure aquifers. Placement of porous clinker within reconstructed alluvial runs and main recharge zones of reconstructed spoil aquifers insures rapid return to steady state conditions and may reduce total dissolved solids. Clinker may be more resistant to dissolution than other, unaltered spoil.

Overburden handled during mining operations may be changed from the reduced condition before mining to an oxidized state during removal and spoil operations. The oxidation liberates toxic elements, such as selenium, and makes those elements available to the ground water or surface water or for plant uptake. Deep burial of the toxic spoil is one suggested remedy with only short-term repercussions from ground water interacting with the oxidized spoil. Return of the ground-water level to a steady state also returns the saturated zone to reducing chemical conditions. In the reduced state, unsuitable elements such as selenium are relatively immobile to dissolution or plant uptake. One method of insuring recharge to deep-burial spoil sites is to use clinker spoil as reclamation material near recharge zones for these sites.

SUMMARY

Spontaneous combustion is the main cause of natural coal burns and, therefore, is responsible for major clinker deposits such as those formed over the last million years in the Northern Great Plains of North America. The conditions conducive to clinker formation include the heat of wetting associated with frequent water level changes in the coal aquifers. By understanding the avenues of subsurface oxidation, the coal exploration geologist and the mining engineer can deal with clinker to find commercial coal, map its extent, and plan efficient coal-deposit development. This paper presents a general discussion of clinker formation and clinker's effect on coal resource development. More significant work requires site specific studies because coal burns and clinker are related to the variability of the area geology. No master deposit model for clinker stratigraphy or handling can be formulated without more detailed site studies.

ACKNOWLEDGEMENTS

I would like to thank Donald A. Coates, Mark F. Middleton, Jack C. Pashin, Janyth S. Tolson, and Margaret M. Walsh for technical discussions and manuscript assistance.

REFERENCES

Coates, D.A., 1980, Formation of clinker by natural burning of coal beds in the Powder River Basin, Wyoming, in L.M. Carter, ed., Proceedings of the Fourth Symposium on the Geology of Rocky Mountain Coal, Golden, Colorado, April 28-29, 1980: Colorado Geological Survey, Resource Series 10, p. 37-40.

Coates, D.A., and C.W. Naeser, 1981, Fission-track ages and landscape development, Eastern Powder River Basin, Wyoming: Geological Society of America, Abstracts with Programs, v. 3, no. 7, p. 428.

Coates, D.A., and C.W. Naeser, 1984, Map showing fission-track ages of clinker in the Rochelle Hills, Southern Campbell and Weston Counties, Wyoming: U.S. Geological Survey Miscellaneous Investigations Series Map I-1462, 1 sheet.

Coŝca, M.A., E.J. Essene, J.W. Geissman, W.B. Simmons, and D.A. Coates, 1989, Pyrometamorphic rocks associated with naturally burned coal beds, Powder River Basin, Wyoming: American Mineralogist, v. 74, p. 85-100.

Davis, R.W., 1976, Hydrologic factors related to coal development in the Eastern Powder River Basin, in R.B. Laudon, ed., Geology and energy resources of the Powder River Basin, 28th Annual Field Conference: Wyoming Geological Association, Guidebook, p. 203-207.

Dunrud, C.R., and F.W. Osterwald, 1980, Effects of coal mine subsidence in the Sheridan, Wyoming area: U.S. Geological Survey Professional Paper 1164, 49 p.

Duzy, A.F., and G.W. Land, 1985, Hot coal--Bulk transport and storage: Mining Engineering, v. 37, no. 2, p. 139-144.

Fullerton, D.S., 1977, Surficial geologic map of the Corral Creek Quadrangle, Campbell County, Wyoming: U.S. Geological Survey Miscellaneous Field Studies Map MF-919, 1 sheet.

Heffern, E.L., 1980, Coal stratigraphy of the Tongue River Member, Northern Cheyenne Reservation, Montana, in L.M. Carter, ed., Proceedings of the Fourth Symposium on the Geology of Rocky Mountain Coal, Golden, Colorado, April 28-29, 1980: Colorado Geological Survey, Resource Series 10, p. 76-80.

Kuchta, J.M., V.R. Rowe, and D.S. Burgess, 1980, Spontaneous combustion susceptibility of U.S. coals: U.S. Bureau of Mines Report of Investigations RI-8474, 37 p.

Mapel, W.J., B.F. Bohor, J.C. Sarnecki, E.A. Magill, Michael Sokaski, and George McIntyre, 1975, Preliminary mineral resources study of the Northern Cheyenne Indian Reservation, Montana: U.S. Bureau of Indian Affairs Administrative Report BIA-3, 79 p.

Miller, C.H., 1979, Seismic, magnetic, and geotechnical properties of a landslide and clinker deposits, Powder River Basin, Wyoming and Montana: U. S. Geological Survey Open-file Report 79-952, 42 p.

Sarnecki, J. C., 1981, The formation of clinker in western coal regions: in Proceedings of the 91st Annual Meeting, Lincoln, Nebraska, April 10-11, 1981: Lincoln, Nebraska Academy of Sciences, p. 34.

Smith, A.C., 1989, Bureau develops spontaneous combustion formula: Coal, v. 26, no. 7, p. 73.

GLACIALLY DEFORMED BEDROCK: IMPLICATIONS FOR COAL EXPLORATION, MINING, AND HIGHWALL STABILITY

Mark M. Fenton and John G. Pawlowicz
Alberta Research Council
Terrain Sciences Department
Box 8330, Station F
Edmonton, Alberta, T6H 5X2
Canada

ABSTRACT

Observations at various Plains coal mines in Alberta have shown that the ramifications of glacially deformed bedrock include 1) loss of recoverable coal, 2) increased ash content, 3) increased uncertainty in reserve estimates, 4) increased variability in pedogenic and geotechnical soil types, 5) increased potential for highwall failure, 6) increased access problems, 7) decreased core recovery, and 8) decreased permeability and drainage potential. One advantage is the disruption of hard layers, making excavation easier. This paper presents two examples to illustrate ramifications 1, 3, 5, and 8.

Glacially deformed bedrock can be detected by airphoto interpretation, pitwall and/or outcrop mapping, coring, downhole geophysical logging, and surface geophysics. Foreknowledge about the location, structure, dimension, and hydrology of glacially deformed bedrock is important to both exploration and mine planning.

INTRODUCTION

During the past seven years, research programs have been undertaken to provide information to the Plains coal industry in Alberta on the implications of glacially deformed bedrock and methods for detection of this material. Foreknowledge about the location, structure, dimension and hydrology of glacially deformed bedrock is important to both exploration and mine planning.

Continental glaciers advanced a number of times during the Pleistocene (Fenton, 1984b). During each advance, the weight and motion of the glacier resulted in local deformation of the underlying material. The products of this deformation are 1) masses of bedrock and/or older glacial and stratified sediment that have been deformed and/or transported more or less intact by the glaciers, and 2) where recognizable, an associated source depression situated up the direction of glacier flow from the deformed material.

Glacially deformed bedrock and sediment have been recognized in many of the glaciated regions of North America and Europe (Aber et al., 1989; Moran et al., 1980), including a considerable portion of the Great Plains (Figure 1) (Fenton, 1984b). Within the Plains, some of the earliest descriptions of glaciotectonism were by Sardeson (1898, 1905, 1906) who interpreted several large anomalous outliers of rock of Cretaceous age in Minnesota to be blocks transported by glacier ice. Later investigations include those of Hopkins (1923), Slater (1927), Byers (1960), Kupsch (1962), Moran (1971), Christiansen and Whitaker (1976), Fenton and

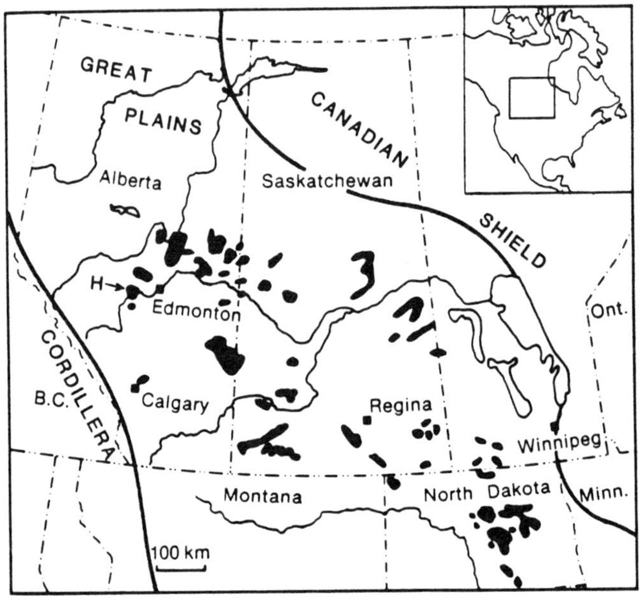

***Figure 1.** Map showing the distribution of glacially deformed bedrock and sediment on the Interior Plains and the location of the Highvale Mine study area (H). Modified from Moran et al. (1980).*

Andriashek (1978), Moran et al. (1980), Fenton (1983a, 1983b, 1984a), Bluemle and Clayton (1984), Fenton et al. (1985a, 1985b), Aber et al. (1989), and Andriashek and Fenton (1989).

The significance of glacial deformation to geotechnical engineering in general and more specifically to mining has been recognized (e.g., Fenton et al., 1986; Sauer, 1978). Some of the best exposures of such deformed bedrock are in open-pit coal mines.

Our observations at various Plains coal mines have shown that the principal deleterious results of glacial deformation are 1) the disruption of both the overburden and coal stratigraphy typified by folding, faulting, and/or displacement, together with fracturing and crushing of the previously flat-lying beds, 2) the removal or repetition of marker beds, and 3) a reduction in the rock strength. Ramifications and/or problems for both coal exploration and mining include

1) Coal recovery: Glacially transported blocks of coal may not be economically recoverable.

2) Coal quality: An increase in the ash content of the coal delivered to the power plant as a result of a) fractures, opened by deformation, allowing oxidation and/or the movement of clastic material into the coal, and b) the inclusion with the mined coal of thin partings which no longer can be removed during the mining process because the coal seams and partings are folded rather than flat-lying.

3) Reserve estimation: Glacial removal of the coal masses increases the complexity of the subcrop pattern which can increase the uncertainty in reserve estimates.

4) Sediment variability: Higher sediment variability, both vertically and horizontally, resulting in higher variability in pedogenic and geotechnical soil types.

5) Highwall stability: Deformed bedrock is weaker than undeformed bedrock, and shear planes both at the base and within the glacial thrust mass may dip into the pit depending on pit orientation. All these conditions increase the potential for highwall failure.

6) Core recovery: Glacial deformation involving crushing without compaction can result in material that is comparatively loose and difficult to contain within a core barrel.

7) Drainage: Pathways present in the undeformed bedrock may be disrupted and/or sealed during deformation.

8) Access problems: Glaciotectonic terrain can be comparatively rugged, making access more difficult (some thrust hills are 200 m (660 ft) high).

One advantage of this deformation is the disruption of hard layers in the overburden. This makes excavation by a dragline easier and more efficient, and reduces the need to blast.

Space limitations restrict this paper to two examples of the above: 1) the effect of glaciotectonism on coal zone continuity, and 2) highwall failures resulting from mining into glacially deformed bedrock. Methods useful for the detection of the deformed bedrock then are discussed briefly.

COAL ZONE CONTINUITY

Two of the characteristic features of glacial deformation are a source depression and adjacent to this, in the direction of glacial flow, a hill formed by the deposition of the mass of glacially deformed material. The Wolf Lake glacial thrust mass, in east-central Alberta, is a typical example of this type of feature (Andriashek and Fenton, 1989). Deformation at this site was caused by a southward flowing glacier. The thrust mass consists of a hill composed of a series of arcuate ridges. The hill is asymmetric, with the steeper face on the south. The hill covers about 42 km^2 (16 $mi.^2$), is about 140 m (460 ft) high, and contains about 6 km^3 (1.4 $mi.^3$) of sediment. Immediately north of the hill is a water-filled depression called Wolf Lake. Fault scarps form the east and west margins of the lake and are parallel to the glacier flow direction. A steep scarp about 10 m (33 ft) high forms the northern side of the lake. The lake varies from about 5 to 25 m (16 to 82 ft) deep and its total volume is less than that of the thrust mass, likely due to partial filling of the lake by sediment subsequent to its excavation by the glacier. The undeformed terrain surrounding the hill and depression is essentially flat.

Glacial thrust hills and associated source depressions range in size from those covering a few hundred square meters to a few exceptionally large examples in which the hill alone may cover 1,000 km^2 (400 $mi.^2$) and reach 300 m (1,000 ft) high. However, at other sites the masses of glacially thrust material either have been truncated and smoothed over by later glacial flow or buried by the debris left by the later flow. Similarily, the source depression may have been filled in by later glacial or nonglacial material. At such sites, there is no geomorphic evidence of presence of either the glacial thrust mass or the source depression.

The source depression is one aspect of glacial deformation of the bedrock that can be overlooked during coal development, particularly when there is no geomorphic evidence for its existance. The assumption that the coal is continuous through a glacial thrust depression can yield overly optimistic estimates of the total tonnages during the exploration phase. The unexpected discovery of the absence of one or more seams during mining can disrupt the mining schedule, with potentially serious economic implications.

Local absence of one or more coal seams, the result of glacial deformation, has been observed in a number of the open pit coal mines in the Interior Plains of Canada. The volume removed may vary from a few cubic meters to more than 1 km^3 (0.2 mi^3). Figure 2 shows a map prepared from borehole and surface geophysical data of the coal subcrop lying below the drape of Quaternary sediments at one of these coal mines (Aber et al., 1989; Fenton and Pawlowicz, in preparation). This area includes a source depression, whose excavation extended deep enough into the bedrock to have removed all of the coal seams, and adjacent low hills situated in the direction of glacial flow from the depression. Geophysical logs from test holes have identified some of the transported coal and overburden within these hills (Figure 3). However, a significant portion of the excavated material has likely been transported much farther because the estimated volume of coal removed from the source depression, about 12,000,000 m^3 (420,000,000 ft^3), exceeds that forming the hill. The detailed coal subcrop pattern (not shown on Figure 2), reveals that within the area between the center of the source depression and the hill, the lowest coal seam subcrops farthest to the north and the highest seam subcrops farthest to the south, indicating that the basal shear zone produced by the glacier at the time of failure, was likely dipping northward at a low angle.

At this site the transported coal remaining within the hills is of no economic use because of the small volume of coal in each hill and the reduction in quality due to fracturing and oxidation. Economic recovery of glacially transported coal masses has been possible at some other sites, however.

Glacial transport of coal increases the uncertainty of reserve estimates made during coal exploration because it increases the complexity of the subcrop pattern. With test holes drilled only at A, B, C, and D the interpretation of the coal subcrop would be the dashed line on Figure 2. The removal of the coal could go undetected without additional drilling such as at holes E, F, and G. The use of a coordinated surface geophysics and drilling program has been found to be the most economical way to avoid such surprise discoveries between test holes (Sartorelli et al., 1986).

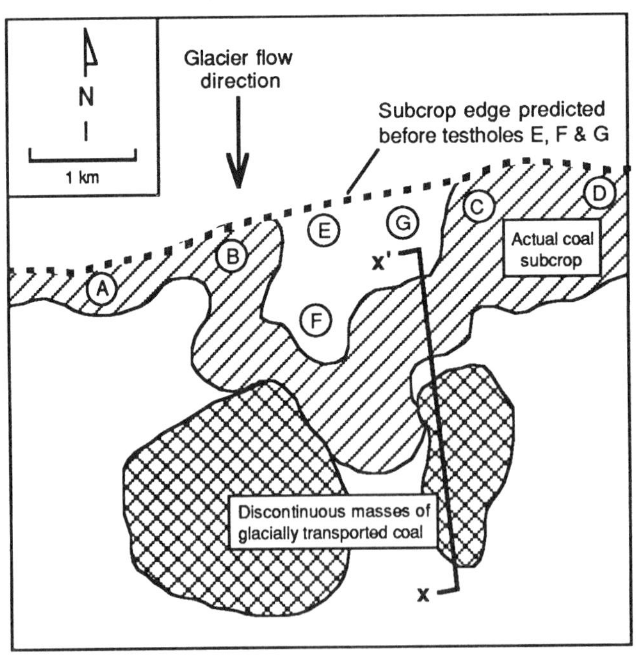

Figure 2. Subcrop map showing predicted and actual subcrop of the coal. Modified from Fenton and Pawlowicz (in preparation).

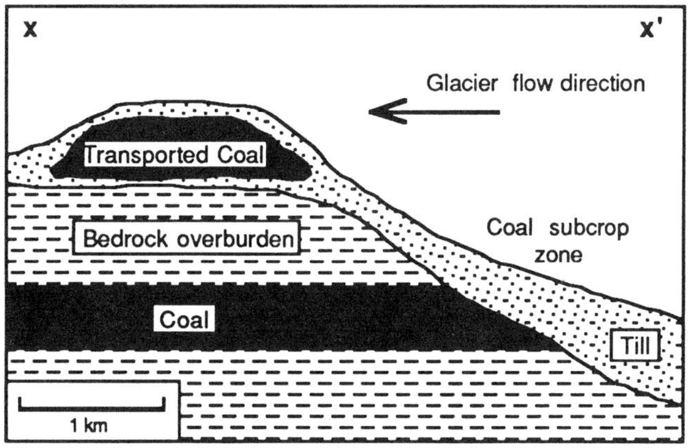

Figure 3. *Schematic cross section X-X', as indicated on Figure 2, showing both the south side of the source depression formed in the bedrock surface by glacial removal of the coal and overburden and the adjacent hills formed by deposition of some of the excavated material. Coal symbol represents the entire multiple seam sequence. Modified from Fenton and Pawlowicz (in preparation).*

HIGHWALL FAILURE

This section provides examples of how knowledge of deformed bedrock can 1) provide an understanding and model for an existing highwall failure, and 2) recognize and predict mine wall segments with high failure potential. The examples are from the Highvale Mine (Figure 1) which is owned by TransAlta Utilities Corporation and which in 1984 was producing 11,400,000 tonnes/yr (12,500,000 tons/yr) (Tapics, 1984). Information for these two sites is taken from Moell et al. (1985), some of which was published in Fenton et al. (1985a and 1986).

Coal is mined from the Ardley Coal Zone, which is Upper Cretaceous to Paleocene in age (Holter et al., 1975) and forms part of the Scollard Member of the Paskapoo Formation (Carrigy, 1970; Irish, 1970). The coal sequence is about 15 m (49 ft) thick and includes six coal seams with a cumulative coal thickness of about 10 m (33 ft). The seams dip southwestward at about 4 m/km (20 ft/mi.). More detailed information on the mine geology can be found in Lyons et al. (1984) and Nikols and Lyons (1985). Regional mapping of the surficial geology indicates that the glaciers advanced southwards, locally folding, faulting, crushing, and transporting masses of bedrock (Andriashek et al., 1979; Fenton 1983c).

Site One: Failure Analysis And Modeling

During 1983 and 1984, a series of highwall failures occurred in Pit 03 between Ramps 3-1 and 3-2 (Figures 4 and 5). Subsequent research focused on determining the conditions prior to and during failure and on identifying the failure mechanisms. Data were obtained through airphoto interpretation, outcrop study, surface geophysical methods, subsurface coring, downhole geophysical logging, and monitoring of a number of piezometer nets.

Geology

The general stratigraphic units are as follows, in descending order:

U1. Thrust bedrock (predominantly mudstone and sandstone).

Figure 4. Highwall at Site 1 before and after failure, 1983. View looking westward from Ramp 3-1. Note that the failure zone extends across the bench and into the adjacent bedrock.

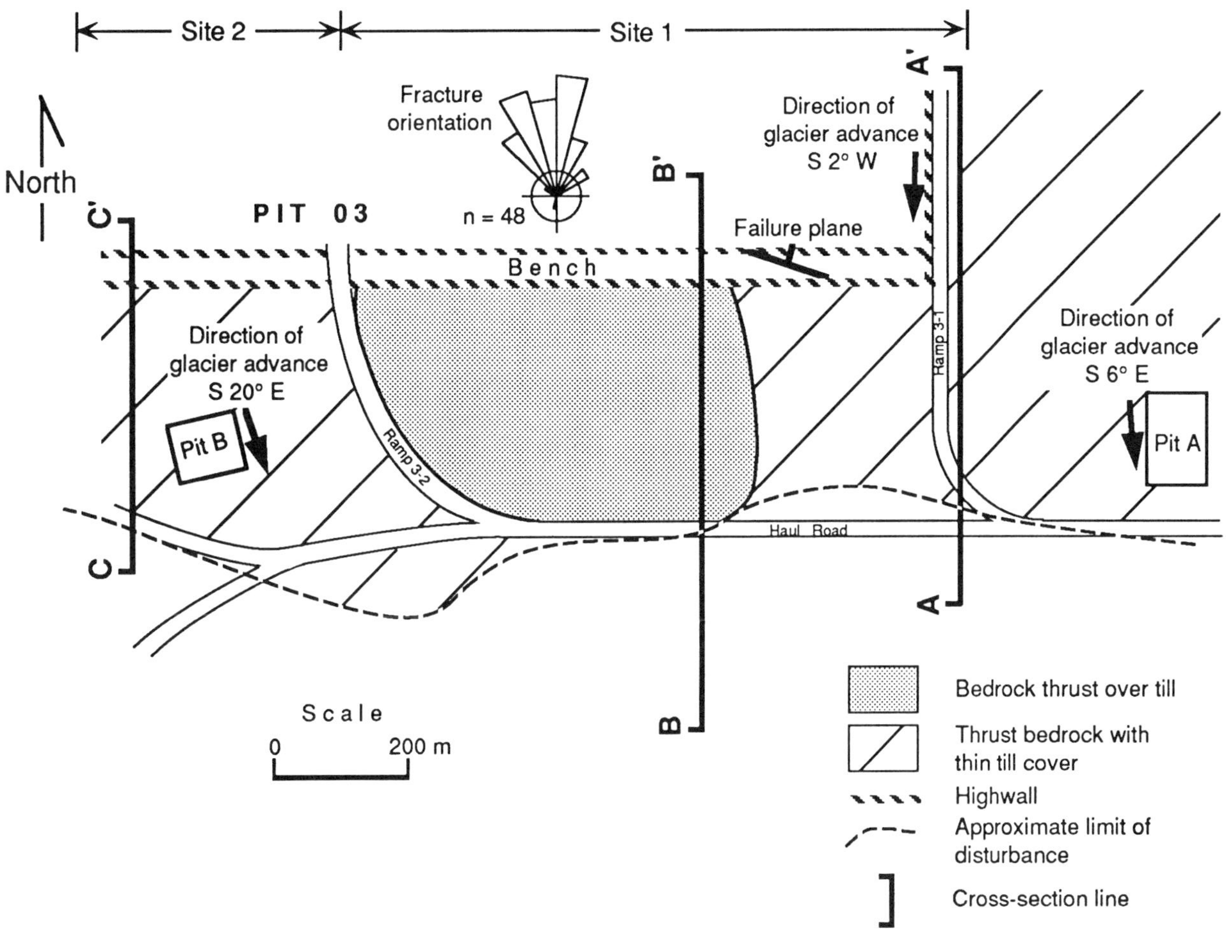

Figure 5. *Map of southeast part of Pit 03 showing areas of glacially deformed bedrock and the sites of geologic structural observations.*

U2. Glacial diamicton (till).

U3. Sandstone.

U4. Mudstone.

U5. Interlaminated mudstone, coal, and bentonite.

U6. Six-seam coal unit.

Portions of the sandstone, mudstone, interlaminated mudstone, coal, and bentonite (U3 to U5), and upper coal seams of U6 had been subjected to glaciotectonism (Figures 5 and 6).

Specifically, a mass of deformed and transported bedrock (U1) overlay the till (U2) in the central third of the area (Figure 5). The till, which both underlay this mass and outcroped beyond it, was grayish brown to dark grayish brown and contained abundant clasts and sheared lenses of bedrock material.

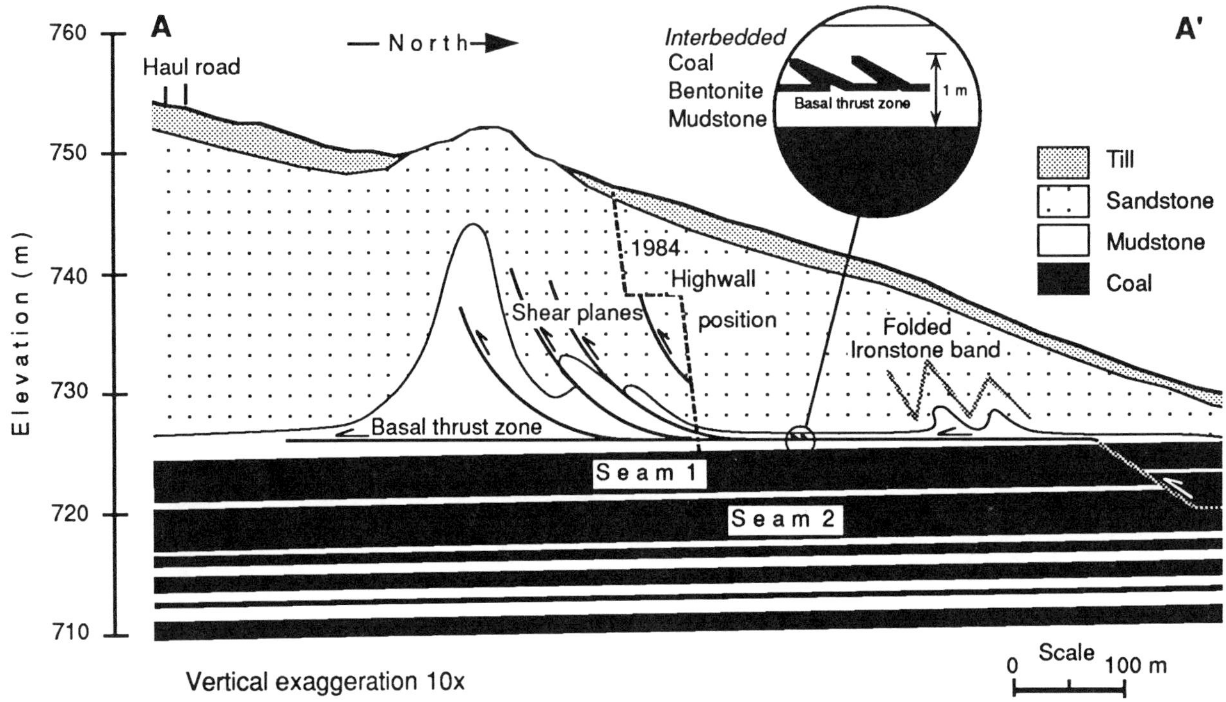

Figure 6. Schematic cross section for Site 1 showing the deformation along Ramp 3-1 and southward along line A-A'. Modified from Moell et al. (1985). See Figure 5 for the location of the section.

The sandstone (U3), underlying the till, was yellowish brown where oxidized, and bluish gray where unoxidized. The generally massive nature of this unit commonly prevented the recognition of small deformation structures; however, observations from the wall of Ramp 3-1 and nearby pits indicated the sandstone had been subjected to large-scale folding and faulting. The underlying mudstone (U4), was gray to dark gray, massive, and highly fractured, with many of the fractures having polished surfaces. The contact with the overlying sandstone was wavy.

The interlaminated mudstone, coal, and bentonite unit (U5) consisted of coal laminae, pale yellow bentonite, and gray mudstone. The contact with the underlying coal was slightly wavy. This unit was strongly deformed with the bentonite and mudstone resembling a melange in many places. The coal laminae near the top of the unit highlighted folds which were generally less than 1 m in height and extended upward into the mudstone. The folds were overturned and the axial plane dipped northward (Figure 6).

Large-scale folds and faults were exposed along Ramp 3-1 in Pit 03 and the smaller Pits A and B (Figures 5 and 6) (Nikols, 1985), allowing the acquisition of detailed data on the glaciotectonic structures (Fenton et al., 1986; Moell et al., 1985). The sites were selected along north-south oriented cuts to permit viewing of the geology in a direction parallel to the general direction of glacial advance and, therefore, parallel to the structural grain of the area. The directions of glacial advance, based on the orientation of major structures, ranged between S20°E and S2°W which was consistent with the direction of about S10°E obtained from the measurement of a number of smaller thrust planes exposed in the highwall (Figure 5) (Fenton et al., 1983). The highwall, although an extensive exposure, was oriented east-west and consequently did not allow the viewing of any of the large structures in a direction parallel to the direction of glacier flow. Almost all the glacially induced thrust planes dipped northward into the pit.

The base of the disturbed bedrock was bounded by a shear zone. This zone rose stratigraphically southward from within Seams 1 and 2 north of the haul road, to a few metres above Seam 1 near the haul road (Figure 6). The shear plane ended at some point south of the haul road as it was absent in those core holes drilled south of, but closest to, the haul road. Where the shear plane immediately overlay Seam 1, the movement had occurred along the 1-m (3-ft) thick layer of interlaminated mudstone, bentonite, and coal (U5) and was well illustrated by a number of small folds within this unit.

Groundwater Conditions

A number of piezometer nests were set during the summer of 1983 (Fenton et al., 1986; Moell et al., 1985). The glacially deformed bedrock was found to be almost completely saturated. Furthermore, the overlying till and underlying bedrock acted to inhibit drainage of this material.

Hydrogeologic cross section B-B' (Figure 7) shows that at site HV83-4 there was both 20 m (66 ft) of saturated overburden and evidence of only very limited drainage in both the overburden and Seam 6. The water level in the piezometer set in the overburden sandstone at this site declined by only 0.6 m (2 ft) between September 1983 and October 1984. Sites farther from the highwall (HV83-6, HV83-1 and HV83-11) showed no decline in water level; in fact, a small increase was detected at HV83-1.

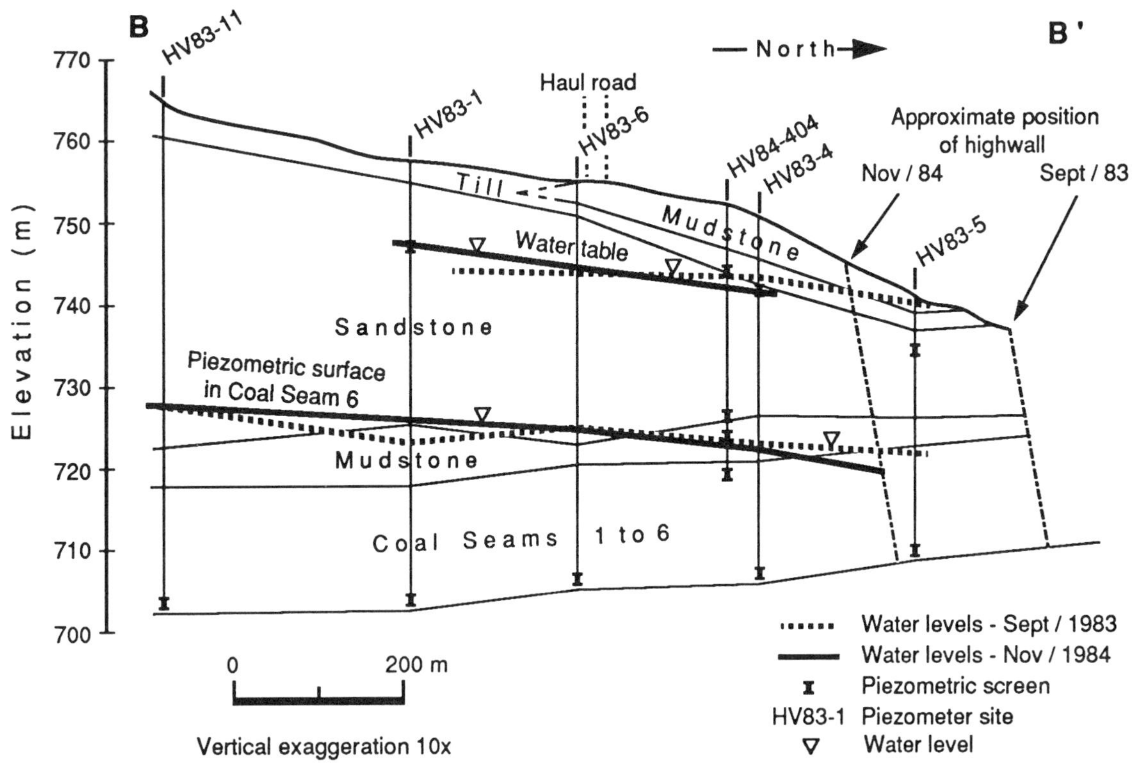

Figure 7. Hydrogeologic cross section for Line B-B' at Site 1 showing for 1983 and 1984 1) the piezometric surfaces in Seam 6 and (2) the water table. Modified from Moell et al. (1985). See Figure 5 for the location of the section.

Failure Cycle: Observations

Observations from March to October, 1984, chronicled an entire "failure cycle" beginning with the excavation of a new bench (Figure 4). An understanding of the type of glacial deformation was essential to the proper understanding and modeling of the failure. The failure was composite, with four mechanisms contributing to the overall result: 1) exfoliation or spalling, 2) block rotation or slump, 3) block slide, and 4) in situ disintegration of all the large, failed blocks (Figure 8).

Both exfoliation and in situ disintegration were volumetrically of minor significance. The exfoliation consists of the falling away from the face, of sheets or slices of sandstone about 1-3 m (3-10 ft) thick. In situ disintegration consists of the gradual disaggregation and, thus, reduction in size of sandstone blocks produced by the other types of failure.

The rotational slumps varied in size, but were generally 50-100 m (160-330 ft) long, 25-50 m (82-160 ft) wide, and roughly semicircular in plan, with the direction of rotation approximately northward. The basal shear zone for each slump was always less than 1 cm (0.4 in.) thick, and either wet if fresh enough, or showed signs of dessication if older. One slump, which occurred on July 1, 1984, was observed within 12 hrs of its initial movement. Here the slump appeared to have moved along a pre-existing, glacially induced, shear zone because 1) the dip direction, of the shear zone, was about N20°E, and the slickensides caused by the rotation were oriented about N25°W (the highwall was oriented approximately east-west), indicating that the block moved diagonally across a pre-existing shear zone rather than down a new shear zone, and 2) the shear zone within the upper portion of the failure was found to intersect undisturbed masses of hangingwall sediment that had not been included in the slide.

Block slides were indicated by vertical fractures. These were a series of cracks that trended approximately parallel to the face and developed over the entire area between Ramps 3-1 and 3-2; the failure zone extended at least 50 m (160 ft) back from the upper edge of the pit face (Figure 4). The fractures were observed to extend to depths of more than 4 m (13 ft) and width increased with the length of time each fracture was open.

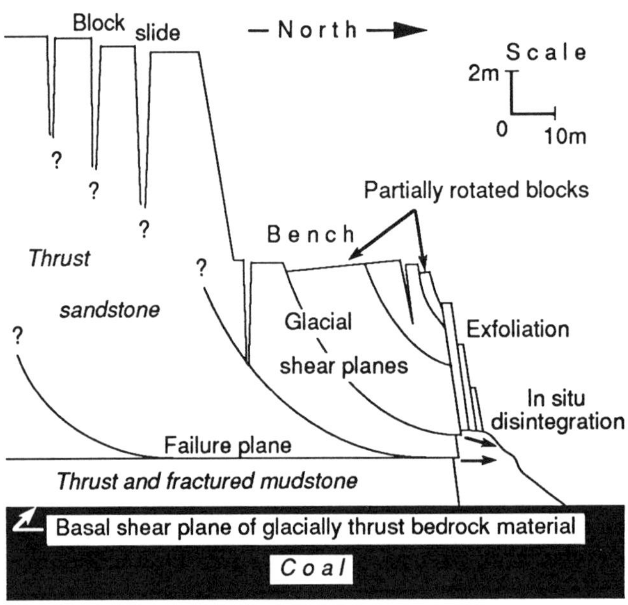

Figure 8. *Schematic illustration of the observational data and model for the highwall failure Site 1. Modified from Moell et al. (1985).*

Failure Model

In summary, the highwall failure was composite, the product of a series of smaller failures of a variety of types consisting of exfoliation, in situ disintegration, rotational slump and block slide (Figure 8). The model involves the initial exfoliation of the newly exposed face and in situ disintegration of the spalled blocks at the base of the cut. This was followed by the development of relatively small rotational slumps and vertical fractures near the outer (northern) margin of the bench, likely the surface manifestation of a combination of early stress release and incipient block slide. With time (about two months for the 1984 Pit 03 failures), this composite failure zone gradually spread southwards across the bench to include the adjacent land surface. This spread is believed to be the result of both the continued development of the small rotational slumps and more importantly, the initiation of the larger-scale rotational slumps and block slides.

Some of the rotational slumps, particularly the large ones, likely were the result of remobilization of preexisting glacial shear zones (Figure 8). Some, particularly the small ones, may have formed spontaneously as a result of the low strength of the sandstone which likely was caused by intergranular glacial deformation. Groundwater conditions, saturation and elevated pore water pressures, likely contributed to both types of slumps.

The block failures are believed to have been caused by a combination of elevated pore water pressures at the contact of the glacially deformed sandstone and fractured mudstone units and the saturated and weakened condition of this deformed sandstone. The dip of this contact toward the pit undoubtedly also contributed to the block sliding. The vertical fractures could have been the result of the block slides, but more likely are due to opening of preexisting joints or perhaps a combination of joints and glacial shear planes.

Site Two: Failure Prediction

This site, located west of Ramp 3-2 (Figure 5), was chosen because, although the highwall in this area was stable throughout 1983 and 1984, existing data from airphotos, surface geophysics, and drilling indicated that this site included glacially deformed bedrock. Detailed drilling and piezometer installation was done to provide additional data along Line C-C' (Figure 9).

The stratigraphy was similar to that at Site 1 east of Ramp 3-2: till over sandstone, mudstone, and the coal sequence (Figure 9). The relationship of the units was different, however. At Site 2, a thickening of the till unit separated a mass of deformed mudstone from a mass of deformed sandstone situated farther to the south (Figure 9).

Figure 9 also shows hydrogeologic conditions in the disturbed overburden. At site H5, the deformed sandstone was absent and the confining till lay directly on a structural high formed by deformed mudstone and underlying bedrock. The till and mudstone at site H5 essentially acted as a plug preventing northward drainage from the deformed sandstone. The groundwater in the northern portion of this sandstone therefore was confined by the overlying till, giving rise to slightly artesian conditions.

Because little drainage of the poorly consolidated, saturated, and deformed sandstone was likely to occur prior to the stripping of overburden south of site H5, the sandstone was considered to have a high potential for failure. No remedial action was taken because the volume of deformed sediment was estimated to be small enough that any failure could be handled as part of normal mining operations, but observation of the highwall continued. Successive mining cuts advanced the highwall southward until the till mass was excavated in 1986. The highwall at this location then failed within a relatively short period of time. Failure continued with each succeeding cut until the undeformed sandstone was reached.

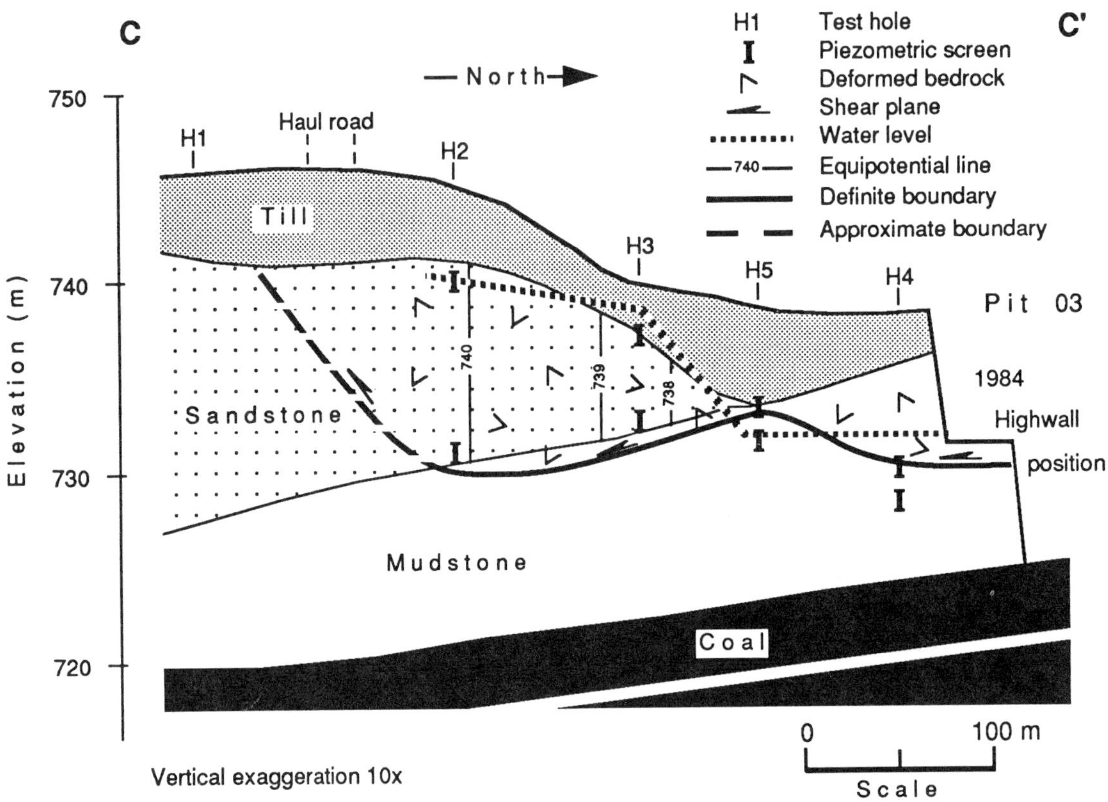

Figure 9. Hydrogeologic cross section for Line C-C', Site 2. Hydraulic head values are in metres above sea level. Equipotential contour interval is 1 m (3 ft). Modified from Moell et al., (1985). See Figure 5 for the location of the section.

A number of solutions to this type of problem have been suggested. These include the construction of a spoil buttress (Khan and Nikols, 1989), an extended operating bench, and drainage of the groundwater (Fenton et al., 1986). Where the existence of preferentially oriented glacial shear zones is recognized, the reorientation of the highwall also might be considered (Fenton et al., 1986).

DETECTION

Useful methods for the detection of deformed bedrock include 1) airphoto interpretation, 2) pitwall and/or outcrop mapping, 3) coring using either rotary core or hollow-stem auger and A-casing, 4) downhole geophysics, and 5) surface geophysics. Airphoto interpretation can identify glacially deformed bedrock where the deformation has left some surface expression. The deformed material generally is indicated by the presence of a large hill or group of smaller hills situated in the direction of glacial flow from a source depression (Andriashek and Fenton, 1989; Bluemle and Clayton, 1984; Fenton, 1983b; Fenton, 1987; Fenton et al., 1985b; and Moran et al., 1980). At some sites, this group of hills extends for several kilometers in the direction of glacial flow and at others the hill(s) is separated from the source depression by many kilometers. At other sites, the masses of glacially thrust material have been either truncated and smoothed over by later glacial flow or buried by the debris left by later flow. Similarily, the source depression may have been filled in by later glacial or nonglacial material. Airphoto interpretation is not effective at these sites. Topographic maps seldom provide the geomorphologic detail needed to recognize many of the deformed masses.

In outcrop, deformed bedrock is most easily recognized in exposures oriented parallel to the direction of glacial flow. Exposures perpendicular to ice flow generally show only more subtle structural changes. Characteristic features include large- to small-scale folds and faults (Figure 10), with an orientation consistent with the local glacial flow direction, and crushing. The crushing breaks the rock into fragments ranging in size from about 10 cm (4 in.) to 5 mm (0.2 in.) in diameter. The degree of crushing can vary within an individual glacial thrust mass. The fragments may have open spaces between them or may have been compacted.

The degree of deformation for a thrust mass that has been deformed in place (i.e., without being transported away from the source area) decreases in the direction of glacial flow and eventually disappears. The sediment forming the distal portion of the mass appears essentially undisturbed. The basal shear zone decreases in thickness and eventually disappears in the direction of glacial flow.

The orientation and structural fabric of each glacial thrust mass is controlled by the local glacial flow direction. If this varies within a mine site, the orientation of the deformation structures such as glacially induced shear planes also will vary. Glacial flow up a small valley, for example, can produce shear planes oriented as much as 90° to the regional flow. A basal shear zone always is present and shear zones within the thrust mass also are common. All types of shear zones have a high failure potential if they dip into the pit. Small faults associated with the vertical loading by the ice have been recognized below the basal shear zone in the Highvale Mine (Fenton et al., 1985b; Moell et al., 1985).

Figure 10. *Roadcut section through hill, exposing glacially deformed bedrock, north of Tomahawk, Alberta.*

Rooting zones or buried soil horizons which often are associated with coal sequences can appear similar to glacially crushed mudstone, although they generally are much thinner than most thrust masses. These rooting zones can be mistaken for a basal shear zone when observed in core or detected with geophysical logs.

In core samples, deformed bedrock commonly is identified by steeply dipping or variably dipping bedding, crushing of the rock, and/or shear zones. If oriented A-casing (Fenton et al., 1987; Fenton et al., in preparation) is taken, the dip direction of the shear zones also can be determined from the core and compared to the glacial flow direction.

Downhole geophysical logs (natural gamma, resistivity, caliper, gamma-gamma density, sonic, dual-spaced neutron, and dipmeter) can be run in rotary and cored drill holes. The density and sonic logs appear to be most effective for the detection of deformed bedrock because they are sensitive to changes in the physical properties of the material (fracturing and density) produced by glacial deformation (Moell et al., 1985). The use of downhole geophysical logs to both detect and to subdivide the glacially deformed bedrock on the basis of geotechnically significant criteria is the subject of a current project (Pawlowicz and Fenton, in press).

Surface geophysical techniques also are effective for the detection of glacially deformed bedrock. Those techniques are discussed elsewhere in this chapter by Henderson et al.

CONCLUSIONS AND RECOMMENDATIONS

Foreknowledge of the location, structure, dimensions, and hydrology of glacially deformed bedrock is important to both exploration and mine planning. The principal deleterious effects of glacial deformation are 1) the disruption of both the overburden and coal stratigraphy typified by folding, faulting, and/or displacement, together with fracturing and crushing of the previously flat-lying beds, 2) the removal or repetition of marker beds, and 3) a reduction in the rock strength.

Ramifications and/or problems for both coal exploration and mining include the following:

1) Loss of recoverable coal.

2) Increased ash content of the mined coal product.

3) Increased uncertainty in reserve estimates.

4) Increased variability in pedogenic and geotechnical soil types.

5) Increased potential for highwall failure.

6) Increased access problems.

7) Decreased core recovery.

8) Decreased permeability and drainage potential.

One advantage of this deformation is the disruption of hard layers in the overburden. This makes excavation by the dragline easier and more efficient, and reduces the need to blast. Glacially deformed bedrock can be detected by a variety of methods: airphoto interpretation, pitwall and/or outcrop mapping, coring, downhole geophysics, and surface geophysics. A program for the detection of deformed bedrock is best initiated as part of an exploration plan or mine feasibility study. The project should begin with an evaluation of the areal and vertical extent of the deformed material using airphoto interpretation, followed by surface geophysical surveys, and finally exploratory coring of the areas of suspected deformed material. After this, individual glacial thrust masses should be investigated in more detail to obtain data on their composition, hydrogeology, and structure. Particular attention should be given to the orientation of the glacial shear planes because they are produced in response to the local glacial flow directions rather than the regional flow direction.

ACKNOWLEDGMENTS

We wish to thank TransAlta Utilities Corporation for allowing the use of data collected at the Highvale Mine. Thanks to Georgia Hoffman and the editors for reviewing the manuscript. Their comments were appreciated and improved the paper. Thanks also to Kathy Gates for typing and editing.

REFERENCES

Aber, J.S., D.G. Croot, and M.M. Fenton, 1989, Glaciotectonic Landforms and Structures: Dordrecht, Netherlands, Kluwer Academic Publishers, 157 p.

Andriashek, L.D, and Fenton, M.M., 1989, Quaternary stratigraphy and surficial geology of the Sand River area 73L, Alberta: Alberta Research Council, Bulletin 57, 154 p.

Andriashek, L.D., M.M. Fenton, and J.D. Root, 1979, Surficial Geology of Wabamun Lake, Alberta: NTS 83G: Alberta Research Council, Map 149.

Bluemle, J.P., and L. Clayton, 1984, Large scale glacial thrusting and related processes in North Dakota: Boreas, v. 13, p. 279-299.

Byers, A.R., 1960, Deformation of the Whitemud and Eastend Formations near Claybank, Saskatchewan: Royal Society of Canada, Transactions, v. 53, ser. 3, sec. 4. p. 1-11.

Carrigy, M.A., 1970, Proposed revision of the boundaries of the Paskapoo Formation in the Alberta Plains: Bulletin of Canadian Petroleum Geology, v. 18, p. 156-165.

Christiansen, E.A., and S.H. Whitaker, 1976, Glacial thrusting of drift and bedrock, in R.F. Legget, ed., Glacial till, an interdisciplinary study: Royal Society of Canada Special Publication 12, p. 121-130.

Fenton, M.M., 1983a, Composition of deformation terrain (glacial thrust terrain): A clue to glacial entrainment and transport (abstr.): Geological Association of Canada, Program with Abstracts, v. 8, p. A22.

Fenton, M.M., 1983b, Deformation terrain mid-continent region: Properties, subdivision, recognition (abstr.): Geological Society of America, Program with Abstracts, v. 15, no. , p. 250.

Fenton, M.M., 1983c, Preliminary Glaciotectonic Map, Highvale Mine: Unpublished Map, Prepared for TransAlta Utilities by Alberta Geological Survey, Alberta Research Council.

Fenton, M.M., 1984a, Deformation terrain Canadian prairies: morphology and sediment facies, in Abstracts and Program of the INQUA Commission on Genesis and Lithology of Quaternary Deposits, Symposium on the Relationship Between Glacial Terrain and Glacial Sediment Facies, Medicine Hat, Alberta, July, 1984: Alberta Research Council, p. 7-8.

Fenton, M.M., 1984b, Quaternary stratigraphy, Canadian Prairies, in R.J. Fulton, ed., Quaternary stratigraphy of Canada--A Canadian contribution to IGCP Project 24: Geological Survey of Canada Paper 84-10, p. 57-68.

Fenton, M.M., 1987, Deformation terrain on the northern Great Plains, in W.L. Graf, ed., Geomorphic Systems of North America: Geological Society of America, Centennial Special Volume 2, p. 176-182.

Fenton, M.M., and L.D. Andriashek, 1978, Glaciotectonic features in the Sand River area northeastern Alberta, Canada (abstr.), in Abstracts of the 5th Biennial Meeting American Quaternary Association, Edmonton, Alberta, September 2-4, 1978: American Quaternary Association, p. 199.

Fenton, M.M., C.E. Moell, J.G. Pawlowicz, G. Sterenberg, M.R. Trudell, and S.R. Moran, 1983, Highwall Stability Project, Highvale Mine study report, December, 1983: Unpublished report prepared for TransAlta Utilities by Alberta Geological Survey, Alberta Research Council, 70 p.

Fenton, M.M., C.W. Langenberg., C.E. Jones, M.R. Trudell, J.G. Pawlowicz, J.A. Tapics, and D.J. Nikols, 1985a, Tour of the Highvale open pit coal mine, in Guidebook for the Petroleum Society of CIM and Canadian Society of Petroleum Geologists Joint Conference, Edmonton, Alberta: Alberta Research Council, Open-File Report 1985-7, 55 p.

Fenton, M.M., C.E. Moell, J.G. Pawlowicz, and C.W. Langenberg, 1985b, Deformation terrain Canadian Prairies: Landforms, lithofacies and processes (abstr.): Geological Society of America, Program with Abstracts, v. 17, no. , p. 287.

Fenton, M.M., C.E. Moell, J.G. Pawlowicz, C.E. Sterenberg, 1987, Using an A-casing sampler for collecting oriented core (abstr.): Geological Association of Canada, Program with Abstracts, Saskatoon, v. 12, p. 41.

Fenton, M.M., C.E. Moell, and J.G. Pawlowicz, in preparation, A-casing, a tool for sampling poorly consolidated sediment and taking oriented cores: Alberta Research Council, Terrain Sciences Department Open-File Report.

Fenton, M.M., and J.G. Pawlowicz, in preparation, Importance of glaciotectonism to coal exploration and development: Alberta Research Council, Terrain Sciences Department Open-File Report.

Fenton, M.M., M.R. Trudell, J.G. Pawlowicz, C.E. Jones, S.R. Moran, and D.J. Nikols, 1986, Glaciotectonic deformation and geotechnical stability in open pit coal mining, in R.K. Singhal, ed., Geotechnical stability in surface mining: Rotterdam, A.A. Balkema, p. 225-234.

Holter, M.E., J.R. Yurko, and M. Chu, 1975, Geology and coal reserves of the Ardley coal zone of central Alberta: Alberta Research Council Report 75-7, 41 p.

Hopkins, O.B., 1923, Some structural features of the Plains area of Alberta caused by Pleistocene glaciation: Geological Society of America Bulletin, v. 34, p. 419-430.

Irish, E.J.W., 1970, The Edmonton Group of south-central Alberta: Bulletin of Canadian Petroleum Geology, v. 18, p. 125-155.

Khan, F.Z., and D.J. Nikols, 1985, Toe buttressing procedure to stabilize slide zones for dragline stripping operations: Canadian Institute of Mining and Metallurgy Bulletin, v. 82, p. 37-47.

Kupsch, W.O., 1962, Ice-thrust ridges in western Canada: Journal of Geology, v. 70, no. 5, p. 582-594.

Lyons, B., S. Argyll, D. Nikols, T. Godfrey, and J. Ellison, 1984, Highvale and Whitewood extended geologic mapping: Prepared for TransAlta Utilities Corporation by MONENCO Ltd., Unpublished Report CPL 8234-6.

Moell, C.E., Pawlowicz, J.G., Trudell M.R., Fenton, M.M., Langenberg, C.W., Sterenberg, G.J., and Jones, C.E., 1985, Highwall stability project, Highvale Mine report of 1984 activities: Unpublished Report Prepared for TransAlta Utilities Corporation by Terrain Sciences Department, Alberta Research Council, 256 p.

Moran, S.R., 1971, Glaciotectonic structures in drift, in R.P. Goldthwait, ed., Till: A Symposium: Columbus, Ohio State University Press, p. 127-148.

Moran, S.R., L Clayton, R.LeB. Hooke, M.M. Fenton, and L.D. Andriashek, 1980, Glacier-bed landforms of the prairie region of North America: Journal of Glaciology, v. 25, no. 93, p. 457-476.

Nikols, D., 1985, October picture and caption: Canadian Society of Petroleum Geology, 1985 Calendar.

Nikols, D., and B. Lyons, 1985, Geology of the Highvale Mine, Presented at the Second District Five Meeting, Hinton, Alberta, September 10-13, 1985: Canadian Institute of Mining and Metallurgy, Preprint, 27 p.

Pawlowicz, J.G., and M.M. Fenton, in press, Detection of bedrock weakened by glacial deformation, in G.L. Hoffman, M.M. Fenton, and J.G. Pawlowicz, eds., Downhole Geophysics Project 1986-1990 Final Report: Alberta Research Council, Open-File Report, Chapter 5.

Sardeson, F.W., 1898, The so-called Cretaceous deposits in southeastern Minnesota: Journal of Geology, v. 6, p. 679-691.

Sardeson, F.W., 1905, A peculiar case of glacial erosion: Journal of Geology, v. 13, p. 351-357.

Sardeson, F.W., 1906, The folding of subjacent strata by glacial action: Journal of Geology, v. 14, p. 226-232.

Sartorelli, A.N., M.S. Pesowski, J.D. Henderson, M.M. Fenton, and D.H. Green, 1986, Compression and shear wave methods applied to the recognition of glaciotectonic deformation, in R.K. Singhal, ed., Geotechnical stability in surface mining: Rotterdam, A.A. Balkema, p. 235-241.

Sauer, E.K., 1978, The engineering significance of glacier ice-thrusting: Canadian Geotechnical Journal, v. 15, p. 457-472.

Slater, G., 1927, Structure of the Mud Buttes and Tit Hills in Alberta: Geological Society of America Bulletin, v. 38, p. 721-730.

Tapics, J.A., 1984, The Highvale Mine--Developments in the next ten years, Presented at the 86th Annual General Meeting of the Canadian Institute of Mining and Metallurgy, Ottawa, Ontario, April 15-19, 1984: Canadian Institute of Mining and Metallurgy, Preprint of Paper 78, 17 p.

Chapter 2

Reserve Estimation

PREFACE

Now that you know where the coal deposits are (or are not), how do you quantitatively define how much coal reserve you have, their overall quality, and the economics of coal extraction and use? Welcome to the newfangled world of reserve estimation! Once upon a time, it was sufficient to simply know there was coal in an accessible location and that you had a market for coal of that rank for heating, steam generation, and/or metallurgical use. Nowadays, things are a bit more complicated. The "easy" (in terms of accessibility and general-use quality) coal reserves are rapidly being depleted, and environmental and other regulatory restrictions/constraints on coal accessibility and use are redefining the dividing line between resources (coal in general in the ground) and mineable reserves (economically and technically mineable and useable).

So, what are coal researchers and the industry doing to identify useable reserves under today's constraints? Palmer's overview paper briefly gives a status report on reserves estimation techniques and active researchers/users of these techniques. He also provides a comparison on where U.S. reserve estimation stands today in comparison to the late 1970s and the beginning of the 1980s when overall estimates first were being formalized.

One of the most recent methods of estimating and investigating coal reserves is the use of geographic information systems (GISs). Such systems use map (i.e., geographically) based data to quantitatively delimit and compare land-use restrictions, quality parameters, coal volumes, etc. to derive standardized and customized numerical and graphical estimates. Treworgy presents an example of use of a GIS for state-wide reserves estimates for Illinois and includes a description of how such estimates can be customized to fit a particular company's needs. Sites et al. discuss how the U.S. Geological Survey and the state geological surveys (Virginia in this paper) are jointly evaluating coal reserves and the various (and often insidious) restrictions on coal resources on a 7.5-min. quadrangle basis. Their bottom line is that there is not as much accessible, useable, quality coal out there as it seem at first glance (i.e., as compared to "resources").

The application of statistical and geostatistical techniques of reserve estimation continues to provide best-constrained estimates of what coal is really in the ground in a mine property. Ghose discusses how such geostatistical estimation is done and describes how drilling costs versus reserves-information gain can be optimized. Drilling a few scattered holes or drilling the property to death do not cut it anymore in today's competitive coal marketplace.

Cross-Reference

Because reserve estimation is so closely tied to exploration and reserve definition, the papers in the previous chapter likewise will be of interest to estimators. If coalbed methane reserves are brought into the question of useability or quality of a coal resource, as is increasingly the case these days, then the papers by Ertekin et al. and by Mitchell and Pappajohn in the Coalbed Methane chapter will be of interest as well. This especially may be true with regards to multiple use and/or premining gas drainage of a coal reserve.

Most of the papers in the Mining chapter describe geologic conditions that affect the technical mineability of a coal seam. The paper by Kane and Powell, in particular, on subsidence will provide insight on evaluating the impact of mining on surface features and, in return, the land-use restrictions on mining imposed by surface structures (surface buildings) and other features (such as railroads, cemeteries, and National Forests).

As pointed out in papers by Treworgy and Sites et al., coal quality is part of the equation for coal useability and reserves. Therefore, the papers in the Coal Quality chapter will be of interest, especially those by Winston, Goodarzi and Van der Flier-Keller, and Hower et al.

It was noted previously that environmental restrictions on coal, in terms of coal use and subsequent emissions (air and water), regulations in general, and land-use restrictions, now are a major consideration in defining reserves versus resources. Papers by Moran et al., Wildeman, Beaver et al., and Goodarzi and Gentzis particularly address topics of interest for this type of restriction on reserves. This is an area where there clearly is an immediate need for further concern on the part of industry and for closer evaluation of coal deposits that currently are called "reserves".

COAL RESOURCES AND RESERVE ESTIMATES IN THE UNITED STATES IN 1991 — A STATUS REPORT

James E. Palmer
Sigma Consultants, Inc.
118 S. 17th St.
Mattoon, IL 61938

Substantial progress has been made in the past decade in both the methodologies of resource/reserve calculations and the applications of such methodologies. Major areas of advancement have occurred in resource/reserve classification, computer applications, and geostatistics. All three areas are to some extent interrelated.

RESOURCE AND RESERVE CLASSIFICATION

As one result of a symposium entitled "The Evaluation of North America Coal Reserves" at the 1979 annual meeting of the Geological Society of America in San Diego, California, a series of papers were published in 1981 (Geological Society of America Bulletin, v. 192, Part I, p. 519-577; includes Palmer, 1981, and Treworgy et al., 1981) defining both the status of resource/reserve estimations in North America and, more notably, the differences and deficiencies between variously applied estimation methodologies. This symposium and the resultant publications were part of the recognition in the early 1980s that existing resource and reserve estimates were lacking in accuracy and meaningfulness for energy planning and coal resource utilization.

As a result of the recognition of this problem, a revised and expanded system of coal resource classification was published by the U.S. Geological Survey (USGS) in 1983 (Wood et al., 1983) which provided full or partial solutions to many of the difficulties or deficiencies of prior classification systems. This system now is used widely in the coal producing states, partly as the result of cooperative agreements between the states and the federal government for mapping and resource estimations. Also, increasing use by various states of the National Coal Resources Data System (NCRDS) of the USGS resulted in greater familiarity by state agencies with the 1983 USGS classification system. Whereas some differences remain in application by certain state agencies, general acceptance of the system has developed during recent years. In addition, where significant differences in resource/reserve numbers between state and federal agencies are recognized due to classification discrepancies, computer programs have been developed to inter-translate coal resource/reserve numbers.

Much of the earlier work of the USGS Branch of Coal Resources and geologists in coal-producing states has been directed toward estimation of total coal resources of the United States (currently estimated at more than 3,600,000,000,000 tonnes (4,000,000,000,000 tons)) and a demonstrated reserve base. Whereas these very large figures have been reassuring to the public, coal geologists have long recognized that they do not realistically reflect the ready availability of coal for mining under current and near-future economic or technical conditions. The actual amount of coal available under existing economic conditions is uncertain, but is drastically less than published values for coal resources and the reserve base.

A new approach to coal resource estimation is being led by the USGS Branch of Coal Resources, the U.S. Bureau of Mines, and the state geological surveys of Kentucky, West Virginia, and Virginia. By taking into account mining methods and costs, and various land-use restrictions to mining, recently completed studies of 7.5-min. quadrangles in the central Appalachians indicate that only about 50% of originally estimated resources are available for mining. Similar restriction-oriented studies of coal resources are underway for Illinois at their state geological survey. As these studies are expanded to other coal-producing states, a much more realistic understanding should be developed of U.S. coal resources and reserves.

COMPUTER APPLICATIONS

Since initial usage in the 1970s of computer applications for data retrieval and storage as part of the U.S. Geological Survey's NCRDS, computer applications in coal resources and reserves have been expanded in numerous interesting ways. For example, a nationwide computer network has been established through which a very large amount of publicly available, detailed coal resources and reserves data can be called upon by private individuals or companies from relatively inexpensive computer workstations. Depending upon quality and detail of input coal data, and capabilities of the user's workstation and printer, such data can be used to produce output maps and data such as

1) location and areal extent of certain features which would limit or preclude mining, or which would be used in mine planning (e.g., power plants, railroad right-of-way, coal loading docks, navigable waterways, reservoir sites, urban areas, active mines, cemeteries, and interstate highways),

2) various characteristics of individual coal seams (e.g., thickness, tonnage calculations, depth, overburden thickness, sulfur content, and development potential or mineability),

3) geologic characteristics of coal seams such as coal outcrops, roof and floor characteristics, suitability for various types of mining (longwall, room and pillar, stoping, etc.), sulfur, moisture, ash, volatiles, fixed carbon, and Btu content, and

4) specialized maps showing, for example, pounds SO_2 per million Btus of heat value and showing high-sulfur areas, summation of combined coal thicknesses of two or more seams with calculation of combined stripping ratios, and state-of-the-art hypothetical ash and pyritic sulfur removal calculations for in-place coal, with calculated resultant coal quality and heat value.

More detailed analyses which can be completed with certain workstations and printers include three-dimensional color visualizations of coal-bearing strata with various layers of strata displayed and/or removed (e.g., glacial till covering of bedrock valleys, roof or floor strata, certain underclay facies, and various other sublayers). Such three-dimensional visualizations from both above and below a coal seam are particularly useful in the recognition of specific conditions in the mine roof or floor and in presentation to management of mining problems and future plans.

By use of the convergence of two or more data sets into a single computer-generated map presentation, it is sometimes possible to recognize important conditions or characteristics of coal seams which might otherwise be overlooked. For example, the plotting of fractures and/or faults from one data set in conjunction with the extent of a soft shale facies from a second data set may indicate severe mining problems in a specific area. Data sets indicating sandstone channels, groundwater, steeply dipping strata, folds, split coal, thick shale partings, mined-out areas, and gassy mine areas are examples of other parameters that can be analyzed through the use of computer-generated convergence maps.

Because the cost of computer printing of even certain elaborate three-dimensional color maps and diagrams usually is less than the cost of conventional color printing, it now is possible to publish maps as needed directly from the computer printer. Such maps easily can be revised and updated as often as necessary according to user requirements.

GEOSTATISTICAL METHODS

Geostatistical applications in coal resource and reserve estimations have attracted substantial interest among coal geologists and mining engineers, but the applications are not yet widely used in the coal industry. Applications of such methods, including kriging, often requires substantial training of users in mathematics and statistics and in coal geology and mining. Others who would use such geostatistical data for decision-making also must have some familiarity with the methodology and its limitations. Mine managers sometimes are unwilling to abandon older "tried-and-true" methods for resource/reserve estimation for the much more complex geostatistical methods. Nevertheless, such methods, at a minimum, provide useful analytical backup for conventional resource/reserve calculations and, in some cases, may allow recognition of weaknesses in conventional resource/reserve databases. When used in conjunction with a planned drilling program for mine development, geostatistical methods can provide precise definition of the amount of drilling data required, thus saving funds that might otherwise be expended on unneeded drill holes.

RETROSPECT

In 1980, I prepared a review for the Oak Ridge National Laboratory of the "Demonstrated Reserve Base of Coal in the United States on January 1, 1979 (Draft) -- A technical review" (Palmer, 1980). In the review (p. 2), it was pointed out that the coal resource database consisted of "a hodgepodge of apples, oranges, and pomegranates that have been aggregated together both accurately and inaccurately depending upon the source, and in many cases, the true identity of even the fruit itself is unknown. It is obvious that the Demonstrated Reserve Base is simply a mathematical manipulation of the resource database, and of necessity, contains all of the inherent weaknesses of the source data."

During the past decade, as the result of numerous cooperative programs between the USGS Branch of Coal Resources and geological surveys of various coal-producing states, through the efforts of some state surveys individually, and with the publication of the newer, detailed coal resource classification system by the USGS, a more standardized approach to resource/reserve classification has been developed nationwide. Finally, by use of computer applications, as previously discussed, much more emphasis now is given to the careful screening of resource data to identify economically, technically, and environmentally mineable coal reserves. This approach represents a giant step away from the nearly useless 1979 reserve base data and should result, in future years, in an accurate assessment of the demonstrated reserve base of coal in the United States.

ACKNOWLEDGEMENTS

Information and observations contained herein were compiled by the author after discussions with the following coal professionals: Harold Gluskoter, U.S. Geological Survey, Reston, Virginia; Thomas Kehn, U.S. Geological Survey (Retired), St. Augustine, Florida; and Colin G. Treworgy, Illinois State Geological Survey, Champaign, Illinois. The information and assistance they provided are gratefully acknowledged.

REFERENCES

Averitt, P., 1975, Coal resources of the United States, January 1, 1974: U.S. Geological Survey Bulletin 1412, 131 p.

Palmer, James E., 1980, Demonstrated reserve base of coal in the United States on January 1, 1979 (Draft) -- A technical review: Oak Ridge, Tennessee, Oak Ridge National Laboratory, Unpublished Report, 17 p.

Palmer, James E., 1981, Coal resources and reserves estimates: Some unresolved problems: Bulletin of the Geological Society of America, v. 92, pt. I, no. 8, p. 519.

Treworgy, C.G., H.H. Damberger, and J.E. Palmer, 1981, Desirable characteristics of a state-wide evaluation of coal resources: Bulletin of the Geological Society of America, v. 92, pt. I, no. 8, p. 574-577.

Wood, G.H., T.M. Kehn, M.D. Carter, and W.C. Culbertson, 1983, Coal resource classification system of the U.S. Geological Survey: U.S. Geological Survey Circular 891, 65 p.

TECHNIQUES FOR ESTIMATING THE COAL RESERVE BASE OF A REGION USING A GEOGRAPHIC INFORMATION SYSTEM: A CASE STUDY FROM ILLINOIS

Colin G. Treworgy
Illinios State Geological Survey
615 E. Peabody Drive
Champaign, IL 61820

ABSTRACT

This paper describes techniques for estimating the reserve base of coal in a region using a class of computer software called a geographic information system (GIS). The term "reserve base", as used in this paper, refers to those coal deposits that are technically, economically, and legally minable at the present time. Factors that influence the minability of coal were identified by interviewing a cross section of experts on mining in Illinois. Though some deposits clearly are unminable, the majority are minable but have varying degrees of attractiveness or likelihood of being mined. To reflect this, deposits are classified by their potential (or likelihood) to be developed as well as their current minability. The GIS provides the ability to catalog all the attributes of a deposit that relate to its minability, to rank the potential of deposits for development, to test the effects of hypothetical economic, legal, or technical developments on the reserve base, and to customize estimates of the reserve base in response to specific issues.

INTRODUCTION

This paper describes techniques for estimating the coal reserve base of a region. The term reserve base, as used in this paper, refers to those coal deposits that are technically, economically, and legally minable at the present time (Wood et al, 1983). Estimates of the coal reserve base in a region (e.g., county, state, coal basin) help to target prospective areas for development and provide a yardstick for measuring the value of a company's or landowner's holdings relative to the total amount of coal available. Utility companies, companies that service the mining industry, and government agencies rely on reserve estimates for long-term planning to identify the character, quantity, and location of coal which may be developed in the future.

Estimating the reserve base for a region can be a challenging task. Because of the large area involved in regional evaluations, there may be many variables and much data to be considered. Field mapping or exploratory drilling to fill in voids in the data generally are impractical within the budget and time constraints of the project. Variations in mining practice among companies or between mining districts may require flexibility in applying criteria to define the reserve base throughout a region. Because criteria may change significantly within the space of a few years, the reserve base estimate must be compiled in a manner that can be revised readily.

Though it is possible to estimate the reserve base of a region using manual methods (e.g., see Treworgy et al., 1978), the process is greatly improved through the use of computers, particularly a class of computer software called a geographic information system (GIS). A GIS permits large amounts of data to be analyzed, tabulated, and displayed in less time and with more flexibility than is possible with manual methods. This capability allows different sets of criteria to be used to estimate the reserve base under various scenarios. Criteria may be adjusted and factors added or updated as conditions change.

Geographic Information Systems

A GIS is designed to analyze and display maps and their associated data. It includes capabilities for encoding, manipulating, analyzing, and displaying large quantities of map data. Individual map features can be extracted, combined with other data, and analyzed based on their location, size, or other attributes. For example, one can calculate the tonnage of all coal greater than 150 cm (60 in.) thick in contiguous blocks of at least 18,000,000 tonnes (20,000,000 tons) and within 8 km (5 mi.) of a railroad. "Buffer zones" may be created around features (e.g., zones of influence around data points and zones of exclusion around abandoned mines or oil wells). Feature categories may be aggregated, or used to create new attributes through mathematical or logical operations (e.g., thickness X area X recovery factor = tonnage of recoverable coal, or tons recoverable / cubic yards of overburden = stripping ratio). The GIS can be used to produce maps and tabulations or digital data for input into other computer programs.

An exhaustive description of the capabilities of a GIS is beyond the scope of this paper. Further details on GISs and their applications can be found in Berry (1987), Tomlinson (1987), Treworgy and Bargh (1986), and Treworgy et al. (1988).

ESTIMATING THE RESERVE BASE OF DEEP-MINABLE COAL IN ILLINOIS

The techniques for estimating the coal reserve base of a region are illustrated by a project conducted by the Illinois State Geological Survey (ISGS) to estimate the reserve base of deep-minable coal in Illinois (Treworgy and Bargh, 1982). Though the details of the factors identified and their application to coal deposits will differ between regions, the concepts used in this project and the techniques used to implement them are applicable to any region or to surface-minable deposits.

The steps for estimating the reserve base of deep-minable coal in Illinois were as follows: 1) identify the factors that influence minability of the coal, 2) compile the data on each factor, and 3) evaluate the minability of coal deposits in conjunction with these factors.

Factors Influencing The Minability of Coal

Factors that influence the minability of coal were identified by interviewing a cross section of experts on mining in Illinois. These experts represented five coal companies and two consulting firms and consisted of eight engineers, four geologists, two managers, and one marketing person. The coal companies operated a total of 13 underground mines in Illinois representing 58% of the state's annual production from underground mines. The mines operated by these companies had annual productions ranging from 730,000 tonnes (800,000 tons) to more than 1,800,000 tonnes (2,000,000 tons). These mine sizes were (and still are) typical of all but a few underground mines in the state. The interviews were conducted with one organization at a time. Several people from the organization representing different fields of expertise were present at most interviews. In addition to discussing a list of factors that influence minability of coal, the participants were asked to comment on potential obstructions to mining as depicted on maps (e.g., areas with various densities of oil wells, different types and configurations of surface land uses, etc.).

A few experts in fields other than mining were contacted to clarify some specific situations. An official at the Illinois Department of Transportation was interviewed to learn the State's policy on mining under roads, and an official of an utility company was contacted to discuss restrictions on mining in the vicinity of nuclear power plants.

The diversity of people interviewed proved to be useful. Though many factors were mentioned repeatedly, some factors were identified by only one or two individuals. In several cases, the companies holding large reserves of thick, shallow coal had more stringent requirements for depth and thickness of a deposit than experts from companies that had small holdings and which were actively exploring for additional reserves. Also, some experts had encountered specific mining problems which were not generally recognized by others. Table 1 shows the criteria that each group used to define minable coal.

There was no consensus among the experts as to the criteria to be used to define minable coal. However, at this stage in the project, identification of the factors that influenced minability was more important than defining specific criteria. Our goal was to build a database that contained the attributes of coal deposits and other factors that were needed to estimate the reserve base under a variety of scenarios of market conditions. For example, though depth was identified as a factor there was no agreement among experts as to what coal was too deep or too shallow to be considered a part of the deep-minable reserve base. It was apparent, however, that only general categories of depth (e.g., 30-m (100-ft) increments) would be needed to evaluate this factor regardless of what specific criteria might later be used.

Acquisition Of Data On Factors Influencing Minability Of Coal

The next task was to compile data for the factors that were identified and enter them into the GIS database. Depending on the level of detail needed and the amount of data available, factors either were mapped from primary source materials, inferred from other data, or left unmapped and omitted from consideration.

Mapped Factors

Maps were compiled for most factors for which ample data were available. Coal thickness was mapped in 0.3-m (1-ft) increments for all seams. These maps also showed mined-out areas, channels, and the 46-m (150-ft) depth line. Coal in Illinois less than 46 m (150 ft) deep is classified as surface-minable and was evaluated in a separate study. Urban areas, infrastructure (railroads and interstate highways), cemeteries, and public lands were compiled from topographic maps and plat books.

Areas densely drilled for oil were delineated using oil and gas development maps maintained by the ISGS. Mining regulations require that drill holes penetrating a seam either must be properly plugged above and below the coal or have a barrier of unmined coal left around them. Most companies assume that oil wells have not been properly plugged and leave a barrier pillar. Because of the size of the barrier required to prevent subsidence (a size that varies somewhat with the mining system), rooms cannot be developed in areas where wells are spaced every 4 hectares (10 acres) (a common spacing in oil fields). Entries cannot be constructed where wells are spaced every 8 hectares (20 acres) or less over large areas.

Inferred Factors

The data available for several factors were insufficient to map for all seams and had to be inferred from other information. Chlorine (Cl) content was available as a statewide map only for the Herrin Coal (Gluskoter and Rees, 1964). Chlorine content of coals in the Illinois Basin has been shown to be controlled primarily by seam depth and regional factors. Because the deep-minable coals in Illinois occur within an interval of a few hundred feet, the Cl content of the Herrin Coal was applied to all coals.

	MC1	MC2	MC3	MC4	MC5	C1	C2
Minimum Thickness (ft.)	4.5	4.5	4	4	3.5	4.5	4.5
Maximum Depth (ft.)	?	600-800	none	?	none	none	500-700
Maximum Intra-seam Partings	10-20% by wt	12" or 15-30% by wt	avoid	<12"	12" or 20% by thknss	12" or 5-30% by wt	20% by thknss
Minimum Tonnage (10^6 tons)	25-30	40-60	30-40	50	40	20	50
Drift to Bedrock Overburden	depends on strata	?	40-50' min rock	?	?	?	1:1 ratio 100' rock minimum
Minimum Interburden (ft.)	30-50	depends	50	25	50-100	60	100 or less
Lower seam Mined-out	nm	ok	ok	nm	nm	ok	ok
Maximum S	3%	1.2 lbs/mil Btu washed	?	2.5-3% washed	most coal ok	?	?
Maximum Cl	?	0.4%	?	?	no max	no max	<0.5%
Towns	nm	nm	nm	40% extraction	nm	nm	nm
Public Land	nm	maybe	?	?	maybe	maybe	maybe
Dams	nm	?	?	ok	ok	nm	nm
Interstates	entries only	entries only	entries only	entries only	ok	entries only	?
Oil Fields (minable well-spacing)	nm	>10 acre spacing	nm	>20 acre spacing	>20 acre spacing	?	?

Table 1. *Criteria used by coal companies and consulting firms to define minable coals in Illinois (MC = mining company; C = consulting firm; nm = not minable; ? = no answer given).*

Rank and moisture content (mineral-matter-free basis) of the Herrin Coal had been mapped statewide (Damberger 1971). Again, because these factors were understood to be regionally controlled, and rank increases relatively slowly with depth, the heating and moisture values of the Herrin Coal were applied to all coals. By assuming an average ash content of 10%, the approximate heating value (Btu/lb) of deposits on an as-received basis was calculated.

Data on sulfur (S) content were too widely spaced to draw S isolines. Instead, areas of moderate to low S content were identified by mapping the geology commonly identified with relatively low-sulfur coals. Illinois coals generally have a S content of 3-5%, except where overlain by thick accumulations of non-marine rock. Statistically, the coal can be expected to have a S content of less than 2.5% (as low as 0.5%) in areas where these nonmarine rocks are 6 m (20 ft) or more in thickness (Gluskoter and Simon, 1968).

Because precise depth to coal was not needed to evaluate minability, maps of coal depth were entered into the GIS only for the Herrin and Springfield coals. The depth to these coals was used to calculate the approximate depths to nine other seams using the average interval between the seams. Where necessary, these figures could be adjusted to accommodate local variations of coal depth or interburden thickness.

Unmapped Factors

Roof and floor conditions, though important, are difficult to determine without geologic and rock mechanics data from closely spaced drill holes. Because extensive data were not available on roof and floor conditions, these factors were not considered. Other factors not considered because of lack of data were partings in seams, ratio of the thickness of the glacial drift to bedrock overburden (important for stability of mine workings), faults, clay dikes, method of extraction (percent recovery), and coal quality parameters other than S, Cl, and rank.

Evaluating The Minability And Development Potential Of Deposits

Once loaded into the GIS, these maps of coal and factors that influence minability are merged so that, for any coal seam at any location, the GIS contains a complete catalog of all the known attributes related to the minability of that seam (e.g., thickness, depth, quality, land cover, etc.). The GIS can categorize and tabulate the resources of coal by any of these attributes (e.g., common categories are political unit, overburden thickness, and S content). The GIS also is used to identify and tabulate those deposits that meet the criteria for a reserve base. The GIS's ability to perform this task relatively rapidly (as compared to manual methods) makes it feasible to modify the criteria and redefine the reserve base as new influences on the minability of deposits are recognized.

In calculating the reserve base of coal in Illinois, we found it useful to go beyond a simple designation of deposits as minable or unminable. There are no standard criteria in the United States for determining a reserve base, and the local experts who we interviewed did not agree on the criteria for defining minable coal. The variation on some of the criteria (e.g., minimum thickness and maximum depth) would result in differences in reserve base estimates of billions of tons. Although some deposits are clearly unminable, the majority are minable with varying degrees of attractiveness or likelihood of being mined. To reflect this, a classification scheme was created to rank the potential of deep-minable coal deposits for development (Treworgy and Bargh, 1982).

The development potential of a deposit is rated by comparing the characteristics of the deposit to the characteristics of deposits currently being mined in the state. Deposits similar to those currently mined are considered to have the highest potential for development. Deposits that are slightly inferior to those being mined, but with characteristics similar to deposits currently being leased, are classified as having a moderate potential for development. Deposits with significantly inferior characteristics have a low potential for development. Deposits that underlie surface features which might be damaged by subsidence or which may be legally restricted from mining have a restricted potential for development.

Currently, the high, moderate, and low development potential categories of deep-minable coal are defined based on the thickness and depth of coal being mined and leased in Illinois in the early 1980s (Figure 1). The high potential category consists of deposits more than 1.4 m (4.5 ft) thick and less than 120 m (400 ft) deep or more than 1.7 m (5.5 ft) thick and less than 300 m (1,000 ft) deep. Moderate potential consists of coal excluded from the high potential category, but more than 1.1 m (3.5 ft) thick and less than 300 m (1,000 ft) deep, or more than 1.4 m (4.5 ft) thick and less than 370 m (1,200 ft) deep, or more than 1.7 m (5.5 ft) thick. All remaining deposits are in the low development potential category. Coal less than 71 cm (28 in.) thick is considered unminable. Deposits with a restricted potential for development underlie cities, dams, public lands, cemeteries, and interstate highways. Deposits located in areas densely drilled for oil and gas also have a restricted potential for development. Tonnage estimates were cross-tabulated by S, Cl, and heating value to show the potential influence of these additional factors (e.g., see Figure 2).

Benefits Of Using A GIS

With the GIS, it was relatively easy to apply the procedures described above and calculate the amount of coal in each category of development potential. The tonnage of coal also may be tabulated by any

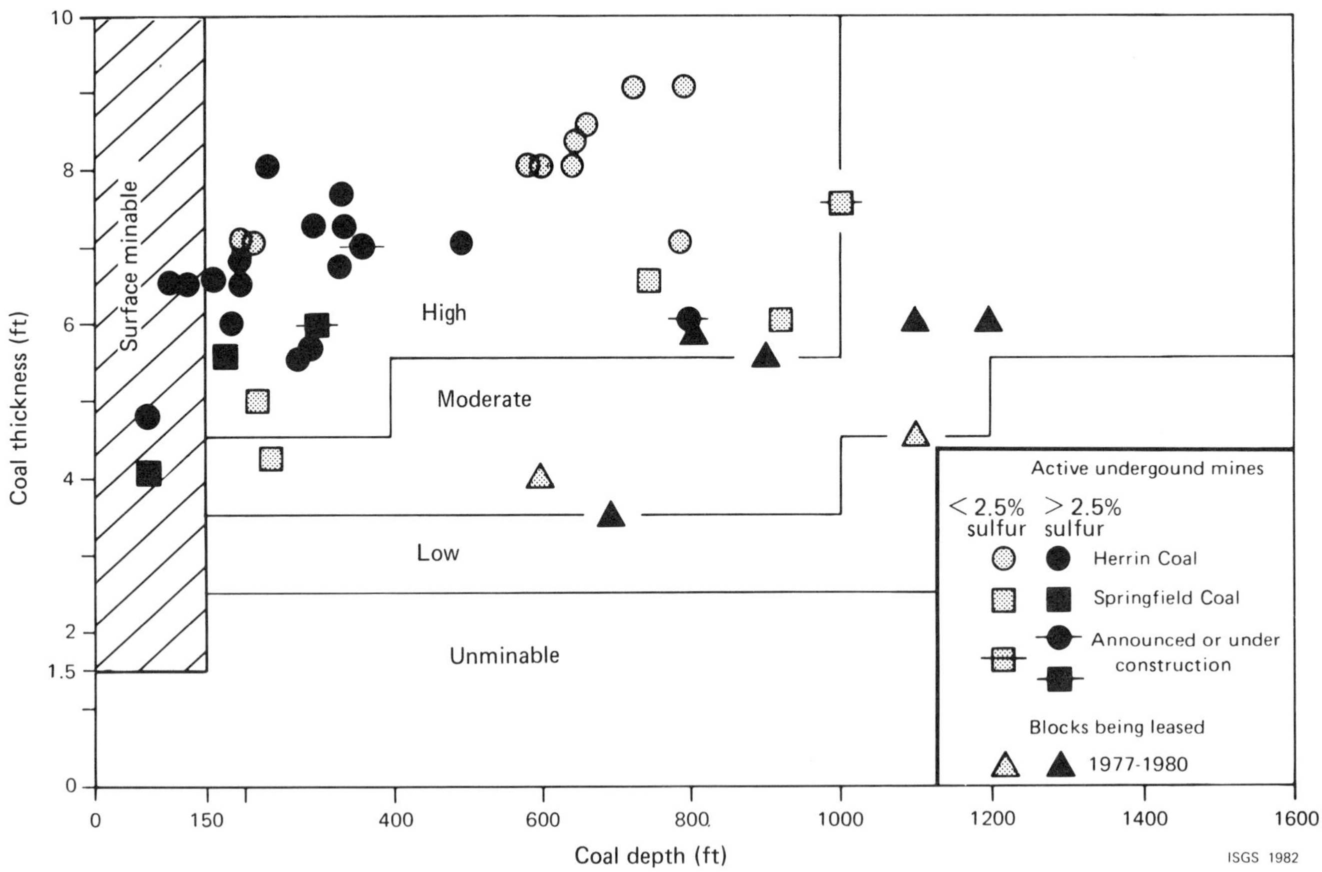

Figure 1. Classification of development potential of deep-minable coal in Illinois based on the thickness and depth of coal being mined or leased in the early 1980s.

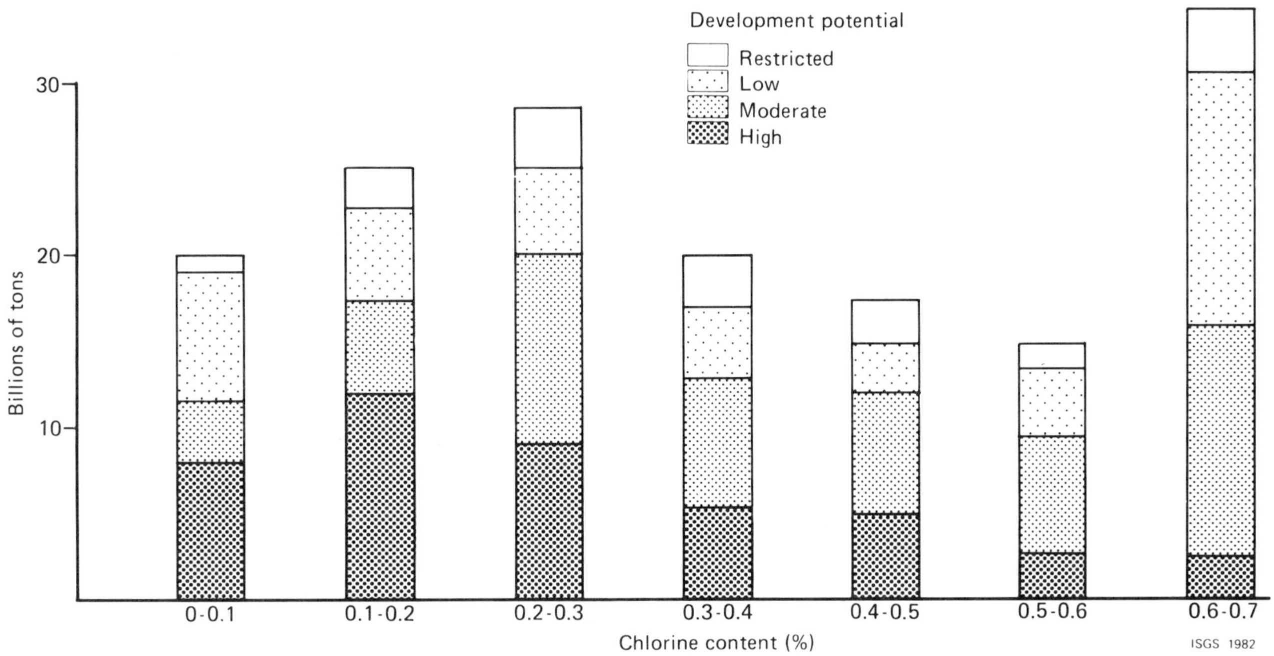

Figure 2. *Deep-minable coal resources of Illinois by Cl content.*

combination of factors in the database and maps can be produced showing the distribution of deposits by these factors or the development potential ranking (e.g., see Figure 3).

Because estimates can be revised relatively quickly with the GIS, the ISGS generally can provide "customized" reserve estimates on request. This capability provides a solution to the different views among experts as to the criteria for minable coal. For reporting purposes, the ISGS considers the high potential category to be equivalent to the reserve base. Users of these data are free to construct their own reserve base estimate according to their own criteria. This can be accomplished either by using the published cross-tabulations of tonnages or with the GIS. For example, a group studying the potential for marketing Illinois coal in Japan and Korea asked how much coal in the reserve base met their specifications for rank and S content. The answer was quickly produced in both tabular and map form using the GIS.

The GIS provides the ability to test the effects of hypothetical economic, legal, or technical developments on the reserve base. For example, a coal company requested an estimate of the tonnage of reserves that met their criteria for longwall mining. The GIS also can be used for siting studies, such as to calculate the reserve base within a certain distance of a proposed plant site or railroad line.

In the future, estimates for the reserve base and development potential will be refined by adding more detailed data for interburden thickness, S content, and coal depth. Additional factors which will be included in the determination of the reserve base and development potential classifications are the ratio of unconsolidated to bedrock overburden thickness, prime farmland, land use, size of mining block, ash content, and proximity to abandoned mines (mines in underlying and overlying seams as well as adjacent mines in the same seam). Also, areas will be classified by the type of mining method that could be used (e.g., longwall or room and pillar) and corresponding recovery factors applied to the reserve base to estimate reserves. The GIS allows these factors to be added incrementally as data or opportunities permit.

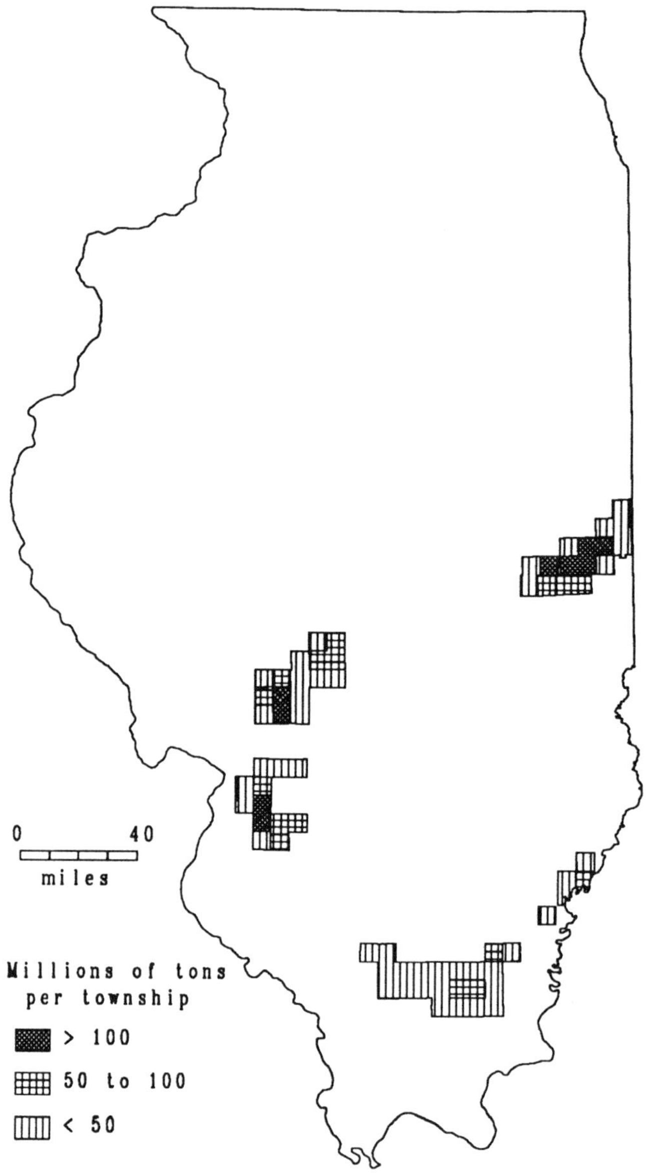

Figure 3. *Map produced by a GIS showing townships in Illinois containing deep-minable coal resources with a high development potential, S content less than 2.5%, and a Cl content of less than 0.4%.*

CONCLUSIONS

Many factors influence the minability of coal. Some of these factors are mappable on a regional basis, others can be inferred, and some factors will require more data than are available and will have to be deferred from consideration. Experts on mining in the region may disagree as to how these factors should be applied to defining the reserve base. These challenges require techniques for estimating the reserve base that are flexible and capable of manipulating large volumes of map data.

A GIS is a powerful tool for making estimates of the reserve base of coal in a region. The GIS makes it feasible to catalog a large number of factors that relate to the minability of a deposit. Using the analytical capability of the GIS, tonnage estimates can be cross-tabulated by various categories and reserve base estimates can be enhanced, such as by ranking deposits by their potential for development. The reserve base easily can be recalculated to reflect an individual user's criteria or for testing the affect on estimates of various scenarios of future economic, legal, or technical developments. Additional factors can be incorporated into the evaluation incrementally as new data become available or conditions change. Estimates of the nation's coal reserve base should improve in the future as state and federal organizations make use of GIS tools (hardware and software) and extensive databases.

REFERENCES

Berry, J.K., 1987, Fundamental operations in computer-assisted map analysis: International Journal of Geographical Information Systems, v. 1, no. 2, p. 119-136.

Damberger, H.H., 1971, Coalification pattern of the Illinois Basin: Economic Geology, v. 66, no. 3, p. 488-494.

Gluskoter, H.J., and O.W. Rees, 1964, Chlorine in Illinois coal: Illinois State Geological Survey, Circular 372, 23 p.

Gluskoter, H.J., and J.A. Simon, 1968, Sulfur in Illinois coals: Illinois State Geological Survey, Circular 432, 28 p.

Tomlinson, R.F., 1987, Current and potential uses of geographical information systems. The North American experience: International Journal of Geographical Information Systems, v. 1, no. 3, p. 203-218.

Treworgy, C.G., and M.H. Bargh, 1982, Deep-minable coal resources of Illinois: Illinois State Geological Survey, Circular 527, 65 p.

Treworgy, C.G., and M.H. Bargh, 1987, Geographic information systems: A computer technology for the coal industry, in Y.J. Wang, R.L. Grayson, and R.L. Sandord, eds., Use of computers in the coal industry: Boston, A.A. Balkema, p. 99-108.

Treworgy, C.G., M.H. Bargh, C. Hindman, and C. Morgan, 1988, Costs and benefits of GIS data management: A case study of a database managed by a state agency, in 1988 Annual Convention Technical Papers: Falls Church, Virginia, American Congress for Surveying and Mapping-American Society for Photogrammetry and Remote Sensing, Vol. 2, p. 186-195.

Treworgy, C.G., L.E. Bengal, and A.G. Dingwell, 1978, Reserves and resources of surface-minable coal in Illinois: Illinois State Geological Survey, Circular 504, 44 p.

Wood, G.W. Jr., T.M. Kehn, M.D. Carter, and W.C. Culbertson, 1983, Coal resource classification system of the U. S. Geological Survey: U.S. Geological Survey Circular 891, 65 p.

USE OF GEOSTATISTICS IN PLANNING AN OPTIMUM DRILLING PROGRAM

S.N. Ghose
AMAX Coal Industries
P.O. Box 6106
Indianapolis, IN 46206-6106

ABSTRACT

Drilling in the coal industry often is conducted in a pattern regulated by spacing between the boreholes. In order to increase the reliability of prediction of deposit characteristics, additional boreholes are drilled. In the absence of a numerical aid, it often is difficult to assess the gain in reliability or marginal benefit achieved by additional drilling. In this paper, a method to optimize drilling with respect to location and the number of boreholes needed is presented. Measurement of precision in the data is expressed by the percent variation at a 95% confidence level which is correlated with the cost of drilling. Fictitious data are added to the initial database at a graduated rate, and precision is determined for each stage of addition. The precision will improve each time a set of additional fictitious data are added, and the cost will increase accordingly. The point of intersection between the cost and precision curves is considered to be the optimum drilling point with respect to spacing and the number of boreholes.

INTRODUCTION

In order to survive in today's exceedingly competitive coal industry, it is necessary more than ever to produce coal at the lowest possible cost. This requires higher productivity without any interruptions in the process of mining. Drilling for reserve determination is an integral part of the coal mining process, which starts with discovery and ends with the exhaustion of a deposit. Information is acquired through drilling to protect the industry from economic, geologic, and safety risks. In a conventional process of mining, there are three distinct stages of progression. Namely, exploration (primary), development (secondary), and extraction (tertiary) stages can be established. Drilling assignments in each stage are distinguished by the function of the drilling. In the absence of verifiable and numerically reliable data on the coal seam, over-drilling in secondary and tertiary stages often occurs. Geostatistics can be used to optimize a drilling program without compromising safety and knowledge of uncertainties. This will ensure an uninterrupted execution of the mining process at a maximum economic advantage.

CHARACTERIZATION OF THE FRAMEWORK

In this section, main characteristics of the geostatistical method and a brief description of the principles related to classical statistics are presented. Only the relevant features associated with coal reserves are discussed.

In classical statistics, as often is the case with empirical methods, no single estimator is universally applicable. Reliability of estimates, projected by linear estimators for a mineability variable (thickness, depth, etc.) in coal, can be reduced by the inhomogeneity and structural anisotropism associated with the deposit. Some degree of bias, caused by asymmetric and skewed data, can be resolved by the use of nonlinear (parametric) and nonparametric estimators. Applicability of an estimator is determined by the magnitude of error between an a priori value and the estimates.

The performance of an estimator is expressed this way:

$$Z^*v = \sum_{i=1}^{n} Y_i \; Z_i$$

where

$E(Zv - Z^*v)$	= (Journel and Huijbregts, 1978)
Zv	= mean sample estimates (a priori value)
Z^*v	= estimated value
Y_i	= weights

If we assume a very simplified linear relationship between the number of boreholes and the precision, then

$$n = Pi^{-a}$$

in logarithmic form, $\ln n = \ln Pi^{-a}$, P = precision = $(Zv - Z^*v)$, Pi = initial precision, n = number of boreholes, and a = percent slope.

The main objective here is to find an optimum n which is compatible with the economics of drilling. Using the following relation, an approximate number of boreholes necessary for a certain degree of precision can be estimated:

$$\sqrt{n} = \frac{st_{0.05}}{0.5CI} = d$$

where

n	$= d^2$ (Waterman and Hazen, 1972)
$t_{0.05}$	= statistical t value for testing at 95% confidence limit
CI	$= \overline{X} \pm S\overline{x}\,(t_{0.05})$ at 95% confidence limit
$\overline{X}$	= sample mean
$S\overline{x}$	= standard error of the mean

Normally, this relation overestimates the number of boreholes necessary for a given degree of precision. This typically is true at the earlier stages of the process. The most likely reasons for overestimation are the imprecise factors of input in the calculation such as

i) Unstable and high standard deviation,

ii) Small initial database (low n), and

iii) Statistical dependence, etc.

This overestimation of the required number of boreholes (*n*) makes this process less applicable for optimization.

Geostatistics, the principles of geostatistics and the main aspects of application, have been discussed thoroughly by Matheron, Journel, Kim, Knudsen, and others (e.g., see Journel and Huijbregts, 1978). In geostatistics, the correlatability of the drilling data is used until the holes are so far apart that one hole does not influence the other or until the data are random in the sense of probability. Therefore, very sparse data in a large area will not provide a meaningful result.

The variogram is the most fundamental part of the geostatistical estimation process. The variogram is a graphical representation of the similarities between observations. Among the various theoretical models available for fitting with the experimental variogram, a spherical model normally is found to be the most applicable for coal deposits. A properly structured variogram of a coal deposit exhibits various relations and estimates which are useful in mining. These are as follows:

i) Extent of similarities between observations

ii) Directions of maximum and minimum homogeneity

iii) Domain of consistency in the database

iv) Inherent inconsistency of the database

The size of the initial database is a very important consideration in the application of geostatistics. An adequate representation of the variabilities in the database and some degree of stability, with a relatively large n, are necessary. However, it would not be useful to assign a standard number of minimum data points needed for every coal deposit. Preliminary descriptive statistics on the database, size of the deposit, and geological consistency should be taken into consideration. Normally, a drilling density of one hole per 130 hectares (320 acres) is a good starting point. However, the size of the reserve also should be taken into consideration. For example, a 91,000,000-tonne (100,000,000-ton), in situ, deep-mineable reserve with a coal thickness of 1.4 m (4.5 ft) would be contained in an area of approximately 52 km^2 (20 mi.^{2}). This will provide 40 drilling data points with a density of two per 2.6 km^2 (1 mi.^{2}). However, if the thickness of the seam is 3 m (10 ft), it will need approximately 23 km^2 (9 mi.^{2}) or 18 drill hole data points for the same amount of reserves. This probably will not be enough data points for a well structured variogram. As is shown later, a significant number of fictitious data points are needed in the system to determine optimum drilling density. Hence, it is necessary that the initial database is stable enough to absorb these fictitious data without distorting the true characteristics of the population.

Ordinary Linear Kriging

At this preliminary stage of evaluation, use of ordinary linear kriging is adequate. This method provides an average estimate for the selected grid (block). Smaller grids to obtain very specific and minute details are rather meaningless if the data are very sparse. If the data are skewed, ordinary linear kriging will not provide a very reliable estimate. It is quite likely that the errors will be relatively large and autocorrelated. Normally, toward the end of the development (secondary) stage, if asymmetry in the data still exists, nonlinear estimators such as disjunctive kriging, lognormal kriging, or indicative kriging should be used for a precise and accurate estimate of tonnage.

METHODOLOGY

Kriging estimates of $\bar{x}$ and variance (s^2) are the two basic quantities needed for drilling optimization. Methodology often is modified by the user according to the need, facilities available, and desired level of accuracy. In this study, software and basic steps to obtain $\bar{x}$ and s^2 concur with the procedure presented by Taylor and Barnes (1983). Additional steps for optimization have been added here. Some practical observations related to coal mining also have been included for convenience. There are four basic parts of this process:

i) Descriptive statistics.

ii) Variogram modelling.

iii) Kriging.

iv) Optimization.

Descriptive Statistics

First, the position statistics for the population need to be determined. Accumulation of a histogram for the population and the Kolmogorov-Smirnov (K-S) test for normality should follow. Bimodality or some degree of skewness often is seen in the data at the initial stages of exploration. These invariably are due to the problem of seam correlation, facies change, depositional irregularities (like split seams, partial washouts, etc.), or structural abnormalities. If necessary based on the above characteristics, the main population can be subdivided into small subpopulations.

If these subpopulations are treated individually, abnormalities can be eliminated. It also is possible that the available data are not truly representative and show a biased picture of the total population. Addition of data from the same source to the database by extending the area of interest or even bringing in data from another location (which is geologically similar) can diminish the skewness or at least indicate how stationary is the dispersion.

Variogram

Variogram modelling at an early stage of drilling with a limited amount of data probably is the most difficult task of all. The most common problems associated with the coal deposits are 1) sagging of the r values with increasing distance h, 2) very high nugget value (normally less than 30% of the sill of the variogram), 3) less pronounced direction of anisotropy, and 4) sometimes clustered or excessively dispersed r values. In most cases, the adjustment techniques suggested by Knudsen and Kim (1978) can overcome the adversities. Also, as suggested earlier, data addition or geological data splitting can improve the variogram. Some degree of data smoothing also can be achieved by transforming it to logarithms (if no zero thickness exists in the set). If the database is very small, then splitting will further reduce the size and the reduced database may become unusable for geostatistical work. In restricted situations like this, interpolated data based on a conservative isopach map may be used. Matching with the true population needs to be verified by evaluating the residuals between the kriging estimates and the true values. In the succeeding drilling program, these locations should be verified by drilling locations.

Grid Selection

Grid or cell (block) size for the interval of estimation should be decided prior to kriging. For each of these cells, an estimated value will be generated by kriging. It is necessary that the grid or cell size be kept constant during the entire process. Obviously, the smaller the cell size the larger would be the percent variation in the estimate. If density of the initial raw data is one point per 130 hectares (320 acres), the cell size for generating estimates should not be less than 16 hectares (40 acres). On the average for the Illinois Basin type of coal deposits, eight empty cells between two adjacent data points for estimation are found to be a good criterion for selection of grids.

Optimization

After the initial operation with the raw data, the mean and percentage variation at a 95% confidence level for the entire deposit is calculated. The input and output files are copied, and the data density and total number of data points are determined. A contour map of percent variation, using each cell, needs to be prepared next for locating zones of weakness and uncertainty in the estimates. Fictitious data points are added to the locations where contour stacking (steep slope) and abnormalities are noticed. Normally, addition of fictitious data is kept to within 25% of the initial raw data. In order to strain the system to its maximum, a lower value at the 95% confidence level is used uniformly. The preceding operations are repeated in succession and fictitious data are added in sequence until the percent variation for each sequence is reduced significantly.

The total number of data points (fictitious and raw), data density, and total variation at the 95% confidence level are determined for each stage of the operation. These quantities are graphed as shown in Figure 1. In order to add a dimension of expenditure, drill holes and cost are plotted against the data density. The point of intersection of these curves represents a very stationary situation which is optimum in terms of drilling density with maximum economic advantage. In other words, the maximum information obtainable at minimum cost.

DISCUSSION

The marginal benefit curve (Figure 1) represents the order of precision after each sequence of drilling. The characteristic of the curve is unique for each coal seam. The change in slope and how quickly a state of stability is established will depend upon the variability in the coal seam. The marginal cost curve is very similar to the marginal benefit curve, but with an inverse disposition.

The slope differentials illustrate a higher degree of instability at the initial stages (at lower borehole density). The probability of geological risk due to uncertainties is significant. Therefore, more data are needed to establish a condition of equilibrium and stability, which is designated as the point of optimum density. This point of equilibrium condition further is complemented by the marginal cost curve. The marginal cost curve, due to its reverse disposition, shows higher economic risk (waste) at the higher drilling density sectors of the curve. It also can be noticed that the degree of variation is diminishing (marginally) at a slower rate beyond this point (optimum drilling) and the required drilling to achieve this is increasing rapidly. Both these curves have been constructed using the same points in the abscissa. This restricts disproportionate variations in the slope of the curve. Normally, for the Illinois Basin type of coal deposit, an overall 10% variation at a 95% confidence level provides a stable base of precision. Beyond this point, local inadequacies are addressed. Henceforth, any additional drilling will help to approach the optimum point.

For this reason, drilling needs to be conducted at a graduated rate of progression. After each sequence, if necessary, the graph needs to be adjusted with respect to the optimum drilling point. The number of boreholes

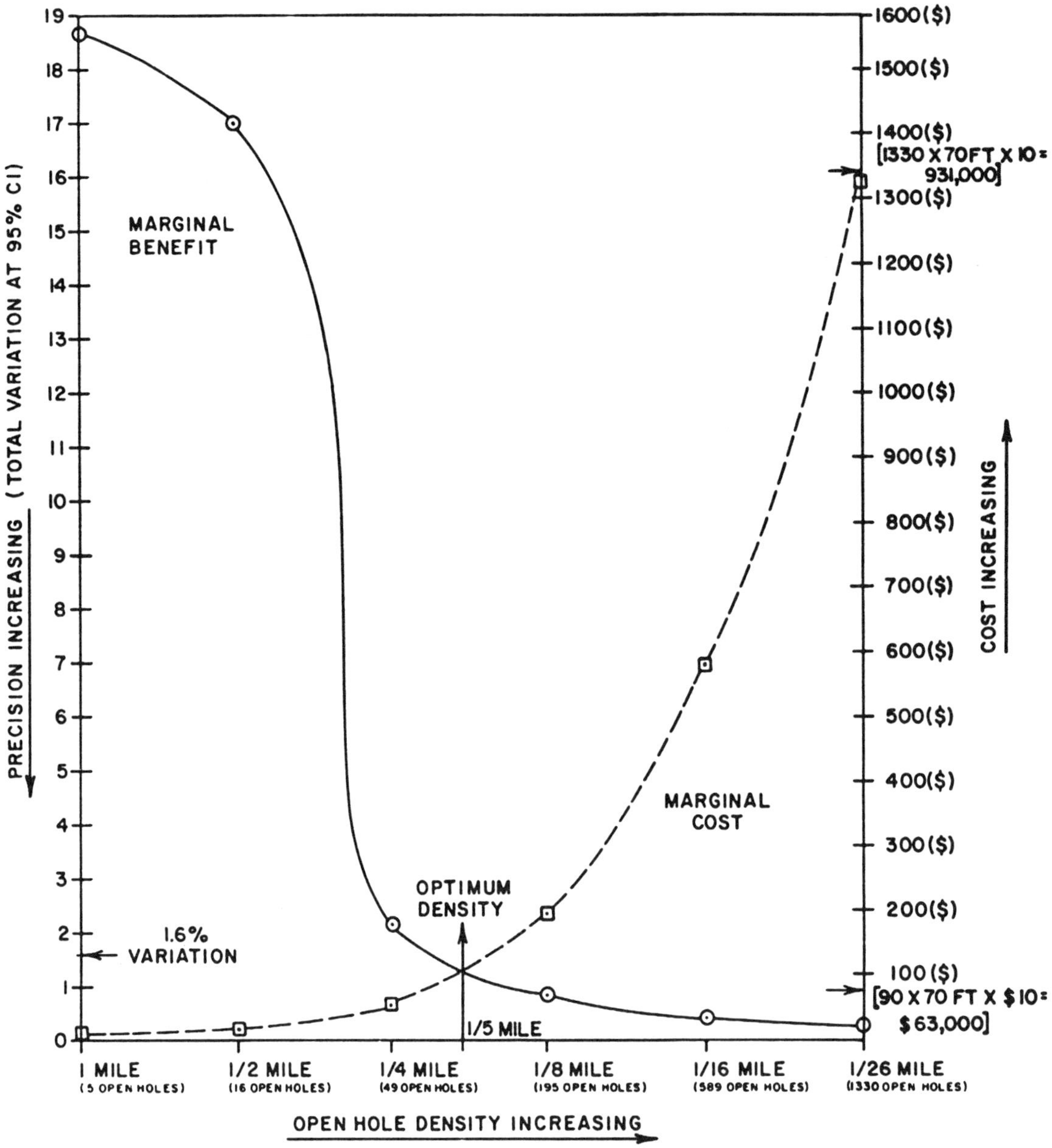

Figure 1. *Determination of optimum borehole density (spacing) by use of marginal-benefit and marginal-cost curves. Cost is calculated as borehole numbers times total depth times cost per foot to drill.*

to be drilled in each sequence is purely an economic consideration. Mobilization and demobilization costs for each sequence are important factors to consider.

The optimization method discussed so far is limited to a single-seam situation. If there is more than one seam in the area, obviously the major seam should be used. In some reserve areas, several seams can be present, but none is consistently dominant over the entire coal field of interest. A proposed mining plan or geological seam plan can help to identify the seam with the greatest potential in a given section and should be used accordingly. Net coal thickness is not influenced by the depth of occurrence or vertical geological succession. Hence, by interchanging the thickness of the seams to be mined, an apparent isotropism can be assumed for the entire area.

In the process of optimization, irrespective of the hole type, each drill hole is considered as a data point. However, in any drilling program, for complete analysis, both open and core holes are drilled. Typically, a 4:1 ratio between the open and core holes is adequate for proving the reserve. Geology and contract specifications may require additional core holes. In general, coring is twice as expensive as open-hole drilling.

Optimization analyses of two geologically very different coal reserves are briefly discussed below. Any significant atypical outcomes of the processes have been substantiated by numerical data and illustrations.

Case 1

This surface minable reserve area of Coal 5 and 6 is located in Southern Illinois. Drilling activities for proving and defining the reserve have been completed and mining also is nearing completion. By recreating the different borehole densities (spacing) in sequence, the drilling program is optimized to demonstrate over-drilling and related cost. Thickness of coal is used as the input variable for its consistency and geological succession. Approximately 2,000 boreholes were drilled, out of which 1,330 were included in the calculations. The remaining boreholes had less than full coal thickness or were drilled to define the crop line. Percent variation contours were drawn for each density to locate the trouble spots. The estimate of precision was established for each sequence and graphically correlated with the cost which is illustrated in Figure 1 (for convenience, only the borehole numbers have been used here).

Case 2

In this example, a reserve area in New Mexico with eight discontinuous coal seams, has been presented for optimizing the drilling program. The reserve is affected by faulting, washouts, and splitting of the coal seams. The prospect is at the initial stage of exploration and presents the typical problems which are encountered at this stage, such as skewness and/or bimodality (in different seams), an unstructured variogram, and large variations in the kriging estimates. These conditions are discussed briefly in succeeding order.

Seam Selection

Due to the complex geology of the area, the seam was selected on the basis of the statistical properties such as position statistics, frequency, K-S statistics, etc. The results are shown below in a tabular form. The coal seams are designated by numbers in vertical succession, such as follows:

6000	Full seam
5990	Lower bench of 6000 seam
5000	Full seam
4990	Lower bench of 5000 seam
6000 + 5990	6000 seam and its lower bench are combined to give the total net coal thickness.

Coal Seam	Combined Benches, Entire Area	Single Seam, Entire Area	Combined Benches, Partial Area	Single Seam, Partial Area
6000		s=1.20, n=79 k-s=9.6		s=1.18, n=35 k-s=31.2
6000+5990	s=1.24, n=77 k-s=27.7		s=0.86, n=33 k-s=60.7	
5990		s=0.94, n=12 k-s=82.8		
5000		s=0.64, n=60 k-s=9.1		s=0.66, n=32 k-s=11.9
5000+4990	s=0.88, n=57, k-s=70.5		s=0.85, n=31 k-s=94.8	
4990		s=0.57, n=21 k-s=70.6		s=0.47, n=17 k-s=95.4
3000		s=2.08, n=95 k-s=0.5		s=2.4, n=41 k-s=3.6
3000+2990	s=1.99, n=91 k-s=1.5		s=2.19, n=40 k-s=7.3	
2990		s=1.36, n=11 k-s=71.0		s=1.48, n=9 k-s=68.3
2980		s=1.37, n=57 k-s=64.5		s=1.27, n=21 k-s=64.4
2980+2970	s=1.39, n=53 k-s=74.3		s=1.26, n=20 k-s=92.4	
2970		s=1.07, n=7 k-s=60.1		s=1.3, n=4 k-s=84.6

Table 1. *Statistical properties of the eight seams for the entire area and geologically homogeneous partial area (n = number of drill holes; s = standard deviation). Kolmogorov-Smirnov (K-S) statistics show the probability of being incorrect if the hypothesis of normal distribution is rejected.*

Based on these statistics, variograms were run to see which seam provided the most organized and structured variogram with maximum data pairs. These analyses indicate that 6000 and its 5990 bench, combined for the entire area, show the best results (Table 1).

Variogram

After routine tests and calibrations, it was found that a distance of 6,100 m (20,000 ft) and 10 intervals (class size) offered the most structured and least scattered variogram. The estimated parameters are c = 0.61, c_o = 0.92, sill = 1.53, range = 3,200 m (10,500 ft), anisotropy direction = 315°, anisotropy ratio = 2.15.

Kriging

Initial kriging was conducted for the entire reserve area with the following specifications: map scale of 2.5 cm = 300 m (1 in. = 1,000 ft), grid cell size = 300 m by 300 m (1,000 ft by 1,000 ft) (9.299 hectares (22.96 acres)), total acreage = 4,771 hectares (11,780 acres), total in situ tonnes (tons) = 32,300,000 tonnes (35,500,000 tons), and average seam thickness = 0.521 m (1.71 ft). The isopach of percent variations at a 95% confidence level is shown in Figure 2.

Optimization

A small portion of the area, which included 101 grid cells totalling 938.8 hectares (2,318 acres), was chosen for optimization. In a sequence of five steps, fictitious boreholes (with thickness data representing the minimum value at 95% confidence level) were added to the locations of maximum variations. Altogether, 100 fictitious boreholes were added to the system for optimization, which is summarized in Figure 3.

As indicated by Figure 3, the expected optimum borehole density is one hole per 20 hectares (50 acres) in order to adequately characterize the reserves. This means that fewer holes probably will not detect all significant variations in the coal seams involved and that a high density of holes will not provide enough additional significant information to justify their drilling cost.

Figure 2. *Isopach map of the percent variation, at the 95% confidence level, for the combined 6000 and 5990 coal seams.*

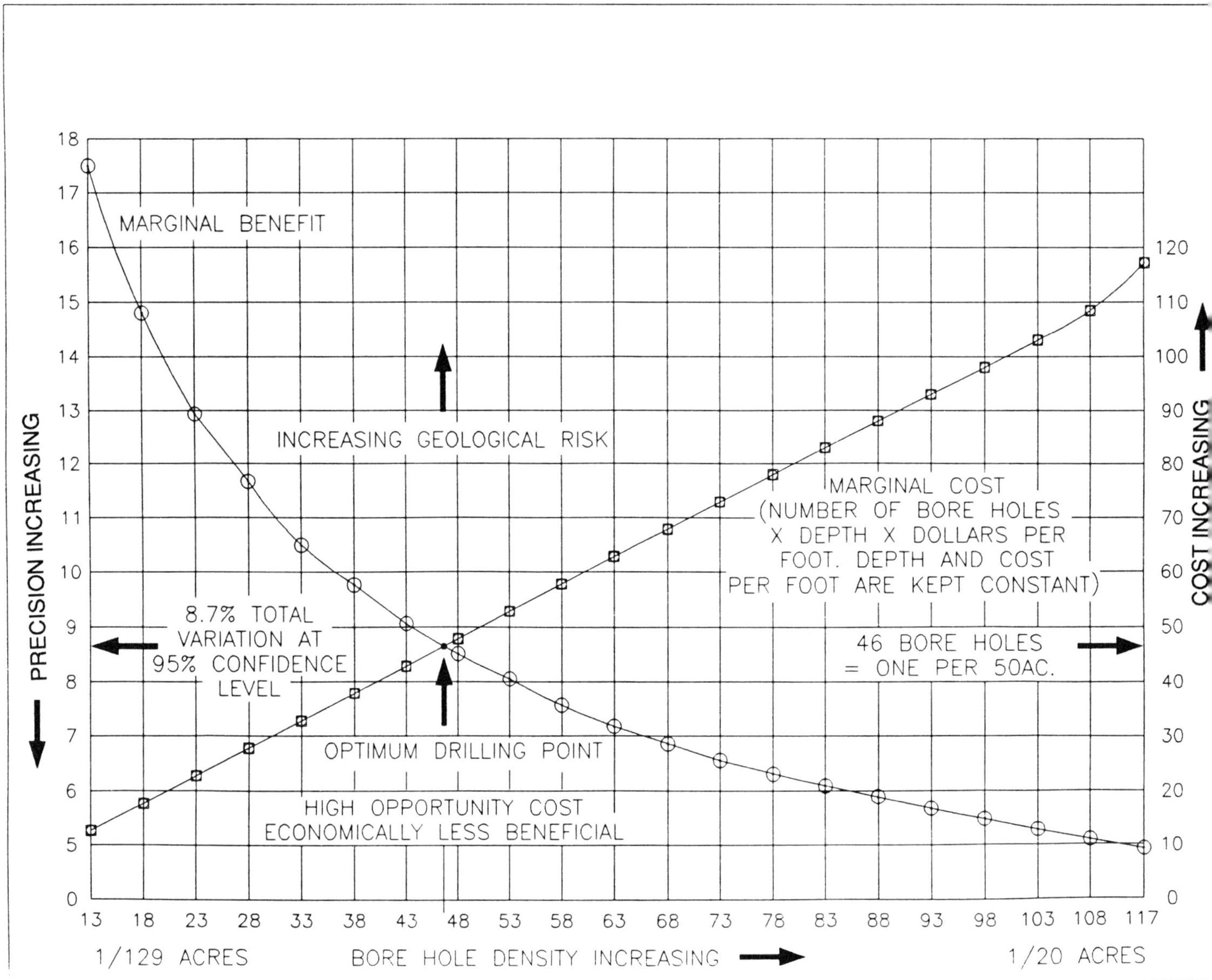

Figure 3. *Graph of marginal-benefit and marginal-cost curves indicating the optimum drilling point for the Standing Rock area.*

CONCLUSIONS

This study shows that, by using geostatistics, the vague or arbitrary way of designing a drilling program can be avoided. A numerical base of precision indicating the number of boreholes necessary and the locations to be drilled can be achieved prior to an elaborate drilling program. Also, uneconomic and excessive drilling can be eliminated within the guidelines of optimization.

REFERENCES

Baafi, E.Y., Y.C. Kim, and F. Szidarovszky, 1986, On nonnegative weights of linear kriging estimation: Mining Engineering, v. 38, no. 6, p. 437-442.

Journel, A.G., and Ch.J. Huijbregts, 1978, Mining Geostatistics: New York, Academic Press, p. 20-24.

Kim, Y.S., P.K. Medhi, and L.S. Roditis, 1987, Performance evaluation of indicator kriging in a gold deposit: Mining Engineering, v. 39, no. 11, p. 947-952.

Knudsen, H.P., and Y.C. Kim, 1978, A short course on geostatistical ore reserve estimation: Tucson, University of Arizona, Unpublished Short Course Notes, p. 118-223.

Mansfield, E., 1982, Microeconomic theory and application: New York, W.W. Norton and Company, p. 142-170.

Neter, J., W. Wasserman, and G.A. Whitmore, 1982, Applied statistics: Boston, Allyn and Bacon, Inc., p. 45-104.

Taylor, H., and R. Barnes, 1983, A short course on geostatistics: Golden, Colorado School of Mines, Mining Engineering Department, Unpublished Short Course Notes.

Waterman. G.C., and S. Hazen, 1968, Development Drilling and Bulk Sampling, in E.P. Pfleider, et al., eds., Surface mining: New York, American Institute of Mining, Metallurgical, and Petroleum Engineers, Inc., p. 85-87.

RESTRICTIONS TO MINING: THEIR EFFECT ON AVAILABLE COAL RESOURCES[1,2]

Roy S. Sites, Elizabeth V.M. Campbell, and Karen K. Hostettler
Virginia Division of Mineral Resources
P.O. Box 3667
Charlottesville, VA 22903

Nancy K. Gardner and Susan J. Tewalt
U.S. Geological Survey
956 National Center
Reston, VA 22092

ABSTRACT

Only a portion of the coal traditionally considered as remaining coal resources is minable because of a variety of restrictions imposed on mining. Two categories of restrictions were examined for impact on coal resources in the Southwest Virginia coal field: land-use and technologic restrictions. Land-use restrictions are limitations imposed upon mining by government regulatory policies and by potential for liability. Technologic restrictions are those imposed upon mining by the limitations of mining technology and by safety concerns. Available coal resource studies were completed for two 7.5-min. quadrangle areas within the southwest Virginia coal field. Although both quadrangles cover portions of the same coal field, coal bed geologic conditions differ between them. The Vansant Quadrangle is located toward the interior of the coal field and contains gently dipping to flat-lying coal beds throughout the entire quadrangle. The Wise Quadrangle, located at the southeastern edge of the coal field, contains coal beds that are folded and faulted, with the majority of the coal beds occurring in the northern half of the quadrangle.

After restrictions to mining were applied, the available coal resources in the Vansant Quadrangle were calculated to be 530,000,000 tonnes (583,000,000 tons) or 67% of the remaining coal resources. Technologic restrictions account for 99.6% of the restricted coal. In the Wise Quadrangle, available coal resources were 470,000,000 tonnes (520,000,000 tons), which also is 67% of the remaining coal resources, with technologic restrictions accounting for 89.6% of the restricted coal.

---

1. Note: The term "coal bed" is used in this paper for resource calculation purposes to signify distinct coal-only units rather than "coal seams" which could include partings and other subunits of non-coal material.

INTRODUCTION

Traditionally, efforts to estimate coal resources by federal and state agencies have concentrated on defining remaining coal resources. These resources are categorized by using coal thickness, overburden thickness, and probability of occurrence. However, not all coal considered as remaining is minable because of a variety of restrictions to mining. To examine the impact of some of these restrictions in different states and to determine what portion of the remaining coal resources realistically is available, the U.S. Geological Survey entered into a cooperative project with the following agencies of three states: the Virginia Division of Mineral Resources, the Kentucky Geological Survey, and the West Virginia Geological and Economic Survey. This report summarizes the results of studies in Virginia of two 7.5-min. quadrangles: the Vansant Quadrangle in Buchanan County and the Wise Quadrangle in Wise and Scott counties. Details of the entire cooperative project are discussed by Carter and Gardner (1989).

Both quadrangles lie near the southeastern edge of the Cumberland Mountains section of the Appalachian Plateaus physiographic province in the Southwest Virginia coal field (Figure 1). The Vansant Quadrangle is located in south-central Buchanan County, and the Wise Quadrangle is located in south-central Wise County and northernmost Scott County. The study areas are characterized by highly dissected topography with generally steep slopes and V-shaped valleys. The maximum relief is as much as 464 m (1,520 ft) in the Vansant Quadrangle and 666 m (2,180 ft) in the Wise Quadrangle.

The available coal resource study undertaken for the Vansant Quadrangle involved evaluating 20 Pennsylvanian-age bituminous coal beds (Figure 2). Fifteen of the coal beds studied are above drainage level. The remaining five are below drainage level and have no traceable outcrop within the quadrangle. All 20 coal beds occur within a stratigraphic interval of approximately 790 m (2,600 ft). This interval thins to around 656 m (2,150 ft) near the northwestern corner of the quadrangle. The deepest coal bed currently mined in the Vansant Quadrangle is the Pocahontas No. 3. This coal bed is from 270 to 700 m (900 to 2,300 ft) below ground and is mined extensively by the longwall method. Only those coal beds found to be more regionally persistent and exceeding 36 cm (14 in.) in thickness were studied.

In the Wise Quadrangle, 18 coal beds were evaluated (Figure 3). All of the coal beds studied are above drainage level locally and have traceable outcrops within the quadrangle. All 18 coal beds are within a stratigraphic interval of approximately 999 m (3,275 ft). The southern half of the quadrangle encompasses the east-west trending portion of the Powell Valley anticline. As a result, the coal-bearing rocks upwarped by the anticline largely were eroded from the southern half of the quadrangle. Only the four oldest coal beds are now exposed to the south, along the flanks of the anticline. As in the Vansant Quadrangle, only those coal beds found to be more regionally persistent and greater than 36 cm (14 in.) in thickness were studied.

The available coal resource study for the two quadrangles included outcrop information from published geologic maps of the quadrangles (Nolde and Mitchell, 1984; Whitlock et al., 1988). Outlines of mined areas were obtained from mine maps acquired from the Virginia Division of Mines. All outcrop and mine line data were compiled on mylar overlays and digitized at a scale of 1:24,000. Additionally, the study made use of coal thickness data from 703 surface locations and 140 drill holes in the Vansant Quadrangle and data from 234 surface locations and 87 drill holes in the Wise Quadrangle. Several detailed stratigraphic sections were constructed from these data within each quadrangle to assure accurate coal bed correlations.

Computer-based geostatistical programs developed by the U.S. Geological Survey (Olson, 1977) were used to calculate resources for each coal bed using categories of coal thickness (greater than 36 cm (14 in.), exclusive of partings), overburden thickness, and occurrence probability. The tonnages of the original coal resources were computed using the boundaries for each coal bed (defined by the quadrangle edge and the outcrop trace), the coal thickness therein, and a coal weight of 13,200 tonnes/hectare-m (1,800 tons/acre-ft).

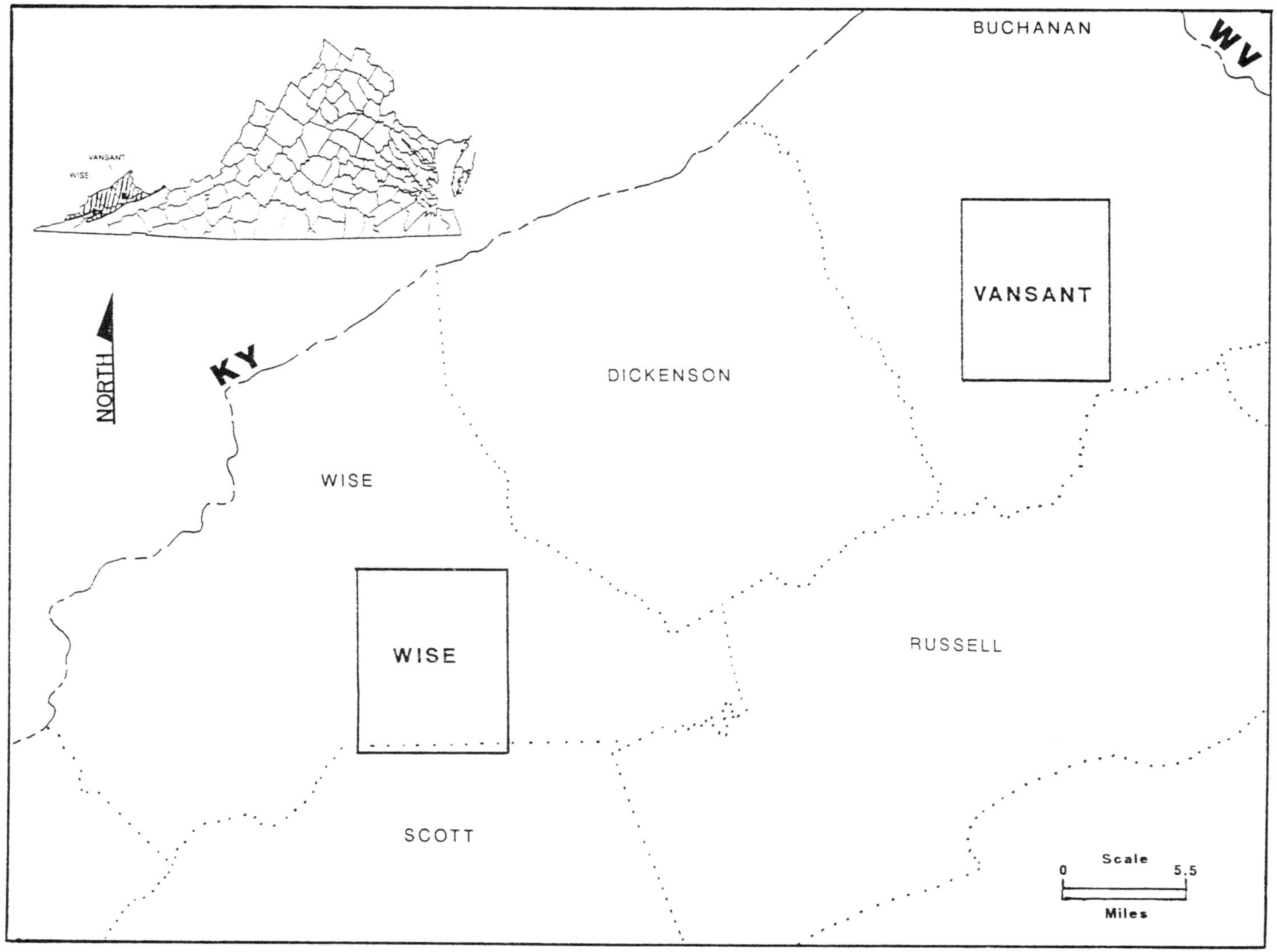

Figure 1. *Location map of the Vansant Quadrangle, Buchanan County, and the Wise Quadrangle, Wise County, used for coal availability studies in the Southwest Virginia coal fields.*

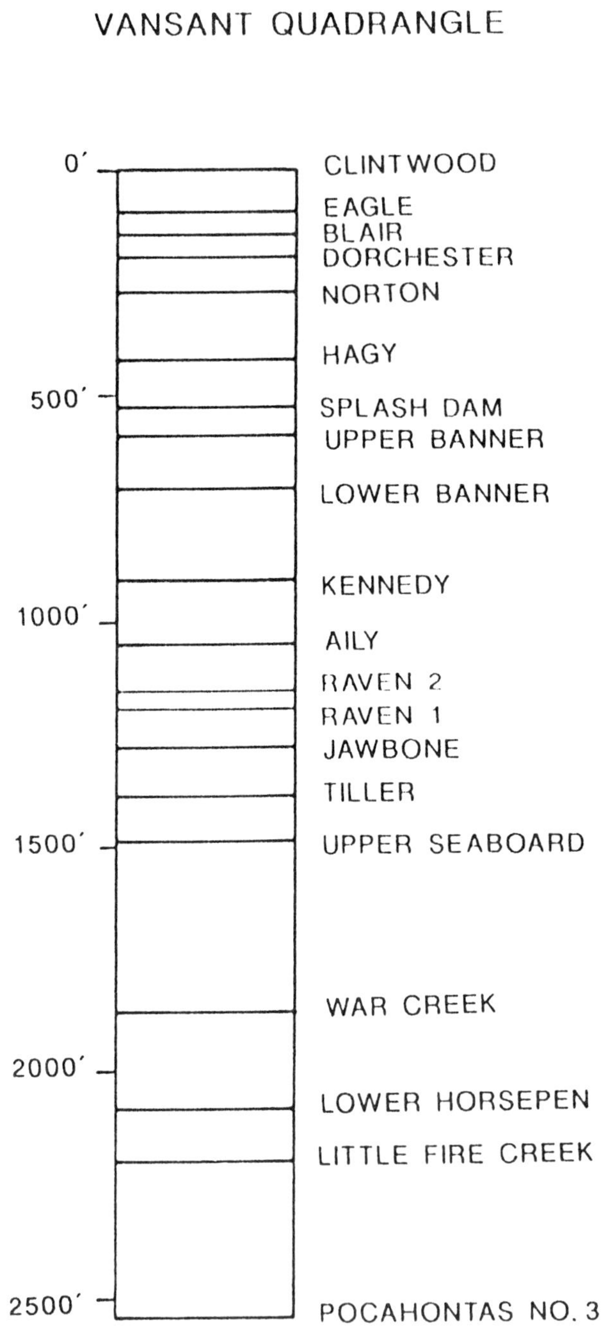

Figure 2. *Coal bed columnar section showing coal beds used for the coal availability study of the Vansant Quadrangle, Buchanan County, Virginia.*

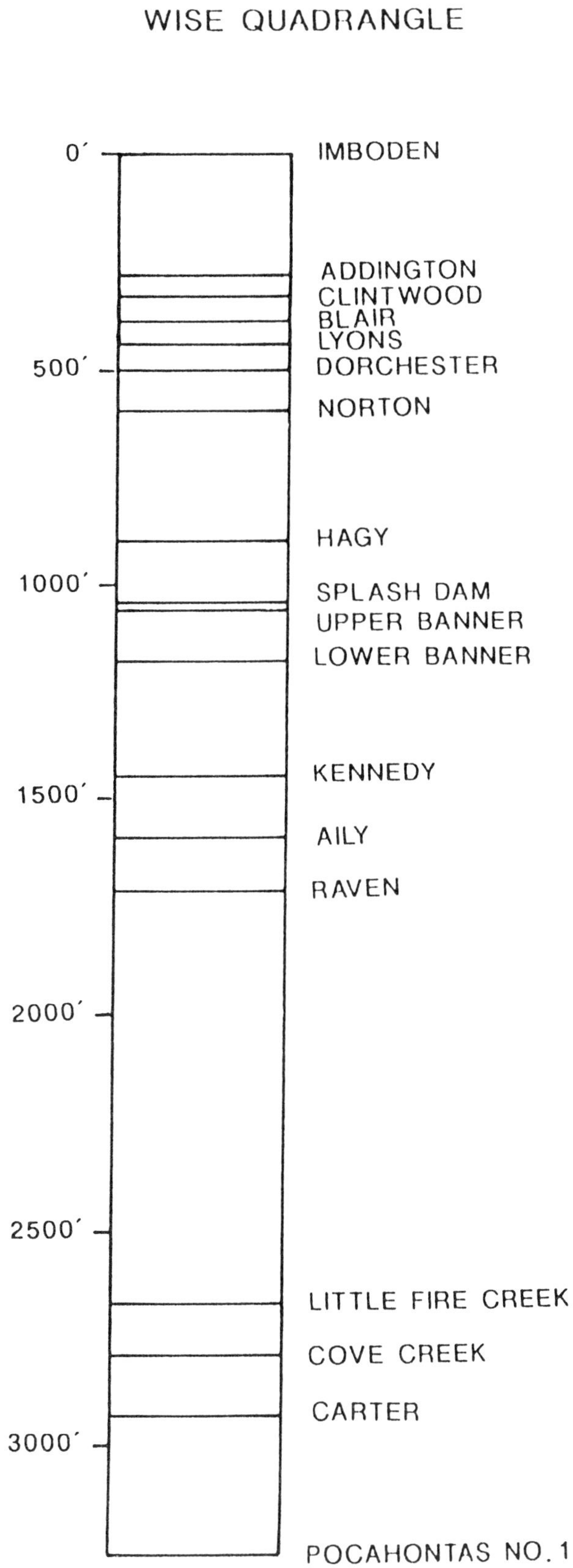

Figure 3. *Coal bed columnar section showing coal beds used for the coal availability study of the Wise Quadrangle, Wise County, Virginia.*

Remaining coal resources were defined by removing all mined coal areas (stripped, augered, and deep mined) from the defined original coal resources of each coal bed. The tonnages of remaining coal resources then were computed, as were the original resources. The tonnages of mined-out coal for each coal bed were determined by subtraction of remaining resources from original resources. Restrictions to mining operations were identified for each coal bed and those areas were removed from each coal bed's remaining resource, thus defining the available coal resources per bed. Tonnages of each type of restriction, exclusive of overlapping areas, then were computed for each coal bed. Available coal resources for each coal bed also were computed in the same manner as the original and remaining resources. The methodology is discussed in greater detail by Eggleston et al. (1990).

RESTRICTIONS TO MINING

The restrictions to mining examined in this study fall into two categories: land-use and technologic. Specific restrictions were identified from discussions with local coal producers and governmental regulators. Land-use restrictions are limitations imposed upon mining by government regulatory policies and by the potential for liability. Land-use restrictions, which primarily affect surface mining and shallow underground mining (depths of 0-60 m (0-200 ft)), applied in both studies were 1) the presence of local population centers, 2) a need for a 60-m (200-ft) buffer zone around oil and gas wells, 3) a need for a 30-m (100-ft) buffer zone around cemeteries, and 4) the need for a 30-m (100-ft) buffer zone along the sides of streams having more than 0.1 m^3/sec (5 ft^3/sec) estimated minimum flow. For the Wise Quadrangle, the presence of Jefferson National Forest lands imposed an additional restriction to surface mining. In both study areas, restrictions to mining because of the presence of highways, pipelines, and powerlines were determined to affect very small areas or to have no affect on mining and, therefore, were not applied.

Technologic restrictions are those imposed upon mining by the limitations of mining technology and by safety concerns. The applied technologic restrictions predominantly affect underground mining. These restrictions were 1) the need for a 60-m (200-ft) buffer zone around all underground mine workings, 2) the occurrence of interburden less than 12 m (40 ft) thick between two unmined coal beds, 3) the existence of previous mining of a coal bed less than 12 m (40 ft) above or below an unmined coal, 4) a coal thickness of less than 71 cm (28 in.) with greater than 300 m (1,000 ft) of overburden, and, 5) for coal beds below drainage level that must be mined through a shaft, a coal thickness of less than 100 cm (40 in.). Coals with overburden thicknesses greater than 300 m (1,000 ft) only are being mined when they are more than 71 cm (28 in.) thick. Also, mine operators in Virginia consider coals below drainage level that are less than 100 cm (40 in.) in thickness not to be economic to mine currently. Buffer zones applied to existing oil and gas wells are common to both land-use and technologic restrictions. Areas of restriction overlap were adjusted to prevent overestimation of restrictions.

RESULTS

Total calculated original coal resources are 916,000,000 tonnes (1,008,000,000 tons) in the Vansant Quadrangle and 734,000,000 tonnes (807,000,000 tons) in the Wise Quadrangle. In the Vansant Quadrangle, 14% of the coal has been mined, leaving 787,000,000 tonnes (866,000,000 tons) of remaining coal resources (Figure 4a). In the Wise Quadrangle, only 4% of the coal has been mined, leaving 701,000,000 tonnes (771,000,000 tons) of remaining coal resources (Figure 4b). After applying the specific restrictions, the total available coal resources were calculated to be 530,000,000 tonnes (583,000,000 tons)(67% of the remaining resources) in the Vansant Quadrangle (Figure 5a) and 470,000,000 tonnes (520,000,000 tons)(67% of the remaining resources) in the Wise Quadrangle (Figure 5b). Figure 6 compares the original, remaining, and

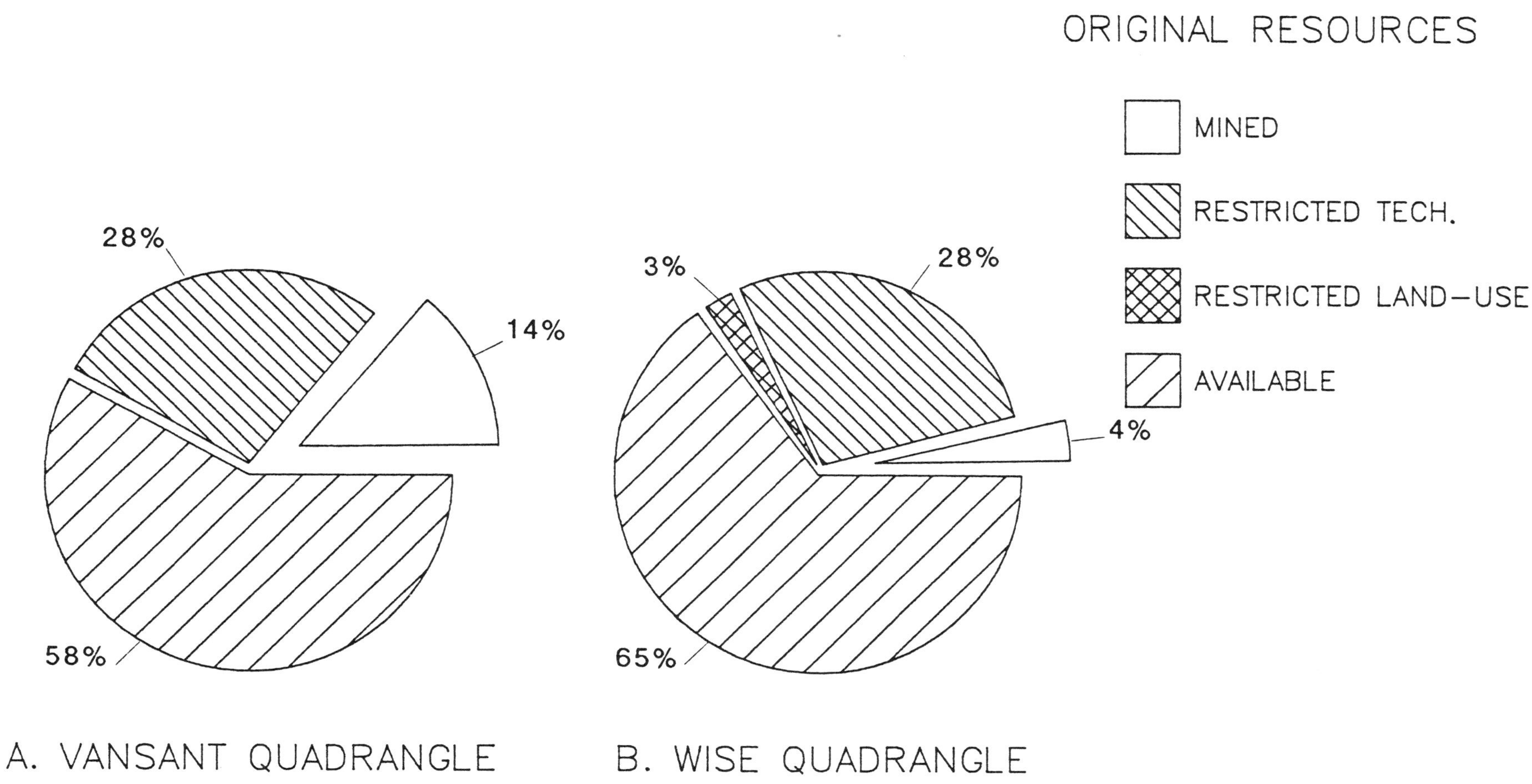

Figure 4. *Pie diagrams of original coal resources showing percentages of mined, restricted, and available coal for the A) Vansant and B) Wise quadrangles. Remaining coal resources are a total of the technologically restricted, land-use restricted, and available coal resources.*

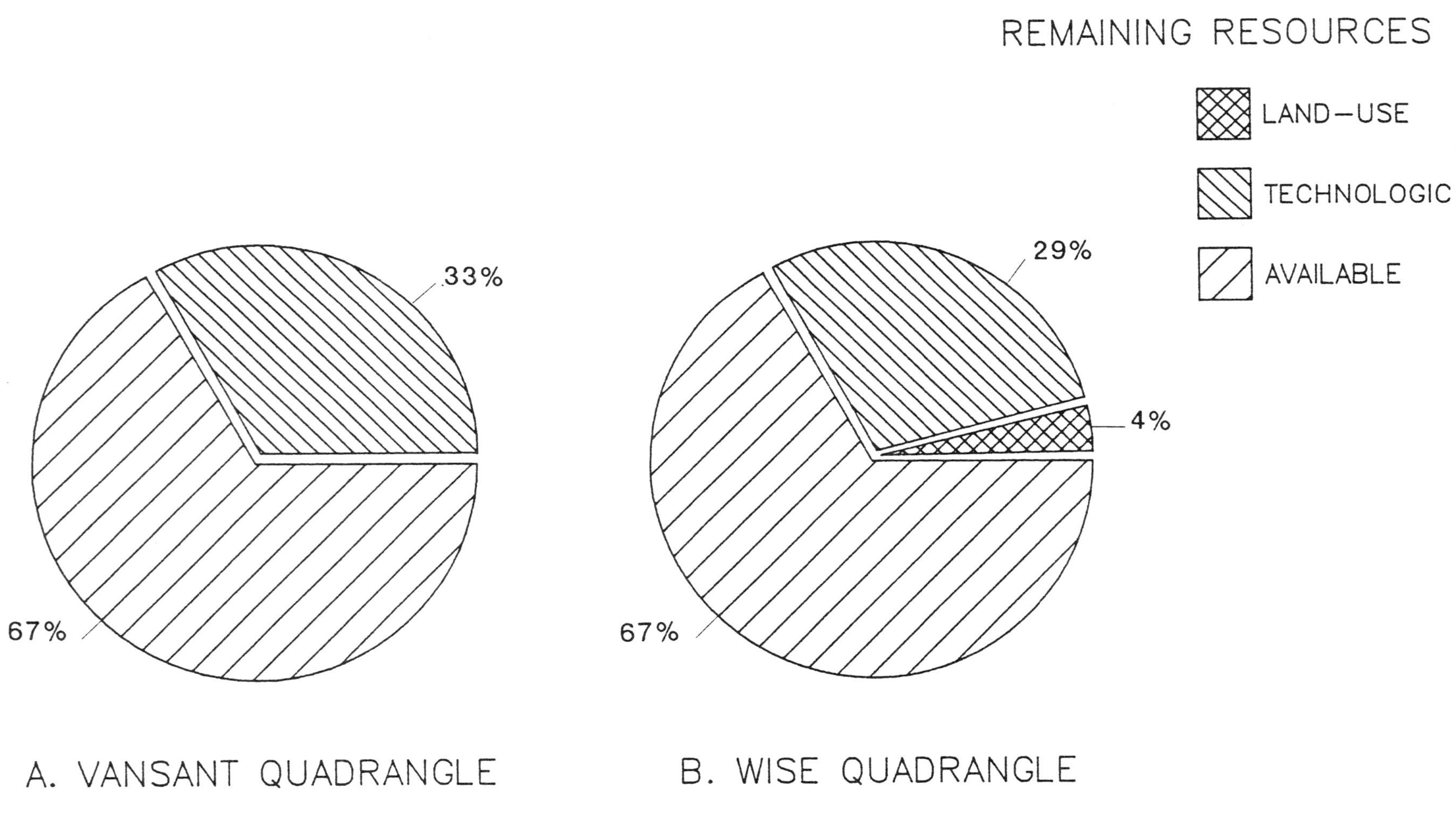

Figure 5. *Pie diagrams of remaining coal resources showing percentages of available coal and percentages affected by land-use and technologic restrictions for the A) Vansant and B) Wise quadrangles.*

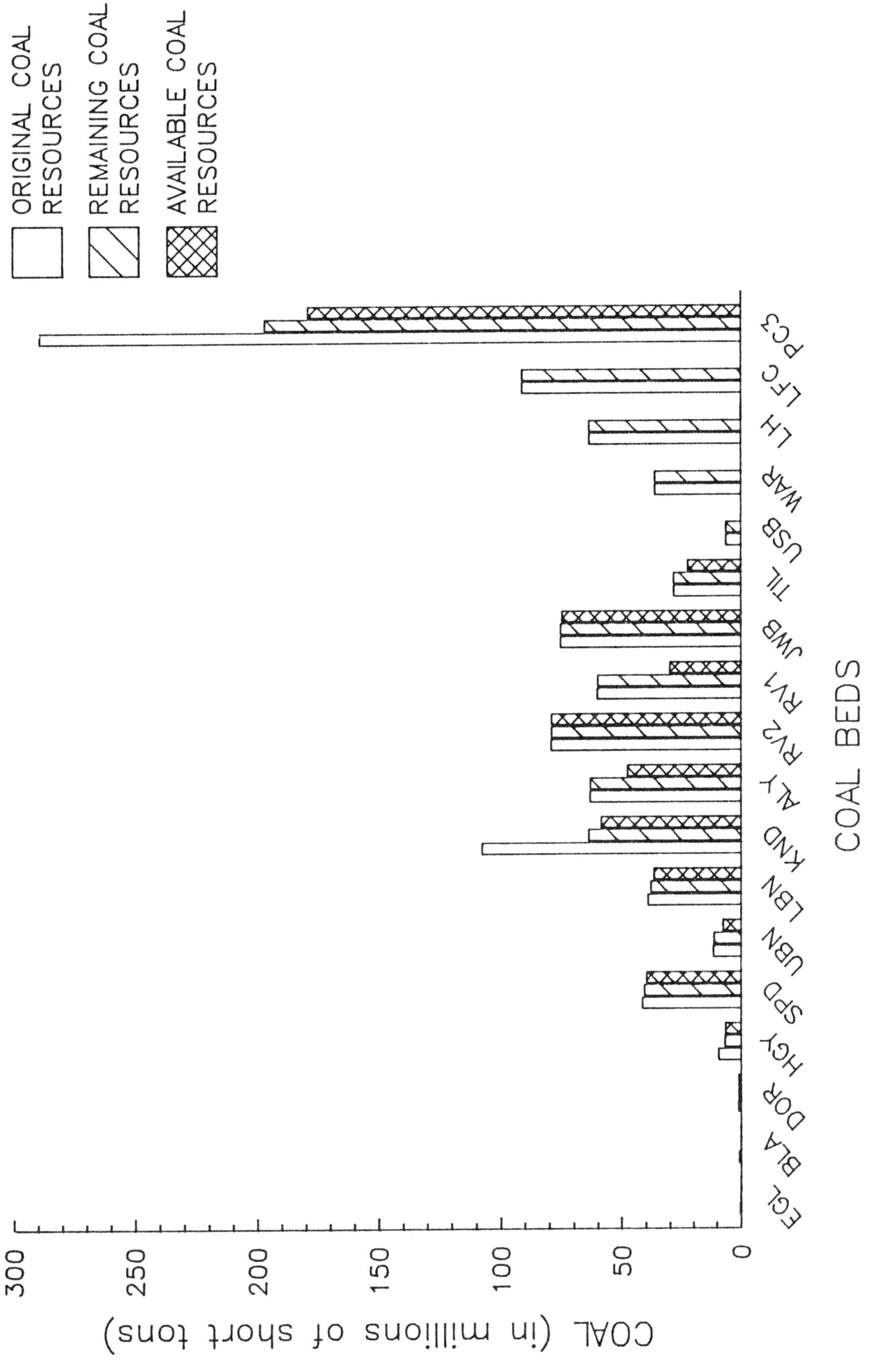

Figure 6. *Bar chart showing original, remaining, and available coal resources for each coal bed addressed in the Vansant Quadrangle study. See Figure 2 for the coal bed names.*

available resources for each coal bed in the Vansant Quadrangle. The Clintwood and the Norton coals were found to be mined out and too thin, respectively. Figure 7 compares original, remaining, and available resources for each coal bed in the Wise Quadrangle.

In the Vansant Quadrangle, the total restricted coal represents 33% of the remaining coal resources (Figure 5a). Land-use restrictions account for only 0.4% of all restricted coal. Local population centers contributed 88% of the land-use restrictions (Figure 8a). The upper five coal beds which occur only along mountain tops are not affected by restrictions because of their limited areal extent. These five coal beds contain 2,360,000 tonnes (2,600,000 tons) of remaining and available resources which account for only 0.5% of the total available coal resource for the quadrangle. The five coal beds below drainage level were not affected by land-use restrictions because most occur at depths greater than 60 m (200 ft). However, because they are below drainage level and require shaft mining, they are affected by the technologic restriction of a 100-cm (40-in.) minimum coal thickness.

Technologic restrictions account for 99.6% of all restricted coal in the Vansant Quadrangle (Figure 5a). The 100-cm (40-in.) minimum thickness restriction for coal beds below drainage level accounts for 73% of the coal excluded by technologic restrictions (Figure 9a). The second most important technologic restriction is the interburden minimum thickness which accounts for the exclusion of 21% of the technologically restricted coal. Only 4,500,000 tonnes (5,000,000 tons)(0.6% of the remaining resource) of the coal from the above drainage-level coal beds are less than 71 cm (28 in.) thick and are covered by more than 300 m (1,000 ft) of overburden. This is only 2% of the coal impacted by technologic restrictions.

In the Wise Quadrangle, the total restricted coal represents 33% of the remaining coal resources (Figure 5b). Land-use restrictions account for 10.4% of all restricted coal. Again, local population centers were the most important cause, accounting for 78% of the coal excluded by land-use restrictions (Figure 8b). National forest lands accounted for the second most important land-use restriction, accounting for 18% of the land-use restricted coal. National forest lands contain 0.6% of the total remaining coal resource. As can be noted, the impact of land-use restrictions is greater in the Wise Quadrangle than in the Vansant Quadrangle because of the restrictions placed on surface mining on national forest lands. Only the uppermost coal bed, the Imboden, was not affected by restrictions. The Imboden, however, contains only 180,000 tonnes (200,000 tons) of remaining and available coal resources and accounts for less than 0.05% of the total available coal resource.

Technologic restrictions account for 89.6% of all restricted coal in the Wise Quadrangle (Figure 5b). The restriction of having a 71-cm (28-in.) minimum thickness when the overburden thickness is greater than 300 m (1,000 ft), accounted for 69% of the coal excluded by technologic restrictions (Figure 9b). The second most important technologic restriction is the interburden thickness minimum which accounts for 24% of the technologically restricted coal. None of the coal beds were entirely below drainage level, therefore the 100-cm (40-in.) minimum coal thickness for below-drainage-level coals applied in the Vansant Quadrangle was not a restriction in the Wise Quadrangle.

FURTHER CONSIDERATIONS

A possible technologic restriction for above-drainage-level coals, not addressed when calculating the available coal resource values for these studies, is the exclusion of coals less than 71 cm (28 in.) thick with 60-300 m (200-1,000 ft) of overburden. Many coal producers consider coal in this category not producible because of the special equipment required and the general difficulty in mining. This restriction could affect

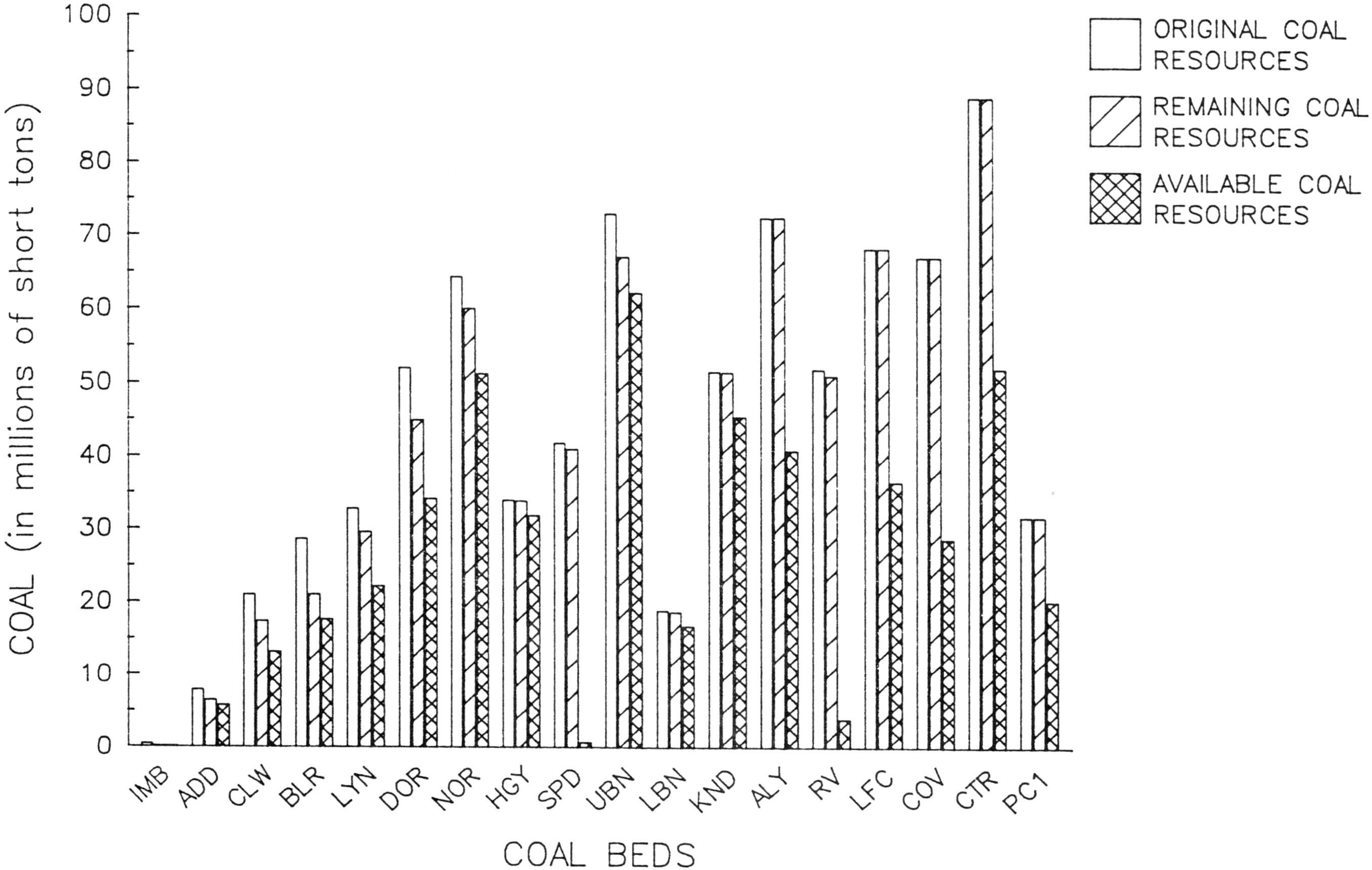

Figure 7. *Bar chart showing original, remaining, and available coal resources for each coal bed addressed in the Wise Quadrangle study. See Figure 3 for the coal bed names.*

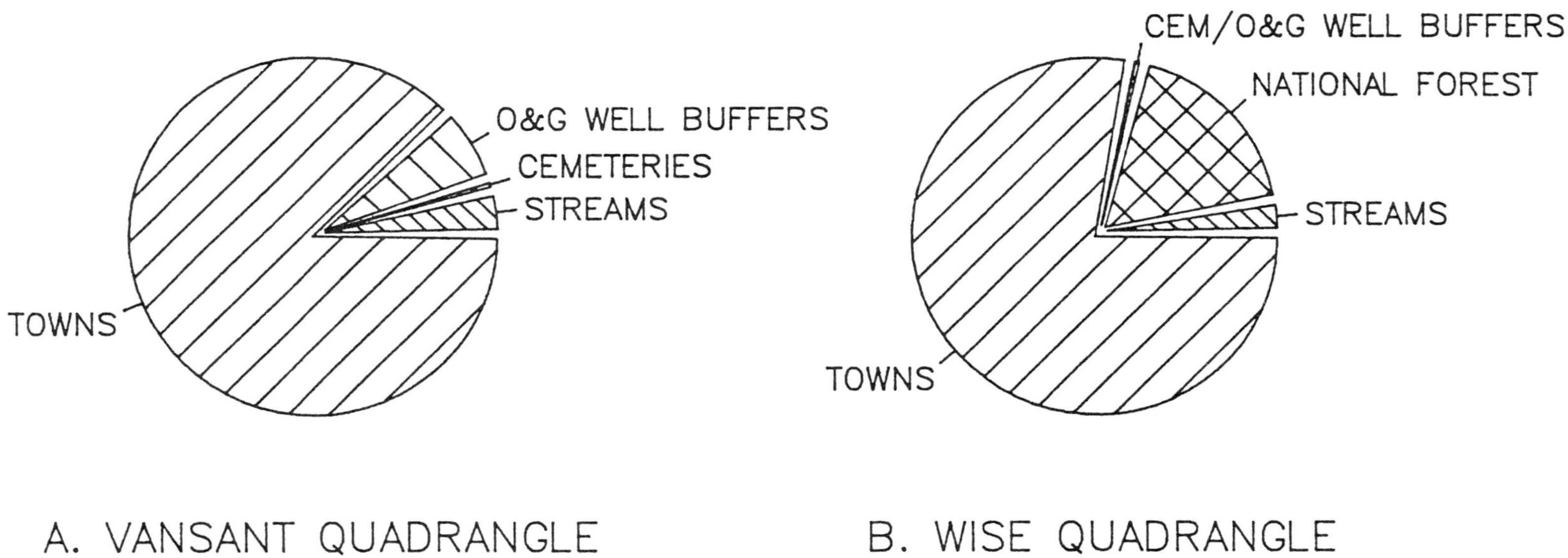

Figure 8. *Pie diagrams of land-use restriction categories showing the relative amounts of each for the A) Vansant and B) Wise quadrangles.*

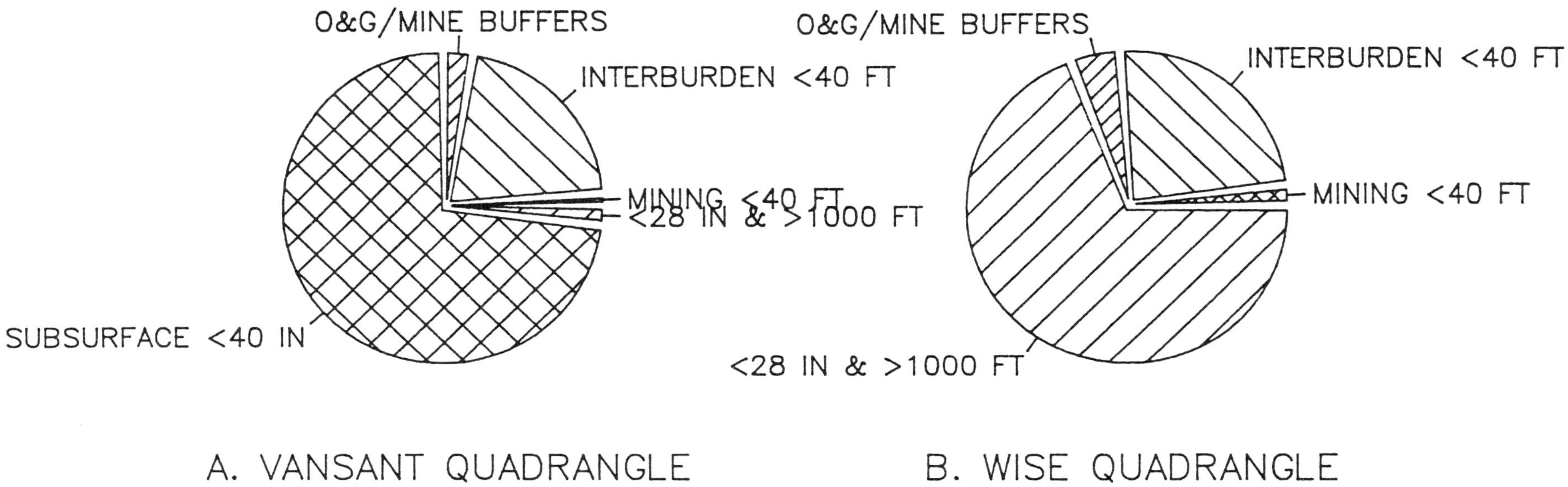

Figure 9. *Pie diagrams of technologic restrictions showing factors considered for the A) Vansant and B) Wise quadrangles. For the Vansant Quadrangle, values of coal affected by mine buffers in areas with less than 60 m (200 ft) of overburden are not included. For the Wise Quadrangle, none of the coal beds were entirely below drainage level; therefore, the "less than 100 cm (40 in.)" restriction did not apply.*

27% more of the remaining coal resources in the Vansant Quadrangle and 17% more of the remaining coal resources in the Wise Quadrangle.

Other categories of restrictions which may be examined as more detailed information becomes available are geologic and geochemical restrictions. Geologic restrictions could include such factors as sand channels, weight percentage of partings, and geologic structures that produce unstable mine roofs. Geochemical restrictions, such as sulfur and ash content, also affect coal minability.

SUMMARY

The available coal resource study for both the Vansant Quadrangle and the Wise Quadrangle found technologic restrictions to be the most significant factor when calculating resource estimates. In the Vansant Quadrangle, 33% of the total remaining coal resources were affected by restrictions, of which 99.6% were technologic. In the Wise Quadrangle, 33% of the total remaining resources were affected by restrictions, of which 89.6% were technologic. Estimates of the available coal resources must take into consideration restrictions to mining such as land-use and technologic restrictions. Remaining resources alone do not give a realistic estimate of the coal available for mining.

REFERENCES

Carter, M.D., and N.K. Gardner, 1989, An assessment of coal resources available for development: U.S. Geological Survey Open-File Report 89-362, 52 p.

Eggleston, J.R., M.D. Carter, and J.C. Cobb, 1990, Coal resources available for development--A methodology and pilot study: U.S. Geological Survey Circular 1055, 15 p.

Nolde, J.E., and M.L. Mitchell, 1984, Geology of the Prater and Vansant quadrangles, Virginia: Virginia Division of Mineral Resources, Publication 52, 28 p., 1 map.

Olson, A.C., 1977, Graphic analysis of resources by numeric evaluation techniques (GARNET): Computers and Geosciences, v. 3, p. 539-545.

Whitlock, W. W., J.A. Lovett, and R.N. Diffenbach, 1988, Geology of the Wise Quadrangle and the coal-bearing portion of the Fort Blackmore Quadrangle, Virginia: Virginia Division of Mineral Resources, Publication 80, 1 map.

Chapter 3
Coalbed Methane

PREFACE

Coalbed methane hardly could be considered a part of "traditional" coal geology. Many practitioners probably would consider themselves more akin to petroleum geologists, especially in the area of exploration and production techniques. Yet, anyone who has ever been involved with coal mining or coalbed methane, especially for the middle-rank coals, understands the inseparable nature of coal deposit geology and methane presence/generation.

This separate chapter is an acknowledgement of the importance of coalbed methane, both to the coal and petroleum industries. The overview paper by Murray provides a general review of the status of knowledge of coalbed methane resources in the U.S. and of the exploration and production techniques involved in exploiting this "unconventional" resource.

The biggest single consideration for coalbed methane exploration and production, other than methane generation in the first place, is the structural geology of the coal deposit. Ertekin et al. provide the general framework of how coal acts as a reservoir and how it releases methane. Moving up the frame-of-reference scale, we come to the paper by Soeder where the impact of depth of the coal deposit on reservoir properties is examined. To broaden the frame a little more, and to consider the problems involved in producing in a tectonically complex area, we come to Mitchell and Pappajohn who detail exploits in an unconventional setting for this unconventional gas resource.

Cross-Reference

Although the papers deal with shallower deposits than is usual for coalbed methane, the three papers in the Exploration and Reserve Definition chapter provide useful geologic and geophysical concepts for coalbed methane exploration as well. Likewise, the papers in the Reserve Estimation chapter involve reserve estimation concepts that are valid for coalbed methane as well. Despite the differences in gas generation, the papers by Beaver et al. in the Underground Coal Gasification and the Environmental Impacts chapters discuss the hydrogeology of coal deposits, a most applicable subject considering that many coalbed methane wells start out as "water wells".

In the Mining chapter, the papers by Collins, Hucka, Molinda and Ingram, and Nelson and Bauer complement the emphasis on structural geology evident in the coalbed methane papers. Finally, the paper by Winston in the Coal Quality Concerns chapter provides a concise integration of basin geology with coalbed methane and other coal quality parameters for the Warrior coal field, one of the two most active areas in the U.S. for coalbed methane development.

COALBED METHANE: NATURAL GAS RESOURCES FROM COAL SEAMS

D. Keith Murray
D. Keith Murray & Associates, Inc.
1658 Cole Blvd.
Golden, CO 80401

INTRODUCTION

Methane trapped, or occluded, in coal seams is a virtually untapped source of clean, sulfur-free, pipeline-quality energy that today constitutes an economically viable exploration and development objective. The production of gas from coal seams can be accomplished by drilling and completing either vertical or horizontal boreholes, utilizing essentially conventional technology, with some modifications, especially in the completion of such wells. In addition, many coalbed methane wells require initial dewatering of the coal by means of a variety of pumping arrangements. Most of the wells drilled for coal degasification require reservoir stimulation, usually through casing perforations, employing various combinations of hydraulic fracturing, such as sand-water, sand-foam, or sand-gel, in order to increase permeability to gas.

The production of gas from coal seams, unlike in situ gasification techniques, is non-destructive to the coal, except possibly by the enlargement of already present cleats or fractures in the coal as a result of hydraulic fracturing. Furthermore, coalbed methane is high in heating value, generally between approximately 900 and 1,050 Btu/scf. Conversely, gas produced from the in situ combustion of coal is low in heat content, typically in the range of 150 to 300 Btu/scf.

The coalification process, by which organic matter (e.g., wood and peat) is converted to coal, generates very large quantities of methane, both biogenic (i.e., formed by bacterial action in the early stages of diagenesis) and thermogenic (formed by thermal reactions during the phase of catagenesis). This gas is stored in high concentrations in both the coal and the associated sediments after the gas-expulsion point in the coal has been attained. Volume-for-volume, high-rank coals are capable of storing several times as much gas as are porous sandstone reservoirs under similar conditions of pressure due to the extremely high internal surface areas of coal — as high as 307 m^2/g (1,500,000 ft^2/lb) (Cervik, 1969).

Until recently, methane from coal seams has been considered an "unconventional" resource because of the unique properties of coal (which constitutes both a source and a reservoir of natural gas). Coal is an extremely complex substance, one that still is incompletely understood. According to VanKrevelen (1961), coal has many attributes: it is a fuel, an organic sediment, a rock, a collection of plant debris, an organic chemical substance, a solid colloid, and a chemical reactant. It is not surprising, then, that such aspects as the reservoir behavior of coal are so difficult to predict and to model, unlike the more conventional sandstone or carbonate rock reservoirs. In geologic basins such as the San Juan of Colorado and New Mexico and the Black Warrior of Alabama, the experience of petroleum industry operators has progressed to the point where coalbed methane now can be considered a conventional resource, albeit one possessing a number of unsolved problems, particularly in the area of production.

An important characteristic of coalbed methane wells, especially those that initially produce water along with the gas, is an increase in gas production with time and a corresponding decrease in water production. This gas production incline may persist for several years before the inevitable decline occurs. Based on limited, but well-documented, production histories, it appears that typical coalbed methane wells will be long-lived (20-40 yr).

When optimum conditions of rank, gas saturation, reservoir temperature, coal permeability, and other critical factors are present, a high rate of success should be experienced in the development of a coalbed methane field or pool. The ubiquitous nature of methane in coal seams generally is independent of structural position in a basin (i.e., whether anticlinal or synclinal), except as structural deformation may affect the permeability of a coal seam reservoir.

The resource base of coalbed methane in the United States alone is indeed immense. Preliminary estimates, based on very incomplete data on certain basins in the 48 conterminous states, place this resource in the range of 400-800 tcf or more of methane in-place. Estimates for individual basins vary from a few tcf to more than 80 tcf.

COAL AS A SOURCE AND RESERVOIR OF NATURAL GAS

Coal seams are both sources and reservoirs of the gas that is formed as a by-product of coalification, which is defined as the process by which vegetal material progressively evolves from peat to lignite to sub-bituminous, bituminous, and anthracite coal. According to VanKrevelen (1961), coalification "...may be defined as the gradual increase in carbon content of fossil organic material in the course of a natural process" (p. 45). The peat-forming process involves biochemical reactions (diagenesis); bituminous and higher-rank coals pass through a geochemical (thermogenic or catagenic) stage. The thermal maturation, or metamorphism, of humic kerogenous (Type III) organic matter (largely oxygen-rich lignin and cellulose) results in a progressive devolatilization of the kerogen in the coal, together with an increase in carbon content, decrease in moisture content, increase in calorific value and in percent vitrinite reflectance, increase in the degree of molecular ordering, and a marked increase in thermally generated methane. The important aspects of calculating the thermal maturity of organic material in sediments (N.V. Lopatin's "Time-Temperature Index" of maturity), of kerogen maturation relative to vitrinite reflectance (as portrayed in the VanKrevelen diagram), and of other source rock modelling techniques are well described in papers by Choate et al. (1986), Meissner (1984), Rightmire et al. (1984), and Waples (1981), in addition to VanKrevelen (1961).

The coalification process, from peat through anthracite, generates very large volumes of methane, with lesser amounts of carbon dioxide and nitrogen. From the wood stage to the low-rank lignite rank, some 34.4 m^3/tonne (1,350 ft^3/ton) of biogenic methane is generated (from Mott's Model, in Francis, 1954). From high-volatile bituminous to anthracite rank, the volume of thermally generated methane may exceed 255 m^3/tonne (10,000 ft^3/ton) of coal (Meissner, 1984, p. 418).

Not all of the methane generated during the coalification process migrates, or is expelled, out of coal seams. Coal has the capacity to retain, store, or absorb methane in varying amounts. Methane storage is achieved by two primary methods: 1) in the microporosity system, wherein the gas is absorbed within or upon the molecular structure of the kerogen in the coal, as well as in the micropores; and 2) in the macroporosity system, by conventional volume storage within the cleats or fractures that almost always are present in the coal (Meissner, 1984). The retention of methane in coal also can be expressed as follows: 1) as sorbed molecules on the internal surfaces or within the molecular structure of the coal; 2) as gas trapped within the matrix (macro- or micro-) porosity, which typically occurs in the 5Å to 500Å-plus size range; 3) as free gas within the cleat and fracture systems; and 4) as gas dissolved in the free water that may exist in the cleats and fractures (Choate et al., 1986). The volumes of methane that can be stored by molecular absorption and in the microporosity system are determined by 1) coal rank, 2) temperature, and 3) reservoir pressure (Meissner, 1984).

It is important to note that lower-rank coals — those below medium-volatile bituminous rank — are characterized by having storage capacity beyond that of generation. Furthermore, expulsion of methane takes place at the point at which generation exceeds storage capacity under conditions of constant temperature and pressure. It can be seen, then, that methane in coals presents problems and paradoxes that are not found in the more "conventional" reservoir rocks. For example, 0.03 m^3 (1 ft^3) of sandstone having 15% porosity and 75% gas saturation, at a depth of 760 m (2,500 ft), can hold 8.4 scf of gas, whereas the same volume of medium-volatile bituminous coal at the same depth can store 22 scf of gas or 2.6 times as much. This phenomenon in part is due to the unique molecular structure of coal, wherein the micropore system appears to behave as a molecular sieve, or a clathrate cage (analogous to the structure of zeolite minerals), in which methane molecules are nested within benzene rings. Another problem involves the dynamic nature of an accumulation of coal-derived methane which results from the ability of coal to act both as a "gas-generating machine", which together with contiguous sandstone reservoir beds contain the critical indigenous elements of source, migration paths, and traps (Meissner, 1984), and as an absorbing "sponge". Under conditions of thermal heating, coals may continue to generate methane, expelling it into the surrounding sediments when the total storage capacity of the coal has been exceeded. On the other hand, in basins undergoing thermal cooling, the coals will tend to reabsorb from the surrounding clastic reservoirs the gas that the coal seams originally generated as their storage capacity is increased during the cooling phase. The thermal heating/high-volume gas generation and cooling/gas reabsorption processes can result in overpressured and underpressured gas accumulations, respectively, which are common in many of the coal-bearing Rocky Mountain basins. From the above, it is obvious that estimating the potential recoverable reserves and resources of methane contained today in a particular coal deposit will be a difficult and elusive task.

COMPOSITION OF COALBED GAS

Gas produced directly from coal seams almost always is of pipeline quality, being composed of from approximately 90% to 95% methane in most instances, with minor amounts of heavier hydrocarbons, CO_2, N_2, O_2, H_2, and He, and with heating values generally between 950 to 1,050 Btu/scf (pure methane has a heating value of 1,012 Btu/scf at 60°F and atmospheric pressure). Analyses of gases recovered from certain relatively deep (greater than 152 m (5,000 ft)), high-rank coals have indicated the presence of ethane and heavier hydrocarbons in concentrations of from 10% to 15%. The presence of H_2S and other sulfur compounds in coalbed gas is virtually unknown, even from high-sulfur coals.

The composition, volumes, and liberation rates of hydrocarbons generated by the coalification process appear basically to be a function of the relative abundance of the various macerals (vitrinite, alginite, exinite, etc.) that are found in coals. Macerals are microscopic components of coal and consist of the remains of the original plant material (Ulery, 1984).

Additionally, the level of thermal alteration has a decided effect upon the composition of the hydrocarbons generated by terrestrial (Type III) organic matter. There is good evidence that autochthonous generation by hydrocarbons may, in some situations, occur at lower levels of thermal maturity (i.e., at vitrinite reflectance levels of less than 0.6% R_o) than generally believed (Snowdon and Powell, 1982). Furthermore, ethane and propane have been observed in immature sedimentary environments and result from diagenetic reactions that may parallel or immediately follow the formation of biogenic methane (Schoell, 1983). In some basins, (e.g., the San Juan Basin of Colorado and New Mexico), significant quantities of high-gravity liquid hydrocarbons (condensate?) are produced in certain wells, indicating the capacity of some types of coals to generate "oil".

PRODUCTION OF COALBED METHANE

Methane has been produced from coal seams through boreholes since the turn of the century. These holes include water wells and degasification holes, ranging from vertical to horizontal, designed to drain as much methane as possible for safety reasons from virgin coal seams prior to mining (see Skow et al., 1980). A number of production case histories are described in Choate et al. (1986), Murray (1981), Rightmire et al., (1984), Shirley (1986), Tew and Mancini (1986), Tilton (1976), Trevits and Finfinger (1986), and TRW Energy Engineering Division (1981).

A paper essential to the understanding of the behavior of coal-gas reservoirs was prepared by Cervik (1969), who observed that gas occurs in coal seams in both an absorbed and a free state. Absorbed gas is stored in the matrix, or micropores, of the coal and desorbs and diffuses through the coal at a rate governed by the diffusion process described by Fick's Law or by other diffusion models, the driving force being a concentration gradient. Once the gas has migrated into the larger pores and into the cleat and fracture system, it then will flow into the wellbore (or mine) according to Darcy's Law, being driven by pressure gradient. These two types of mass transport are interdependent. The majority of coalbed reservoirs are at essentially hydrostatic pressure, and they depend on a system of fractures and cleats for most of their permeability. The relative permeabilities to both gas and water are critical to the initial production of methane from coalbed reservoirs. In water-saturated coals, water must be removed (usually by pumping) in order to upset the equilibrium that exists between the methane absorbed within the micropores and that existing in the fracture system. Once a pressure gradient has been established, methane will first diffuse into the fracture system and then from the fractures into the wellbore, where the pressure has been lowered to less than hydrostatic. Ultimately, the productivity of a coalbed gas well will be dependent largely upon the ability to lower reservoir pressure and water saturation (if present) in the coalbed reservoir. A multiwell pattern is necessary in order to create drainage boundaries or areas of interference. Kissell and Edwards (1975) have demonstrated that by lowering the water saturation in the fractures and cleats in the coal, the effective permeability to gas is increased (i.e., more space is made available to the gas phase), resulting in an <u>increasing</u> rate of gas production and a corresponding <u>decrease</u> in rate of water production. Such "negative declines" have been observed in a number of coalbed methane wells in productive areas such as the San Juan and Black Warrior basins.

Drilling and production activities involving coalbed methane wells in the most active areas in the United States — the San Juan, Black Warrior, Piceance, and Raton Basins — are summarized in the issues of the <u>Quarterly Review of Methane from Coal Seams Technology</u> (published by the Gas Research Institute since 1983).

RESOURCES OF COALBED METHANE

A wide range of estimates pertaining to total in-place coalbed methane resources in the conterminous United States have been published since 1978. These estimates vary from a minimum of 72 tcf to a maximum of 860 tcf, with estimates of <u>recoverable</u> resources ranging from 10 to 487 tcf. All of these estimates must be considered as very preliminary because they are based on incomplete data regarding both the magnitude and character of U.S. coal resources below 915 m (3,000 ft) in depth and on the in-situ gas content of most of the coal seams involved. If one uses the data presented in Averitt (1975), the following estimate could be made: Estimated remaining coal resources in the United States, as of January 1, 1974, including total identified and hypothetical resources remaining in the ground beneath 0-1,830 m (0-6,000 ft) of overburden, are 3,601,000,000,000 tonnes (3,968,000,000,000 tons). If only 50% of this resource has an insitu methane content of 5 m^3/tonne (200 ft^3/ton), then the total insitu coalbed methane resource of the United States, including Alaska, could be on the order of 400 tcf. This probably is a conservative figure for several reasons: 1) The total remaining U.S. coal resource in the ground is believed to be considerably more than 3,600,000,000,000 tonnes (4,000,000,000,000 tons), based on recent resource evaluations and on the fact that thick coal seams are known to exist below the 1,830-m (6,000-ft) depth cutoff used by Averitt (1975) (in fact, coals occur at depths greater than 3,050-4,600 m (10,000-15,000 ft) in some basins in the Rocky Mountain region); and 2) the gas content of many deposits of coal in the subsurface (principally, below 150-300 m (500-1,000 ft)) exceeds 5 m^3/tonne (200 ft^3/ton), as shown by Rightmire et al. (1984) and Diamond et al. (1986). Published maps (as in Averitt, 1975; and Rightmire et al., 1984) also do not accurately represent the rank of all the coal deposits known to exist in many of the Western U.S. basins, especially of those coals occurring below mineable depths. For example, from sample data from wells drilled for oil and gas, it is known that the rank of coals may increase considerably with depth. The low-rank bituminous coal known from analyses of outcrop or coal mine samples may be low-volatile bituminous, or even approaching anthracite, from the same stratigraphic interval at a depth of 3,000 m (10,000 ft).

PROBLEMS OF ASSESSMENT OF COALBED METHANE RESOURCES

The following are examples of some of the problems that complicate the assessment of this very large resource:

1. How can the gas resource present in low-permeability ("tight") sandstone reservoirs be separated from that stored in or derived from coal seams, particularly where these two types of reservoirs are intimately interbedded (as in the case of the Mesaverde Group in the Rocky Mountain region)? Most of the gas generated by the coalification process today is not present in the coals themselves; much of this coal-derived gas now may be trapped in other reservoir beds that are in close proximity to the coals.

2. If situations exist where it can be demonstrated that coal seams continue to recharge contiguous producing clastic reservoirs as a pressure differential between the two is established, how should such possibilities be addressed in unproven areas (as described in Meissner, 1984, and Rightmire et al., 1984)? Wyman (1984) believes that at least 50% of the gas in certain coal seams in the Lower Cretaceous sequence in the "Deep Basin" of Western Canada can be recovered from the adjacent sandstones and conglomerates by means of diffusion from the coal matrix and Darcy flow through open fractures.

3. How does one evaluate an area in which coal seams occur within the window of active gas generation? Welte et al. (1984, p. 47) describe the Elmworth gas field, located in the "Deep Basin" of northwestern Alberta, as being in "...a dynamic situation where gas is continually being generated in the center part of the Deep Basin and lost toward the surface and the more porous edge. In the inner core of the gas-generating rock column diffusion processes seem to be the predominating mode of transportation." The dynamic nature of coalbed "gas machines" is an extremely complex, yet very important, phenomenon that demands considerably more research.

RATIONALE FOR COALBED METHANE EXPLORATION

Exploration for coalbed methane should include aspects of both coal geology and petroleum geology, as well as a nonconventional approach to reservoir engineering. Pre-drilling activity should evaluate the geology of the entire coal-bearing sequence, including any interbedded low-permeability ("tight") gas-bearing sandstones.

Exploration studies should address such topics as the physical and chemical nature of the coal (rank, chemistry, depositional environment, etc.), the thermal history and hydrodynamics of the region of interest, thickness of the coal seams and of the overburden, geologic structure and tectonic features such as fracture patterns and igneous activity, coalbed gas desorption data in the study area, and a petrographic analysis of available coal cores and well cuttings.

CONCLUSIONS

Coal constitutes one of the richest known sources of hydrocarbons. The coalification process generates very large volumes of methane — more than 204 m^3/tonne (8,000 ft^3/ton) of high-rank coal — which is both biogenic and thermogenic in origin. Coal is an extremely complex organic substance and possesses the unique capability of being both a source and a reservoir of natural gas. Consequently, the reservoir behavior of coal, its production characteristics, and other aspects so important to the natural gas industry are difficult to predict and to model. In this sense, coalbed reservoirs can be considered "unconventional".

However, the growing successful experience of operators in basins such as the San Juan in Colorado and New Mexico and the Black Warrior in Alabama has advanced to the point where coalbed methane now may be treated as a conventional resource. Gas from coals is an attractive exploratory objective for reasons that include the relatively shallow depths of most deposits of coal, the ubiquitous occurrence of gas in coal, and the very large gas generation and storage capacities of the higher rank coals. Furthermore, coal gas wells typically experience an <u>increase</u> in gas production with time, and a corresponding <u>decrease</u> in water production, if any. The production <u>incline</u> of some coalbed methane wells is expected to persist for several years, based on the few well-documented production histories which are available.

Methane trapped in coal seams is a virtually untapped source of clean, pipeline-quality energy that today is an attractive, economically viable objective for the gas-producing industry. Based on very preliminary studies, trillions of cubic feet of <u>recoverable</u> methane are believed to exist in many of the coal-bearing areas in the United States, as well as in both Western and Eastern Canada. The Potential Gas Committee (1988) estimates that the most likely <u>recoverable</u> resource, in all resource categories, of coalbed methane in selected basins in the conterminous United States is at least 90.1 tcf.

During 1990, as much as 50% of all gas well completions in the United States will be coalbed methane wells. By the end of 1990, possibly 7,000 coalbed methane wells will have been drilled in the U.S., primarily in the Black Warrior, Greater Green River, Piceance, and San Juan basins. These wells, when in full production, are expected to supply approximately 6% of our domestic natural gas consumption, an amount equivalent to our current imports from Canada (G. Warfield Hobbs IV, personal communication).

REFERENCES

Averitt, Paul, 1975, Coal resources of the United States, January 1, 1974: U.S. Geological Survey Bulletin 1412, 131 p.

Cervik, J., 1969, Behavior of coal-gas reservoirs: U.S. Bureau of Mines Technical Progress Report 10, 10 p.

Choate, Raoul, J.P. McCord, and C.T. Rightmire, 1986, Assessment of natural gas from coalbeds by geologic characterization and production evaluation, in D.D. Rice, ed., Oil and Gas Assessment: American Association of Petroleum Geologists, Studies in Geology No. 21, p. 223-245.

Diamond, W.P., J.C. La Scola, and D.M. Hyman, 1986, Results of direct-method determination of the gas content of U.S. coalbeds: U.S. Bureau of Mines Information Circular IC 9067, 95 p.

Francis, Wilfred, 1954, Coal, its formation and composition: London, Edward Arnold Publications, Ltd., (especially) p. 317 (Mott's Classification) and p. 359-363 (including Table VIII.II, Mott's Table).

Kissell, F.N., and J.C. Edwards, 1975, Two-phase flow in coalbeds: U.S. Bureau of Mines Report of Investigations RI 8066, 16 p.

Meissner, F.F., 1984, Cretaceous and lower Tertiary coals as sources for gas accumulations in the Rocky Mountain area, in J. Woodward, F.F. Meissner, and J.L. Clayton, eds., Hydrocarbon Source Rocks of the Greater Rocky Mountain Region: Denver, Rocky Mountain Association of Geologists, p. 401-431.

Murray, D.K., 1981, Methane from coal: Coal and Energy Quarterly, no. 29, p. 30-38.

Potential Gas Committee, 1988, Potential Supply of Natural Gas in the United States (as of December 31, 1988): Golden, Colorado, Potential Gas Agency, p. 123-132.

Rightmire, C.T., G.E. Eddy, and J.N. Kirr, eds., 1984, Coalbed methane resources of the United States: American Association of Petroleum Geologists, Studies in Geology Series #17, 378 p.

Schoell, M., 1983, Genetic characterization of natural gases: American Association of Petroleum Geologists Bulletin, v. 67, no. 12, p. 2225-2238.

Shirley, Kathy, 1986, Coalbed methane coming of age: AAPG Explorer, v. 7, no. 12, p. 1 and 8-9.

Skow, M.L., A.G. Kim, and M. Deul, 1980, Creating a safer environment in U.S. coal mines--the Bureau of Mines Methane Control Program, 1964-79: U.S. Bureau of Mines Impact Report, 50 p.

Snowden, L.R., and T.G. Powell, 1982, Immature oil and condensate-Modification of hydrocarbon generation model for terrestrial organic matter: American Association of Petroleum Geologists Bulletin, v. 66, no. 6, p. 775-788.

Tew, B.H., and E.A. Mancini, 1986, Energy resources of northwestern Alabama--Guidebook for the 1986 Annual Meeting of the American Association of Petroleum Geologists, Energy Minerals Division, Atlanta, Georgia, June 18-20, 1986: Tuscaloosa, Geological Survey of Alabama, 44 p.

Tilton, J.G., 1976, Gas from coal deposits, in Natural gas from unconventional geologic sources: Washington, National Academy of Sciences, p. 206-229.

Trevits, M.A., and G.L. Finfinger, 1986, Results given of studies concerning methane extraction from coalbeds: Mining Engineering, v. 38, no. 8, p. 805-808.

TRW Energy Engineering Division, 1981, Coalbed methane production case histories: Morgantown Energy Technology Center-U.S. Dept. of Energy, Report DOE/METC/08089-115, various pages.

Ulery, J.P., 1984, Effects of petrographic characteristics on coalbed gas content, in J.R. Pederson, ed., Research 1984: Washington, U.S. Bureau of Mines, p. 15-16.

VanKrevelen, D.W., 1961, Coal--typology, chemistry, physics, constitution: Amsterdam, Elsevier Publishing Co., 514 p.

Waples, D.W., 1980, Time and temperature in petroleum formation: Application of Lopatin's method to petroleum exploration: American Association of Petroleum Geologists Bulletin, v. 64, no. 6, p. 916-926.

Welte, D.H., R.G. Schaefer, W. Stoessinger, and M. Radke, 1984, Gas generation and migration in the Deep Basin of Western Canada, in J.A. Masters, ed., Elmworth--Case study of a Deep Basin gas field: American Association of Petroleum Geologists, Memoir 38, p. 35-47.

Wyman, R.E., 1984, Gas resources in Elmworth coal seams, in J.A. Masters, ed., Elmworth--Case study of a Deep Basin gas field: American Association of Petroleum Geologists, Memoir 38, p. 173-187.

STRUCTURAL PROPERTIES OF COAL THAT CONTROL COALBED METHANE PRODUCTION

T. Ertekin and W. Sung[1]
Petroleum & Natural Gas Engineering Section
Mineral Engineering Department
102 Mineral Science Bldg.
Pennsylvania State University
University Park, PA 16802

H.I. Bilgesu
Department of Petroleum and Natural Gas Engineering
West Virginia University
Morgantown, WV 26506

ABSTRACT

This paper provides a review of structural properties of coal that control coalbed methane production. Coalbed methane reservoirs are characterized by a dual-porosity nature. Gas molecules stored in the micropore structure by adsorption are subject to desorption from the coal grain surfaces and diffusional transport to a well-defined natural fracture network composed of face and butt cleat systems. Using the results generated by numerical models, typical production performance characteristics of horizontal and vertically fractured wells are generated. Furthermore, the roles of coal seam properties, such as seam thickness, sorption time constant, and permeability anisotropy in determining the methane production potential of the coal seams are examined.

INTRODUCTION

Unconventional natural gas has been defined as gas produced from sources other than those historically exploited by the oil and gas industry. This unconventional resource is a pipeline-quality, high-BTU-content gas recovered from reservoirs such as geopressured aquifers, Devonian shales, and coal seams. Methane drainage from coal seams is an increasingly important technology because it provides an effective means of increasing mine safety and recovering an otherwise lost valuable energy source. Estimates of technically recoverable gas contained in domestic coal seams range from 300,000,000,000,000 scf to 800,000,000,000,000 scf (Deul and Kim, 1975; Olson, 1979; Wise, 1979). Coalbed methane is high in quality and requires little or no processing other than water removal (demisting) before transmission.

---

1. Present Address: Hangdang-dong, Sungdong-ku, Hanyang University, College of Engineering, Mining and Mineral Engineering Department, Seoul, South Korea.

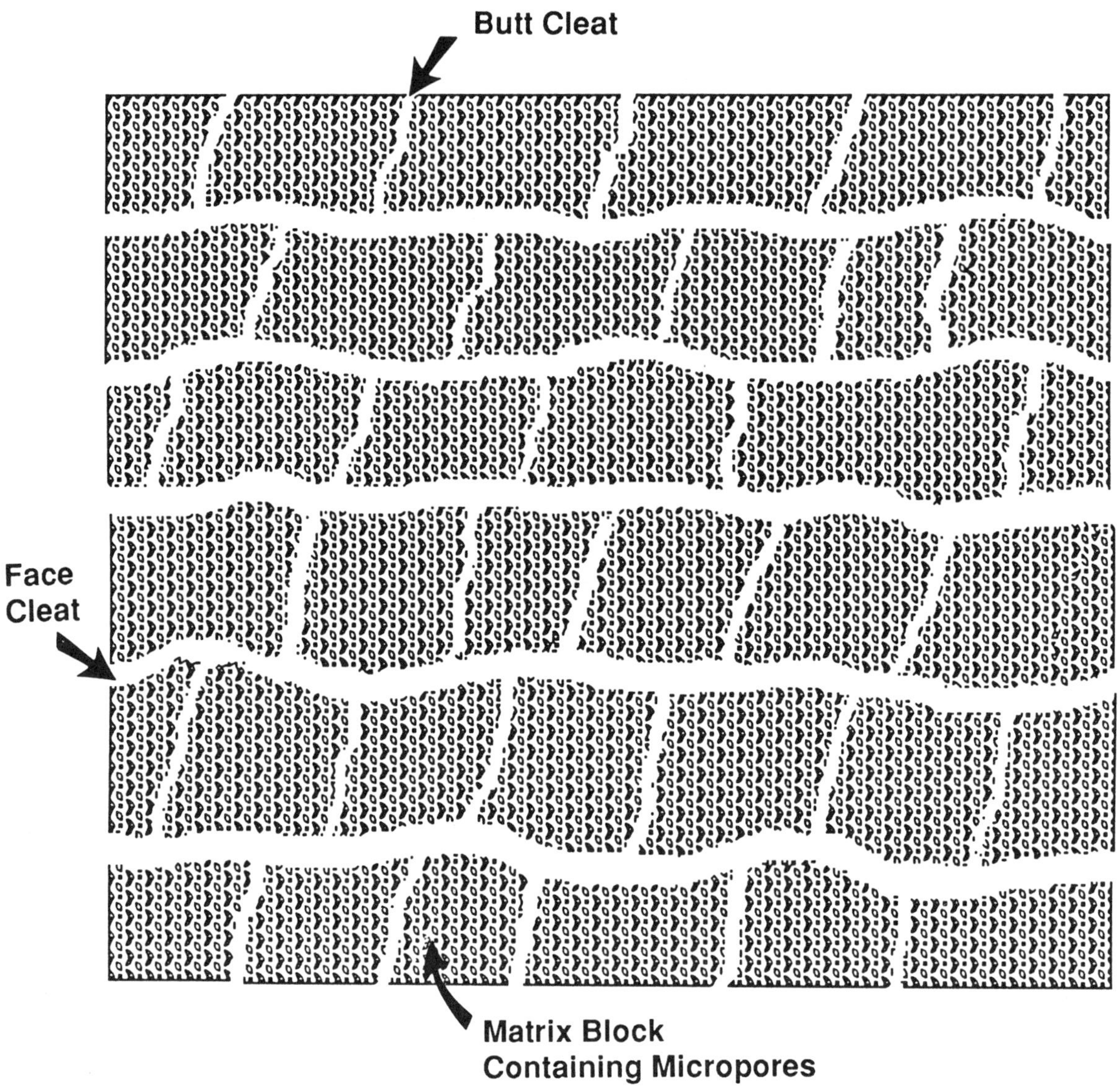

Figure 1. *Schematic drawing of a coal seam cleat system.*

The study of the reservoir characteristics of coal seams is crucial in understanding the transport of gas and brine through coal seams. The coal matrix is heterogeneous and characterized by two distinct porosity systems: macropores (commonly known as the cleat or joint network) and micropores (primary porosity matrix). Although the role of the primary porosity is similar in coal seams and conventional dual-porosity reservoirs, there are two major differences. Whereas the gas stored in the matrix of a conventional double-porosity reservoir is in free state, and obeys the real gas law, the major portion of the gas stored in the micropores of a coal seam is in the adsorbed state and, thus, obeys an entirely different set of laws. The second difference is caused by the size of the micropores. The principle transport mechanism through the primary porosity is considered to be a diffusion process in coal seams, whereas in a conventional dual-porosity reservoir, laminar flow in the matrix is the dominant mechanism.

This paper presents an overview of the general characteristics and potentials of coal seams as unconventional gas reservoirs by addressing 1) reservoir characteristics of coal seams, 2) fluid flow dynamics in coal seams, and 3) field development strategies. Most of the analyses have been conducted with the aid of single-

(King et al., 1986) and multi-well (Sung et al., 1986), multidimensional, two-phase finite-difference numerical models which have been developed and applied to the coal seam degasification process. In all of these model formulations, partial differential equations representing the flow of water and gas were derived from mass balances applied on a finite element in the reservoir. Two- and three-dimensional rectangular and cylindrical coordinate systems were imposed on the coal seam and used to transform the continuous-flow equations into discrete form by application of Taylor series expansions.

In the development of the models, the coal seam was considered to be anisotropic and heterogeneous with respect to fluid flow and storage properties. Water velocity was described by Darcy's Law, and gas velocity was calculated as the sum of the velocities defined by Darcy's Law and Fick's Law. The system of nonlinear equations generated by the finite-difference approximation was linearized with the aid of a generalized Newton-Raphson iteration procedure. Entries to the Jacobian matrix are supplied by numerical differentiation and the resulting system of equations is solved using a direct solver equipped with a block-band algorithm.

The models have options to accommodate a number of vertical wells as well as horizontal wells. In the case of vertical wells, the model is able to handle the simulation of post-fracture performance. Furthermore, options exist to include a number of impermeable geological barriers within the system. The models were verified through field applications (King et al., 1986; Sung et al., 1986) and by performing comparisons against existing analytical and/or numerical solutions (King et al., 1986).

RESERVOIR CHARACTERISTICS OF COAL SEAMS

Coal seams are characterized by a natural fracture network commonly referred to as cleat. The cleat system consists of two nearly perpendicular fissures, the more predominant of which is called the face cleat. The butt cleat is less continuous and often terminates against the face cleat. The physical relationship between the matrix and fracture systems is shown schematically in Figure 1. As shown in this figure, mutual strike angles for the cleat systems vary between 75° and 85°. The macropore system is made up of the volume created by cleating. Although in most analytical and numerical treatments the fracture spacing is assumed to be uniform, there still is a fair amount of variation in cleat spacing which tends to follow a log-normal distribution. The aperture of the macropores is very stress dependent and may vary from angstroms to millimeters. There do not appear to be any transitory pores between the micropore and macropore systems (van Krevelin and Schayer, 1967).

The total effective porosity of coal to water generally is less than 5%, whereas the effective porosity to free gas in the same coal seam may be as high as 10% (Patching, 1970). Fracture or macropore space frequently accounts for less than 2% of the seam bulk volume (Dabbous et al., 1976), and the gas storage mechanism in the macropores is the same as in conventional sandstone or carbonate reservoirs.

The primary porosity of a coal seam is defined by the micropore structure of the coal matrix. Micropore space accounts for as much as 85% of coal's total porosity. Small pore diameters ranging from 5Å to 10Å give the coal matrix a large internal surface area of 100-300 m^2/g (1,100-3,200 ft^2/g) as evaluated from the adsorption isotherms of carbon dioxide using the Dubinin-Polanyi equation (Marsh, 1965; Thimons and Kissel, 1973).

In addition to gas stored in the free state in the pore volumes of the fracture and matrix systems, coal seams also store methane gas as an adsorbed mono-molecular layer film on the coal grain surface. The dense mono-molecular layer packing of the adsorbed methane molecules and the large surface area of the coal grains in the matrix elements together make adsorption the predominant gas storage mechanism (Thimons and Kissel, 1973). The unsteady-state nature of the sorption phenomenon further complicates the reservoir engineering characteristics of coal seams. The sorption characteristics are dependent upon the chemical properties (e.g., rank) of individual coals. In general, the sorption capacity increases with the fixed carbon content of a coal and with depth. Figure 2 shows the interrelationship between sorptive capacity, rank, and depth of burial (Kim, 1977).

The quantity of gas adsorbed also is a function of pressure. Figure 3 shows the equilibrium sorption isotherms, the functional relationships between the volume of gas adsorbed versus pressure, for several coals in the United States (Kissel et al., 1973). The shape of isotherms suggests that gas readily is adsorbed onto the coal grain surface at low pressure levels. As pressure increases, the adsorption rate begins to monotonically decline as gas molecules compete for increasingly less available space on the surface of the coal grains. The crowding of the gas molecules becomes severe at high pressure levels, and the sorption rate approaches zero.

The matrix sorption phenomenon can be formulated using three different approaches: equilibrium sorption, quasi-steady-state sorption, and unsteady-state sorption. In the equilibrium sorption approach, the amount of gas adsorbed on the micropore walls is a function only of pressure in the fracture. Thus, as the fracture pressure declines over a discrete time interval, sorbed gas is instantaneously released into the fracture. Several models (Airey, 1968; O'Shaughnessey, 1980; Owili-Eger, 1973; Price and Abdalla, 1972) based on this approach have been developed to describe the release of methane gas from underground mine workings.

The quasi-steady-state sorption model follows a slightly more rigorous development. In this formulation, it is assumed that the desorption rate is directly proportional to the difference between gas concentration at the external matrix surface and the average concentration within the matrix. Models developed by Boyer et al. (1982), Pavone and Schwerer (1984), King et al. (1986), Sung et al. (1986), and Remner et al. (1986) are based on this approach.

The unsteady-state sorption model relates the desorption rate to the concentration gradient across the coal matrix. During early desorption times, the concentration gradient is large, and the unsteady-state model predicts larger desorption rates than the quasi-steady-state model. At later times, the matrix concentration gradient assumes some degree of uniformity and the unsteady-state and quasi-steady state models predict similar sorption rates. Smith and Williams (1982), Price and Ancell (1978), and Ancell et al. (1980) all have developed models using the unsteady-state matrix formulation.

Continuous, unfractured coal is relatively impermeable to gas and water. The in situ permeability of coal seams predominantly is caused by the cleat. The dependence of the permeability of the seam on cleat has two important consequences. The first is that the permeability is anisotropic, and, most of the time, with the greatest permeability in the direction parallel to the face cleat. The second consequence is the dependence of the permeability on confining stress. Increasing the confining stress causes crack (cleat) closure, thus reducing the absolute permeability. Permeability reductions of two orders of magnitude have been observed, with the greatest relative changes occurring in less permeable coals (Dabbous et al., 1976; Mordecai and Morris, 1974; Somerton et al., 1975). This phenomenon becomes of practical importance when the state of stress around a wellbore or mine face is considered.

Table 1 lists some of the reservoir properties of major U.S. coal seams. Although permeabilities of U.S. coal seams tend to be high ($k>1.0$ md), this is not the case for all coal seams. Jones et al. (1982) report samples of European coals with permeabilities as low as 0.003 md.

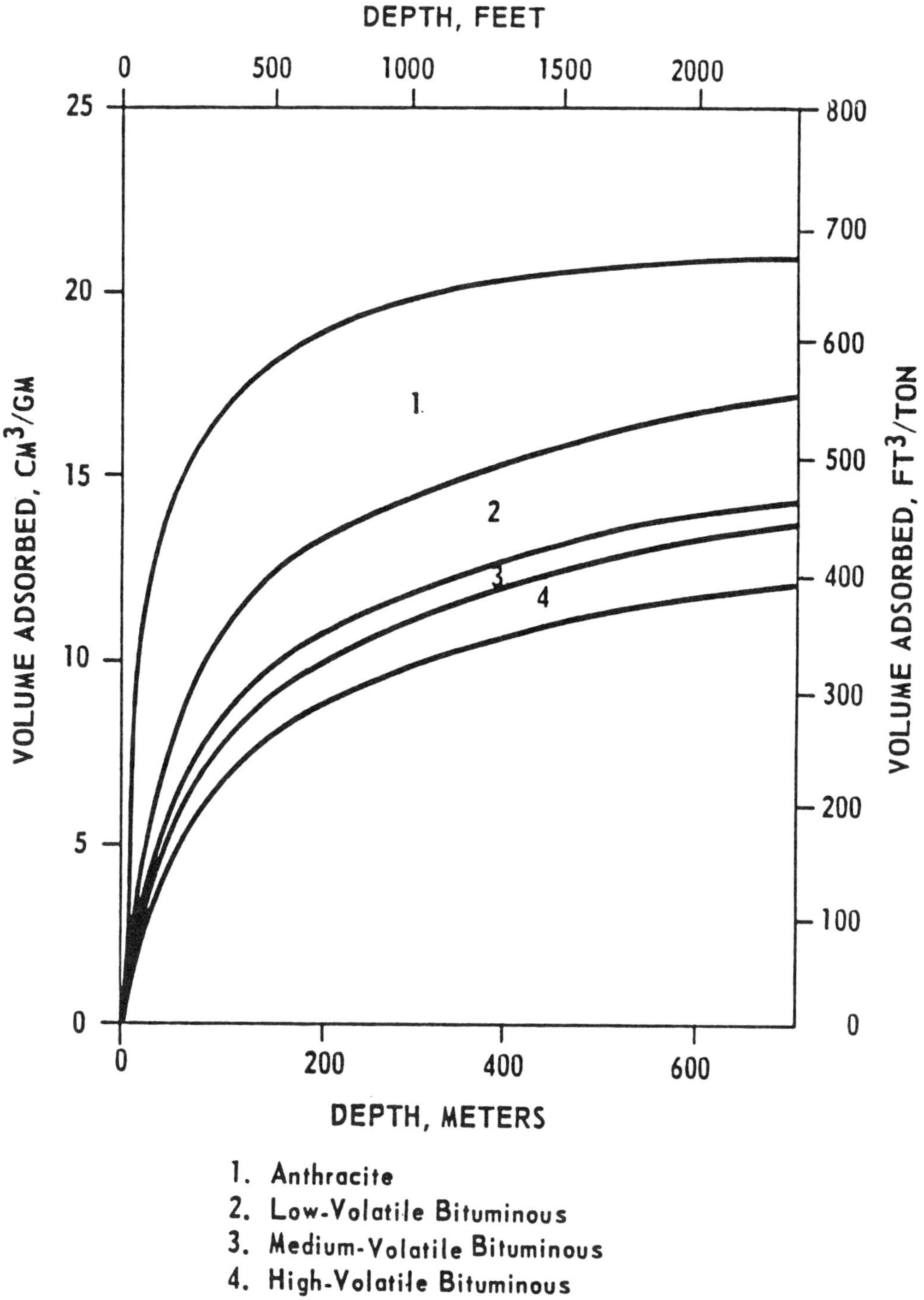

Figure 2. *Variation of methane content with coal rank and depth (from Kim, 1977).*

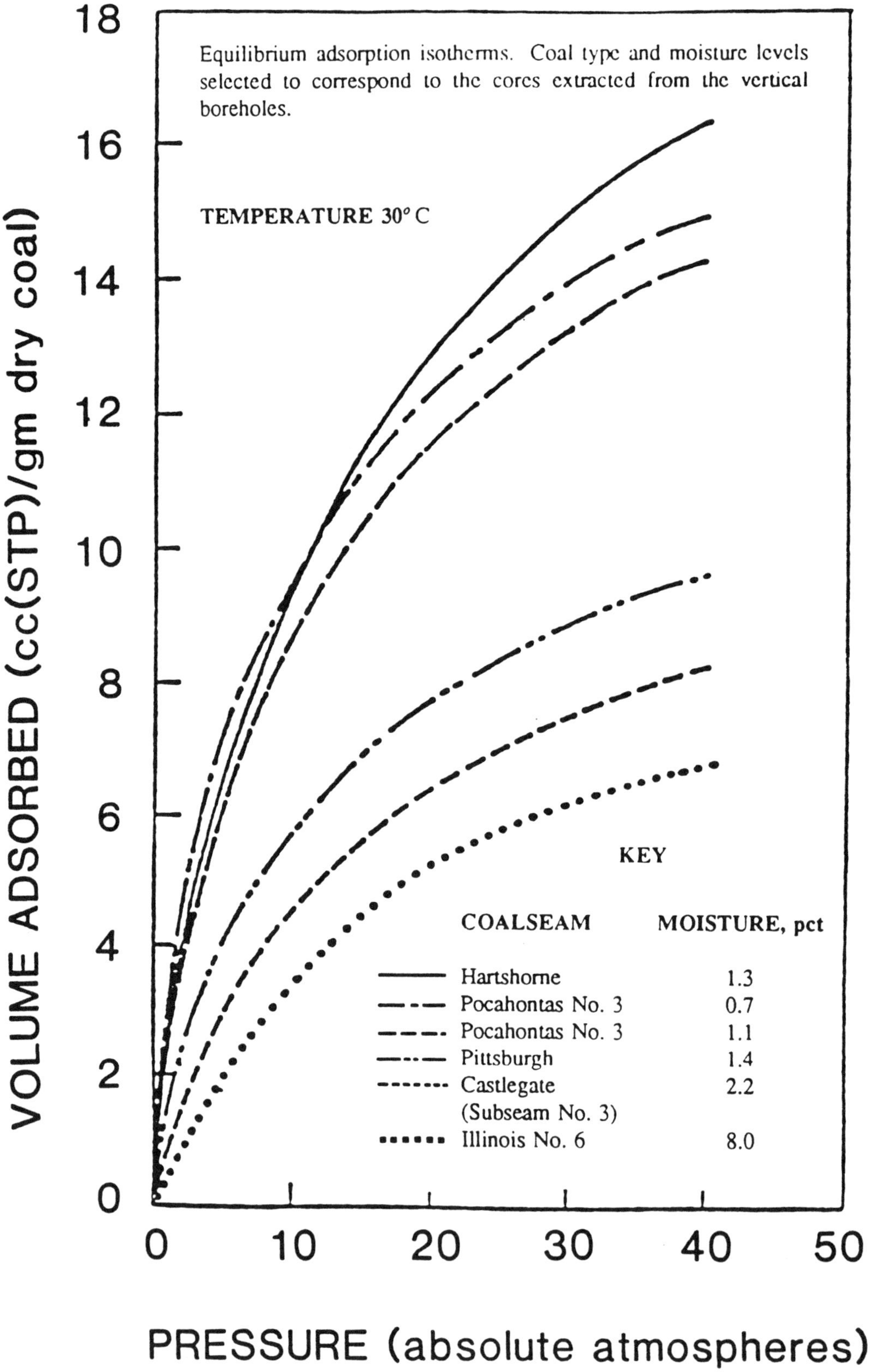

Figure 3. *Equilibrium sorption isotherms for several U.S. coals (from Kissell, 1973).*

Gas Source	Rank	Geologic Era	Overburden (Feet)	Adsorbed Gas (ft^3/ton)	Virgin Reservoir Pressure (psi)	Permeability (md)
Pittsburgh	High-Volatile Bituminous	Pennsylvanian Series (Carboniferous)	500-900	218	275	10
Pocahontas #3	Low-Volatile Bituminous	Pennsylvanian Series (Carboniferous)	1600	430	650	2 to 9
Mary Lee	High-Volatile Bituminous	Pennsylvanian Series (Carboniferous)	2100	512	750	7
Sunnyside	High-Volatile Bituminous	Mesaverde Group (Cretaceous)	2500	160	620	--

Table 1. *U.S. coal seam characteristics (from Makismovic et al., 1977).*

FLUID FLOW DYNAMICS IN COAL SEAMS

The application of Darcy's Law to gas flow in these low-permeability formations requires a correction for the Klinkenberg effect (gas slippage across the capillary walls of the pore channels). This correction takes the form of a slippage factor (b) in the Klinkenberg equation (Klinkenberg, 1941):

$$k = k_\infty \left(1 + \frac{b}{p}\right)$$

where p and k represent average pressure and permeability to the gas at that average pressure, respectively, and k_∞ represents the absolute (liquid) permeability. The slippage factor (b) has values in the range of 70-700 kPa (10-100 psia).

Rose (1948) and, more recently, Sampath and Keighin (1982) conducted gas flow experiments in cores partially saturated with water. Their results show that slippage factor decreases with increasing water saturation. During depletion, a gas reservoir undergoes changes (both in time and location) in pressure and saturation. The effect of slippage, therefore, varies throughout the life of the reservoir. To date, no detailed experimental investigation has been conducted regarding dynamic behavior of gas slippage. This is surprising because of the large reserves of gas trapped in tight formations.

Ertekin et al. (1986) formulated a pressure-, composition-, and saturation-dependent slippage factor which is similar to the approach of Adzumi (1937) for slip through capillary tubes. This approach is based on the premise that, in tight formations, gas flow is under the simultaneous influences of pressure and concentration fields. In this approach, the driving mechanisms exerted by these fields are assumed to be acting in parallel (that is, the individual velocities are additive). Away from the wellbore the flow through the pressure field obeys Darcy's Law (Darcy, 1856) (macroscopic flow), and as such, all assumptions inherent to this law apply. The flow through the concentration field obeys Fick's Law (Fick, 1855) (random molecular flow). In this approach, it is assumed that gas can be assigned a constant diffusivity coefficient and molecular weight.

That is, chromatographic separation of the gas will not occur because of the individual diffusivities of the gas constituents. This ensures a diffusivity coefficient and molecular weight independent of time.

When a coal seam is in virgin condition, the volume of free gas is small when compared to the volume of the gas in the adsorbed state. Again, for coal seams in virgin condition, it is assumed that the cleat system initially is fully saturated with water. As water is removed from the macropores, gas is desorbed from the micropore surfaces caused by the decline of reservoir pressure and then diffuses towards the macropore structure. This process results in a nonuniformly distributed source mechanism over the macropore structure. Once in the fractures, coalbed gas is transported simultaneously with water to the low pressure areas. Gas and water transport in the fracture system is governed by Darcy's Law (see Figure 4).

FIELD DEVELOPMENT STRATEGIES

The use of vertical wells for the drainage of methane from coal seams prior to mining operations was suggested by Rice (1928). In the ensuing years, many methods of methane production have been employed with varying degrees of success. Two of the more promising methods are methane drainage via vertical boreholes from the surface and horizontal boreholes from the surface or from the shaft bottom.

Of these two methods, the use of vertically oriented wells is more common (Elder and Deul, 1975; Kim, 1978). In this technique, degasification wells are drilled and completed several years prior to mining. Vertical wells generally are fractured hydraulically to increase the permeability of the coal seam in the vicinity of the wellbore.

Vertical wells also are used to drain methane from previously mined areas, when a gas-bearing overburden is present (Kuuskraa et al., 1978; Moore and Zabetakis, 1972). In this technique, wells are drilled into the porous, but not necessarily permeable, roof rock. After mining, usually using the longwall method, the unsupported roof collapses into the mine, creating a highly permeable rubble (gob). Gas is released during the collapse of the roof and travels through the broken roof or gob to the drainage wells. Hydraulic stimulation of these gob degasification wells also has been attempted (Makismovic et al., 1977); however, unsatisfactory results were obtained.

Although the use of vertical drainage wells has the advantages of simplicity, low drilling costs, and low maintenance costs, there are several disadvantages associated with this technique. Because of the permeability anisotropy which exists in virgin coal seams, it is expected that only a relatively narrow volume, parallel to the direction of the greatest permeability, will be drained (Davis, 1971). For vertical wells to be successful, therefore, the hydraulically induced fracture should be oriented perpendicular to the direction of the greatest permeability. Because controlling the propagation of the fracture (in terms of its orientation and penetration length) in hydraulic fracturing is not possible, this perpendicularity requirement cannot be satisfied. If the seam to be produced also is mineable, the hydraulic fracture must be fully contained within the formation. A poorly designed stimulation treatment could cause irreparable damage to the roof rock, rendering the seam unmineable in the vicinity of the well.

Horizontal boreholes for coal seam degasification were implemented for the first time in the United States in 1958 (Thakur and Poundstone, 1980). The drilling of horizontal drain holes several years ahead of mining appears to be feasible mostly because of the reasonably precise directional control, nature of the operation, and the economics involved. Several coal companies have reported that significant amounts of methane have been produced over extended periods of time through horizontal boreholes. Mine safety and economics

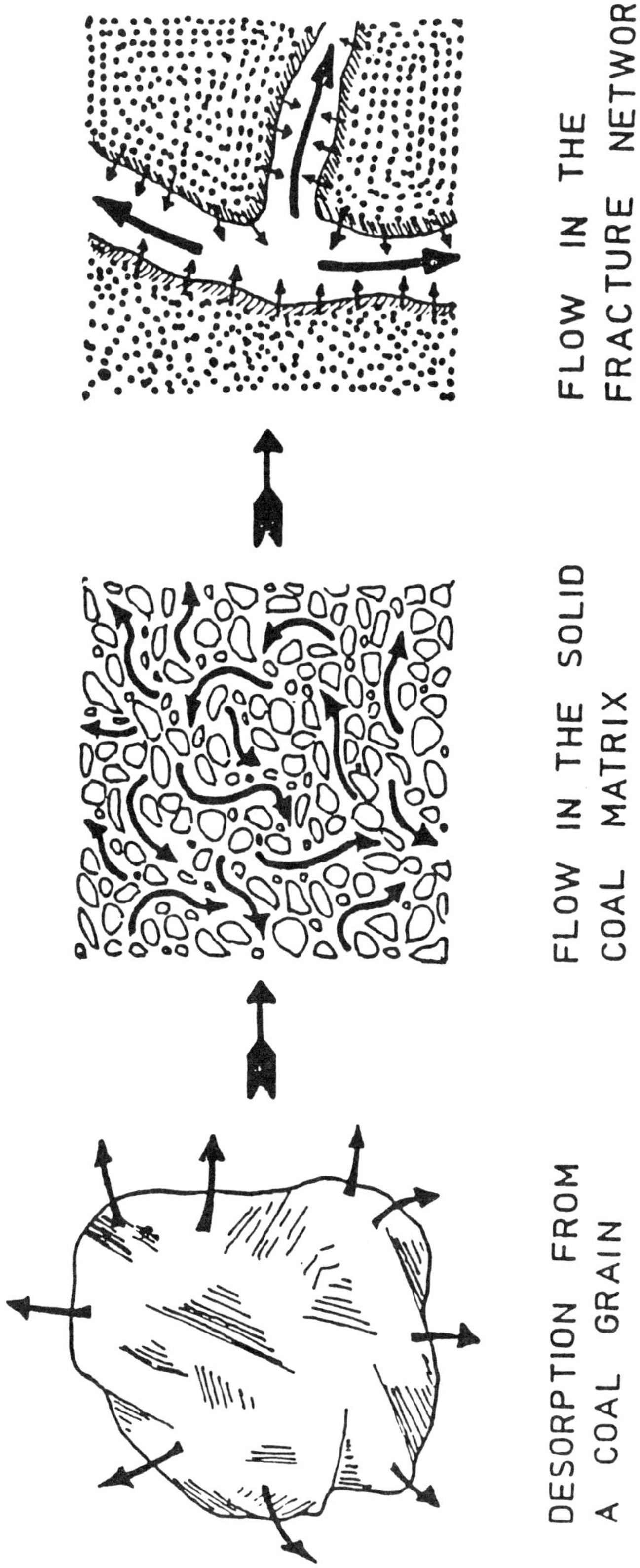

Figure 4. *Schematic drawing of methane flow dynamics in a coal seam.*

also have been improved considerably by potential reduction of methane emissions into mines (von Schonfeldt et al., 1980).

When producing methane from a shaft bottom, several 33-610-m (100-2,000-ft) long, horizontal boreholes are drilled radially outward (Fields et al., 1973, 1975; Kim, 1978; Satriana, 1980). Gas production from these horizontal boreholes ranges from 330 to 1,480 scf/day per linear meter (100 to 450 scf/day per linear foot) of hole (Kim, 1978; Satriana, 1980). The major advantage of this technique is that direction of the boreholes can be selected, so that at least one borehole is guaranteed to be oriented perpendicular to the direction of the greatest permeability.

Typical Production Performance of Coal Seam Drainage Wells

Typical gas production from a coal seam reservoir via either vertical or horizontal drainage holes is quite different from that of a conventional natural gas reservoir because cleat systems of the coal generally are assumed to be fully saturated with water in virgin conditions. During the dewatering process, reservoir pressure is lowered. This in turn allows gas to desorb from the micropore surfaces.

As shown in Figure 5, initially a large amount of water is produced from the natural fracture network, and this causes an early peak in the gas production rate. This rate declines rapidly for a short time period during which gas is depleted from the cleat spaces. After reaching a brief minimum, the gas production rate again increases steadily until it reaches a second maximum. This period is known as the negative decline period, and it is a characteristic of gas production from coal seams. This steady increase in the gas production rate is caused by an increase in relative permeability to gas because of the desorption of large volumes of gas from the micropores and its subsequent diffusion into the fracture network. Therefore, this desorbing gas creates a localized, high gas-saturation region in the area enclosed by the pressure transient. The desorbed gas provides an internal pressure maintenance for the reservoir, and at the same time it decreases the relative permeability to brine because of the increased gas saturation. This creates a competing effect with respect to water production.

Because water removal strongly influences the pressure decline and, consequently, the desorption rate, a unique production mechanism is observed. The gas production rate, after reaching a maximum, starts to decline as the withdrawal of the gas from micropores begins. The time of occurrence and the magnitude of the early production rate peak, as well as the duration and magnitude of the negative decline period can be shifted, both in time and magnitude, by the individual or coupled effects of coal seam thickness, permeability, porosity, and sorption characteristics. Therefore, the magnitude of the early peak was found to be a function of the ability of the micropore matrix to yield gas to the macropore system and of the coal seam's conductivity to water. This leads to the following observations: 1) the magnitude of the early peak increases, and takes place later, as the permeability and the thickness of the coal seam increase; 2) the magnitude of the second peak-production rate decreases as porosity and sorption time constant increase; and 3) the time at which the second peak-production rate occurs is independent of gas content or sorption pressure (Kim, 1977; Sung and Ertekin, 1987b).

Effects of Coal Seam Properties on Gas Drainage Efficiency

It is important to understand the role of some key coal seam characteristics on methane drainage efficiency for optimizing the gas production from a coal seam. In this discussion, results from Ertekin et al. (1988), Kolesar and Ertekin (1986), Remner et al. (1986), and Sung and Ertekin (1987a) are reviewed.

First, the effect of the absolute permeability of a coal seam on gas drainage efficiency is studied. In general, as the hydraulic conductivity of the coal seam increases, variations in the production characteristics of

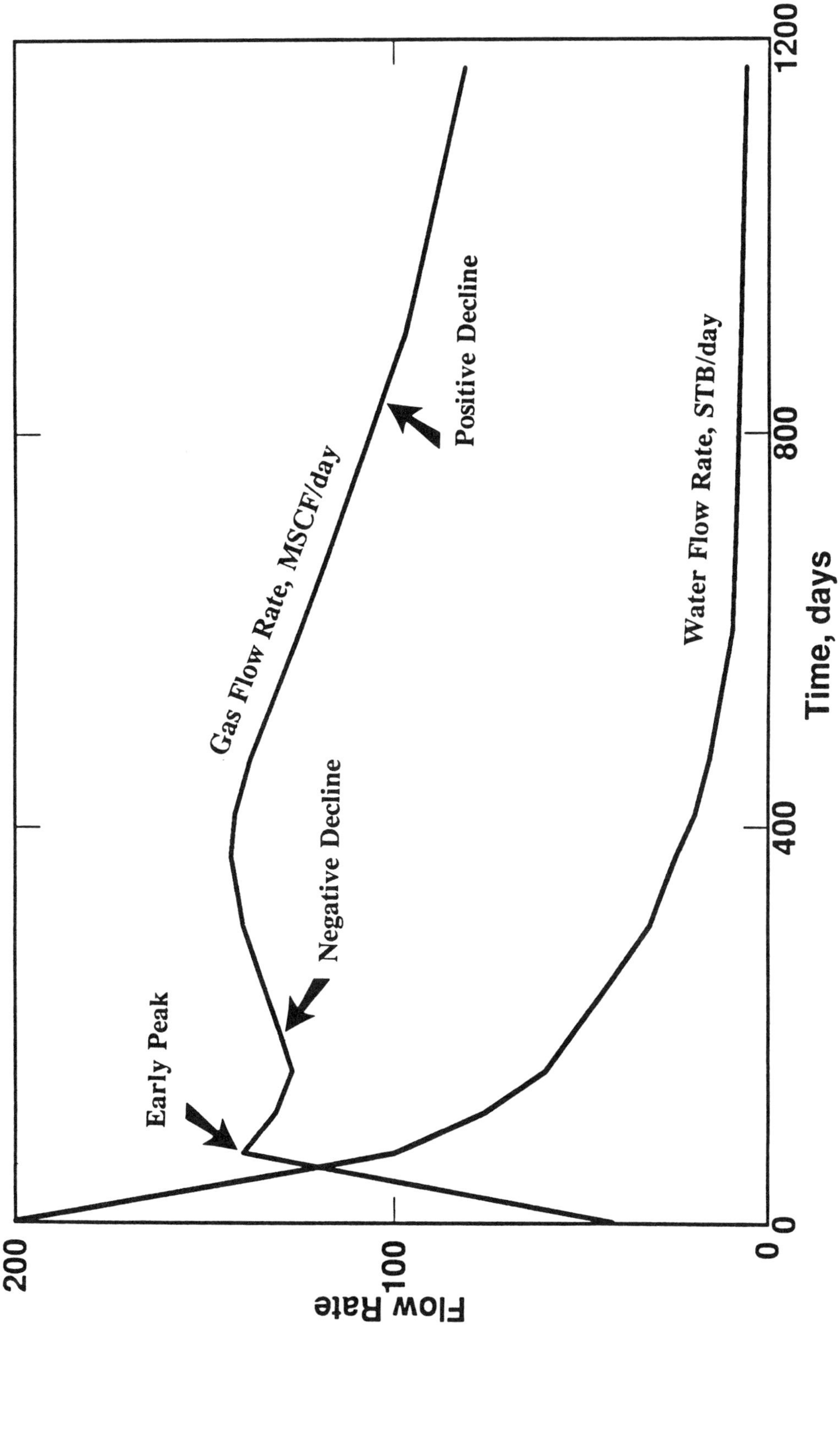

Figure 5. *Typical production performance of a horizontal drainage well.*

drainage wells during the course of gas and water production become more drastic. This is simply because more permeable systems will interact with the wellbore more effectively and drain a larger coal volume than less permeable systems, and because the ability of the coal seam reservoir, in terms of its conductivity, to drain large areas will enhance the desorption rate. When a well is placed in production and water is removed, a zone of reduced pressure will be created. With time, this low pressure zone expands. Gas desorption will occur only in this reduced pressure zone. Because the pressure-transient wave travels faster in more permeable coals, more gas will tend to desorb in a given time if everything else remains constant.

In the application of vertical or horizontal methane drainage wells, the thickness of the coal seam needs to be taken into consideration. The results of numerical model studies as shown in Figures 6 and 7 indicate that the effect of coal seam thickness is greater on production from a vertical, hydraulically fractured well than in a horizontal borehole. This is expected because, in the case of hydraulically created vertical fractures, if fractures are 100% contained within the coal seam, the exposed fracture surface area increases as the thickness of the coal seam increases. On the other hand, the corresponding surface area for a borehole of given diameter increases only as its length is increased (Figure 8). From the numerical exercises conducted on this parameter (Ertekin et al., 1988; Sung and Ertekin, 1987a), it is concluded that the superiority of horizontal boreholes over vertical wells becomes more obvious for thinner reservoirs.

Kolesar and Ertekin (1986) and Remner et al. (1986) investigated the effect of the sorption time constant (τ in sec) which regulates the rate at which gas is released from the micropores into the macropores as given in the following equation:

$$\tau = \frac{R^2}{D}$$

where R is the external radius of the matrix element (in centimeters) and D is the diffusivity coefficient (in cm^2/sec). For a typical coal, the value of D ranges between 10^{-5} to 10^{-10} cm^2/sec, whereas the value of τ may vary between 10^6 to 10^8 sec (Thimons and Kissel, 1973). As shown in Figure 9, a decrease in the value of the sorption time constant accelerates the effects of desorption on the well production or, stated in a slightly different manner, a coal with a smaller time constant has a more active diffusion mechanism. This observation can be explained further by examining the impact of fracture spacing and of diffusion constant on the diffusion characteristics of coals. A coal with a greater fracture density has a greater matrix surface area than a coal with a lower fracture density. Accordingly, diffusion is faster in coals with short diffusion times because of greater fracture density. Small diffusion times also are characteristic of coals with large diffusion constants. A coal with a large diffusion constant (D) will have a high rate of methane diffusion out of the solid coal matrix. This results in improved pressure maintenance in the fracture system and a more gradual decline in the production rate of the well.

Field Development

Optimum field development in a coal seam degasification process depends mainly on the coal seam characteristics, well spacing, and water production (which facilitates desorption of methane gas and, hence, leads to the flow of gas).

Well spacing in coal seams has a much more significant role in obtaining maximum recovery than in conventional gas reservoirs. The additional pressure reduction caused by more densely drilled wells increases the desorption rate, thus supplying more free gas. This results in an increased gas flow rate per well. Furthermore, it is speculated that desorbed gas, because of its compressibility, acts as a shielding element by diminishing the transients caused by the other wells. Therefore, the presence of additional wells in the pattern yields more uniform pressure and saturation distributions in the coal seam. A uniform pressure

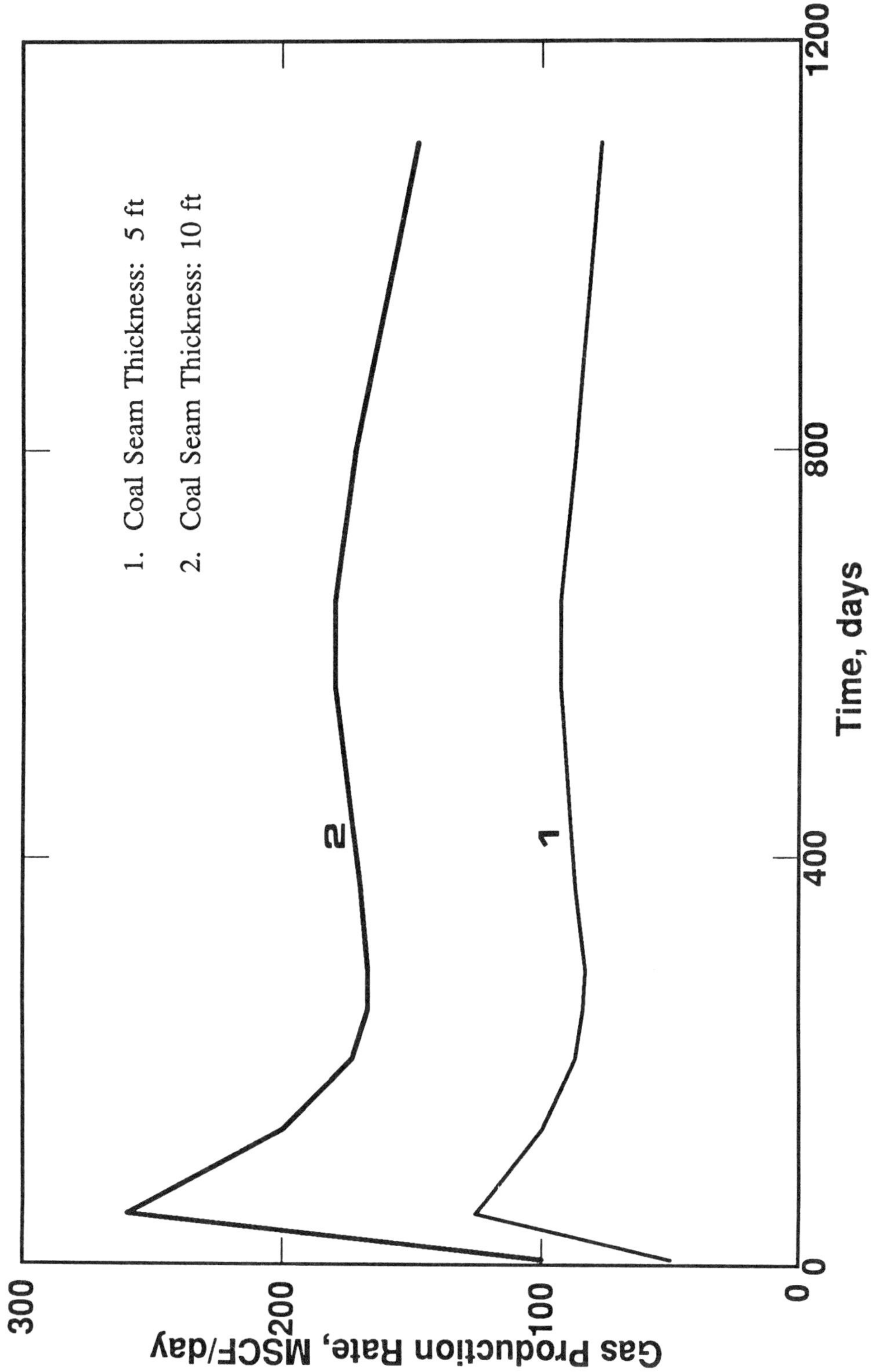

Figure 6. *Gas production rates from vertically fractured wells (fracture height matches the formation thickness).*

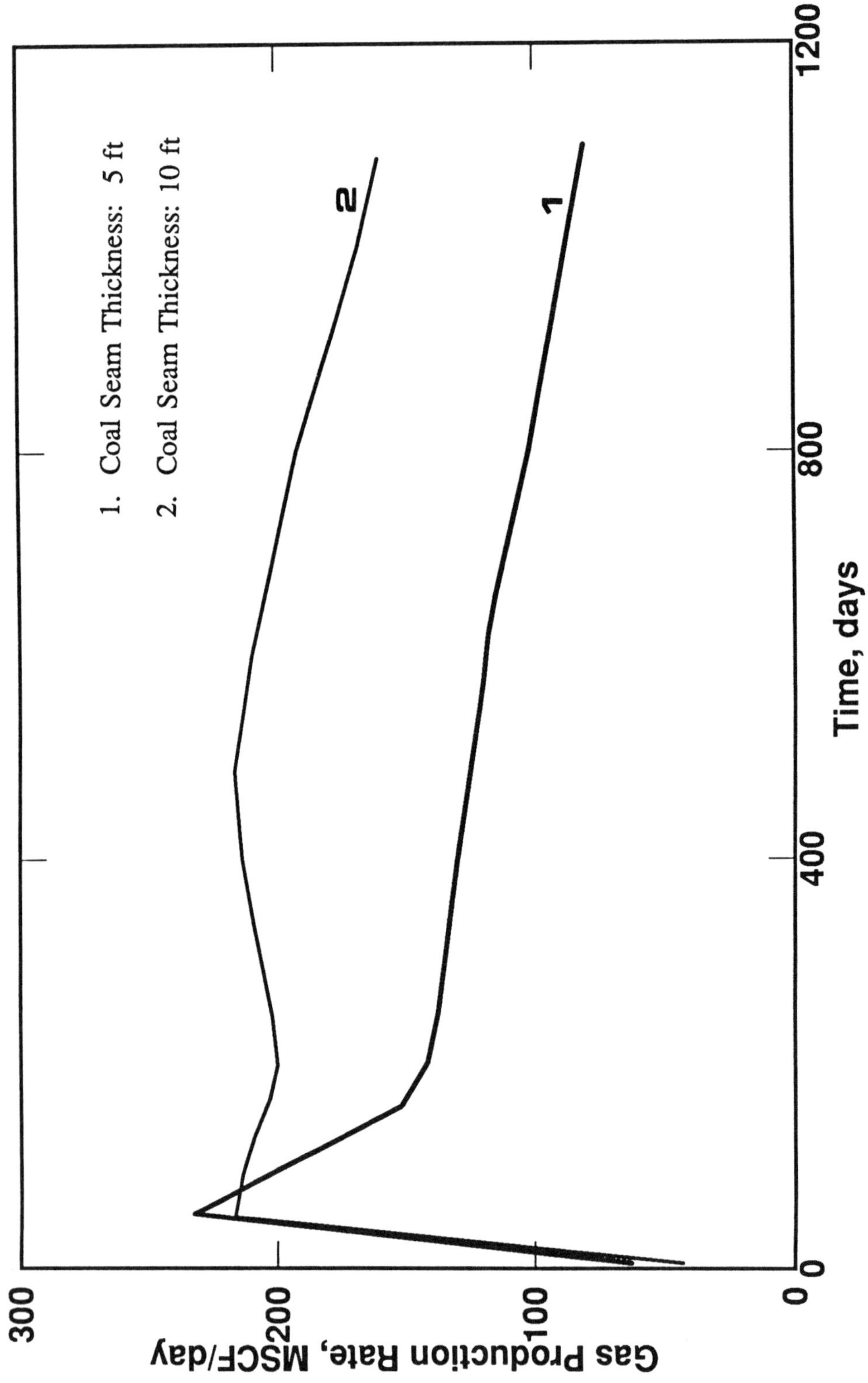

Figure 7. *Gas production rates from horizontal boreholes completed in different coal seams with different thicknesses.*

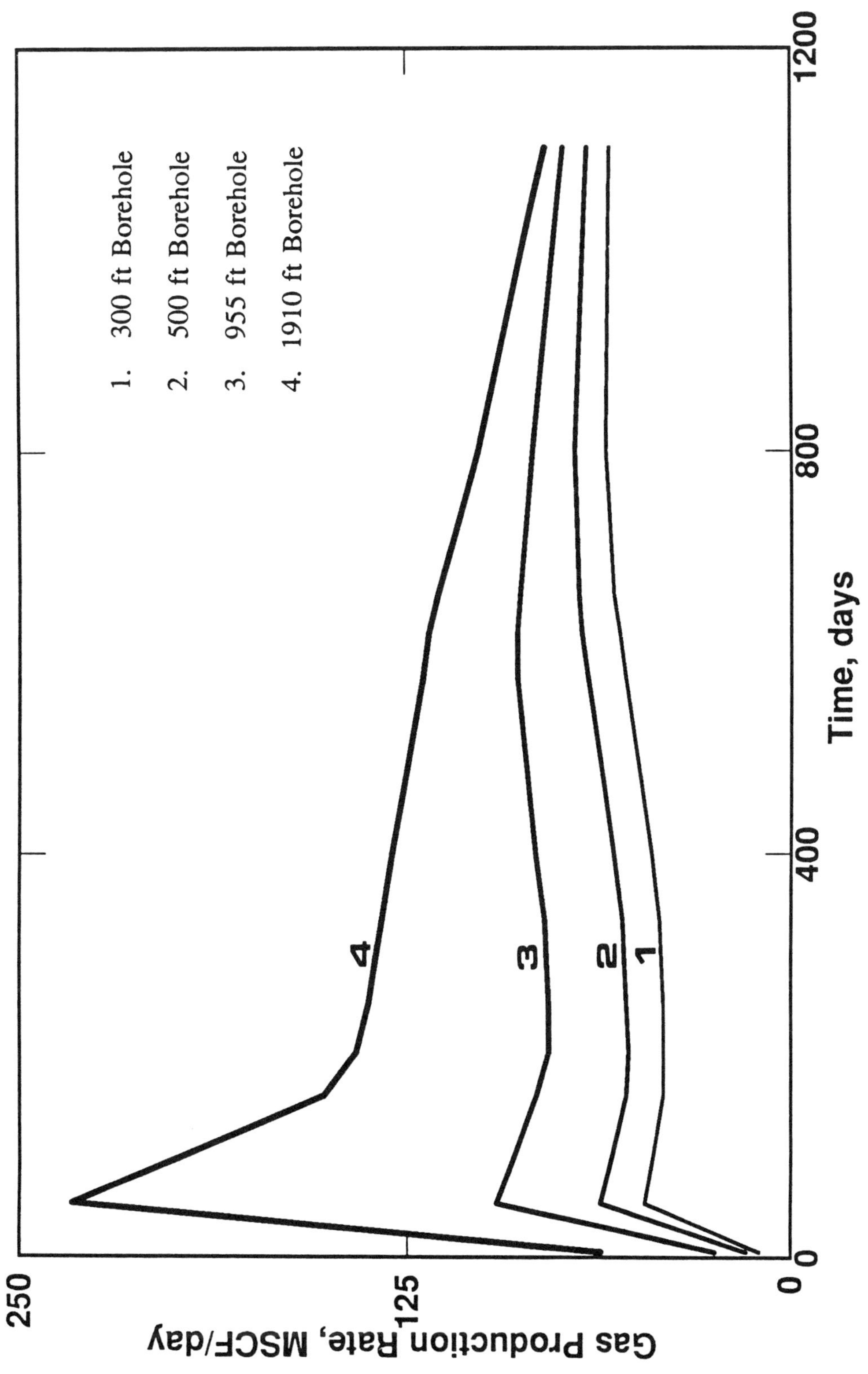

Figure 8. *Variation of methane drainage rate with horizontal borehole length.*

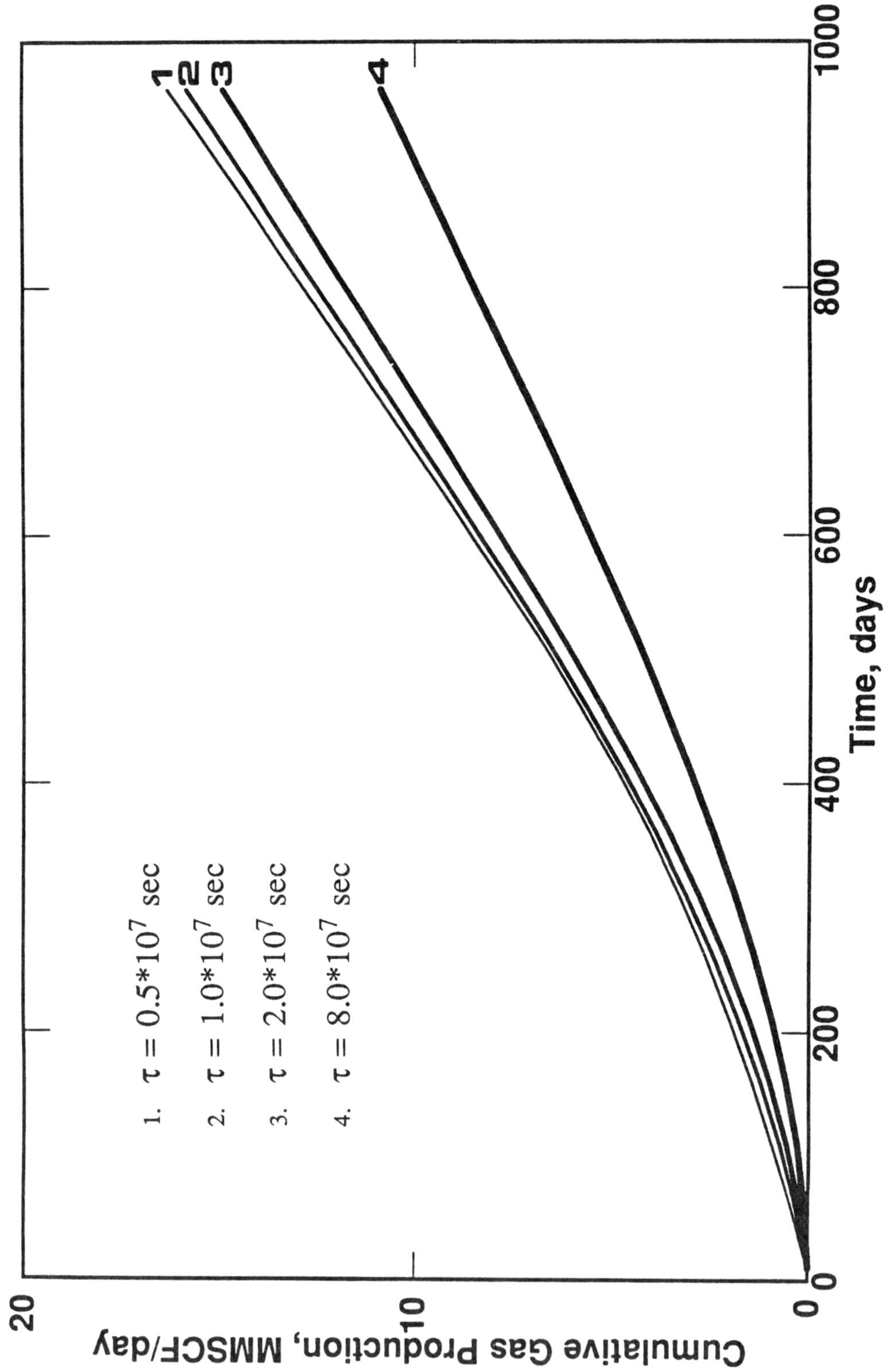

Figure 9. *Effect of sorption time constant on cumulative gas production.*

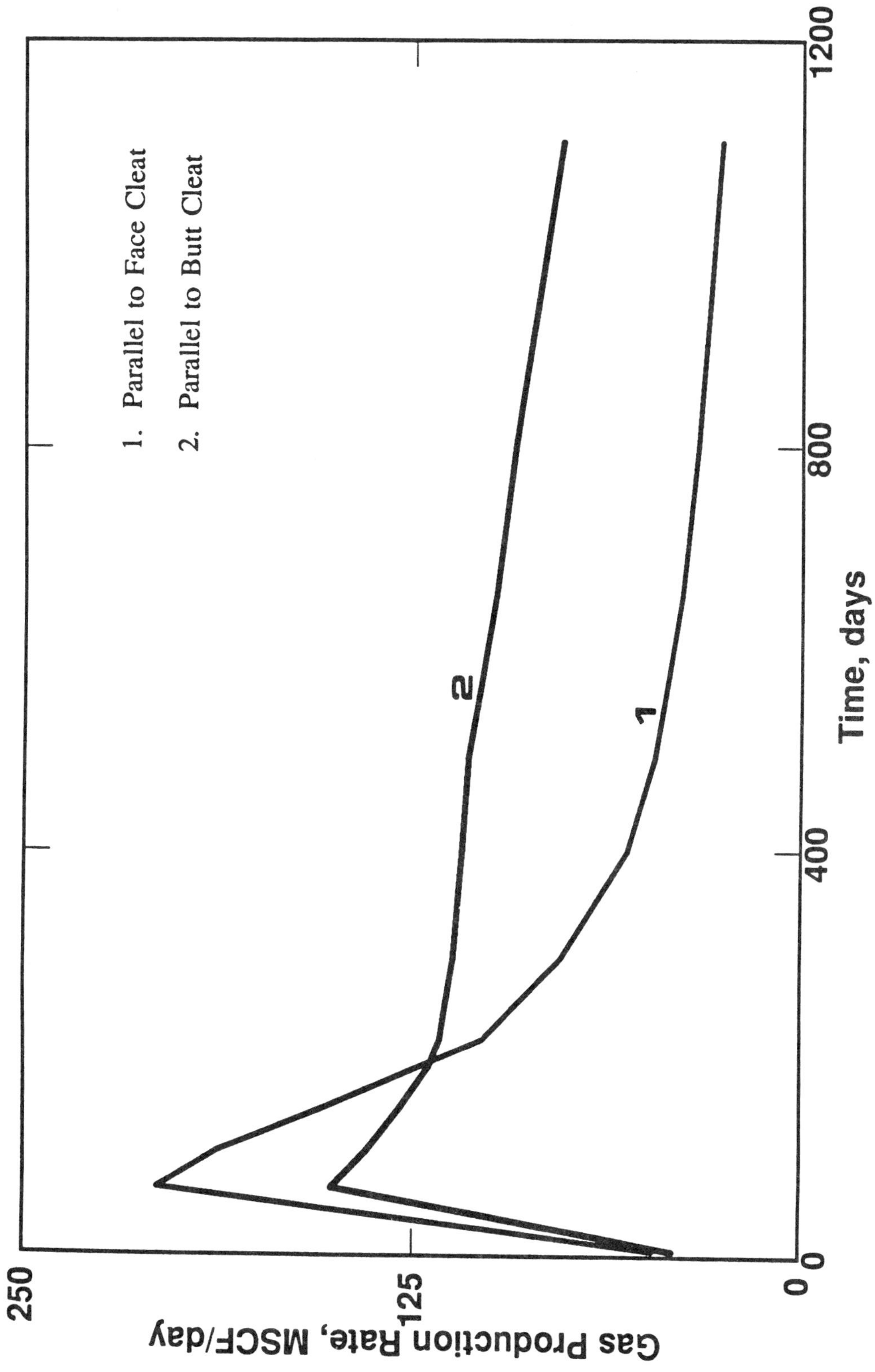

Figure 10. *Comparison of the production performances of two horizontal drain holes with different orientations with respect to cleat systems of a coal seam with a permeability contrast of 0.1.*

distribution during the depletion of a coal seam is very important because it enhances the existence of a continuous free gas phase (Sung and Ertekin, 1987b). In a sparsely developed well field, infill drilling will have a strong impact on the ultimate gas recovery because the drilling of additional wells on established units represents an attempt either to increase the recovery by facilitating the desorption of more gas (which presumably would not be produced by the original wells) or to prevent gas migration to other wells on other leases. The degree of success of an infill drilling program in a coal seam depends on the uniformity of pressure and saturation distributions created by the original wells. If infill wells can enhance the quality of the existing continuous free gas saturation, then they are expected to contribute significantly to the field production (Sung and Ertekin, 1987b).

As mentioned earlier, coal seams often have a well-defined, natural fracture network. The face and butt cleats create a distinct permeability anisotropy. This strong permeability anisotropy infers well arrangements whether wells should be positioned parallel to face cleat or butt cleat direction (in the case of vertical wells, this discussion refers to the relative distances between the wells placed parallel to the face cleat and butt cleat directions). When more drainage wells are drilled parallel to the face cleat direction, water is produced rapidly from the more discontinuous butt cleats creating a more effective pressure decline thereby yielding more gas during the early production period. As production continues, drainholes drilled to intersect face cleats which have greater continuity effectively degas a greater volume of coal (see Figure 10). Therefore, although initial production during the early period is larger when more wells are placed parallel to the face cleat direction, maximum recovery requires drainage wells placed perpendicular to the face cleat direction (Ertekin et al., 1988; Sung and Ertekin, 1987b).

SUMMARY

Very large volumes of natural gas are locked in unconventional gas reservoirs. Coal seams, as one of the unconventional gas reservoirs, often are not being fully developed because of the lack of understanding of the dynamics of fluids in the reservoirs. The presence of a fracture-induced permeability anisotropy, together with the strong sorption characteristics of coal, underscores the important differences between this resource and conventional gas resources. With appropriate research and development, unlocking of this vast resource no doubt will help tremendously in the quest of earth scientists to alleviate the technological problems associated with the production of these untapped resources.

REFERENCES

Adzumi, H., 1937, Studies on the flow of gaseous mixtures through capillaries--II: The molecular flow of gaseous mixtures: Bulletin of Chemical Society of Japan, v. 12, No. 6, p. 285-291.

Airey, E.M., 1968, Gas emission from broken coal: An experimental and theoretical investigation: International Journal of Rock Mechanics and Mining Sciences, v. 5, p. 475-494.

Ancell, K.L., S. Lambert, and F.S. Johnson, 1980, Analysis of the coalbed degasification process at a seventeen well pattern in the Warrior Basin of Alabama, in Proceedings of the Society of Petroleum Engineers/Department of Energy Symposium on Unconventional Gas Recovery, Pittsburgh, Pennsylvania: Houston, Society of Petroleum Engineers, paper No. 8971.

Boyer, C.M., H.L. Morrison, A.M. Pavone, and F.C. Schwerer, 1982, Methane modeling--Predicting the inflow of methane gas into coal mines: U.S. Department of Energy, Final Contract Report No. DE-AC22-80PC30123.

Dabbous, M.K., A.A. Reznik, B.G. Mody, P.F. Fulton, and J.J. Taber, 1976, Gas water capillary pressure in coal at various overburden pressures: Society of Petroleum Engineers Journal, v. 16, no. 5, p. 261-267.

Dabbous, M.K., A.A. Reznik, J.J. Taber, and P.F. Fulton, 1974, The permeability of coal to gas and water: Society of Petroleum Engineers Journal, v. 14, no. 6, p. 563-572.

Darcy, H., 1856, Les Fontaines Publiques de la Ville de Dijon, Dalmont, Paris.

Davis, J.G., 1971, An outline of Consol's research program on methane control as related to the Pittsburgh Seam, in Proceedings of the West Virginia Coal Mining Institute, Bluefield, West Virginia, April 23-24, 1971: Morgantown, West Virginia, Morgantown Printing Co., p. 61-68.

Deul, M., and A.G. Kim, 1975, Coalbeds: A source of natural gas: Oil and Gas Journal, v. 73, no. 24, p. 47-49.

Elder, C.H., and M. Deul, 1975, Hydraulic stimulation increases degasification rate of coalbeds: U.S. Bureau of Mines Report of Investigations RI 8047, 17 p.

Ertekin, T., G.R. King, and F.C. Schwerer, 1986, Dynamic gas slippage: A unique dual-mechanism approach to the flow of gas in tight formations: Formation Evaluation, v. 1, no. 1, p. 43-52.

Ertekin, T., W. Sung, and F.C. Schwerer, 1988, Production performance analysis of horizontal drainage wells for the degasification of coal seams: Journal of Petroleum Technology, v. 40, no. 5, p. 625-632.

Fick, A., 1855, Ann. Physik: v. 59, p. 170.

Fields, H.H., S. Krickovic, A. Sainato, and M.G. Zabetakis, 1973, Degasification of virgin Pittsburgh Coalbed through a large borehole: U.S. Bureau of Mines Report of Investigations RI 7800, 27 p.

Fields, H.H., J.H. Perry, and M. Deul, 1975, Commercial-quality gas from a multipurpose borehole located in the Pittsburgh Coalbed: U.S. Bureau of Mines Report of Investigations RI 8025, 14 p.

Jones, A.H., U. Ahmed, M.A. Abou-Sayed, and B. Sakashita, 1982, Fractured vertical wells versus horizontal boreholes for methane drainage in advance of mining U.S. coals, in A.J. Hargraves, ed., Proceedings of the Symposium on Seam Gas Drainage with Particular Reference to the Working Seam, Parkville, Victoria, Australia: Parkville, University of Wollongong.

Kim, A.G., 1977, Estimating methane content of bituminous coalbeds from adsorption data: U.S. Bureau of Mines Report of Investigations RI 8245, 22 p.

Kim, A.G., 1978, Methane drainage from coalbeds: Research and utilization, in Methane gas from coalbeds--Development, production, and utlization: U.S. Department of Energy, Morgantown Energy Research Center, Report SP-78/1, p. 12-17.

King, G.R., T. Ertekin, and F.C. Schwerer, 1986, Numerical simulation of the transient behavior of coal seam degasification wells: Formation Evaluation, v. 1, no. 2, p. 165-183.

Kissell, F.N., C.M. McCulloch, and C.H. Elder, 1973, The direct method of determining methane content of coalbeds for ventilation design: U.S. Bureau of Mines Report of Investigations RI 7767, 17 p.

Klinkenberg, L.J., 1941, The permeability of porous media to liquids and gases, in Drilling and production practice: New York, American Petroleum Institute, p. 200-213.

Kolesar, J.E., and T. Ertekin, 1986, The unsteady state nature of sorption and diffusion phenomena in the micropore structure of coal, in Proceedings of the Society of Petroleum Engineers Unconventional Gas Technology Symposium, Louisville, Kentucky, May 18-21, 1986: Houston, Society of Petroleum Engineers, p. 289-314.

Kuuskraa, V.A., J.P. Brashear, T.M. Doscher, and L.E. Elkins, 1978, Enhanced recovery of unconventional gas: U.S. Department of Energy, Final Contract Report No. EF-77-C-01-2705.

Makismovic, S.D., C.H. Elder, and F.N. Kissell, 1977, Hydraulic stimulation of a surface borehole for gob degasification: U.S. Bureau of Mines Report of Investigations RI 8228, 17 p.

Marsh, H., 1965, The determination of surface areas of coals--Some physiochemical considerations: Fuel, v. 44, p. 253-268.

Moore, T.D., and M.G. Zabetakis, 1972, Effect of surface borehole on longwall gob degasification (Pocahontas No. 3 Coalbed): U.S. Bureau of Mines Report of Investigations RI 7657, 9 p.

Mordecai, M., and L.H. Morris, 1974, The effects of stress on the flow of gas through coal measure strata: Mining Engineer, v. 133, no. 164, p. 435-443.

Olson, M.L., 1979, Development of coalbed methane as an energy source: Energy Sources, v. 4, no. 4, p. 353-365.

O'Shaughnessy, S.M., 1980, The computer simulation of methane flow through strata adjacent to a working longwall coalface: Nottingham, England, The University of Nottingham, Unpublished Ph.D. Dissertation.

Owili-Eger, S.C., 1973, Simulation of quantity and quality control in mine ventilation: University Park, The Pennsylvania State University, Unpublished M.Sc. Thesis, 259 p.

Patching, T.H., 1970, The retention and release of gas in coal--A review: Canadian Institute of Mining and Metallurgy Bulletin, v. 63, no. 703, p. 1302-1308.

Pavone, A.M., and F.C. Schwerer, 1984, Development of coal gas production simulators and mathematical models for well test strategies: Gas Research Institute Final Contract Report No. 5081-321-0457, 510 p.

Price, H.S., and A.A. Abdalla, 1972, Mathematical model simulating flow of methane and water in coal mine systems: U.S. Bureau of Mines, Contract Report No. S0110752.

Price, H.S., and K.L. Ancell, 1978, The feasibility of methane from coal, in Methane Gas from Coalbeds--Development, Production, and Utilization: U.S. Department of Energy, Morgantown Energy Research Center, Report SP-78/1, p. 37-73.

Remner, D.J., T. Ertekin, W. Sung, and G.R. King, 1986, A parametric study of the effects of coal seam properties on gas drainage efficiency: Reservoir Engineering, v. 1, no. 6, p. 633-646.

Rice, G.S., 1928, Safety in coal mining: U.S. Bureau of Mines Bulletin 277, 141 p.

Rose, W.D., 1948, Permeability and gas-slippage phenomenon, in Drilling and Production Practice: New York, American Petroleum Institute, p. 209-217.

Sampath, K., and C.W. Keighin, 1982, Factors affecting gas slippage in tight sandstones of Cretaceous age in the Unita Basin: Journal of Petroleum Technology, v. 34, no. 11, p. 2715-2720.

Satriana, M., 1980, Unconventional natural gas: Resources, potential, and technology: Noyes Data Corporation, p. 148-182.

Smith, D.M., and F.L. Williams, 1982, Diffusional effects in the recovery of methane from coalbeds, in Proceedings of the Society of Petroleum Engineers/Department of Energy Symposium on Unconventional Gas Recovery, Pittsburgh, Pennsylvania: Houston, Society of Petroleum Engineers, paper No. 10821.

Somerton, W.H., L.M. Soylemezoglu, and R.C. Dudley, 1975, Effect of stress on permeability of coal: International Journal of Rock Mechanics and Mining Sciences and Geomechanics, v. 12, no. 6, p. 129-145.

Sung, W., and T. Ertekin, 1987a, Performance comparison of vertical and horizontal hydraulic fractures and horizontal boreholes in low permeability reservoirs: A numerical study, in Proceedings of the Society of Petroleum Engineers/Department of Energy Symposium on Low Permeability Reservoirs, Denver, Colorado, May 18-19, 1987: Houston, Society of Petroleum Engineers, p. 185-194.

Sung, W., and T. Ertekin, 1987b, An analysis of field development strategies for methane production from coal seams, in Proceedings of the Society of Petroleum Engineers 62nd Annual Technical Conference and Exhibition, Dallas, Texas, September 27-30, 1987: Houston, Society of Petroleum Engineers, p. 231-242.

Sung, W., T. Ertekin, and F.C. Schwerer, 1986, The Development, testing and application of a comprehensive coal seam degasification model, in Proceedings of the Society of Petroleum Engineers Unconventional Gas Technology Symposium, Louisville, Kentucky, May 18-21, 1986: Houston, Society of Petroleum Engineers, p. 457-472.

Thakur, P.C., and W.N. Poundstone, 1980, Horizontal drilling technology for advance degasification: Mining Engineering, v. 32, no. 6, p. 676-680.

Thimons, E.P., and F.N. Kissel, 1973, Diffusion of methane through coal: Fuel, v. 52, no. 4, p. 274-280.

van Krevelin, D.W., and J. Schayer, 1957, Coal science aspects of coal constitution: Princeton, New Jersey, Elsevier Scientific Publishing Co., 352 p.

von Schonfeldt, H.A., B.R. Pothini, and G.N. Aul, 1980, Advance methane control and its impact on gas emissions in the Pocahontas #3 coal seam, in Proceedings of the Society of Petroleum Engineers/Department of Energy Symposium on Unconventional Gas Recovery, Pittsburgh, Pennsylvania: Houston, Society of Petroleum Engineers, paper No. 8969.

Wise, R.L., 1979, Methane recovery and utilization from coalbeds, in Proceedings of the Society of Petroleum Engineers 54th Annual Technical Conference and Exhibition, Las Vegas, Nevada, September 23-26, 1979: Houston, Society of Petroleum Engineers, paper No. 8357.

THE EFFECTS OF OVERBURDEN STRESS ON COALBED METHANE PRODUCTION

Daniel J. Soeder[1]
Institute of Gas Technology
3424 South State Street
Chicago, IL 60616

ABSTRACT

The ability of methane gas to diffuse from the matrix of a coal seam and move to a production well via the natural fracture (cleat) system is strongly influenced by changes in stress on the coal seam. Reducing the pore pressure in coal by withdrawal of liquids and gas results in an increase in the net overburden pressure or effective stress on the seam. Linear, fracture-like features such as cleats respond to elevated values of effective stress by narrowing down. This constriction commonly leads to a decrease in formation permeability of cleated, coalbed methane reservoirs during the normal course of production and drawdown. Stresses introduced into coal during hydraulic fracturing also can adversely affect cleat permeability. Recent core analysis and petrographic work has indicated that increased stress preferentially closes the smaller cleats and microcleats in coal, although the larger cleats remain open. The increased flowpath tortuosity due to closure of interconnecting microcleats results in lower coal seam permeabilities. Assessing and predicting stress-related phenomena in coal are important to understanding the production geology of coalbed methane.

INTRODUCTION

Methane-bearing coals are considered to be a significant gas resource. Although coal is a porous medium, permeabilities are usually quite low, and the pore structure is considerably more complex than that usually found in conventional clastic reservoirs. The porosity of coal characteristically consists of a microporous matrix, macroporosity in fossilized plant fragments or "phyterals", and fracture porosity by way of the cleat system (Wyman, 1984). The "face cleats" and "butt cleats" in coal are the relatively more highly and less well-developed fracture sets, respectively. Coals also contain subhorizontal "bedding-plane joints" which are less common than cleats, but are important to fluid movement because they intersect much of the near-vertical cleat system. In a typical virgin coal seam, the cleats contain almost no methane, but are filled with water. The methane resides in the phyteral pores and micropores, as well as in an adsorbed state on carbon complexes in the matrix (Holditch et al., 1989). The seam is produced by withdrawing water from the cleat system until the point on the relative permeability curve is reached where gas begins to flow, after which gas and water are co-produced from the well. Gas desorbs continuously from the coal matrix to replace that produced from the cleat system, and water production gradually declines over time.

---

1. Current Address: U.S. Geological Survey, P.O. Box 327, MS 721, Mercury, NV 89023.

The adsorption potential of coal is awesome, allowing it to contain very large amounts of gas. Wyman (1984) points out that 0.45 kg (1 lb) of coal can have up to 46,500 m^2 (500,000 ft^2) of adsorptive surface area, and estimates that the Canadian Elmworth coals in Alberta can contain up to 14 m^3/tonne (500 ft^3/ton) of free and adsorbed methane. This compares with recent estimates of the gas resource potential of the Menefee Coal in the San Juan Basin by Crist et al. (1989), where 38 tcf of gas may be contained in a total of 131,000,000,000 tonnes (145,000,000 tons) of coal. Crist and her co-workers calculated an average gas content of 7.5 m^3/tonne (262 ft^3/ton) of methane for the Menefee Coal. These examples show that the total coalbed methane resource in North America is significant, with potentially recoverable coalbed methane estimated to be on the order of hundreds of tcf.

Although high gas-in-place values in a coal seam are certainly desirable for coalbed methane production, large methane contents alone do not necessarily guarantee commercial gas production. One additional factor is critical to production from any tight gas resource: reservoirs must have a sufficient number of high-permeability pathways through which the matrix gas can move to a wellbore. Permeable pathways in a coal seam consist of cleats, and coalbed methane production is intimately related to the nature and structure of the cleat system. Because cleats are narrow, flat, near-vertical fractures, they are sensitive to closure with increased normal stress. Net or effective stress is increased by producing gas from the seam and reducing the internal pore pressure, which acts to offset part of the overburden pressure and horizontal stresses in a virgin reservoir. In a study on flow through microfractures in tight sands, Ostensen (1983) noted that the crack-like flowpaths were propped open by an irregular distribution of bumps or "asperities" on the fracture faces. Mechanically, these asperities are rather poor supports, and he found that flowpaths through fractures were more susceptible than open, grain-supported pores to narrowing in response to stress. The same effect often occurs in the coal cleat system to an even greater degree, because coal generally is more easily deformed than sandstone.

COAL PERMEABILITY UNDER STRESS

A number of laboratory experiments have been carried out over the past few years to try to improve understanding of the effects of confining stress on permeability in coal. As part of a project funded by the Gas Research Institute (GRI) to determine the amount of formation damage done to coal by drilling muds and fracturing fluids, the Institute of Gas Technology (IGT) measured the liquid permeability of coal samples at net confining pressures that simulated effective stress from normal dewatering and production drawdown (Rose and Foh, 1984). "Effective stress", as used in this paper, is a laboratory approximation of actual reservoir net stress using standard equipment and techniques. It is a best-judgment, averaged value for the combination of vertical stress and directional horizontal stress experienced by a sample in the ground. Effective stresses imposed on laboratory samples at IGT generally are selected midway between the net overburden stress gradient and the hydraulic fracture gradient, when such data are available. We feel that this is more representative than applying just a straight net overburden pressure. It would, of course, have been better to apply true, oriented biaxial, or even triaxial, stresses to the cores in the laboratory, but equipment to do this was not available, and these measurements were not the main objective of Rose and Foh's (1984) research project.

With respect to the laboratory analyses themselves, a number of workers over the years have questioned whether "restored" stress states in the laboratory can possibly be representative of "in situ" stress in the reservoir, given the unavoidable stress cycling and hysteresis that occur when a core is removed from the ground. Core analysis data in this paper are compared only against other core analysis data, and not with well test permeabilities or other bulk reservoir analyses, in order to avoid this problem. Thus, if hysteresis exerts an effect on the core data, it is a common effect to all samples. Problems such as stress cycling and hysteresis on cores are important questions that should be investigated, but not at the expense of foregoing core analysis altogether, which is the only way some types of data can be obtained. Although bulk reservoir tests are critical to understanding the production potential of a gas reservoir, they are less useful for investigating individual effects, such as change in permeability with effective stress or water saturation. These analyses are much better suited to a laboratory environment, where the variables can be controlled and the desired effect isolated and measured. The data then can be compared in a relative sense, or related back to the reservoir by incorporation into numerical models and by correlation to defined geological properties.

Figure 1 shows the horizontal permeability parallel to the face cleat direction in the Cameo Coal from the Piceance Basin as a function of effective or net confining stress. Gas permeabilities of typical Devonian shale and tight sand samples under stress are included for comparison. The figure demonstrates that the permeabilities of both the coal and shale are severely reduced under stress, whereas the sandstone permeability is more mildly affected. This is true even in sandstone samples which contain flowpath apertures and crack-like pore morphologies similar to the coal. Rose and Foh (1984) concluded that the strong stress-dependence of permeability in coal would be an important factor in nearly all phases of coalbed methane production and economics.

The theoretical treatment of microcrack permeability in tight gas sandstone by Ostensen (1983) predicts a power law relationship between permeability and effective stress on the rock matrix:

$$K \propto P^{-\alpha}$$

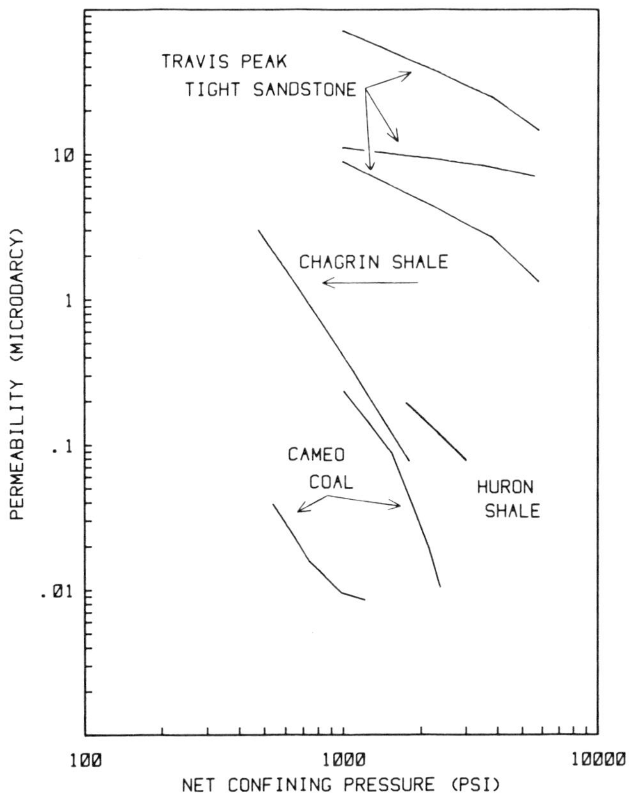

Figure 1. Permeability as a function of net confining pressure (effective stress) for a variety of tight gas reservoir samples.

where K is the single-phase permeability at net confining pressure P. For this type of relationship, a plot of the $\log_{10}$ of permeability versus the $\log_{10}$ of net stress should be a straight line with a negative slope of value α. Higher values of exponent α indicate a steeper slope and a greater reduction of permeability with effective stress (Ostensen, 1983). Values of α measured on tight sandstones at IGT typically range from less than 1 to 1.5, whereas those for Devonian shale and coals range from 2 to 3 or higher.

Mathematical modeling of coalbed methane reservoir performance has shown that relationships between effective stress and permeability must be included to accurately describe production (McKee et al., 1987). McKee and his colleagues determined that their theoretical mathematical relationships for coal porosity, permeability, and density as a function of effective stress fit both laboratory and field data quite well, with a correlation coefficient (r^2 value) above 0.95 for the laboratory and as high as 0.92 for the field data. They concluded that much more realistic predictions of coalbed methane reservoir performance could be obtained by including an effective stress parameter in the numerical simulation.

The physical manifestation of high effective stress on the pore morphology in coal can be deduced mathematically from the Klinkenberg (1941) correction for gas slippage in tight rocks. Work at IGT on other low-permeability formations with crack-like flowpaths has shown that the characteristic width or flowpath aperture ($\overline{w}$) of the cracks can be calculated from parameters measured during core analysis:

$$\overline{w} = \frac{16\,c\mu}{b}\left(\frac{2RT}{\pi M}\right)^{0.5}$$

where

b	= empirical constant from Klinkenberg equation
c	= Adzumi constant (assumed to be 0.9)
μ	= viscosity of the gas
M	= molecular weight of the gas (in Kg/Kmole)
C	= gas constant = 8,314 Joule/(Kmole x T)
T	= temperature (Kelvin)

Once the value for characteristic flowpath aperture is known, the tortuosity (τ) of the flowpaths also can be determined:

$$\tau = \overline{w}\left(\frac{\varphi}{12K_{\infty}}\right)^{0.5}$$

where

φ	= porosity of flowpaths (total porosity of sample is generally used, but it gives a minimum limit value for tortuosity)
K_{∞}	= Klinkenberg-corrected gas permeability (i.e., permeability to an ideal gas at infinite pressure)

Flowpath tortuosity in this usage describes the total distance a gas molecule must travel to traverse the axial length of a cylindrical core plug. Tortuosity is expressed in multiples of plug lengths with the straightest, most direct path having a value of 1. Higher tortuosity values generally indicate poorer pore-to-pore interconnections in a sample, and higher tortuosities usually correlate with lower permeabilities. Full details concerning the derivation and use of the above equations for crack-like flowpaths can be found in Randolph et al. (1984).

The work cited earlier on stress-dependent measurements of permeability in coal by Rose and Foh (1984) unfortunately does not lend itself to direct analysis using the above equations because the measurements were made with water, instead of gas. Nevertheless, work on other tight gas reservoir samples with crack-like flowpaths has indicated a consistent response to increased effective stress: the characteristic flowpath aperture increases, as does the tortuosity. This was found on laboratory measurements of shale (Soeder, 1988a), tight gas sands (Soeder and Randolph, 1987), and igneous rocks (Soeder, 1988b). These studies led to the conclusion that tight rocks containing crack-like flowpaths react to elevated levels of stress in a similar manner: the smallest crack apertures are preferentially closed down under higher effective stress. Because the smallest size cracks are closed and, therefore, eliminated as gas flowpaths at high effective stress, the characteristic width of flowpaths through the remaining openings is larger. However, closure of the small cracks under stress also reduces the number of interconnections between the larger flowpaths in the sample, increasing the flowpath tortuosity and reducing the permeability. Flowpath tortuosity in a Devonian gas shale, for example, increased from four plug lengths to over 50 plug lengths when net confining pressure was doubled (Soeder, 1988a), and permeability was reduced by nearly 70%. The characteristic flowpath width in this shale was seven times larger at the high confining pressure, indicating a significant closure of smaller, interconnecting pores.

PETROGRAPHY OF COAL PORES AND FLOWPATHS

Observations of cleat and pore morphology in coal can be carried out with an optical reflectance microscope if the void spaces are first impregnated with an epoxy plastic containing fluorescent dye (Soeder, 1990). The technique employed at IGT utilizes a low-viscosity, endothermic casting epoxy stained with rhodamine-B dye (also known as Red Dye #19). The rhodamine fluoresces very brightly under incident light and can be observed by using the proper illumination and filter configuration.

The lamp housing and filters for rhodamine-B epifluorescence can be obtained readily from most major microscope manufacturers. The beam splitter and filters are designed to illuminate the specimen with light wavelengths concentrated around the "exciter frequency" or the frequency which will cause the dye to strongly fluoresce. A schematic diagram of the microscope optics arrangement used to observe fluorescent-dyed epoxy in pores and fractures is shown in Figure 2.

Because the fluorescent wavelengths of most materials are different from the exciter wavelengths, the beamsplitter and filters in the optical path prevent the exciter frequency from reaching the eyepiece while passing the fluorescent wavelengths. Green light, for example, elicits a strong reddish-orange fluorescent response from the rhodamine dye. The filter system is configured such that the green light illuminates the specimen, but only the reddish-orange light is permitted to reach the microscope eyepiece. The effect is to produce an image in which only the rhodamine-dyed epoxy areas can be seen, while the rest of the coal and mineral grains are "blacked-out".

Epoxy impregnation on coal and other low permeability samples at IGT is performed using a vacuum/pressure method, in which the air is first evacuated from both the sample pores and the epoxy. The sample is submerged in the epoxy while still under vacuum, and atmospheric pressure then is introduced to push the epoxy into the sample pores. Under normal conditions, it is necessary to place the sample in a hydraulic chamber, where pressures of tens of MPa (several thousand psi) can be applied to overcome capillary pressures and push the epoxy into small pores. This is done in a uniform manner in small steps to prevent a differential pressure from building between the pores and the surface of the sample (i.e., elevated pressures inside and outside the sample are equal), and we have never observed a coal sample fracturing as a result of impregnation. High pressure is required for the impregnation of coal because epoxy is a non-wetting fluid, and capillary forces prevent it from entering pores in the "water-wet" coal. Impregnation by capillary imbibition of a dyed, water-based compound also was tried successfully, without utilizing any external pressure to force

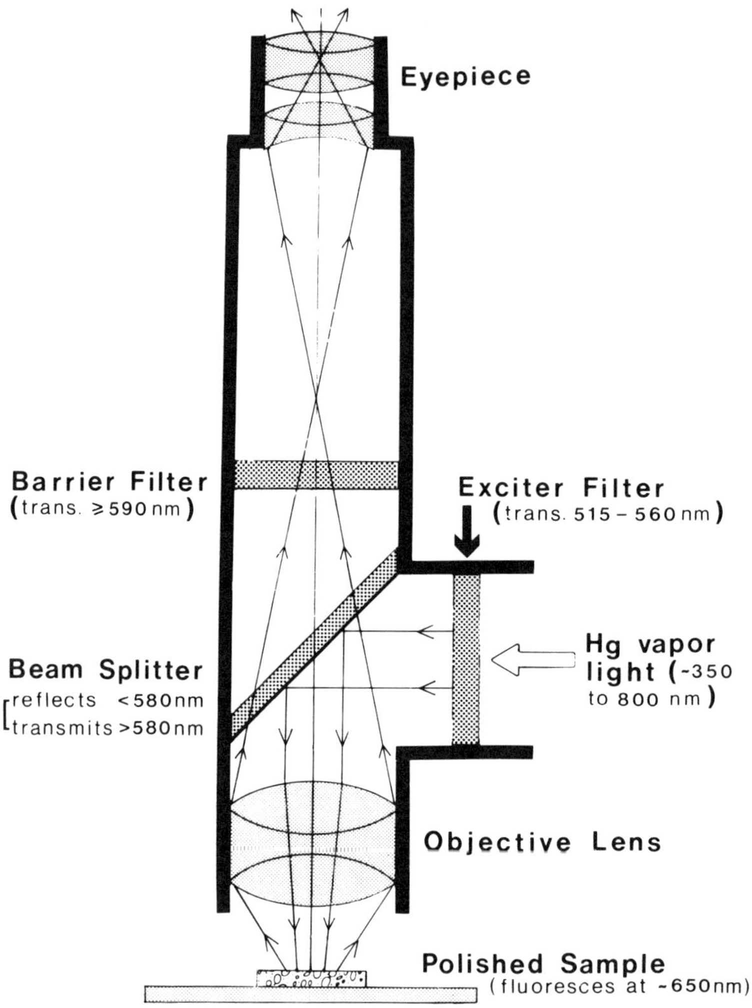

Figure 2. *Schematic drawing of epifluorescence petrographic-microscope configuration used to observe rhodamine-dyed epoxy in the pores and cleats of coal.*

the liquid into the pores. It was abandoned in favor of the epoxy, however, because it did not cure as well, was not as hard, and produced no visible difference in the size and numbers of cleats and microcleats visible in the coal. Drying the coal at elevated temperatures appears to be far more harmful to the pore structure than pressure impregnation (Rose and Foh, 1984), and this is carefully avoided when preparing samples at IGT. Drying, if needed, is performed at room temperature in a dessicant chamber.

An apparatus also has been developed at IGT for impregnating pores while a sample is confined under "restored" effective reservoir stress. This assembly is shown schematically in Figure 3. It consists of a vertically-mounted, standard Hassler-type coreholder containing the sample, with a vacuum line on the lower end and a plastic tube containing epoxy under a gas pressure head at the upper end. The air is evacuated from the sample pores and the epoxy simultaneously is forced in under pressure while the rock is confined. Confining pressure is maintained until the epoxy is cured. Epoxies of sufficient tensile strength are used to ensure that the sample will hold together when the confining pressure is released.

Figure 4 shows two photomicrographs (A and B) of rhodamine-impregnated, gas-productive Pratt Coal from the Pottsville Formation in the Warrior Basin of Alabama. A single horizontal core, 2.5 cm (1 in.) in diameter by 5.1 cm (2 in.) long, was sliced in two in a direction perpendicular to the cylinder axis to produce a pair of

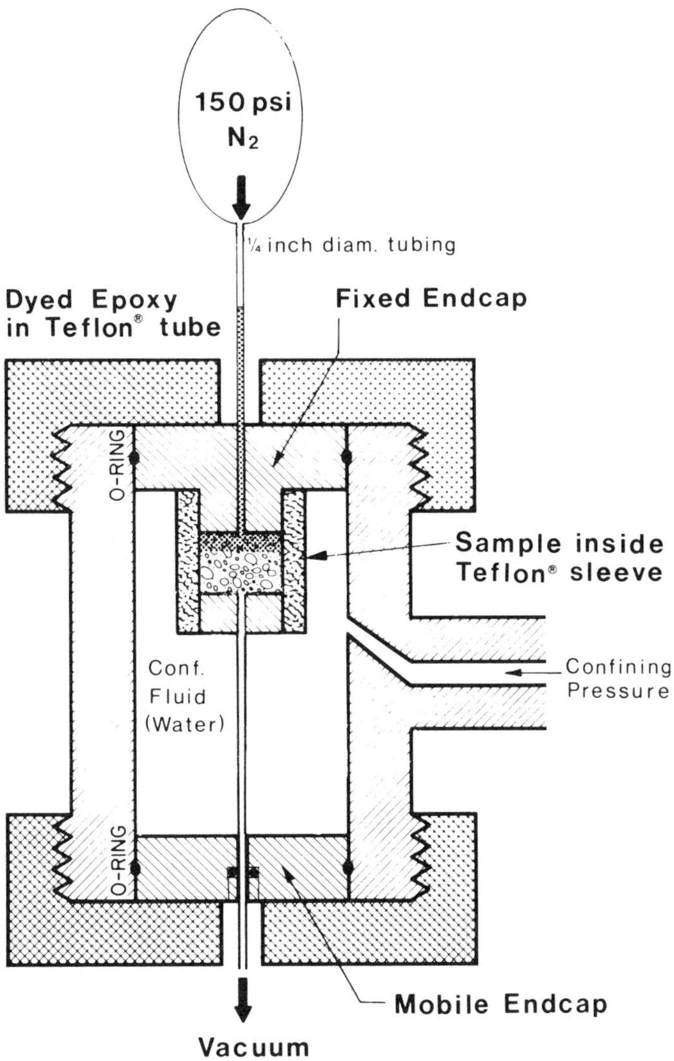

Figure 3. *Schematic drawing of apparatus used to impregnate samples with rhodamine-dyed epoxy under elevated net confining stress.*

very similar, 2.5-cm (1-in.) long samples which had the same orientation from the same bedding plane. The plane of the photographs is oriented approximately perpendicular to face cleats and bedding, and more or less parallel to the plane of the butt cleats. Sample B was impregnated with rhodamine-dyed epoxy under 6.9 MPa (1,000 psi) net confining stress, whereas sample A was impregnated at zero net stress.

The fluorescent dye clearly delineates phyteral porosity (P), cleats (C), and microcleats (mc) in sample A, the unstressed coal (Figure 4A). However, in the photomicrograph of sample B (Figure 4B), impregnated at 6.9 MPa (1,000 psi) net stress, it is apparent that many of the microcleats are not present, although the larger cleats and phyterals appear unaffected. This is graphic evidence that the closure of small flowpaths and the increase in flowpath tortuosity under elevated stress are real, observable phenomena, and not just mathematical abstractions. To my knowledge, these are the first photomicrographs showing changes in flowpath morphology under stress that clearly support the conclusions derived from mathematical calculations of permeability data. It is very important to recognize that stress-dependence of permeability in coal is a real phenomenon which must be considered when modeling production decline curves and reservoir flow properties for coalbed methane wells.

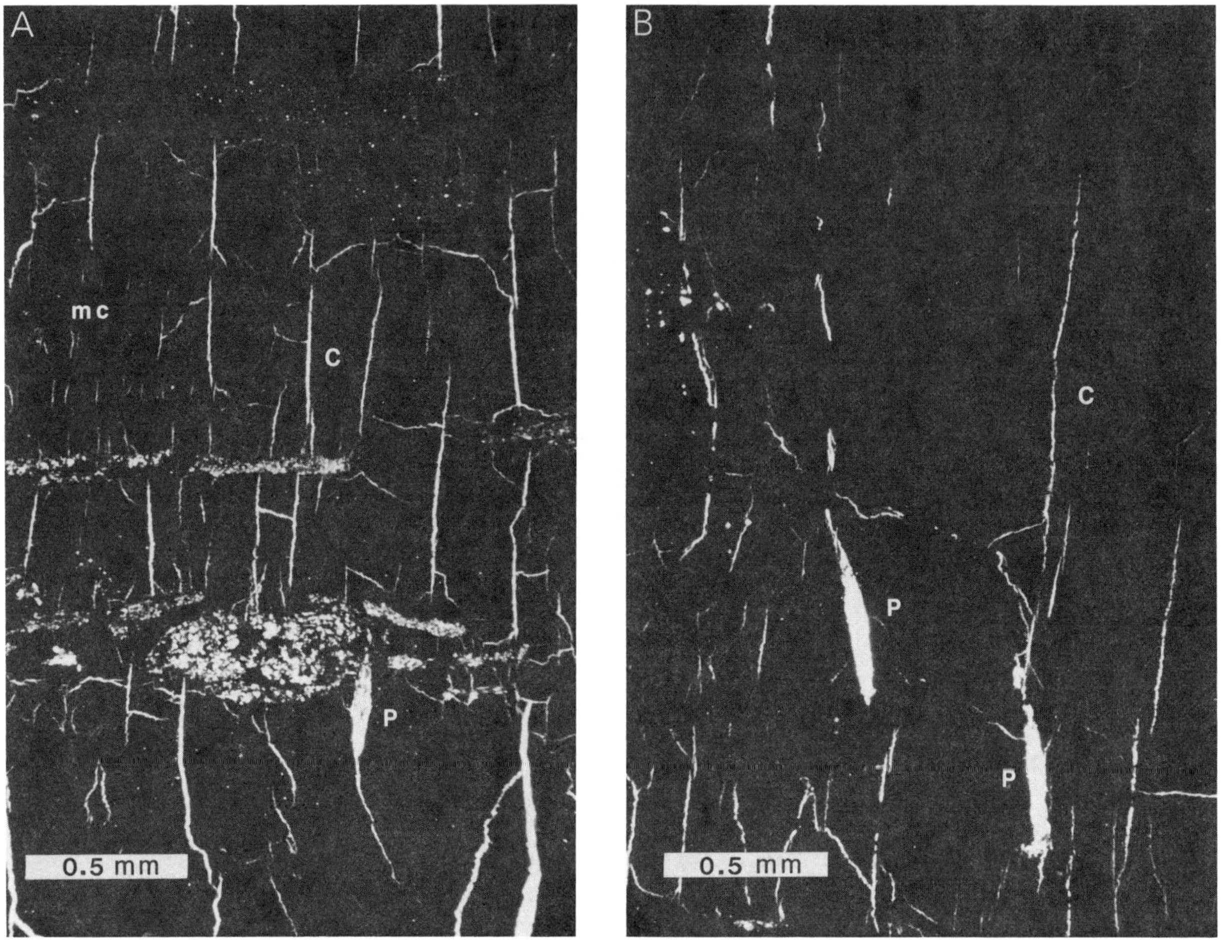

Figure 4. Photomicrographs of rhodamine dye epifluorescence in Pratt Coal of the Pottsville Formation (from the Oak Grove Mine near Birmingham, Alabama). A) Sample A was impregnated with rhodamine-dyed epoxy while unconfined and shows details of phyteral porosity (p) and cleats (c) as well as microcleats (mc). B) Sample B, taken from the same bedding plane about 2 cm (0.8 in.) away from sample A, was impregnated under 6.9 MPa (1,000 psi) net confining pressure. Many of the cleats are narrowed but still visible, and the phyteral pores appear undisturbed, but smaller cleats are all but absent from sample B.

MATRIX SHRINKAGE

Based on studies of gas containment and transport in Australian coals, Gray (1987) concluded that physical shrinkage of the coal matrix during gas desorption makes an important contribution to reducing the effective stress on a seam. Essentially, his data indicate that the reduction in coal volume by gas desorption may be significant enough to compensate for cleat closure under increased stress during drawdown. In fact, according to Gray, large amounts of matrix shrinkage caused by desorption may actually lead to an increase in coalbed permeability during drawdown, rather than a decrease. This proposition of "negative effective stress" on a coal seam is highly controversial among many workers in the coalbed methane area, primarily because no one knows precisely all of the different factors influencing the stress fields in coalbed methane reservoirs. Not the least of these is the inhomogeneity of the coal itself, which varies on a small scale and is extremely difficult to quantify.

Ting (1987) reported some striking relationships between optical anisotropy in vitrinite and the adsorption capacity of coal. Strongly anisotropic coals adsorbed and desorbed significantly less gas than weakly anisotropic coals of the same rank. The degree of optical anisotropy in vitrinite is proportional to the amount of structural deformation and tectonic stress experienced by the coal (Levine and Davis, 1984). These results suggest that the most significant desorption, and therefore the largest amount of matrix shrinkage, occurs in weakly anisotropic coals exposed to low degrees of tectonic stress. These findings imply that significant matrix shrinkage in moderately-stressed coals may indeed offset drawdown stress and that "negative effective stress" is possible in certain cases.

Still, questions remain concerning the volume of matrix shrinkage reported by Gray (1987). The data used in the calculations primarily consist of a series of measurements made using CO_2 as the adsorbing gas. Other work is cited by Gray (1987) which used methane desorption and resulted in much smaller amounts of shrinkage. Measurements using methane under controlled temperature and pressure conditions have been reported by Harpalani and Zhao (1989), but the results are inconclusive, and fail to clearly demonstrate whether their measured trend of increased permeability in coal at lower pore pressure is entirely due to matrix shrinkage, only to the Klinkenberg effect of "gas slippage", or some combination of the two. Further experiments performed with a variety of adsorbing and non-adsorbing gases under constant net confining pressure may resolve this issue.

STIMULATION

Because coal is a low permeability material, coalbed methane reservoirs generally require stimulation treatments to produce economical quantities of gas. Stimulation is a straightforward engineering procedure designed to increase the effective contact area of the wellbore inside the target formation. Tight formations in contact with a large wellbore area can produce as readily as conventional formations flowing into a standard diameter well. Two common methods used to increase the wellbore contact area in a coalbed methane reservoir are horizontal drilling and hydraulic fracturing. Horizontal drilling involves keeping the wellbore inside the coal seam for a considerable distance by drilling within it, rather than through it. Horizontal boreholes have been used for decades to drain methane from coal seams at the mine working face. However, drilling horizontal boreholes from a vertical well on the surface was, until recently, a difficult-to-impossible, expensive process. Developments in "whipstock" type drilling strings, downhole motors, and better drill bit navigation allow modern drillers to kick off from a vertical well and turn the hole horizontal, hitting the target formation with an accuracy of a few feet. A favorable feature of horizontal drilling in a coal seam is the high degree of control over direction of penetration, which enables the borehole to intersect the major cleat set and provide the most efficient methane drainage.

Classic hydraulic fracturing techniques involve pressuring-up a water-filled vertical well until the rock strength of the target formation is exceeded, and a near-vertical crack grows outward from the wellbore perpendicular to the least principal stress direction. The fracturing fluid generally carries solid "proppant", consisting of a coarse sand or other granular material which goes into the crack to hold it open after pressure is released. Hydraulic fracturing originally was developed as a stimulation treatment for low permeability gas sands and petroleum reservoirs, and certain inherent difficulties are encountered when it is applied to coalbed methane.

The major problems involved in hydraulic fracture stimulation of a coalbed methane well are caused by the fact that coal is a relatively weak, fractured material, which tends to be more compressible than inorganic rocks (Holditch et al., 1989). The low mechanical strength of coal, and the associated lack of stress contrast commonly found between coal seams and surrounding strata, often cause hydraulic fractures to form somewhat different morphologies than those in tight sands. Holditch et al. (1989) identified four distinct types of hydraulic fracture morphology in coal: 1) horizontal fracture in a shallow seam caused by low overburden pressure; 2) a single vertical fracture through multiple, stacked thin seams; 3) compound vertical fracture (T-shaped) in single, thick seams which follows the pre-existing cleat system, and 4) a compound, uncontained vertical fracture through a thick seam and upper and lower bounding strata because of a lack of stress contrast. Improved understanding of the nature of stress and mechanical properties in coal seams has led to better knowledge of the effect stimulation treatments have on the existing cleat system, resulting in better-designed hydraulic fractures. From a capital cost standpoint, fracture stimulations tend to be less expensive than drilling deviated or horizontal wells from the surface, although the financial competitiveness of horizontal drilling is improving rapidly. Technologically advanced hydraulic fracturing of coal may significantly improve the economics of coalbed methane recovery, if the potential damage caused by introducing high stresses into the reservoir can be avoided.

SUMMARY AND CONCLUSIONS

The coalbed methane resource in the U.S. represents a major supply of future natural gas. Favorable production potential of individual seams requires the presence of two important factors: significant volumes of gas in-place, and a well-developed cleat system through which gas can move to a well bore. Because of the nature of cleats as flat, narrow, poorly-supported cracks, the width or aperture of the cleats is reduced at higher levels of effective stress. Effective stress is increased during production and drawdown by lowering the gas pressure in the seam: overburden pressure and tectonic stresses remain constant, therefore the net confining pressure increases. Mathematical calculations and petrographic observations indicate that the smallest cleat sizes in coal are closed off completely under increased stress, whereas the larger cleats only narrow down somewhat. Loss of the smaller cleats decreases permeability by reducing the number of interconnections between pores and major cleats. This leads to higher values for flowpath tortuosity, resulting in the lower permeabilities. This seems to be a common mechanism of permeability reduction under stress for nearly all rocks tested by IGT which contain microfracture flow paths in a tight matrix. Reduction of cleat aperture under stress may be offset to a greater or lesser degree by shrinkage of the coal matrix resulting from gas desorption, although the effectiveness of this process is far from well understood.

Coalbed methane reservoirs require stimulation to produce economical amounts of gas. This is commonly accomplished by either drilling horizontal boreholes through the coal, or by driving a vertical hydraulic fracture into the seam from a wellbore. Hydraulic fracturing generally is less expensive than horizontal drilling, but because of the stresses it introduces into the coal cleat system, it can be less effective. Recent improvements in the understanding of hydrofracture morphology in coal have resulted in more effective fracture treatment designs, allowing for efficient recovery of coalbed methane with improved economics.

REFERENCES

Crist, T.E., C.M. Boyer, and B.S. Kelso, 1989, A geologic and coalbed methane resource analysis of the Menefee Formation in the San Juan Basin, Southwestern Colorado and Northwestern New Mexico, in Proceedings of 1989 Joint Rocky Mountain Regional/Low Permeability Reservoirs Symposium, March 6-8, Denver, Colorado: Richardson, Texas, Society of Petroleum Engineers, p. 153-160.

Gray, I., 1987, Reservoir engineering in coal seams: Parts 1 and 2: SPE Reservoir Engineering, v. 2, p. 28-40.

Harpalani, S., and X. Zhao, 1989, An investigation of the effect of gas desorption on coal permeability, in Proceedings of the 1989 Coalbed Methane Symposium, April 17-20, Tuscaloosa, Alabama: Tuscaloosa, University of Alabama, p. 57-64.

Holditch, S.A., J.W. Ely, M.E. Semmelbeck, R.H. Carter, J. Hinkel, and R.G. Jeffrey, Jr., 1989, Enhanced recovery of coal bed methane through hydraulic fracturing, in Proceedings of 1989 Joint Rocky Mountain Regional/Low Permeability Reservoirs Symposium, March 6-8, Denver, Colorado: Richardson, Texas, Society of Petroleum Engineers, p. 43-53.

Klinkenberg, L.J., 1941, The permeability of porous media to liquids and gases, in Recommended practices for drilling and production: New York, American Petroleum Institute, p. 200-213.

Levine, J.R., and A. Davis, 1984, Optical anisotropy of coals as an indicator of tectonic deformation, Broad Top coal field, Pennsylvania: Geological Society of America Bulletin, v. 95, no. 1, p. 100-108.

McKee, C.R., A.C. Bumb, and R.A. Koenig, 1987, Stress-dependent permeability and porosity of coal, in Proceedings of the 1987 Coalbed Methane Symposium, November 16-19, Tuscaloosa, Alabama: Tuscaloosa, University of Alabama, p. 183-193.

Ostensen, R.W., 1983, Microcrack permeability in tight gas sandstone: Society of Petroleum Engineers Journal, December, p. 919-926.

Randolph, P.L., D.J. Soeder, and P. Chowdiah, 1984, Porosity and permeability of tight sands, in Proceedings of the 1984 SPE/DOE/GRI Unconventional Gas Recovery Symposium, May 13-15, Pittsburgh, Pennsylvania: Richardson, Texas, Society of Petroleum Engineers, p. 57-66.

Rose, R.E., and S.E. Foh, 1984, Liquid permeability of coal as a function of net stress, in Proceedings of 1984 SPE/DOE/GRI Unconventional Gas Recovery Symposium, May 13-15, Pittsburgh, Pennsylvania: Richardson, Texas, Society of Petroleum Engineers, p. 253-259.

Soeder, D.J., 1988a, Porosity and permeability of eastern Devonian gas shale: SPE Formation Evaluation, v. 3, no. 2, p. 116-124.

Soeder, D.J., 1988b, Laboratory measurements of permeability in igneous and pyroclastic rocks (abstr.): Geological Society of America Abstracts with Programs, v. 20, no. 7, p. A310.

Soeder D.J., 1990, Applications of fluorescence microscopy to the study of pores in tight rocks: AAPG Bulletin, v. 74, no. 1, p. 30-40.

Soeder, D.J., and P.L. Randolph, 1987, Porosity, permeability and pore structure of the tight Mesaverde Sandstone, Piceance Basin, Colorado: SPE Formation Evaluation, v. 2, no. 3, p. 129-136.

Ting, F.T.C., 1987, Optical anisotropism and its relationship with some physical and chemical properties of coal: Organic Geochemistry, v. 11, no. 5, p. 403-405.

Wyman, R.E., 1984, Gas resources in Elmworth coal seams, in John A. Masters, ed., Elmworth: Case study of a deep basin gas field: American Association of Petroleum Geologists Memoir 38, p. 173-187.

COALBED METHANE PRODUCTION POTENTIAL IN COMPLEX GEOLOGIC SETTINGS

Terry E. Mitchell and Steven P. Pappajohn
GeoTrends Inc.
6627 S. 191st Place, Suite F-100
Kent, WA 98032

ABSTRACT

The volume of proven coalbed methane reserves being added to the natural gas resource base in the United States has continued to increase substantially. Most coalbed methane drilling activity, however, remains confined to the Warrior and San Juan basins, both of which are characterized by well-documented uniform structural and stratigraphic relationships relative to the coal-bearing strata in other prospective basins. The coalbed methane production potential of tectonically deformed basins also should be demonstrated so that this resource may contribute more fully to the nation's energy supply. One of the first attempts to produce coalbed methane in a relatively complex geologic setting is underway in Western Washington where five closely spaced test wells were drilled into multiple coal seams of the Eocene Puget Group. Preliminary test results indicate the potential for an average per-well production rate of 500 mcfd from an in-place reserve base of greater than 5.5 bcf/km^2 (14 bcf/mi.^{2}).

INTRODUCTION

The volume of the coalbed methane resource in the coal basins of the continental United States (Figure 1) has been estimated by the U.S. Department of Energy (DOE), the Gas Research Institute, and others to be on the order of 800 tcf (Morz et al., 1983). The number of coalbed methane production wells being drilled has increased significantly in recent years to the degree that, in certain regions where hundreds of new wells are proposed, the level of development activity is reminiscent of the "boom times" experienced by the petroleum industry circa 1980 (Stremel, 1988). Interest in coalbed methane resource development has surged specifically because of a growing awareness of the increased level of economic viability of such programs. Economic interest in coalbed methane has increased recently because the effectiveness of drilling, production, and reservoir engineering procedures, which were pioneered by the coal industry to extract the volatile gas from their mining operations, have benefited from the incorporation of engineering technology from the petroleum industry. Development interest also has been intensified by the extended availability of lucrative federal tax credit incentives for coalbed methane production.

Initial Development Confined to Uncomplicated Geologic Settings

The majority of coalbed methane production programs in the United States are located in the Warrior Basin of Alabama and the San Juan Basin of Colorado and New Mexico, not because these coal basins are the only prospective areas for such development, but because they are the sites of the initial coalbed methane projects of the coal and petroleum industries, respectively. Consequently, initial risk to development has been

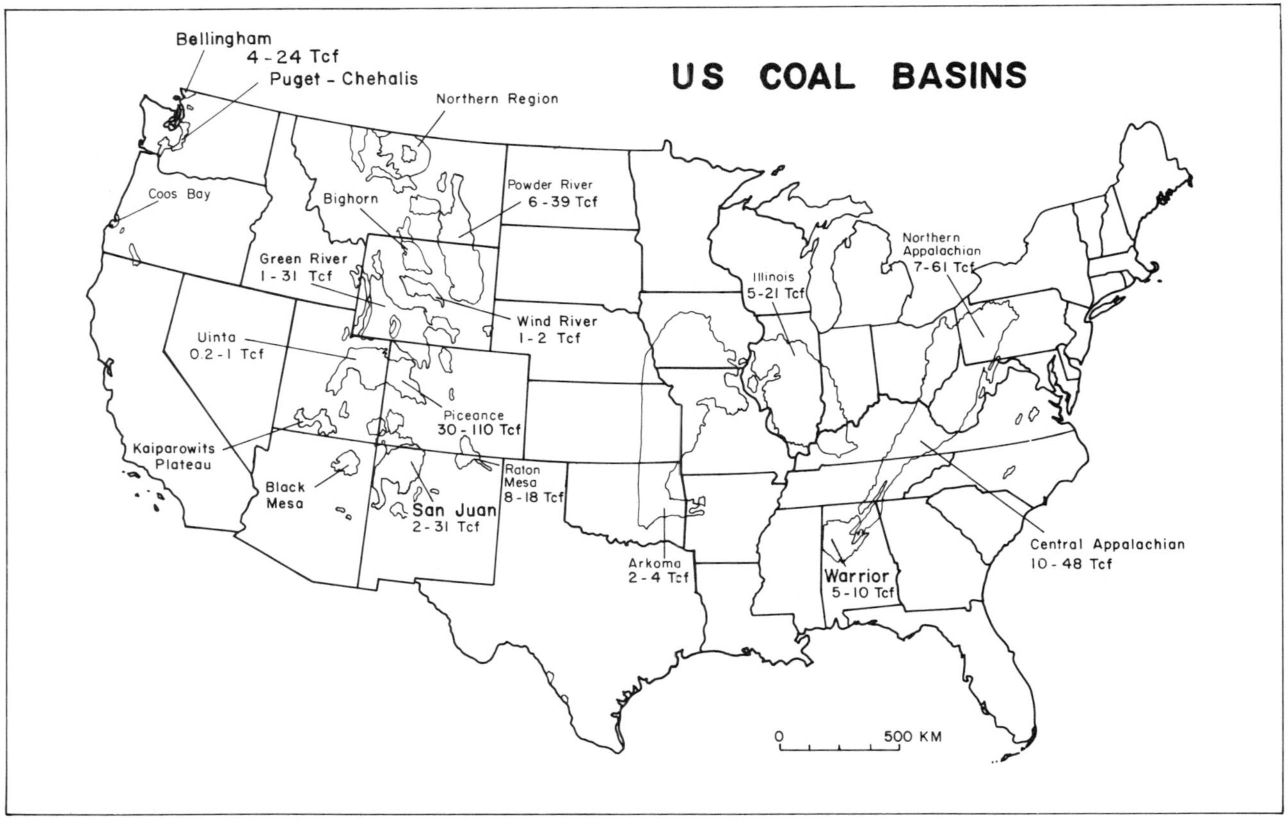

Figure 1. *Coalbed methane reserve estimates in the continental United States (after Robertson and Rightmire, 1986, and Kilso and Wicks, 1988). Dark stippling indicates the basins where coalbed methane exploration/production has been active. Sedimentary basins not shown have not been evaluated for coalbed methane potential.*

reduced in the Warrior and San Juan basins because prior extensive work has lead to the definition of the parameters necessary to establish successful production programs.

In the rush to capitalize on the investment potential of coalbed methane development prior to the January 1, 1991 expiration of the "nonconventional fuels tax credit" (which now has been extended to 1993), most producers conservatively have initiated operations primarily in the only two basins where successful production techniques already have been established. Little interest has been shown for entering new areas where a potentially risky "engineering learning curve" must be dealt with. Industry appears to be particularly reluctant to initiate coalbed methane development in regions such as Western Washington where, despite excellent reserve and market potential, greater degrees of geologic complexity are exhibited relative to the more predictable and thoroughly studied structural and stratigraphic relationships of the coal-bearing intervals in the Warrior and San Juan basins (Culbertson, 1964; Fassett and Hinds, 1971).

Benefit of Establishing Production in Complex Geologic Settings

Energy forecasters generally have acknowledged that the United States likely will begin to experience an excess in demand over supply of natural gas during the early 1990s as a consequence of the current trend of depleting domestic reserves at a faster rate than they are being added to by new discoveries. The potential supply problem could escalate further as a result of the ongoing increase in environmental awareness that problems of atmospheric pollution can be reduced by supplementing relatively cleaner-burning natural gas for many current uses of oil, gasoline, and coal.

The present reluctance of most operators to initiate coalbed methane programs in complex geologic settings could delay the ability of the nation to fully benefit from the resulting potential additions to the nation's proven reserves of natural gas. The problem likely will become intensified following the expiration of the federal tax credit program for coalbed methane production. Unless coalbed methane production projects are proven successful in the more challenging geologic settings prior to the termination of the tax incentives for undertaking such economic risks, these potential reserve additions could remain undeveloped and, therefore, not capable of contributing in a timely manner to anticipated national increase in the demand for natural gas.

INITIAL PRODUCTION EFFORT IN A COMPLEX GEOLOGIC SETTING

A DOE report by the Gas Research Institute (GRI) concluded that up to 24 tcf of methane may be present in coal fields in the western portion of the state of Washington, a part of the tectonically-deformed, geologically-complex Pacific Northwest (Choate and Johnson, 1980). This resource estimate is based upon the inferred presence of up to 55,000,000,000 tonnes (60,000,000,000 tons) of coal in the region, and on the assumption that 440 standard cubic feet of methane per tonne (scf/tonne) (400 scf/ton) represents a reasonable average gas-content value for these reserves.

All of the approximately 100 bcf of natural gas that is consumed per year in Washington, where petroleum production has yet to be established, is transported into the region via one interstate pipeline company. A majority of the potential coalbed methane reserves in Washington are present along the western foothills of the north-trending Cascade Mountains, which are bordered on the west by the most densely populated region in the state (Figure 2). Conservatively assuming that only 12 tcf of coalbed methane is present in logistically developable regions that are in close proximity to existing gas markets and, furthermore, that only 50% of the methane is recoverable, the resulting 6 tcf of reserves represents a significant resource base worthy of development in this petroleum-import dependent state.

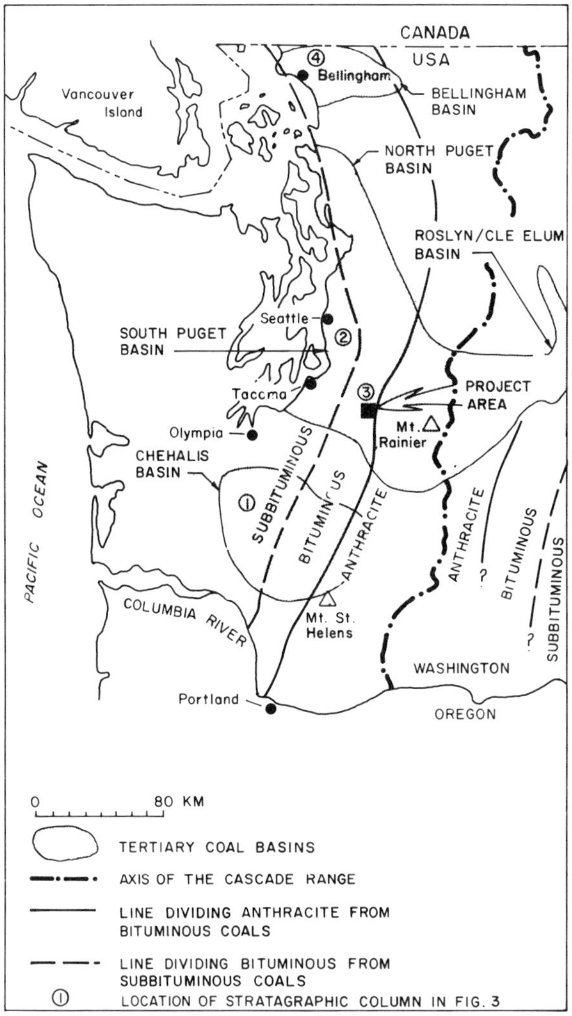

Figure 2. *Generalized geographic position of Eocene-age coal basins and regionally symmetric thermal maturation trends relative to the axis of the Cascade Range (after Choate and Johnson, 1980). Note that the circled numbers represent the locations of the stratigraphic sections shown in Figure 3.*

Depositional and Tectonic Influences on Regional Methane Potential

The geology of Western Washington evolved in response to dynamic interactions between converging oceanic and continental plate systems during the Tertiary, which resulted in complex structural and stratigraphic relationships throughout the region (Armentrout and Suek, 1985; Engebretson et al., 1985; Snavely et al., 1980). The depositional and tectonic evolution of Western Washington resulted in several physical characteristics beneficial to coalbed methane resource development potential, most notably the thickness and the degree of thermal maturation of the prospective gassy, coal-bearing intervals.

Prolific rates of fluvial-deltaic sedimentation in rapidly subsiding depositional basins along a north-south belt, roughly parallel to the present trend of the younger Cascade Mountains, during the Eocene resulted in the creation of a sequence of thick (up to 6,000 m (20,000 ft) in places), non-marine, arkosic, coal-bearing strata

(Johnson, 1982). Coal seams ranging in thickness from a few centimeters to greater than 7 m (23 ft) are present throughout the resulting Eocene stratigraphic section (Vine, 1968). The laterally discontinuous nature of individual coal seams across distances of several kilometers or greater, which resulted from deposition in delta plain environments, also has been influenced by subsequent tectonic deformation (Buckovic, 1979). Despite the inability of explorationists to correlate individual seams for great distances in the predictable manner that is possible in the Warrior and San Juan basins, the Eocene sedimentary section contains numerous enough coal seams throughout its known thickness interval to assure that an adequate volume of gas-bearing coal can be penetrated at reasonable drilling depths (300-1,200 m (1,000-3,900 ft)) in most prospective areas, independent of structural complexity.

The current geographic distribution and thermal maturation trends of the coal basins in Washington are a direct result of post-Middle Eocene igneous activity and tectonic uplift of the Cascade Range in response to episodic changes in plate-motion dynamics (Choate and Johnson, 1980; Hammond, 1979; and Walsh and Phillips, 1983). The western portion of the north-trending belt of coal basins, which formed in progradational, near-shore to delta-plain depositional environments, was separated geographically from its eastern non-marine, fluvial facies as a consequence of intense post-Eocene igneous activity and tectonic uplift of the north-trending Cascade Mountains system. A prolific outpouring of flood basalt to the east of the Cascades during the Miocene buried a majority of the fluvial facies, but coal deposits in the western portion of the state were buried only by sporadic volcanic and glacial activity.

Igneous intrusion of the coal basins superimposed a regional aureole of thermal maturation upon them that resulted in the trend of increased rank that is apparent in coal deposits located progressively closer to the crest of the Cascade Mountains (Walsh and Phillips, 1983). Among the coal deposits on both the eastern and western sides of the Cascades, there are parallel belts that matured to the high- to medium-volatile bituminous rank, which represents optimal conditions for the generation and retention of methane (Kim, 1978) and, hence, viable coalbed methane exploration objectives. These deposits reflect an anomalous level of thermal maturity more typical of coal of pre-Tertiary age and, therefore, are anticipated to exhibit a correspondingly high methane content.

Post-Eocene tectonic uplift of the Cascades has resulted in prolific rates of precipitation from Pacific storm systems along the western flanks of these mountains. The resulting high water table and a corresponding high hydrostatic head may have provided subsurface reservoir pressures to assure the optimal retention of occluded methane in the coal basins of Western Washington (Choate and Johnson, 1980).

Prospective Coalbed Methane Formations in Washington

Many regions along the western foothills of the Cascades are prospective for coalbed methane exploration in Washington. In general, regional exploration objectives include the Puget Group and its undesignated stratigraphic equivalents in the central portion of Western Washington, the slightly older Chuckanut Formation to the north, and the slightly younger Skookumchuck Formation to the south (Figure 3). The widespread presence of these thick coal-bearing stratigraphic intervals enhances the likelihood of developing numerous coalbed methane production programs throughout the region. Should an initial test drilling program in Pierce County prove economically successful, then Washington eventually could become the focus of additional programs that could lead to the first significant coalbed methane development activity outside of the San Juan and Warrior basins.

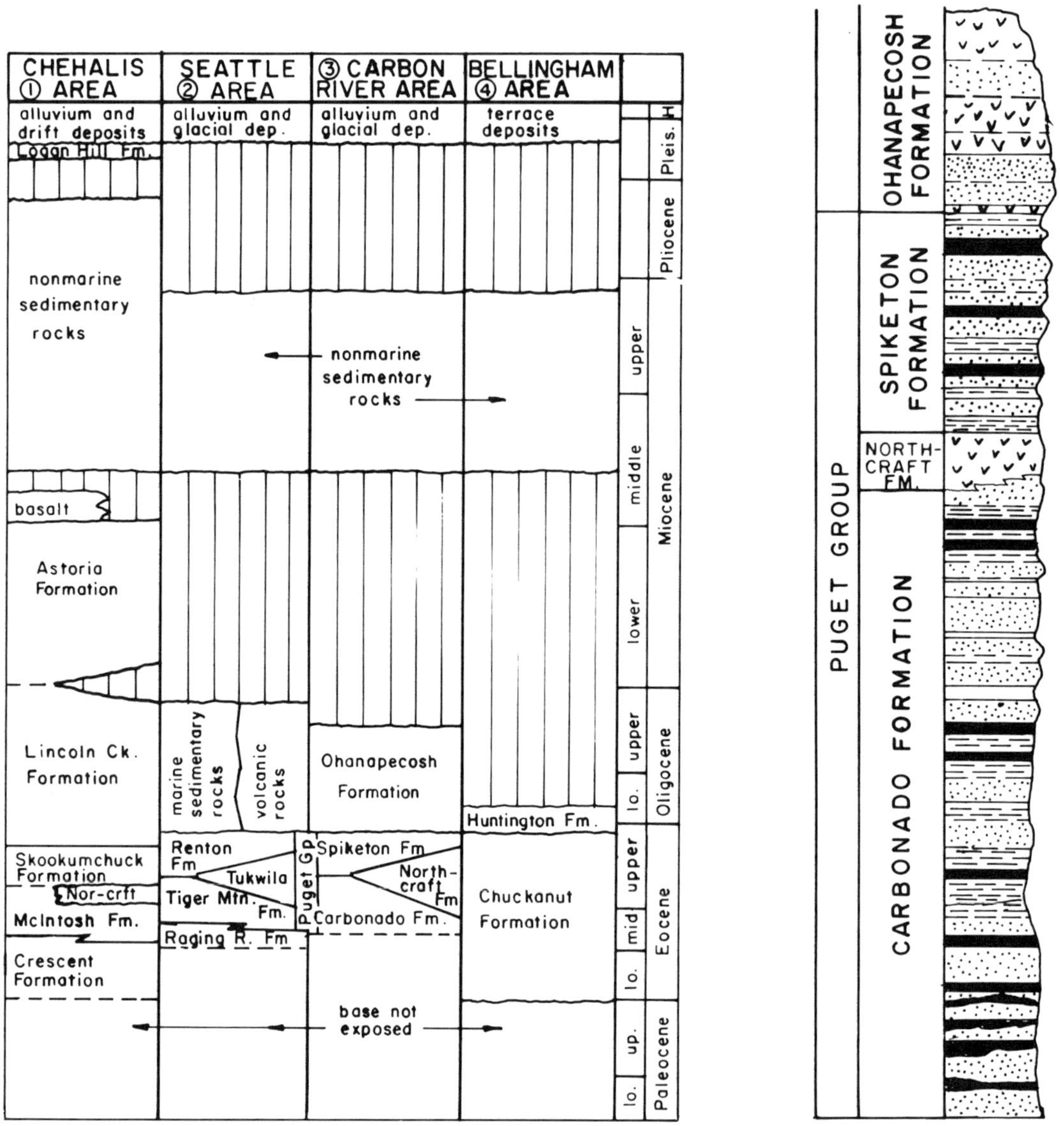

Figure 3. *Generalized time-stratigraphic columns depicting the Eocene-age, nonmarine, coal-bearing units — the Chuckanut and the Skookumchuck formations and the Puget Group — which comprise the primary coalbed methane exploration objectives in Washington State (modified from Armentrout et al., 1983). Note that the locations for the stratigraphic sections depicted are shown on Figure 2.*

Washington Drilling Program

Exploration Objective

The primary exploration objective of the initial coalbed methane drilling program in Washington reflected the need to gain additional geological and engineering data with which to evaluate the production potential in this relatively unexplored and geologically complex region. The main geological objective of the drilling program in Pierce County was to penetrate coal seams in the lower portion of the Puget Group, and to complete five closely spaced wells (approximately 65-hectare (160-acre) centers) in one or more seams that exhibited sufficient thickness (2 m (6.6 ft) or greater) and gas content (minimum of 275 scf/tonne (250 scf/ton) of coal) to represent viable development prospects.

Primary engineering objectives included the need to design effective completion procedures and stimulation treatments needed to establish producibility in multiple coal seams with dips anticipated to be in the range of 45°-70°. Potential complications in the initial drilling program were minimized by using relatively conventional engineering procedures common to the petroleum industry, including the use of a rotary rig using a mud system, cased and perforated completions, and sand-propped, hydraulic fracture-stimulation treatments. "Exotic" practices, such as directional drilling, drilling with air instead of a mud system, etc., await additional experience along the engineering learning curve in this frontier exploration area.

Project Area Selection

Among the many potential project areas appropriate for initiating coalbed methane exploration and development activity in the Pacific Northwest, the region near the historic mining town of Carbonado (in T18N, R6E) in Pierce County, Washington (Figure 4) was selected based on favorable criteria including the following: 1) the local presence of the largest known volume of high-quality coking coal in the western coastal states, which suggested that other coal seams in the stratigraphic section were thermally mature and likely to be very gassy (as also is indicated by the high number of methane-related mining accidents in this district); 2) the availability, from previous drilling and mining operations in the general area, of geologic data suggesting the widespread presence of thick, multiple beds of gas-bearing bituminous coal; 3) proximity of the project site to gas delivery lines and end-user markets in the Seattle-Tacoma area approximately 10 to 20 km to the west; 4) the logistical ease of access to potential drillsites in this rugged, heavily forested region; and 5) the presence of large, unleased tracts of prospective mineral acreage comprised of timber company lands characterized by little, if any, cultural development, minimal potential title problems, and low acquisition costs.

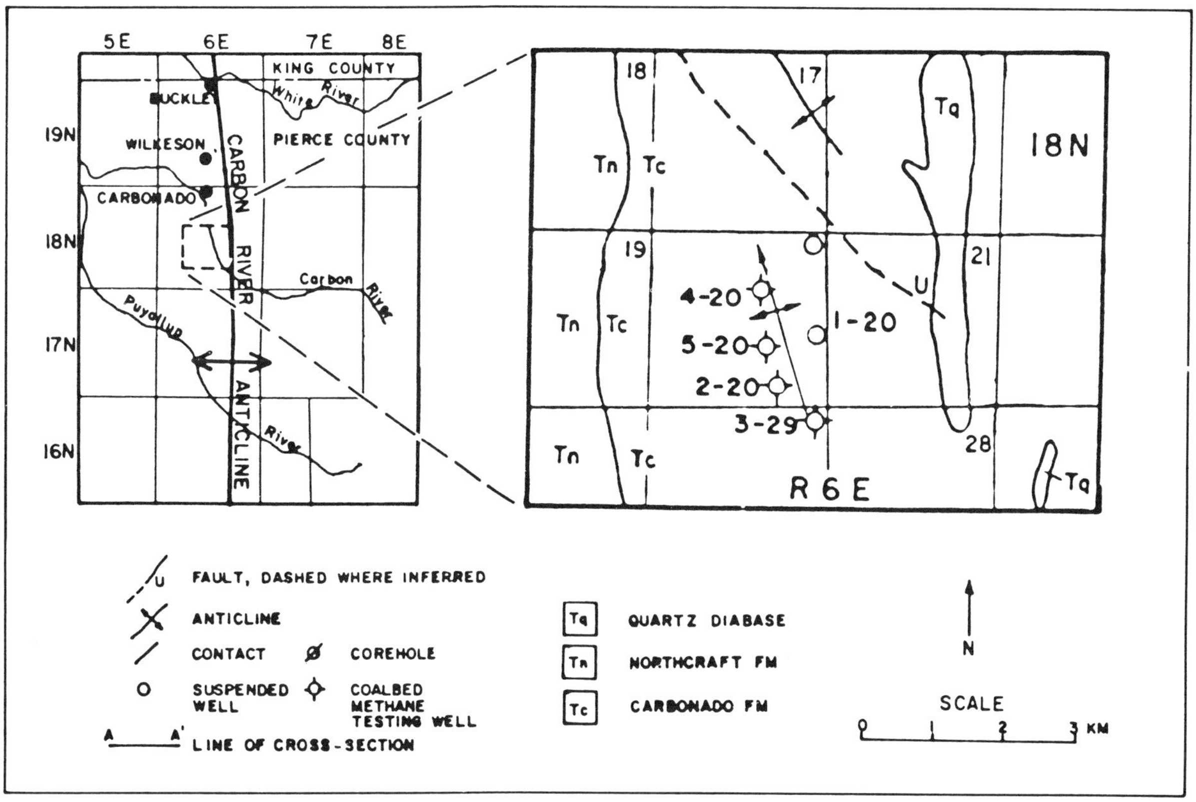

Figure 4. *Basic geology of the drill-site area in T18N, R6E, Pierce County, Washington (modified from Gard, 1968).*

Geologic Setting

The project area in Pierce County is traversed by a regional structure known as the Carbon River Anticline (Gard, 1968), which extends along a north-plunging trend for a distance of greater than 48 km (30 mi.) from its southern terminus in the vicinity of the Nisqually River drainage to just beyond the region of the White River drainage. The anticline provides a window through the post-Eocene volcanic deposits into the sequence of interbedded arkosic sandstones, siltstones, and coals of the Puget Group along a band approximately 8-10 km (5-6 mi.) wide trending parallel to the crest of the structure. Superimposed upon the anticline, at least in the few areas that have been mapped, are numerous smaller-scale folds and faults at various angles to the anticlinal axes. In most places, however, bedrock exposures are covered by Pleistocene to Recent volcanic and glacial deposits and thick vegetation, which makes surface studies extremely difficult.

The stratigraphic section selected for the project consists of the known lower portion of the Puget Group anticipated to be present in the unmapped vicinity of Section 20, T18N, R6E. The Puget Group in this portion of Pierce County is subdivided into the coal-bearing Carbonado Formation and the younger Spiketon Formation, which are separated stratigraphically by volcanic deposits of the Northcraft Formation, all of which are of Eocene age. Coal seams of the Wilkeson Series of the Carbonado Formation initially constituted the primary exploration target for the drilling program. Subsurface information of specific significance to the project included mining data from the Wilkeson and Melmont mining districts, which are located approximately 8 km (5 mi.) north and 3 km (2 mi.) east, respectively, of the drillsite area. Additional subsurface data were derived from the geophysical logs from the 605-m (1,990-ft) deep Utah International C-83-1 exploratory coal corehole in the northeastern-most corner of the drill site section. The C 83-1 data suggested the presence of an interbedded sequence of sandstone, siltstone, and coal with easterly dips of 60°-70°, which was correlated as being stratigraphically equivalent to a portion of the Wilkeson coal series of Daniels (1914) and Gard (1968) (Figure 5).

Despite the paucity of outcrop and subsurface data, the crest of a small (less than several kilometers) asymmetric anticline trending north-northwest was anticipated to be present in the proposed drillsite area in Section 20. This expectation was based on evaluation of the dipmeter log from the nearby corehole, as well as by interpretation of the trends of other structural features in the area. The first well drilled in the five-well program, the Carbon River 1-20, was located on a structurally high position on the east flank of the postulated anticline (Figure 4). The four wells drilled subsequently, the Carbon River 2-20, 3-29, 4-20, and 5-20, were located updip to the west of the 1-20 well in a position closer to, and along the axial trend of, the anticline.

The wells were drilled in structurally high locations to take advantage of the possibility of encountering certain conditions favorable to methane production potential, the presence of which were evident in mining records. Such conditions include 1) greater degree of fracturing near the axial plane of a fold (enhanced reservoir permeability), 2) lower hydrostatic head in updip locations (facilitates reservoir dewatering), and 3) slightly overpressured reservoir conditions adjacent to fault planes, anticlinal axes, and plunging noses of folds (entrapment of associated entrained gas and correspondingly high gas/water ratios, sufficient enough to create economic rates of methane production during the initial stages of dewatering).

Drilling Program

Well design consisted of drilling 91.5 m (300 ft) of 31.12-cm (12.25-in.) hole and setting 22.03-cm (8.675-in.) casing and cementing the casing to the hole annulus for the length of the hole. Soft insert bits were used to drill a 20.00-cm (7.875-in.) hole to an average depth of 1,170 m (3,840 ft) into which 11-cm (4.5-in.) production casing was run to total depth. This casing was cemented to approximately 150 m (490 ft) above the shallowest potentially productive coal seam.

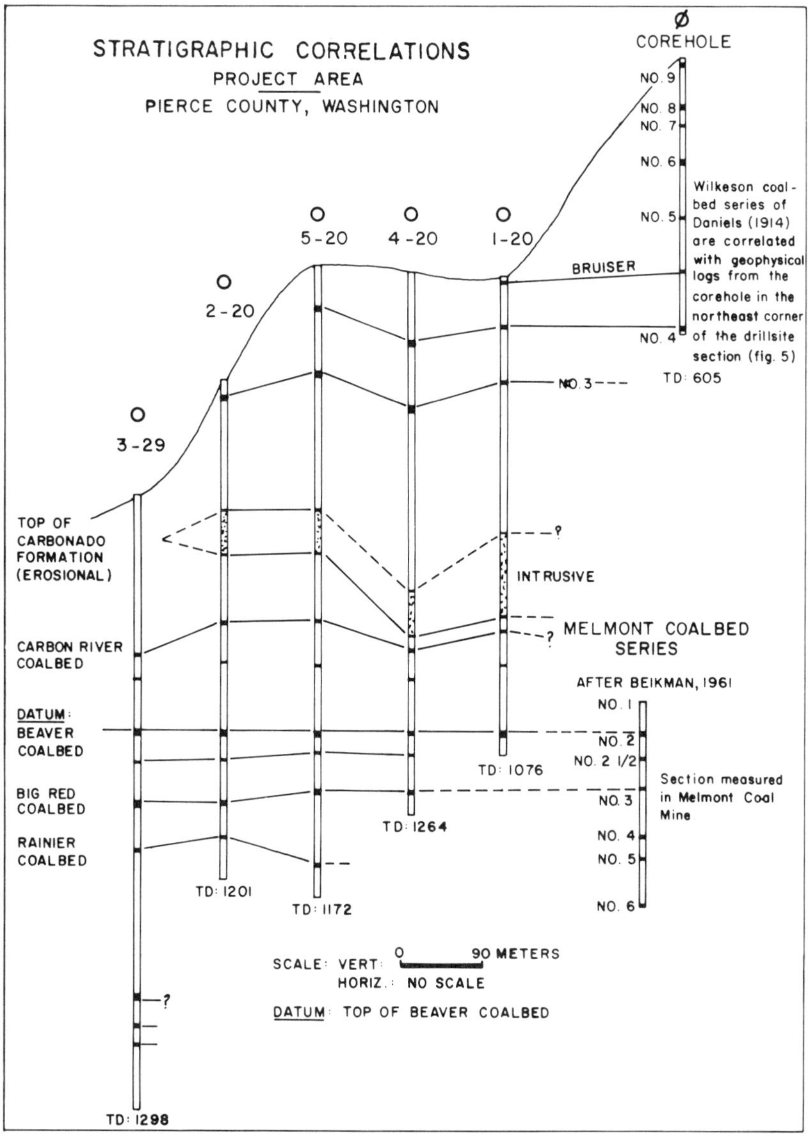

Figure 5. *Inferred stratigraphic relationships between the coal seams in the drill-site area and the Wilkeson and Melmont mining districts. The coal section from the Melmont coal mine shown at lower right is taken from Beikman (1961).*

The drilling of each wellbore was conducted as expediently as possible (rates ranging from 6 to 13 min./m (1.8 to 4.0 min./ft)) to minimize formation damage in the targeted coal seams. A stiff, three-point, packed-hole bottom assembly was used in an attempt to control wellbore deviation while drilling with optimum weight-on-bit into the steeply dipping stratigraphic section. The tungsten carbide insert cutters normally used in similar bottom-hole assemblies were replaced with blank rollers and a rubber, non-rotating, sleeve-type stabilizer in order to achieve adequate stabilization with minimal wear.

Drilling Problems

All five wells generally were drilled in a trouble-free manner, considering the nature of the logistical and geological constraints discussed previously. Despite the use of the stiff, three-point bottom hole assembly, severe deviation at optimum weight-on-bit and rotary speeds could not be prevented unless essentially all the weight was taken off the bit. Rapid drilling into the variable lithologies of the interbedded coal-bearing section resulted in wellbore deviations being greater than planned (up to 36°). Wireline surveys were required to track hole deviation because it quickly exceeded the capabilities of the Teledrift instrument originally planned to be used for this purpose. The wells were allowed to continue to deviate as total depth was approached in order to penetrate the maximum amount of the steeply dipping stratigraphic section.

Lost circulation was expected in light of the tectonically disturbed nature of the region, but it was encountered only in the 2-20 well and, to a lesser extent, in the 3-29 and 4-20 wells. In the 2-20 well, the second well drilled, total returns of drilling mud and cuttings were lost at 833.5 m (2,735 ft) after tripping for a new bit and while drilling just below a 2-m (6.6-ft) thick coal seam. Despite spotting several lost-circulation material (LCM) pills, returns could not be recovered until a thixotropic cement plug had been set across the loss zone, which was identified by a temperature log to be at a depth of 819 m (2,690 ft). The third well drilled, the 3-29 well, also experienced lost returns while the driller was trying to unplug the bit jets at 1,078.5 m (3,538.6 ft), but this was readily cured and definitely was operator-induced rather than naturally occurring. In the 4-20 well, the fifth well drilled, partial returns were lost just below the surface casing and total returns were lost at 404.5 m (1,327 ft). Although the loss seemingly was cured, the rest of the well was drilled with a persistent slight fluid loss that could not be stopped with LCM, but did not warrant the rig time to justify setting a cement plug.

Drilling Results

The 1-20 well was drilled less than 1 km due south of the C 83-1 corehole to a total depth of 1,076 m (3,530 ft). The suite of geophysical logs that were run in the 1-20 well included focused-electric, high-angle dipmeter, gamma ray density, and seam-thickness tools. Comparison of the density logs both from the 1-20 well and the nearby corehole indicated an overlap in the upper and lower portions, respectively, of the two wellbores of an approximately 90-m (300-ft) thick section, inclusive of the Brusier and the No. 4 coal seams of the Wilkeson Series (Figures 5 and 6).

Comparisons of the mud log and density logs from the 1-20 well with measured sections and other geologic data from the mining records of the Wilkeson District suggested a possible correlation of an additional 250-300-m (820-1,000-ft) thick coal-bearing section in the 1-20 well with the remainder of the known stratigraphic sequence of the Wilkeson Series that was not penetrated in the shallower C 83-1 corehole. A similar comparison of the 1-20 well-log responses with mine data from the Melmont District suggested a correlation between the lower 90-180 m (300-590 ft) of this well with the uppermost section of the Melmont Series. If the correlations are correct, the 1-20 well provides the first definitive information confirming the subsurface stratigraphic relationships between the coal seams of the Wilkeson and Melmont districts.

The analyses of geologic data from the 1-20 well led to the decision to locate subsequent wells up-dip to the west of this location to penetrate, at a relatively shallower depth, a more complete section of the stratigraphic equivalent of the Melmont Series, which replaced the Wilkeson Series as the primary production objective for subsequent completions. Although less was known geologically about the Melmont Series relative to the Wilkeson Series, mine records indicated that it exhibited physical characteristics conducive to production potential similar to the Wilkeson Series. These characteristics include the presence of multiple, gas-bearing bituminous coal seams of adequate thickness to justify completion in subsequent wells.

Drilling of the five wells confirmed the presence of an asymmetric anticline with a north-northwest trend and a plunge to the north in the drill site section (Figure 6). Even though the surface locations of the four wells drilled along the axial trend of the anticline lie on the eastern side of the anticlinal axis, all of the coal seams that subsequently were completed for production are situated along the western flank of the structure. The

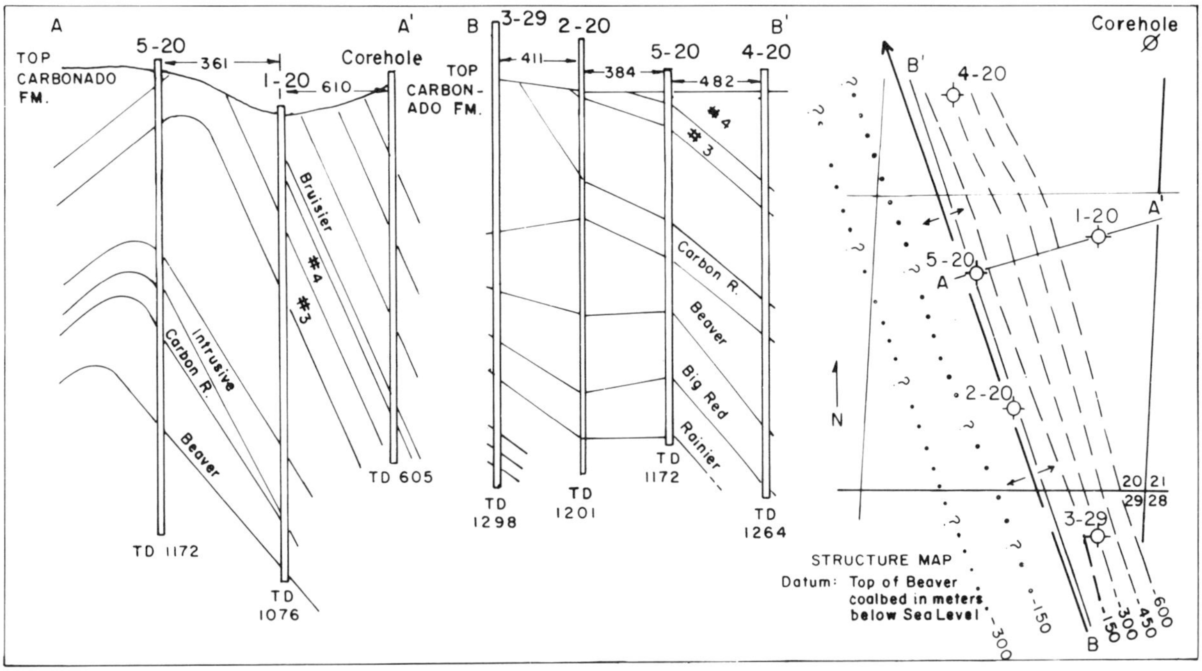

Figure 6. Structural map and geologic cross sections of the drill site area. The Beaver coal seam, the uppermost zone completed for production, serves as the datum for this structure map. The corehole is projected into the plane of cross section A-A' from its location in the northeast corner of the drill site section. The 4-20 and 3-29 wells are projected into the plane of cross section B-B'.

axial trend of the anticline may be comprised of a high-angle reverse fault with minor west-side-up displacement (not depicted on Figure 6).

Among the engineering objectives of the drilling program was the need to determine the most effective interval to space wells to optimize reservoir drainage. Proper well spacing assists the process of reservoir dewatering/pressure drawdown, which is necessary to initiate desorption of the occluded methane at economic rates from the matrix of the coal. The four wells drilled subsequently to the 1-20 well were located at average intervals of 400 m (1,300 ft), which was estimated to represent approximately 65-hectare (160-acre) drainage centers in the steeply dipping reservoir system. The 65-hectare (160-acre) drainage pattern was chosen as the most appropriate starting point to initiate reservoir testing based on our prior experience at the Amoco Cedar Hill coalbed methane project in the San Juan Basin. All four wells were situated near to, and on trend with, the anticipated position of the crest of the anticline, instead of a more-conventional stand-up/lay-down combination. "Stand-up/lay-down" refers to the arrangement of wells in a wellfield, "stand-up" being a north-south pairing of wells and "lay-down" being an east-west pairing. This was done to take advantage of the potentially more favorable reservoir characteristics that were anticipated to be present in such a structural position, as discussed previously.

Completion Program

The four test wells that were drilled along the axial trend of the anticline averaged 1,190 m (3,900 ft) in depth, and each penetrated an aggregate of over 30.5 m (100 ft) of bituminous coal comprised of beds from several centimeters to greater than 7.5 m (25 ft) in thickness. The coal seams that were completed for production in the four wells, although correlative to the No. 2, No. 2 1/2, No. 3 and, possibly, the No. 6 coal seams of the Melmont Series, are designated respectively herein as the Carbon River, Beaver, Middle, Big Red, and Rainier coal seams (Figures 5 and 6). Several seams that were penetrated in the 3-29 well below the Rainier (No. 6?) represent coal seams which were not encountered in the nearby mining districts. The 3-29 well was not completed in the newly discovered coal seams because they were situated too deep to be penetrated in the offsetting wells along the trend of the north-plunging anticline.

Casing opposite the entire thickness interval of each completed coal seam was perforated with 13 jet shots per meter (4 jet shots per foot). Following limited attempts to swab and test the wells without artificial stimulation, it became apparent that they would require such treatment in order to overcome effective and\or inherent permeability problems in the coal seam reservoir. Therefore, the completed zones were fractured using an 18-kg (40-lb) cross-link gelled KCl water and 12\20 sand proppant. Treatment sizes varied from 32,000 L (8,500 gal.) of gelled fluid with 3,400 kg (7,500 lb) of sand to 230,000 L (60,000 gal.) of fluid and 34,000 kg (75,000 lb) of sand. The fracture treatments were pumped down 6.032-cm (2.375-in.) tubing between a retrievable bridge plug and packer because all of the perforating and breakdown tests on the wells had already been performed.

Final completion consisted of anchoring 6.032-cm (2.375-in.) tubing just above the top set of perforations, connecting a 4.44-cm by 5-cm (1.75-in. by 2-in.) tubing barrel pump via a string of 2.22-cm (0.875-in.) sucker rods to a 160-size pumping unit driven by size-three Sargent Econo-pac electric motors and controllers. Power for the motors was supplied by individual diesel-powered generators because the remoteness of the well sites made the use of regional electric power impractical. (Of importance to subsequent mechanical and logistical difficulties experienced in trying to maintain continuous multiwell production testing operations is the noteworthy fact that the four pumps used in the project represented the first such equipment to be introduced into Washington since a failed attempt to establish conventional hydrocarbon production in the state in 1957.)

Coal, Gas, and Water Analyses

The gas content of the completed coal seams averaged 465 scf/tonne (423 scf/ton), based on calculations made from numerous desorption measurements made while drilling. Measurements of the vitrinite reflectance of samples of six of the completed intervals exhibited a range in values from 0.84 $\%R_o$ to 1.18 $\%R_o$. The vitrinite reflectance values are in agreement with prior classifications of the Melmont Series coal seams as ranging in rank from high-volatile A to medium-volatile bituminous, a level of maturity capable of generating a significant volume of methane.

Analyses of the coalbed gas from the wells indicated it was of pipeline quality. Gas samples were comprised of slightly less than 99% methane, 1.0% nitrogen, and less than 1% carbon dioxide. Calculation of heating values ranged from 853 to 1,003 Btu.

The results of extensive analyses of water produced during the limited pump testing of the four completed wells demonstrated that its quality was sufficient to meet state and federal requirements for surface discharge into the Carbon River, which drains the Carbon Glacier on nearby Mount Rainier. Among some of the key constituents of coal seam water considered to be of particular environmental concern, water from the test wells had an average pH of 8.4 and contained 1,620.0 mg/ml total dissolved solids, less than 1.0 mg/ml chloride, 1,170.0 mg/ml alkalinity (as $CaCO^3$), 12.4 mg/ml magnesium, and 647.0 mg/ml sodium. The economics of producing coalbed methane in the project area obviously are aided by the absence of need to incur significant capital expenditure for water disposal systems.

Production Potential

The average gas content value of 465 scf/tonne (423 scf/ton) of the coal seams in the project area represents approximately 5.5 bcf/km^2 (14 bcf/mi^2.) of in-place methane reserves. This is in close agreement with the maximum value of 440 scf/tonne (400 scf/ton) used in the GRI/DOE study, which suggested that up to 24 tcf of coalbed methane reserves are present in Western Washington.

The reservoir system in the immediate drill site area appears to be normally pressured to slightly underpressured and, therefore, dewatering operations will have to be conducted prior to achieving economic rates of gas flow, which is typical of the history of a majority of other productive coalbed methane development programs. The specific reservoir conditions of the Carbon River drill site area, however, are in no way indicative of the production potential of the up to 55,000,000,000 tonnes (60,000,000,000 tons) of prospective coal elsewhere in the region. Considering the geologic complexity of the coal basins, it is very likely that local conditions of reservoir overpressuring will be encountered in other locations. This could result in many wells that are capable of producing as prolifically as the best wells in the San Juan Basin.

The Carbon River wells underwent extremely intermittent production testing on pump over a 5-month period during which they were flowed to atmosphere and not against backpressure. On the average, initial gas-water ratios were on the order of 1:8. This value is representative of similar initial results in a majority of other coalbed methane projects that eventually established economic levels of production as additional offset wells were drilled to dewater the reservoirs and liberate increasing volumes of gas. It was not possible to document changes in the gas-water ratio during the sporadic test interval because frequent mechanical and logistical problems (parted rods, generator breakdowns, lack of readily available parts and service in the region, etc.) and administrative/budgetary "distractions" (e.g., commitment of the small staff to drill development wells in the San Juan Basin) prevented simultaneous multiwell pumping from ever occurring for an adequate length of time. Consequently, it was not possible to accomplish an accurate assessment of the production potential prior to exhaustion of the limited available capital for the project.

Computer projections of potential peak-production rates, however, range from 200 mcfd to 800 mcfd per well in a worst-case/best-case scenario, based on the data from the drilling and preliminary testing programs. Therefore, assuming an average production rate of 500 mcfd for the 125 wells that could be drilled on 65-hectare (160-acre) centers on the prospective approximately 8,100 hectares (20,000 acres) in just the one township of the project area, the resulting production of 62.5 mcfd represents significant potential in a state where all gas supplies are imported. Furthermore, considering the fact that the project township represents only 20% of the prospective coal-bearing acreage in Western Washington, the potential for producing coalbed methane in this region should be of obvious enough significance not to be discounted based solely on the incomplete results of the present study. On the contrary, the encouraging results of the Carbon River wells should justify additional effort to explore for and develop the coalbed methane reserves in the region.

SUMMARY

The primary result to date of the first multiwell coalbed methane drilling project in Washington is the formal documentation of gas content and in-place reserve values, which substantiate the validity of the reserve estimates of 24 tcf in the GRI/DOE study of Choate and Johnson (1980). Conservatively assuming that, of the estimated 24 tcf of in-place coalbed methane reserves in Western Washington, only 75% logistically are assessable and, furthermore, that only 50% of these are recoverable, the resulting 9 tcf of reserves could represent a significant local source of clean-burning natural gas for many years to come in this energy-import dependent, but rapidly growing, region. Perhaps even more importantly, energy consumers, particularly large industrial end users, could benefit from the introduction of a local source of gas as an alternative to the present somewhat uncompetitive situation of being supplied entirely by only one interstate pipeline system.

As environmental pressure continues to mount for cleaner-burning fuels and the supply/demand ratio ultimately begins to decrease, the role of coalbed methane in natural gas development in the United States likely will escalate accordingly. Although the initial attempt at establishing coalbed methane production in the relatively unexplored Pacific Northwest remains in an early stage of engineering development, the encouraging results to date justify similar programs be undertaken in this and other geologically complex regions to realize the potential benefit of developing this unconventional energy resource. Industry should be encouraged to expand coalbed methane production programs to more than just two of the many prospective coal basins in this country. Hopefully, the preliminary, but encouraging, results of the project in Washington will assist in this effort.

REFERENCES

Hull, D.A., J.D. Beaulieu, and W.W. Rau, 1983, Correlation of Cenozoic stratigraphic units of western Oregon and Washington: State of Oregon Department of Geology and Mineral Resources Oil and Gas Investigation 7, 90 p.

Armentrout, J.M., and D.H. Suek, 1985, Hydrocarbon exploration in western Washington and Oregon: American Association of Petroleum Geologists Bulletin, v. 69, p. 627-643.

Buckovic, W. A., 1979, The Eocene deltaic system of west-central Washington, in Armentrout, J.M., R.M. Cole, H. Ter Best Jr., eds., Cenozoic paleogeography of the Western United States: Society of Economic Paleontologists and Mineralogists Pacific Section, Pacific Coast Paleogeography Symposium 3, p. 147-163.

Byrer, C.W., T.H. Morz, and G.L. Covatch, 1987, Coalbed methane production potential in U.S. basins: Journal of Petroleum Geology, p. 821-834.

Choate, Raoul, and C.A. Johnson, 1980, Geologic overview, coal and coalbed methane resources of the Western Washington coal region: Lakewood, Colorado, TRW Energy Systems Group report for U.S. Department of Energy Morgantown Energy Technology Center Contract DE-AC21-78MC08089 (distributed under Gas Research Institute Contract 5011-321-0101), 262 p.

Culbertson, W.C., 1964, Geology and coal resources of the coal-bearing rocks of Alabama: U.S. Geological Survey Bulletin 1182-B, 79 p.

Daniels, J.Y., 1914, The coal fields of Pierce County, Washington: U.S. Geological Survey Bulletin 10, 146 p.

Engebretson, D.C., Allen Cox, and R.G. Gordon, 1985, Relative motions between oceanic and continental plates in the Pacific Basin: Geological Society of America, Special Paper 206, 59 p.

Fassett, J.E., and J.S. Hinds, 1971, Geology and fuel resources of the Fruitland Formation and Kirtland Shale of the San Juan Basin, New Mexico and Colorado: U.S. Geological Survey Professional Paper 676, 76 p.

Gard, L.M., Jr., 1968, Bedrock geology of the Lake Tapps Quadrangle, Pierce County, Washington: U.S. Geological Survey Professional Paper 388-B, 33 p.

Hammond, P.E., 1979, A tectonic model for evolution of the Cascade Range, in Armentrout, J.M., R.M. Cole, H. Ter Best Jr., eds., Cenozoic paleogeography of the Western United States: Society of Economic Paleontologists and Mineralogists Pacific Section, Pacific Coast Paleogeography Symposium 3, p. 219-238.

Johnson, S.Y., 1982, Stratigraphy, sedimentology, and tectonic setting of the Chuckanut Formation, northwest Washington: Unpublished Ph.D. Thesis, University of Washington, Seattle, 221 p.

Kim, A.G., 1978, Experimental studies on the origin and accumulation of coal bed gas: U.S. Bureau of Mines, RI 8317, 18 p.

Kilso, B.S., and D.E. Wicks, 1988, A geologic analysis of the Fruitland Formation coal and coal-bed methane resources of the San Juan Basin, Southwestern Colorado and Northwestern New Mexico, in Fasset, J.E., ed., Geology and coal-bed methane Resources of the northern San Juan Basin, Colorado and New Mexico: Denver, Rocky Mountain Association of Geologists, p. 69-90.

Morz, T.H., J.G. Ryan, and C.W. Byrer, 1983, Methane recovery from coalbeds--A potential energy source: U.S. Department of Energy Morgantown Energy Technology Center, Publication DOE/METC/83-76.

Robertson, R.L., and C.T. Rightmire, 1986, Commercial development of coalbed methane: Chemical Engineer Progress, February, p. 48-53.

Snavely, P.D. Jr., W.C. Wagner, and D.L. Lander, 1980, Interpretation of the Cenozoic geologic history, central Oregon continental margin: Cross-section summary: Geological Society of America Bulletin, Part 1, v. 91, p. 143-146.

Stremel, K., 1988, Coming out of the closet: Oil and Gas Investor, v. 8, p. 38-44.

Vine, J.D., 1968, Depositional environment of coal-bearing Paleogene strata, Western Washington (abstr.): Geological Society of America, Special Paper 101, p. 230.

Walsh, T.J., and W.M. Phillips, Rank of Eocene coals in Western Washington and Central Washington State: A reflection of plutonism?: State of Washington Department of Natural Resources Division of Geology and Earth Resources, Open File Report 83-16, 21 p.

Chapter 4

Underground Coal Gasification

PREFACE

Underground coal gasification (UCG) was focused upon as a topic for a separate chapter, albeit a short one, because of the promise this form of coal resource utilization holds for the extensive coal deposits in the U.S. and elsewhere which are too deep, too thin, etc. to extract any other way. Many seams are likely to stay that way regardless of advancements in other extraction technologies.

Surface coal gasification, as pointed out by Beaver et al. in this chapter, is a controlled process and, therefore, not particularly subject to the vagaries of the geologic setting. The geologic concerns for surface gasification fall into the areas of mining, because the coal must be extracted to use in the gasification chamber, and coal quality, because the quality of the gasification process and products rely on the characteristics of the coal. Therefore, surface gasification will not be covered explicitly in this book.

The overview paper in this chapter will provide you with a review of past UCG attempts and tests in the U.S. and of the basic technology involved in different UCG configurations. This paper also briefly reviews the geologic controls on UCG success.

The paper by Beaver et al. goes into greater depth on the results of the Rocky Mountain 1 (RM1) test and the insights it provided for future UCG work, especially in the area of hydrogeology. With recently increased environmental awareness on the part of the public and coincident concern about groundwater supplies, the hydrogeology of potential UCG sites and prevention of contamination of groundwater during UCG will be of paramount concern if UCG becomes widespread. Environmental geologists, take note!

It is true that, at least in the U.S., UCG work has cooled-off considerably and is now done with a relatively low level of national interest. However, continued concern about energy independence, both in the U.S. and in other coal-bearing countries, and "alternative" fuels and energy sources likely will bring UCG back to a state of high interest before long. The results presented by Beaver et al. show that this coal utilization technique can be a "clean" technology when the influence of geology on the UCG process is understood.

Cross-Reference

The papers by Sarnecki (in the Exploration chapter) and Goodarzi and Gentzis (in the Environmental Impacts chapter) discuss clinker and the natural coal burning which produces it. UCG has similarities to the natural burning of coal, and insights on subsurface results of the UCG process may be gained from these papers.

Because the UCG process is a form of coal extraction, the papers on subsidence (Kane and Powell) and roof stability (e.g., Sames and Moebs, Greb, and Collins) in the Mining chapter will provide information on how geology can impact UCG cavity stability. Likewise, papers in the Mining chapter (e.g., Weisenfluh and

Ferm and Molinda and Ingram) and the Coal Quality Concerns chapter (e.g., Hower et al. and Winston) on the ways depositional environment and geologic structure impacts coal seams can provide insight on conditions that could control UCG progress or quality. Some of the waste concerns for UCG sites are touched upon by Beaver et al. in the Environmental Impacts chapter.

UNDERGROUND COAL GASIFICATION

Robert L. Oliver[1]
Layne Geosciences, Inc.
N 4140 Duplainville Rd.
Pewaukee, WI 53072

G.F. "Pete" Dana
1719 Downey St.
Laramie, WY 82070

ABSTRACT

Underground coal gasification (UCG) technology has been tested in several areas of the United States over the past two decades. Each experiment attempted to demonstrate the technical and/or economic feasibility of various technologies. Evaluations of past experiments have shown that geologic and hydrologic conditions are a critical factor in the results of any UCG test. Some conditions are potential "stoppers" if they are present. Therefore, complete geologic and hydrologic characterization of potential UCG sites is essential to successful UCG projects.

Past experiments have proven UCG to be technically and environmentally feasible, especially with proper site selection and characterization. Further development depends on demand and other economic factors. Probably the greatest demand for UCG utilization will be in the industrial Midwest and East where only limited testing has been done on the thin, bituminous coal resources.

INTRODUCTION

Underground coal gasification (UCG) has the potential to recover 1,270,000,000,000 tonnes (1,400,000,000,000 tons) of U.S. coal resources which presently are considered uneconomical to recover by conventional mining technology (Frye, 1985). Coal gasification converts coal into liquid and gaseous products. The gaseous products contain combustible components which have heating values ranging from 100 to 450 Btu/scf. The product gases can be used directly to generate steam or electricity, or the products can be used as feedstock to produce liquid fuels, synthetic natural gas, or specialty chemicals such as ammonia.

The major setbacks to development of UCG at the present time are economic unattractiveness and environmental impacts. Locating UCG modules in areas with appropriate geologic and hydrologic characteristics can improve economic attractiveness by reducing environmental impacts and increasing process

---

1. Current Address: Atec Environmental Consultants, 11356 Mathis Avenue, Dallas, TX 75229.

efficiency. Understanding the UCG process and its interaction with the hydrogeologic environment, coupled with knowledge gained by evaluating past tests, has lead to the development of a set of desired characteristics for a UCG operation.

Process Technologies

UCG is coal gasification conducted in situ, so it does not require mining, gasification reactors, or ash disposal. However, UCG does require some coal seam preparation. Figure 1 shows a simplified UCG process consisting of a vertical well pair. Before gasification can proceed, the permeability of the coal between the well pair must be enhanced. This is normally done through reverse combustion linking, directionally drilled linking, or a combination of both.

In reverse combustion linking, coal at the base of one well is ignited (ignition well) and high-pressure air is injected in the other well (injection well). Air permeates through the coal from the injection well and intersects the combustion zone at the base of the ignition well. The combustion zone at the base of the ignition well will expand and follow the oxygen source back to the injection well, creating a high-permeability channel between the two wells. This channel often migrates toward the top of the coal seam at the same time.

In directionally drilled linking, a borehole is drilled between the vertical well pair. This requires directional drilling techniques which are expensive. This is the preferred method for establishing the link, however, because the borehole can be placed near the base of the coal seam, thereby increasing resource recovery. Because of the difficulties in accomplishing direct mechanical linking, reverse combustion linking is used after directional drilling to complete the mechanical link.

Other configurations also have been developed and tested. Figure 2 shows the steeply dipping bed (SDB) UCG configuration. In this configuration, a slant production well is drilled near the bottom of a steeply dipping coal seam. A slant injection well is drilled below the coal seam and the well enters only at the bottom of the module to be gasified where it intersects the production well. The module acts as a packed-bed reactor providing high resource recovery and minimal roof interaction.

Another configuration, termed the controlled retracting injection point (CRIP), was developed by Lawrence Livermore National Laboratory (Figure 3). A horizontal injection well is drilled at the base of a coal seam that intersects a vertical and a horizontal production well. The vertical production well is used during process start-up, after which the horizontal production well is used. A liner is inserted in the injection well, and a mobile igniter-burner is inserted inside the liner. The ignitor-burner can be retracted and used to burn off the liner, thereby exposing unaltered coal to the process. In this fashion, multiple modules can be formed using the same well pair.

In all configurations, UCG is conducted by injecting air and/or a steam-oxygen mixture into one well (injection well) and producing the product gases from another well (production well). The coal is consumed laterally away from the injection well toward the production well. As the coal is consumed, a cavity forms that eventually exposes the roof rock to the gasification area. Gasification is terminated when the roof rock interaction is sufficient to degrade gas quality beyond a set limit.

Past Experiments

Although UCG has been in development since the early 1900s, research in the United States did not begin in earnest until 1972 with the onset of four experiments conducted by the U.S. Department of Energy (DOE) at a site near Hanna, Wyoming. The resource at the Hanna site is a flat-lying, subbituminous coal, approximately 9 m (30 ft) thick, lying at a depth of 49-122 m (160-400 ft). The tests used typical well-pair configurations (Figure 1) and air injection. Hydraulic fracturing, to enhance coal seam permeability, and

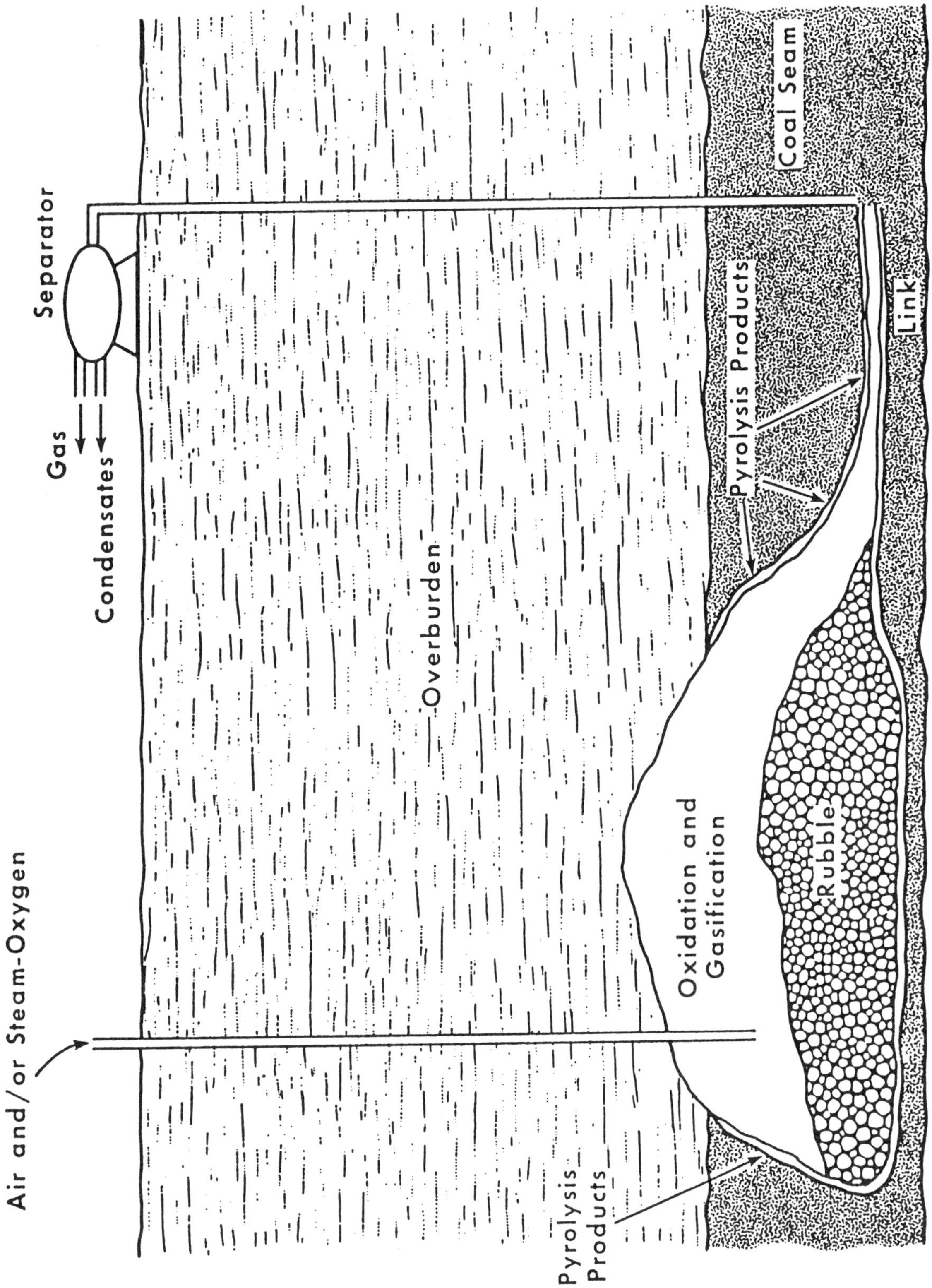

Figure 1. *Typical configuration for vertical well-pair gasification.*

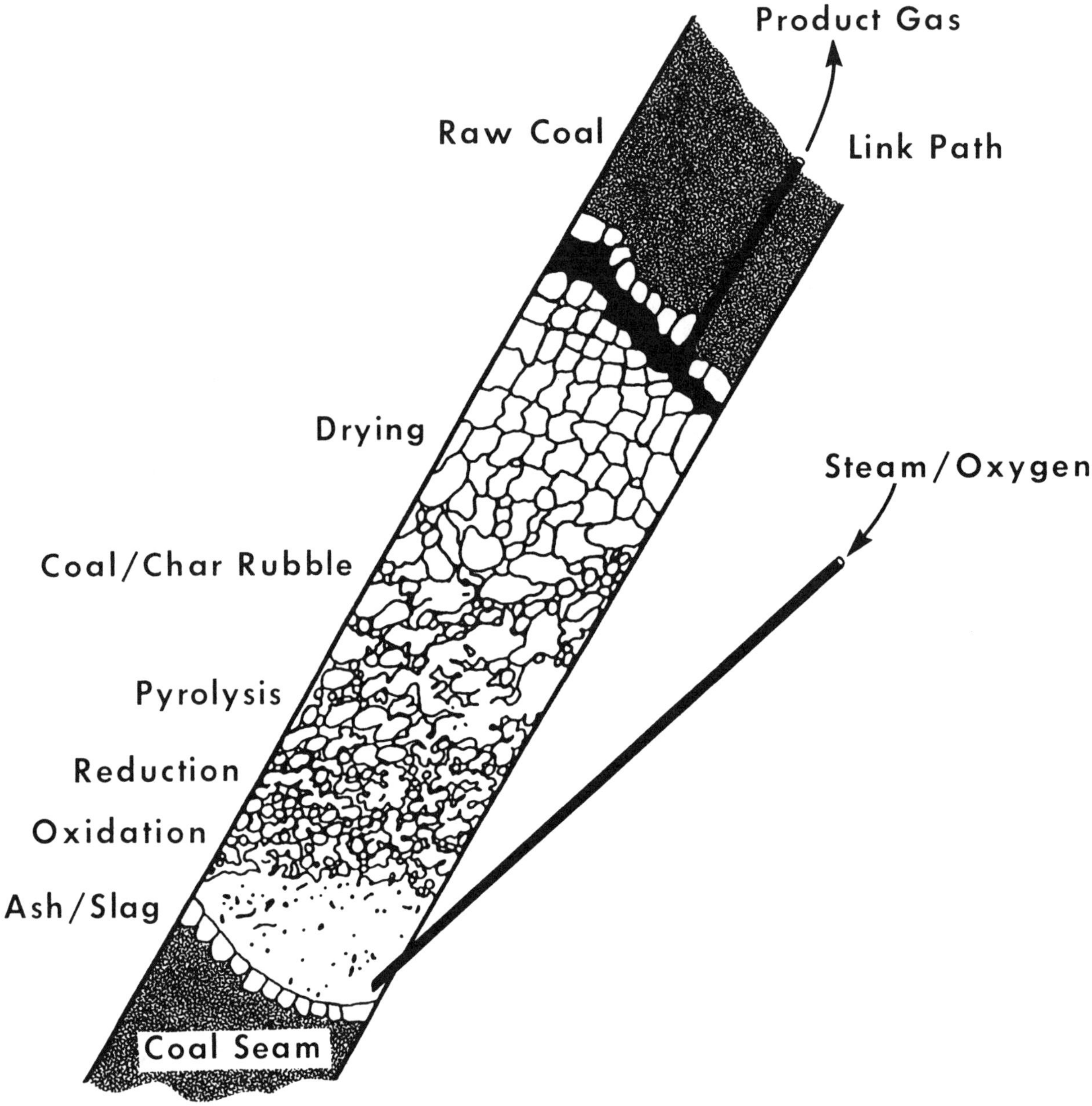

Figure 2. *Typical configuration for steeply dipping bed (SDB) gasification.*

forward combustion linking were unsuccessfully attempted during the first test. The second experiment pioneered reverse combustion linking and was successful in producing excellent gas quality. The third test was located in an area where the coal seam was very dry, so insufficient hydrogen was available to drive the reactions and resulted in low gas quality. Undetected faulting in the area of the fourth test caused linking and gas-flow problems. Proper site selection and characterization could have minimized the problems experienced during the third and fourth tests.

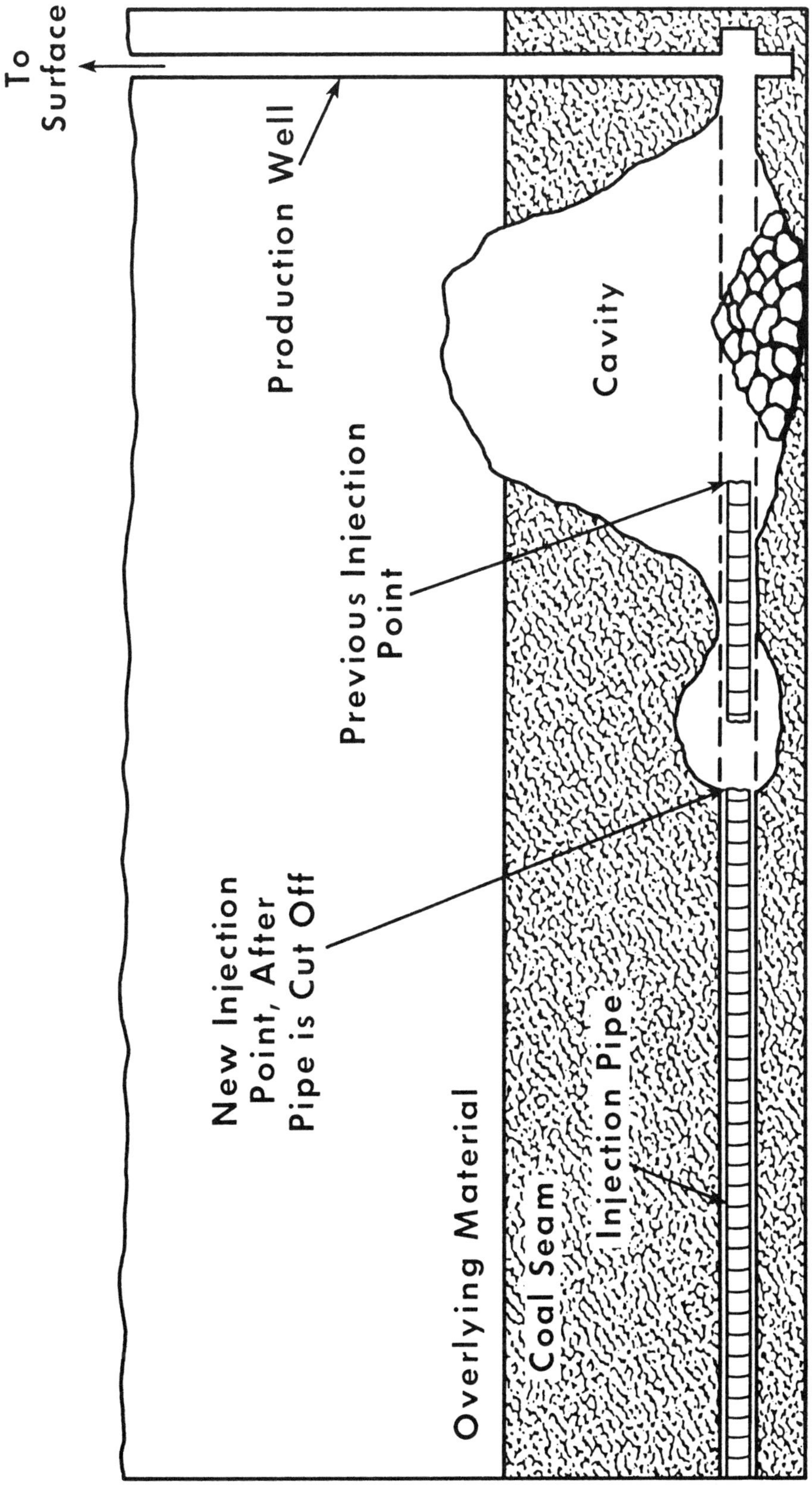

Figure 3. *Typical configuration for controlled retracting injection point (CRIP) gasification.*

Another significant series of tests was conducted by the Lawrence Livermore National Laboratory from 1976 to 1979 at the Hoe Creek site near Gillette, Wyoming. The resource is a flat-lying, subbituminous pair of coal seams overlain by as little as 20 m (66 ft) of wet, incompetent material. The first test unsuccessfully attempted explosive fracturing and forward combustion. The second test successfully used reverse combustion linking and air and steam/oxygen mixtures for the injection stream. The third test pioneered the use of directionally drilled linking. The second and third tests experienced operational difficulties because of extensive overburden collapse which caused water influx and gas loss problems. These problems could have been avoided by locating the tests in a more desirable hydrogeologic environment through adequate site selection and characterization.

The DOE conducted a 275-m (900-ft) deep test at a site near Pricetown, West Virginia. The resource is a high-swelling bituminous coal approximately 1.8 m (6 ft) thick. The test used reverse combustion linking and air injection. After 12 days of gasification, the test was terminated because of well casing separation and subsequent plugging. Swelling of the coal reduced air flow, leading to the plugging problems.

Gulf Research and Development Company (Gulf) conducted two tests at a site near Rawlins, Wyoming from 1979 to 1981. The resource is a steeply dipping (63°), subbituminous coal approximately 7 m (23 ft) thick which outcrops at the surface. The tests used SDB technology (Figure 2). The first test used a combination of air, steam, and oxygen injection gases, and the cavity was about 122 m (400 ft) below ground level. The second test was conducted at a depth of approximately 183 m (600 ft) using steam and oxygen as the only injection gases. During one 19 day period, the product gas heating value averaged 356 Btu/scf, making this the most successful UCG experiment reported in the United States. Part of the reason for the success of the test is that a UCG cavity formed in a steeply dipping bed converts the coal more efficiently because of reduced roof intéraction.

Lawrence Livermore National Laboratory conducted a test for the DOE at a site near Centralia, Washington in 1983. The resource is a moderately dipping (20°), high-ash subbituminous coal approximately 11 m (35 ft) thick. The upper 6 m (20 ft) of the seam was used for the test which pioneered the CRIP technology (Figure 3). Only one of the planned CRIP maneuvers was accomplished because of control problems experienced during drilling of the injection well; however, feasibility of the concept was demonstrated.

The most significant test conducted to date is the Rocky Mountain 1 (RM1) Experiment located near the Hanna test sites. The test was conducted during the winter of 1987-1988 with the majority of the funding provided by DOE and the Gas Research Institute. The RM1 Experiment was the first test to have adequate, pre-burn site characterization. The test consisted of two closely spaced modules operating simultaneously. One module used the CRIP technology and the other module used the previously untested Extended Link Well (ELW) technology. Operational measures also were taken to reduce environmental impacts, including operating at cavity pressures below hydrostatic head and venting and cooling the cavities rapidly after test shut-down.

Other tests also have been conducted in the United States. These include 1) Gorgas, Alabama, conducted by DOE, 2) Colonial Mine in Western Kentucky, conducted by Gulf, 3) Reno Junction, Wyoming, conducted by ARCO Coal Company, and 4) Texas lignites, conducted by Basic Resources, Inc. and Texas A&M University.

GEOLOGIC AND HYDROLOGIC INTERACTIONS

Geologic and hydrologic interactions with the UCG process have been documented by postburn studies of past experiments (e.g., Cooke and Oliver, 1985; Dana et al., 1985; Ethridge et al., 1983; Oliver, 1986; Oliver et al., 1987; Strickland and Jennings, 1979). Geologic and hydrologic interactions strongly affect linking and process efficiencies. They also can cause environmental damage which is expensive to remediate.

To prevent or minimize these impacts, the geologic and hydrologic regime of a potential site must be identified. The characterization should include a critical evaluation for faults and fracture systems at each potential site. The evaluation also should be made to determine if any detected faults or fracture systems intersect the surface because of the potential for process gas loss to the atmosphere. The lithology of the overburden should be identified to predict roof collapse and potential gas loss. Major partings should be located so that linking problems can be avoided. An evaluation of coal, overburden, and underlying rock permeabilities, aquifer flow systems, directional permeabilities, and the interrelationships of water-bearing strata also should be conducted. A site with proper characteristics then can be selected and appropriate process designs developed so that the process attains the best possible results and economic return.

Linking in UCG technologies that use vertical well pairs can be substantially hindered or even impossible if faults cross the link path (Oliver et al., 1987). Faults at Hanna IV offset the permeable horizon along which the link was established (Figure 4). For processes using directionally drilled links, faults or folds make the drilling of directional process wells difficult because of the limitations in making rapid wellpath changes. At the Tono I site (Dana et al., 1985), the position of the borehole moved up within the coal seam because of the change in dip (Figure 5). This reduced resource recovery because the CRIP process only consumes coal that is above the level of the directional injection well (Oliver, 1986). Linking also may be hindered if thick partings are present along the link horizon or at the ignition point.

Process efficiencies (or gas quality) can be reduced because of geologic and hydrologic conditions. Extensive gas loss can result from abundant fractures in the coal or overburden, fractures that extend from the coal into the overburden, or abundant unconsolidated material in contact with the coal. These conditions provide high-permeability pathways through which process gas can migrate out of the reactor cavity. This significantly reduces the amount of process gas recovered from the system. Any of these factors also may enhance the introduction of non-coal material into the gasification cavity through increased roof collapse.

Thick sequences of unconsolidated overburden material caused extensive roof collapse during the Hoe Creek III experiment (Ethridge et al., 1983). As a result, the experiment experienced gas loss and reduced gas quality. In addition, high permeability in the unconsolidated overburden allowed process gas to move into the overburden strata causing a significant groundwater contamination problem.

Process efficiencies also can be reduced by excessive influx of groundwater into the cavity. This can occur if permeability of the coal is high or if the overburden material is wet and highly permeable. Groundwater leakage from underlying strata into the coal can occur if upward vertical hydraulic gradients are present. Water also may be introduced into the reactor cavity as a result of roof collapse and subsequent connection of the cavity to overlying water-bearing zones. The result of excessive water in the reactor cavity is a reduction in gas quality.

Environmental impacts of UCG resulting from these interactions are groundwater contamination, interconnection of water-bearing units, and disruption of natural groundwater flow. Groundwater contamination can be extensive if the target coal seam is an aquifer with high permeability. Further contamination of adjacent aquifers can occur if the aquifers become interconnected through roof collapse (Figure 6). Water quality of the overlying water-bearing unit at the Hanna site was impacted through this mechanism (Cooke and Oliver,

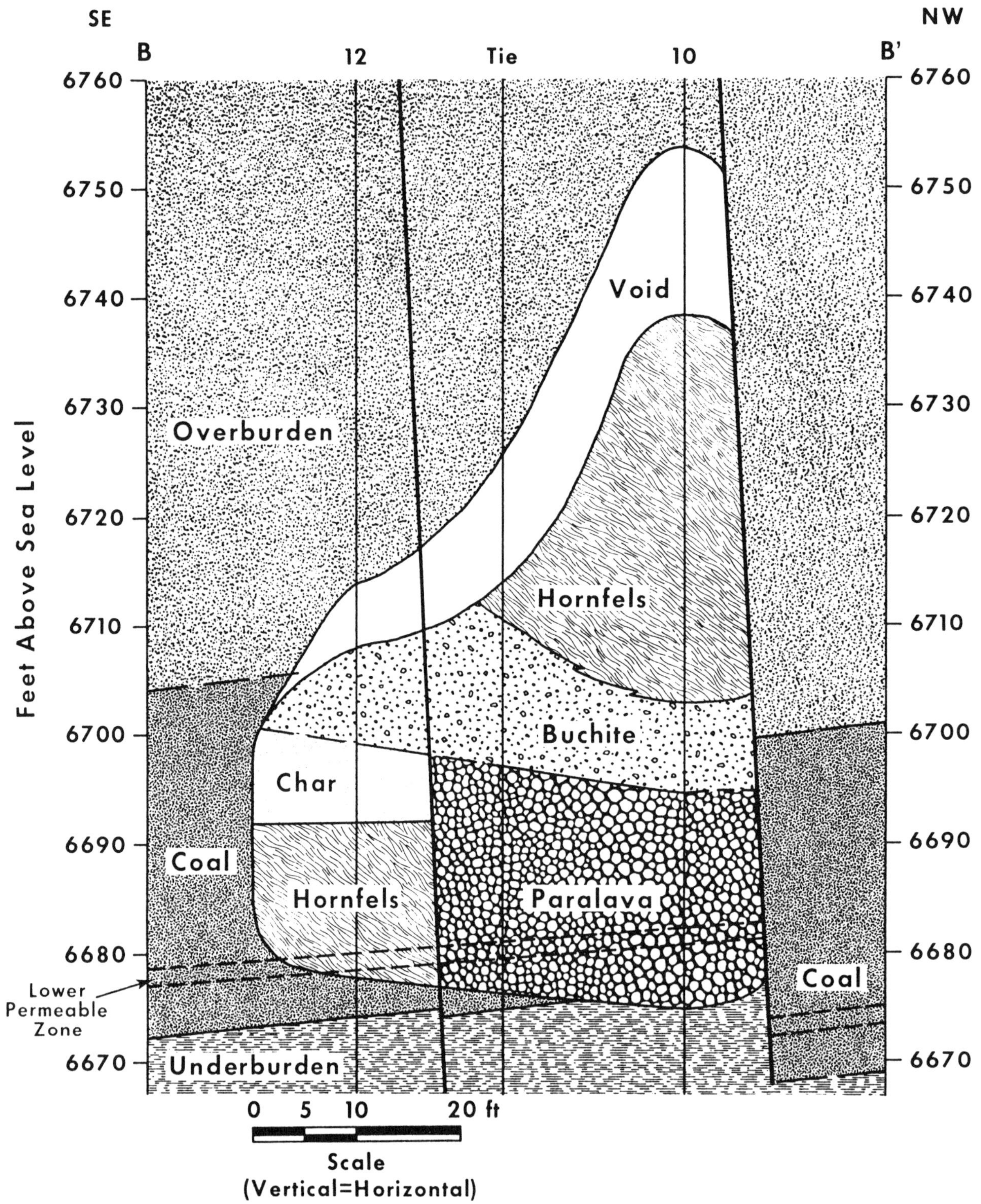

Figure 4. Offset of the permeable link zone by faulting.

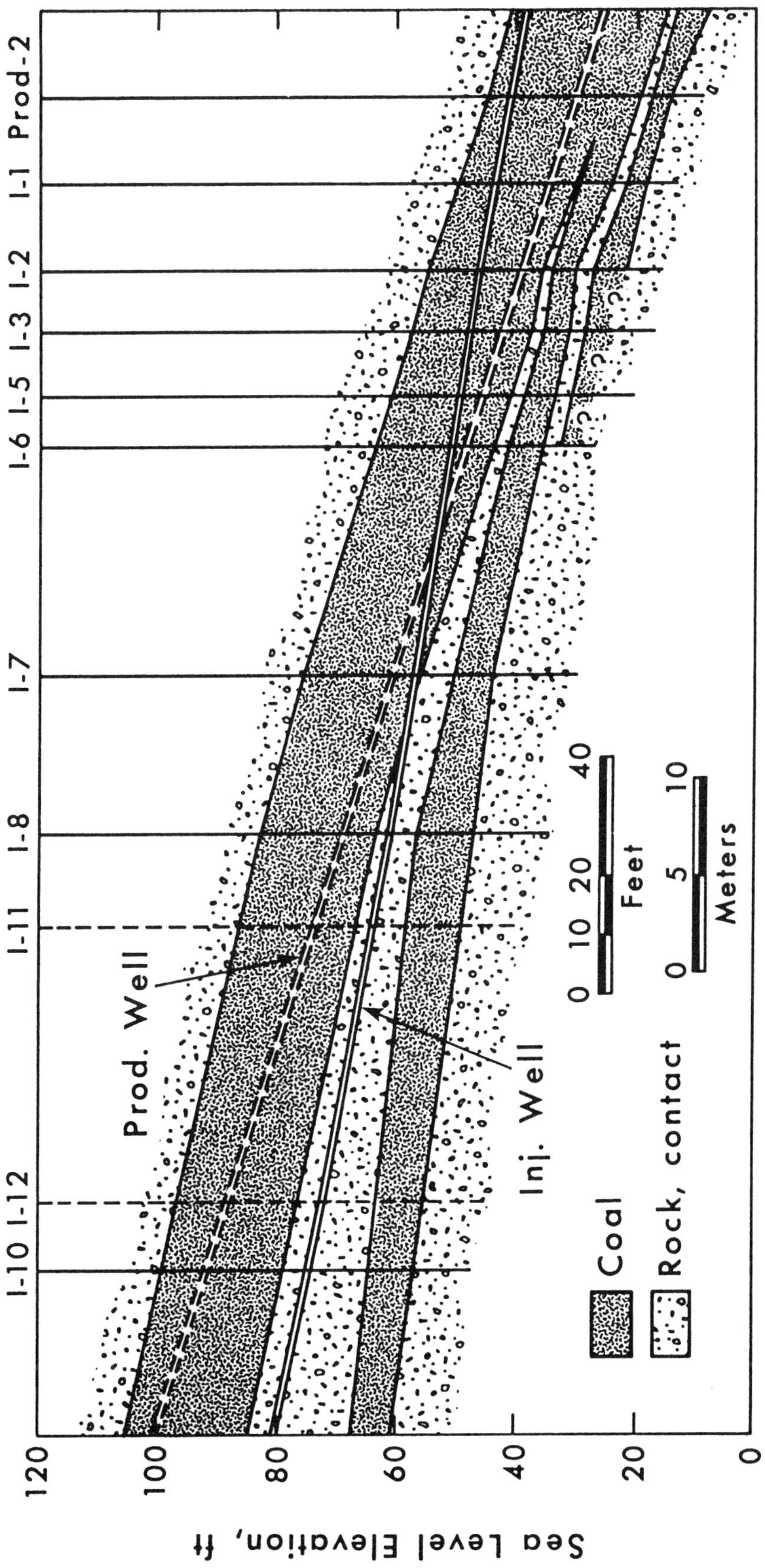

Figure 5. Change in vertical location of horizontal process wells due to coal seam dip variation.

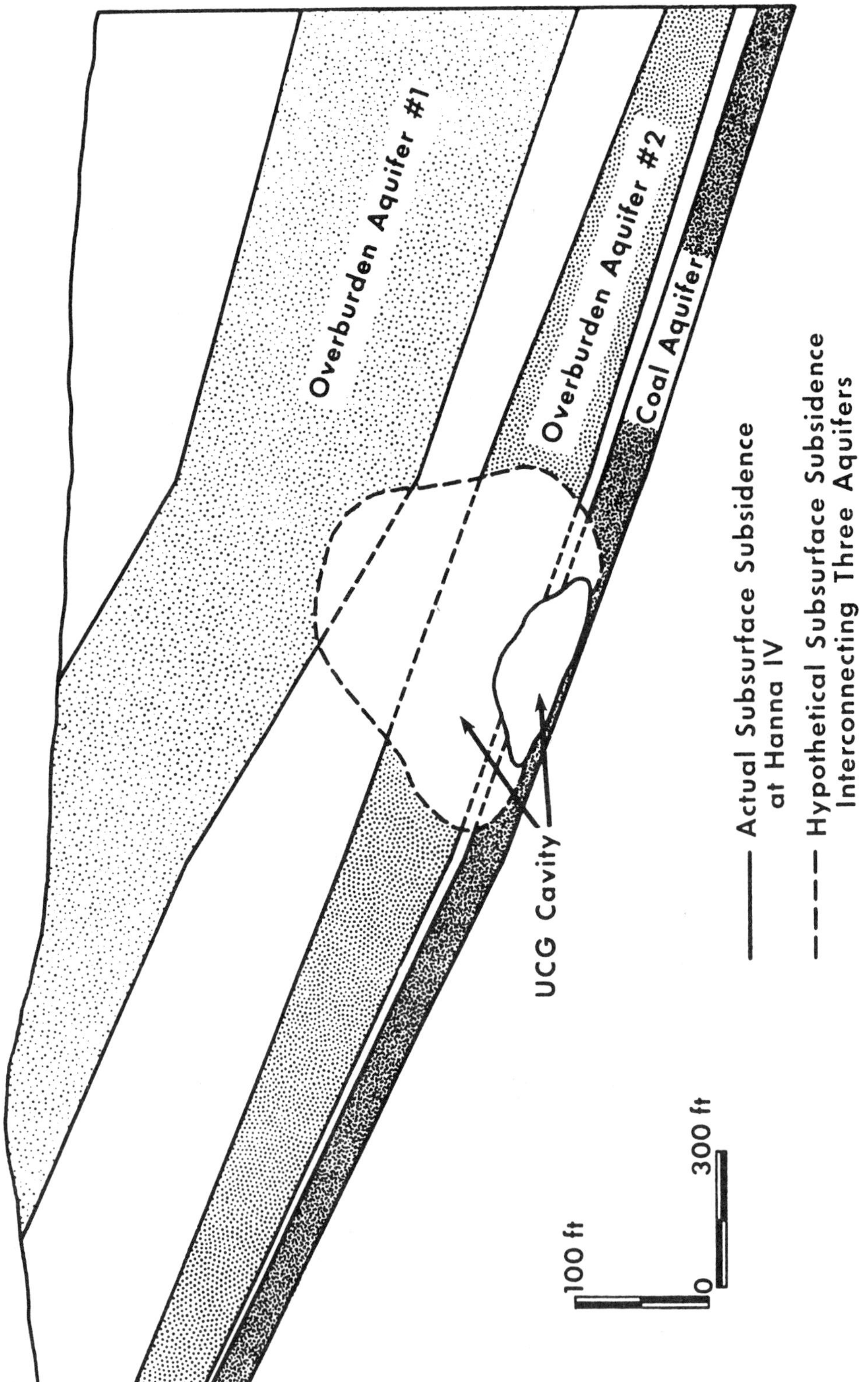

Figure 6. *Interconnection of aquifers caused by roof collapse.*

1985). Contaminated water also can migrate into other aquifers through preexisting or induced fracture systems, or process gases can leak into overlying or underlying aquifers which have high permeabilities.

DESIRED SITE CHARACTERISTICS

The economic potential of commercial UCG operations depends on a number of factors, some of which are unique to thermal in situ processes such as UCG. Because the UCG process is conducted in subsurface environments, the geologic and hydrologic characteristics of the subsurface environment are most important to process efficiency and environmental impacts. The major influence on the success of UCG operations has been shown to be the selection of a site with favorable geologic and hydrologic characteristics and sufficient characterization to permit adequate design and well construction.

Ideal geologic and hydrologic characteristics for a UCG site have been developed based on past test results (Table 1). Included in Table 1 are values that define coal seam, overburden, and hydrologic characteristics that are most suitable to a successful UCG operation. An ideal site would contain a relatively continuous, dry coal seam of moderate thickness and consistent dip, low permeabilities in the boundary strata, and competent, relatively homogeneous overburden material.

Parameter	**Value**
Coal Thickness (m)	1.5-15
Thickness Variation (% of seam thickness)	<25
Dip[1] (degrees)	0-70
Dip Variation[2] (degrees/31 m)	<2
Single Parting Thickness (m)	<1
Total Partings Thickness (% of seam thickness)	<20
Fault Displacement (% of seam thickness)	<25
Fault Density (number of faults/31 m)	<1
Coal Rank[3]	≤Bituminous
Coal Moisture (wt %)	<15
Coal Sulfur (wt %)	<1
Overburden Thickness (m)	92-460
Thickness of Consolidated Overburden[4] (m)	>15
Coal Seam Permeability (md)	50-150
Immediate Overburden Permeability[5] (md)	<5
Distance to Nearest Overlying Water-Bearing Unit (m)	>31
Coal Aquifer Characteristic	Confined
Nearest Producing Well Completed in Coal Seam (km)	>1.6
Available Coal Resources[6] (10^6 m^3)	15.4

[1] Depends on technology.

[2] If using directionally drilled process wells.

[3] If bituminous, free-swelling index should be low.

[4] Thickness should be directly above or near top of coal seam.

[5] 15 m (50 ft) directly overlying coal seam.

[6] For 20-year long plant operation.

Table 1. *Desired underground coal gasification site characteristics.*

Seam Thickness

Although the coal seam thickness is not critical, a relatively thick seam and a fairly constant seam thickness generally are more desirable. Coal seams that are less than 1.5 m (5 ft) thick are more susceptable to excessive heat loss than thicker seams because of rapid exposure of the inert roof rock (Bartel et al., 1980). Additionally, water influx per unit volume of consumed coal increases with decreasing seam thickness (Krantz and Camp, 1980).

Seams thicker than about 15 m (50 ft) may have substantial roof stability problems when gasified because of the large volume of material removed. Depending on bulking factors and competence of the overburden rock, this coal removal may lead to subsidence at the surface. Subsidence to the surface is less of a problem in thinner seams.

A coal seam that changes thickness over short distances may cause operational difficulties and complicates the drilling of directional process wells. A commercial venture will require approximately 2.6 km^2 (1 mi^2) of a fairly continuous, 6-m (20-ft) thick coal seam to operate a moderate-sized plant for about 20 yrs.

Partings

Partings within the coal seam can cause operational problems if they are located near ignition points or along linking paths. Single partings, greater than about 1 m (3 ft) in thickness, may totally stop gasification and will, in all cases, cause a substantial decrease in gas quality during the period of time that they are part of the material being gasified. If the coal seam contains more than 20% parting material, gas quality may be lowered to the point that gasification of that seam would be economically infeasible.

Depth

Depth to the coal should be greater than 92 m (300 ft) to prevent surface subsidence, but no more than 460 m (1,500 ft) until such time as the technology has been proven at greater depths. The greater depths currently cause operational problems because of high lithostatic pressure which restricts gas flow.

Seam Dip

The process that is used will determine what dip of the coal seam is most desirable. Dip variation is especially significant to processes that will require directionally drilled link paths. Processes that use vertical process wells also may experience linking and override problems if the dip changes substantially over short distances. However, process designs and module locations can be modified to minimize the negative effects of existing dips.

Faults

Faults can cause significant operational and environmental problems, and UCG sites should be chosen to avoid them if possible. Linking across fault zones is difficult, if not impossible, and faults may act as conduits for groundwater or gas transport through which significant contamination or gas loss can occur. Drilled links reduce the effects of faults on linking, but the faults create problems in drilling the link. Small-scale faults that displace less than 25% of the coal seam thickness are acceptable if they are adequately identified. In any case, fault density should not be greater than one fault every 31 m (100 ft). This density will allow sufficient space to correct drill hole locations and to place modules between faults.

Coal Rank

The major impact of coal rank is the related swelling characteristics of the coal. Coals with a rank of bituminous or higher generally swell when heated which closes off gas flow pathways. Coals with a rank of subbituminous or lower generally shrink when they are heated, making them easier to link and gasify. Anthracite is considered undesirable for gasification because it is difficult to ignite and tends to occur in thin seams.

Coal Quality

The sulfur content of the coal is important because excessive sulfur emissions will require expensive clean-up processes. Moisture in the coal is significant because of water influx and its related impact on the UCG process. Some moisture is beneficial to the process; however, too much moisture can create operational problems.

Overburden

It is crucial to have a well-consolidated overburden section immediately above the coal seam that will inhibit roof collapse. Roof collapses propagating 9-12 m (30-40 ft) into consolidated overburden have been noted in previous tests (Oliver, 1986; Oliver et al., 1987). Consequently, the consolidated overburden section should be at least 15 m (50 ft) thick. In addition, the immediate overburden should be dry and relatively impermeable so that it will inhibit water influx and gas loss. To prevent possible interconnection of water-bearing strata with the gasified coal seam by roof collapse, no water-bearing zone should be within 31 m (100 ft) above the target coal seam. It also is desirable to have impermeable strata immediately below the target coal seam.

Coal Permeability

The coal seam should have a moderate permeability so that gas can flow through it to propagate the process. Very low permeabilities make gasification difficult, whereas very high permeabilities will increase the amount of process gas lost into the coal seam. In addition, the coal seam should be a confined aquifer so that contaminants produced by the process are contained within the coal seam.

FUTURE DEVELOPMENT

The future for UCG is unknown at this time. The basic technology has been proven and the RM1 test established environmental acceptability. Site characteristics have been established so that appropriate sites can be selected to maximize economic return. Commercialization will depend on several factors including energy costs, demand, and the ever-changing regulatory requirements.

The major coal resource areas of the U.S. are the western subbituminous coals and the eastern bituminous coals. Great quantities of coal resources are available in the West, where most of the testing has been conducted. However, these resources are very thick and close to the surface and will probably be most economically recovered through conventional mining techniques well into the foreseeable future. Lack of demand and high costs to transport the gasification product to high-demand areas further hinder development.

The eastern bituminous coals are located in high-demand areas and are thin and often deep below the surface. These resources are the most likely for UCG utilization. Only limited testing has been conducted on these resources, however, and they will be more difficult to gasify because of their swelling characteristics. In addition, they have high sulfur contents which present additional environmental problems, especially in light of the ongoing acid rain legislation. More testing in these resources should be conducted, with an emphasis on developing techniques to improve gas flow through the coal and to reduce sulfur emissions.

REFERENCES

Bartel, L.C., T.L. Dobecki, and R. Stone, 1980, Results of the characterization of an in situ coal gasification site in the state of Washington, in Proceedings of the Sixth Annual Underground Coal Gasification Symposium, Afton, Oklahoma, July 13-17, 1980: U.S. Department of Energy, Report #800716, p. I-41 to I-48.

Cooke, S.D., and R.L. Oliver, 1985, Summary of the groundwater monitoring program at the Hanna, Wyoming Underground Coal Gasification Test Sites: U.S. Department of Energy, Technical Note #DOE/METC-85/4006, 40 p.

Dana, G.F., R.L. Oliver, and J.R. Covell, 1985, Evaluation of the postburn drilling, coring, and geotechnical surveying program at the WIDCo-CRIP UCG experimental test site, Centralia, Washington, in Proceedings of the Eleventh Annual Underground Coal Gasification Symposium, Denver, Colorado, August 11-14, 1985: U.S. Department of Energy, Report #DOE/METC-85/6028, p. 9-36.

Ethridge, F.G., L.K. Burns, W.G. Alexander, G.N. Craig II, and A.D. Youngberg, 1983, Overburden characterization and postburn study of the Hoe Creek site and comparison with the Hanna, Wyoming site: U.S. Department of Energy, Report #DOE/LC/10705-T1, 185 p.

Frye, K.N., 1985, Underground coal gasification challenges and potential, in Proceedings of the Eleventh Annual Underground Coal Gasification Symposium, Denver, Colorado, August 11-14, 1985: U.S. Department of Energy, Report #DOE/METC-85/6028, p. 1-8.

Krantz, W.B. and D.W. Camp, 1980, A water-influx model for UCG, in Proceedings of the Sixth Annual Underground Coal Gasification Symposium, Afton, Oklahoma, July 13-17, 1980: U.S. Department of Energy, Report #800716, p. III-21 to III-31.

Oliver, R.L., 1986, Final results of the Tono I UCG cavity excavation project, Centralia, Washington: Lawrence Livermore National Laboratory, Contract Report #8568605, 23 p.

Oliver, R.L., D.J. Sinks, and G.L. Berdan, 1987, Postburn evaluation of the Hanna IVA and B UCG experiments, Wyoming: U.S. Department of Energy, Report #DOE/FE/60177-2289, 86 p.

Strickland, R.F., and J.W. Jennings, 1979, Analysis of geological limitations to underground coal gasification: In Situ, v. 3, no. 3, p. 209-226.

THE ROLE OF HYDROGEOLOGY IN UNDERGROUND COAL GASIFICATION WITH AN EXAMPLE FROM THE ROCKY MOUNTAIN 1 (RM1) TEST, CARBON COUNTY, WYOMING

F.W. Beaver, G.H. Groenewold, C.R. Schmit, and D.J. Daly
Energy and Environmental Research Center
University of North Dakota
Box 8213, University Station
Grand Forks, ND 58202

R.L. Oliver
ATEC Environmental Consultants
11356 Mathis Avenue
Dallas, TX 75229-3157

ABSTRACT

A thorough understanding of the hydrogeology of an underground coal gasification (UCG) site, complemented with understanding potential interactions between the hydrogeologic system and the UCG process, is fundamentally important to the design and conduct of UCG operations that are both technically successful and environmentally acceptable. Field experience to date suggests that the required hydrogeologic insight must include 1) the occurrence, flow, and quality of fluids (groundwater and gas) moving through the three-dimensional matrix of geologic materials that serves as the UCG "reactor" and 2) the textural, mineralogical, chemical, and structural character of the geologic materials.

The hydrogeologic aspects of the Rocky Mountain 1 (RM1) UCG site were characterized to an extent unprecedented in UCG research and development. Subsequently, during the course of the test program, events occurred and observations were made that required responses and explanations, the nature of which clearly justified the need for comprehensive prior hydrogeologic characterization. For example, an episode of gas migration during the start-up phase of the test, which affected both groundwater flow and quality, could not have been addressed adequately without the detailed knowledge that the hydrogeologic characterization activities had provided about both the local groundwater and the geologic materials through which it flowed. Knowledge of the fractured nature of the primary aquifer, which was also the target coal for gasification, was particularly useful in this case, as well as in explaining observations of groundwater drawdown patterns and the overall sensitivity of the hydrogeologic system to test-related pressure conditions.

The RM1 test demonstrated the intimate relationship between a site's hydrogeologic system and the UCG process active within that system. In so doing, the RM1 test dramatically underscored the significant part that hydrogeologic considerations must play in planning and operating successful UCG tests and commercial ventures in the future.

INTRODUCTION

Underground coal gasification (UCG) is the process by which coal is ignited in-place underground, thereby producing heat that, in turn, pyrolytically decomposes and gasifies additional coal, resulting in the production of a combustible gas. Air/oxygen and/or steam are injected into a coal formation that has been ignited, and product gases including methane, carbon monoxide, hydrogen, nitrogen, carbon dioxide, and water are produced (Figure 1). The gases are retrieved by means of appropriately placed horizontal and/or vertical wells and can be processed and utilized at the surface.

UCG differs substantially from conventional above-ground gasification. In surface gasification, the process is contained in an engineered reactor vessel within which the operating conditions can be closely monitored and controlled by the operators. In essence, the environment within the reactor is not only controllable, but largely predictable. However, surface gasification requires the mining and, in some cases, preparation of coal, as well as the management of the refractory residuals and other waste streams. Both mining and waste management impact the economy of surface gasification and, at the same time, have significant environmental implications.

In contrast, UCG utilizes the geologic environment as the gasification reactor, thus requiring that the process be designed to accommodate the geological and hydrogeological constraints inherent in the site. The "control" of the gasification environment, typical of engineered surface settings, must be replaced with "understanding and accommodating" the natural setting which serves as the reactor. The advantages of the

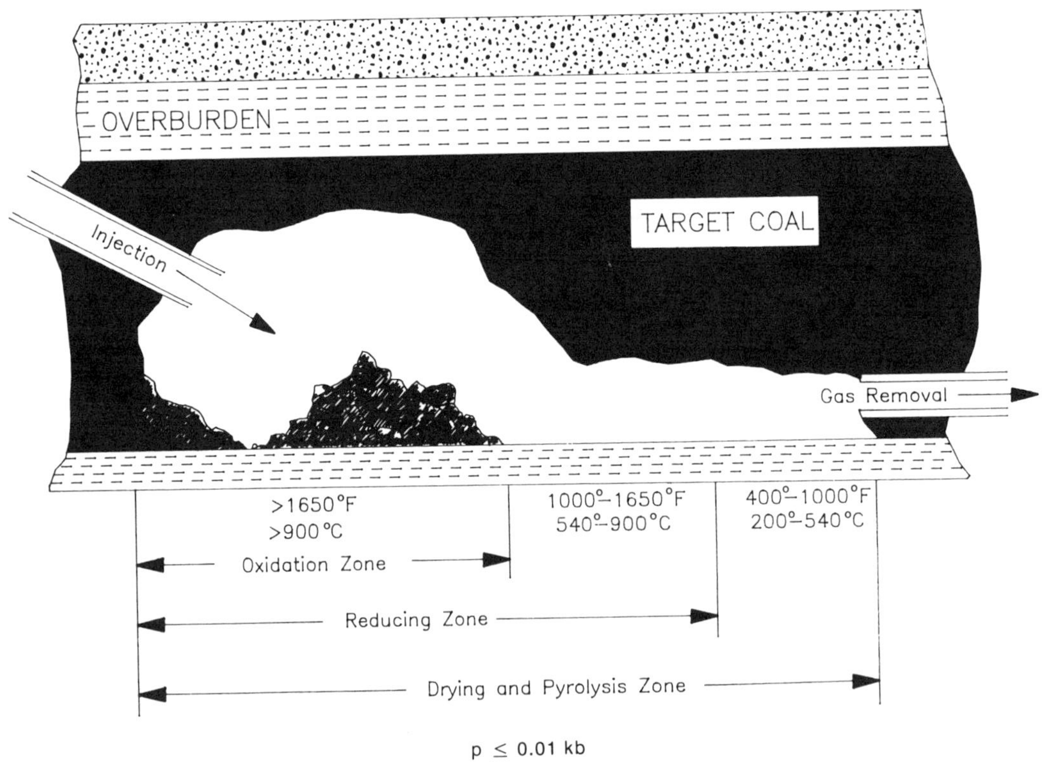

Figure 1. *Schematic illustration of the underground coal gasification process (modified from Dockter, 1986).*

UCG process include not having to mine the coal and handle the by-products, in addition to access to presently unmineable reserves.

Previous UCG research has shown that UCG process applications can have adverse effects on the local hydrogeologic system, including the occurrence, flow, and quality of subsurface water. Furthermore, it has been clearly demonstrated that an ability to identify, prevent, and/or mitigate potential groundwater contamination problems must be a key aspect of any UCG program (Daly et al., 1987). At the same time, it is well known that groundwater plays an active role in the gasification process and, on occasion, can adversely affect that process (Thorsness and Creighton, 1983).

Because the local hydrogeologic system and the UCG process interact, the development of a basic understanding of the hydrogeology of a UCG site must be a fundamental part of the pretest activities of any UCG program. This understanding is accomplished through detailed characterization of that site, and this, in turn, must be followed by monitoring of the hydrogeologic system before, during, and after the active gasification phase. The evolving knowledge of that system gained through thorough site characterization and ongoing monitoring may be used to reduce or prevent impacts either on the hydrogeologic system by the process or vice versa. Moreover, in those situations where impact prevention is not possible, knowing the site well enables an appropriate response to observed impacts.

These principles were clearly demonstrated during the course of the Rocky Mountain 1 (RM1) UCG test which involved both the characterization of the site to an extent unprecedented in UCG research and development and extensive groundwater monitoring. This evaluation proved fundamental to an understanding of operations as well as environmental protection, both during the active burn and in the post-burn monitoring phase. It is the intent of this paper to provide a brief history of UCG, a description of the UCG process, and an illustration of the role of hydrogeology in UCG as based on the experience in the RM1 test. The section on the RM1 test focuses on the hydrogeologic investigation portion of the environmental program and provides an analysis of the interaction of the process operations and the setting during the period from the fall of 1987 to the fall of 1988, the period that included active gasification ("burn") and the subsequent groundwater restoration activities.

DEVELOPMENT OF UNDERGROUND COAL GASIFICATION

Historical Perspective: 1868-1985

Consideration of in situ coal gasification began at least as early as 1868 (Arthur D. Little, Inc., 1972), and speculative interest continued in the U.S., England, and the U.S.S.R. until the 1930s when the first field test was conducted in the U.S.S.R. By the eve of World War II, gas was being produced from at least one UCG facility in the U.S.S.R., and, following the war, testing continued into the 1960s when emphasis shifted to conventional natural gas. The end of World War II saw the initiation of UCG research activities in a number of other countries, most notably Great Britain and the United States, although the scope of the Soviet effort has yet to be approached elsewhere. Unfortunately, little substantive information is available on Soviet UCG research and development.

The first field test in the U.S. was conducted at Gorgas, Alabama, by the U.S. Bureau of Mines in 1946, but it was not until the early 1970s that the UCG research and development effort in the U.S. was focused and accelerated (Figure 2, Table 1). Beginning in 1972, a variety of federal and private entities conducted a number of field tests generally designed to evaluate the UCG process as applied to low-rank coals in Texas, Wyoming, and finally Washington (Dockter, 1986). In the wake of the test (Tono I) that took place at Centralia, Washington, in 1983, the status of UCG in the U.S. could be summarized as follows:

Figure 2. *Locations of UCG tests in the conterminous United States during the period 1969 to 1991 (refer to Table 1).*

1) The process, when applied to subbituminous coals, could be controlled in such a manner as to efficiently produce acceptable quantities of low- to medium-grade gas.

2) Reverse "linking" was the best available method to enhance permeability and, thus gasification, but drilled linkage offered substantial promise, particularly as a means of linking effectively over substantial distances.

3) The use of horizontal process wells, either exclusively or in conjunction with vertical wells, offered the potential for both increased process efficiency and increased coal utilization.

4) The use of injected oxygen, rather than air, produced superior process results and yielded higher quality gas, though optimum steam to oxygen ratios remained to be determined.

5) The economics of UCG were becoming increasingly favorable as progress continued in refining and applying the process.

6) The future development of the technology was increasingly linked to identifying, characterizing, and effectively responding to the potential UCG-related environmental impacts.

The last point was particularly critical because it required addressing commonly held perceptions that UCG was a high-impact technology and that the impacts might very well be insurmountable. This view had arisen

Location	Sponsor	Experimental Designation	1965	1970	1975	1980	1985
Colonial Mine, KY	Gulf	--	□				
Hanna, WY	LETC (DOE)[1]	I		△			
		II-IA			△		
		II-IB			△		
		II-II			△		
		II-III			△		
		III			△		
		IV-A(a)			△		
		IV-A(b)			△		
		IV-B(a)			△		
		IV-B(b)			△		
Hoe Creek, WY	LLNL (DOE)	I			△		
		II			△		
		II			▲		
		II			△		
		III			△		
		III			▲		
Fairfield, TX	Basic Resources	--			○		
College Station, TX	Texas A&M	--			○		
Tennessee Colony, TX	Basic Resources[2]	--			○		
Reno Junction, WY	ARCO Coal Co.	--			△		
Rockdale, TX	Texas, A&M	--			○		
						○	
Princeton, WV	METC (DOE)	--			□		
Rawlins, WY	Gulf/DOE[3]	I			△		
		I			▲		
		II				▲	
Centralia, WA	LLNL (DOE)	Tono 1				▲	
Hanna, WY	DOE/GRI	Rocky Mountain 1					▲

[1] II and III used combination of air and O_2/steam.
[2] Steam/O_2 in final phase.
[3] I used combination of air and O_2/steam.

Coal	Injection Gas: Air	Injection Gas: O_2/Steam
Bituminous	□	
Subbituminous	△	▲
Lignite	○	

Table 1. *Summary of underground coal gasification tests in the conterminous United States, 1969 to 1991.*

because the tests conducted through 1983 were process-oriented and typically were designed, sited, and conducted without the benefit of detailed site-characterization and environmental-monitoring programs. The potential effects of site geology, including topography, structure (e.g., faults), stratigraphy, sedimentology, mineralogy, and hydrogeology (groundwater moving through earth materials) largely were ignored, and as a result, a number of environmental impacts occurred that were inherently major and perhaps serious enough to threaten further development of the technology. In view of this, the RM1 UCG test addressed the potential for UCG impacts as an integral part of test design.

Rocky Mountain 1 Field Test

The RM1 test, near Hanna, Carbon County, Wyoming, occurred adjacent to the site of the earlier Hanna UCG tests (Figure 2) and was designed as a simultaneous test of two UCG technologies, controlled retracting injection point (CRIP) and the extended linked well (ELW), in the Hanna No. 1 coal seam, which lies approximately 107 m (350 ft) beneath the land surface (Bloomstran et al., 1987). Specifically, the test program was designed to respond, in large part, to critical process and environmental research needs through the following:

1) the use of horizontal holes as part of the two innovative UCG technologies being evaluated in an effort to increase both process efficiency and coal utilization;

2) the evaluation of the effects of variation in the ratio of injected steam and oxygen on product-gas quality;

3) the evaluation of the effects of oxygen injection rate variations on product-gas quality;

4) the collection of data that would be useful in evaluations of the economics of UCG; and

5) the conduct of an extensive environmental research and monitoring program that would involve a) the detailed pre-test characterization of the site, b) the detailed evaluation of UCG-related environmental impacts, and c) the incorporation of environmental considerations in the test design to prevent or mitigate these impacts, particularly those involving groundwater.

Furthermore, the RM1 program was designed to be managed and conducted by a multidisciplinary group of scientists and engineers representing a variety of entities. These organizations included the U.S. Department of Energy, the Gas Research Institute, Amoco, Union Pacific Resources, the Lawrence Livermore National Laboratory, Stearns-Roger, Energy International, the Project Construction Corporation, the Western Research Institute, and the University of North Dakota Energy and Environmental Research Center.

The RM1 UCG environmental program consisted of site characterization, site monitoring, and groundwater restoration activities. The hydrogeologic investigation portion of the program, as integrated into the RM1 test design, consisted of pre-test site characterization and instrumentation, followed by detailed monitoring of the character of groundwater at the site during and after the test burn. Pre-test characterization of the RM1 site, including the characterization of groundwater occurrence, flow, and quality, as well as the characterization of the flow matrix, is reviewed by Mason et al. (1987a and b). Groundwater quality during and shortly after the RM1 test burn is discussed in Mason et al. (1987a), whereas Moody (1990) provides an evaluation of the effects over the period of the test and the subsequent 2-yr period. Post-test groundwater-quality data continue to be collected and evaluated.

In order to minimize impacts to groundwater, the results of the hydrogeologic characterization were used to determine operational parameters and the formulation of the groundwater restoration plan (Riley et al., 1987). These activities, carried out in the period following the burn, involved venting the cavities, allowing them to cool and fill with groundwater, then pumping out the cavities, and finally treating and disposing of the contaminated water. These activities, summarized and evaluated in Boysen et al. (1990) and Moody (1990), affected the hydrogeologic system of the site.

METHODOLOGY

Water-level measurements were made on a quarterly basis by Western Research Institute (WRI) personnel before the test in the 24 groundwater-monitoring wells at the RM1 site and by WRI and Project Construction Corporation personnel on a more frequent basis during and after the test. The construction and installation of these wells are discussed in detail by Mason et al. (1987b) (Figure 3). The collected data were reduced by North Dakota Mining and Mineral Resources Research Institute personnel to yield an evolutionary sequence of potentiometric surface maps for the site depicting variations in hydraulic head in the Hanna No. 1 coal seam in response to test and post-test events. This sequence provides a history of groundwater flow conditions at the RM1 site during that period. A comprehensive hydrograph incorporating the water-level data from most of the groundwater-monitoring wells also was developed. Figure 4, a schematic representation of the comprehensive hydrograph, provides a history of well responses from January to September of 1988 and can be used to establish relationships to selected operational events. A number of potentiometric surface maps, along with the schematic hydrograph, will be used to illustrate the hydrogeologic effects of significant RM1 events and will serve as the basis for the following discussion.

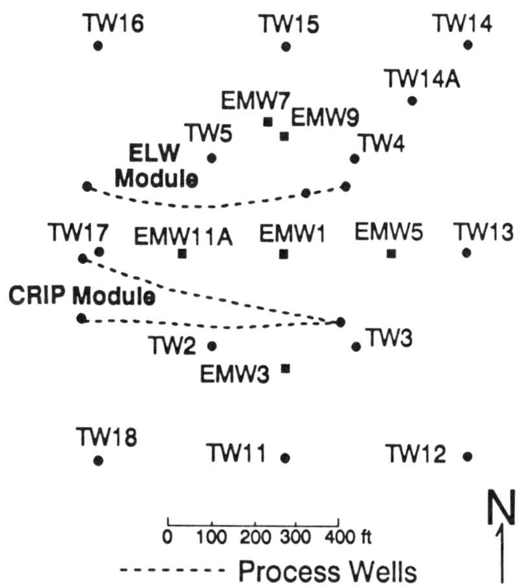

Figure 3. Location of groundwater-monitoring instrumentation and the extended linked well (ELW) and controlled retracting injection point (CRIP) modules at the Rocky Mountain 1 UCG test site (modified from Mason et al., 1987b). The dashed lines indicate the locations of horizontal wells which define the working areas of the modules.

RESULTS AND DISCUSSION

Potentiometric Surfaces

The Hanna No. 1 Coal is a confined aquifer in the test area, and contours on potentiometric surface maps represent lines of equal hydraulic head for the Hanna No. 1 coal seam. Because groundwater flows from regions of higher head to regions of relatively lower head, these maps can be used to establish flow patterns at the time that a set of water levels is taken. On this basis, pre-test water-level data collected by WRI indicated a gentle head gradient and, therefore, groundwater flow to the northwest across the RM1 site in the Hanna No. 1 aquifer (Figure 5). This flow direction was along the strike of the northeast-dipping coal seam. The baseline hydraulic head was taken to be the value at the approximate center of the site, or about 2,906 m (6,910 ft) above mean sea level (Figures 4 and 5).

Initiation of the burn in early November of 1987 resulted in changes in the baseline areal flow system and, therefore, the original flow pattern in response to the effects of the local RM1 hydrogeologic anomaly that developed because of test-related changes in conditions at the gasification centers. During the start-up phase of the test, operating pressures at the gasification points were substantially higher than the hydraulic head in the surrounding area. At the same time, the groundwater-monitoring wells in the southwest quadrant of the site also were characterized by high pressures, as well as by the presence of product gas. Apparently, the high pressures at the gasification centers initiated movement of product gas updip (that is, to the southwest through fractures in the coal aquifer) and the gas, being less dense than water, displaced the

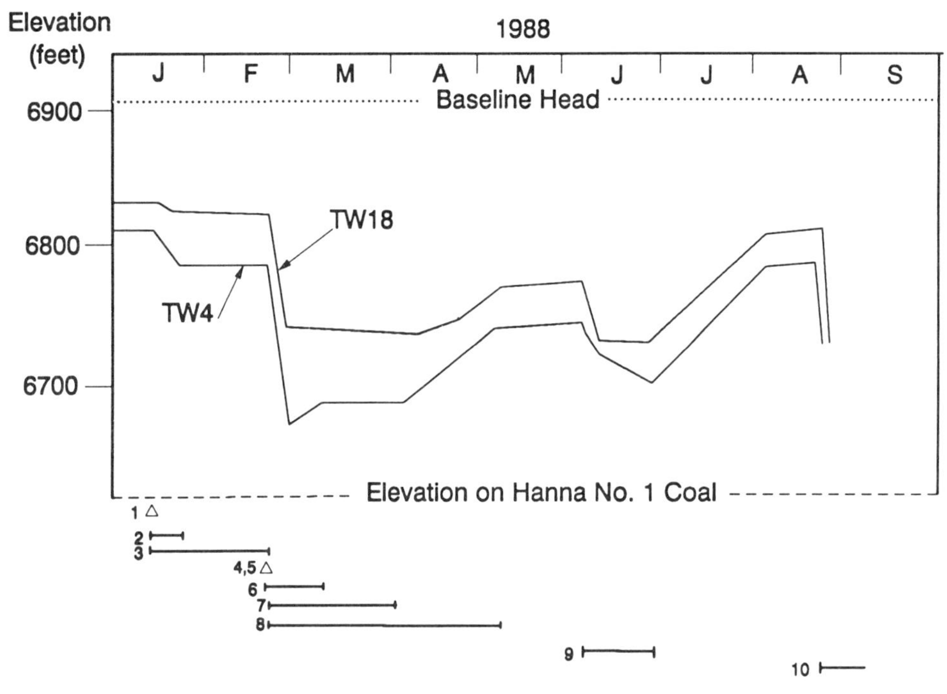

Figure 4. Schematic hydrograph for the Rocky Mountain 1 UCG test and post-burn monitoring period, January to September of 1988 (based on Boysen et al., 1990). 1) 1/16 -- end of ELW gasification, 2) 1/16 - 1/25 -- Steam flushing and controlled venting of the ELW cavity, 3) 1/16 - 2/26 -- Controlled venting of the ELW cavity (pressure maintained at approximately 620 kPa (90 psia)), 4) 2/26 -- End of CRIP gasification, 5) 2/26 - 2/27 -- Simultaneous, controlled, depressurization of CRIP and ELW UCG cavities, 6) 2/26 - 3/7 -- Steam flushing and controlled venting of the CRIP cavity, 7) 2/27 - 3/2 -- Venting of ELW cavity, 8) 2/27 - 5/12 -- Venting of CRIP cavity, 9) 6/14 - 7/1 -- Venting of CRIP cavity using groundwater-restoration well, 10) 8/26 -- Pumping initiated for cavity cleanup.

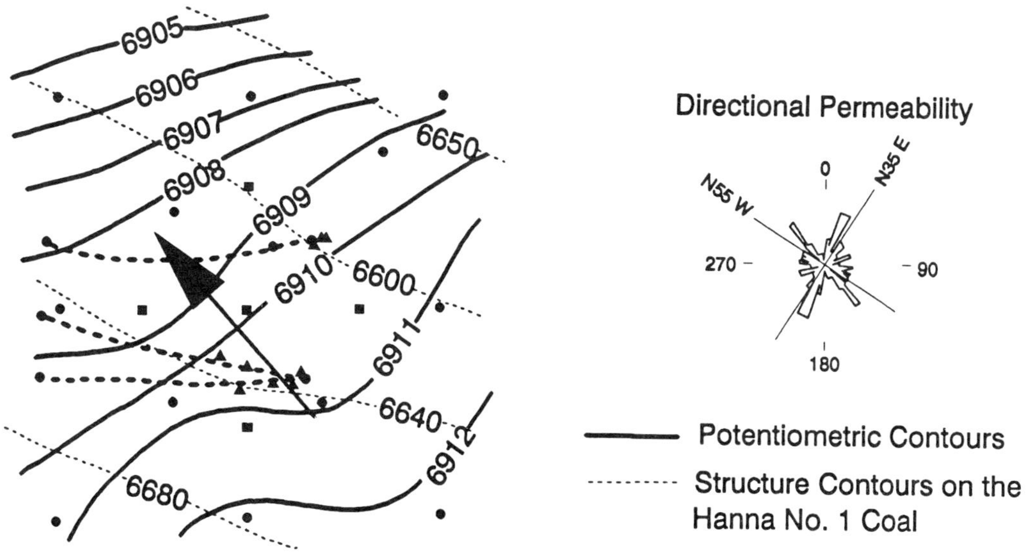

Figure 5. Baseline potentiometric surface (October 1986), structure, and fracture permeability for the Hanna No. 1 Coal (based on Mason et al., 1987b). Arrow indicates direction of groundwater movement in the Hanna No. 1 Coal. Directional permeability from Glass and Roberts (1980). Dashed lines for gasification modules are as in Figure 3.

water at the top of the coal aquifer (Figure 6). As migration progressed, gas was detected in monitoring wells screened in the full interval of the coal, resulting in the observed conditions. Both the dip of the coal and its highly fractured nature had been established by WRI during the pre-test characterization of the site. Given this knowledge of the geologic setting and the pressure relationships that were known to exist, this episode of gas migration was a predictable if, in fact, unanticipated event, which highlights the importance of adequate site characterization.

The measurement of water levels in many of the manually monitored groundwater wells prevented the development of potentiometric surface maps for this event. However, some indication of aquifer condition can be gained from Figure 7, which allows the comparison of cavity injection pressures and water levels in EMW7, a groundwater-monitoring well close to the ELW system (Figure 3) which was instrumented by WRI to record the water level on a continuous basis. In general, the water-level and pressure records included in Figure 7 correlate quite closely. Note in particular the response in EMW7 to ELW shutdown in mid-January. This correspondence is indicative of good hydraulic interconnection in the cavity area, a condition that was not unexpected in view of the known extent of fracturing in the coal aquifer. Indeed, good hydraulic communication was not limited to the cavity area, but rather characterized the entire site as indicated by the rapid site-wide response of the hydrogeologic system to a variety of test and post-test events that followed.

In late November, after start-up, operating pressures were reduced and maintained at less than 80% of the baseline hydrostatic pressure in the Hanna No. 1 aquifer in the area of gasification. This was in accordance with the established operations plan. In response to this reduction in pressure, the gas "bubble" related to start-up contracted to the cavity areas. At the same time, a head gradient became established toward the developing gasification cavities, resulting in groundwater flow in those directions (Figure 8).

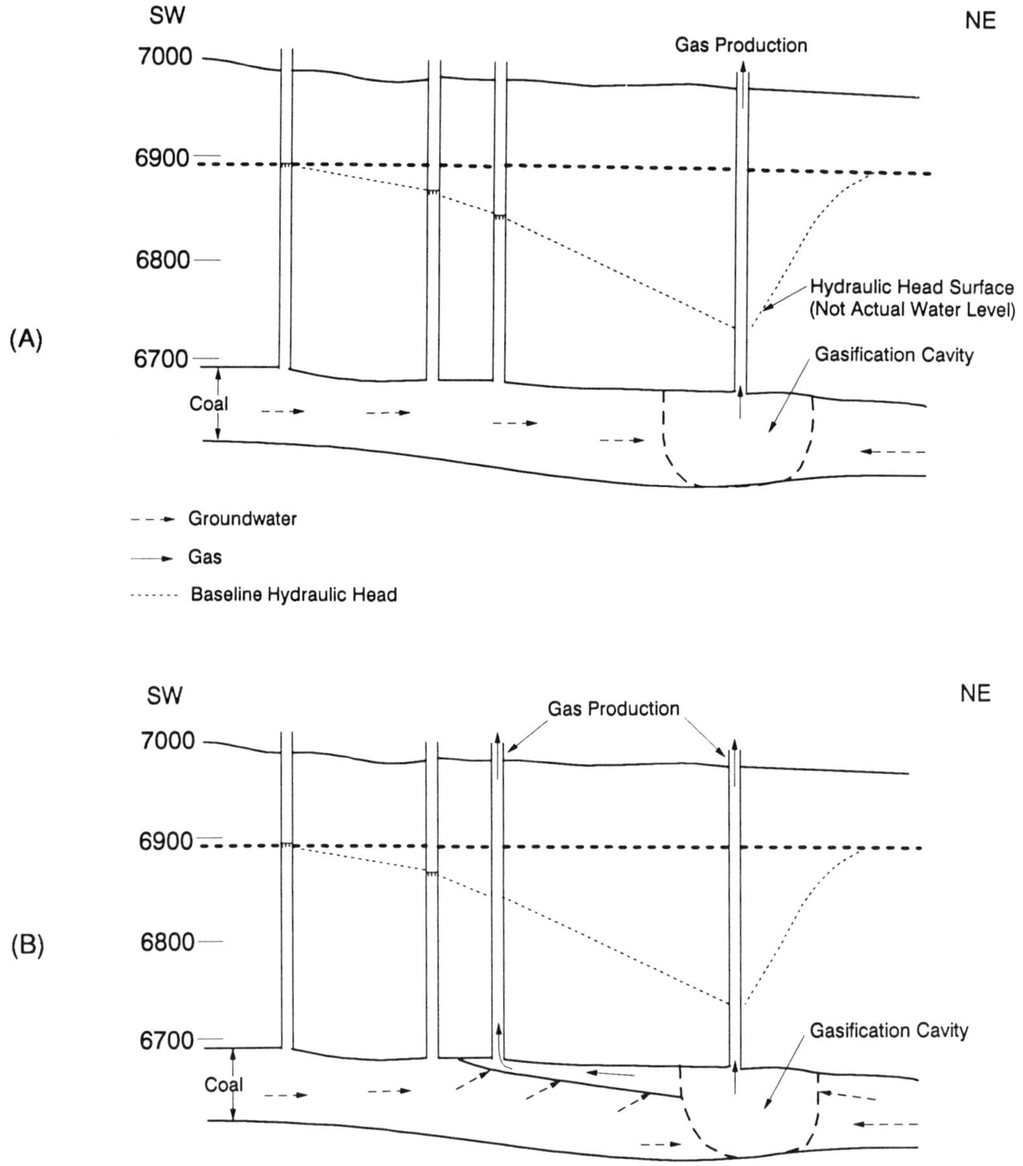

Figure 6. *Schematic diagram of hydrogeologic relationships under high- and low-pressure operating conditions during the Rocky Mountain 1 UCG test. A) Under typical operating conditions (cavity pressure less than hydrostatic), groundwater moves toward the cavity where it is consumed in the process and converted to steam, resulting in a loss of head. B) Under elevated operating pressures, product gas migrated updip along fractures and was detected in groundwater-monitoring wells in the southwest portion of the site.*

A select sequence of potentiometric surface maps (Figures 8 and 9) can be used to document the formation, development, and eventual disappearance of the cone of depression in the potentiometric surface. This feature represents a hydraulic head gradient in the coal aquifer that was maintained throughout the test, and for a substantial time after the test, by keeping the operational gasification cavity pressures lower than the fluid pressure in the aquifer distant from the cavities. Significant aspects of the head-contour maps include the elevation differences between the innermost and outermost contours of the cone of depression, the shape of the zone, and the locus of the bottom of the cone.

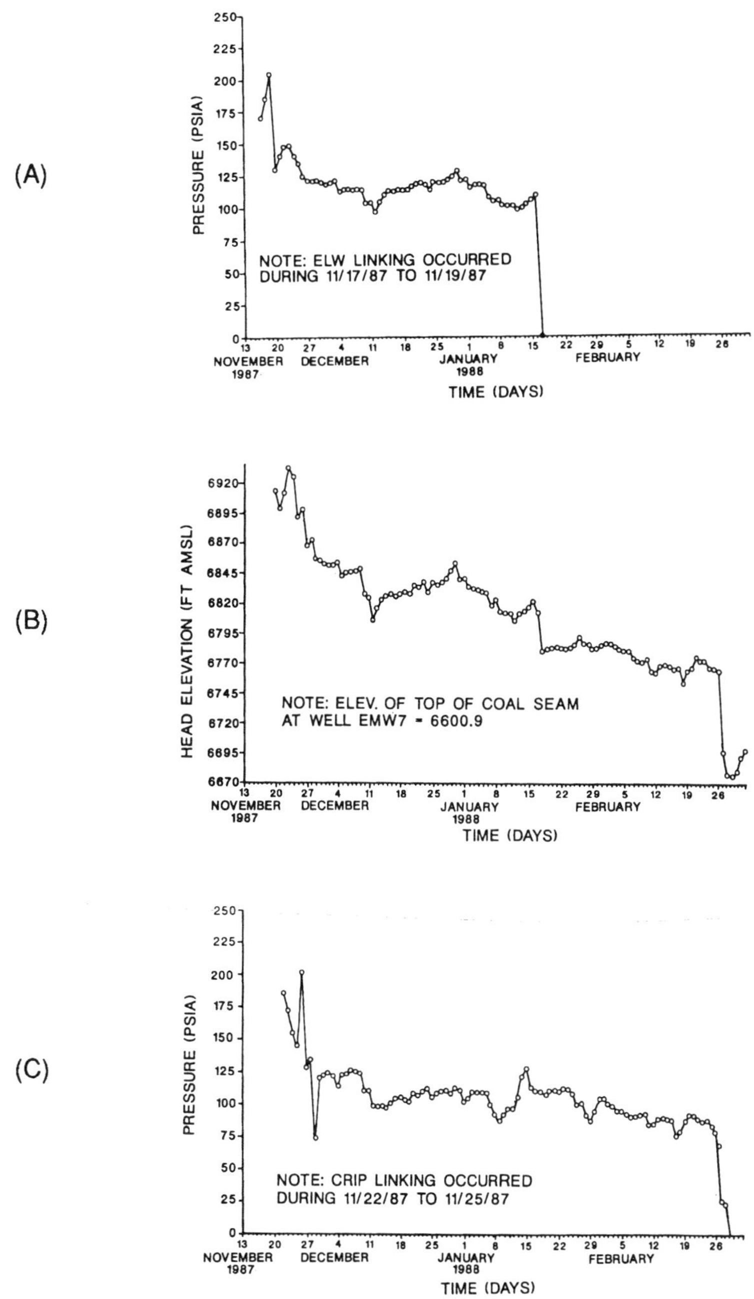

Figure 7. *Comparison of test injection pressures and hydraulic head at the Rocky Mountain 1 UCG test site (modified from Moody, 1990): A) ELW injection pressures, B) hydraulic head in groundwater-monitoring well EMW-7, and C) CRIP injection pressures.*

The general character of the potentiometric surfaces representing the gradual development of the cone of depression during the early stages of the test is illustrated by Figure 8a. The operation of the two gasification centers at reasonably comparable pressures is reflected in the inclusion of both cavities in the deepest portion of the cone. Termination of the ELW burn in mid-January and subsequent venting resulted in even further deepening of the cone of depression and a shift of the locus to the ELW cavity (Figure 8b).

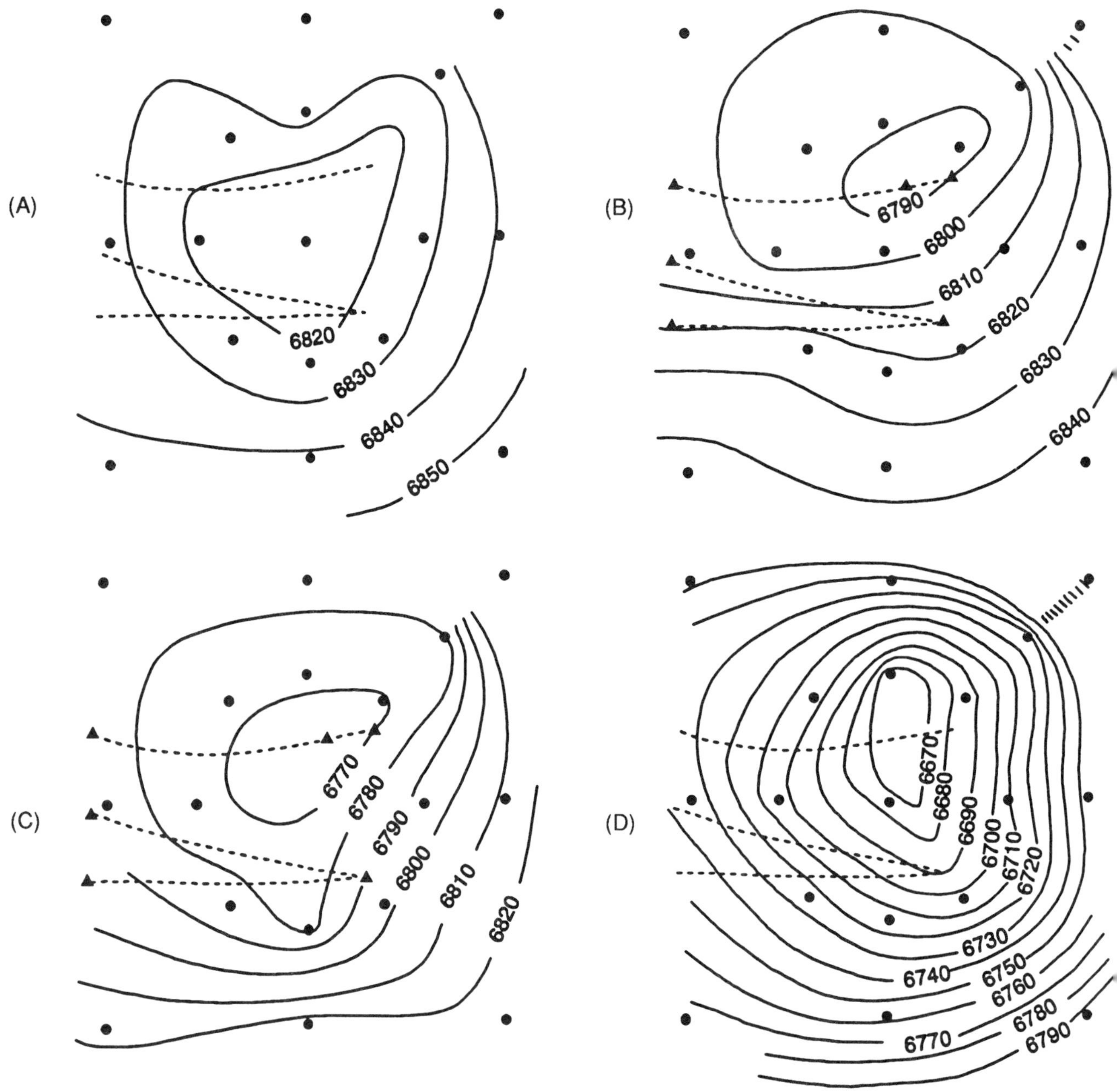

Figure 8. *Potentiometric surface maps indicating hydraulic head at the Rocky Mountain 1 UCG test site (elevations are in feet above mean sea level): A) January 5, 1988, B) February 2, 1988, C) February 24, 1988, and D) February 27, 1988. Dashed lines for gasification modules are as in Figure 3.*

At the same time, indications of a northeast-southwest trend coinciding with one of the two major fracture directions determined during site characterization also began to be manifested (Figure 8b). This correspondence of drawdown and fracture trends, which at times was very strong, provides further support for the control of fluid flow at the RM1 site by fractures in the coal aquifer.

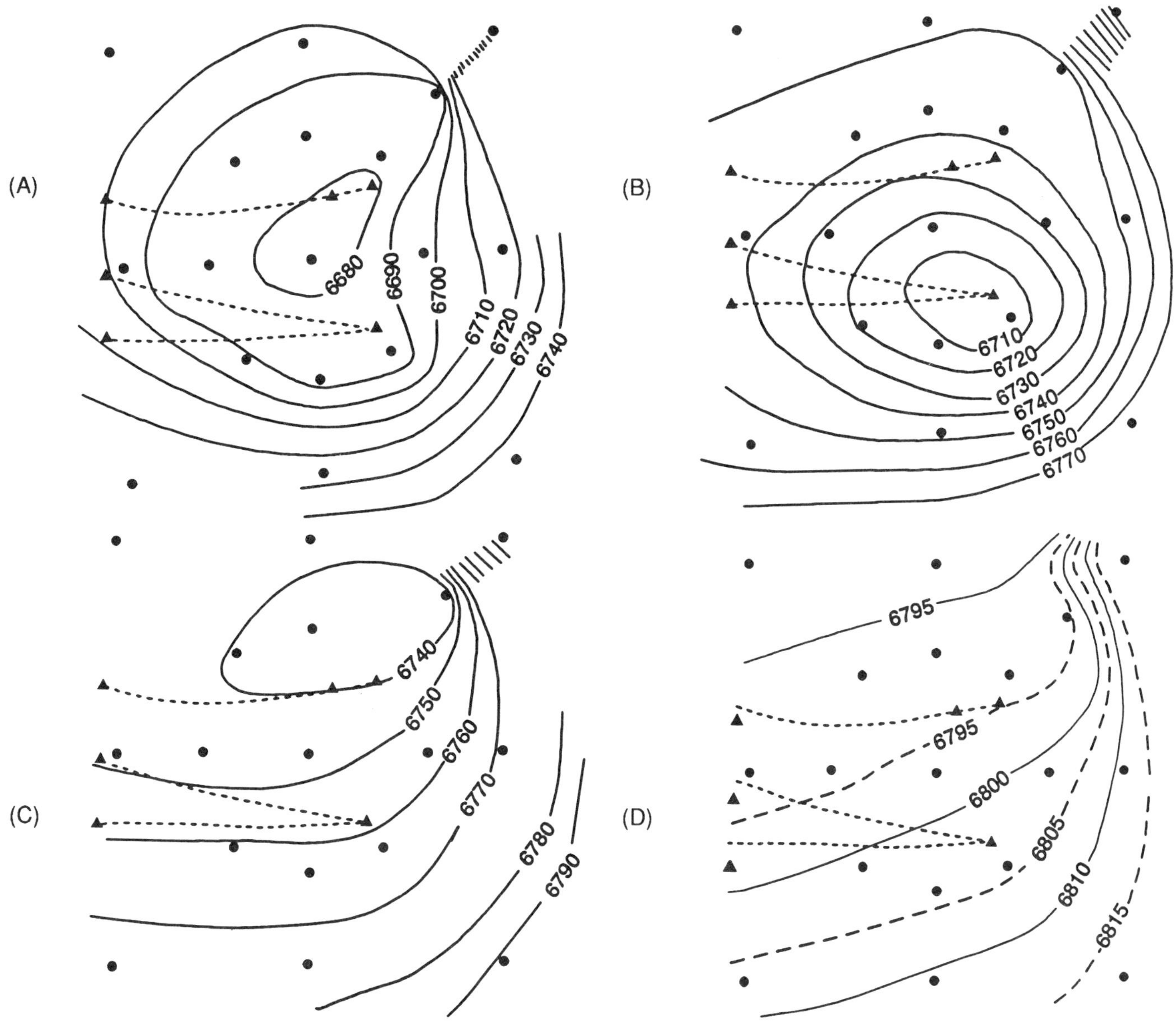

Figure 9. *Potentiometric surface maps indicating hydraulic head at the Rocky Mountain 1 UCG test site (elevations are in feet above mean sea level): A) April 14, 1988, B) June 15, 1988, C) July 20, 1988, and D) August 2, 1988. Dashed lines for gasification modules are as in Figure 3.*

The venting and cooling process that followed ELW shutdown lasted until February 26, 1988. During that time, venting was controlled so that the pressure in the ELW cavity was held at or very near CRIP-operating pressure in order to prevent or minimize movement of contaminants between the cavities. Throughout this period, the focus of the cone, which continued to deepen slowly, remained over the ELW cavity (Figure 8c) as a result of water loss as steam during venting.

By early February, a steep gradient had begun to develop between monitoring wells TW14 and TW14A in the northeast quadrant of the site (Figure 8b). This gradient continued to steepen throughout the test and during the period of cavity venting and cooling that followed (Figure 8c and d). Pre-test site characterization by WRI (Mason et al., 1987a and b) revealed an offset in the Hanna No. 1 coal seam and accompanying units in that area, although the orientation of the fault never was totally resolved. The hydraulic conductivity of this feature was not adequately determined, so it has been impossible to definitively establish whether it enhanced water movement or impeded flow; either situation could have resulted in the observed steepening gradient. However, the relatively low cavity-inflow rates that were observed tend to suggest that this feature functioned as a flow barrier.

The test burn was terminated on February 26, 1988, with cessation of oxygen injection to the CRIP system followed by the depressurization of both cavities by venting. In response to the simultaneous venting of the two cavities, the head levels throughout the site dropped significantly, which resulted in a much deeper cone of depression, although it remained focused on the ELW cavity (Figure 8d). The water levels at this time were the lowest yet recorded and were as close as 16 m (50 ft) to the top of the coal aquifer. This dramatic drop in hydraulic head occurred because the cavities were depressurized to or near to atmospheric pressure, thus creating a much greater hydraulic gradient from the undisturbed portions of the coal aquifer toward the cavities.

Cavity depressurization and the accompanying decrease in water head levels was followed by a short period during which water levels generally rose at gradually decreasing rates before starting to drop very slowly once again. This slow decrease in levels persisted through mid-April, and its effect on the potentiometric surface is illustrated in Figure 9a. At the cessation of venting, the hydraulic head began to increase at a constant, if moderate, rate. In response to this rise in hydraulic head, which signaled the initiation of the recovery of the hydrogeologic system in the area, the cone of depression became shallower through early May, after which it remained reasonably stable into early June. This latter period of apparent equilibrium will be discussed further in the next section.

Post-burn drilling associated with the installation of pumps for "cavity-cleanup" activities occurred in early June 1988. When the drilling associated with these activities penetrated the CRIP cavity, a substantial release of steam resulted. This "geyser" was accompanied by another significant, rapid reduction in hydraulic head over the site, which resulted in a deepened cone of depression once again, the focus of which now shifted to the CRIP cavity (Figure 9b). Over a period of several weeks following the initial drop, water levels generally fell at a gradually decreasing rate, although in some areas, levels began to rise very slowly toward the end of this period. Significant venting stopped about July 1, at which time head levels began to rise at a rate higher than any previously observed, resulting in a rapidly shallowing cone of depression (Figure 9c). This trend indicates that, once again, the hydrogeologic system was in recovery. The fact that the rate of recovery was notably higher apparently indicates that recovery was no longer impeded by cavity-related phenomena such as the compression of a large gas bubble.

Water levels continued to rise steadily, and by August the cone of depression had disappeared. In its wake, flow off-site to the northwest was reestablished (Figure 9d). Head levels remained approximately 30 m (100 ft) lower than the baseline potentiometric surface (Figure 4).

RM1 Hydrograph

A detailed hydrograph incorporating available hydraulic head data from most of the groundwater-monitoring wells at the RM1 site was constructed as the data were received. A schematic representation of that hydrograph illustrating significant changes associated with the test and post-test events is included here (Figure 4) to serve as a reference for a discussion of those changes based on consideration of the more detailed hydrograph and to supplement the preceding discussion which was based on the potentiometric-surface maps.

In general, the detailed hydrograph revealed that, with the exception of Monitoring Well TW14 which was on the far side of the apparent flow barrier discussed above, and EMW8, which was completed in the overburden, the remaining monitoring wells, all screened in the coal, quickly responded to system stimuli in basically the same manner and essentially at the same time, with only small differences in the magnitude of the response. Likewise, the recovery patterns and rates generally were similar for those wells. This site-wide phenomenon, which was mentioned earlier, provides particularly strong evidence for good hydraulic communication throughout the RM1 site and corresponds to what would be expected in view of the highly fractured nature of this coal aquifer system.

As has been noted, there was a gradual lowering of the water levels across the site during the burn as water flowed toward the cavity and became involved in the gasification process. Superimposed on this general test-long pattern of decline was the effect of the shutdown and venting of the ELW module in mid-January. When this took place, the water levels in the monitoring wells closest to the ELW cavity (signified by TW4 on Figure 4) all began to drop at an accelerated rate relative to the CRIP-related wells (represented by TW18 on Figure 4), resulting in the two distinct sets of individual hydrographs.

When the test phase of the RM1 program was completed in late February, water levels dropped as much as 30 m (100 ft), as close as 16 m (50 ft) to the top of the coal aquifer. As discussed previously, this extreme drop in water levels was the result of venting the cavities to atmospheric pressure subsequent to shutdown, thereby creating a relatively larger hydraulic gradient from the undisturbed portions of the coal aquifer toward the burn cavities. A brief period followed during which the levels increased slightly, before beginning slowly to fall in response to cavity venting and cooling. This trend persisted until mid-April, at which time venting effectively ceased, and the head began to rise significantly across the site. During the second week in May 1988, this rate of rise began to level off, and a period followed that was characterized by reasonably stable head over the site.

Core samples of thermally altered material were obtained from immediately east of the ELW cavity in early June, at which time large quantities of water were introduced into the system from drilling activities. The result can be seen in a general, slight rise in water levels at that time. Immediately following this coring, a hole was drilled for the installation of a pump in the CRIP cavity in preparation for cavity-water cleanup activities. When the cavity was penetrated, a "geyser" resulted, which prompted an immediate and significant drop in head. Although the magnitude of this reduction in head was less than that related to the start of venting and cooling activities, the water levels once again fell as close as 16 m (50 ft) to the top of the coal aquifer.

Substantial quantities of steam continued to be emitted from the CRIP pump hole for several weeks, during which time the head generally continued to fall at increasingly slower rates, although slow rises eventually began in some areas. When significant steam release ceased, the head began to rise at a relatively rapid, constant rate. This pattern was significantly different than the pattern that was observed following the initial venting and cooling period, which was characterized by a slower rate of head recovery that gradually leveled off. This earlier pattern almost certainly represents the effects of gases that were trapped in the cavities and were increasingly compressed by inflowing groundwater until they could be compressed no further, resulting in water levels that remained more or less static. The presence of these gases trapped under high-pressure was demonstrated by the geyser eruption observed when the CRIP cavity was penetrated after head stability had persisted for several weeks. During the "post-geyser" period, the rate of head rise was greater, apparently because the trapped gas largely had been vented, thus allowing the cavities to finally fill with water in an unhindered manner. Under those circumstances, the only space available to inflowing groundwater was the actual well bores of the monitoring wells, and, consequently, water levels rose faster.

The period through late July 1988 coincides with the disappearance of the cone of depression as cavity-filling neared completion. Subsequent cavity cleanup pumping operations revealed that the hydrogeological system responds quickly to pumping stress and that it recovers fully. Water quality, although not the specific subject of this paper, has essentially returned to baseline levels, indicating that contaminants essentially were steam-stripped from the cavity during venting and cooling (Moody, 1990).

CONCLUSIONS AND RECOMMENDATIONS

A UCG system is typified by a complex set of interactions between the process and the hydrogeologic environment because the environment is the reactor. Process considerations cannot be separated from environmental considerations. Unlike conventional surface coal gasification facilities wherein the reactor environment is totally engineered and controlled, in situ gasification is controlled in large part by the geologic and hydrogeologic character of the setting in which it is placed. As a result, the design, conduct, and evaluation of any UCG operation must evolve from a fundamental knowledge of the potential gasification site and an understanding of the process-environment interactions that may result. It follows, therefore, that thorough site characterization is absolutely essential for successful UCG implementation.

Groundwater is not used only in the UCG-related chemical reactions, but also is converted to steam by the available heat, both processes together resulting in the imposition of an anomaly on the baseline hydrogeology which, over the course of its development, approximates a pumping operation. Observations made during the RM1 test suggest that not only is the volume of water used substantial, but the resulting impacts on the hydrogeologic system in the fractured coal medium are felt rapidly over a substantial area. It is estimated that approximately 2% of the water available in the coal-seam aquifer was consumed during the RM1 test. Although this is a small percentage of the total available, it represents a sizeable volume of water, the utilization of which substantially depleted the hydraulic head on the site and affected head beyond the permit boundaries to a lesser, though no less significant, extent. These observations, which are derived from a UCG test that was small-scale and of short duration, have important implications, particularly for the longer-term, larger-scale UCG tests and commercial UCG operations which eventually will be conducted. Specifically, it is evident that great care must be taken to design, site, and conduct such operations so as to ensure an adequate supply of water in order to maintain a functional hydraulic head without adversely affecting the water resources beyond the permit area.

The strategy employed during the RM1 test to maintain a head gradient (and, therefore, groundwater flow) toward the cavities by keeping operating pressures at some fraction of the hydrostatic pressure in the coal-seam aquifer was reasonably successful. It must be kept in mind, however, particularly in longer-term operations, that the hydrostatic pressure at the gasification center is not fixed, but declines as groundwater is utilized. Consequently, in any UCG operation that attempts to maintain conditions similar to RM1, it will be necessary to provide for ongoing monitoring of the constantly changing hydrostatic head so that the operating pressure can be adjusted accordingly. This could mean that UCG could significantly deplete the head in coal seams where there is little recharge and/or minimal hydraulic conductivity, which underscores the need for adequate hydrogeologic characterization before UCG sites even are chosen. In settings where variations in these characteristics are found to occur, it should be possible to site and conduct UCG operations so as to take the best advantage of recharge and flow systems. Nevertheless, in longer-term, large-scale tests or commercial UCG operations, artificial recharge may have to be considered.

The pressure at which the UCG reactor operates is also a primary consideration in controlling the movement of product gases and associated potential groundwater contaminants. If the system pressure is maintained at a level that results in the confinement of the gases to the approximate vicinity of the cavity, there will be minimal potential for contaminant transport by means of an expanding gas "bubble". In the RM1 case, maintenance of the pressure gradient toward the cavities for most of the test and for an extended period after the test seems to have had this result. However, during this short test initiation phase when the operating pressure was allowed to exceed the hydrostatic pressure, product gases predictably migrated updip through the coal fracture system. Although the maximum extent of this migration was not determined, there is some evidence that limited contamination did occur before the lowering of the operating pressure caused the gas "bubble" to recede.

Gas migration and groundwater flow are components of the multiphase flow that moves through the confined flow system in response to hydraulic head. Both the hydraulic gradient and coal structure, therefore, must be considered in the design, siting, and conduct of UCG operations in order to optimize gas production, while minimizing contaminant production and migration. The effects of gas migration in an updip direction may be mitigated by updip gasification, which would allow the consumption of any contamination transported by prior updip movement of gas. Gasification in a down-dip direction, on the other hand, although promoting reduced coal recovery and the potential for coal override, would move away from zones contaminated by any updip gas migration that might occur. The post-operation remediation of such zones, though not a certainty, would, if required, increase the costs of that operation and, perhaps, influence its long-term acceptability.

Both pre-burn aquifer air pressurization tests and hydrogeologic observations made during and after the test indicated that good hydraulic communication existed between all points in the RM1 UCG system. This situation can be attributed to the extensive directional fracturing of the coal aquifer, which provided a pervasive network of flow paths that were neither homogeneous nor isotropic. A system such as this (i.e., one in which flow is controlled by fracture permeability) does not readily allow the application of numerical techniques to predict groundwater movement simply because the assumptions of isotropy and homogeneity inherent in, and required by, presently available numerical solutions do not reflect the reality of the situation. Because fracture flow is typical of coal aquifers, which in turn are typical UCG targets, efforts to apply or modify existing groundwater flow models to UCG systems generally are inappropriate. Emphasis should be placed on the conceptual interpretation of the available real data collected during site characterization to select reasonable UCG sites and then design operations that are appropriate to those sites. Likewise, the same process should form the basis for interpreting the data and observations resulting from those operations.

The hydrogeologic system at and near the RM1 site also responded quickly to operations-related hydraulic stimuli. Moreover, the system typically readjusted quite rapidly, as was demonstrated repeatedly in responding to test-related stimuli such as operating-pressure changes. It generally was observed that, within one day, the effect of a particular event of that type was felt across the RM1 site and was reflected back in a manner somewhat analogous to resonance damping. The specific nature of these responses could be determined more precisely if continuous recording, downhole pressure transducers were used as an integral part of the groundwater-monitoring program at UCG sites. An entire well field could be networked into one data logger or microcomputer, and with the appropriate software, data could be continuously collected and reduced. Although the basic products would continue to be potentiometric surface maps and hydrographs, the increased detail of the data set and the timeliness of its treatment would enhance an operator's ability to utilize that data in the day-to-day decision-making processes focused on the achievement of efficient, economical, and environmentally acceptable in situ gasification in any setting and on any scale. In the same way, continuous data collection and reduction would allow for better control of the scheduling and conduct of post-gasification restoration activities, for which a timely knowledge of the extent and rate of cavity-filling by groundwater is very important.

Other on-site monitoring techniques which may prove valuable include microseismic arrays and geothermal sensors. Seismic information could be used to track locations of horizontal boreholes during drilling and also could help refine and understand UCG cavity development through the detection of events such as rock falls in the cavity that produce vibrational signals. Geothermal monitoring, used for the first time in the RM1 test, provides data useful in delineating water movement in the cavity area (Gosnold et al., 1988).

REFERENCES

Arthur D. Little, Inc., 1972, A current appraisal of underground coal gasification: Cambridge, Massachusetts, Report ADL-C-7367-1, 280 p.

Bloomstran, M.A., Olson, T.E., and Davis, B.E., 1987, Design, operation, and test plan for the Rocky Mountain 1 underground gasification test, in J.W. Martin and S.P. Barone, eds., Proceedings of the Thirteenth Annual Underground Coal Gasification Symposium, Laramie, Wyoming, August 24-26, 1987: U.S. Department of Energy, Report DOE/METC-88/6095, p. 14-22.

Boysen, J.E., Covell, J.R., and Sullivan, S., 1990, Results from venting, flushing, and cooling of the Rocky Mountain 1 UCG cavities: Chicago, Illinois, Gas Research Institute, Report GRI 90/0156, 75 p.

Daly, D.J., Groenewold, G.H., and Schmit, C.R., 1987, The evolution of the design of environmental research associated with UCG, in J.W. Martin and S.P. Barone, eds., Proceedings of the Thirteenth Annual Underground Coal Gasification Symposium, Laramie, Wyoming, August 24-26, 1987: U.S. Department of Energy, Report DOE/METC-88/6095, p. 309-316.

Dockter, L., 1986, Underground coal gasification in the United States: An overview, in Proceedings of the Twelfth Annual Underground Coal Gasification Symposium, Saarbrucken, Federal Republic of Germany, August 24-28, 1986: U.S. Department of Energy, Report DOE/FE/60922-H1, p. 2-13.

Glass, G.B., and Roberts, J.T., 1980, Coals and coal-bearing rocks of the Hanna coal field, Wyoming: Geological Survey of Wyoming, Report of Investigation 22, 43 p.

Gosnold, W.D., Beaver, F.W., and Schmit, C.R., 1988, Groundwater temperature at the Rocky Mountain 1 underground coal gasification site, Carbon County, Wyoming, in T.C. Bartke and V.L. Hill, eds., Proceedings of the Fourteenth Annual Underground Coal Gasification Symposium, Chicago, Illinois, August 15-18, 1988: U.S. Department of Energy, Report DOE/METC-88/6097, p. 263-273.

Mason, J.M., Oliver, R.L., and Moody, C.G., 1987a, Geology and groundwater hydrology of the proposed Rocky Mountain 1 underground coal gasification site, Hanna, Wyoming, in J.W. Martin and S.P. Barone, eds., Proceedings of the Thirteenth Annual Underground Coal Gasification Symposium, Laramie, Wyoming, August 24-26, 1987: U.S. Department of Energy, Report DOE/METC-88/6095, p. 317-333.

Mason, J.M., Oliver, R.L. Schrieber, J.D., Moody, C., Smith, P., and Healy, M.J., 1987b, Geohydrology of the proposed Rocky Mountain 1 underground coal gasification site, Hanna, Wyoming -- Volume 1: Denver, Colorado, United Engineers and Constructors, Inc., Stearns-Roger Div., Unpublished Report, 100 p.

Moody, C.G., 1990, Changes in groundwater quality and subsurface hydrology: Chicago, Illinois, Gas Research Institute, Report GRI 90/155, 161 p.

Riley, C.J., Masson, C.A., Zbranek, V., and Covell, J.R., 1987, Groundwater restoration program for the Rocky Mountain 1 UCG test, in J.W. Martin and S.P. Barone, eds., Proceedings of the Thirteenth Underground Coal Gasification Symposium, Laramie, Wyoming, August 24-26, 1987: U.S. Department of Energy, Report DOE/METC-88/6095, p. 334-342.

Thorsness, C.B., and Creighton, J.R., 1983, Review of underground coal gasification experiments at Hoe Creek, in W.B. Krantz and R.D. Gunn, eds., Underground coal gasification: The state of the art: American Institute of Chemical Engineers, Symposium Series, v. 79, No. 226, p. 15-43.

Chapter 5
Mining

PREFACE

The largest single coal resource extraction method and initial means by which coal resources are utilized is the mining process. It has been such for a long, long time, and there is no reason to doubt that it will remain so for quite a while to come. As noted by Platt in the Introduction to this book, the 910,000,000 tonne/yr (1,000,000,000 ton/yr) production (by mining) mark was reached in the U.S. in 1990. Whereas this level may not be sustainable forever, the number is an indication of the importance of coal mining (and consumption) to the U.S. and world energy supply in general.

Because underground mining still is one of the most hazardous professions, much applied coal research and mine site work is related to identifying and mitigating hazardous geological conditions. Such conditions as roof falls, floor heave, and water and gas inflows also can severely impact mine productivity and equipment. The overview paper by Weisenfluh and Ferm is aimed at providing a review and case study of how coal depositional environment influences mine safety and productivity.

Papers by Sames and Moebs and by Greb follow through on this theme by providing examples from the Eastern U.S. of roof control and hazard recognition through application of geological concepts. Structural geology and its effects on mining are the basis for the papers by Hucka and by Molinda and Ingram. Nelson et al. provide a case study from the Eastern U.S. of how both depositional and structural geology impact, and can be planned for in, a longwall mine.

Stress and its effects on mining are brought up first by Sames and Moebs. Collins carries the concept further by discussing stress in a mine from a mine operator's practical point of view. His suggestions should be useful to small and large operators alike. Nelson and Bauer go on to discuss techniques for dealing with stress in a coal field known for residual tectonic stress-related problems, the Illinois Basin.

Finally, what happens after you remove the coal? Eventually, the mine will at least partially, if not completely, collapse, sometimes by design (such as in longwall mines and pillar retreat areas). The geological controls and constraints on the process of subsidence over coal mines are covered by Kane and Powell.

Cross-Reference

As noted in the preface to the Exploration and Reserve Definition chapter, the papers there tie in intimately with the concept of coal mining. The paper by Sites et al. in the Reserve Estimation chapter deals with the restrictions to mining imposed by cultural or technological factors.

One of the big concerns in coal mining is the presence of methane in the coal and its release, sometimes catastrophically, into the mine. Clearly, the papers on Coalbed Methane, particularly those by Ertekin et al. and Soeder, are applicable to this problem.

If the coal quality is not appropriate, then mining will be economically difficult or require a degree of footwork (i.e., blending) to bring the mine product into compliance. Changes in quality over a mine property, especially those undetected by an inadequate drilling program, can cause serious economic problems for the mine operator. Papers in the Coal Quality Concerns chapter apply to this consideration, particularly the papers by Winston and Hower et al.

Surface mining was not intentionally excluded from this chapter; that simply is how the articles were submitted. Nonetheless, surface mining is represented in this book through papers by Henderson et al. and by Fenton and Pawlowicz in the Exploration and Reserve Definition chapter. Reclamation of surface mines also is addressed by Moran et al. in the Environmental Impacts chapter.

Also in the Environmental Impacts chapter, papers that deal with the aftermath of coal mining in the form of liquid wastes (i.e., acid and/or toxic drainage) are those by Wildeman and by Klusman and Machemer. A unique, but potentially significant, problem for air and water quality is discussed by Püttmann et al. Finally, the considerations of coal waste disposal in open pit mines are discussed by Beaver et al.

APPLICATION OF DEPOSITIONAL MODELS TO MINING PROBLEMS

Gerald A. Weisenfluh and John C. Ferm
Department of Geological Sciences
University of Kentucky
Lexington, KY 40506

ABSTRACT

Increasing involvement of geologists in operational aspects of the coal industry has led to applications of depositional process models for solving mining problems. Procedures for developing these models include synthesizing extant borehole and mine records with direct observations to assemble a three-dimensional construct of the coal seam and adjoining rock bodies, establishing relationships between the geologic parameters and mining conditions, and comparing these data to examples of modern depositional systems. The latter step enables the geologist to make predictions about characteristics of the coal and rock which may have gone unrecorded in existing data.

INTRODUCTION

Historically, the role of the geologist in the coal industry has been that of an explorationist. In this capacity, widely spaced borehole and outcrop data are assembled in order to determine the approximate limits, tonnage, quality, and general mining characteristics of a particular coal seam. The tasks of mine planning and oversight of development and operational problems have been the realm of the mining engineer who applies the principles of his trade (mechanics, hydraulics, etc.) to the extraction, transportation, and processing of the coal. This division of labor had been successful for the most part because the target seams were thick, widespread, and relatively uniform in thickness. In the past 15 yrs, however, the seams that are being mined have become increasingly complex, and the constraints on coal quality, extraction, and reclamation procedures are more numerous and stringent. This situation has led to a greater involvement of geologists in the operational aspects of the coal industry.

The kinds of problems that have arisen are numerous, but most involve seam thickness and continuity, roof and floor control, or coal quality. In each case, a detailed understanding of the three-dimensional characteristics of the seam and associated rock bodies is required although available data, exploration drill logs and a few on-site observations, generally are not sufficient to generate such detailed information. Hence, the primary goal of the geologist is one of generating projections from existing control points into areas where no data are available.

Whereas the mining geologist has many tools that can assist in making such predictions, one that has proved very useful is the concept of depositional environments or depositional process models. This concept assumes that the coal and the adjoining rocks are a product of a depositional process such as the accumulation of peat in a swamp or the sediment infilling of a channel or lake. The predictive value of the process concept arises from the documentation of depositional processes and their products in modern coal-forming environments, recognition of these products in the ancient rock record, and application of the resulting analogies to make predictions about coal and rocks.

A simple, small-scale example of the link between modern processes, analogous rock units, and mining applications is the material generally found lying directly beneath coal seams. This rock type (known as underclay, seat earth, or fireclay) is characterized by chaotic grain orientation, abundant small-scale polished surfaces (slickensides), and carbonaceous films believed to be remnants of fossil roots. These characteristics of disrupted bedding diminish downward from the base of the coal. Inspection of modern alluvial soils (Birkeland, 1974) shows that they have similar properties, and the inference is made that the underclays represent ancient soil profiles (Huddle and Patterson, 1961).

Application to mining problems could proceed in two directions. The first is direct and obvious. The disoriented grain fabric and particularly the slickensides yield a rock which is almost without strength and, if found in the immediate roof of an underground mine, can be the source of serious roof problems. This occurrence is quite common and can arise, in a depositional sense, in two ways. It may represent a failed attempt to establish a subsequent swamp after burial of the original peat by detrital sediment or a separation of the upper part of a seam by a wedge-shaped body of sediment (a "seam split"). The second application involves the question of position of this material throughout the mine area. Does it remain directly on top of the coal (and continue to present problems) or does it rise far enough above the seam to no longer be a source of difficulty? This requires reconstruction of depositional events from immediately following the cessation of the main coal swamp to the formation of the next following soil zone. It is this kind of inference that draws most heavily upon the knowledge of modern depositional processes.

Numerous other process models have been applied to coal-bearing rocks in the mining industry (Hylbert, 1984; Krausse et al., 1979; Milici and Gathright, 1985; Moebs, 1977). Whereas these depositional concepts (or at least their names) generally are familiar in the industry at large, their application often is missed or neglected by the average operating man. Consequently, another important task of the geologist is to transmit the information in terms which are meaningful to those involved in the day-to-day operation of the mines.

The objectives of this paper are to provide a general format for generating depositional models, to illustrate by example the kinds of predictions that depositional process models can yield and which can be applied to mining situations, and to suggest means by which this information can best be conveyed to operating personnel.

AN APPROACH FOR GENERATING PROCESS MODELS

In most instances, the mining problems for which the advice of the geologist is sought are "after the fact" situations. For example, a specific area of a mine has encountered an unusually high frequency of roof falls or headings have experienced abrupt reduction in coal thickness and attempts to reorient the entries have met with similar misfortune. The approach to such problems does not differ from other types of geological investigation and consists of three basic steps: observation and data synthesis, development of associations, and construction of a predictive depositional model. Although these are standard geological procedures, some aspects are unique to coal mining and will be described below.

Observation And Synthesis

The basic task in this step is assembly of data with the objective of developing, with the greatest possible precision, a three-dimensional construct of the coal seam and associated rocks, or at least one that comprises a testable hypothesis. Such a model may include only the coal seam, but, in most cases, rock layers lying above and below the seam are included, with special emphasis on the nature and rate of lateral change of rock and coal characteristics.

In underground mines, and in particular those that are large and well developed, a vast amount of data generally is available in the form of mine maps that are prepared as part of basic engineering procedures. These maps show the elevation at the base of the seam as surveyed approximately every 50 m (160 ft), and data of this kind are often sufficient to detect the presence of geologic structures associated with paleochannels or small faults within the seam. Seam height and, in some cases, individual bench thicknesses are measured with similar frequency and are useful for establishing seam stratigraphy and delineating thickness reductions related to seam splitting, tapering, or "washouts". Mine maps often show areas where the entries are irregularly distributed or reduced in number, providing an indication of difficult mining situations. Many companies record on the map the locations of roof falls, weak or cribbed roof, as well as other information about mining conditions. In addition, there are records of boreholes both within and beyond the mined area, and these may be accompanied by analyses of the coal seam. These drill-hole analyses, plus analyses of in-mine samples, can provide some general quality characterization information about the coal. All of this extant data, if relatively complete, constitutes an important database describing the coal body, but often little information is included about the rock above or below the seam.

In surface mining, the situation is reversed in that information about strata overlying the seam is abundantly available in the highwall. In some cases, elevation at the base of the seam has been surveyed and, as a rule, borehole data proximal to the mining face are more abundant than in underground mines. The quality of the records in both cases, however, generally is low. Whereas surface mining, unlike underground workings, provides an excellent record of rocks overlying the seam at one locality, it lacks the three-dimensional component of underground data unless records have been kept systematically as mining proceeded.

Despite this large body of existing data, it is, in most cases, not sufficient to solve a particular mining problem. Small features such as minor displacements go unrecorded, and data concerning overlying and underlying rock are especially incomplete. This usually can be remedied by direct observation. In underground mines, this often is difficult and time consuming as "roof rock" is rarely exposed except in roof falls or at overcasts. In surface mines, the reverse is true and highwalls provide excellent sites for observation of the character and lateral variation of rock overlying the seam.

Synthesis of existing data and those arising from direct observation lead to the next step of development of associations between geologic parameters and mining conditions.

Association Of Attributes

Although the combination of existing data and those arising from direct observation rarely is sufficient to develop a three-dimensional construct in adequate detail to solve specific problems for an entire mine, it is possible at least to develop associations which can be helpful. These associations are of two general types. The first is the association of a particular geologic attribute with a difficult mining situation. As previously cited, many underclays or seat earths lack physical strength and, where found directly above the seam, can lead to serious problems. In these cases, operating personnel can be informed of the rocks' properties and measures taken to observe the occurrence of such rock elsewhere in the mine. Similarly, where channel sandstones have cut into soft shale and the two rocks occur side by side, the shales, in many cases, are densely slickensided parallel to the channel margin and can be a source of difficulty in roof control. Operations personnel can be alerted to this situation and, when such contacts are found in the roof, extra support can be applied. Associations of this type are somewhat simplistic and do not necessarily involve inferences derived from depositional models, but can be very helpful to those for whom production is the prime goal.

The second type of association is that which links a geological characteristic which is known to create mining problems with some other geological property that can be readily determined or is available in existing data. If, for example, boreholes are closely spaced throughout the property, the margin of the sandstone (inferred to be flanked by slickensided shale) can be mapped, and operations personnel can be alerted to expect roof problems in such areas. Similarly, if it can be shown that coal quality deteriorates as a seam changes in thickness or splits, then borehole data can demonstrate the approximate area where such quality changes can be anticipated.

Whereas the latter type of association involves establishment of relationships between rock properties without any recourse to their mechanism of formation, it nonetheless has genetic implications. In the slickenside/channel sandstone example, the erosional contact between the sandstone and shale implies a depositional event and, to some degree, incorporates a depositional model. Similarly, the change in coal thickness or splitting of a coal seam records a depositional event which has left an imprint on the quality and thickness of the coal. Hence, while purely objective, associations mark a transition between data synthesis and a full-scale predictive process model.

Predictive Depositional Models

Though the associations described above have their origin in depositional processes and can be very helpful in solving mining problems, they differ from predictive depositional models in that they depend on data that can be observed directly. In contrast to associating attributes, predictive depositional process modeling is based on the knowledge of a recognizable product of a depositional event which permits prediction of characteristics which may not be derivable from the limited available data.

An example of this type of modeling could occur in the case of the sandstone juxtaposed with slickensided shale. If this sandstone could be shown to be of fluvial origin and to be a point bar deposit of a meandering stream, then the opposite margin of the sandstone body could be expected to be flanked by a clay-filled plug in the abandoned channel, and perhaps, slump deposits from the cut bank of the channel. Such deposits would produce a very poor roof in an underground mine. However, their presence could go undetected by widely spaced drilling or their importance might not be recognized from a sketchy driller's log. The primary point is that the recognition of a meandering channel deposit leads to predictions that are not found in the existing data.

This unique aspect of predictive depositional modeling arises from the knowledge of depositional processes and, perhaps, represents one of the most promising tools for the mine geologist. Two examples of such predictive modeling are discussed below: the Pratt Seam in Northern Alabama and the Cedar Grove Seam in Southern West Virginia. Both examples address problems of seam continuity and roof control in underground mines.

ROOF CONTROL IN THE PRATT SEAM

A study in Northern Alabama was undertaken because an unusually large number of roof falls had occurred in a mine in the Pratt Seam. Operating personnel wanted to know the cause of the falls, where similar problems could be encountered in the future, and what actions should be taken to stabilize the roof in existing workings.

Examination of the mine workings and drill hole records indicated that the immediate roof was comprised of 30 cm (12 in.) to 2 m (7 ft) of dark shale or siltstone overlain by 1-7 m (3-20 ft) of flat-bedded sandstone with ripple cross-laminations. The coal seam consists of two benches that are each about 1 m (3 ft) thick with a parting of fireclay that is 5 cm (2 in.) thick. Major roof falls were found in two geologic settings (Figure 1). The first setting is along the margins of narrow depressions in the seam which are 20-40 m (70-130 ft) wide and 2-3 m (7-10 ft) deep (called "swags" by the miners). Within these swags, the upper bench of coal was partially to completely removed and replaced by smooth-textured shale. Falls had occurred on the margins of the swags where the overlying sandstone was close to the top of the coal with little or no intervening shale. The swags were inferred to have formed as narrow sluggish creeks or channels which drained the coal swamp during its late stages. This inference was based on their linear character, U-shaped cross section, replacement of the coal by organic-rich mud, and intertongueing of this mud with the uppermost part of the seam.

The second type of fall occurred where the overlying sandstone thinned to 1-2 m (3-7 ft) thick, as shown on the right side of Figure 1. The character of the bedding within this tapering edge of the sandstone body was flat, and the abrupt contacts between the sandstone and the overlying and underlying shales were the planes of weakness along which the roof failed. These bedding characteristics led to the interpretation that this sandstone was deposited, not in a confined channel, but on a relatively flat surface as a crevasse splay.

In both cases, modified bolting procedures were developed to improve the roof support in problem areas. Along the swag margins, additional bolts of greater length were added between positions of the normal rectilinear grid. The edge of the splay sandstone was supported with conventional point-anchor devices, ensuring that the bolt lengths were sufficient to anchor at least 50 cm (20 in.) above the top of the sandstone. Moreover, bolting crews could be trained to recognize these conditions by recording the proximity and thickness of sandstone in the course of drilling bolt holes. In practice though, the monitoring program was difficult to consistently maintain throughout the mine, and it became necessary to identify specific areas where these problems were most likely to occur.

This situation led to the development of a predictive process model that included both the swags and the tapered edge of the overlying sandstone, both features being associated with roof falls. In order to generate the model, three maps were prepared. One, shown in Figure 2, depicted the distribution of swags and tapered edges of the sandstone and was used to make projections in the immediate mine area. The other maps

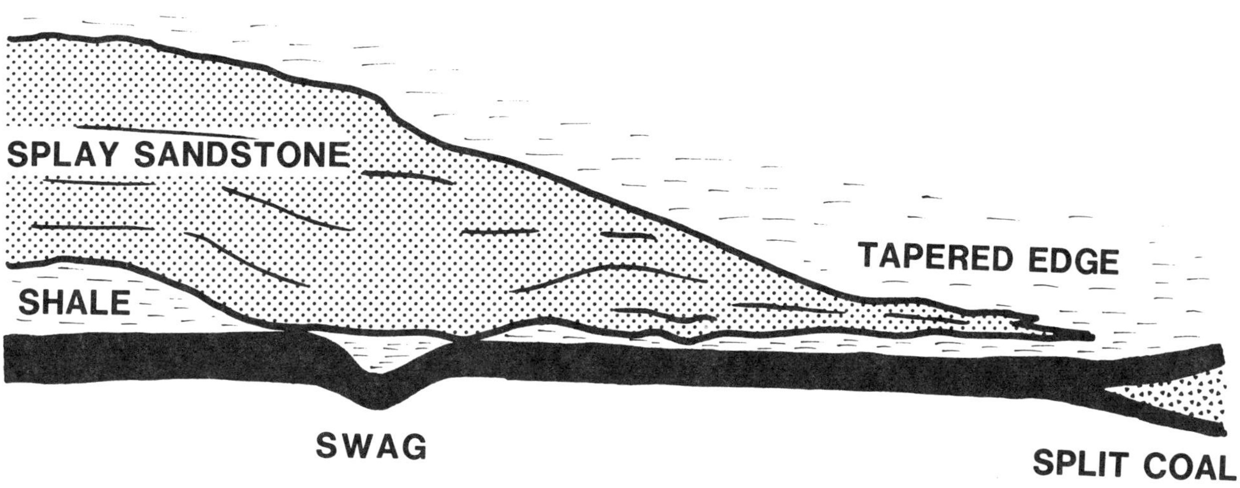

Figure 1. Reconstruction of splay sandstone and associated sediments overlying the Pratt Seam showing the character of "swags" and tapered edge of sandstone which are a source of roof problems.

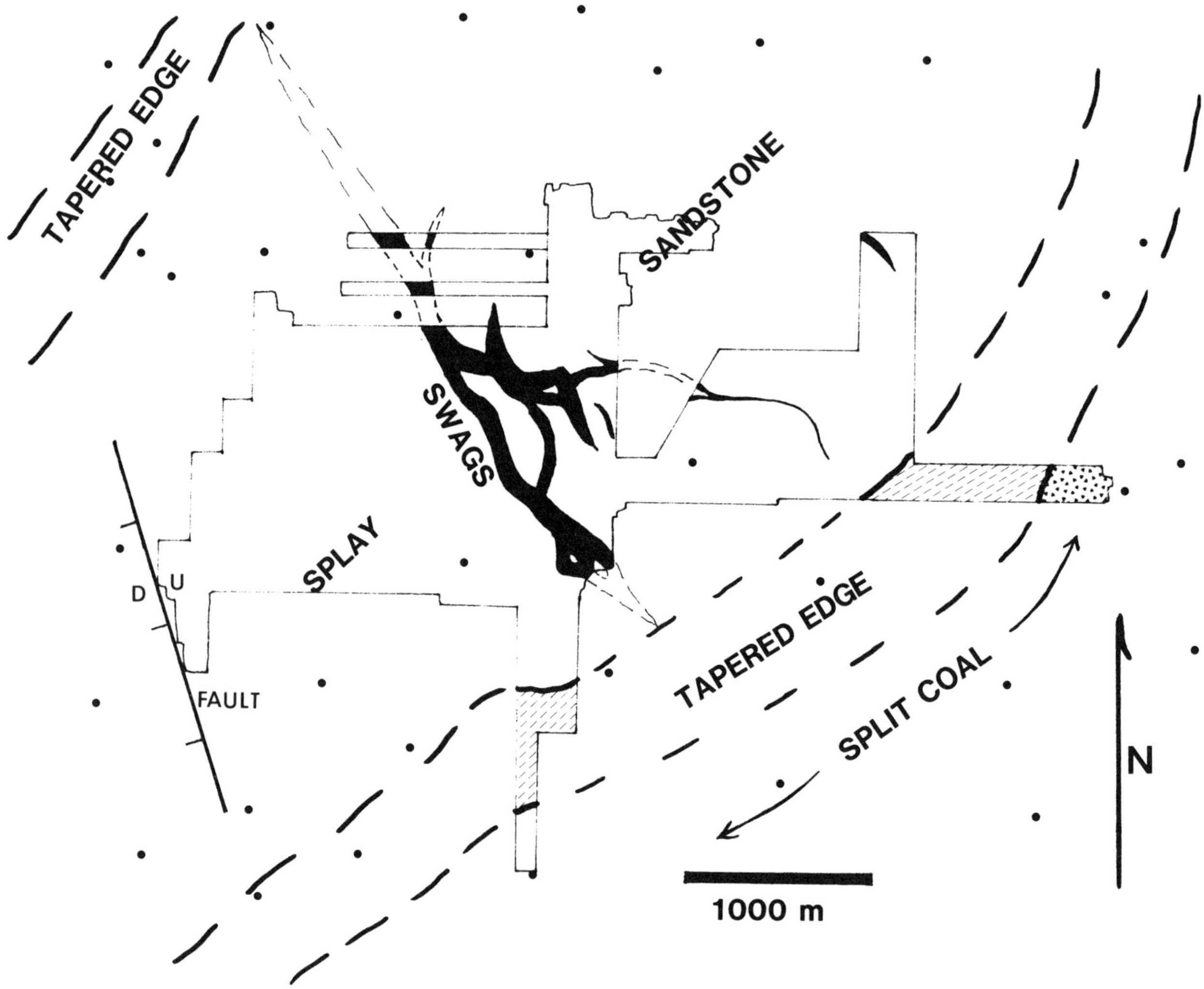

Figure 2. *Distribution of swags and tapered edges of splay sandstone mapped within the mine in the Pratt Seam and their projections (dashed) based on nearby borehole records (dots).*

(Figure 3a and 3b) place the mine data in a more regional context and in relation to normal faults which are believed to have influenced the distribution of peat and adjoining sediments (Weisenfluh and Ferm, 1984).

The distribution of swags and the geometry of the splay sandstone, based on in-mine mapping and borehole records, are shown on Figure 2. The swags are most abundant in the central part of the mine (Figure 2) and trend to the northwest. They commonly bifurcate in either direction along their trend and one of the resulting branches usually terminates within 100 m to 200 m (300 ft to 700 ft). Projections of these mapped channels into the northwestern part of the mine were accurate to within 30 m (100 ft) and aided in the planning of longwall panels in this area. The tapered edge of the splay sandstone was observed in the eastern and southern mains and, using nearby borehole records, the opposing edges of the sandstone body were defined approximately 1,000 m (3,000 ft) west and 3,000-5,000 m (10,000-16,000 ft) north of the mine.

In order to extend these projections, which are proximal to the mine, to include the entire coal seam, remaining borehole data were used to generate a fault-control process model. The premise of this depositional model is that the associated normal faults were active at the time of peat accumulation and produced topographic

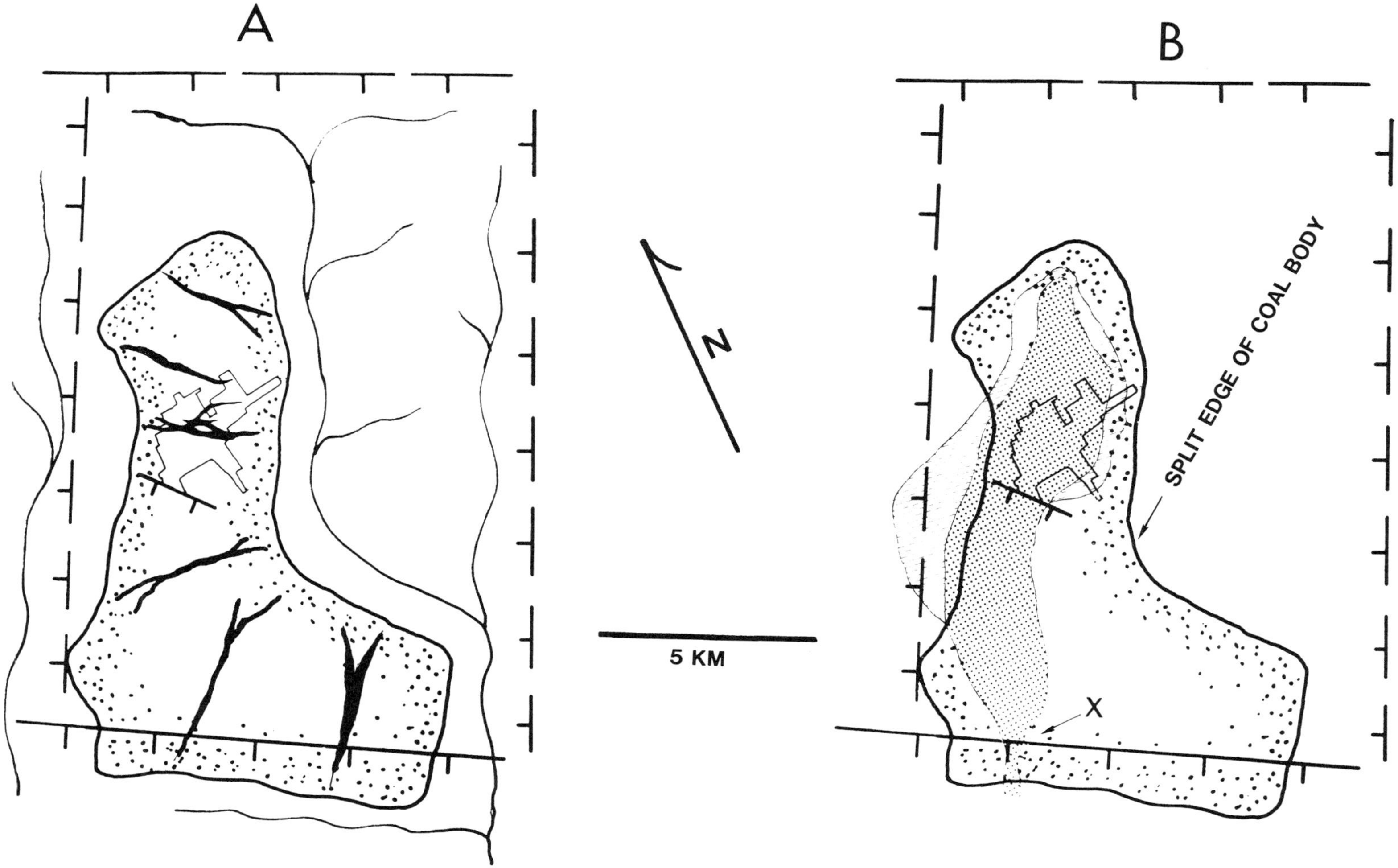

Figure 3. *Fault-control depositional model for the Pratt Seam. Faults solid where known from borehole data and surface exposures and dashed where data are sparse. Tick marks are on the downthrown side of the faults. The mineable Pratt coal body, delineated by the area of merged benches (1.5-2 m (5-7 ft) thick), is indicated by the large stippled pattern. A) Inferred locations of swag channels relative to faults. B) Position of splay sandstone (small stippling) and tapered edges (dashed pattern) based on borehole data.*

irregularities which affected the distribution of peat and sediments. The upthrown edges of the blocks were topographically high, protected from detrital influx, and most favorable for peat development. The downthrown sides attracted sediment resulting in splits within the seam. Figure 3 shows the location of the mine within the northeastern extension of an arcuate coal seam and the position of known and inferred faults within the region. These faults presently display up to 50 m (160 ft) of offset at the Pratt horizon, but throw at the time of deposition of the seam was probably only 3-6 m (10-20 ft). The main body of the coal seam occupies the higher, southwestern and northwestern sides of a fault block which is roughly 12 km by 18 km (7.5 mi. by 11 mi.) in extent. According to this reconstruction, the "swag" channels draining the surface of the bog would flow in a southeastern or northeastern direction away from the topographically highest part of the block. The precise position of the channels cannot be predicted or easily detected by existing borehole data in the area, but their presence and approximate vector can be inferred.

The other feature related to roof control problems in the present mine is the feathered edge of the splay sandstone overlying the coal seam. The position of this feature, based on borehole data, is virtually the same as the thicker part of the underlying seam (Figure 3b) suggesting that the distribution of the splay was largely controlled by differential compaction of the underlying peat. The latter will compact more than laterally equivalent detrital rocks and, as the swamp was buried, the area underlain by the thickest peat would generate a topographic low into which the flow of the splay would be attracted. Hence, the position of the feathered edge can be expected as areas of split coal are approached.

Although the effects of the faults on the splay sandstone appear to have been small relative to those of differential compaction, it would be anticipated that the sandstone deposit would become more narrow and incised as the highest part of the fault block was approached (X on Figure 3). This is supported by the apparent lack of the "tapered edge" facies as well as a restricted area of sandstone in the southwestern part of the block. Although none of the existing boreholes indicate significant reduction in coal thickness in this area, it would be expected that the "incised" splay would locally erode the coal seam.

Effects of fault-controlled deposition also can be applied to other kinds of mining problems. Kane and Powell (this volume) suggest that the vertical stacking of certain lithologies (mostly sandstones) due to contemporaneous fault movements have a direct effect on the style of overburden subsidence after mining has taken place.

CUTTING THROUGH A "CHANNEL" IN THE CEDAR GROVE SEAM

The operational problem of cutting through a "channel" occurred in the Cedar Grove Seam in Southern West Virginia where sandstone was reported to have replaced the coal. It was known that a substantial body of reserves lay beyond the no- or low-coal area (a "want"), so the problem was one of devising the best way through the area of low coal (Figure 4). Mining on one side of the "want" and drilling on the other had shown that the zone was about 2,000 m (6,600 ft.) wide and extended entirely across the property. Direct examination of the area where some probing had been done showed that a 10-15-cm (4-6-in.) thick parting, which occurred throughout the mine, had rapidly thickened and at the end of the entries was about 60 cm (24 in.) thick. The thick parting consisted (from bottom to top) of shale, sandstone, and underclay, and the two benches of coal split by these rocks were slightly thinner than elsewhere in the mine. Some drill holes within the want area were available and showed a maximum thickness of 6 m (20 ft) of rock between the benches, but the quality of the core logging was insufficient to indicate bedding and other details.

The splitting relationships of the parting and the general dimensions of the rock body suggested a splay deposit similar to those observed in large road cuts and surface mines in the region and in modern deltaic

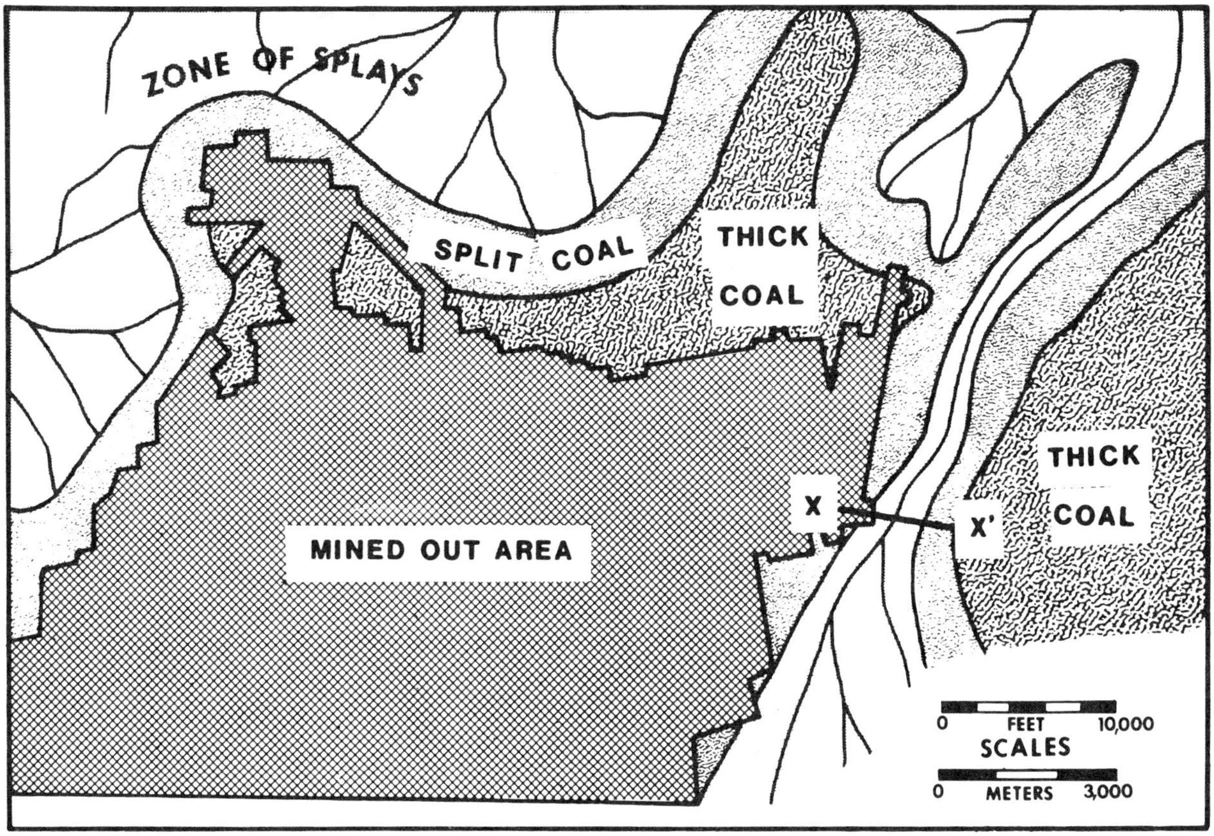

Figure 4. Detailed diagram of mine property showing location of mined-out Cedar Grove Coal and position of cross section X-X' (from Ferm, 1978).

deposits (e.g., Baganz et al., 1975; Saxena, 1976). On modern deltas, splays form when the banks of a major channel are breached and a mound of mud and sand is deposited beyond the edge of the levee (Saxena, 1976; Staub and Cohen, 1979). As additional layers of sediment are added, one or more small channels cut into the previously deposited mud and sand as a hydraulically more efficient means of moving water over the splay surface. These channels carry sediment to the edge of the mound where, if enough material is deposited, the newer material, in turn, is cut through by the channel. The mound, which was originally roughly fan shaped in outline, may continue to grow as a linear, lobate deposit until the breach in the levee is healed and no further water or sediment is provided to the splay. At this time, the surface of the splay may become revegetated. Such events are particularly common on delta plains where levees are low, and the surrounding backswamps are marshy and sodden.

Given this knowledge of splay formation, the low/no coal area was reconstructed as Figure 5a. In this particular case, the splay was built onto a peat deposit now represented by the lower bench of coal and, after splay construction ceased, the swamp returned and more peat, now forming the upper bench, was accumulated. Note on the figure that a channel was to be expected near the thickest part of the deposit, and there was some possibility that this channel would cut into the lower bench of coal. Also, it was expected that the material filling the channel would be coal or organic-rich shale which would make poor roof rock.

Because the problem was one of driving a rock tunnel, options were proposed in an effort to find a mining solution which maximized coal thickness while minimizing roof problems. Two alternatives for tunneling,

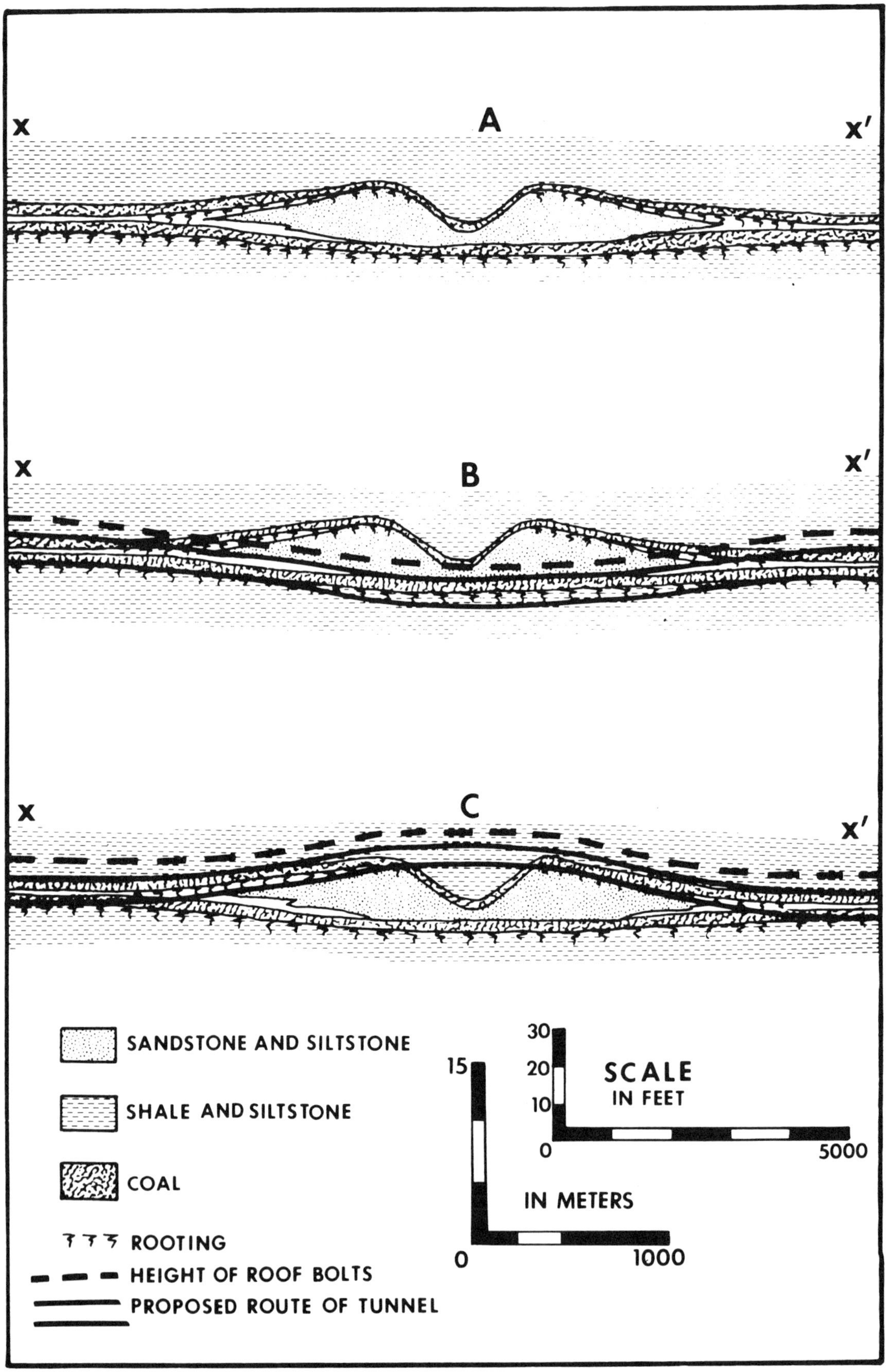

Figure 5. Depositional model for the Cedar Grove split. A) Cross section of splay deposit that crosses the mine property and splits the seam. Location of cross section shown on Figure 4. B) Location of proposed tunnel under splay deposit. Dashed line is height to which roof bolts are driven. C) Location of proposed tunnel over splay deposit (from Ferm, 1978).

shown as Figures 5b and 5c, were constructed following the upper or lower bench of the seam as coal is easily cut and followed in tunneling. The disadvantage of following the lower bench lay in possible roof problems on the margins of the deposit where roof bolts would be anchored in coal or underclay of the upper bench or in the abandoned channel where coal or soft shales would be encountered. There also was a possibility that the channel would have cut into the lower bench, further complicating the problem.

The upper bench route seemed a better alternative because the roof rock there was similar to that elsewhere in the mine where there had been few roof problems. The disadvantages would be that the coal would be expected to thin on the crest of the splay deposit and that very steep dips (up to 30°) could be encountered in the central channel area. As shown in Figure 5c, this problem could be avoided by driving horizontally or only slightly downward where the crest of the deposit was reached rather than following the coal into the channel.

As it turned out, the upper bench route was chosen, and two channels instead of one were found. Neither channel was as deep as was predicted, and grading across the channel area solved the problem of a potentially very rolly floor.

The main point in these two examples is that both followed the approach described in the "Approach for Generating Process Models" section. Mine map and existing borehole data were incorporated with direct observations in order to create a rough model of the three-dimensional characteristics of the coal and overlying rock bodies. Associations were generated between geologic characteristics and the operating problem to be solved. Finally, knowledge of modern depositional processes were applied to the data and predictions were made that, in at least these cases, were tested by further mining.

CONCLUSIONS

In reading the case studies above, one would be inclined to believe that depositional models are solving all the operational problems in the mining industry. This is far from the truth. Some mining problems, such as insitu stress (Nelson and Bauer, this volume) and tectonic-related faulting (Krauss et al., 1979; Milici et al., 1982), involve geologic processes that have modern analogs, but these processes are not observable directly. Solutions to these problems entail extensive observation and sophisticated physical measurements, and the geologic models generated are grounded, at least in part, in theoretical bases. These problems interact with those of depositional origin so that, even in applications of this kind, the methods described in this paper may be applied.

Also, some modern processes are not well understood. A great deal of research effort is being directed toward using peat swamp depositional models to predict variations in chemical and petrographic properties of coal seams. Whereas much is known about the surficial characteristics of swamps, only recently has detailed information about the internal portions of swamps been collected in such a way that it can be related to ancient coal seams (Esterle, 1990).

As shown by some papers in this chapter, a start has been made, but there are major problems facing full utilization of the methods described in this paper. The first is simple inertia. Both explorationists and management are satisfied with their respective roles and see little reason to change. Moreover, the use of concepts such as depositional environments that appear to be remote from practical mining problems naturally can be expected to be viewed with some suspicion. The second problem, and one of at least equal importance, lies on the geological side of the fence. There are few geologists actually assigned to operations problems in industry and fewer still who are engaged in research whose results can be transferred to application. Given such constraints, the following topical "wish list" would seem to be appropriate.

Take The Sting Out Of Routine Tasks

Techniques are now available that reduce the time-consuming tasks that formerly consumed the efforts of the industry geologist. Powerful computer hardware now has been simplified to the level of a household appliance, and a large variety of software suitable for manipulation of large amounts of data is readily available. Such systems should reduce the time formerly required in simple repeated calculations and in map preparation. Tasks such as core logging and seam description, which are critical in interpreting depositional environments, can be speeded up for geologists or can be reduced to a technical-level operation, not requiring highly trained personnel, with the use of photographic guides (Ferm and Smith, 1981; Ferm et al., 1985; Ferm and Weisenfluh, 1981). Freeing geologists from direct involvement in such routine tasks will allow more time for direct field observation which, as has been shown above, is absolutely essential in solving mining problems. Also, these core-logging guides provide for standardized descriptions which have enhanced applications to reclamation problems (Weisenfluh and Ferm, 1989) and to general lithologic analysis (Dickenson, 1984).

Getting Straight With Depositional Processes

This is a devil with two tails — one toward the mining geologist and the other to the researcher. Realizing that the efforts of industry geologists are mostly consumed in dealing with immediate tasks, some time could be given to building a mental bridge between the rocks that are observed every day and the probable processes that brought them about. This can be done in two ways: 1) becoming familiar with summarized information now found in standard texts describing sedimentary processes and their products (e.g., Galloway and Hobday, 1983; Morgan, 1970), and 2) trying to apply this information to some large nearby outcrops or areas of work where data from closely spaced drill holes (300-600 m (1,000-2,000 ft) apart) are available.

The major effort in this direction, however, should be carried out by the research geologist who has two major tasks. The first is seeking among those rock or coal bodies, where the data are most complete and the depositional setting known, those criteria for recognizing the environment which can be used where data are sparse and/or of poor quality. The second is the study of modern depositional settings. There is a substantial amount of data assembled by geomorphologists and civil engineers about fluvial, deltaic, and shoreline settings, including swamps and peat bogs (e.g., Cohen et al., 1984; Coleman and Gagliano, 1965; Frazier, 1967; Frazier and Osanik, 1961; Nummedal, 1982; Staub and Cohen, 1979; Welder, 1959). However, many studies of modern environments deal only with the surface morphology, and the complete shape and internal characteristics of the deposits are only poorly known. This suggests that cooperative efforts between groups dealing with surface and solid body characteristics would yield promising results.

Focus The Geophysical Applications

The applications of geophysical techniques, particularly seismic, to mining problems have been in place for over 10 yrs (Hasbrouck and Hadsell, 1976; Buchanan, 1979; Lepper and Ruskey, 1979). These techniques are still considered experimental, though, and will be enhanced greatly with a better knowledge of the discontinuities or rock properties that are being investigated. Old, tried-and-true geophysical methods, such as gravity or magnetics, recently have been used to detect basement faults like those in the Pratt example and to relate them to characteristics of the overlying sedimentary cover (e.g., Root, 1989). These and other geophysical techniques show great promise if used in conjunction with depositional process models.

Wave The Calculator Not The Arms

Perhaps the greatest single problem facing the application of geology to mining problems is the method of thought used by geologists and engineers. Engineers, whose basic concepts are derived from physics, are accustomed to thinking in terms of precise measurements and relationships. In contrast, geologists, realizing that predictions of the type that they are asked to make cannot achieve this level of precision, are inclined to provide answers that are vague and unsatisfying to engineers with whom they must work. The solution would seem to lie in the area of probabilistic answers which satisfy the concept of precision yet allow for uncertainties. Some progress in this direction has been made (e.g., Kim et al., 1980; Agterberg, 1989), but for the most part, the field remains open.

Show And Tell

Despite the large and increasing number of journals, basic information transfer from research to application has been modest. The primary difficulty appears to lie in the problem of reducing complex results to simple easily grasped concepts. Some notable examples, such as papers by Greb and by Sames and Moebs (both in this volume) and an audio-visual presentation of roof fall information (Greb and Cobb, 1987) illustrate what can be achieved, but more effort should be pointed toward quantitative expression of results for the engineering audience.

It should be evident that, whereas the use of predictive depositional models shows considerable promise in operational mining problems, the potential is greater than the results so far. The above "wish list" is long, the number of geologists involved are few, and the issues complex. Nonetheless, such a situation suggests that much can be achieved with only modest effort. The means are within our grasp.

REFERENCES

Agterberg, F.P., 1989, Computer programs for mineral exploration: Science, v. 245, p. 76-81.

Baganz, B.P., J.C. Horne, and J.C. Ferm, 1975, Carboniferous and Recent Mississippi lower delta plains--A comparison: Gulf Coast Association of Geological Societies, Transactions, v. 25, p. 183-191.

Birkeland, P.W., 1974, Pedology, weathering and geomorphological research: New York, Oxford University Press, 285 p.

Buchanan, D.J., 1979, The location of faults by underground seismology: Colliery Guardian Annual Review, p. 419-427.

Cohen, A.D., D.J. Casagrande, M.J. Andrejko, and G.R. Best, 1984, The Okefenokee Swamp--Its natural history, geology, and geochemistry: Los Alamos, New Mexico, Wetland Surveys, 709 p.

Coleman, J.M., and S.M. Gagliano, 1965, Sedimentary structures--Mississippi River deltaic plain, in G.V. Middleton, ed., Primary sedimentary structures and their hydrodynamic interpretation: Society of Economic Paleontologists and Mineralogists, Special Publication 12, p. 133-148.

Dickinson, W.W., 1984, Analysis of lithologic descriptions using a simple numeric code: Journal of Sedimentary Petrology, v. 54, p. 638-641.

Esterle, J.S., 1990, Trends in petrographic and chemical characteristics of tropical domed peats in Indonesia and Malaysia as analogues for coal formation: Lexington, University of Kentucky, Unpublished PhD. Thesis, 269 p.

Ferm, J.C., 1978, New concepts and techniques in predicting seam discontinuities, in R.M. Voelker, ed., Proceedings of the Coal Seam Discontinuities Symposium, November 3-4, 1976, Pittsburgh, Pennsylvania: Pittsburgh, D'Appolonia Consulting Engineers, p. 1-9.

Ferm, J.C., and G.C. Smith, 1981, A guide to cored rocks in the Pittsburgh Basin: Lexington, University of Kentucky, 109 p.

Ferm, J.C., G.C. Smith, G.A. Weisenfluh, and S.B. Dubois, 1985, Cored rocks in the Rocky Mountain and High Plains coal fields: Lexington, University of Kentucky, 90 p.

Ferm, J.C., and G.A. Weisenfluh, 1981, Cored rocks of the Appalachian coal fields: Lexington, University of Kentucky, 93 p.

Frazier, D.E., 1967, Recent deltaic deposits of the Mississippi River--Their development and chronology: Gulf Coast Association of Geological Societies, Transactions, v. 17, p. 287-311.

Frazier, D.E., and A. Osanik, 1961, Recent peat deposits--Louisiana Coastal Plain, in E.C. Dapples and M.E. Hopkins, eds., Environments of coal deposition: Geological Society of America, Special Paper 114, p. 63-85.

Galloway, W.E., and D.K. Hobday, 1983, Terrigenous clastic depositional systems: New York, Springer-Verlag, 423 p.

Greb, S.F., and J.C. Cobb, 1987, Geology of roof falls--A training guide: Mine Safety and Health Administration, Instructional Program AV28.

Hasbrouck, W.P., and F.A. Hadsell, 1976, Geophysical exploration techniques applied to Western United States coal deposits, in W.L. Muir, ed., Coal Exploration: Proceedings of the First International Coal Exploration Symposium, London, England, May 18-21, 1976: San Francisco, Miller Freeman Publ., Inc., p. 256-287.

Huddle, J.W., and S.H. Patterson, 1961, Origin of Pennsylvanian underclay and related seat rocks: Bulletin of the Geological Society of America, v. 72, p. 1643-1660.

Hylbert, D., 1984, Geologic structures and mine roof falls in selected coal beds within Appalachia: Appalachian Development Center, Research Report No. 8, 82 p.

Kim, Y.C., F. Martino, and I.K. Chopra, 1980, Application of geostatistics to a coal deposit: Littleton, Colorado, Society of Mining Engineers--AIME, Preprint 80-90, 12 p.

Krausse, H.-F., H.H. Damberger, W.J. Nelson, S.R. Hunt, C.T. Ledvina, C.G. Treworgy, and W.A. White, 1979, Roof strata of the Herrin [No. 6] coal member in mines of Illinois--Their geology and stability: Illinois State Geological Survey, Mineral Note 72, 54 p.

Lepper, M.C., and F. Rusky, 1976, High-resolution seismic reflection techniques for mapping coal seams from the surface: U.S. Bureau of Mines, Technical Progress Report 101, 17 p.

Milici, R.C., and T.M. Gathright II, 1985, Geologic features related to coal mine roof falls--A guide for miner training: Virginia Division of Mineral Resources Publication 55, 13 p.

Milici, R.C., T.M. Gathright II, B.W. Miller, and M.R. Gwin, 1982, Geologic factors related to coal mine roof falls in Wise County, Virginia: Virginia Division of Mineral Resources, Appalachian Region Commission Report CO-7232-80-1-302-0206, 103 p.

Moebs, N.N., 1977, Roof rock structures and related roof support problems in the Pittsburgh Coalbed of Southwestern Pennsylvania: U.S. Bureau of Mines, Report of Investigations RI8230, 30 p.

Morgan, J.P., and R.H. Shaver, 1970, Deltaic sedimentation--Modern and ancient: Society of Economic Paleontologists and Mineralogists, Special Publication 15, 312 p.

Nummedal, D., 1982, Sedimentary processes and environments along the Louisiana-Texas coast: Geological Society of America, Annual Meeting Guidebook, 92 p.

Root, S.I., 1989, Effect of the Transylvania fracture zone on evolution of the western margin of the Appalachian Basin, in M.J. Bartholomew, D.W. Hyndman, D.W. Mogk and R. Mason, eds., Characterization and comparison of Precambrian through Mesozoic continental margins, 8th International Conference on Basement Tectonics: Dordrecht, Holland, D. Reidel Publishing Co.

Saxena, R.S., 1976, Modern Mississippi Delta depositional environments and processes: American Association of Petroleum Geologists/Society of Economic Paleontologists and Mineralogists, Field Trip Guidebook, 125 p.

Staub, J.R., and A.D. Cohen, 1979, The Snuggedy Swamp of South Carolina--A back barrier estuarine coal-forming environment: Journal of Sedimentary Petrology, v. 49, p. 133-144.

Weisenfluh, G.A., and J.C. Ferm, 1984, Geologic controls on deposition of the Pratt Seam, Black Warrior Basin, Alabama, U.S.A., in R.A. Rahmani and R.M. Flores, eds., Sedimentology of Coal and Coal-Bearing Sequences: International Association of Sedimentologists, Special Publication 7, p. 317-330.

Weisenfluh, G.A., and J.C. Ferm, 1989, Reclamation characteristics of coal-bearing rocks on federal lands in Northern Alabama: Southeastern Geology, v. 29, p. 221-234.

Welder, F.A., 1959, Processes of deltaic sedimentation in the Lower Mississippi River: Coastal Studies Institute, Louisiana State University, Technical Report 12, 90 p.

GEOLOGIC DIAGNOSIS FOR REDUCING COAL MINE ROOF FAILURE

Gary P. Sames and Noel N. Moebs
U.S. Bureau of Mines
P.O. Box 18070
Pittsburgh, PA 15236

ABSTRACT

Geologic diagnosis, a method proposed by the U.S. Bureau of Mines to determine the cause of persistent mine roof failure, uses commonly available geologic data and observations on the character and distribution of roof falls. Successful diagnosis requires an understanding of a site's geology, diligent maintenance of a mine roof fall map, and recognition of characteristic patterns or modes of failure. Roof falls are categorized as related to either geology or stress. Geology-related roof failure is linked to geologic discontinuities and rock character. Stress-related roof failure is recognized by patterns of failure. Subtypes further divide the nature of roof failure in each category until an appropriate technique for reducing the incidence of failure can be recommended. This framework enables the user to begin to sort out roof control problems and, by experimenting within the constraints of the approved roof support plan, arrive at effective, systematic control measures.

BACKGROUND

Many methods of classifying mine roof for predicting roof falls are available. Most are based on knowledge of local geologic features. Wier (1970), for example, describes six kinds of roof falls: shale dusting or slaking, sandstone rolls, concretions, slabbing, clay seams, and massive. Although appropriate for the specific area of Indiana studied, they are of limited usefulness outside of the area studied. Hylbert (1978) proposed classifying roof falls based on the structural and lithologic character of the roof rock at a mine in Eastern Kentucky. This method proved useful in projecting trends and anticipating problem areas in advance of mining. Patrick and Augenbaugh (1979) devised a classification based on roof fall geometry. This simplified system of dome-, arch-, minor-, and sloughing-type roof falls enables the reporting of falls independent of local conditions. However, whereas the cause of failure sometimes can be inferred, the ability to provide an accurate analysis of failure is limited if geometric data are used alone.

The geologic diagnosis approach to roof fall classification proposed by the U.S. Bureau of Mines requires more information than the classification methods of other investigators. However, it is meant to provide a less regionally specific method, supply a sound basis for determining the underlying causes of roof failure, and indicate some appropriate means of reducing the rate of failure.

GEOLOGIC DIAGNOSIS

The subsurface environment surrounding a coal mine is complex, and a complete and accurate understanding of bolted mine-roof failure often is difficult to obtain and describe. Even so, a good qualitative evaluation of roof failure sometimes can provide a basis for modifying either mine design or support procedures to alleviate the conditions leading to failure. Most operators are aware of the both subtle and pronounced changes in roof conditions that sometimes occur in a mining operation and the difficulty in either predicting these changes or responding appropriately to prevent roof failure. Once failure occurs though, a knowledge of the site's geology and recognition of a general characteristic of individual falls or a pattern of multiple falls can lead to understanding their cause and, therefore, an appropriate remedial action.

The purpose of geologic diagnosis is to provide the mine operator with a guide for determining the probable cause of roof failure based on either geologic discontinuities and rock character, or on the occurrence, character, and distribution of individual falls in a mine. Once a probable cause is determined, suggestions can be offered on how to avoid additional failures of the same type.

Geologic diagnosis of roof failure is based chiefly on the careful combining of exploration and in-mine data on rock character and roof lithology with a recognition of certain patterns of distribution or modes of failure. This requires diligent use of commonly available geologic data and documentation of a significant number of falls. Occasional, widely distributed falls are insufficient for analysis and in most mines would not indicate a need for action. The geologic data and roof fall documentation necessary for successful geologic diagnosis should include the following:

1) A mine map showing all roof falls. Indicate location, date, approximate shape, and dimensions of the falls. Geologic notes on roof rock composition, condition, and presence of any anomalous structures also should be included. This map provides information for determining roof geology and failure trends.

2) A mine map with surface topography superimposed for correlating roof falls to topographic features.

3) A map showing positions of superjacent or subjacent mining.

4) A map showing immediate roof lithology and thickness, structures such as clastic dikes or paleochannels, and major geologic features (faults, folding, etc.).

The information gathered in this type of mapping is extremely useful in the diagnosis and prevention of roof failure. The roof fall patterns and geologic relationships revealed are used to attribute failure to either geology-related or stress-related conditions.

ROOF FALL CATEGORIES

Geologic diagnosis groups roof falls into one of two main categories: Type G, geology related; and Type S, stress related. These two broad categories are intended to guide mine operators to a preliminary assessment using the geologic and roof fall information gathered on the mine maps. Each category then is further divided into a number of subtypes to identify the true nature of the failures. Remedial action is possible once the cause of failure is established. The geologic associations and roof fall patterns identified with both categories and their subtypes are described and schematically illustrated in the following sections.

Type G: Geology-Related Category

The Type G category of roof failure attributed to geologic discontinuities and rock character is divided into the following five subtypes: G1 (low rock strength); G2 (water sensitivity); G3 (bedding-plane spacing); G4 (minor structures); and G5 (major structures).

Subtype G1--Low Rock Strength

Rock character is the primary identifier of G1 roof failure. This subtype includes all roof rock that is relatively soft, usually sparsely laminated, low in RQD (rock quality designation), point load, and compressive strength. The physical properties for subtype G1 roof rock are commonly less than the following values: 0.3 point load index, 20 shore hardness, and 17 MPa (2,500 psi) compressive strength. This includes claystone and "drawslate" that commonly overlie coal seams. These rocks in general are not self-supporting for normal entry widths and tend to fall from the roof soon after coal extraction.

Areas of low rock strength can be identified early in the exploration process in drill core logs and by simple point load tests. An isopach map of roof lithology prepared from core logs can delineate areas of low rock strength. This subtype commonly occurs between coal splits. The roof of the lower coal seam will consist of the underclay of the upper coal seam. In ongoing mining operations, roof rock that must be supported quickly after exposure to prevent mostly small, continual falls near the face generally belongs to this subtype.

Bolted headers or metal straps are useful in supporting small pockets of roof that tend to fall between bolts. Tension bleed-off in mechanical bolts is a problem in soft rock types due to anchor slippage. If possible, mechanical bolts should anchor in a more competent overlying rock. Full-column resin bolts with metal straps have been found to better support G1 roof if anchorage occurs within the soft rock type. Under a thick interval of claystone, posts and crossbars may be necessary if failure occurs above the bolting horizon.

Subtype G2--Moisture Sensitivity

Rock character is the primary identifier of G2 roof failure. Roof fall patterns also can indicate moisture sensitivity. Moisture-sensitive rock suffers a significant loss of strength from exposure to prolonged high levels of humidity and fluctuations in humidity that occur in the mine environment. Moisture sensitivity results in a progressive softening or slaking that can gradually disintegrate rock to a mud-like unconsolidated condition. The slaking process is one of moisture absorption and expansion during periods of high humidity, and drying in periods of low humidity (Haynes, 1975). In a humid mine atmosphere, slaking of moisture-sensitive rock is a gradual process often measured in months or years. Roof failure occurs by attrition as the dislodging of small fragments exposes more roof and leads to larger falls.

The principal types of moisture-sensitive roof rock are poorly laminated claystone containing many slickensides, and extremely fissile, often highly fossiliferous shales. Simple water immersion tests or exposure outdoors will determine the relative moisture sensitivity of roof rock samples and the extent to which problems may occur underground. Moisture-induced roof fall patterns are recognized by numerous falls and severe slaking near shaft bottoms or drift mouths and along intake air courses.

Techniques developed to prevent or control moisture-induced roof failure include supplementary support, leaving top coal, applying sealants, and air tempering. Supplementary support usually is required to reduce the failure of exposed moisture-sensitive roof. Experience with full-column resin bolts has shown their ability to anchor in soft rock and prevent mine air circulation in open bolt holes. However, as disintegration due to moisture progresses, a larger area of roof than that immediately above the bolt plates requires support. This can be provided by bolted headers, straps, or wire mesh. Continued long-term deterioration in main haulage and airways eventually requires massive, expensive support maintenance.

Leaving the top 10-15 cm (4-6 in.) of coal and bolting it to the roof is a widely accepted method of controlling moisture-induced roof failure. The coal serves as a moisture barrier. Top coal works well when it adheres strongly to the roof, but requires careful mining to avoid exposing the shale.

A coating of gunite, shotcrete, coal tar, or a polymeric sealant on the rock surface is very effective in controlling deterioration. However, their high cost limits the widespread use of coatings in mines.

Air tempering stabilizes air moisture content before circulating it throughout the mine. Intake air is stabilized by coursing it through old workings or a series of entries specifically designed for that purpose (Sames, 1985). Low air velocity through the tempering entries allows it to reach mine temperature and humidity (Figure 1). The air then is directed to working areas of the mine where interaction between it and the moisture-sensitive roof is minimized.

Subtype G3--Bedding-Plane Spacing

Rock character is the primary identifier of G3 roof failure. Weakly bonded, thinly laminated strata are prone to roof separation and sag. Weak bonding usually results from an abundance of mica flakes, clay, or coal along bedding planes. Thinly laminated roof rock can consist of sandstone, interbedded sandstone and shale, or shale, and is identified easily in drill core or in underground exposures. The more closely spaced the bedding planes or laminations, the more difficult it is to form a competent beam in the immediate roof.

Roof falls attributed to closely spaced and poorly bonded laminations usually occur at intersections, where roof spans are greatest. Occasional, seemingly random falls may occur where installed roof support is inadequate. Subtype G3 falls of immediate roof tend to develop first as sag when mechanical bolts are not well anchored in a competent overlying stratum. Sag occurs as slippage along the bedding planes. This often results in the development of a tension fracture along the span's center line caused by the bending moment. As sag continues, the fracture at mid-span extends, leading to general collapse (Moebs, 1977).

Mechanical bolts must anchor in a competent overlying stratum to create an effective beam to support roof prone to G3 failure. If the immediate roof is too thick to accomplish this with mechanical bolts, full column resin bolts provide effective bolt-length anchorage and resistance to slippage. Subtype G3 roof failure that does not respond to longer, mechanically anchored bolts or full-column resin bolts sometimes can be supported effectively with roof trusses. Roof trusses put the roof in compression and exert some upward thrust at the mid-span, counteracting the tendency to slip and sag. Beyond these measures, posts and crossbars or narrower entries become necessary.

Subtype G4--Minor Structures

Subtype G4 roof failure is recognized by a minor structure exposed in the roof fall, or lying adjacent to it, and by some patterns of failure. Minor structures include any geologic feature that constitutes a discontinuity in the normal roof rock. These include slickensides (Figure 2) (Ellenberger, 1979; Moebs and Ellenberger, 1982), kettlebottoms (Figure 3) (Chase and Sames, 1983) and dinosaur footprints (Sames and Laird, 1988), clastic dikes (Figure 4) (Chase and Ulery, 1987), paleochannels (Figure 5) (Moebs, 1977), joints (Figure 6) (Moebs and Ellenberger, 1982; Sames and Laird, 1988), pinchouts (Figure 7) (Moebs and Ellenberger, 1982), concretions (Figure 8), hillseams (weathering-enlarged tension joints near outcrop; Figure 9)(Sames and Moebs, 1989), minor faulting (Figure 10), and slumps. None of these structures are readily identified in core, and all disrupt the formation of an effective roof beam through rock bolting. It is impractical to describe the individual effects of the above structures on mine roof stability here. More detailed information on each structure is available in the references cited.

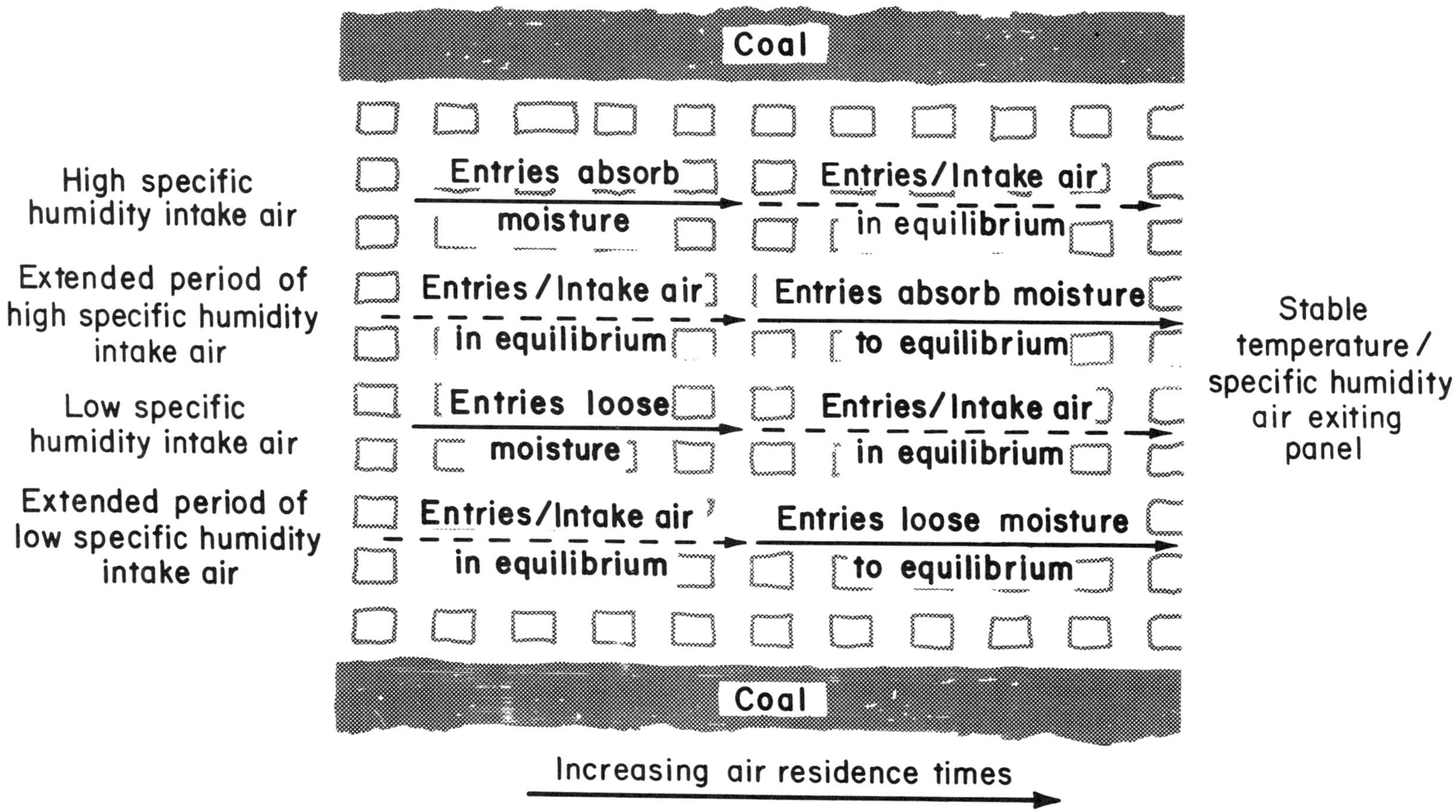

Figure 1. *Moisture gain and loss characteristics of an air-tempering panel.*

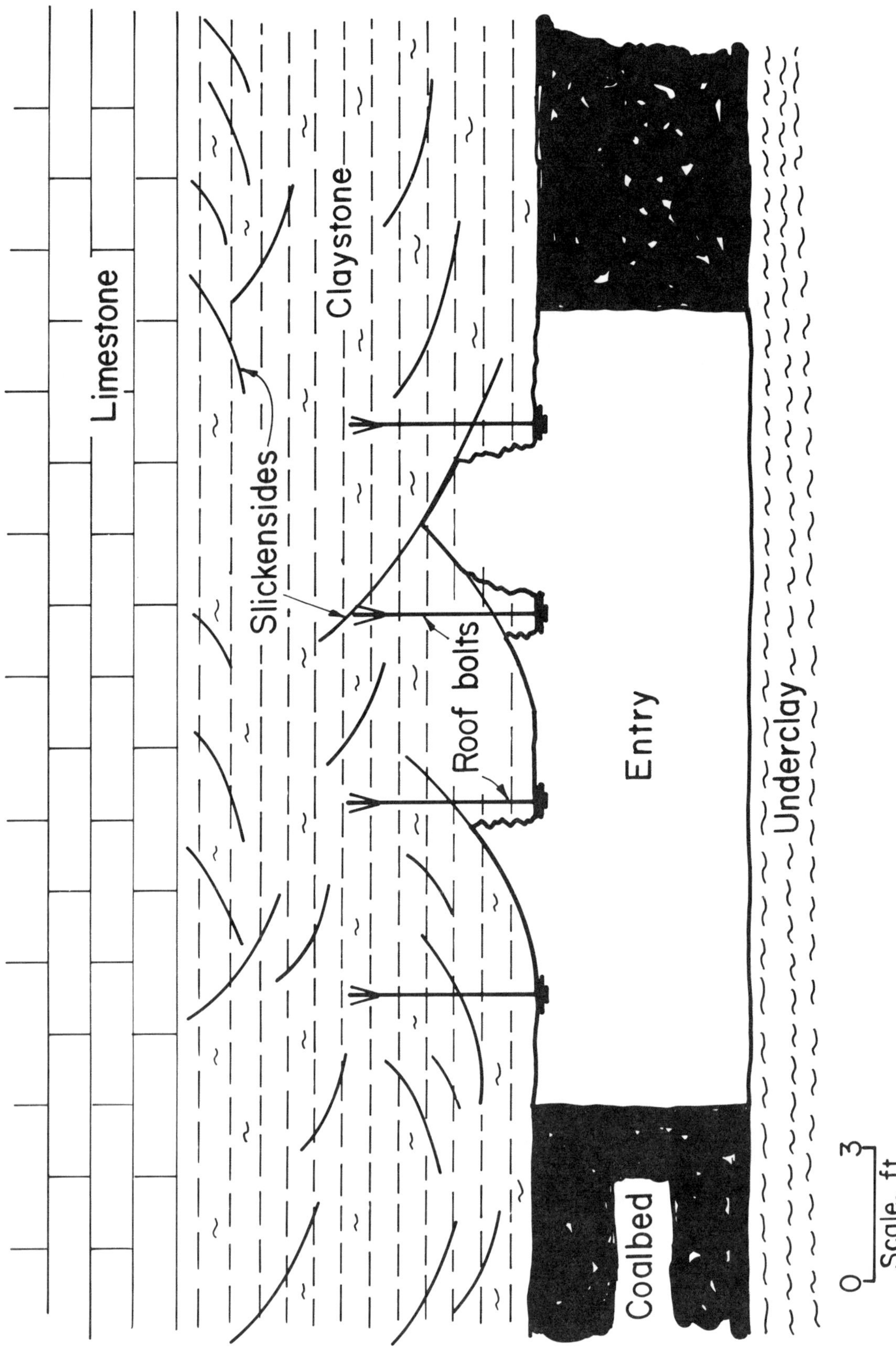

Figure 2. *Slickenside-induced roof failure.*

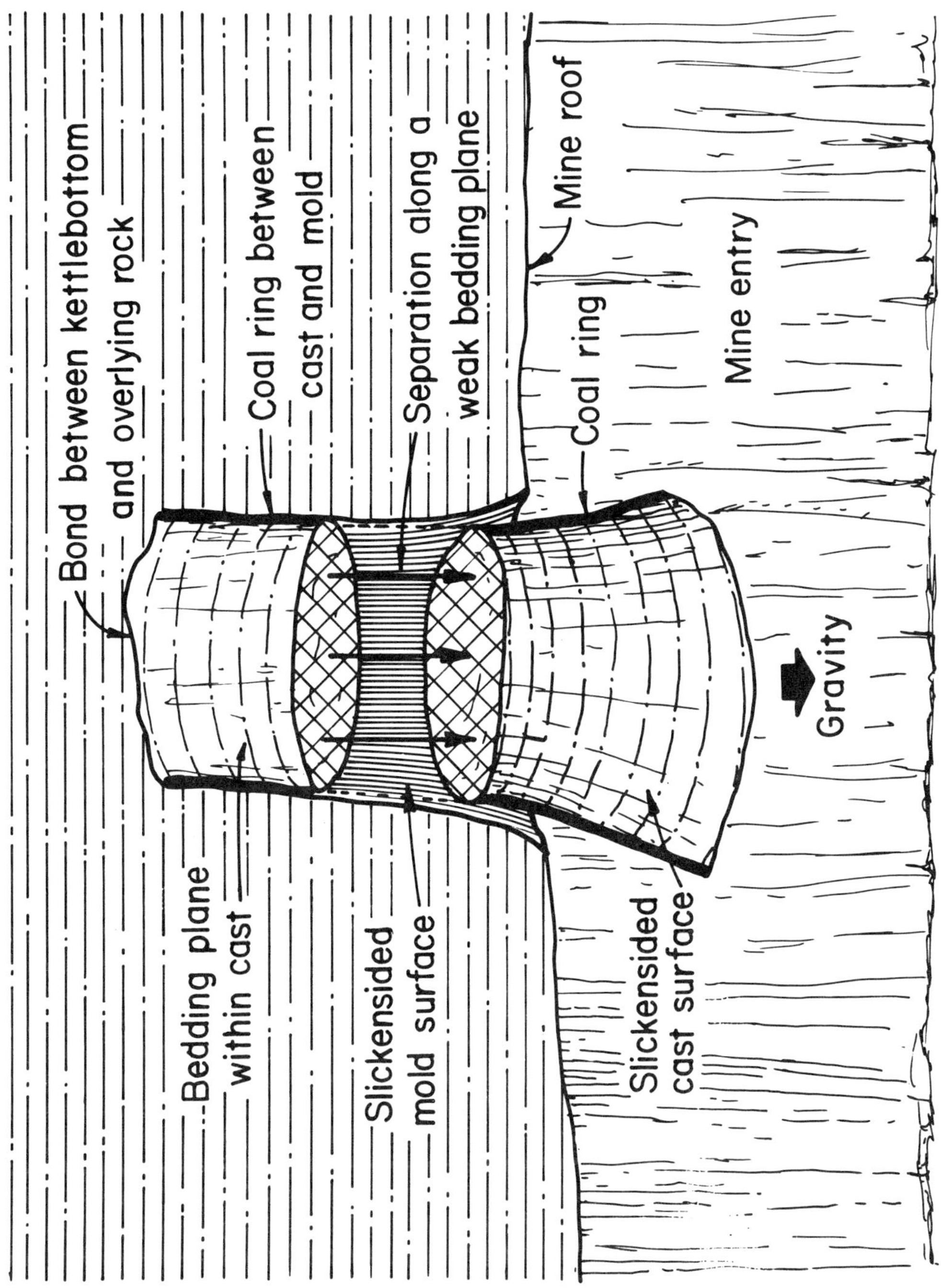

Figure 3. *Kettlebottom characteristics and mode of failure.*

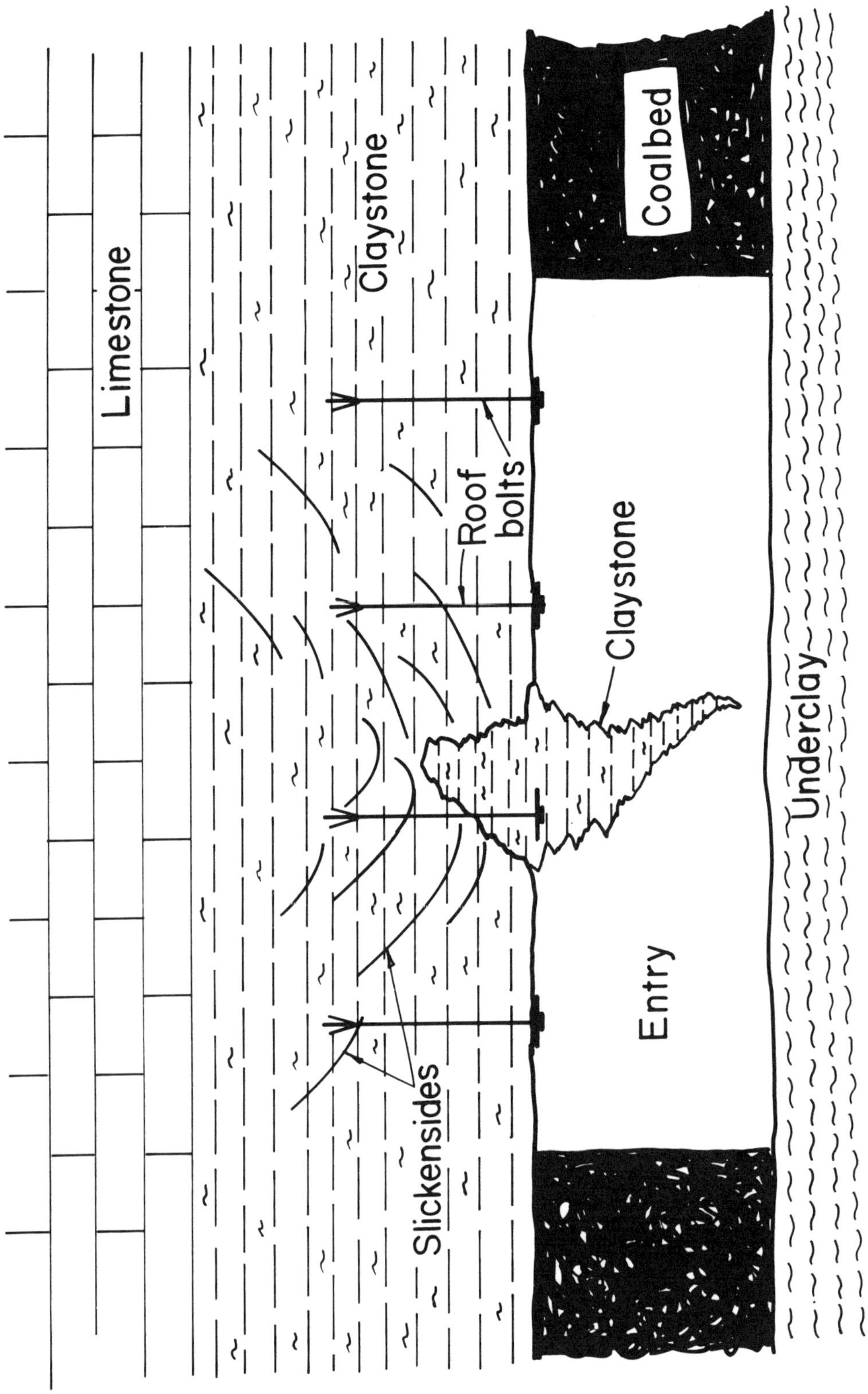

Figure 4. *Effects of a clastic dike on mine roof.*

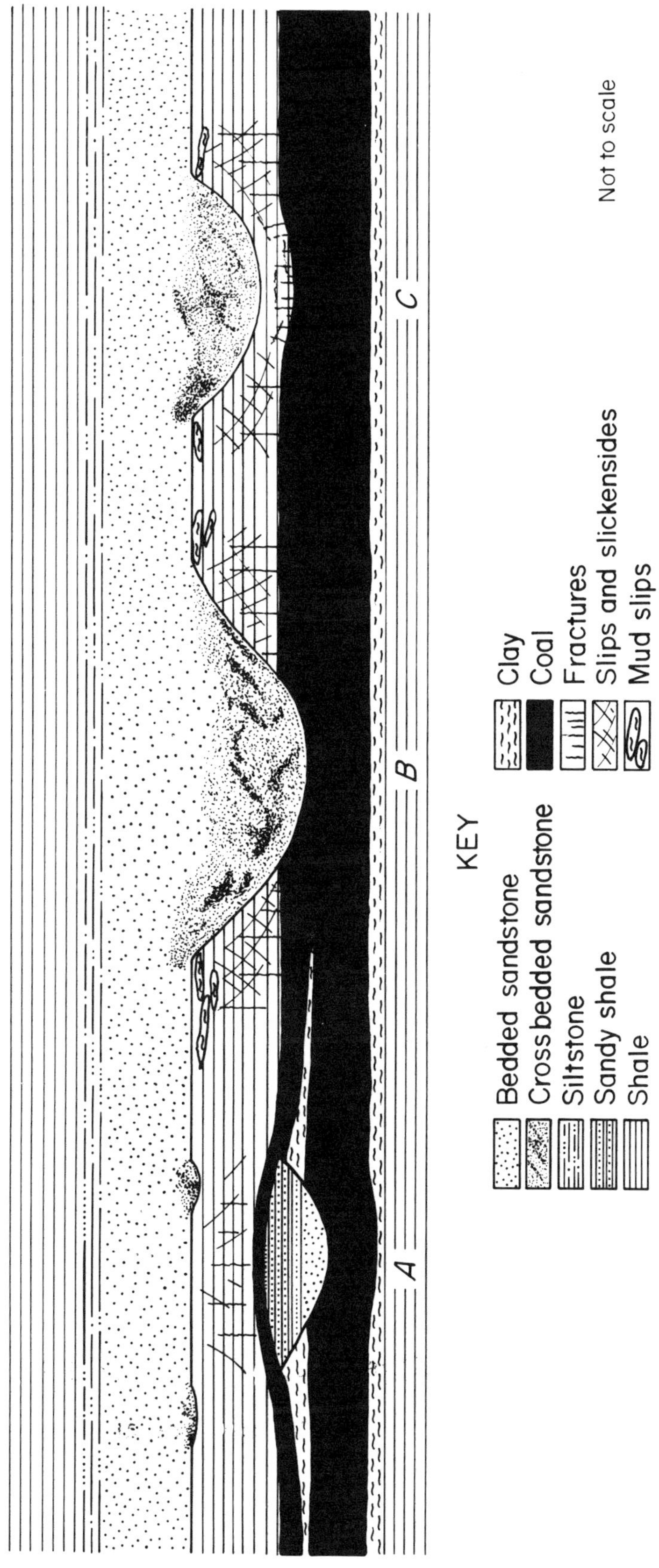

Figure 5. *Paleochannel characteristics that affect roof stability.*

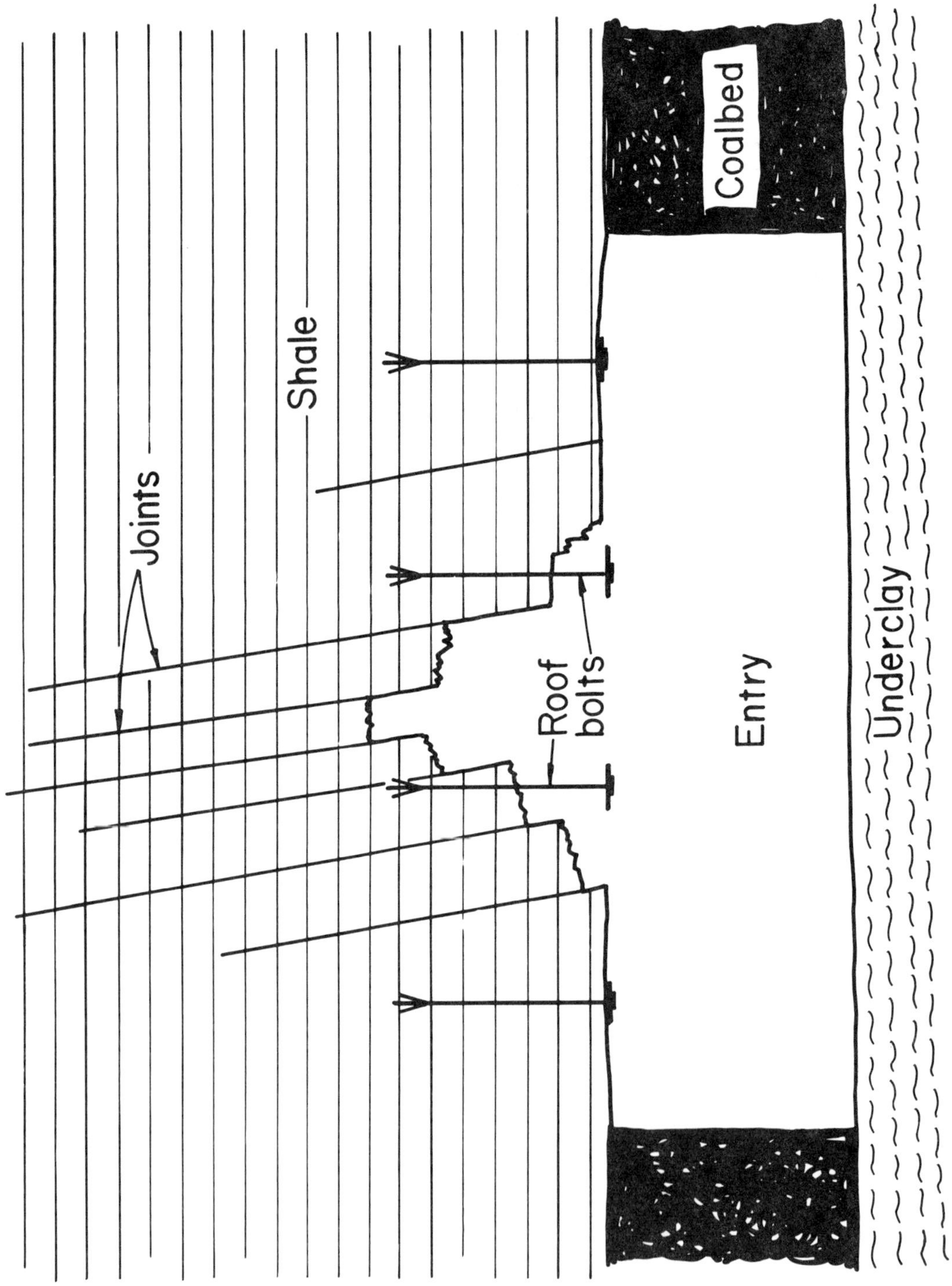

Figure 6. *Roof failure associated with jointing.*

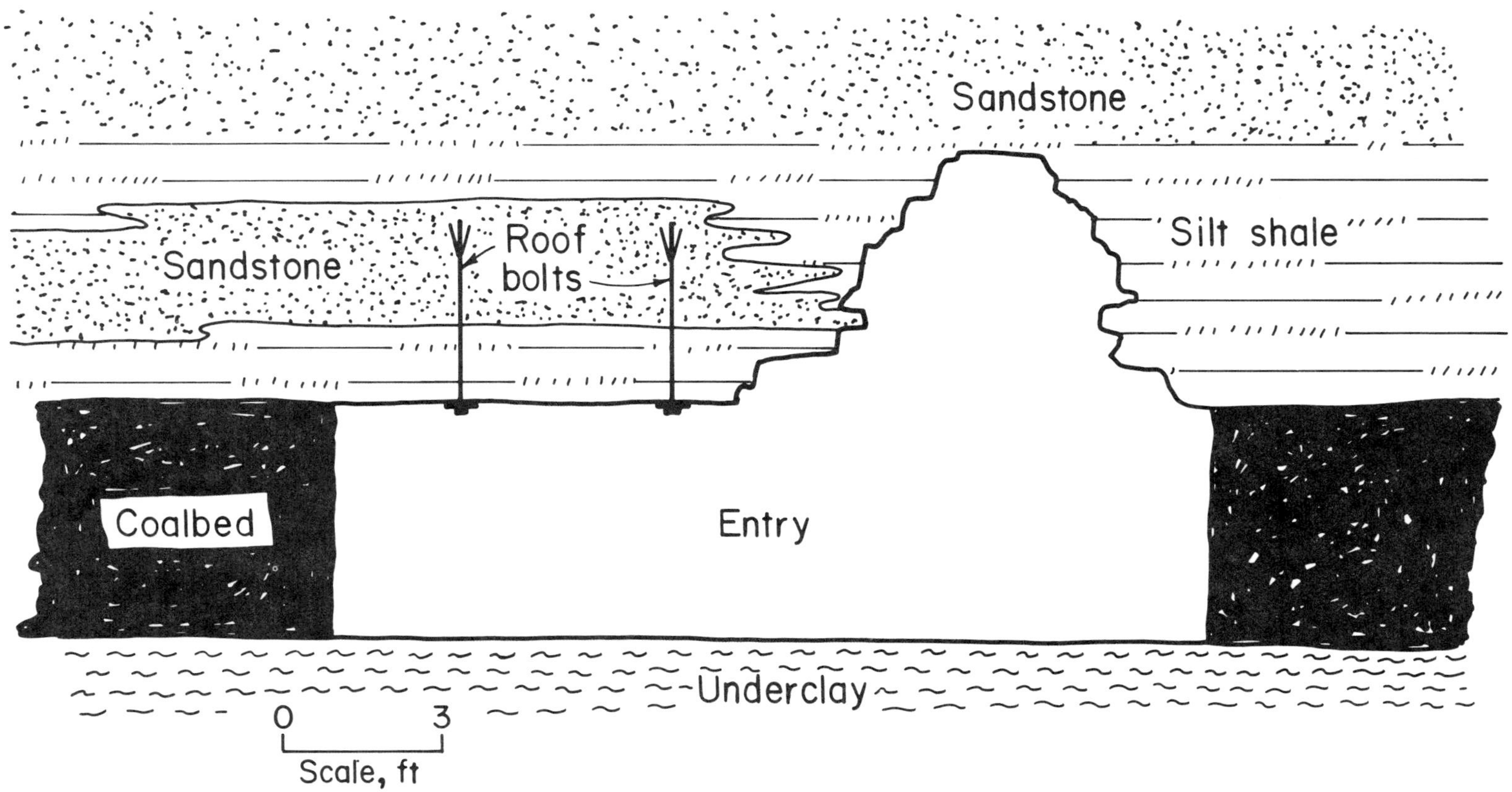

Figure 7. *Roof failure associated with pinchouts.*

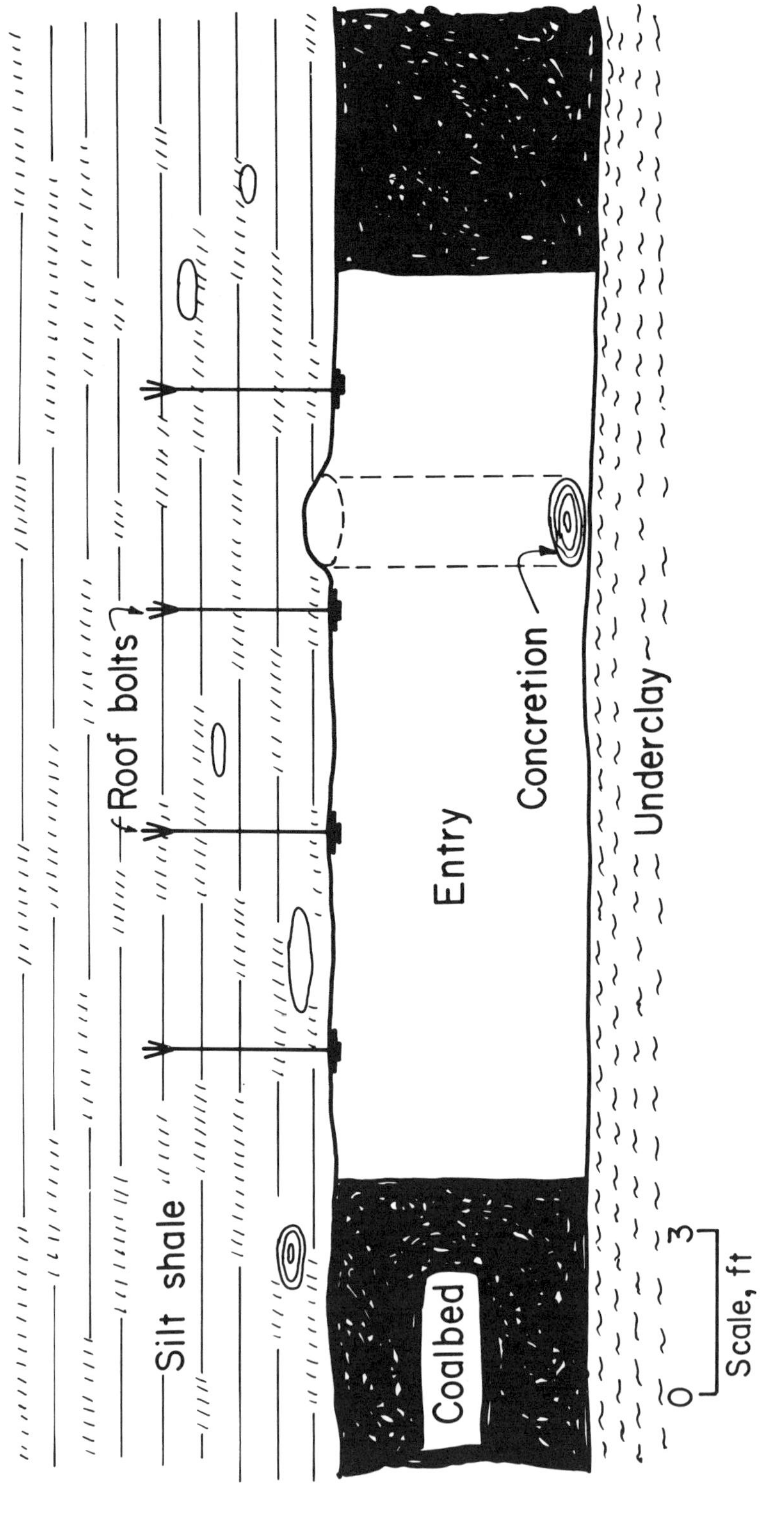

Figure 8. *Roof failure associated with concretions.*

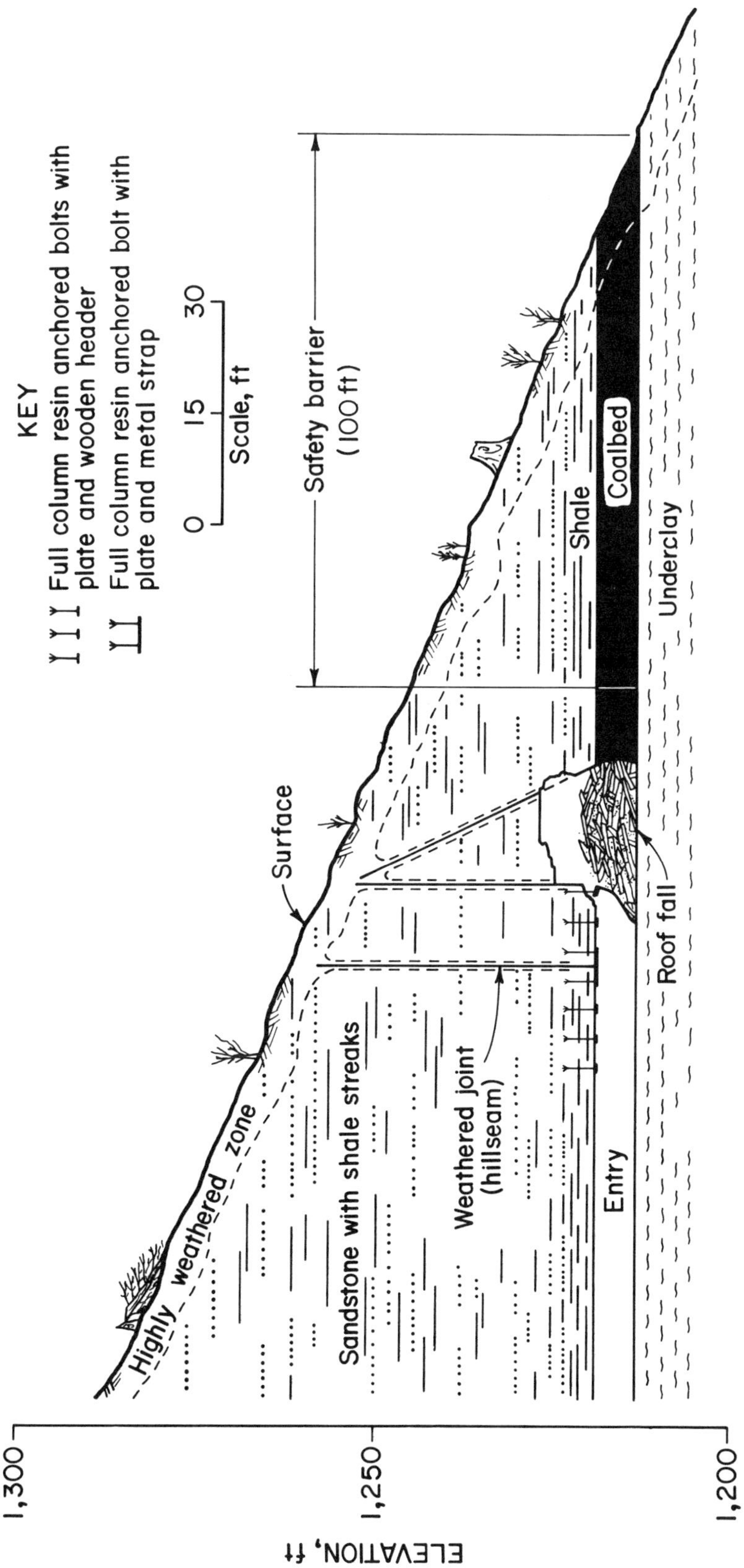

Figure 9. *Roof failure associated with hillseams.*

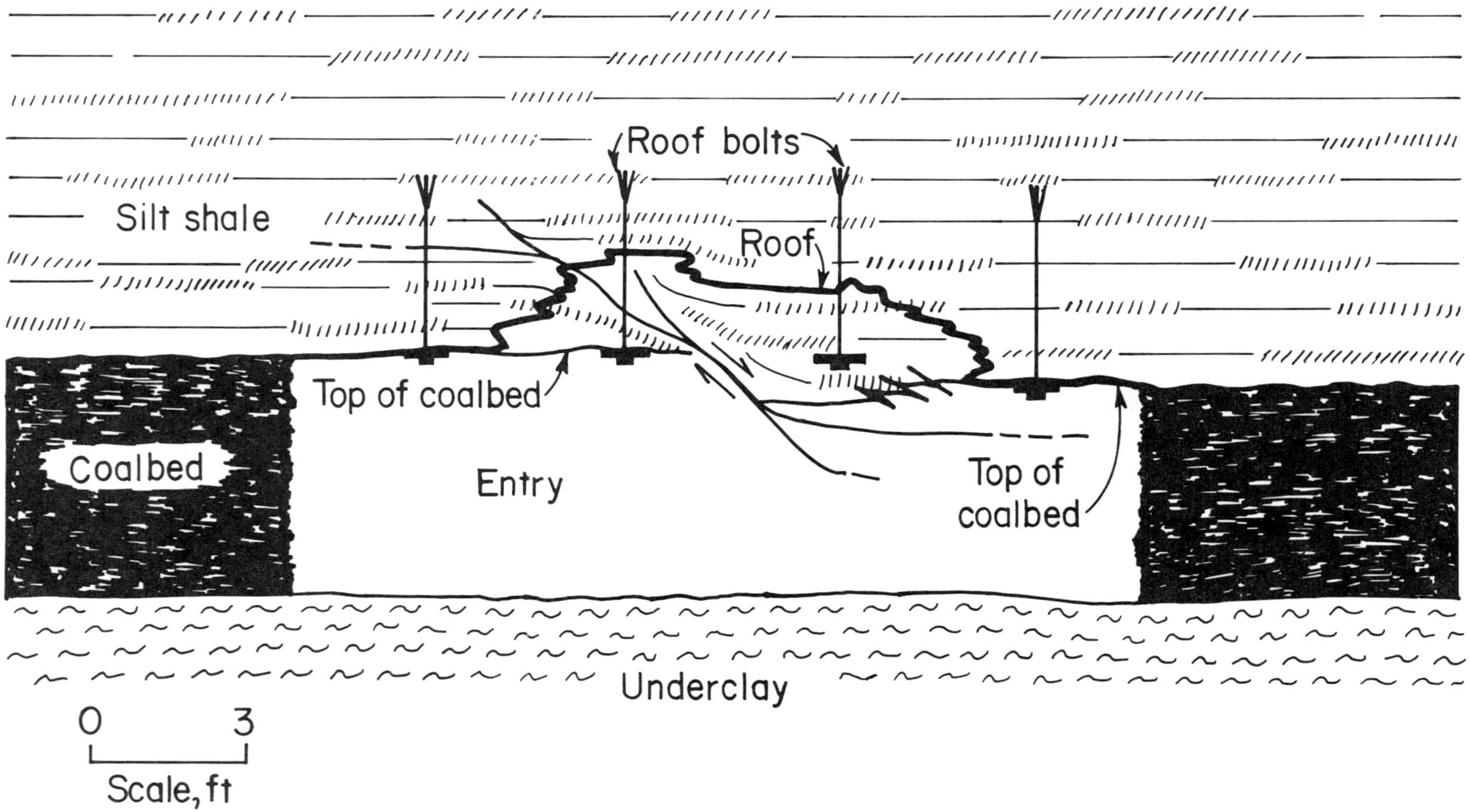

Figure 10. *Effects of minor faulting on roof stability.*

The identity of each of these minor structures becomes apparent during careful examinations of roof falls. The correlation between a structure and persistent roof failure then is established by systematic mapping of both the roof falls and structure. Most of the structures, including paleochannels, clastic dikes, slickensides, slumps, joints, hillseams, pinchouts, and minor faults, tend to be linear. Directional trends can be projected once several occurrences are established on a mine map. Kettlebottoms, dinosaur footprints, and concretions tend to occur in limited areas, and when one is encountered others should be expected.

It also is impractical to attempt to describe all the various support options possible for controlling roof instability caused by each of the mentioned structures. Most of the references cited above also contain support information, but knowing the true nature of a structure causing persistent roof failure allows the tailoring of supplementary support and mine planning to local conditions.

Subtype G5--Major Structures

Subtype G5 includes roof failure associated with tectonic faults and folds that occur along the eastern limits of the Appalachian coal region, and in the Coosa and Warrior basins of Alabama, the Illinois Basin, and some Western U.S. coal fields. Most major tectonic structures are well known, and plans are usually developed for operating in or near them well in advance of mining. They are, therefore, outside the scope of this paper. However, roof instability associated with major structures in Illinois (Nelson, 1981), Western U.S. coal fields (Sames and Laird, 1988), and Southwest Virginia (Molinda and Ingram, 1989) is documented.

Type S: Stress-Related Category

The Type S category of roof failure attributed to stress is divided into the following three subtypes: S1 (valley effect); S2 (regional stress); and S3 (induced stress).

Subtype S1--Valley Effect

Roof fall pattern and failure mode are the primary indicators of S1 roof failure. It is one of the most common and easily recognized types of roof failure in the northern Appalachian coal region. Subtype S1 failure occurs beneath stream valleys in areas of high topographic relief. Roof failure is the result of high lateral compressive stresses (Moebs, 1977). Differential gravitational loading between a stream valley and adjacent topographic highs concentrates the lateral stresses (Figure 11).

A map of surface topography, showing valleys of 30 m (100 ft) or more relief, reveals valley effect failure when superimposed on the roof fall map. The roof falls will occur beneath the valleys to a depth of about 180 m (600 ft). Below 180 m (600 ft), the effect diminishes rapidly (Moebs, 1977). Subtype S1 roof failure generally begins within a few days of mining as a cutter, or shear, where the roof and rib intersect (Figure 12). Failure usually initiates between intersections, but often progresses along an entry parallel to the valley for many feet. Even the most competent roof rock is subject to this type of failure.

The immediate installation of support commonly is recommended to prevent any yielding of the roof under S1 failure conditions. Angled bolts installed to intersect the potential shear plane and anchor above the pillar is an effective supplementary support (Figure 13). Tensioned angle bolts near the roof-rib intersection redistribute the shear stress in the corner (Hill, 1986). The reduction of shear stress reduces the probability of failure. A drawback of angle bolting is the need for tilt-head bolters for on-cycle installation. Roof trusses are sometimes successful where angle bolting alone is not. Again, on-cycle installation is important and requires a bolter able to drill inclined holes. Posts and crossbars or cribs often are necessary to control the roof when S1 failure already has formed.

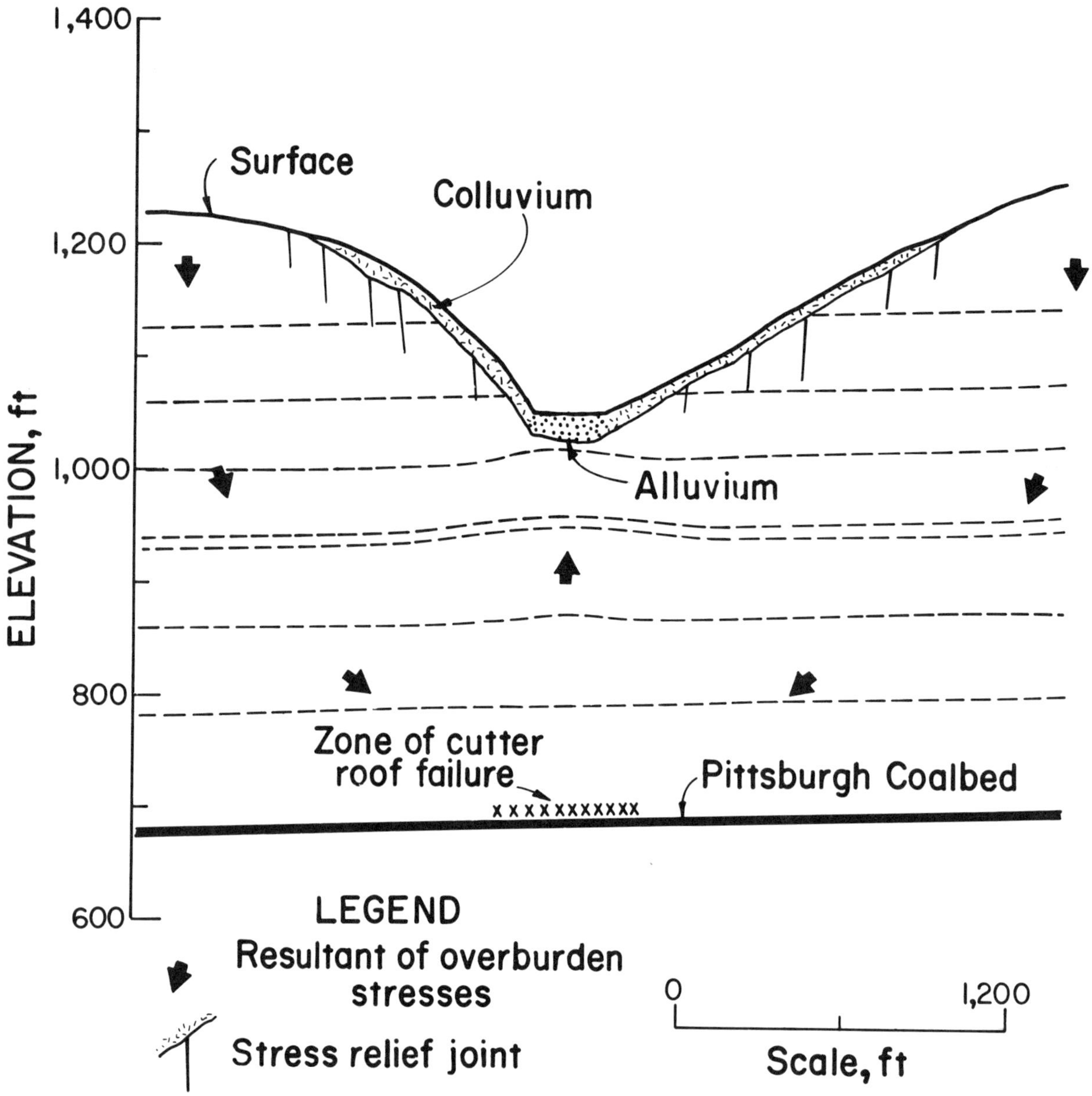

Figure 11. *Concentration of lateral stress by differential loading (valley effect).*

Subtype S2--Regional Stress

Roof fall pattern and failure mode also are the primary indicators of S2 roof failure. S2 failure is caused by remnant, high lateral compressive stress related to past tectonic activity. It is similar to S1 failure, but is not limited to areas beneath valleys, and generally occurs in deeper mines well below present surface drainage elevations.

S2 failure may appear to occur at random on the roof fall map, but with sufficient roof fall mapping often will show preferred directional trends. A preferred north-south trend is common in the northern Appalachian coal region. S2 failure also is characterized by the development of a shear or cutter along the rib line within a few

Figure 12. *Early cutter, or shear, formation along roof and rib intersection.*

weeks of mining. Falls generally begin between intersections and may zig-zag around pillars or along cutters across intersections to follow the preferred trend. When the high horizontal stress field is nondirectional (little difference between the maximum and minimum principal horizontal stress), a clear, preferred directional trend may not be recognizable (Hill, 1986).

In addition to the support recommended for controlling cutter roof already mentioned under subtype S1, mine design changes often can lessen the effects of regional stress. Design changes include entry reorientation, sacrifice entries, and yielding and softened pillars.

Subtype S2 failure will occur in entries perpendicular to the maximum stress when the horizontal stress field is strongly directional. Reorientation of the main headings at an oblique angle to the maximum horizontal stress direction can offset the influence of the stress field (Hill, 1986). Orienting the main headings parallel to the main stress field and staggering pillars isolates roof failure to crosscuts.

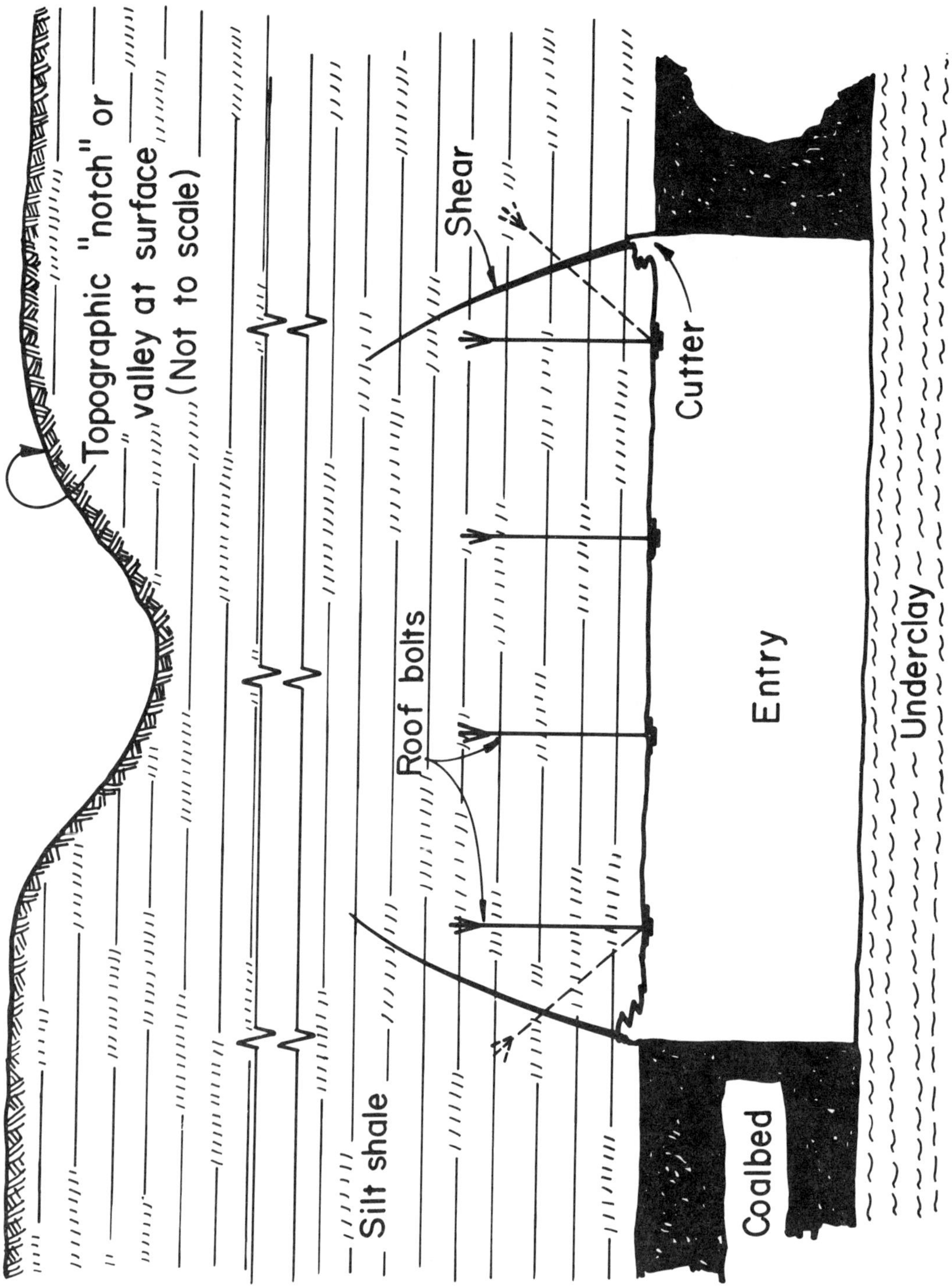

Figure 13. *Intersection of potential shear plane by angled bolts.*

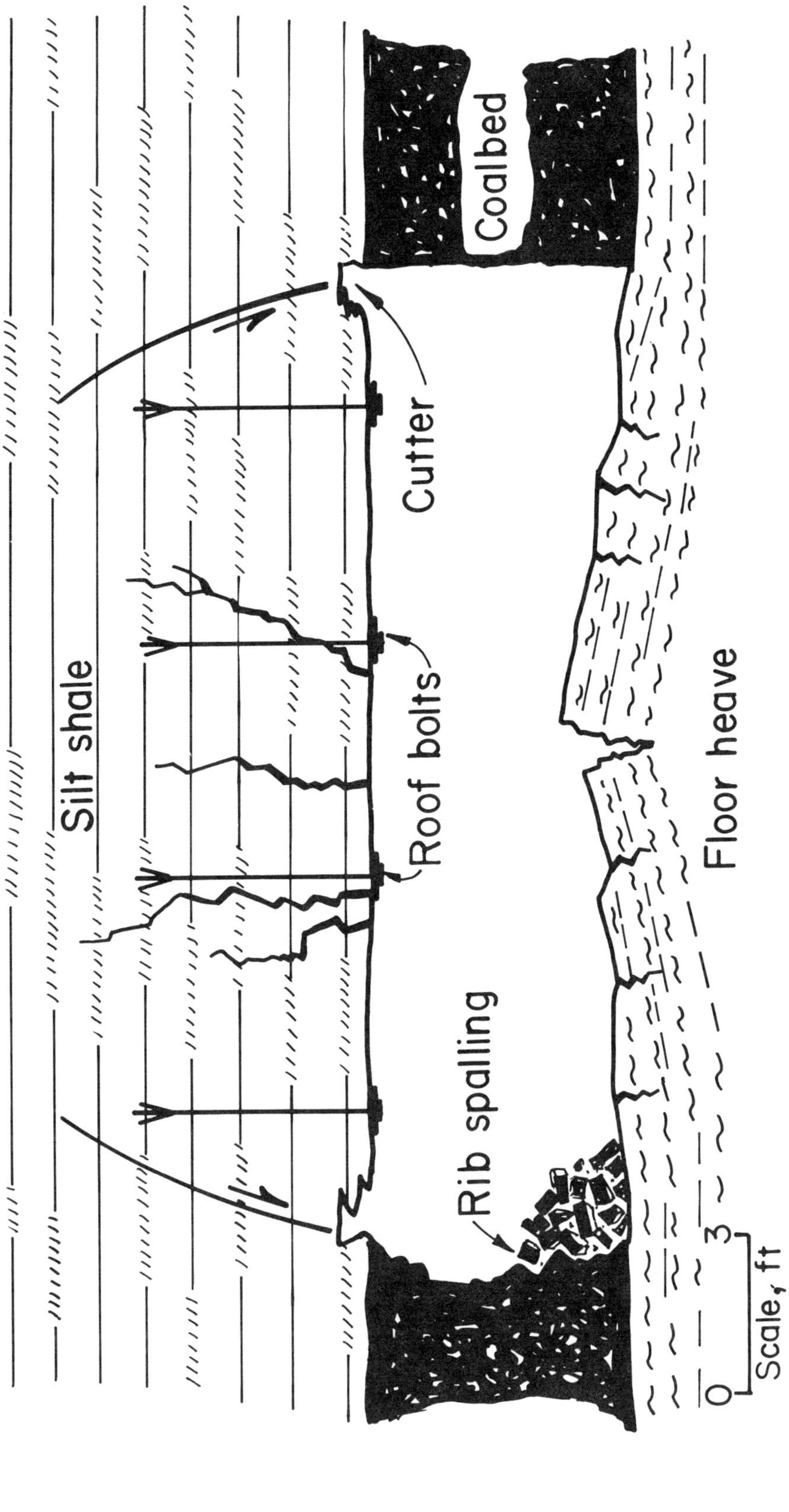

Figure 14. *Effects of multiple-seam mining on ground control.*

Sacrifice entries are driven and the roof allowed to fail to relieve horizontal stress (Hill, 1986; Kripikov, 1982). Adjacent entries then are driven into the destressed area.

Yield pillars transfer roof loads to abutments, reducing overall roof and floor stresses (Kripikov, 1982). Pillar softening, by drilling 15-cm (6-in.) diameter holes into the pillars and face during entry advance, is a technique for reducing stress concentrations in the corners of entries (Kripikov, 1982; Wang, Ropchan et al, 1974).

Subtype S3--Induced Stress

Subtype S3 roof failure is indicated by failure mode and pattern. It almost always is related to superjacent or subjacent mining (multiple-seam mining), or to pressure override from an adjacent panel which is not fully caved. Depending on mining sequence in multiple-seam operations (overmining, undermining, or simultaneous mining), several different manifestations of induced stress can occur.

Subtype S3 failure resulting from pressure override generally results in roof instability within a few hundred feet of the pillared panel. This relationship will show clearly on the mine roof fall map. The remedy for S3 roof failure due to pressure override is improved pillar extraction and caving or oversized pillars (Moebs and Stateham, 1986). Control measures for cutter roof resulting from pressure override are the same as those discussed for S1 failure.

Subtype S3 problems associated with multiple-seam mining include pillar crushing and failure, rib spalling, floor heave, roof cutters, and roof fracturing (Figure 14) (Haycocks and Karmis, 1983). The relationship between these problems and either overmining or undermining become apparent when the current operating mine map is overlaid by the map of any superjacent or subjacent mines.

Because of economics, ownership, and availability, a mine operator may have little control over coal seam mining sequence. Minimizing problems that already are occurring due to multiple-seam mining may require changes in the mining plan. Changes can include reducing percentage extraction, decreasing roof span width, staggering pillars, or changing heading directions (Chekan and Matetic, 1988). Changes in the roof control plan also may be necessary to contend with changing roof conditions caused by overmining or undermining (Chekan and Matetic, 1986). Extraction-sequencing guidelines are available for advance development of mining plans for multiple coal seams (Bieniawski, 1982; Britton, 1980).

CONCLUSION

Geologic diagnosis for categorizing roof falls is based on causative factors. The data collection required by this method tailors the subsequent analysis to the individual mine setting. This allows the roof specialist of a mining company to sort out ground control problems and arrive at several alternatives for their control. Although it is impractical to describe each subtype in detail here, once a subtype is suspected as the causative factor, references are available with in-depth descriptions and remedial actions for each. Geologic diagnosis, if pursued conscientiously, can contribute to accident prevention, improved roof stability, and reductions in cleanup and repair costs.

REFERENCES

Bieniawski, Z.T., 1982, Improved design of room-and-pillar coal mining: Department of Energy Grant DE-FG01-78ET-11428, Pennsylvania State University, Report DOE/ET/11428T1, 165 p.

Britton, S.G., 1980, Mining multiple seams: Coal Mining and Processing, v. 17, no. 12, p. 64-70.

Chase, F.E., and G.P. Sames, 1983, Kettlebottoms: Their relation to mine roof and support: U.S. Bureau of Mines Report of Investigations RI 8785, 12 p.

Chase, F.E., and J.P. Ulery, 1987, Clay veins: Their occurrence, characteristics, and support: U.S. Bureau of Mines Report of Investigations RI 9060, 19 p.

Chekan, G.J., and R.J. Matetic, 1986, Strata interactions in multiple-seam mining--Two case studies in Pennsylvania: U.S. Bureau of Mines Report of Investigations RI 9056, 17 p.

Chekan, G.J., and R.J. Matetic, 1988, Loading characteristics of pillars in multiple-seam mining operations: U.S. Bureau of Mines Report of Investigations RI 9173, 19 p.

Ellenberger, J. L., 1979, Slickenside occurrence in coal mine roof of the Valley Camp No. 3 Mine near Wheeling, W. Va.: U.S. Bureau of Mines Report of Investigations RI 8365, 17 p.

Haycocks, C., and M. Karmis, 1983, Ground control mechanisms in multi-seam mining: U.S. Bureau of Mines Open-File Report 7-84, 342 p.

Haynes, C.D., 1975, Effects of temperature and humidity variations on the stability of coal mine roof rocks: U.S. Bureau of Mines Open-File Report 8-77, 385 p.

Hill, J.L., III, 1986, Cutter roof failure: An overview of the causes and methods for control: U.S. Bureau of Mines Information Circular IC 9094, 27 p.

Hylbert, D.K., 1978, The classification, evaluation, and projection of coal mine roof rocks in advance of mining: Mining Engineering, v. 30, no. 12, p. 1667-1676.

Kripakov, N.P., 1982, Alternatives for controlling cutter roof in coal mines, in Peng, S.S., and J.H. Kelley, eds., Proceedings of the Second Conference on Ground Control in Mining, Morgantown, West Virginia, July 19-22, 1982: Morgantown, West Virginia University, p. 142-151.

Moebs, N.N., 1977, Roof rock structures and related roof support problems in the Pittsburgh Coalbed of Southwestern Pennsylvania: U.S. Bureau of Mines Report of Investigations RI 8230, 30 p.

Moebs, N.N., and J.L. Ellenberger, 1982, Geologic structures in coal mine roof: U.S. Bureau of Mines Report of Investigations RI 8620, 16 p.

Moebs, N.N., and R.M. Stateham, 1986, Coal mine roof instability: Categories and causes: U.S. Bureau of Mines Information Circular IC 9076, 15 p.

Molinda, G.M., and D.K. Ingram, 1989, Effects of structural faults on ground control in selected mines in Southwest Virginia: U.S. Bureau of Mines Report of Investigations RI 9289, 25 p.

Nelson, W.J., 1981, Faults and their effect on coal mining in Illinois: Illinois State Geological Survey, Circular 523, 40 p.

Patrick, W.C., and N.B. Aughenbaugh, 1979, Classification of roof falls in coal mines: Mining Engineering, v. 31, no. 3, p. 279-283.

Sames, G.P., 1985, Coal mine air tempering: Effectiveness, design, and roof support: U.S. Bureau of Mines Report of Investigations RI 8955, 20 p.

Sames, G.P., and R.B. Laird, 1988, Geologic conditions affecting coal mine ground control in the western United States: U.S. Bureau of Mines Information Circular IC 9172, 30 p.

Sames, G.P., and N.N. Moebs, 1989, Hillseam geology and roof instability near outcrop in Eastern Kentucky drift mines: U.S. Bureau of Mines Report of Investigations RI 9267, 32 p.

Wang, F.D., D.M. Ropchan, and M.C. Sun, 1974, Proposed technique for improving coal mine roof stability by pillar softening: Transactions of the Society of Mining Engineers of AIME, v. 255, p. 59-63.

Weir, C.E., 1970, Factors affecting coal mine roof rocks in Sullivan County, Indiana: Proceedings of the Indiana Academy of Sciences 1969, v. 79, p. 263-269.

CLEAT AND JOINT SYSTEM EVALUATION IN COAL SEAMS AND ZONES OF THE BLACKHAWK FORMATION AND FERRON SANDSTONE MEMBER OF THE MANCOS SHALE FORMATION FOR MINE PLANNING AND SITE CHARACTERIZATION IN THE BOOK CLIFFS, WASATCH PLATEAU, AND EMERY COAL FIELDS, UTAH

Brigitte P. Hucka
Utah Geological Survey
2363 South Foothill Drive
Salt Lake City, UT 84109-1491

ABSTRACT

Cleats in coal and joints in the roof rocks of coal mines are important features to be considered during mine planning and design. Knowledge of their presence and orientations is important for evaluating roof stability and for predicting methane and water problems. Cleat and joint systems were examined in 21 active mines located in the Book Cliffs, Wasatch Plateau, and Emery coal fields of Utah. The coal-bearing strata include the Blackhawk Formation and the Ferron Sandstone Member of the Mancos Shale Formation, both units being Cretaceous in age.

The frequency of cleat and joint distributions and orientations in the seams and outcrops was investigated by use of stereographic projection and rose diagrams. Eighty-four rose diagrams and Schmidt stereograms were prepared from approximately 4,000 field measurements. Further, the cleat and joint planes were studied to determine the extent of healing by calcite or other accessory minerals. Minerals in fillings cause changes in cohesion along cleats. Analytical data suggest that cleats in coal and joints in outcrops are the results of tectonic forces. Further, cleats and joints maintain the same pattern in the Book Cliffs coal field and differ in the more tectonically disturbed Wasatch Plateau and Emery coal fields. The relationship of cleats and joints to faults was addressed in the study as well. The results, presented on three regional maps, will help the mining industry in planning and designing mine structures and in degasification of gassy coal mines.

INTRODUCTION

The objective of this study was to provide mine operators of selected mines with complete information on cleat and joint systems and other discontinuities in coal seams and in areas surrounding the mines. Regional maps prepared from the results should assist coal operators in designing mine structures, allowing them to consider the influence of cleat and joint systems on mine stability and potential migration of water and methane gas into mine workings.

The stratigraphic interval where cleat and joint measurements were completed included 16 coal seams and zones in the Cretaceous Blackhawk Formation and Ferron Sandstone Member of the Cretaceous Mancos Shale Formation. The area of study covers approximately 41,200 km^2 (1,590 mi.^{2}) (Doelling, 1972) and extends over three main coal-producing fields in Utah: the Book Cliffs, Wasatch Plateau, and Emery coal fields.

The distribution of cleats in coal and the joint systems in adjacent rocks are important in the coal mining industry. Cleat orientation is one of several factors that influence mining efficiency and, hence, coal-mine layout. The density, orientation, and aperture of coal cleats also greatly influence the permeability to gas or water, and the orientation of joint systems influences roof stability. It is known that sudden gas and coal outbursts (ejections of significant volume of coal accompanied by natural gas) are, among others, directly related to predominant cleat and joint systems (Ammosov et al., 1963). Therefore, a thorough knowledge of the cleat and joint systems is essential in the planning of safe and efficient coal mines. Consequently, data on coal cleats and on joints in surrounding rocks should be collected from outcrops before coal mine development and should be updated with in-mine measurements as mining progresses.

JOINTS AND CLEATS

Joints are fractures along which little or no movement has occurred. In coal, such fractures are known as cleats. The presence of cleats in coal seams has been known for centuries (Kendall et al., 1933), but only in the past century have studies focused on the origin of cleats. Both tectonic and non-tectonic origins have been proposed (Hofer, 1915). Ammosov et al. (1963) accepted both concepts and termed cleats of non-tectonic origin as "endogenetic" and those formed by tectonic forces as "exogenetic." The endogenetic cleats are thought to form by contraction of the coal during consolidation and coalification. A brief but thorough description of the theories of the origin of cleats is presented by McCulloch et al. (1974).

The fracturing of rocks in the laboratory and in natural settings has been discussed by several researchers (e.g., Friedman, 1964; Price, 1966; Paterson, 1978). According to Friedman (1964), rock fracturing involves the separation of rocks into two or more parts after the rock loses its cohesion and resistance to load and releases stored elastic strain energy. In Figure 1, extension fracturing occurs normal to the least principal stress (σ_3) and parallel to the maximum compressive stress (σ_1). Macroscopically, σ_3 may be either tensile (negative) or compressive (positive).

The most prominent cleats are known as the "face" or "main" cleats or "master-joint" cleats (Kendall et al., 1933). They usually are orthogonal to the "butt" cleats, which previously were termed "cross-cleats" or "board cleats". Vertical cleats in a coal seam may be continuous with joints in the immediate roof rock. When such joint systems are oriented perpendicular to bedding planes, they form blocks which may be supported in the roof only by inter-block friction (Figure 2) or by compressional horizontal stresses. Such blocks must be prevented from falling into mine workings by artificial supports.

The cleat directions in coal also could influence the mine layout plan. Coal mine entries are driven at different orientations with the cleat/joint system, depending on mine roof and coal face stability, ease of coal cutting, etc. For example, in 3-4-m (10-13-ft) thick seams, with coal separating with ease from the roof, a main cleat system dipping into the working space may enable the release of large blocks of coal (slabbing) thus constituting a safety hazard and resulting in decreased productivities. In the case of unstable mine roofs where joint systems in roof rocks coincide in orientation with cleats in coal, entries oblique to the cleats may improve the roof control (Figure 3).

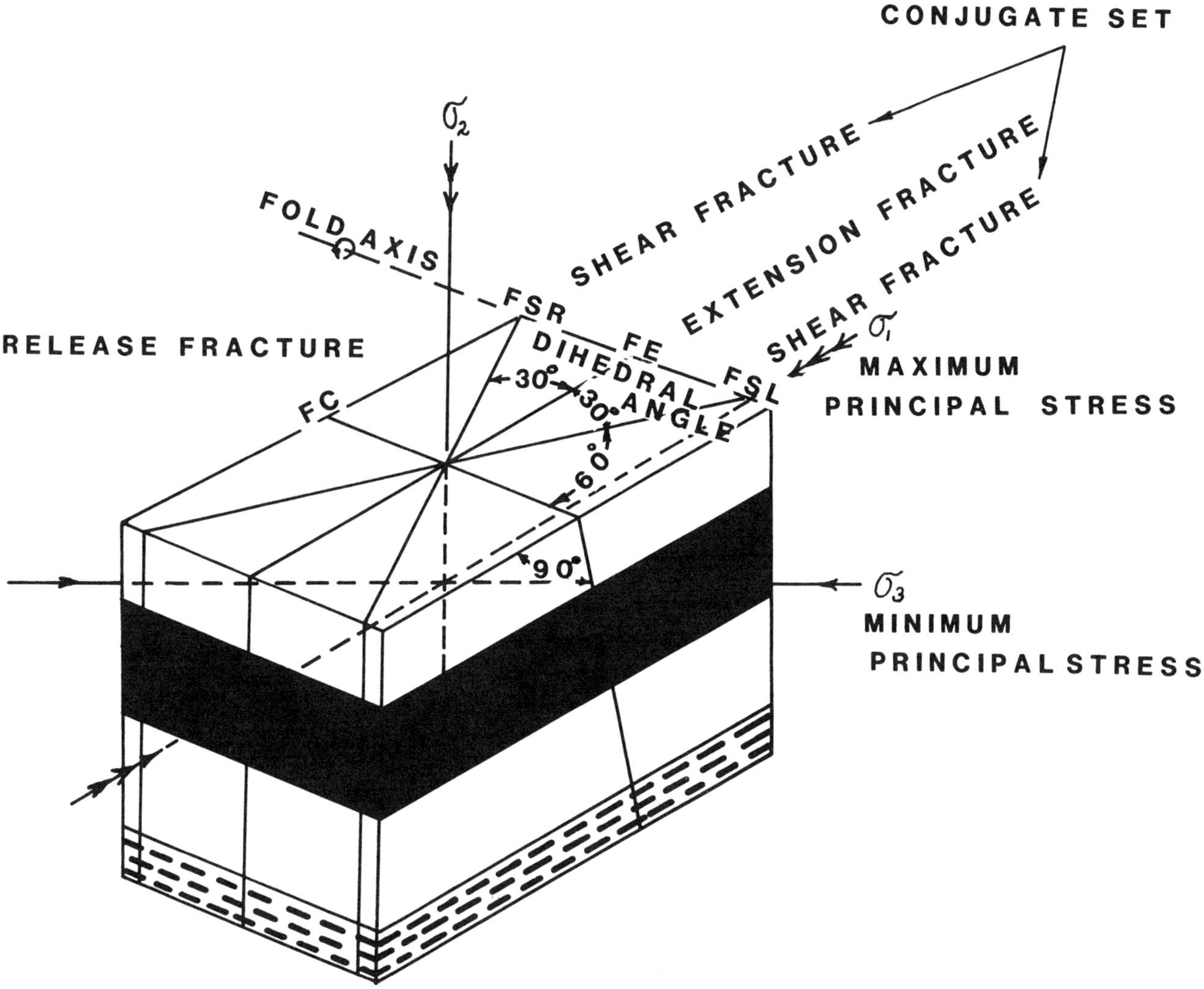

Figure 1. *Fractures in rocks (modified from Friedman, 1964).*

COLLECTION AND PRESENTATION OF DATA

Mine operators in Utah started collecting the cleat and joint systems information only recently, mostly because of a need to improve stability at working faces (M. Bunnell, oral communication, 1988) and for methane drainage tests (C. Allen, oral communication, 1988). As a consequence, the Utah Geological and Mineral Survey, as part of its program to help the mining industry, initiated and funded a 1-yr study (1987-1988) on selected geological factors affecting coal mining (Hucka, 1989).

The data for the structural analysis of cleats and joints were obtained from 16 coal seams and zones in 21 active mines in the Book Cliffs, Wasatch Plateau, and Emery coal fields of Utah (Figure 4). The coal-bearing strata of interest were the Blackhawk Formation and the Ferron Sandstone Member of the Mancos Shale Formation. The results of field measurements were analyzed to establish relationships among 1) joint systems in the areas surrounding the mines, 2) cleats in the coal seams, and 3) joints in the roof strata. The

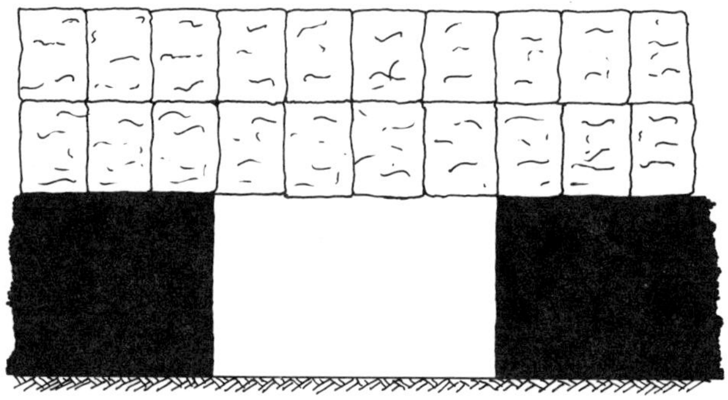

Figure 2. Joints in roof rocks in an entry. Blocks in the roof are supported by friction and keying between blocks.

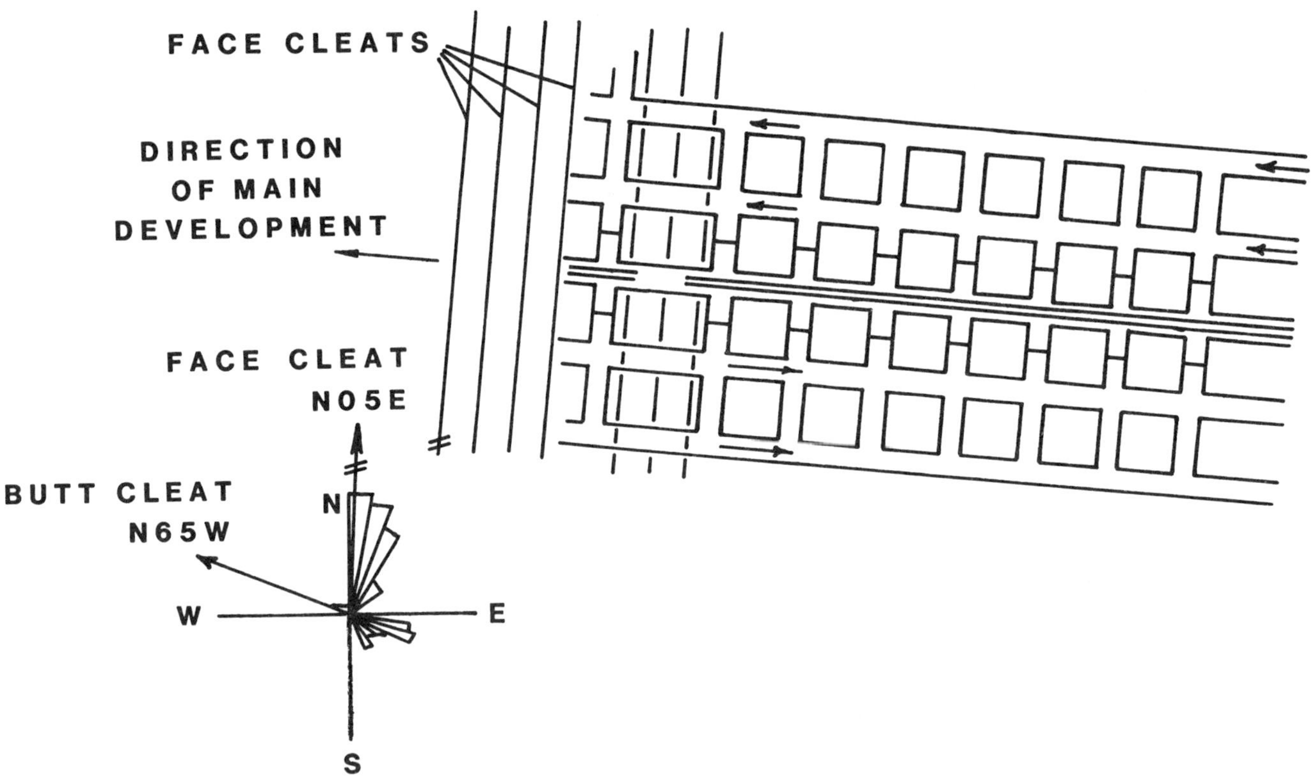

Figure 3A. Idealized mine development plans related to the direction of face cleats. A) Plan preferred by mine operators.

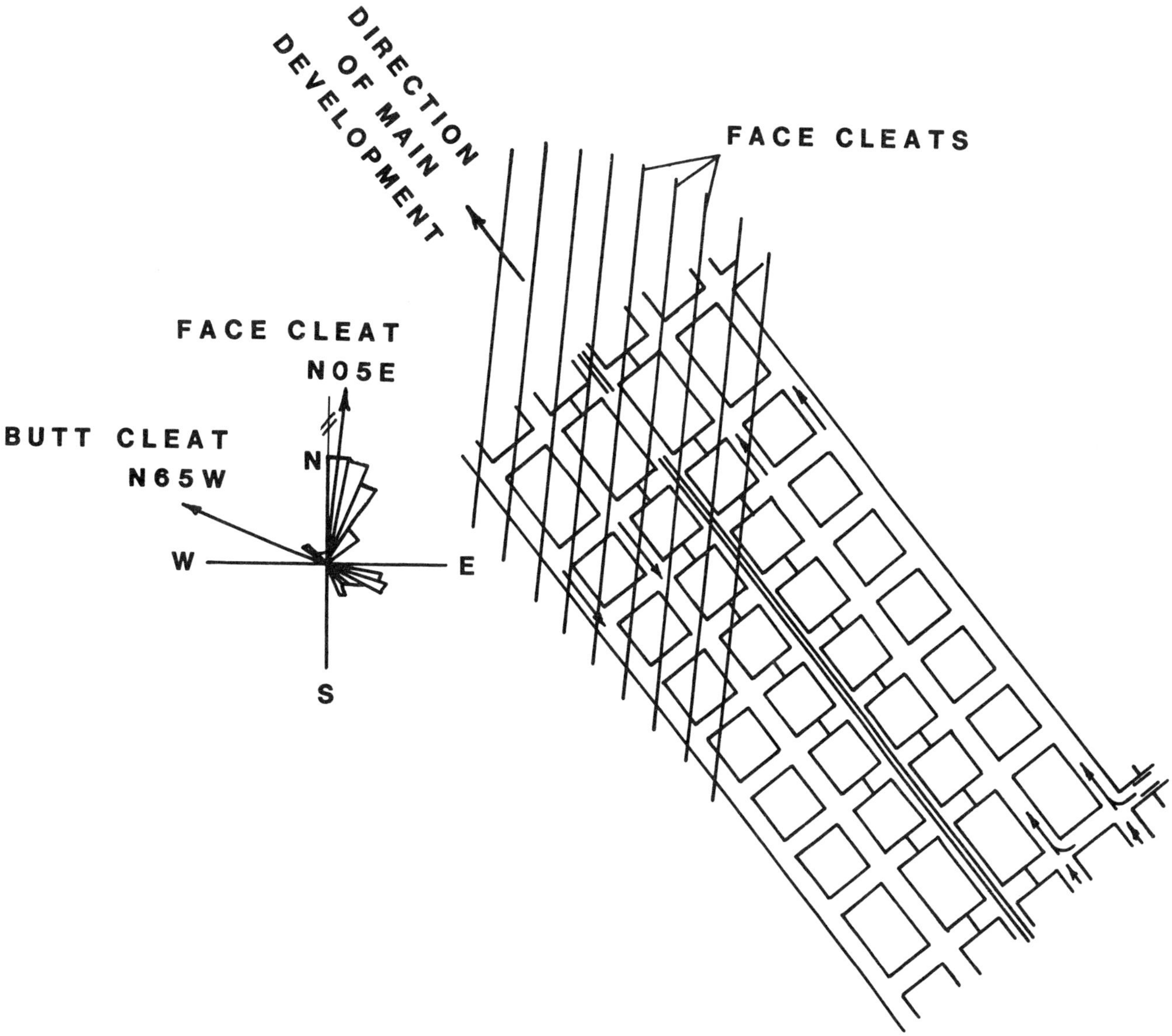

Figure 3B. Idealized mine development plans related to the direction of face cleats. B) Plan that is more common if cleats coincide with roof joints.

cleats and joints were measured using a geological stratum compass (COCLA) which enabled measurement of the azimuth and the dip in a single operation.

At each mine, an average of 100 cleat readings were taken. A similar number of joint readings were taken in the strata within 1.6 km (1 mi.) of the mine.

Coal seams were characterized megascopically and microscopically. The megascopic study consisted of determining the relative proportions of the lithotypes (vitrain, clarain, durain, and fusain) and the mineral accessories and by describing the rock partings. The microscopic analysis consisted of determining the percentage of microscopic constituents (macerals) of the coal. The maceral composition of each seam was determined in white and ultraviolet light on a Leitz Ortholux II microscope at 500X magnification and with a

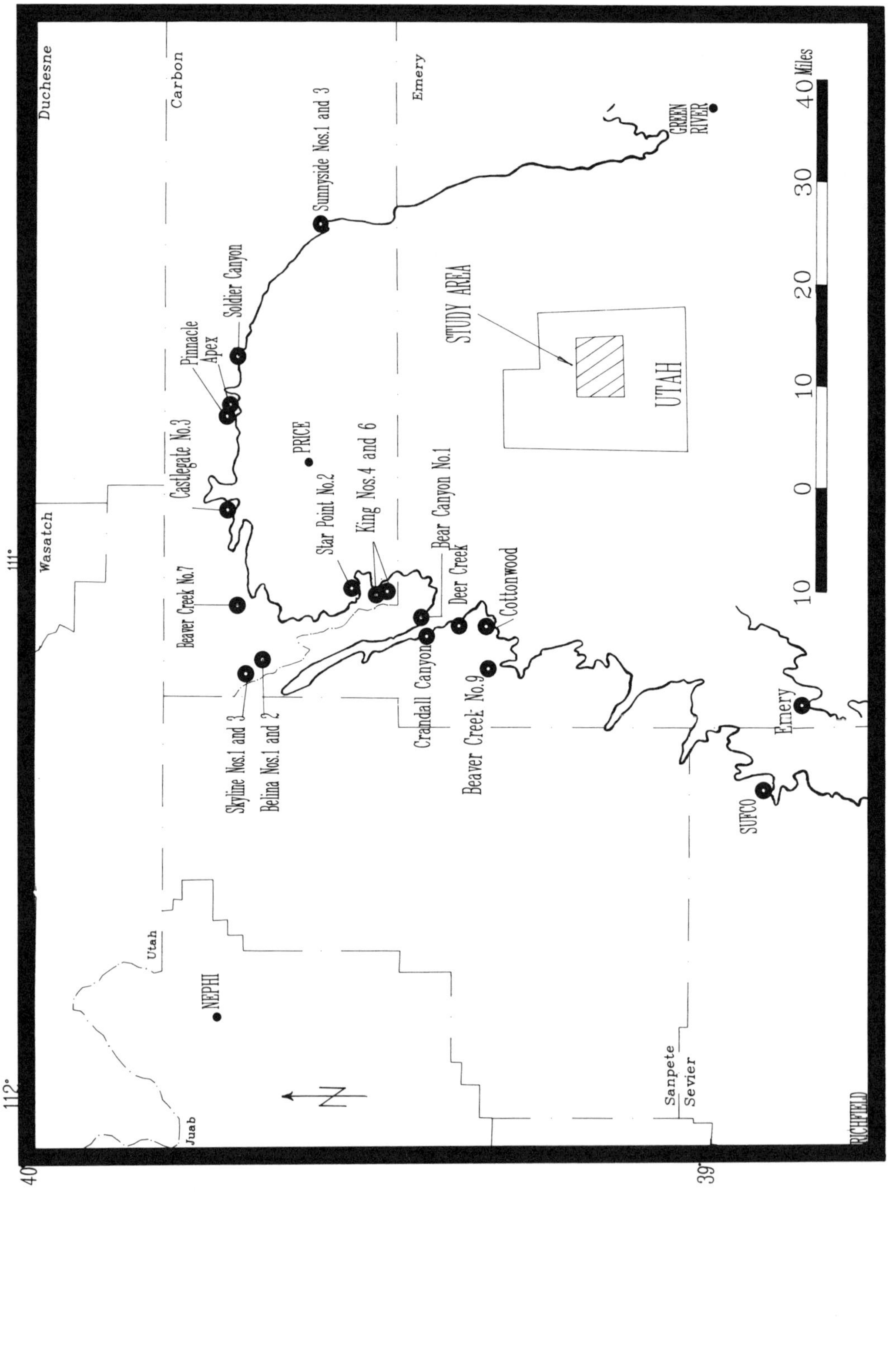

Figure 4. *Locations of mines where cleat and joint orientations were measured.*

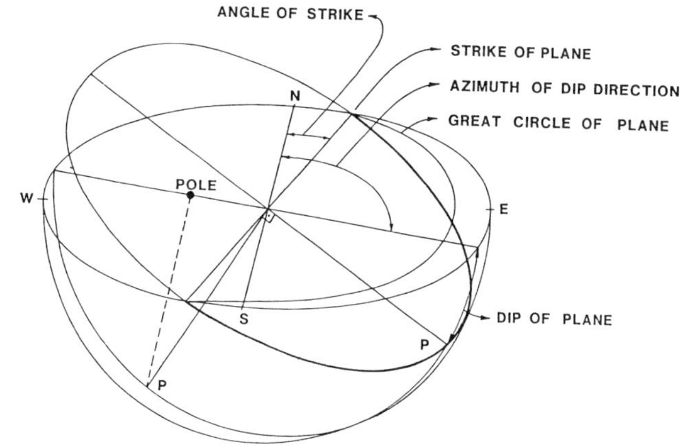

Figure 5. *A schematic of the stereographic projection.*

50X oil immersion objective. The average maceral percentages were based on 1,000-point counts of prepared coal pellets. Four coal pellets were examined from each coal seam (channel sample as well as samples from the top, middle, and bottom of the seam).

Additionally, reflectance measurements of the vitrinites were taken with an MPV (microscope photometer variable) compact photometer to determine coal rank. All coal seams under investigation are ranked high-volatile bituminous (A-C).

Cleat and joint data are presented in two graphic forms: rose diagrams and stereograms. The disadvantage of the former representation is that the rose diagram shows the direction and the density of the cleats, but not their dip. The stereogram most often used for structural data is based on the Schmidt stereographic projection (named after the man who pioneered its use in structural geology; Schmidt, 1925). The stereographic projection uses the lower hemisphere as a reference sphere in which planes and lines pass through the center and intersect the sphere (Figure 5). These intersections are projected onto the equatorial plane of the sphere. A face of the cleat represented by a plane is then plotted either in the form of a great circle or as a pole (the projection of the point at which the normal to the plane cuts the reference hemisphere). The resultant projection is called a stereogram.

To facilitate identification of the cleat and joint patterns, equal area diagrams of the poles were contoured. In constructing contour diagrams, the computer program MICRONET was used to develop a hemisphere on the surface of which the density of cleats and joints is shown by contours indicating percentage concentration of poles. The contour intervals on the plots were chosen at 1%, 6% and 10%. Well-defined patterns of cleats and joints are apparent in the plots (Figure 6).

RESULTS

Eighty-four rose diagrams and stereograms were prepared to analyze approximately 4,000 measurements. Two systems of cleats (face and butt), characterizing the coal, intersect at an angle of 90°. The same applies to the joint systems (main and minor) in the outcrops. Face cleats and main joints generally are perpendicular

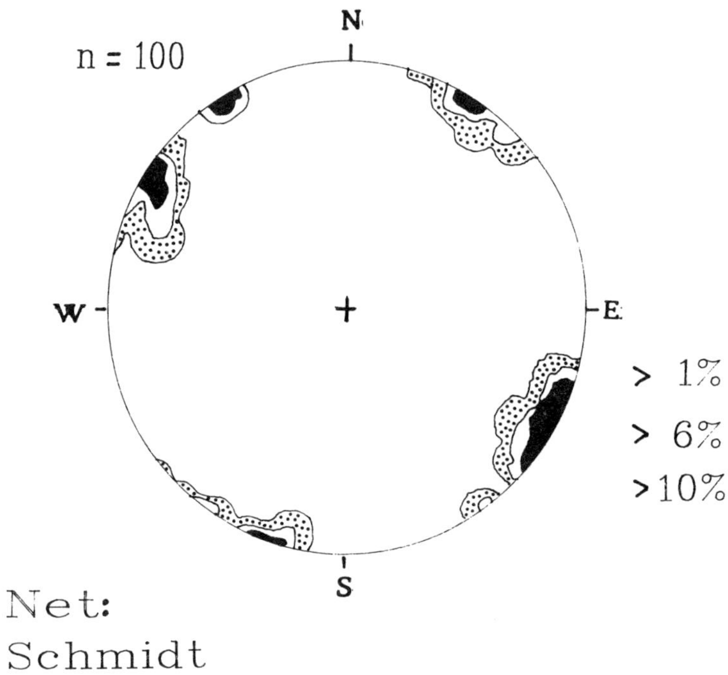

Figure 6. *An example of a contoured stereogram for 100 data points (contour values 1%, 6%, and 10% of poles per 1% area) as output from MICRONET.*

to the direction of the bedding planes. Locally, bimodal trends of face and butt cleats and bimodal trends of main and minor joints are present. The main fracture systems in coal and in the other outcropping rocks generally maintain their orientations in overlying beds.

Further, the orientations of cleats generally coincide with the orientations of regional joints. Regional maps of cleat and joint orientations are shown in Figures 7-9 and are summarized in Table 1.

Cleats

The predominant face and butt cleat orientations are northwest and northeast, respectively. The dip of cleats ranges from 80° to 90°.

There are two prevailing systems of face and butt cleats (Figure 10):

1) The face cleat of the dominant system is oriented N65°W, and the butt cleat is oriented N35°E. This system of cleats occurs especially in the mines of the Book Cliffs coal field (Figure 7), which is not as tectonically disturbed as the Wasatch Plateau coal field (Figure 9).

2) The face cleat of the second system is oriented N5°E, and the butt cleat is oriented N75°W, a rotation of 70° from the dominant system. The differences in the face and butt cleat orientations were noticed especially in the vicinity of faults.

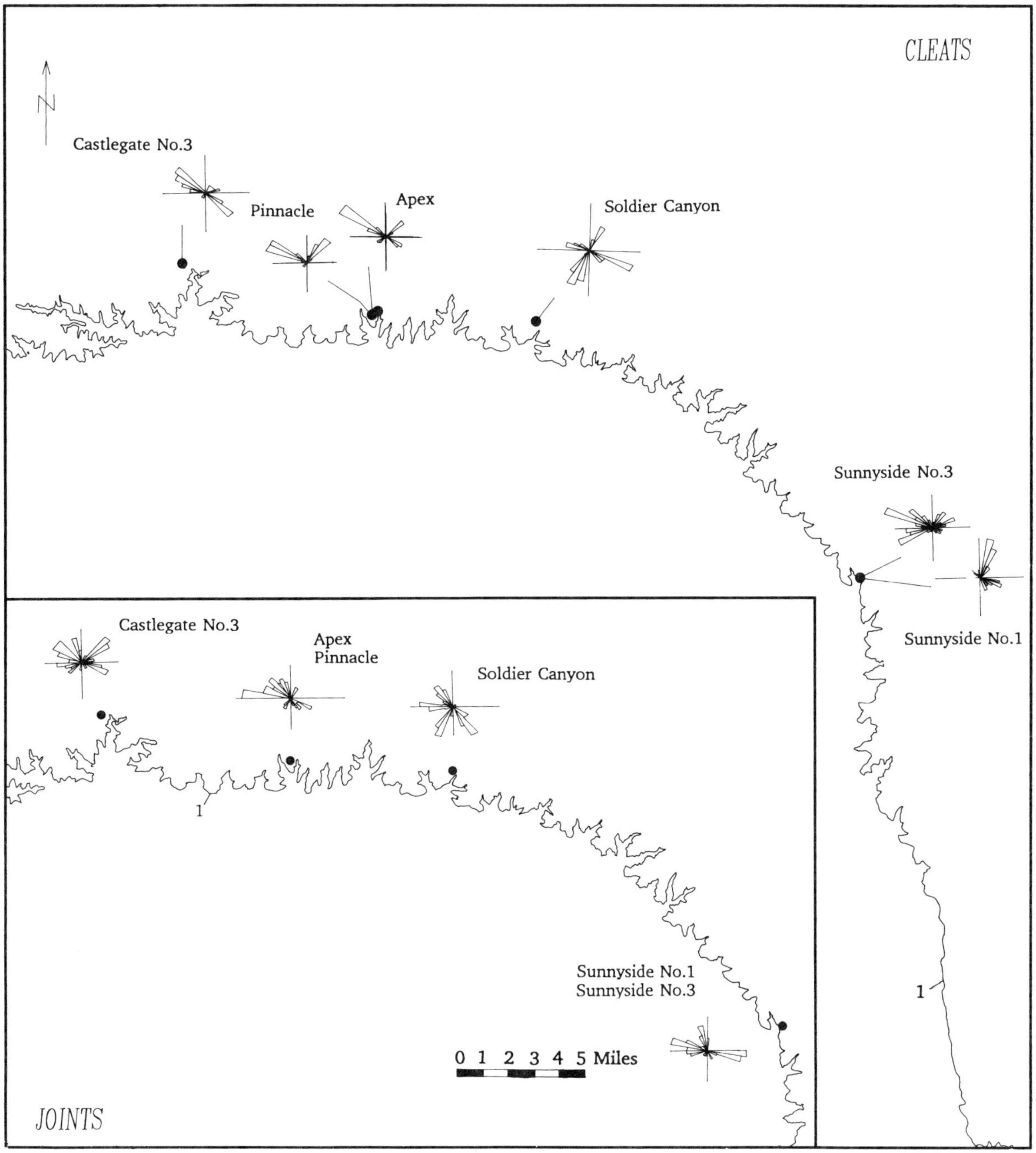

Figure 7. *Cleat and joint orientations in coal seams and surrounding rocks of the Blackhawk Formation in the Book Cliffs coal field (• = mine location; 1 = coal outcrop).*

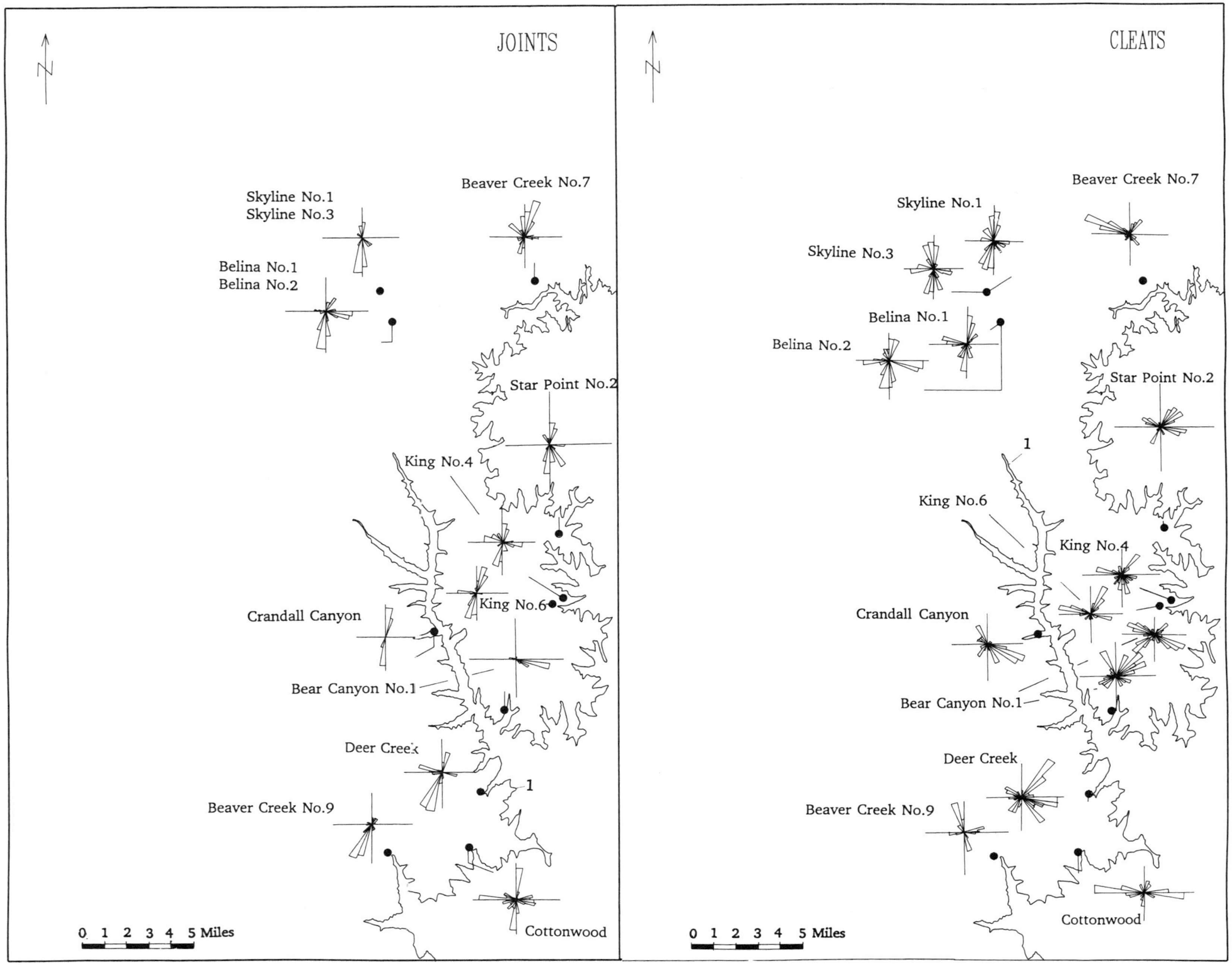

Figure 8. *Cleat and joint orientations in coal seams and surrounding rocks of the Blackhawk Formation in the Wasatch Plateau coal field (• = mine location; 1 = coal outcrop).*

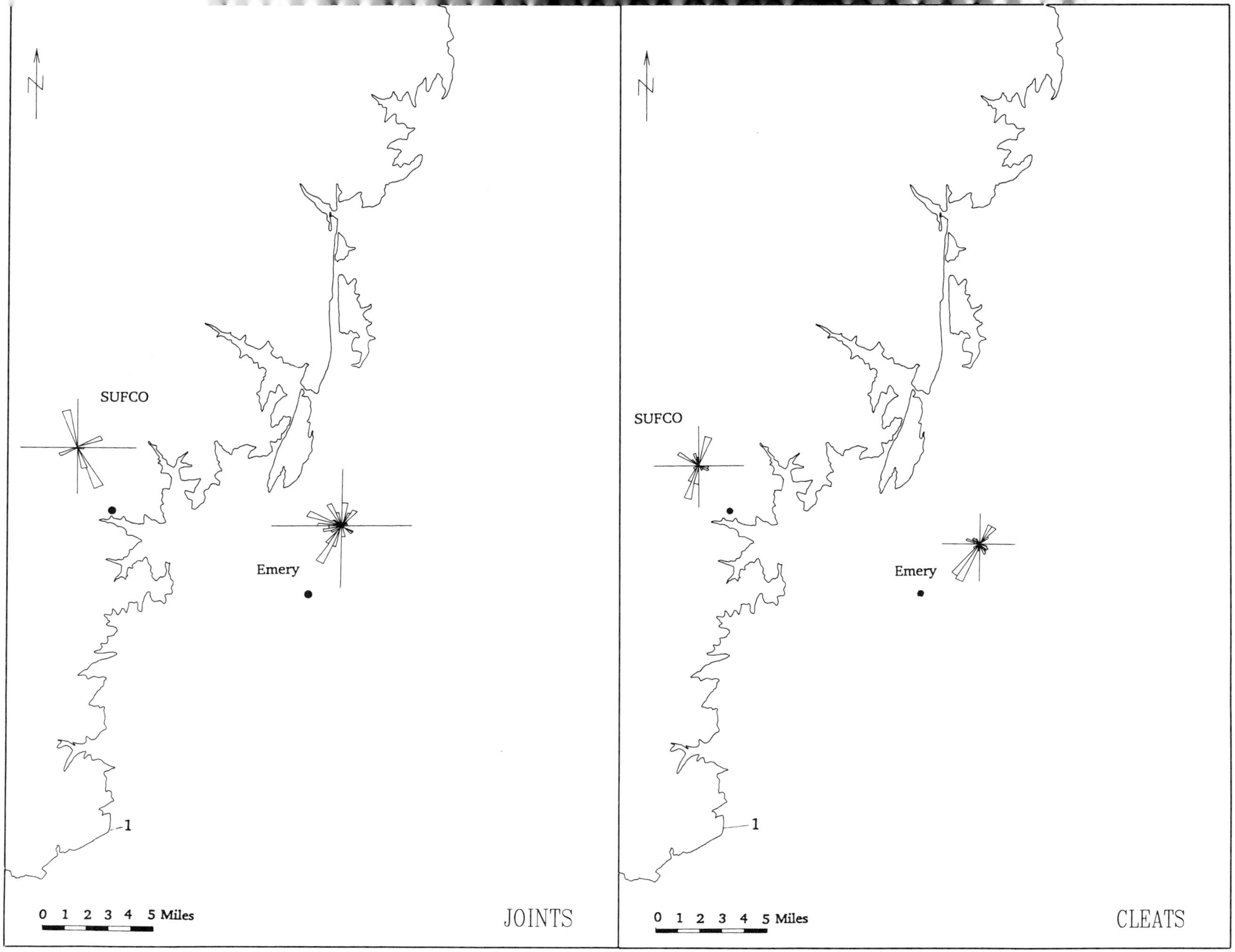

Figure 9. *Cleat and joint orientations in coal seams and surrounding rocks of the Blackhawk Formation and the Ferron Sandstone Member of the Mancos Shale Formation in the Emery and Wasatch Plateau coal fields (• = mine location; 1 = coal outcrop).*

Mine	1. Cleat: Face 2. Joint: Main	Butt Minor	Secondary Secondary	
Emery	1. N25°E	N65°W	N25°W	
	2. N25°E	N65°W	N25°W	
Southern Utah Fuel	1. N15°E	N55°W	N15°W	N85°W
Company No. 1 (SUFCO)	2. N15°W	N65°E		
Beaver Creek No. 9	1. N15°W	N75°E		
	2. N25°E	N45°W		
Cottonwood	1. N85°W	N15°E		
	2. N85°W	N5°E		
Deer Creek	1. N35°E	N75°W	N35°W	
	2. N15°E	N75°W		
Bear Canyon No. 1	1. N65°E	N25°W	N65°W	N25°E
a)				
b)	1. N65°W	N45°E		
	2. N75°W	N25°E		
Crandall Canyon	1. N65°W	N35°E	N25°W	
	2. N65°W	N35°E		
	2. N5°E			
King No. 6	1. N75°W	N25°E		
	2. N15°E	N85°W		
King No. 4	1. N75°W	N35°E	N15°W	
	2. N15°E	N85°W		
Star Point No. 2	1. N25°E	N75°W		
	1. N5°E	N85°W	N25°W	
Belina No. 1	1. N5°E	N80°W		
Belina No. 2	2. N5°E	N85°W		
	2. N5°E	N85°W		
Skyline No. 1	1. N15°E	N85°W	N25°W	
Skyline No. 3	2. N5°E	N75°W		
	2. N5°E	N65°W		
Beaver Creek No. 7	1. N65°W	N35°E		
	2. N15°E	N45°W		
Castlegate No. 3	1. N55°W	N30°E		
	2. N65°W	N35°E		
Pinnacle	1. N65°W	N35°E		
Apex	1. N55°W	N45°E		
	2. N85°W	N35°E	N45°W	
Soldier Canyon	1. N65°W	N25°E		
	2. N35°W	N85°W	N25°E	
Sunnyside No. 3	1. N65°W	N35°E	N65°E	N45°W
Sunnyside No. 1	1. N5°E	N65°W		
	2. N75°W	N15°W	N5°E	

Table 1. *Measured cleat and joint systems in Book Cliffs, Wasatch Plateau, and Emery coal fields. All measurements are given as the mode (±5°) which are the cleat or joint strike measurement values with the greatest frequency.*

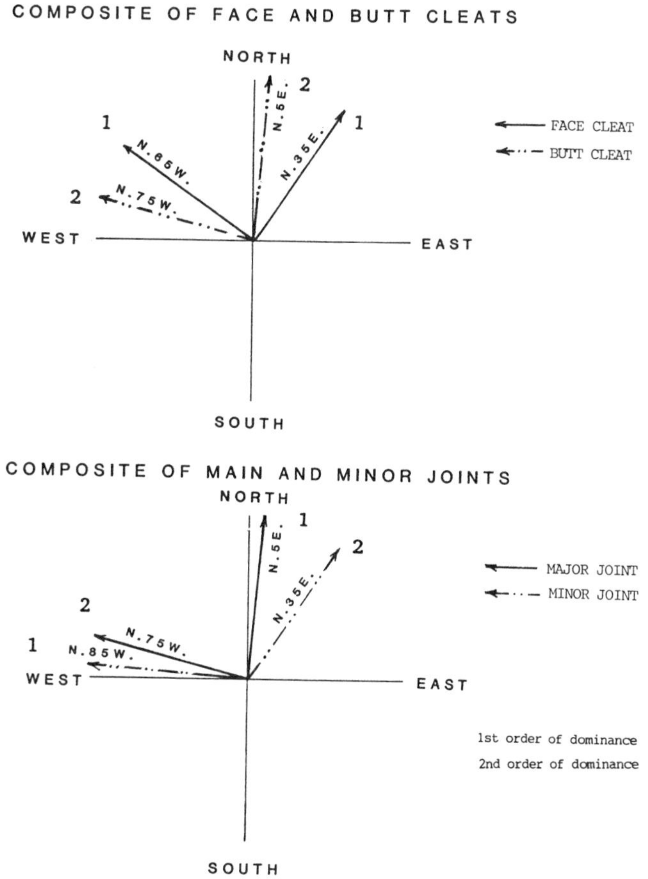

Figure 10. *Composite cleat systems in coal and joint systems in outcropping rocks in the Book Cliffs, Wasatch Plateau, and Emery coal fields.*

Data from several mines also indicate a pattern of conjugate shear fractures with two shear sets (with a dihedral angle of 30°-50°). The dihedral angle can be as great as 60° (Muehlberger, 1961). Because shear fractures tend to form approximately 30° to the principal maximum stress (σ_1), the directions of the maximum and minimum stresses responsible for the cleat or joint can be ascertained. For example, if a set of conjugate shear fractures with two shear sets of N15°E and N15°W is present, it can be concluded that the maximum stress at the time the fractures formed acted in approximately a north-south direction with the minimum stress normal to it. Fractures approximately perpendicular to the shear fractures then would be the strain energy release fractures.

The cleats either cross the whole seam or are limited to the individual stratiform petrographic components of the coal. In a few locations, the cleats continue into the roof rocks as joints with the same orientation. Face cleats, as a rule, are better developed than butt cleats. The face cleats, usually almost vertical, also are better developed in the bright components of the coal (clarain and vitrain) and are not well developed or may be absent in the dull components of the coal (such as durain). Variations also exist within the clarain and vitrain components where the cleats are more highly developed in the vitrain (sometimes causing the coal to disintegrate upon mining).

The cleat surfaces in most cases are smooth or striated. Traces of shearing, bending, and slickensides along cleat planes were observed. In two cases, microfaulting within the coal seam with small dislocations also were observed. These dislocations are filled with pulverized coal, and the presence of associated accessory hematite also was noted.

Mineral accessories associated with cleats included calcite and pyrite as fissure fillings. These fillings appear as thin films or coatings, in tabular form, or as disseminated grains on the cleat surfaces.

The amount of mineral matter in the cleats may be important because it may influence the permeability of the fractures. Resin, in some cases in large amounts, was found either in fractures or disseminated throughout the seams.

Spacing of cleats, in most cases, averaged 12-20 per linear foot (2.5-1.5 cm (1.0-0.6 in.)). Local deviations from the average spacing include the following:

1) in the case of large cleats crossing the entire thickness of the coal seam, spacing was random and ranged from a few inches to several feet; and

2) in the case of individual coal lithotypes (e.g., vitrain), a network of densely spaced cleats with an average spacing of 120 per linear foot (0.25 cm (0.10 in.)) was observed.

Quantitative data on spacing suggest that spacing is related to lithology. The difference between spacing of cleats in coal and that of joints in the adjacent rocks (shale, sandstone, siltstone, etc.) is of an order of magnitude; where the spacing in coal averages less than an inch, the spacing in sandstone is several feet. Further, spacing also is proportional to layer thickness. For example, two successive beds consisting of a thin bed of siltstone and a thick bed of sandstone showed joints closely spaced in the siltstone, whereas in the sandstone the spacing was moderate to wide.

The aperture of cleats proved to be dependent on the thickness of the overburden strata. In mines where overburden thickness exceeded 460 m (1,500 ft), the aperture of the cleats was very small and, on some occasions, fractures lacked mineralization. The same phenomenon was observed in gassy mines where the apertures were small as well.

Joints

Joint systems on the surface were measured with a view toward establishing whether there are any correlations between the cleats in the coal seams and joints in the surface rocks. Based on their pattern, the joints can be classified as either systematic or nonsystematic, but predominantly are the former. Similarly as with the cleats in the coals, two prevailing systems (main and minor joint systems) were present in the rocks. However, in some locations three sets of joints were present. Carbonate minerals constituting the primary filling in the cleats were absent in the joints. In some localities, plumose structures (features of extensional fracturing) were present on the jointing planes in the sandstones.

The same orientation pattern as that for cleats was found for the joint systems in the surface rocks in the Book Cliffs coal field (Figure 7), suggesting that the forces responsible for the formation of cleats and joints were the same. However, this does not hold for the joint systems in the Wasatch Plateau coal field. This field is characterized by complex geologic structure where the difference between joint and cleat orientations can be as much as 90° (Figures 8 and 9). Such a divergence suggests a change in the direction of maximum and minimum stresses, probably caused by more than one deformational event.

Lithotypes

The megascopic whole-seam summaries of the lithotype distribution indicated that the coals under study are predominantly clarains. The resulting microscopic data indicate in most cases coals with high vitrinite content. Further, the maceral composition of the coal appears to control the amount of fractures in a particular seam. Coals with higher vitrinite content show higher density of cleats. The rank of the bituminous coals on the basis of the reflectance measurements range from high-volatile A to C.

SUMMARY AND CONCLUSIONS

The cleats in the coals and the joint systems in the strata of the Blackhawk Formation and Ferron Sandstone Member of the Mancos Shale Formation are mostly the result of tectonic forces. This conclusion is based on the morphology of the cleat and joint surfaces (such as smooth, planar, or slickensided surfaces), regular spacing, presence of pulverized coal along the cleat planes (gouge zone), and plumose structures. All these factors reflect shear and extension. In addition, cleat and joint orientations are continuous over a large areal extent, which suggests they reflect regional stresses.

The regional trends indicate a relationship between cleats in coal and joints in surface rocks. This especially is reflected in the Book Cliffs coal field where the orientations of the butt cleats and the minor joints roughly parallel the axial trend of the fold structure, whereas face cleats and major joints are roughly perpendicular to it.

The surface joint data in the Book Cliffs coal field coincide with the results of cleat data in coal. Therefore, the knowledge of joint orientations in surface rocks could be a reliable estimate of cleats and joints in subsurface coal seams for planning of new mine developments. However, knowledge of potential cleat orientations in the subsurface becomes less reliable for the Wasatch Plateau coal field where faulting is widespread and seems to be one of the factors responsible for the divergence in orientation of cleats in coal and joints in surface rocks.

Furthermore, the results indicate that cleats are influenced by the presence of faults. In 11 mines located close to major faults, a rotation of as much as 90° in orientation of the two main fracture systems in coal seams and outcrops occurred. This divergence in orientation from the regional trend is characteristic of the Wasatch Plateau coal field. However, it is uncertain that the faults are solely responsible for the change in direction of the cleat or joint orientations. Further analysis is being planned to confirm or reject this concept and to ascertain the validity of findings.

For safety reasons, the orientation and spacing of joint systems in mine roofs is of great importance for mine operators. Mine operators prefer to design faces of main entries parallel to face cleats to increase the efficiency of continuous-miner machines. However, if these cleats are continuous with mine roof joints, they may, in combination with prominent bedding planes, outline unstable blocks which are not even partly supported by coal pillars (Figure 11). If the entries are oriented at an angle to the face cleats (and to the joints), the cutting productivity may decrease, but safety is enhanced.

Furthermore, when longwall mining thick seams at great depths, the cleats oriented parallel to the working face often cause an excessive spalling of coal from the face because of overburden strata pressures. The detached slabs of coal must be cleared, mostly manually, from the face conveyor, thus decreasing productivity.

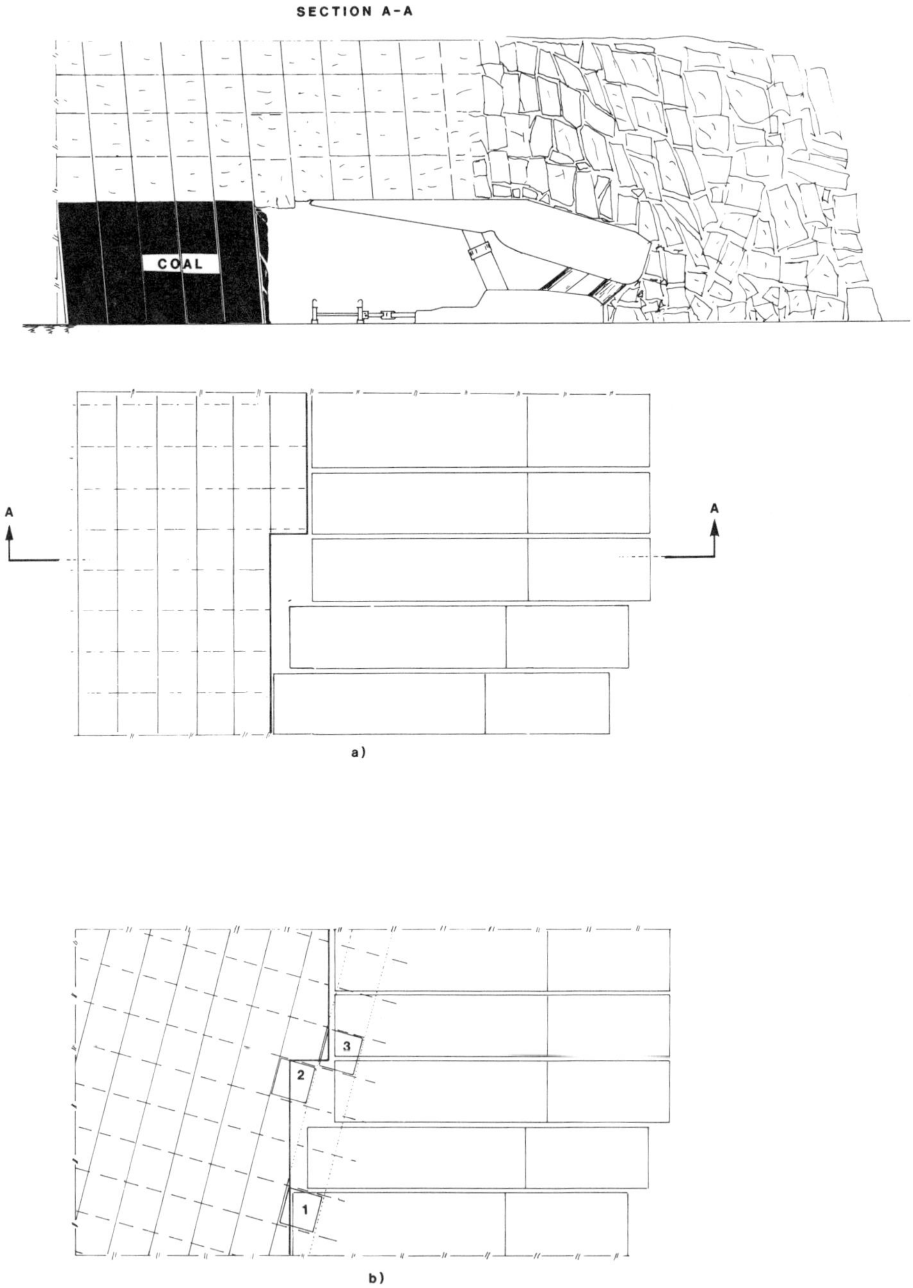

Figure 11. *An example of longwall face orientation compared to the main joint system in the roof A) with the joint system parallel to the longwall face and B) with the longwall face oriented at an angle of 15° to the direction of the main joint system and where detached blocks 1-3 are partly supported either by the shields or by the coal face.*

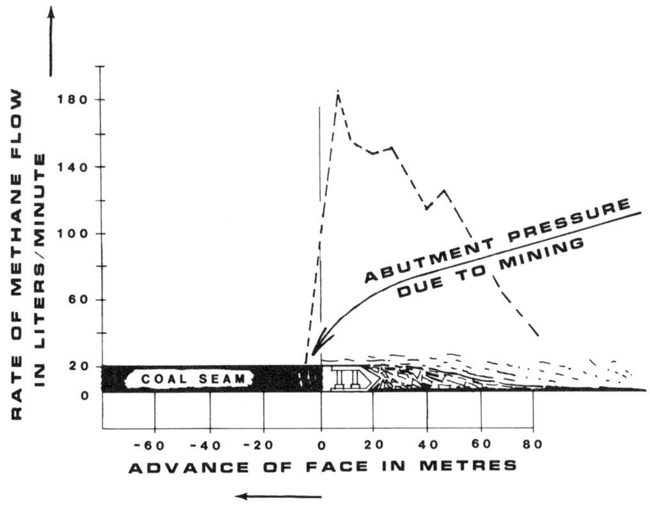

Figure 12. Relation between released methane and advance of a longwall face (modified from Hucka, 1984).

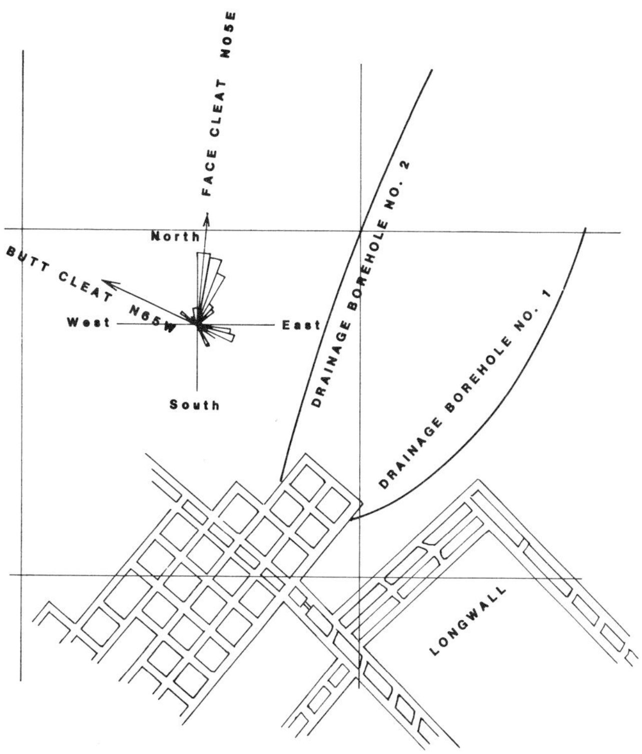

Figure 13. Orientation of drainage holes at the Sunnyside Mine No.1, Utah (modified from Perry et al., 1978).

In some coal mines, longwall faces purposely are oriented at angles of 15°-30° to the face cleats. Such a design results in improved roof stability because the roof rock blocks are supported at least partly by the longwall self-advancing support system (Figure 11b).

When mining a gassy coal seam by use of longwall technology, so-called abutment pressure is created as a result of the mining activity. The abutment pressure acts inside the intermediate strata ahead of the working face, thus causing a temporary closure of cleats that otherwise would interconnect the coal pores and fractures. In other words, the permeability of the coal is decreased. When the face advances, this pressure is released and the cleats are reopened. Additional fractures induced during this process, together with the reopened cleats, increase the permeability and thus free the flow of methane, as shown in Figure 12 (Hucka, 1984).

The importance of detailed knowledge of cleat and joint systems also can be demonstrated by the underground methane drainage tests undertaken by the U.S. Bureau of Mines at the Sunnyside mines (Perry et al., 1978). It is known that the face and butt cleats provide a directional permeability for the flow of methane. It also was found that the flow of methane could be up to 12 times greater when the boreholes are drilled perpendicular to the direction of the face cleats. Such holes intersect larger, better-developed face cleats, yielding more gas than the holes drilled perpendicular to the butt cleats (Rightmire et al., 1984). Drainage tests on coalbed methane at the Sunnyside mines started in 1977 (Perry et al., 1978). Orientations of N5°E for the face cleat and N65°W for the butt cleat (Lower Sunnyside Seam) were measured by the author at the longwall face and are shown in Figure 13. During the tests, a production of gas up to 0.125 mcf per day from a borehole 137 m (450 ft) long was achieved. The hole (No.1) was drilled at an angle of 16° to the face cleat direction. However, the yield of methane might have been up to 15% higher if the borehole had been drilled perpendicular to the face cleat (Hucka et al., 1989).

Resource Enterprises, Inc., Salt Lake City, Utah is presently engaged in extensive coalbed methane production from the Soldier Canyon Mine (Rock Canyon Seam). The orientation of their drainage holes is based on prior detailed study of cleat and joint systems in the seam and the surrounding outcropping rocks. A daily production of gas up to 0.013 mcf was gained from a borehole 520 m (1,700 ft) long. The first borehole was drilled at an angle of 44° to the direction of the face cleat. The production improved strongly when a second, 412 m (1,350 ft) long, was drilled perpendicular to the face cleat. The second borehole yielded 0.2 mcf of gas per day (Schwoebel, 1987). Thus, the results confirm that it has been possible to significantly optimize the production of gas by using geologic data collection prior to methane drainage testing.

In conclusion, the various potential mining applications of knowledge about cleats and joints can be summarized as follows:

1) Mine roof control (defining potential instabilities caused by geologic structures).

2) Improvements in the productivity of continuous-miner sections (reduction in slabbing and safety hazards).

3) Degasification of gassy coal seams (determining the permeability pathways of coal seams).

4) Designing water drainage systems (in the case of excessive water occurrence where the inflow is affected by cleats and joints).

ACKNOWLEDGEMENTS

The author wishes to thank mine personnel for their time and help during the field investigations. Many thanks also go to my colleagues A. Keith, S. Sommer, and T. Thompson, for their assistance during underground cleat and surface joint measurements. Special thanks are extended to Dr. H. Doelling for reviewing the manuscript.

REFERENCES

Ammosov, I.I., and I.V. Eremin, 1963, Fracturing in coal: Jerusalem, Israel Program for Scientific Translations, 107 p.

Doelling, H.H., 1972, Central Utah coal fields: Sevier-Sanpete, Wasatch Plateau, Book Cliffs, and Emery: Utah Geological and Mineral Survey, Monograph Series No. 3, 570 p.

Friedman, M., 1964, Petrofabric techniques for the determination of principal stress directions in rocks, in W.R. Judd, ed., Proceedings of the International Conference on the State of Stress in the Earth's Crust, June 13-14, 1964, Santa Monica, California: New York, American Elsevier Publishing Company, p. 451-552.

Hofer, H., 1915, Schwundspalten (Schlechten, Lassen): Mitteilungen der Geologischen Gesellschaft, v. 7, nos. 1 and 2.

Hucka, B., 1989, Analysis and regional implication of cleat and joint systems in selected coal seams, Emery, Sanpete, Sevier, and Summit counties, Utah: Utah Geological and Mineral Survey, Special Studies Series No. 74, 47 p.

Hucka, V.J., 1984, Methane gas from underground coal mines: A clean source of energy, in F.A. Curtis, ed., Proceedings of the Global Energy Forum '84, Saskatchewan, Canada, May 14-19, 1984: New York, Pergamon Press, p. 31-37.

Hucka, V.J., and D.M. Bodily, 1989, Methane drainage from gassy Western U.S. coal seams, in H. Majima, ed., Proceedings of the International Symposium on Today's Technology for the Mining and Metallurgical Industries, Kyoto, Japan, October 2-4, 1989: Kyoto, Mining and Minerals Institute of Japan, p. 453-460.

Kendall, P.F., and H. Briggs, 1933, The formation of rock joints and the cleat of coal: Proceedings of the Royal Society of Edinburgh, v. 53, p. 164-187.

McCulloch, C.M., M. Deul, and P.W. Jeran, 1974, Cleats in bituminous coalbeds: U.S. Bureau of Mines, Report of Investigations RI 7910, 25 p.

McCulloch, C.M., S.W. Labert, and J.R. White, 1976, Determining cleat orientation of deeper coalbeds from overlying coals: U.S. Bureau of Mines, Report of Investigation RI 8116, 19 p.

Muehlberger, W.R., 1961, Conjugate joint sets of small dihedral angle: Journal of Geology, v. 69, p. 211-219.

Paterson, M.S., 1978, Experimental rock deformation--The brittle field: New York, Springer-Verlag, 254 p.

Perry, J.H., G.N. Aul, and J. Cervik, 1978, Methane drainage study in the Sunnyside Coalbed, Utah: U.S. Bureau of Mines, Report of Investigation RI 8323, 11 p.

Price, N.J., 1966, Fault and joint development in brittle and semi-brittle rock: New York, Pergamon Press, 176 p.

Rightmire, C.T., G.E. Eddy, and J.N. Kirr, 1984, Coalbed methane resources of the United States: American Association of Petroleum Geologists, Studies in Geology Series #17, 378 p.

Schmidt, W., 1925, Gefugestatistik: Tschermak Mineralogie und Petrologie Mitteilungen, v. 38, p. 392-423.

Schwoebel, J.J., 1987, Coalbed methane: Conversion of a liability to an asset: Mining Engineering, v. 39, no. 2, p. 270-274.

ROOF FALLS AND HAZARD PREDICTION IN EASTERN KENTUCKY COAL MINES

Stephen F. Greb
Kentucky Geological Survey
University of Kentucky
228 Mining and Mineral Resources Bldg.
Lexington, KY 40506-0107

ABSTRACT

Different types of roof falls in underground coal mines are characterized by different patterns of roof rocks. Types of falls in Eastern Kentucky mines include weakly bonded shale falls, shale-with-coal falls, weakly bonded stackrock falls, slickensided-shale falls along rolls, weakly bonded sandstone falls, rotated-bedding falls, and falls compounded by joints and fractures. Construction of detailed roof rock maps and geologic models can be used to predict the patterns of strata, and thus areas of potential roof falls, in advance of mining.

INTRODUCTION

Many types of roof falls have been encountered in Appalachian coal fields (McCulloch et al., 1975; Horne et al., 1978; Moebs and Ellenberger, 1982; Hylbert, 1984; Millici and Gathright, 1985; Greb and Cobb, 1987). Documentation of roof falls in the Eastern Kentucky coal field by the Kentucky Geological Survey, mine inspectors, and mining personnel indicates several common types of roof falls: weakly bonded shale falls (Figure 1a), shale-with-coal falls (Figure 1b), weakly bonded stackrock falls (Figure 1c), kettlebottom falls (Figure 1d), slickensided-shale falls beneath rolls (Figure 1e), weakly bonded sandstone falls (Figure 1f), rotated-bedding falls (Figure 1g), and falls along joints or fractures (Figure 1h). Other falls, such as those due to stress fields created by mining, are not discussed in this paper.

TYPES OF ROOF FALLS

Weakly Bonded Shale Falls

Weakly bonded shale falls are defined as roof falls in which 90% to 100% of the fall debris and roof-scar material is composed of shale (Figure 1a). The falls often are domed or arch-shaped. They tend to be small (e.g., spalling falls), but may continue after bolting. These falls occur because of irregular carbonaceous concentrations (often plant fossils) along shale laminae (Figure 2) which disrupt or weaken the bonding between the laminae. Breaks along laminae may be accentuated by shale swelling when the shale is exposed to moisture (Aughenbaugh and Bruzewski, 1976). Roof bolts themselves may act as points of weakness, with many falls resulting from shale breaking at a bolt (Figure 3).

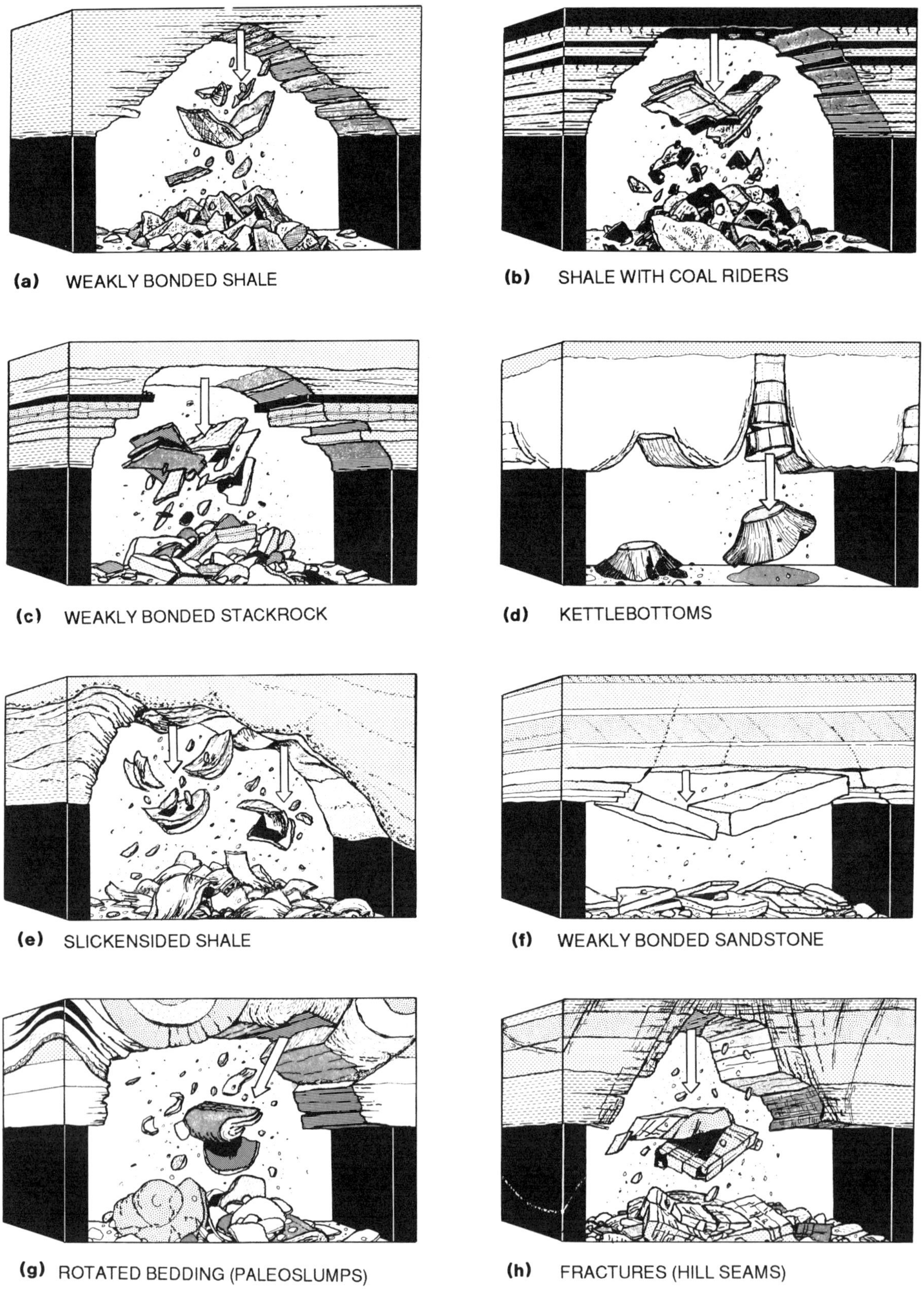

Figure 1. *Types of roof falls discussed in this report.*

Figure 2. Photograph of shale falls along plant fossil concentrations (black) in a mine in the Blue Gem Coal.

Shale-With-Coal Falls

Shale-with-coal falls are defined as roof falls in which shale and coal are the only rock types in the fall debris and roof-scar material (Figure 1b). Falls are domed or arched where the roof is shale-dominated, and flat-topped where the fall is beneath a thick, competent coal rider seam (Figure 4). Persistent, and thick coal riders are often left in the roof for support, but thin, discontinuous coal "streaks", "rash", "stringers", and rider seams often cause falls (Horne et al., 1978; Hylbert, 1984; McCulloch et al., 1975; Moebs, 1977). Causes for these falls include poor bonding between the thin coals and shales, ancient rooting structures beneath the coals which disrupt bedding, slickensides in clays beneath coals, and shale/coal contacts that concentrate moisture, which may promote shale swelling and continued falls. Coal streaks and underclays often can be recognized in cores, from drilling rates, through the use of borescopes, or by observing rock- and coal-dust layering in the dust box of the roof bolting machine after a test hole is drilled.

Weakly Bonded Stackrock Falls

Weakly bonded stackrock falls are defined as roof falls in mixed-rock roofs with appreciable amounts of thin, interbedded shale, sandstone, and sometimes coal (Figure 1c). Fall shapes tend to be flat-topped where the fall roof is at the level of a competent sandstone bed, and domed, arched, or irregularly shaped where the fall roof is dominated by shale (Figure 5). Shales in stackrock conditions contain all the weaknesses that weakly bonded shale and shale-with-coal falls have. Laminated sheet sandstones may be weakened by irregular concentrations of mica, shale, or clay. Rooting structures in thin sandstones beneath coal riders also may disrupt bedding. Problems arise when thin sand sheets or wedges rapidly pinch out into surrounding shales.

The stacking of thin sheet sandstones is common in modern levee and crevasse-splay (flood) deposits (Horne et al., 1978; Hylbert, 1984; Moebs, 1977; Moebs and Ellenberger, 1982; Moebs and Ferm, 1982; Nelson, 1983). In plan view, stackrock strata deposited in crevasse-splays occur in lobate patterns adjacent to sandstone cutouts and rolls that were deposited in ancient stream channels. In section, stackrock strata often are wedge-shaped, with the thickest part of the wedge toward the sandstone roll (ancient channel) and the

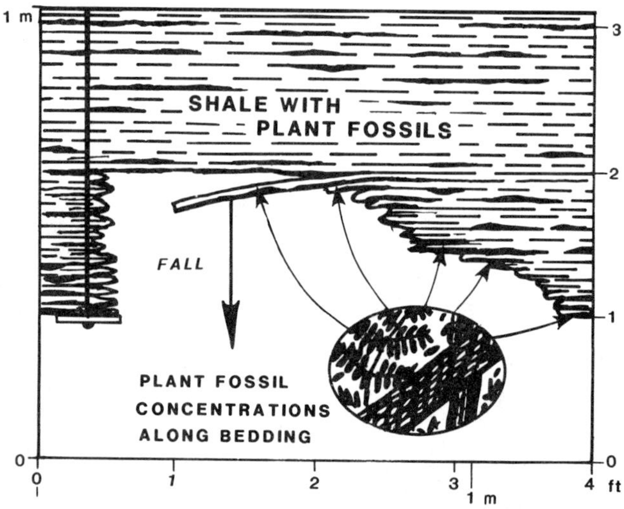

Figure 3. Diagram of a weakly bonded shale fall at a roof bolt in a mine in the Fireclay Coal.

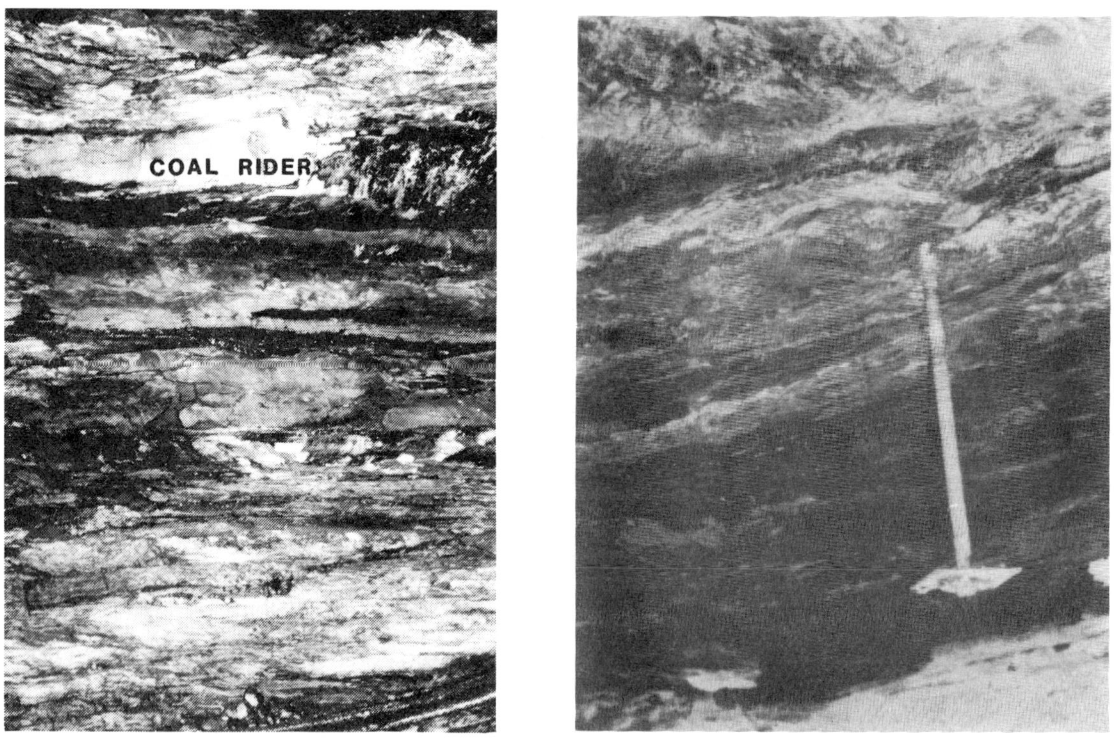

Figure 4. Two photographs of shale-with-coal falls in a mine in the Hazard No. 8 Coal. Light zones are layers of rooted and slickensided underclay.

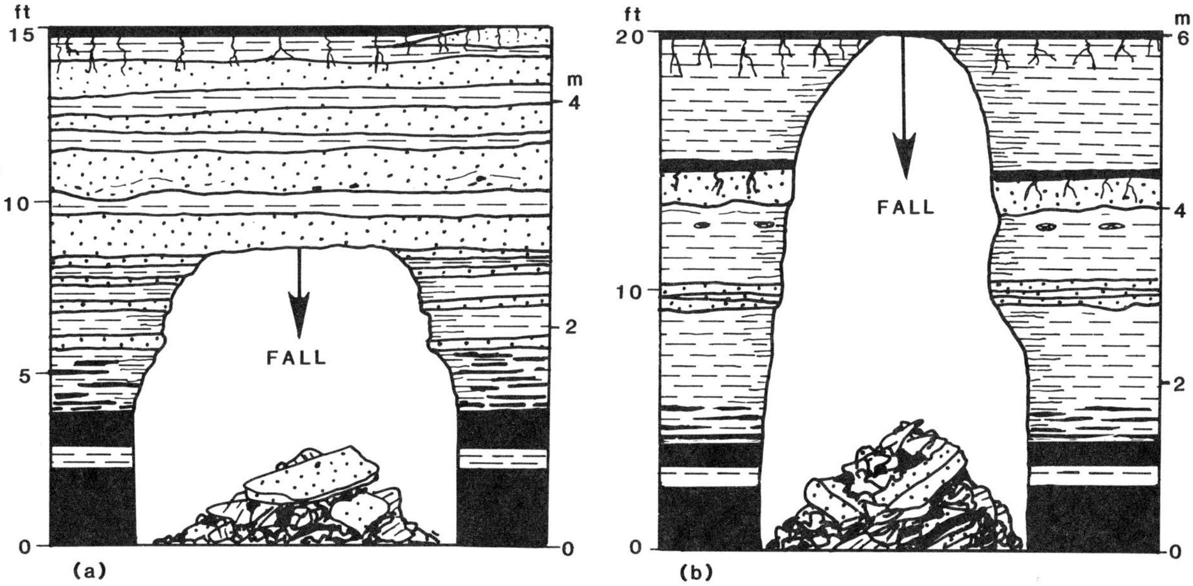

Figure 5. *Comparison of two falls in the same stackrock strata from a mine in the Hazard No. 8 Coal. Note that fall height is controlled by the competence of the sandstone layers. A) Where the second sandstone is competent, the roof fall is flat topped. B) Where the sandstone fails, the fall continues up into the shales and has a dome shape.*

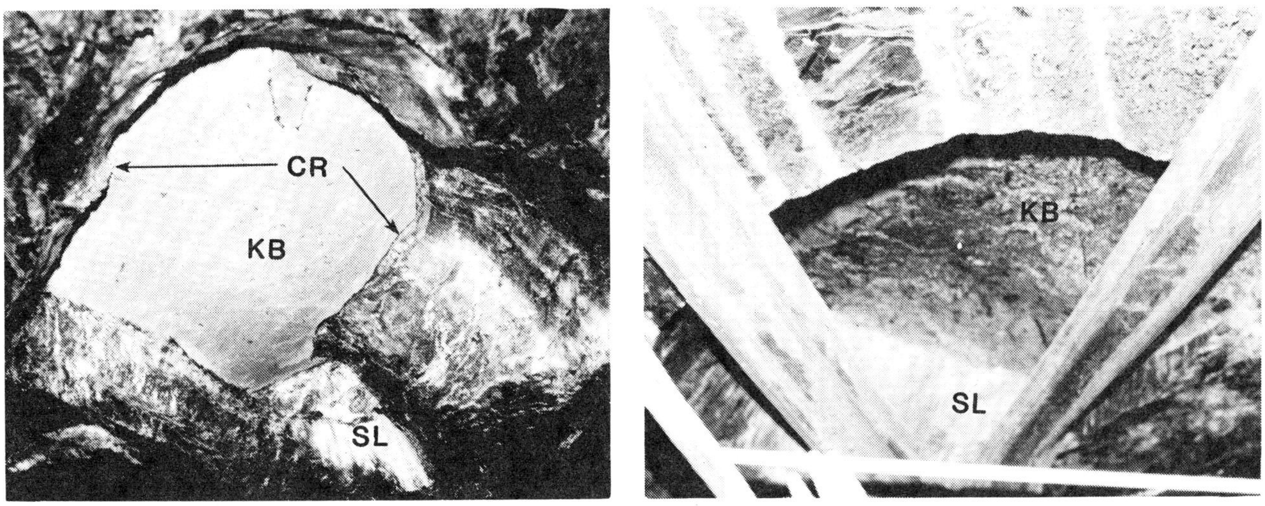

Figure 6. *Photographs of kettlebottoms (KB) from a mine in the Fireclay Coal. Note bounding slickensides (SL) and coal rings (CR).*

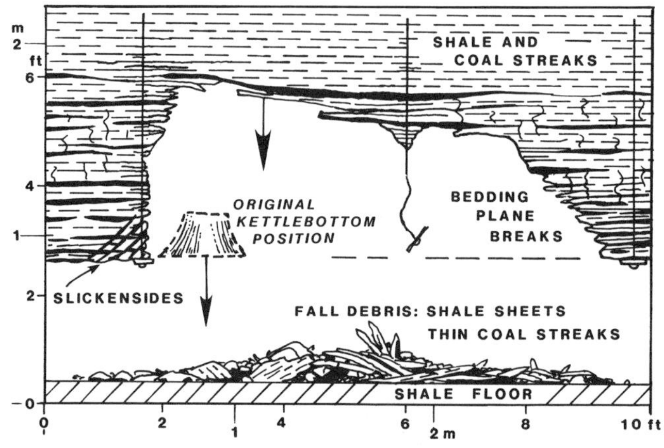

Figure 7. Diagram of a kettlebottom fall in a mine in the Blue Gem Coal where adjacent shale and coal roof also was dislodged during the fall.

thinnest parts of the wedge (where roof falls are common) toward adjacent shaly strata (floodplain/marsh deposits).

Kettlebottom Falls

Kettlebottom falls are isolated rock masses in the roof that are tubular, conical, or bell-shaped (Figure 1d). Falls are irregularly shaped to circular, depending on the shape of the kettlebottom (Figure 6). Most falls are less than 2 m (7 ft) across. Fall height depends on the height of the kettlebottom and the amount of material adjacent to the kettlebottom that is dislodged during the fall (Figure 7). These falls occur because kettlebottoms often are slickensided, surrounded by a thin coal ring that is poorly bonded to the surrounding roof and conical or bell-shaped so that roof pressures can easily "squeeze" them out of the roof. Also, kettlebottoms vary in height, so they are difficult to bolt uniformly (Chase and Sames, 1983).

Slickensided-Shale Falls

Slickensided-shale falls (beneath rolls) are defined as falls in which (1) the fall debris is mostly shale, (2) slickensides are noted in the fall debris or fall scar, and (3) a rock roll or cutout (Figure 8) occurs in the roof scar at the top of the fall (Figure 1e). The rolls and cutouts are elongate bodies of sandstone, shale, sandy shale, siltstone, or in rare cases limestone which truncate the coal seam (cutouts) or occur just above the coal (roll). Slickensides developed in the shales around the rolls and cutouts because muds forming the shale were more compactable than the sediments in the rock that makes up the roll or cutout. Slickensides disrupt or break across bedding, forming natural slip planes in the roof. Slickensided-shale falls are common where sandstone rapidly cuts through shale and where there are isolated shale pockets beneath sandstones with undulatory bases (Figure 9). Slickensides often parallel the trend of the sandstones in the roof (Ingram and Chase, 1987; McCabe and Pascoe, 1979; McCulloch et al., 1975; Moebs, 1981; Nelson, 1983).

Many rolls and cutouts in coal mines were formed by ancient stream-channel processes (Greb and Cobb, 1987; Ingram and Chase, 1987; Horne et al., 1978; Hylbert, 1984; McCabe and Pascoe, 1979; McCulloch et al., 1975; Moebs, 1981; Moebs and Ellenberger, 1982; Nelson, 1983). These paleochannels appear lenticular in cross section and elongate in plan view. They tend to be straight or slightly curvilinear, so once a paleochannel margin has been determined, a trend can be extended in advance of mining. Some of the

Figure 8. Photograph of a cutout in a mine in the Hazard No. 8 Coal. In this case, the cutout was less than 70 m (230 ft) wide. Slickensided-shale falls were common along the cutout margin.

(A)

(B)

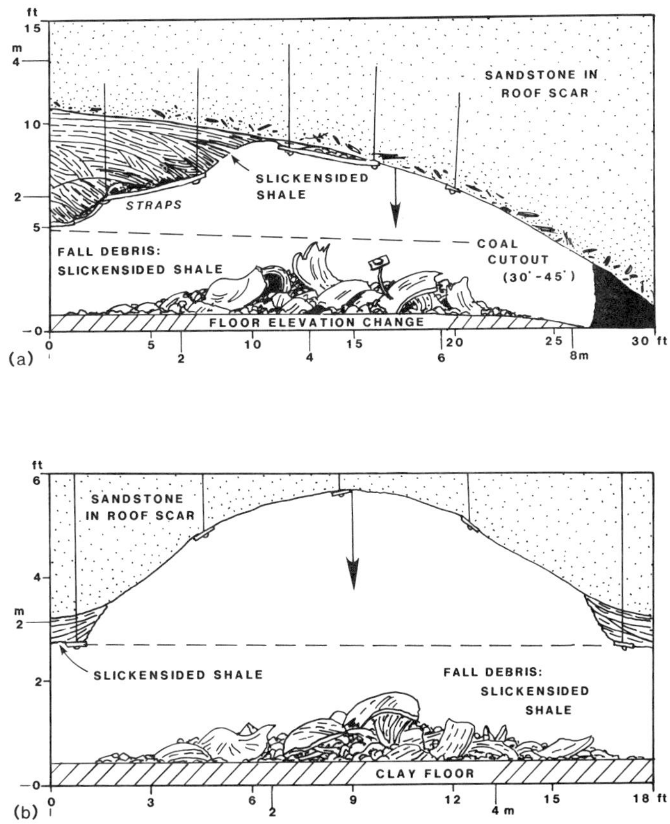

Figure 9. Two types of slickensided-shale falls associated with sandstone rolls: A) fall from beneath a channel cutout from a mine in the Hazard No. 8 Coal, and B) fall in a pocket of shale beneath the undulatory base of a sandstone from a mine in the Blue Gem Coal.

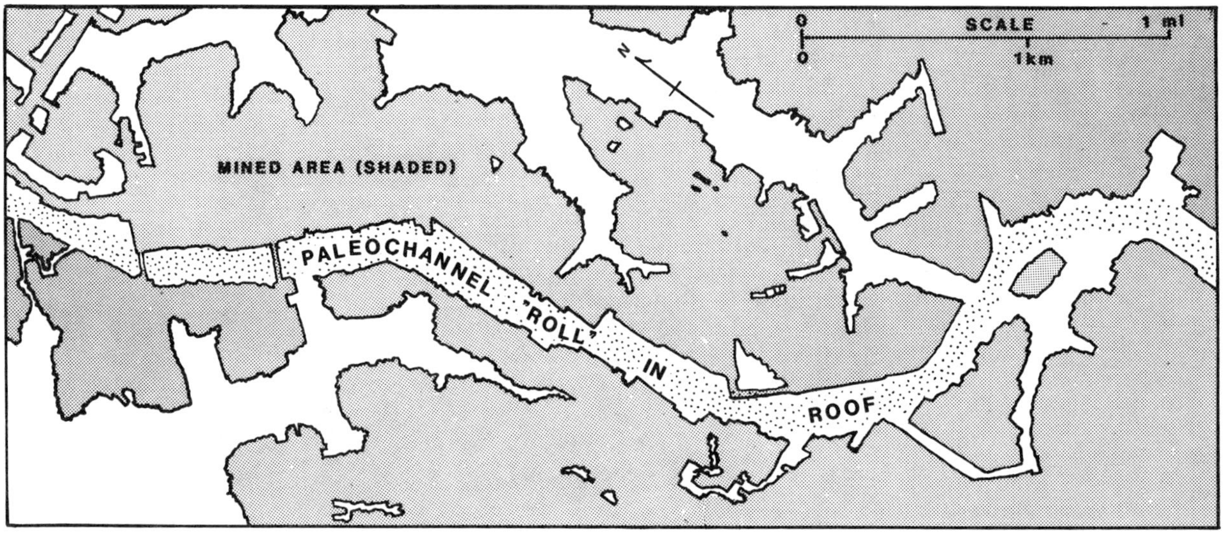

Figure 10. Map of a paleochannel from mines in the Upper Elkhorn No. 3 Coal illustrates the typical sublinear trend of these features. Both continuous and discontinuous paleochannels have been documented in coal mines. Hylbert (1984) mapped a wide variety of paleochannels ranging in width from 65 to 650 m (210-2,130 ft) in the Upper Elkhorn No. 3 Coal. Narrow, discontinuous channels are the most difficult to detect in advance of mining.

paleochannels can be traced for only short distances, whereas others can be traced for many miles (Figure 10).

Weakly Bonded Sandstone

Weakly bonded sandstone falls are defined as roof falls in which the fall material and roof scar are more than 90% sandstone (Figure 1f). These falls are often flat-topped, and large block-like slabs of sandstone fall as single units. Many of the bedding weaknesses of sandstones in the stackrock-type fall also apply to the weakly bonded sandstone fall. In fact, stackrock strata may grade laterally into thick, weakly bonded sandstone. Thick sandstones often are good roofs in coal mines, but where concentrations of shale, mica, and coal occur along bedding, breaks may occur (Cross, 1954; Greb and Cobb, 1987; Horne et al., 1978; McCabe and Pascoe, 1979; McCulloch et al., 1975; Moebs and Ferm, 1982). Breaks in sandstone cores may help to differentiate good sandstone roof from weakly bonded sandstone strata. Sandstone roof falls are uncommon, but where they have occurred, they have been large. Sandstone roof falls can occur within paleochannel deposits as well as in the thick parts of crevasse-splay deposits near channels.

Rotated-Bedding Falls

Rotated-bedding falls are falls in which the roof scar reveals high-angle bedding that appears folded or rolled (Figure 1g). Rotated bedding may occur in any rock type, but is most hazardous in mixed-rock or shaly conditions where slickensides are abundant. Fall shapes tend to be irregular, conforming to bed weaknesses or large slickenside planes. Falls may be large. Rotated and folded bedding 1) disrupts the original rock bedding and prevents formation of a competent beam during bolting, 2) results in oversteepened bedding which may fall when undermined, and 3) often contains deformed or high-angle bedding which commonly is bounded by slickensides or isolated blocks of competent rock (such as sandstone) surrounded by mixed-rock types between slickenside planes (Greb and Cobb, 1987; Hester and Brant, 1981). Rotated-bedding falls easily are overlooked because they are difficult to discern underground. High-angle, deformed beds

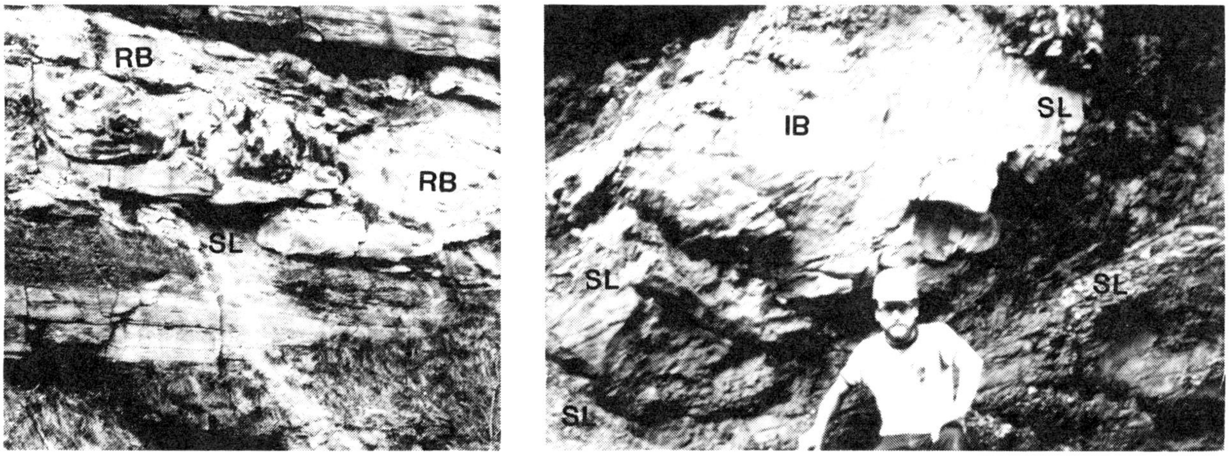

(a) (b)

Figure 11. Photographs of paleoslump deposits from coal-bearing strata in the Eastern Kentucky coal field. Note the folded and rotated bedding (RB) in A, and the isolated and rotated sandstone block (IB) in B (photograph by James Cobb). Note also the slickenside planes (SL) bounding the blocks.

sometimes can be seen in cores and highwall exposures prior to underground mining (Figure 11). In these cases, weak roof conditions can be anticipated.

Rotated bedding and folding in mine roofs often is the result of paleoslumping. Paleoslumps are formed from slumped material such as the deposits that result from collapse of a channel bank. In fact, paleoslumps are most common along palechannel margins and within paleochannel deposits (Hester and Brant, 1981; Horne et al., 1978; McCabe and Pascoe, 1979; Moebs, 1981; Nelson, 1983).

In addition to the types of falls previously discussed, many large falls occur because of a combination of factors (for example, slickensided shale along a sandstone roll in which the shale also contains abundant coal streaks). One of the most common combinations is fractures and joints in combination with one of the previously discussed types of falls.

Fracture And Joint Falls

Fracture and joint falls are roof falls that occur along cracks in the roof (Figure 1h). These falls tend to be sharp-faced along the fracture and irregular away from the fracture (Figure 12). Fracture and joint falls in shale roofs (Horne et al., 1978; Hylbert, 1984; Millici and Gathright, 1985; Nelson, 1983; Overbey et al., 1973) and sandstone roofs (Horne et al., 1978; Moebs and Ferm, 1982; Millici and Gathright, 1985; Overbey et al., 1973) may accentuate weaknesses already inherent in the roof rock. Some fractures result from roof sag and related mining pressures. Others are related to regional stress fields and faulting.

Many fractures in the Eastern Kentucky coal field are caused by the process of overburden unloading ("valley stress relief"). When material is removed from an area by natural stream erosion in a valley (Figure 13) or man-made removal in a entry portal or along a contour strip bench (Figure 14), the unloading of that material relieves confining pressures on the underlying/surrounding compressed and compacted rocks which subsequently relax and propagate fractures and joints called "hillseams" (Greb and Cobb, 1987; Hylbert, 1984;

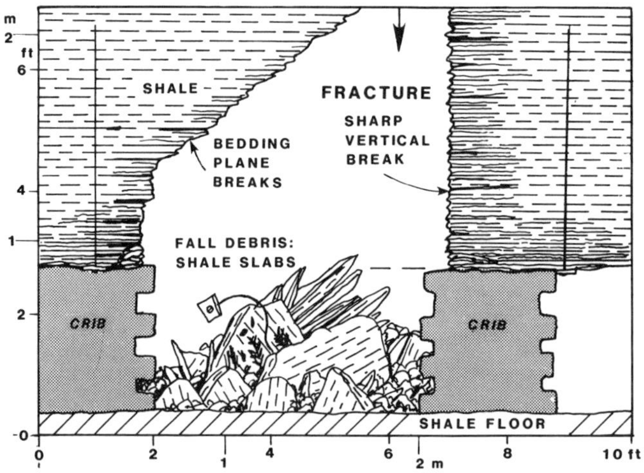

Figure 12. Diagram of a fracture fall from a mine in the Blue Gem Coal. Fractures combine with weakly bonded shale in this fall.

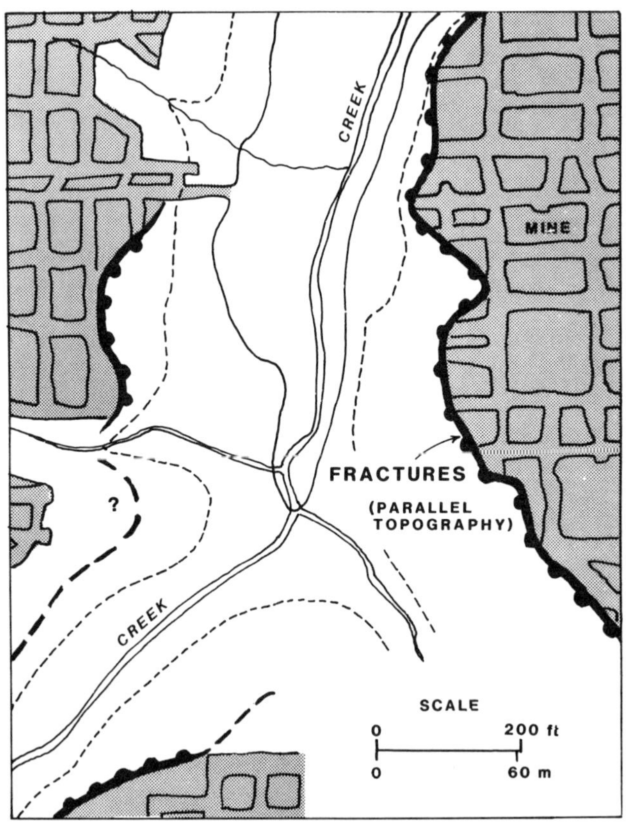

Figure 13. Plan view of fractures that parallel topography from a mine in the Upper Elkhorn No. 3 Coal. These fractures are called "hillseams" in Eastern Kentucky and are caused by overburden unloading.

Figure 14. Photograph looking up at the roof of a fall in an entry portal in a hillside in a mine in the Hazard No. 8 Coal. Note the fracture (FR) at the crest of the fall. Fractures combine with stackrock roof conditions in this fall.

Overbey et al., 1973). Hillseams are one of the most common causes of roof falls in Eastern Kentucky coal mines. Hillseams may be anticipated by making topographic overlay maps to determine narrow and rapidly steepening valleys in the overburden which commonly are associated with fractures. This method is part of the basis of lineament studies (Hylbert, 1984; Jansky and Valane, 1983; Overbey et al., 1973).

GEOLOGIC ASSOCIATIONS

The critical point of the previous roof fall descriptions is that roof falls can be defined by the type of rock in which the fall occurs (i.e., sandstone, shale with coal, etc.) and that these rocks occur in patterns. The advantage of this type of description over simple measurement of fall angles and shape is that patterns of strata can be projected in advance of mining. Thus, specific types of roof falls can be predicted that are known to occur in association with specific patterns of rock strata.

When roof rocks are mapped, patterns of strata become evident. The patterns of strata are related to the manner in which the strata were originally deposited. Comparison between the patterns of strata in mine roofs and the patterns of sediment deposited in modern coal-forming environments can help in determining specific environments of deposition for the strata. This type of comparison is used to construct a geologic model. Geologic models can be used to infer changes in strata that pertain to rock type, shape, bedding, and

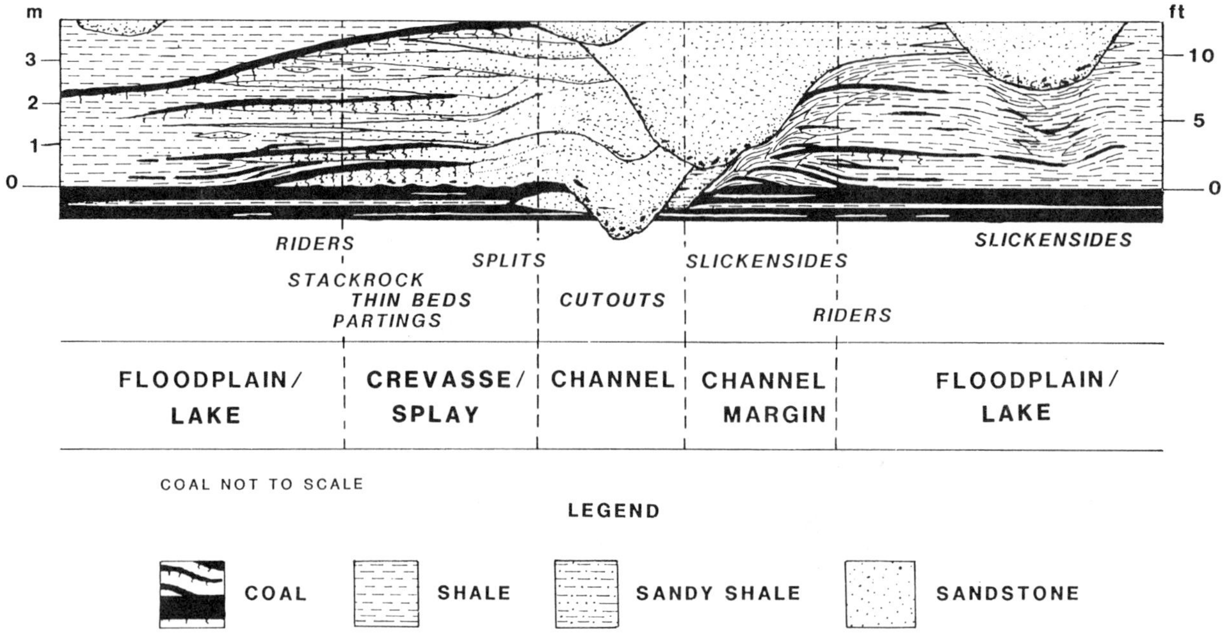

Figure 15. *Example of a geologic model in which characteristic lateral associations of strata and falls in a mine are related to the manner in which the strata originally were deposited. Geologic models are powerful tools for roof-hazard prediction.*

thickness between known data points and, thus, to roof falls and their prediction (Greb, 1990; Horne et al., 1978; Ingram and Chase, 1987; McCulloch et al., 1975; Moebs, 1981; Nelson, 1983; Overbey et al., 1973).

For example, a geologic model for a mine in the Hazard No. 8 coal (Figure 15) illustrates two common rock-pattern associations in Eastern Kentucky coal mines: (1) shale grading laterally to a wedge of stackrock strata that becomes sandier and thicker toward a lenticular sandstone body, and (2) shale sharply truncated by a lenticular sandstone body. The two rock patterns in the Hazard No. 8 example are related to the deposition of sediment in ancient stream channels, floodplains, and lakes (Greb, 1990). In this example, paleochannels are associated with coal thinning, partings, and splits. These features have been associated with paleochannels in many coal seams (Horne and Ferm, 1978; Ingram and Chase, 1987; McCabe and Pascoe, 1979; McCulloch et al., 1975; Moebs, 1981; Nelson, 1983). Because these features are commonly associated with paleochannels, they are incorporated into the geologic model. Thus, even if a paleochannel is not indicated directly within a drill core, paleochannels (and the potential for slickensided-shale falls along the channel margin) can be inferred when cores or mining encounter thinning coal, partings, and splits. In addition to indicating the possible position of a paleochannel, the depositional model can be used to infer the lateral changes in roof rocks away from the paleochannel into unmined areas.

ROOF-HAZARD PREDICTION

Roof-hazard prediction is a dynamic process (Figure 16) that combines geologic knowledge of rock patterns and depositional models with engineering knowledge of the best methods to support those rocks in a mine roof on a continuing basis for the life of a mine. Because the types of roof falls change with changes in roof geology, roof-support plans must accommodate the geologic changes accordingly.

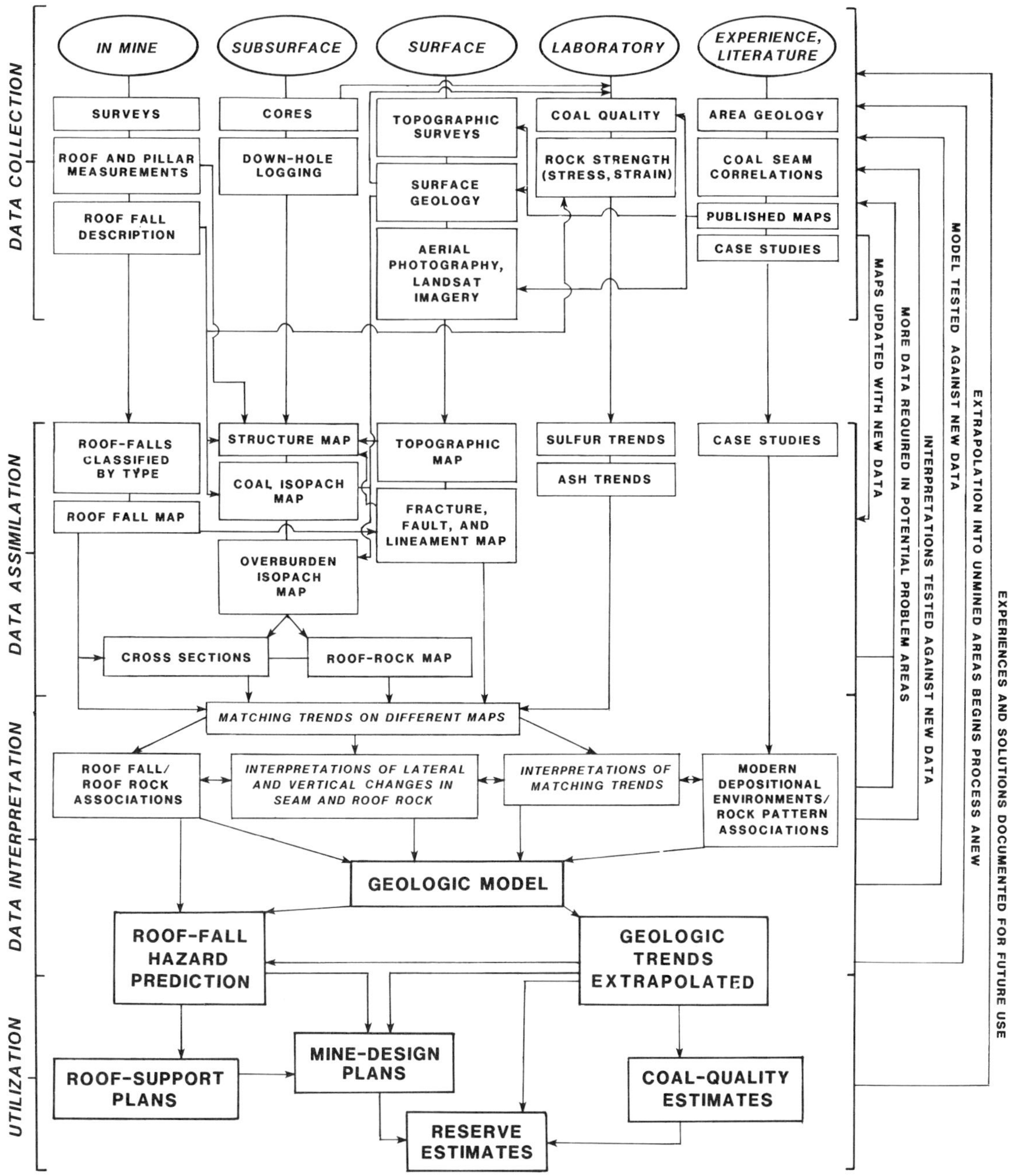

Figure 16. *Flow chart illustrating important parts of a roof-hazard prediction program. Note that the process is dynamic, with new data requiring updating of maps and interpretations. Also note that data are not goal specific, and the same data can be utilized for several purposes.*

The first step in any mining operation is data collection (Figure 16). Data are needed to understand the local geology (coal-thickness variations, common rock-pattern associations, fault trends, etc.). Information on specific areas, or specific seams may be available from mine inspectors, geologic surveys, university mining and geology departments, consultants, and federal mining agencies that work in the coal fields. In the Eastern Kentucky coal field, geologic quadrangle maps at a scale of 1:24,000 are of particular value (Figure 16).

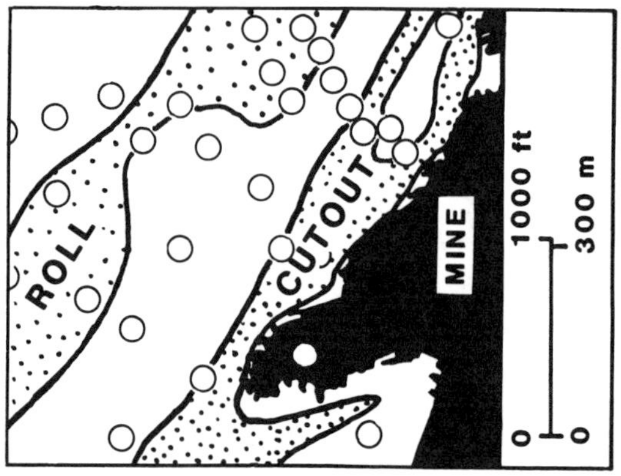

Figure 17. Plan view of cutout channels in a mine in the Hazard No. 8 Coal showing spacing of boreholes needed to determine the multiple paleochannel trends pictured in Figure 8. Slickensided-shale roof falls were common along the cutout channels.

Exploration drilling commonly is used to determine coal thickness, coal quality, and elevation (or dip) of the seam so that initial reserve estimates and mine designs can be made. Maps made from the data (isopach maps, coal-quality maps, structure maps, etc.) also can be used to infer roof conditions (Figure 16). Areas of thin to absent coal on isopach maps, linear topographic lows on structure maps, and increasing sulfur values on coal-quality maps may indicate paleochannels and slickensided-roof problems (Horne et al., 1978; Hylbert, 1984; Ingram and Chase, 1987; McCabe and Pascoe, 1979; McCulloch et al., 1975; Moebs and Ellenberger, 1982; Nelson, 1983). For example, some mines in the Blue Gem Seam of southeastern Kentucky find an increase in sulfur content where sandstone rolls are within 1 m (3 ft) of the top of the seam (Rimmer et al., 1985). These rolls also are associated with slickensided-shale falls (Figure 9).

Drilling density must be sufficient to determine variation in both seam and roof conditions. Whereas achieving the required density of drilling may appear an unnecessary expense in an already tight market, the data collected from additional coring can be used to predict more accurate reserve estimates, coal-quality estimates, possible barriers to mining, and potential roof hazards. The costs of a few additional cores pales in comparison to clean-up and liability costs in roof fall areas. Establishing the necessary spacing of boreholes is difficult because the rate of change in roof rocks is highly variable. In areas affected by narrow paleochannels, a borehole spacing of less than 75 m (200 ft) has been needed to determine the trends of cutouts and rolls (Figure 17). Determining these trends is critical for roof-hazard prediction because falls commonly parallel the trends of cutouts and rolls. If paleochannel trends can be inferred in advance of mining (through sufficient drill hole spacing), then the paleochannel margins can be approached at an oblique angle so that slickenside planes do not parallel entries or crosscuts (Ingram and Chase, 1987; McCulloch et al., 1975; Moebs and Ellenberger, 1982).

Cores should be used not only for determining coal thickness and quality, but for determining potential breaks in roof strata (Figure 18). This determination requires that detailed descriptions of cores be made (especially within 10 m (33 ft) of the top of the coal), noting fossils, breaks, clayey and shaly laminae, slickensides, fractures, bedding, sharp or gradational contacts between different rock types, and unusual features. With detailed descriptions of cores, the first inferences about roof conditions in the area of the borehole can be made. For example, thin sandstones exhibiting the features in Figure 18c and 18d may indicate weakly

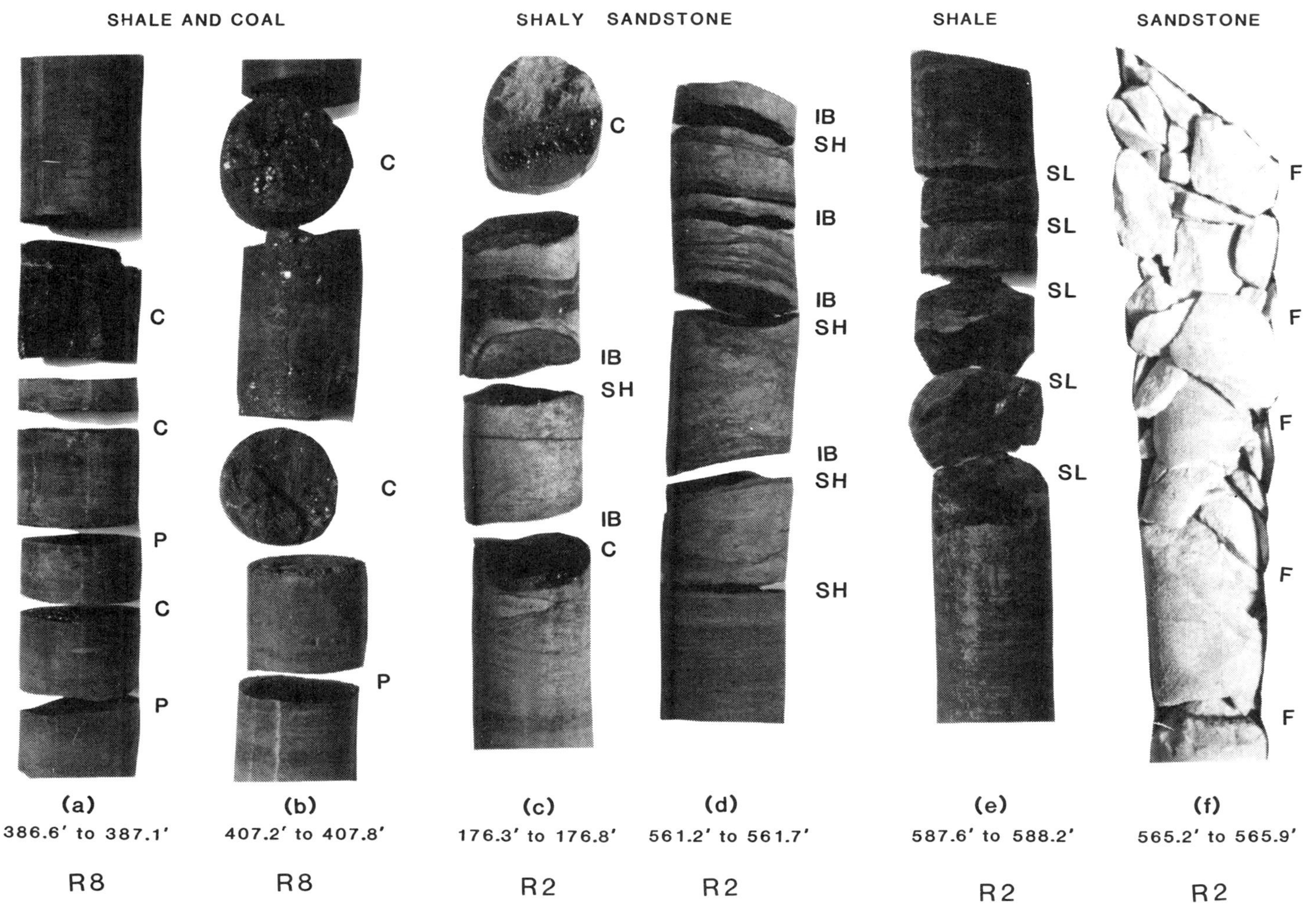

Figure 18. *Examples of some rock weaknesses that can be seen in cores. Samples are from two cores in Clay County, Kentucky. Core R8 is located at Carter coordinate 1-F-70, 2375 FEL x 4375 FSL, and core no. R2 is located at Carter coordinate 7-J-71, 2750 FEL x 2220 FSL. Note the numerous breaks in the sections of cores shown. Breaks occur along thin coal streaks (C), carbonaceous plant-fossil concentrations (P), shaly laminae in sandstone (SH), irregular bedding in sandstone (IB), slickensides (SL), and fractures (F). Noting these features in core descriptions can aid in planning appropriate roof support for areas projected to have similar strata.*

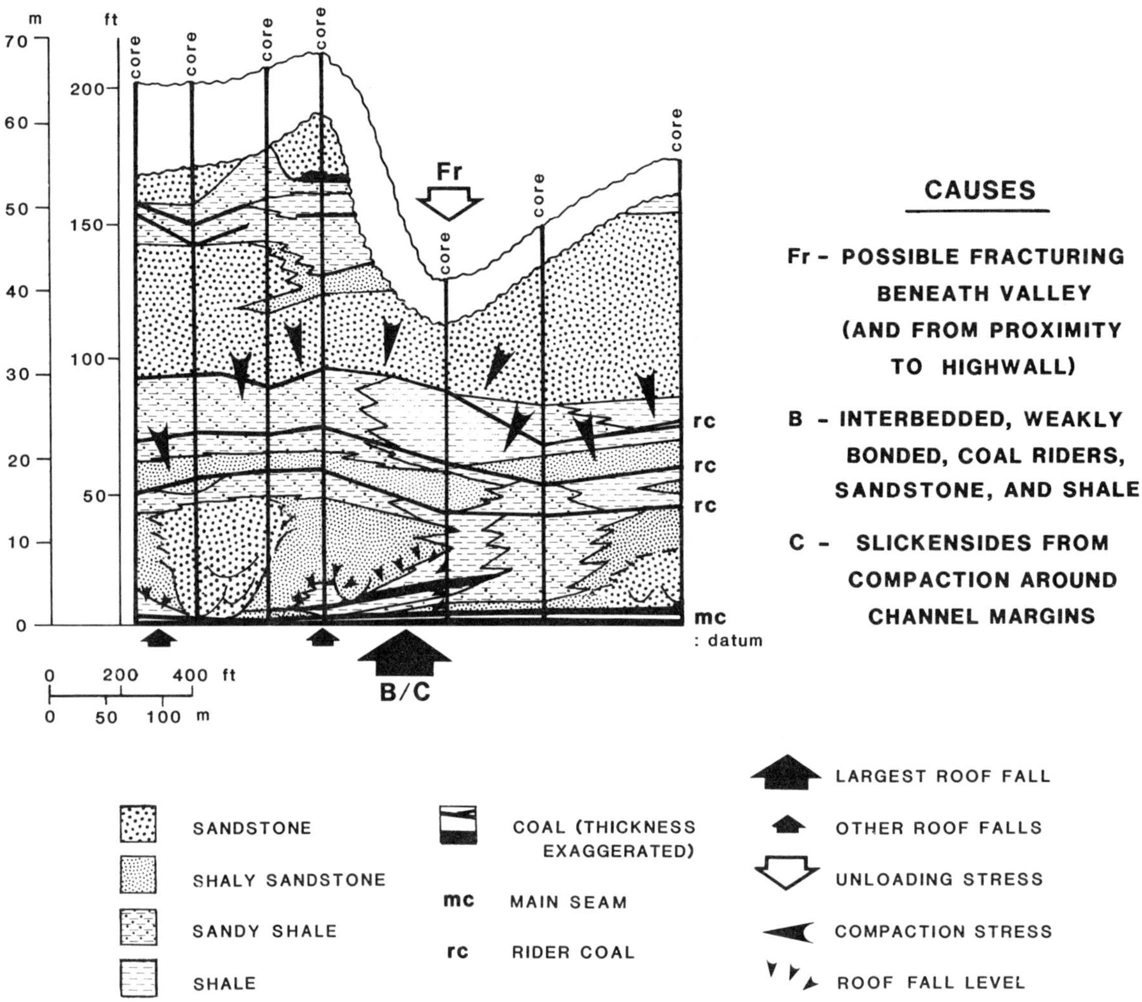

Figure 19. *Cross section illustrating a combination of factors that led to a large fall in a mine in the Hazard No. 8 Coal. Cross sections can help to define potential areas of unstable roof in advance of mining and may be useful in defining the height of the first competent rock layer.*

bonded stackrock strata. Lateral changes in the strata and bed weaknesses must be inferred from nearby boreholes.

Lateral changes between cores should be shown on cross sections so that zones of weak roof can be anticipated (Figure 16). For example, a section through a mine in the Hazard No. 8 Seam (Figure 19) shows a lenticular sandstone roll that cuts through adjacent stackrock strata. In addition, numerous coal riders conform to the geometry of the sandstone (a warning of differential compaction and slickensides along sandstone margins), and overburden rapidly thins beneath a narrow valley (a warning of possible fractures). This combination of factors resulted in a large roof fall. Abrupt changes in roof rocks often are associated with large falls.

Roof falls provide critical data that can be used to document roof changes and roof weaknesses (Figure 16). As falls occur, they should be analyzed geologically. Documentation of falls should include some of the features illustrated in Figure 20. Notes on falls also should include a sketch or photo (where it is safe to take one) of the fall. When falls are documented in this manner, they can be compared against case studies of

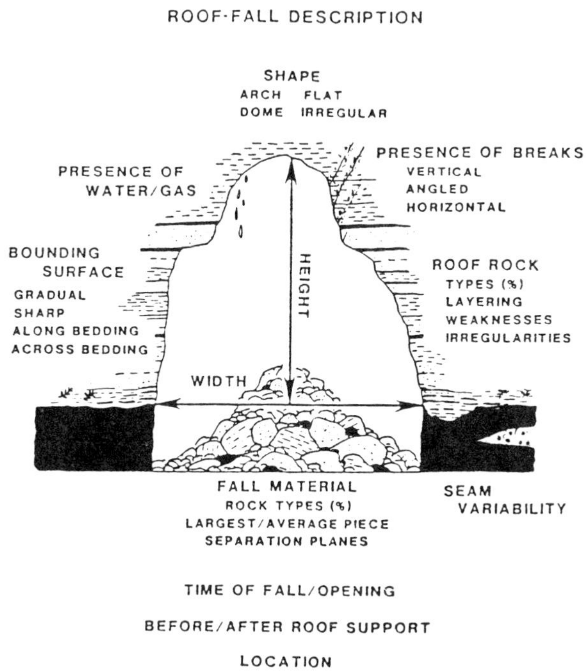

Figure 20. *Features that can be noted in a roof fall.*

roof falls in the literature so that solutions to specific problems can be found (see the References section of this paper).

Different types of falls should be noted on mine map overlays (Figure 16). If falls are not categorized, specific trends may be masked. For example, a mine might have roof falls caused by kettlebottoms and slickensided-shale beneath sandstone rolls. Kettlebottom falls often occur in clusters. In contrast, slickensided-shale falls along sandstone rolls often have linear patterns. If the falls are not distinguished by type prior to mapping, then the grouping of falls may yield no useful trend. Not only are the trends masked, but the causes for each of the falls are obscured, thereby making solutions to the problems more difficult. If fall types are categorized and mapped separately, then specific roof fall trends can be determined and appropriate changes in mine design and the roof support plan can be made in advance of mining.

ACKNOWLEDGEMENTS

I acknowledge the support and funding of the Kentucky Geological Survey, specifically Margaret L. Smath, James C. Cobb, Robert C. Holladay, John D. Kiefer, and Donald C. Haney. Also, the final form of the illustrations in Figure 1 was aided by the reviews of Frank Chase and Tom McLoughlin during the production of "Geology of Roof Falls" for MSHA (Greb and Cobb, 1987). Finally, I acknowledge the mine operators and inspectors who cooperated in this study.

REFERENCES

Aughenbaugh, N.B., and R.F. Bruzewski, 1976, Humidity effects on coal mine roof stability: U.S. Bureau of Mines, Open-File Report 5-78, 164 p.

Chase, F.E., and G.P. Sames, 1983, Kettlebottoms: Their relation to mine roof and support: U.S. Bureau of Mines Report of Investigations RI 8785, 12 p.

Cross, A.T., 1954, The geology of the Pittsburgh Coal: West Virginia Geological and Economic Survey Report of Investigation 10, p. 32-99.

Greb, S.F., 1991, Heterogeneity in seam and roof related to mineability prediction: Hazard No. 8 Coal, a case study, in J. Platt, M. Miller, and J. Price, eds, New geologic perspectives on Central Appalachian Basin low-sulfur coal, Coal Decisions Forum: Palo Alto, California, Electric Power Research Institute, in press.

Greb, S.F., and J.C. Cobb, 1987, Geology of roof falls: A training guide: Mine Safety and Health Administration, Instructional Programs No. AV28.

Hester, N.C., and R.A. Brant, 1981, Paleoslumps: A coal mine roof hazard, in J.C. Cobb, D.R. Chesnut, N.C. Hester, and J.C. Hower, eds., Coal and coal-bearing rocks of Eastern Kentucky: Guidebook and roadlog for Coal Division of Geological Society of America Field Trip No. 14: Kentucky Geological Survey, Series 11, p. 120-126.

Horne, J.C., J.C. Ferm, F.T. Carrucio, and B.P. Baganz, 1978, Depositional models in coal exploration and mine planning in the Appalachian region: American Association of Petroleum Geologists Bulletin, v. 62, p. 2379-2411.

Hylbert, D., 1984, Geologic structures in selected coal beds within Appalachia: Morehead, Kentucky, Morehead State University, Appalachian Development Center, 82 p.

Ingram, D.K., and F.E. Chase, 1987, Effects of ancient stream channel deposits on mine roof stability: A case study: U.S. Bureau of Mines Report of Investigations RI 9092, 33 p.

Jansky, J.H., and R.F. Valane, 1983, Correlation of LANDSAT and air photo linears with roof control problems and geologic features: U.S. Bureau of Mines Report of Investigations RI 8777, 22 p.

McCabe, K.W., and W. Pascoe, 1979, Sandstone channels: Their influence on roof control in coal mines: Mine Safety and Health Administration, Informational Report 1096, 24 p.

McCulloch, C.M., P.W. Jeran, and C.D. Sullivan, 1975, Geologic investigations of underground coal mining problems: U.S. Bureau of Mines Report of Investigations RI 8022, 30 p.

Millici, R.C., and T.M. Gathright, II, 1985, Geologic features related to coal mine roof falls--A guide for miner training: Virginia Division of Mineral Resources Publication 55, 13 p.

Moebs, N.N., 1977, Roof rock structures and related roof support problems in the Pittsburgh Coalbed of Southwestern Pennsylvania: U.S. Bureau of Mines Report of Investigations RI 8230, 30 p.

Moebs, N.N., 1981, The geologic character of some coal wants at the Westland Mine in Southwestern Pennsylvania: U.S. Bureau of Mines Report of Investigations RI 8555, 25 p.

Moebs, N.N., and J.L. Ellenberger, 1982, Geologic structures in coal mine roof: U.S. Bureau of Mines Report of Investigations RI 8620, 15 p.

Moebs, N.N., and J.C. Ferm, 1982, The relation of geology to mine roof condition in the Pocahontas No. 3 Coalbed: U.S. Bureau of Mines Information Circular IC 8864, 8 p.

Nelson, W.J., 1983, Geologic disturbances in Illinois coal seams: Illinois State Geological Survey Circular 530, 47 p.

Overbey, W.K., Jr., C.A. Komar, and J. Pasini, III, 1973, Predicting probable roof fall areas in advance of mining by geologic analysis: U.S. Bureau of Mines Technical Progress Report 70, 17 p.

Rimmer, S.M., T.A. Moore, J.S. Esterle, and J.C. Hower, 1985, Geological controls on sulfur content of the Blue Gem coal seam, southeastern Kentucky: Morgantown, West Virginia, Appalachian Basin Industrial Associates, v.9, p. 212-224.

GEOLOGICAL MODELING TECHNIQUES FOR EVALUATION OF PRODUCTIVITY-RELATED LONGWALL MINING ROOF CONDITIONS: A CASE STUDY

J. Scott Nelson, Ronald H. Mullennex, and Marshall S. Miller
Marshall Miller and Associates, Inc.
P.O. Box 848
Bluefield, VA 24605-0848

ABSTRACT

A longwall mining operation in the Lower Elkhorn Seam in Pike County, Kentucky intermittently experienced severe roof problems at its longwall face from 1980 to 1984. Mapping and analysis of mine productivity and "roof delay" trends showed particularly strong relationships between the low-productivity zones and three geological factors: (1) roof transition from sandstone to shale; (2) presence of a rider coal in the roof; and (3) steep, small-scale synclinal "trough" areas. Principal ground control difficulties associated with these zones included 1) insufficient gob support of the massive main roof, 2) resultant strong front-abutment stresses, and 3) breakage of the weak, thin immediate roof in the face area.

A geologic-hazard rating system was developed based on the relative mining impacts of multiple geologic factors. Composite mapping of these factors generated a hazard prediction model which demonstrated a positive correlation of actual post-evaluation mining conditions to those projected by the model.

INTRODUCTION

This case study presents a technique for analysis of mining delays and subsequent productivity losses related to geological conditions. The principal values of this type of geological analysis result from increased capabilities to 1) estimate the suitability of in-use face equipment for geological and ground control conditions projected within future longwall panels, 2) rationally select new face equipment, 3) anticipate potentially difficult ground control conditions prior to the occurrence of downtime events, and 4) optimize mine design and ground control strategies according to the mine-site geological conditions.

Previous work has been done in predictive composite geological hazards mapping by Overby et al. (1973), Dunham et al. (1976), Ealy et al. (1979), and Miller (1982, 1984). These techniques were used in the current study, with special emphasis on incorporating the derived relationships between geological conditions and longwall productivity based on correlation between longwall-performance mapping and geologic mapping. This approach allowed for a high level of confidence in the predictive capabilities of the geological model for the mine site.

MINE LOCATION AND GEOLOGICAL SETTING

As shown on Figure 1, the subject mine is operated by the Kentucky Carbon Corporation, a Pike County, Kentucky, subsidiary (Kentucky Carbon) of Carbon Industries, Inc. Kentucky Carbon mining operations extract the Lower Elkhorn Seam, which long has been a principal deep-mining coal seam in Eastern Kentucky. Previous workers, including Hunt et al. (1937), McKay et al. (1969), Outerbridge (1968), Rice et al. (1977), and Trent (1965), have demonstrated that the strata on the subject property are of Middle and Lower Pennsylvanian age and belong to the Breathitt Formation. The rock strata consist chiefly of interbedded sandstone, siltstone, shale, and coal.

Within the study area, the strata generally exhibit a gentle northwest dip, and no faults are known to be present. Local steep, small-scale synclinal and anticlinal undulations are present and have had a negative impact on previous mining efforts.

PREVIOUS MINING HISTORY AND EXPERIENCE

Previous mining on the Kentucky Carbon property has been most extensive in the eastern portion of the property where conventional room and pillar mining, with a high extraction percentage (up to 75%), successfully was employed. In the western portion of the property, the Lower Elkhorn Seam occurs below drainage, providing areal distribution sufficient for application of longwall mining. Initial longwall development commenced in 1976 with a chock roof-support system working on 98-120-m (320-400-ft) long faces and 1,040-1,100-m (3,400-3,600-ft) panel lengths. In 1979, the chock support system was abandoned in favor of shield supports rated at 550 tonnes (600 tons) with a "two plus two" leg arrangement. Mining conditions in the initial longwall area (under predominantly sandstone roof) were reported by Kentucky Carbon personnel to have been generally favorable. Face widths in the second longwall area ranged from 130 to 150 m (430 to 480 ft), whereas panel lengths ranged from 920 to 1300 m (3,000 to 4,200 ft). Difficult mining conditions encountered intermittently during this period resulted in excessive delays and lost production.

A central purpose of the investigation was to determine what geological factors controlled the occurrence and areal distribution of poor-roof/low-productivity zones for longwall mining within the study area. Roof control problems characteristic of these low-productivity zones are shown on Figure 2. The weak shale immediate roof here was too thin to provide good gob support for the thick sandstone main roof. The main roof cantilevered into the gob, breaking the friable immediate roof due to the resultant load transfer imposed by the main roof. Therefore, the caving line moved forward to the face area, causing the fractured roof material to fall out immediately after the shearer passed. The fractured shale also was being forced backward into the gob and downward through any gaps which existed between the shields.

The degradation of the immediate roof into a rubble overlying the shields resulted in a number of further difficulties, including 1) the accumulation of broken rock above and below the shields which caused a loss of support resistance, further aggravating control of the main roof, 2) the presence of roof cavities above the shields which caused the shields to be differentially loaded, leading to shield overextension, misalignment of shields, or depression of canopy tips below the top of the seam, and 3) broken roof rock falling at the face, which caused increased clean-up time delays, material handling costs, equipment wear, and plant reject. Production time lost due to either preventative or remedial response to these problems resulted in a reduced rate of face retreat and, subsequently, a longer time for front abutment pressures to further weaken the immediate roof in advance of the face. The synergistic relationship between poor roof, lost production time, and slower face retreat was a primary factor resulting in production rates occasionally as low as 180-360 clean tonnes (200-400 clean tons) per shift in the most difficult roof zones.

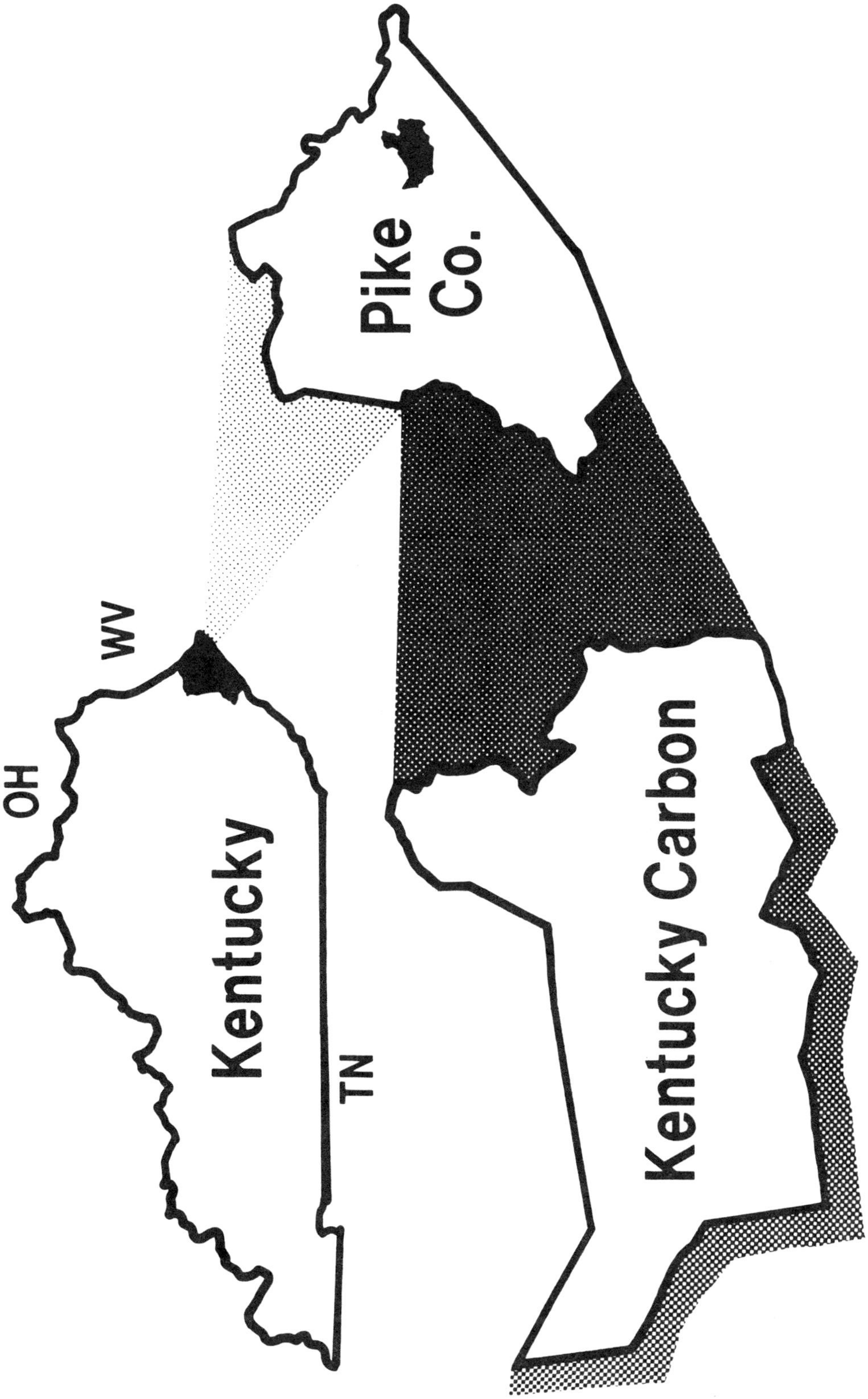

Figure 1. *General location map of the study area and the Kentucky Carbon property.*

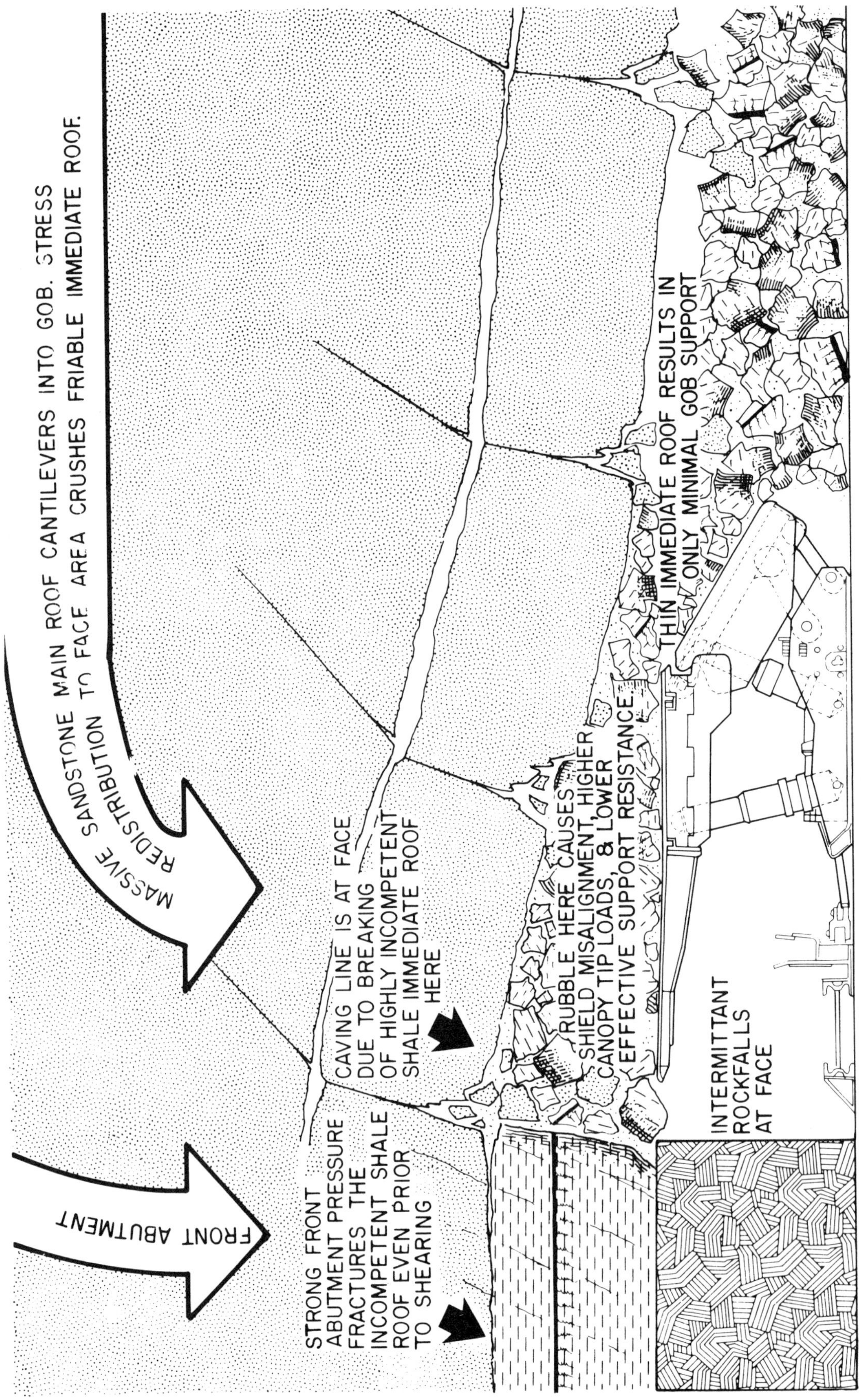

Figure 2. *Roof control problems in the "transition zone" of the rider roof area.*

In order to verify a correlation between poor-roof/low-productivity zones and specific geological conditions, exhibits showing average monthly production (in clean tons per shift) and percentage of available work time lost each month due to "roof delays" (see Figure 3) were developed from mine management's longwall performance records. This mapping immediately demonstrated the presence of a difficult-roof zone traversing the Calloway No. 1 longwall area.

FACTORS CONTROLLING ROOF CONDITIONS

A number of geological factors had a critical role in causing the poor roof zones encountered in the Calloway No. 1 longwall area. Mapping of the individual geological factors — both stratigraphic and structural — served two critical purposes: 1) it allowed for comparison and correlation of the individual geologic trends to known poor roof areas encountered underground; and 2) it allowed projection of these trends into future reserve areas.

Rider Seams

As shown on Figure 4, a thin rider seam overlies the Lower Elkhorn Seam throughout much of the area west of the No. 3 longwall panel in the Calloway No. 1 area. This area is referred to as the "rider roof area". The rider is 3-13 cm (1-5 in.) thick, ranges from 0.2 m to almost 6 m (0.5 ft to almost 20 ft) above the top of the seam, and is located within the shale immediate roof. All available data demonstrate the absence of rider coals south of the Calloway No. 1 mains. This second area is referred to as the "rider-free area". A second, thin, distal rider of very local distribution also is known to occur in the area where the shale roof interval is very thick.

The rider coal and its weak, rooted seat earth form a zone of low strength and low cohesiveness in the roof strata. Where the rider was present in the immediate 0.3-1.5 m (1-5 ft) of roof strata, its weakening effects definitely facilitated breakage of the immediate roof at the longwall face. As the rider became progressively higher above the top of the Lower Elkhorn Seam, its negative effects on mining conditions accordingly diminished.

Shale Immediate Roof

Throughout most of the study area, the Lower Elkhorn is immediately overlain by a shale immediate roof ranging from 0 m to over 6 m (0 ft to over 20 ft) in thickness (see Figure 5). This unit is in turn overlain by a competent, massive sandstone main roof as mentioned previously. The shale roof contains up to two thin and rather lenticular rider coals and their associated seat earths. Where the shale roof is less than about 0.46 m (1.5 ft) thick, it was reportedly too thin to have any significant impact on longwall advance and production. Likewise, where it exceeds about 3 m (10 ft) in thickness, it generally provides sufficient strength and gob support to resist fracturing in advance of the longwall shields.

However, in the "transition" zone where the shale ranges between approximately 0.46 m and 3 m (1.5 ft and 10 ft) in thickness, it tended to break above the shield supports and fall immediately at the longwall face due to 1) increased slickensiding of the shale caused by differential compaction beneath the overlying sandstone, 2) poor strata cohesion associated with the proximal rider coal and the bedding plane surface at the sandstone-shale contact, and 3) stress-induced fracturing of the shale resulting from the load transfer effect of the sandstone, which tended to "beam" for some distance over the gob area. As the thickness of the shale immediate roof increased beyond 3 m (10 ft), gob support increased, the weak roof zone became distal from the seam, slickenside intensity diminished, and roof conditions generally improved correspondingly.

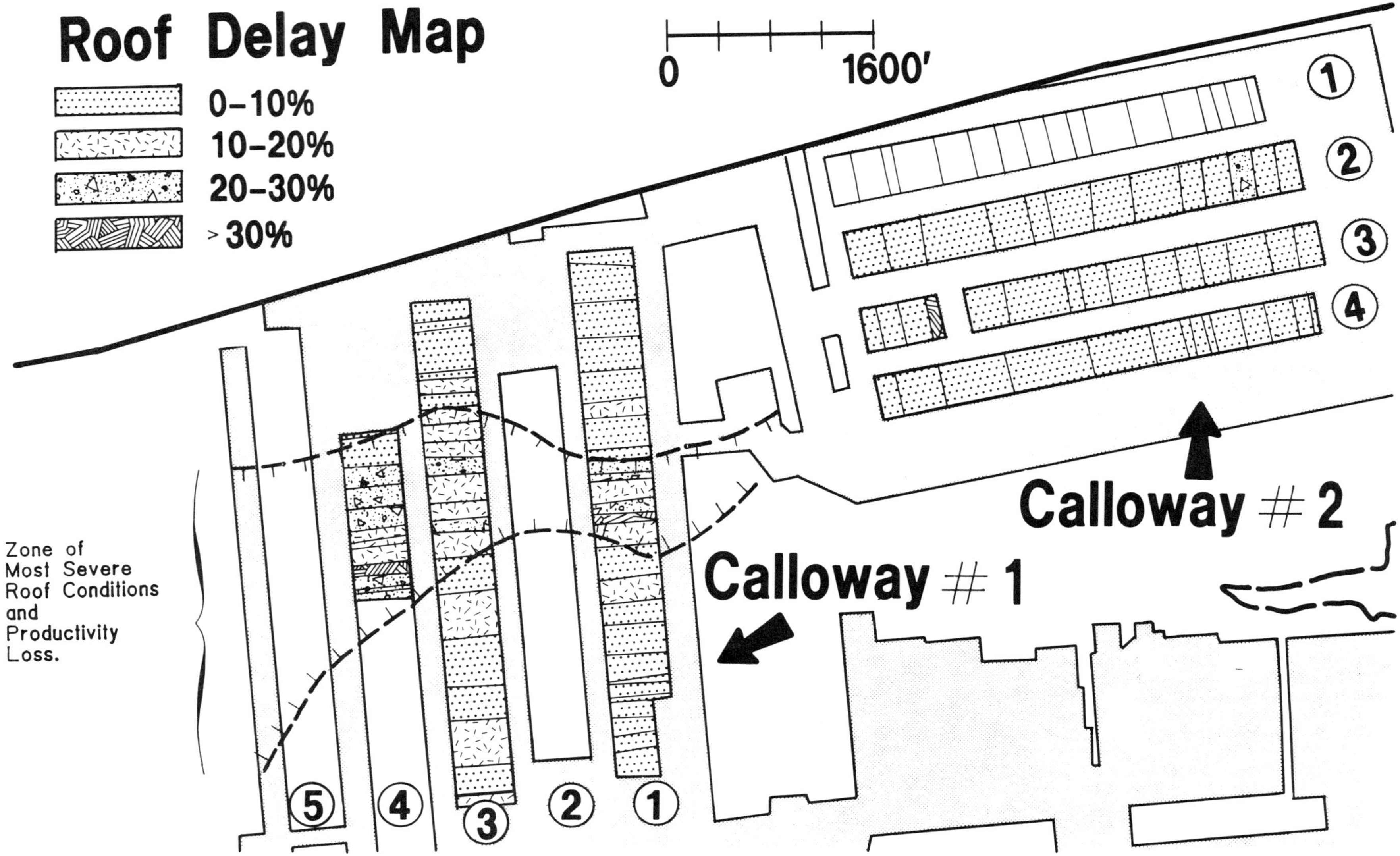

Figure 3. *Roof delay map of the Calloway Mine property showing percentage of available work time lost each month due to roof control difficulties. Note the excellent conditions reported for the Calloway No. 2 area and the trend of difficult conditions traversing the Calloway No. 1 longwall area.*

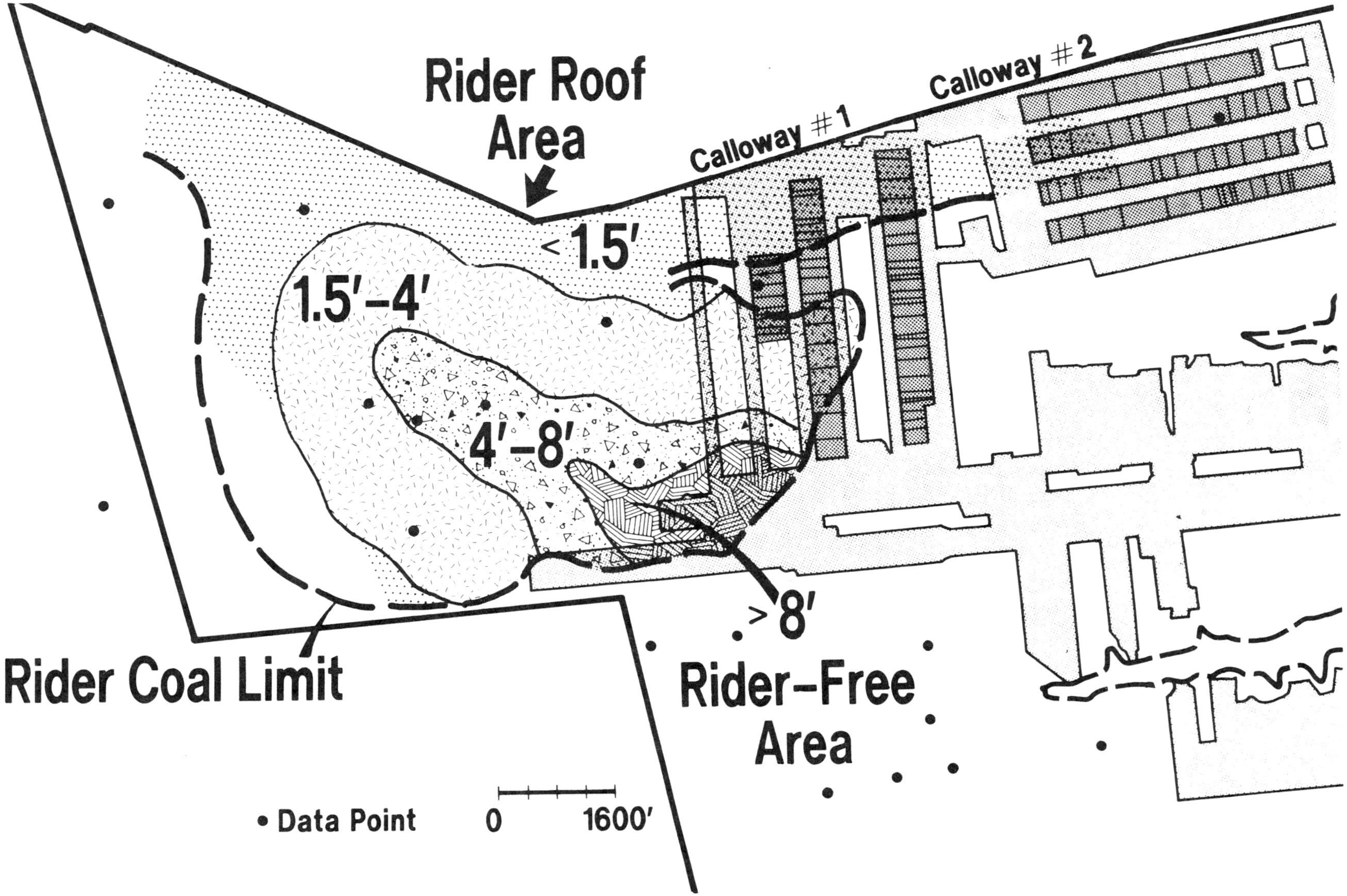

Figure 4. *Isopach of the interval from the top of the Lower Elkhorn Seam to the base of the first overlying rider coal. Note the absence of rider coals throughout most of the previously mined area.*

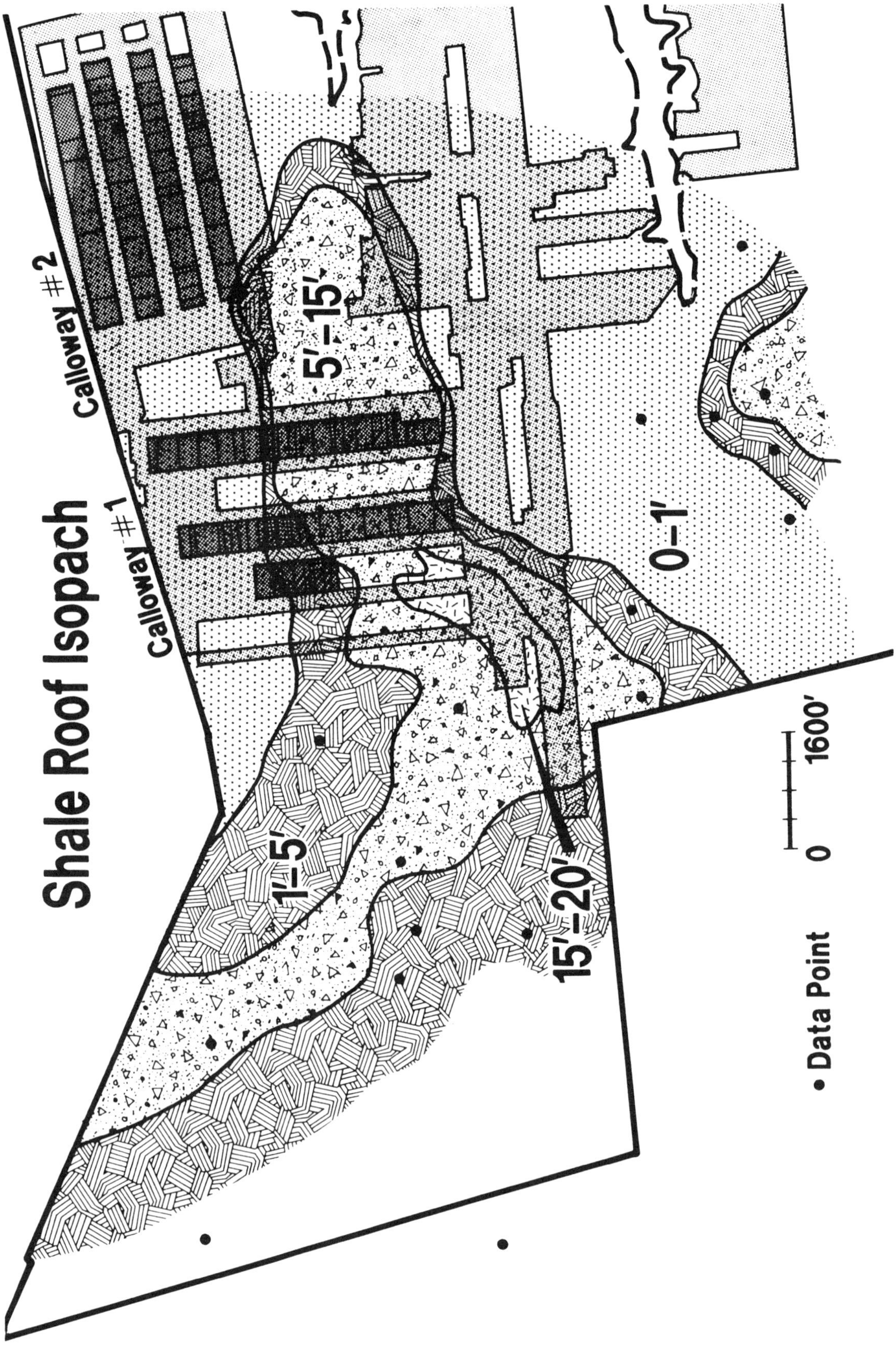

Figure 5. *Isopach map of the shale immediate roof.*

Sandstone Main Roof

Overlying the Lower Elkhorn Seam throughout the study area is a thick- to massive-bedded sandstone. As shown on Figure 6, the thickness of the sandstone unit ranges from about 3 m to over 12 m (10 ft to over 40 ft). Comparison of the shale interval and sandstone unit isopach maps (Figures 5 and 6) demonstrates an approximate inversely proportional relationship of the thickness of these two units.

Figure 7 illustrates the interrelationship between the sandstone main roof and shale immediate roof. This thickness ratio map was generated to define areas in which differential compaction features are likely to occur and, more importantly, to define areas exhibiting an adverse ratio of strong main roof to weak immediate roof. As this map shows, the northern parts of the Calloway No. 1 longwall panels lie in a sandstone-dominated roof area (sandstone-shale ratio 24:1). The shale immediate roof is less than 0.46 m (1.5 ft) thick in this northern area and is overlain by sandstone generally greater than 9 m (30 ft) in thickness. This "thin-shale" area typically did not pose significant problems for longwall advance, simply because the volume of shale available to fall was relatively small. In contrast, the southern limits of these panels lie under dominantly shale roof overlain by relatively little sandstone. In this area, characterized by a sandstone-shale ratio of less than one, abutment stress fracturing of the immediate roof was not a significant problem.

The "transition" zone, however, represented by the area lying between 24:1 and 1:1 sandstone-shale ratios on Figure 7, demonstrated the highest potential for hindrances due to breaking of the shale at or in advance of the longwall face. This zone, particularly when combined with synclinal undulations in the mined seam and presence of the rider coal, represents the anticipated most severe problem area and relates very closely to areas where previous roof problems have been most extreme. The schematic diagram on Figure 8 graphically illustrates the distribution and causes of this "poor roof" zone.

Geotechnical Factors

Prior to this study, geotechnical data for future reserves in the subject area were very limited. To address that deficiency, nine fully cored holes were drilled and full suites of pertinent data were acquired for each hole. Analysis of the geotechnical data generated for the shale immediate roof demonstrated that overall strata cohesiveness and strength characteristics of the rider-free area south of the Calloway No. 1 mains were more favorable than those of the rider roof area.

Roof strata cohesiveness was evaluated by determination of the Rock Quality Designation (RQD), a method developed by Deere (1964). A summary of RQD data for the two areas under consideration is presented in Table 1.

As shown in Table 1 and in Figure 9, the cohesiveness of roof strata in the rider-free area south of the Calloway No. 1 mains is consistently in the good (RQD = 75-90%) to excellent (RQD = 90-100%) range. In the rider roof area, strata cohesiveness ranges from very poor (RQD = 0-25%) to excellent with overall fair to poor conditions prevalent, a condition most likely attributable to the rider coals and associated features. Such features include 1) the poorly cohesive bedding contacts of the rider seam and its seat earth and 2) numerous slickensides caused by differential compaction of the soft, clay-rich shale and seat earth lithologies beneath the overlying massive sandstone.

Analysis of available roof rock compressive strength data also suggests a similar contrast in shale roof strength between the two areas (see Figure 9). A summary of roof rock strength test data for the study area is presented in Table 2.

It is emphasized that there is considerable variation in the mean compressive strength values from borehole to borehole, that the quantity of data is limited, and that some of the very weak roof rocks could not be tested.

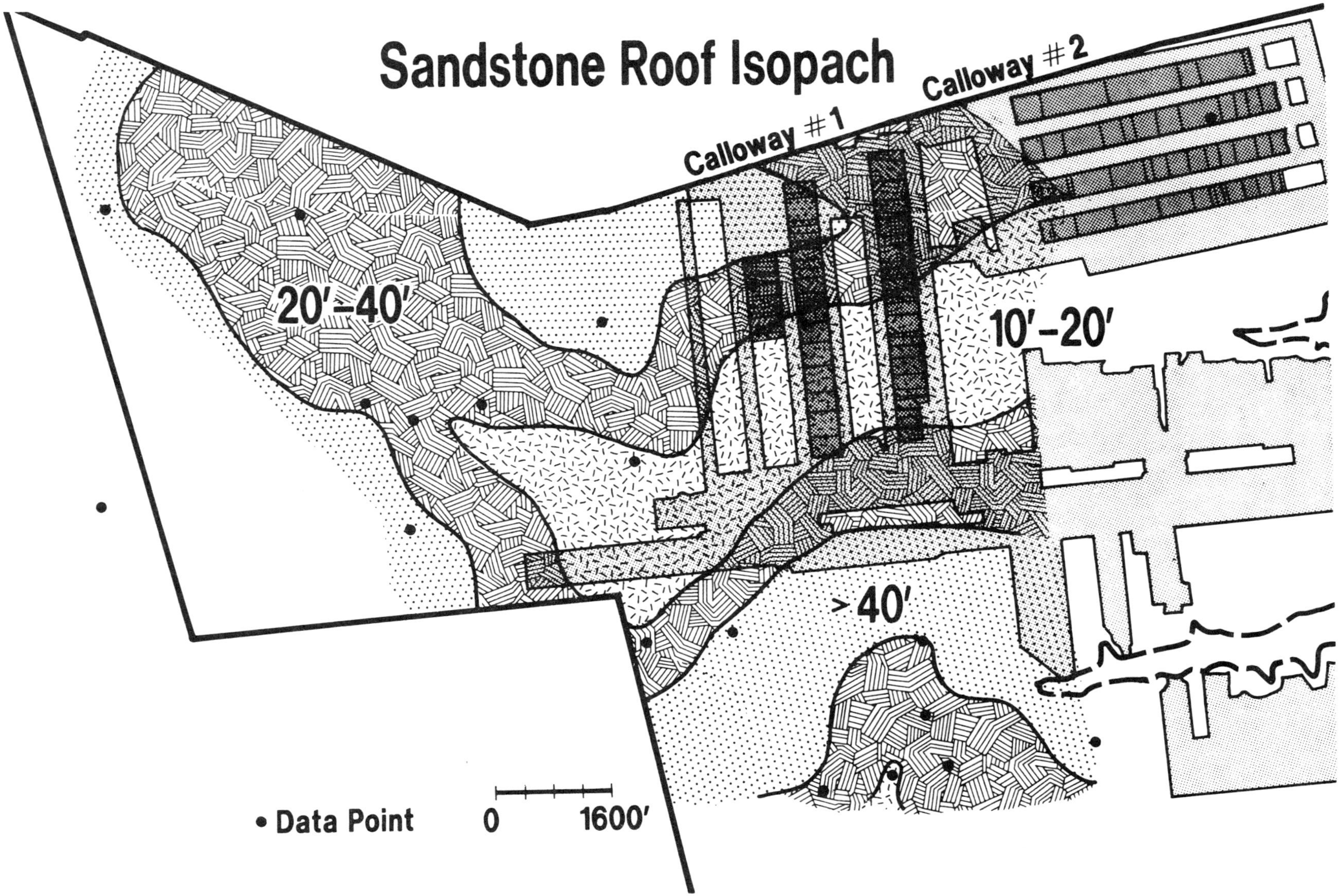

Figure 6. *Isopach map of the sandstone main roof.*

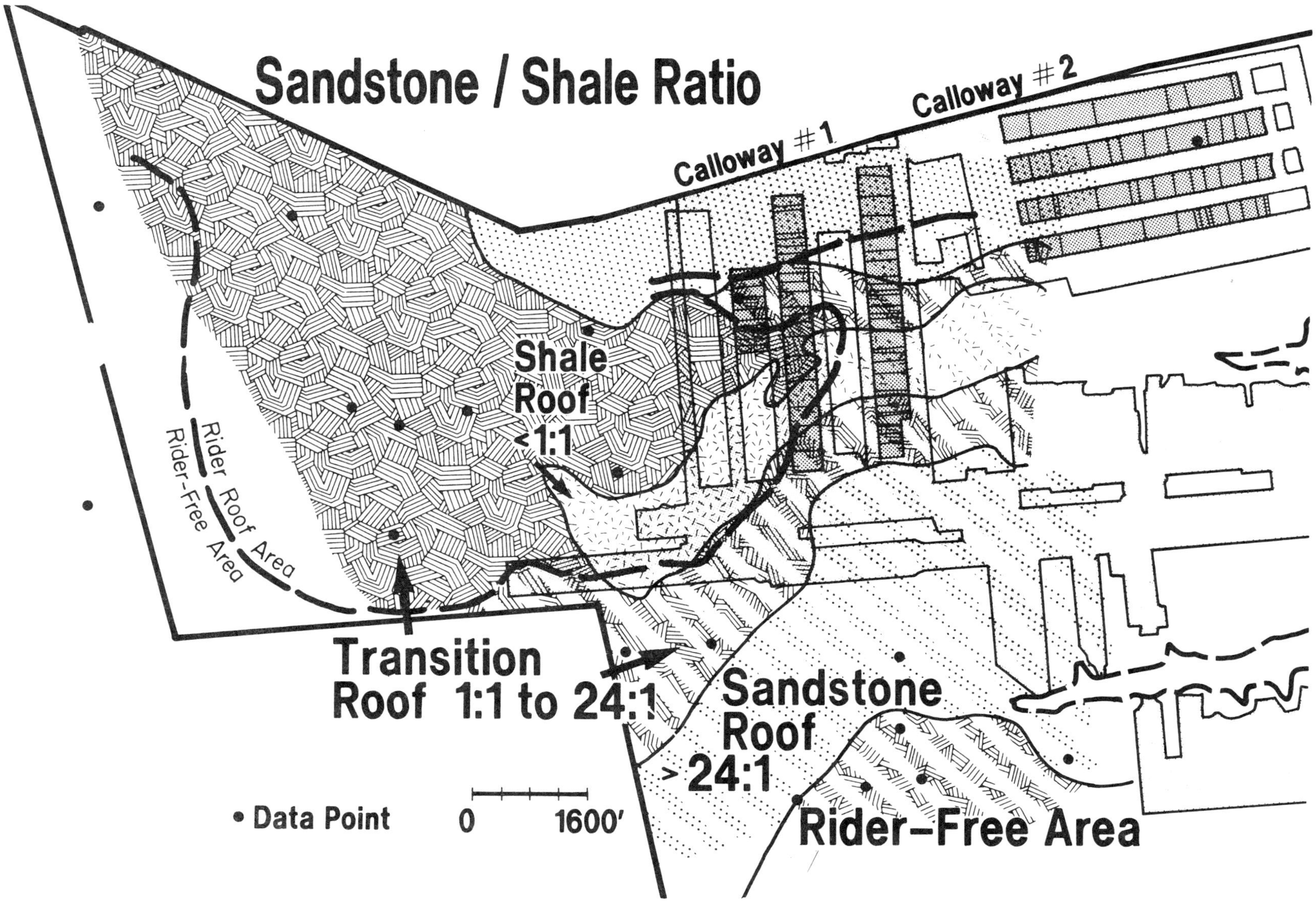

Figure 7. *Ratio of sandstone main-roof thickness to shale immediate-roof thickness.*

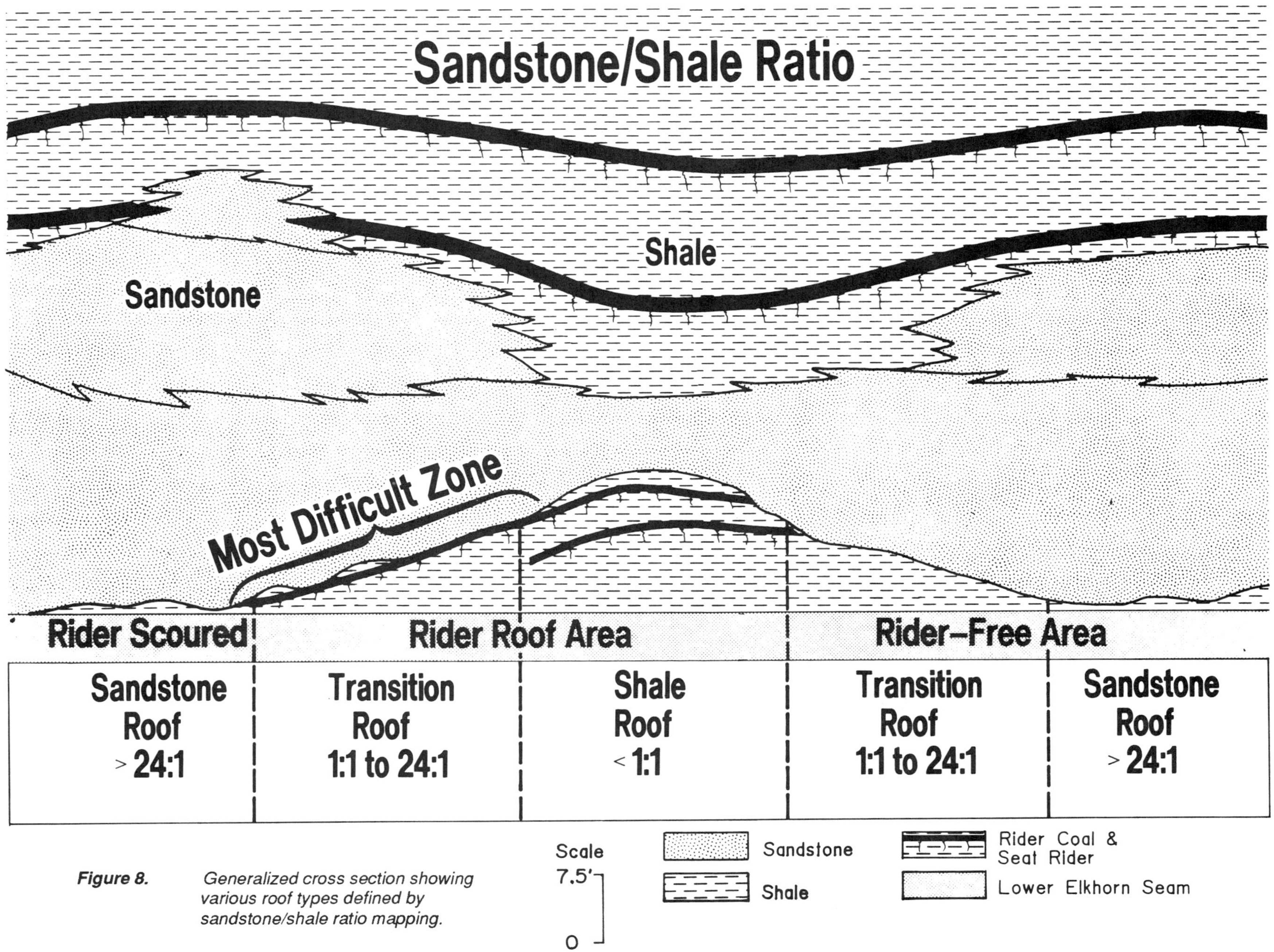

Figure 8. *Generalized cross section showing various roof types defined by sandstone/shale ratio mapping.*

Average RQD Values (%)		
Rider Roof Area		
No. Samples	Range	Average
5	24-93	63
Rider-Free Area		
No. Samples	Range	Average
3	88-95	92

Table 1. *Shale immediate-roof strata cohesion based on the RQD method.*

However, the overall averages suggest shale roof strata of medium strength in the rider-free area and of very low to low strength in the rider roof area. This contrast again most likely is related to presence of the rider coal itself because the soft, clay-rich shale and seat earth lithologies associated with this rider are frequently rooted and lack the laminar bedding features normally characteristic of stronger shale rock types.

The conclusion which generally can be derived from the geotechnical data is that shale immediate-roof strata cohesiveness and strength are anticipated to be more favorable where the rider seam is absent. Significantly less breakage of the shale immediate roof in the face area of a future longwall panel developed in the rider-free area therefore could be anticipated, even in "transition" zone sandstone/shale ratio conditions.

Abrupt Changes In Bed Attitude

Elevational contours on the base of the Lower Elkhorn Seam are presented on Figure 10. In the eastern part of the study area, where most previous mining has occurred, the bedding structure is characterized by a relatively uniform regional northwest dip. Westward in the Calloway No. 1 longwall area, the structural configuration of the mined coal seam changes to a series of small abrupt synclinal "troughs" with associated local occurrences of relatively steeply dipping strata. These features probably represent depositionally (paleotopographically) low areas and/or the result of differential compaction of sediments. Tensional stresses in the roof strata were observed to be associated with these small "troughs", resulting in relatively weaker strata than in adjacent monoclinal or anticlinal areas. Overlay of the structure contour map and the roof delay map (Figures 10 and 3, respectively) demonstrates a strong correlation between the occurrence of these "troughs" and increased delays in longwall mining due to poor roof conditions.

Associated zones of steeper dip, or "rolls", also correspond to delays encountered in longwall development, especially where the longwall face crosses the "roll" at an oblique angle. This circumstance is particularly a problem in poor roof areas where shield overextension is caused by roof cavities. Subsequent rotation and misalignment of the shields in these poor roof areas is compounded by a differentially and steeply pitching floor.

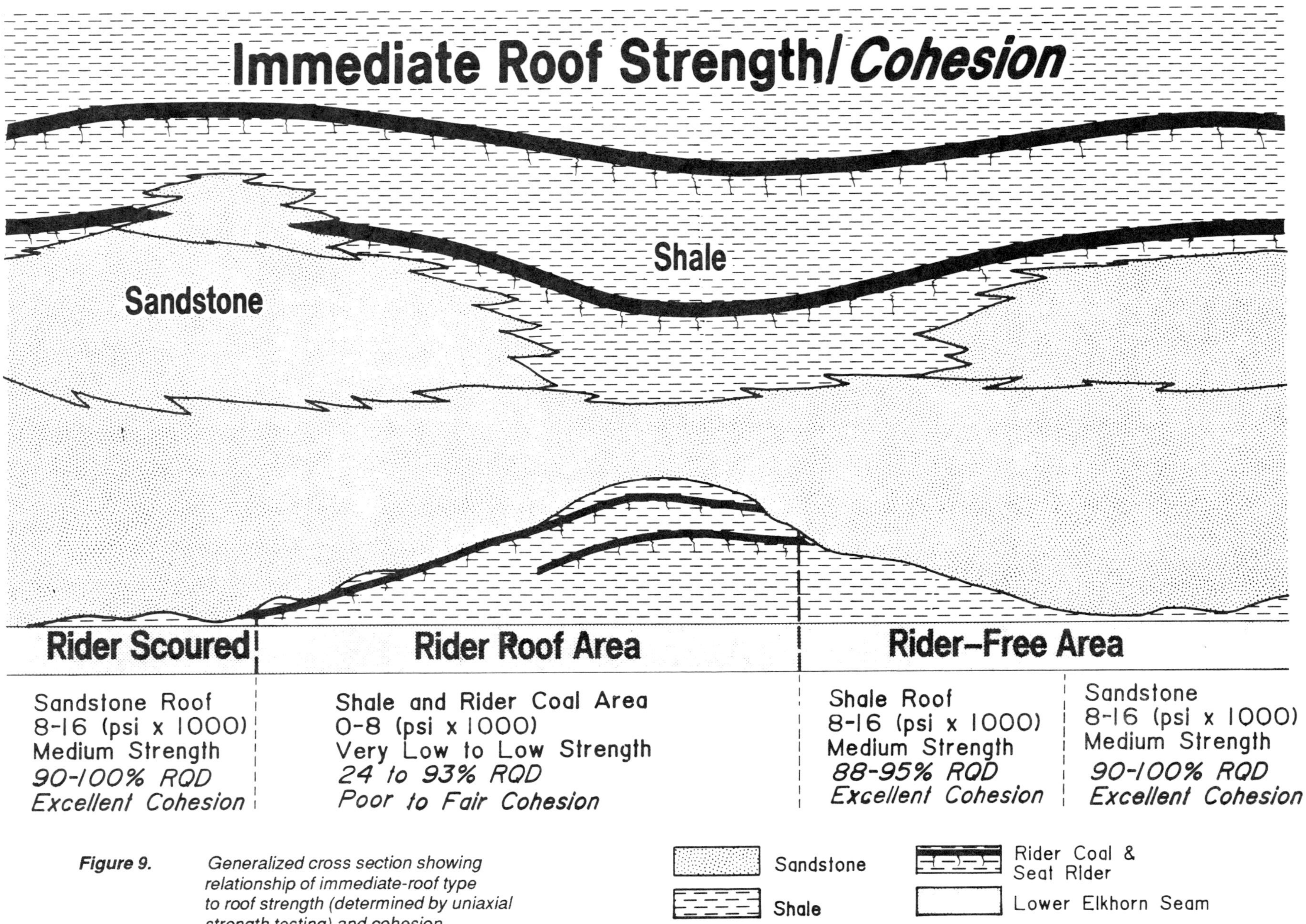

Figure 9. *Generalized cross section showing relationship of immediate-roof type to roof strength (determined by uniaxial strength testing) and cohesion (determined by RQD).*

Mean Compressive Strength Values (values in kPa (psi))		
Rider Roof Area		
KC-4-84	KC-6-84	Average
14,300 (2,074)	49,823 (7,226)	32,062 (4,650)
Rider-Free Area		
KC-1-84	KC-3-84	Average
100,030 (14,508)	55,519 (8,052)	77,776 (11,280)

Table 2. *Strengths for shale immediate-roof strata (KC numbers refer to drill holes). Data source is U.S. Bureau of Mines research on the Kentucky Carbon Corporation property.*

Influence Of Lineaments And Overburden

Fracture systems in the area were delineated using satellite remote sensing imagery analyzed on video enhancement equipment for increased resolution, definition, and classification of linear features (lineaments). Lineaments detected in this study are shown on Figure 11. As the lineaments are representative of fractures or fracture zones in the strata, mining in these areas may be expected to encounter weakened roof material and increased ground water transmission as documented in previous studies by Peng and Haddad (1981), Miller (1982), and Jansky and Jeran (1982). Previous mining experience has demonstrated this to be true for some of the more pronounced lineaments on the subject mine site, especially in thinner overburden areas.

Thickness of overburden above the Lower Elkhorn seam also is presented on Figure 11. Overburden mapping primarily was used in this study to define low-cover areas which potentially could exhibit ground control difficulties related to valley-floor stress relief.

ASSESSMENT OF COMPOSITE GEOLOGIC FACTORS

A considerable variety of geologic factors interact to produce the roof conditions encountered in the Calloway No. 1 area. Whereas each of these factors may be important at any given location, the factors that most strongly correlate to mapped roof delay areas were judged herein to be of a more critical nature and were more heavily weighted in the composite geologic mapping process. As shown on Figure 12, the previously identified "transition" roof zone demonstrates the most positive correlation to difficult ground control areas for the longwall sections. The presence of synclinal troughs and steep-dip zones also can be related to generally poorer mining conditions. Table 3 presents the geologic hazard rating system employed in this investigation.

Composite mapping (Figure 13) of all rated factors yielded overall hazard prediction values ranging from 1 (best anticipated working conditions) to 15 (worst anticipated working conditions). The exhibit readily demonstrates 1) a generally favorable geological setting for the Calloway No. 2 longwall area (predicted

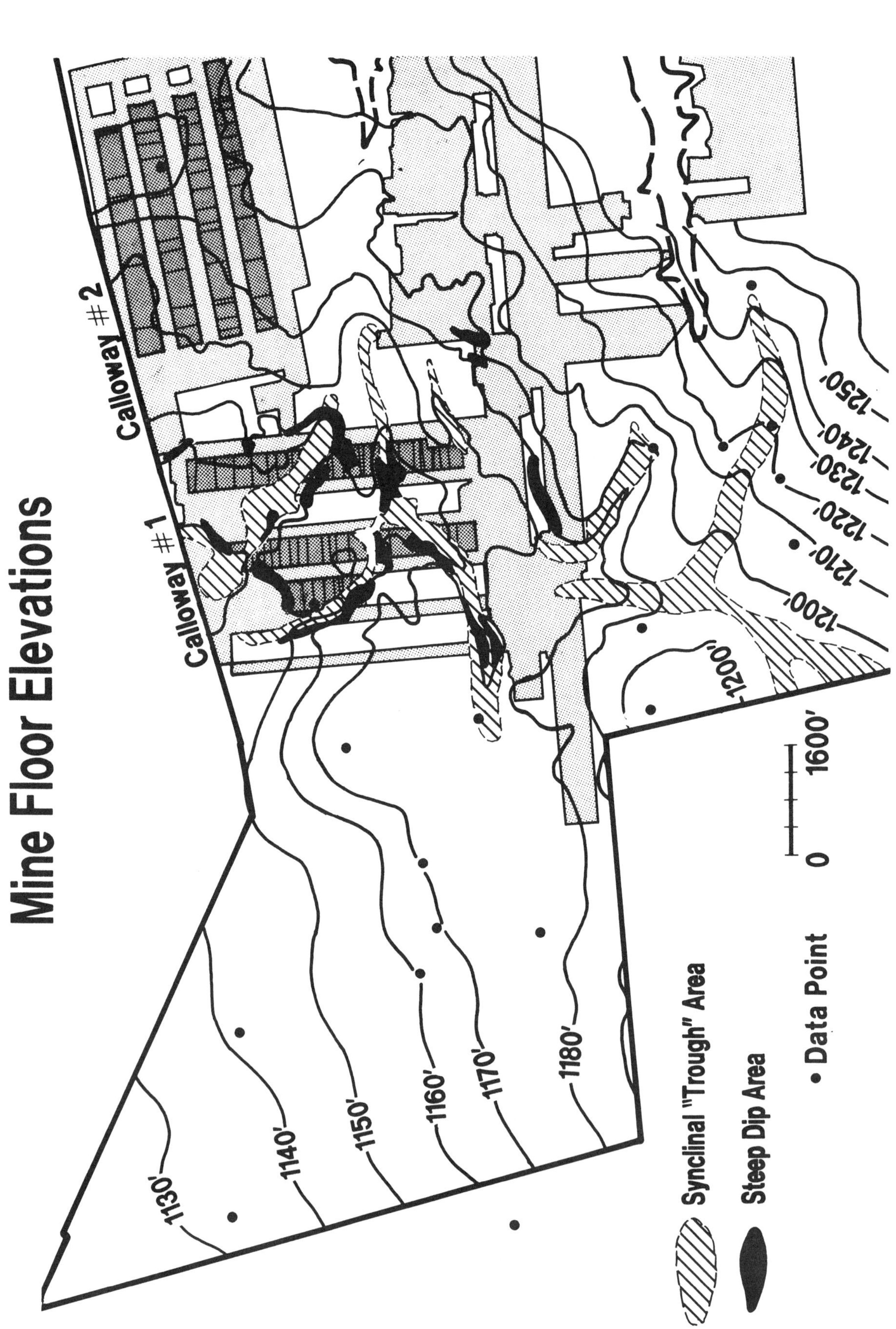

Figure 10. *Structure contour map of the mine floor. The synclinal trough areas probably represent depositionally (paleotopographically) low areas.*

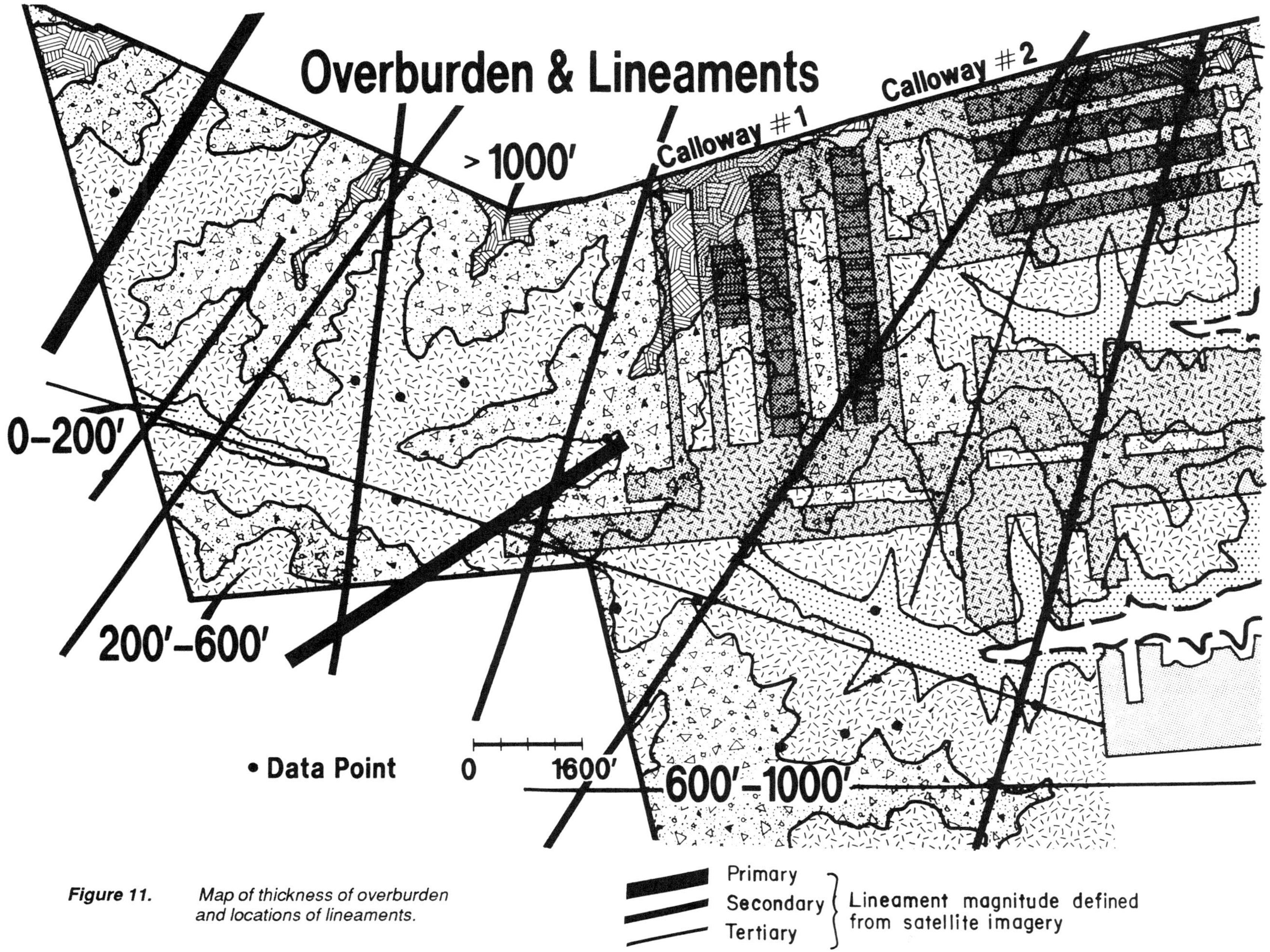

Figure 11. *Map of thickness of overburden and locations of lineaments.*

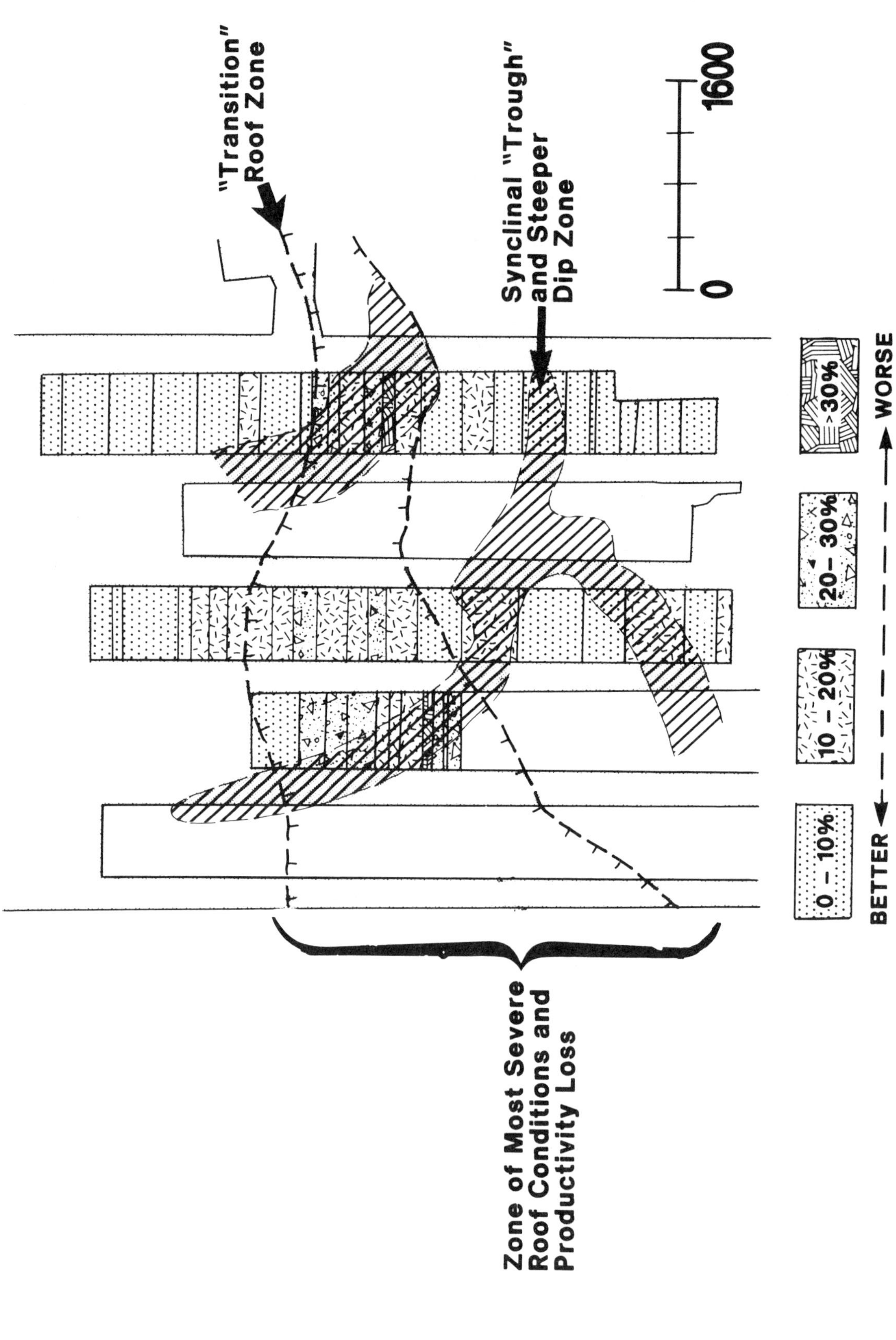

Figure 12. *Map showing the comparison of roof-delay frequency and the areal distribution of "transition" roof and synclinal-roof/steep-dip zones.*

Geological Factor	Value	Geological Factor	Value
Presence of Rider Coal	3	Synclinal Trough Area	5
Shale Roof Thickness	4	Steep-Dip Area	2
1.5' - 5'	3		
10' - 15'	2		
> 15'	1		
< 1.5'	1		
Overburden Thickness:		Lineaments	
0 - 100'	7	Primary Lineament	3
100' - 200'	5	Secondary Lineament	2
200' - 300'	3	Intersection Area	3
300' - 400'	2	Intersection Area Involving	
400' - 600'	1	Primary Lineament	4
> 600'	0		

Table 3. *Hazard Rating System.*

conditions basically similar to those reported to have been experienced during actual mining of the area), 2) an increasingly difficult geologic setting for the Calloway No. 1 longwall area as mining advances westward in the rider roof area (this condition also basically was substantiated by mining experience up to the date of this study), and 3) an overall favorable geological setting for the rider-free roof area south of the Calloway No. 1 mains.

In order to further verify the predictive capability of the composite geological modeling techniques employed, average composite hazard values were generated for the same monthly reporting periods used in roof delay mapping. This was done on a weighted average basis by planimetric survey of the longwall area mined during each monthly reporting period. Results of this comparison are presented on Figure 14 which demonstrates 1) a positive correlation between the predicted average-composite hazard value and the actual roof delay time incurred during mining and 2) the strong influence of "transition" roof conditions and mine floor-elevation changes in defining longwall mining ground control conditions for the subject property.

POST-EVALUATION MINING EXPERIENCE

Subsequent to this evaluation, mining efforts on the Calloway No. 1 longwall panel No. 4 demonstrated a positive correlation of predicted and actual mining conditions, according to operating company personnel. Mine management elected to limit further mine development in the rider roof area and to phase its operations into the rider-free area. As projected in the evaluation, improved ground control conditions have been realized by continuing mine development (both advance mining and pillar extraction) in the new operational area.

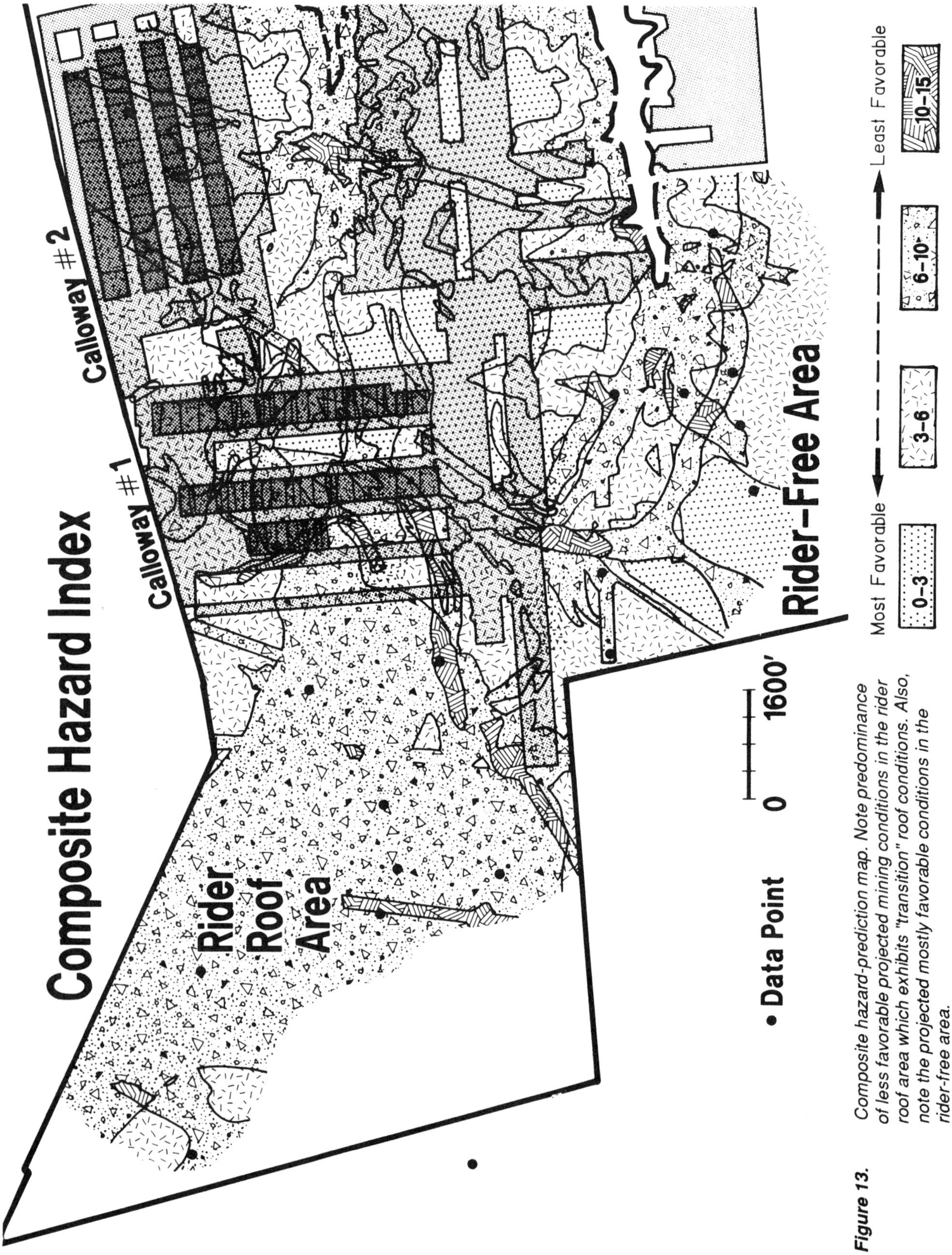

Figure 13. *Composite hazard-prediction map. Note predominance of less favorable projected mining conditions in the rider roof area which exhibits "transition" roof conditions. Also, note the projected mostly favorable conditions in the rider-free area.*

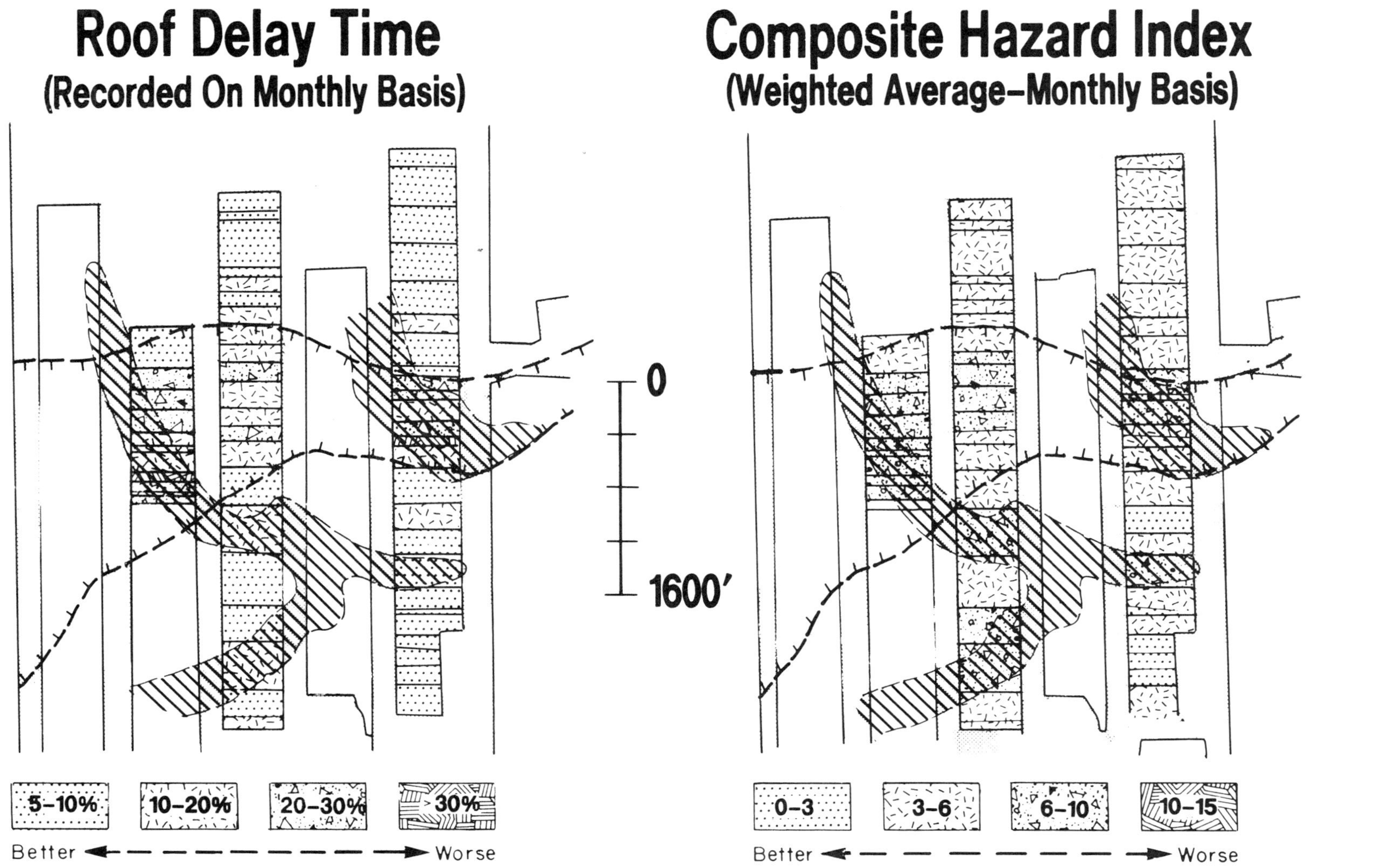

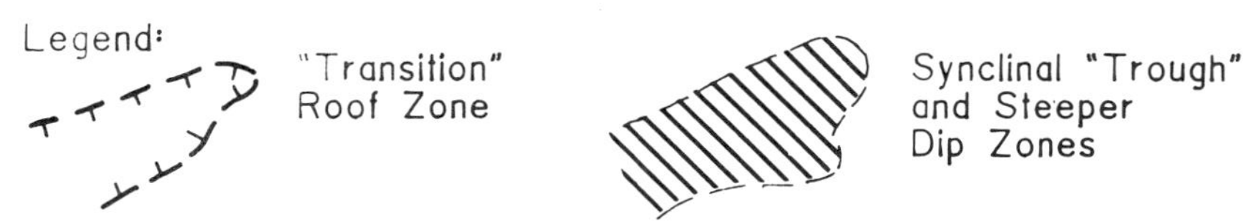

Figure 14. *Map showing a comparison of the recorded longwall section roof-delay time and the composite-hazard value derived from the geologic modeling and analysis. Note the correlation between actual and predicted hazards and the strong influence of the "transition" roof and structural "trough" zones.*

CONCLUSIONS

Results of this investigation confirm that, for the subject property, a number of geological factors interact and result in a variety of roof conditions which exert a strong influence on potential productivity of longwall mining operations. Analysis of mining experience on the subject property demonstrates that "transition" roof, synclinal "troughs", and presence of a rider seam combine to cause the most severe interruptions of mine productivity.

These factors can be identified in advance of mining and, through proper geologic analysis, can be mapped and delineated to afford a predictive model for use in mine planning. Potential benefits include increased capability to 1) anticipate and respond to areas of difficult ground control conditions prior to the occurrence of down-time events, 2) optimize mining design and ground control strategies according to mine-site geological conditions, 3) estimate the suitability of in-use face equipment for future working conditions at the mine, and 4) rationally select face equipment which is properly matched to anticipated working conditions at new longwall installations.

ACKNOWLEDGEMENTS

We thank Carbon Industries, Inc. for permission to publish the findings of this study which initially was presented at the 1988 Longwall USA Conference in Pittsburgh, Pennsylvania.

REFERENCES

Deere, D.U., 1964, Technical description of rock cores for engineering purposes: Rock Mechanics and Engineering Geology, v. 1, no. 1, p. 17-22.

Deere, D.U., and R.P. Miller, 1966, Engineering, classification, and index properties of intact rock: Air Force Weapons Laboratory, Technical Report No. AFNL-TR-65-116.

Dunham, K.R., A.G. Thurman, and R.D. Ellison, 1976, The use of geological/geotechnical investigation as an aid to mine planning, in Proceedings of the 18th U.S. Symposium on Rock Mechanics, Keystone, Colorado, June, 1976: Golden, Colorado School of Mines Press, p. IC 4-1-6.

Ealy, D.L., R.E. Mazurak, and E.L. Laugrand, 1979, A geological approach for predicting unstable roof and floor conditions in advance of mining: Mining Congress Journal, v. 65, no. 3, p. 17-22.

Hunt, C.B., G.H. Briggs, Jr., A.C. Munyan, and G.R. Wesley, 1937, Coal deposits of Pike County, Kentucky: U.S. Geological Survey Bulletin 876, 92 p.

Jansky, J.H., and P.W. Jeran, 1982, Determination of geologic features associated with linears: U.S. Bureau of Mines Report of Investigations RI 8660, 8 p.

Jansky, J.J., and K.W. McCabe, 1979, Evaluation of remote sensing techniques for planning mining operations, in Fun-Den Wang and G.B. Clark, eds., Proceedings of the Mini-Symposium on Underground Coal Mine Design and Planning, New Orleans, Louisiana, February, 1979: New York, Society of Mining Engineers-AIME, p. 27-38.

McKay, E.J., and D.C. Alvord, 1969, Geologic map of the Lick Creek Quadrangle, Pike County, Kentucky: U.S. Geological Survey Geologic Quadrangle Map GQ-716.

Miller, Marshall S., 1982, Predictive geological mapping of mining conditions in Appalachian coal fields (abstr.): American Association of Petroleum Geologists Bulletin, v. 66, no. 8, p. 1172.

Miller, Marshall S., 1984, Composite geologic and linear mapping for defining safe, high productivity in the Appalachian coal fields, in Sutherland, Lucas, M. Karmis, D. Forshey, and Faulkner, eds., Proceedings of the Fifteenth Annual Institute on Coal Mining Health, Safety and Research, Blacksburg, Virginia, August, 1984: Blacksburg, Virginia Polytechnic Institute and State University, p. 195-205.

Outerbridge, W.F., 1968, Geologic map of part of the Jamboree Quadrangle, Pike County, Kentucky: U.S. Geological Survey Geologic Quadrangle Map GQ-775.

Overby, W.K., C.A. Komar, and J. Posini, 1978, Predicting probable roof fall areas in advance of mining by geological analysis: U.S. Bureau of Mines Technical Progress Report TPR 70, 15 p.

Peng, S.S., and M.R. Haddad, 1981, A study of "linear" as a tool for predicting underground instability: Morgantown, West Virginia University, Department of Mining Engineering, Technical Report 80-7, 63 p.

Rice, C.L., R.G. Ping, and J.L. Barr, 1977, Geologic map of the Belfry Quadrangle, Pike County, Kentucky: U.S. Geological Survey Geologic Quadrangle Map GQ-1369.

Trent, V.A., 1965, Geology of the Matewan Quadrangle in Kentucky: U.S. Geological Survey Geologic Quadrangle Map GQ-373.

EFFECTS OF STRUCTURAL FAULTS ON GROUND CONTROL IN SELECTED COAL MINES IN SOUTHWEST VIRGINIA

Gregory M. Molinda and David K. Ingram
U.S. Bureau of Mines
P.O. Box 18070
Pittsburgh, PA 15236

ABSTRACT

Two right lateral, strike-slip faults related to the Pine Mountain overthrust sheet near Buchanan County, Virginia were investigated by the U.S. Bureau of Mines to determine their effects on ground control and to develop recognition criteria for prediction of the locations of the faults. Both faults are overprinted simultaneously with thrust faulting resulting in offsets of the coal seam averaging 2 m (6 ft). Associated with the faulting are severe roof conditions including 6-m (20-ft) high roof falls occurring through an average 9-m (30-ft) wide zone of disturbance. Recognition criteria, resulting from field mapping, include characteristic overthrusting, coal seam swag, bedding plane slippage, and fault gouge. Ground control strategies include mine designs that minimize fault exposure and the use of combination roof bolt systems or "super bolts" (2.5-cm (1.0-in.) diameter resin anchor combination bolts) to reduce the shearing effects of lateral movements in the mine roof.

INTRODUCTION

Roof and rib stability at a number of coal mines in Buchanan Co., Virginia, have been severely affected by faulting related to the Pine Mountain overthrust sheet. Coal seams and roof sequences have been disrupted for several miles along the trace of these faults as well as vertically in the geologic section. More than five mineable coal seams are affected. Recognition that the coal seam discontinuities encountered during mining were part of a large crustal deformation and not local sedimentary readjustments is critical to prediction and avoidance of the discontinuities. The objective of this investigation was to describe the nature and extent of the structures and to document recognition criteria to aid in this prediction.

GEOLOGIC DESCRIPTION OF FAULTS

The Pine Mountain overthrust sheet is a detached section of crust about 201 km (125 mi.) long and about 40 km (25 mi.) wide which has moved in a northwesterly direction as a slab up and over younger strata for many miles (Miller and Fuller, 1954). The rectangular block is located at the junction of Virginia, Kentucky, and Tennessee in the Southern Appalachian coal field (Figure 1). The Pine Mountain Fault originated as a relatively shallow bedding-plane failure which, when bedding resistance became excessive, tore diagonally towards the surface and finally was exposed at Pine Mountain. This movement was due to regional compression oriented northwest-southeast. Rupture of the crust bounding the block on the northeast side occurred progressively northwestward along the trace of the Russell Fork Fault (Figure 2). The crust to the northeast of the Russell Fork Fault was splintered by other tear faults caused by the overthrusting of the Pine Mountain sheet. These faults, which include the Keen Mountain, Pistol Gap, and Jess Fork faults, did not isolate overthrust blocks, but are oriented roughly parallel to the direction of overthrust in the study area. During the formation of these tear faults, the Pine Mountain overthrust block was rotated counterclockwise (Miller, 1974; Wentworth, 1921). As a result of this rotation, the bounding tear faults, including the Keen Mountain and Pistol Gap faults, show some secondary compressional overthrusting. Two of these tear faults, the Keen Mountain Fault and the Pistol Gap Fault, and their effects on roof control are the subject of this paper (Figure 2).

Structures like the Keen Mountain and Pistol Gap faults are large enough to be seen readily on side-looking airborne radar (SLAR) images (McLoughlin, 1986). The Pistol Gap Fault also has been identified as a lineament on Landsat satellite imagery (T.F. McLoughlin, personal communication, 1987). Such imagery can help locate the fault zone, but the width or depth of the fault zone can be accurately determined only by field mapping. To what depth these faults extend is unknown, but the effect of the Keen Mountain Fault on ground control can be documented to depths greater than 460 m (1,500 ft) (Elder et al., 1974). The following discussion describes the mechanics of fault disturbance and the effects on ground control in several coal mines in the area.

INTERPRETATION OF EVENTS IN THE FORMATION OF THRUST FAULTS IN THE STUDY AREA

Evidence of both strike-slip movement and thrust faulting at underground exposures of the Keen Mountain and Pistol Gap faults indicate two types of fault disturbance. Strike-slip movement, indicated by horizontal slickensides on vertical fault planes at all underground exposures of these two faults, is likely a result of the original northwest-southeast compressional field which created the Pine Mountain overthrust sheet. Whereas the Pine Mountain Sheet became completely detached from the surrounding strata by the compression, the rock strata on the northeast boundary of the sheet were only partially isolated by the development of several tear faults, including the Keen Mountain, Jess Fork, Pistol Gap, and Little Paw Paw faults (Figure 2). These faults may represent the first phase of detached sheet formation adjacent to the Pine Mountain Sheet, even though complete detachment did not occur. The movement on the faults is right lateral which is consistent with the northwest-southeast compression direction.

Figure 3 is an interpretation of the two types of faulting believed to have occurred on the Pistol Gap and Keen Mountain faults. Diagram A on Figure 3 shows simple strike-slip movement as a result of northwest-southeast compression. Diagram B on Figure 3 shows simultaneous bedding plane faults and deformation as a result of compression due to counterclockwise rotation of the Pine Mountain block. This compressional failure is responsible for the bulk of the current roof control problems observed in local underground mines.

A northeast-southwest compressional stress orientation caused thrust faulting to occur along weak interfaces (bedding planes and coal seams). The Pistol Gap and Keen Mountain faults are the discontinuities at which compressional stress was relieved by vertical strata failures, as seen in the Chaparal Mine and the Falcon Fuel Mine. These tear fault zones or zones of weakness are the natural locations for the relief of regional stress in the form of rock failure.

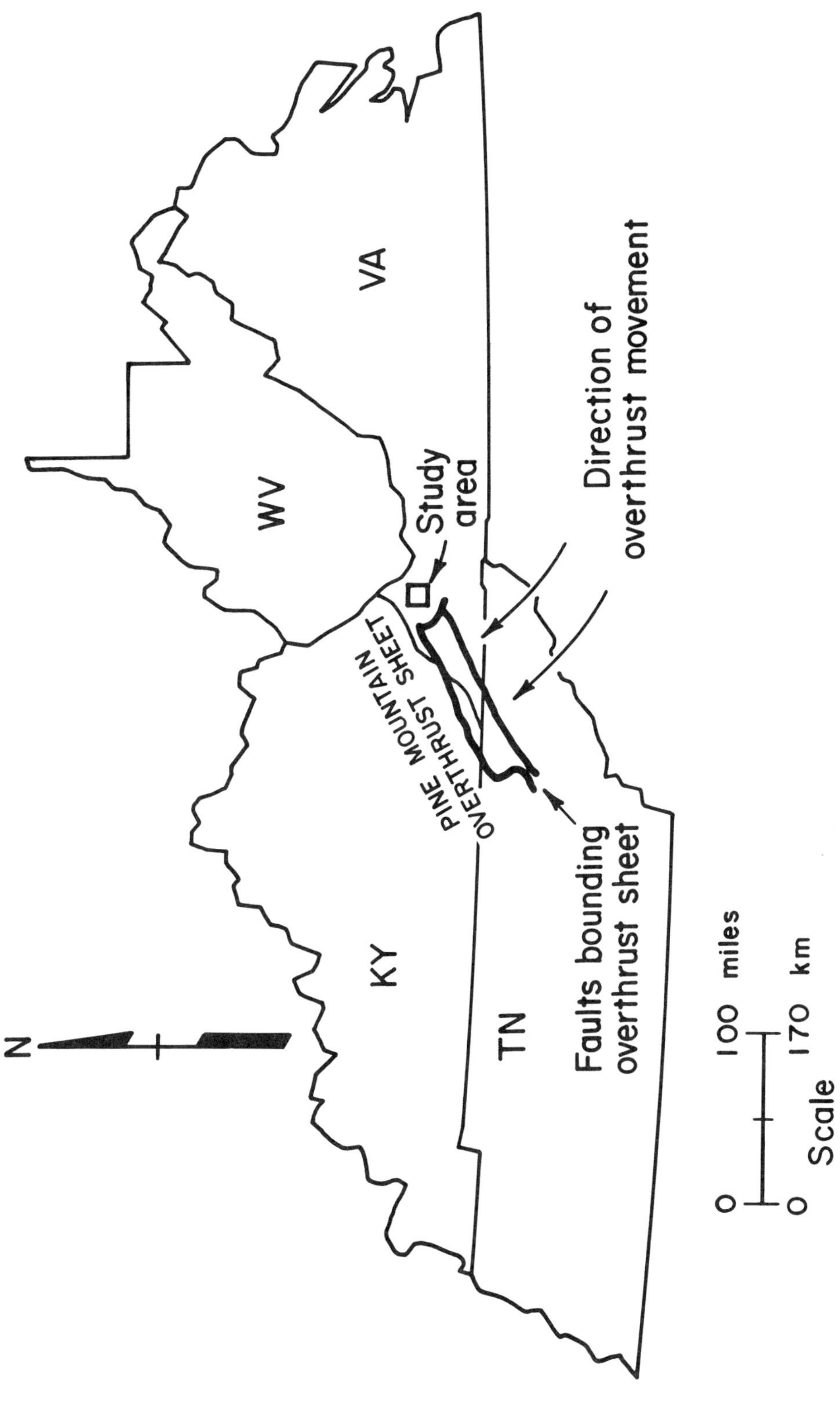

Figure 1. *Location of Pine Mountain overthrust and study area.*

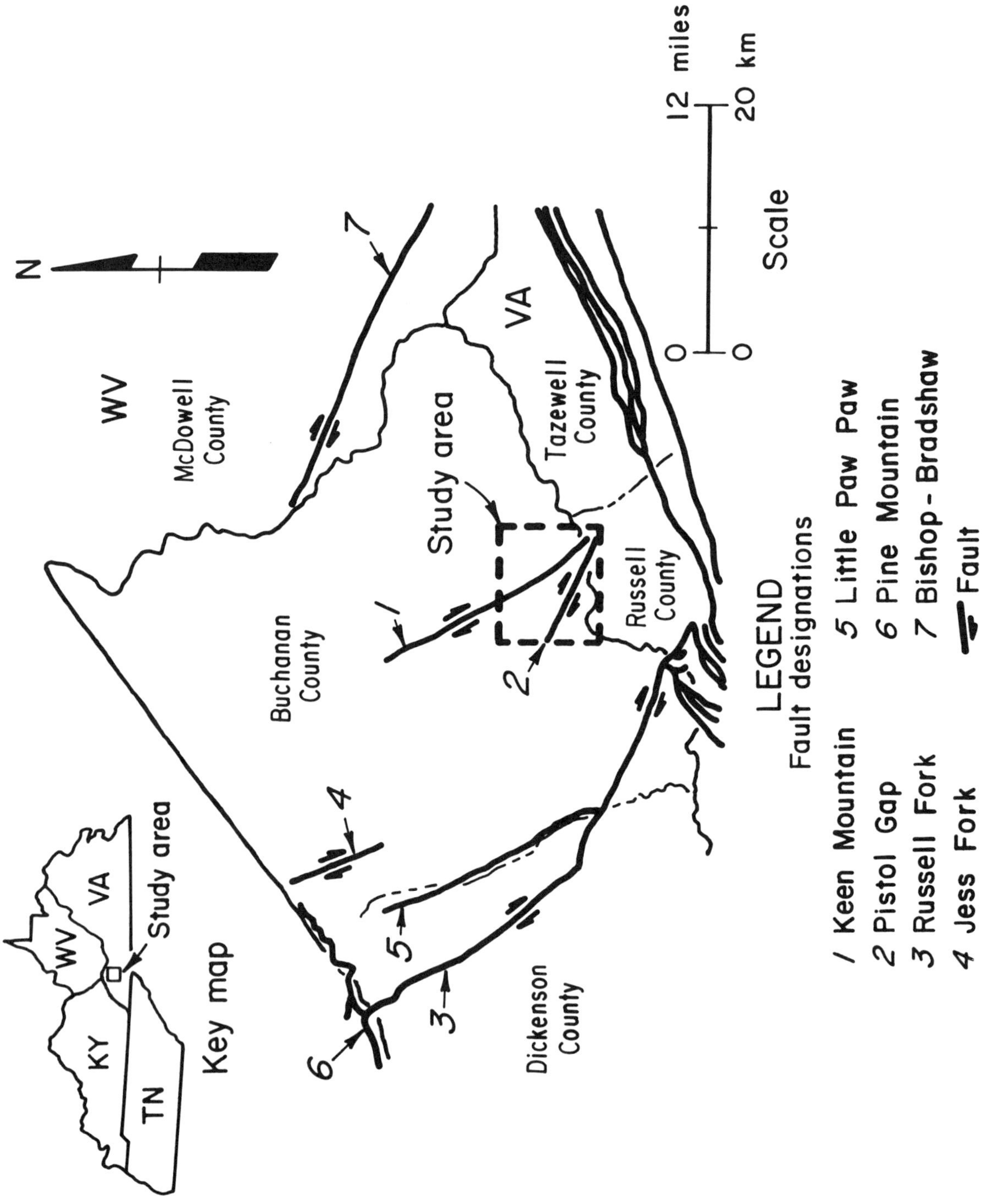

Figure 2. *Structural map of Buchanan County, Virginia with study area indicated.*

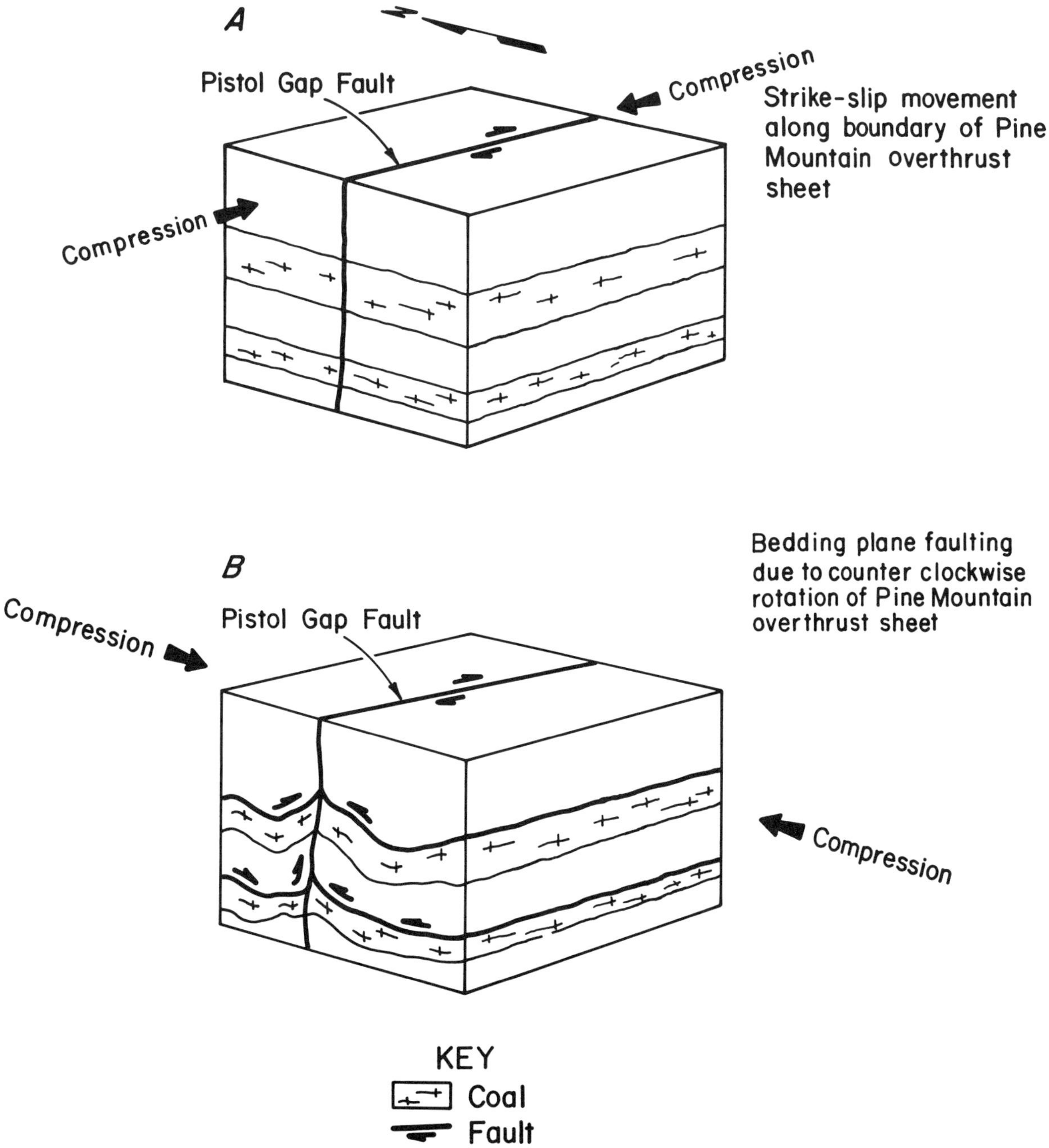

Figure 3. *Block model showing sequence of coal seams subjected to lateral forces, causing coal seam buckling and bedding-plane and vertical failure at Pistol Gap and Keen Mountain faults.*

SELECTED COAL MINES EXPERIENCING DISRUPTION DUE TO FAULTING

Seven heavily mined coal seams in the study area are affected by the Keen Mountain and Pistol Gap faults (Figure 4). Following are examples of the deformation encountered near the faults in several underground mines and of the resulting roof damage.

Falcon Fuel Mine

The Falcon Fuel Mine, which is developed in the Jawbone A&B coal seam, has been severely affected by the Keen Mountain and Pistol Gap fault zones (Figure 5). The mine first encountered the Pistol Gap Fault zone at 850 m (2,800 ft) inby the portal (Figure 6, location C). The operators pulled back because of adverse roof conditions and began driving a four-entry development to the southwest (Figure 6, location D). After approximately 120 m (400 ft), the development was halted as the mine again intersected the Pistol Gap Fault zone. The operators did not know at the time that the two discontinuities were the same fault zone.

Mapping at location C indicates that compressive forces oriented northeast-southwest have caused downbuckling of the coal seam and a thrust fault-type failure in the coal and roof strata. Figure 7 (cross section A-A') shows that the fault runs along bedding at the roof and coal interface from A' to A until it reaches the failure zone where it turns up into the roof. In fact, numerous faults occur along bedding in the roof shales, much like two decks of playing cards being shuffled into one deck. Figure 8 shows the coal seam thrust upward at B-B', resulting in fracturing that caused a 1.2-m (4-ft) high roof fall. In addition, the coal seam is squeezed out to only a few inches thick by bedding-plane movement. Fault gouge and extreme deformation indicate compression. In Figure 7 (cross section A-A'), the slippage, both at the roof/coal interface and at the floor/coal interface, has completely truncated the coal seam. Highly polished slickensides on roof shales indicate slippage along the roof/coal interface (Figure 9). Slippage occurred along bedding within the roof shale as well as within the coal seam.

At cross section B-B' (only 15 m (50 ft) away from A-A'), the nature of the disturbance is different (Figure 7). Here the compressive forces have dragged the coal seam into the roof and partially severed it. A series of three main faults have isolated two fault blocks, which have been uplifted as horsts. The coal seam and adjacent strata have been severely disturbed between the fault planes (coal fragmentation, fault gouge, and cleat rotation). The mine operator notes that roof bolts have been sheared in the vicinity of the faults indicating active bedding-plane movement since bolt installation. The movement which has caused bolts to be sheared appears to be mining-induced and not the result of active tectonic compression. The entire mine is topographically isolated and, therefore, not currently subject to active horizontal stresses related to the Pine Mountain overthrust sheet.

In cross sections B-B' and C-C', the coal seam has been overturned because of compression. Also, at all three places where the fault is exposed at location C (Figure 7), horizontal (strike-slip) movement is indicated by slickensides. This movement is attributed to the original crustal shearing of the Pine Mountain block. Roof conditions in the three accessible entries were poor for nearly 60 m (200 ft) about the fault zone, with falls averaging 2.4 m (8 ft) high and extending up to 6 m (20 ft) high (Figure 10). Supplementary roof support included heavy crossbar sets, posts, I-beams, and 2.4-m (8-ft) long resin bolts. Figure 7 (cross sections A-A' and B-B') also shows that the fault zone is accompanied by a "swag" (this mining term will hereafter be used to indicate a depression in the coal seam caused by downbuckling) zone up to about 23 m (75 ft) wide where the coal seam is downbuckled approximately 2.4 m (8 ft). This swag feature, in these cross sections, is characteristic of compressional stress fields and may serve to warn miners when found elsewhere that they are approaching a fault.

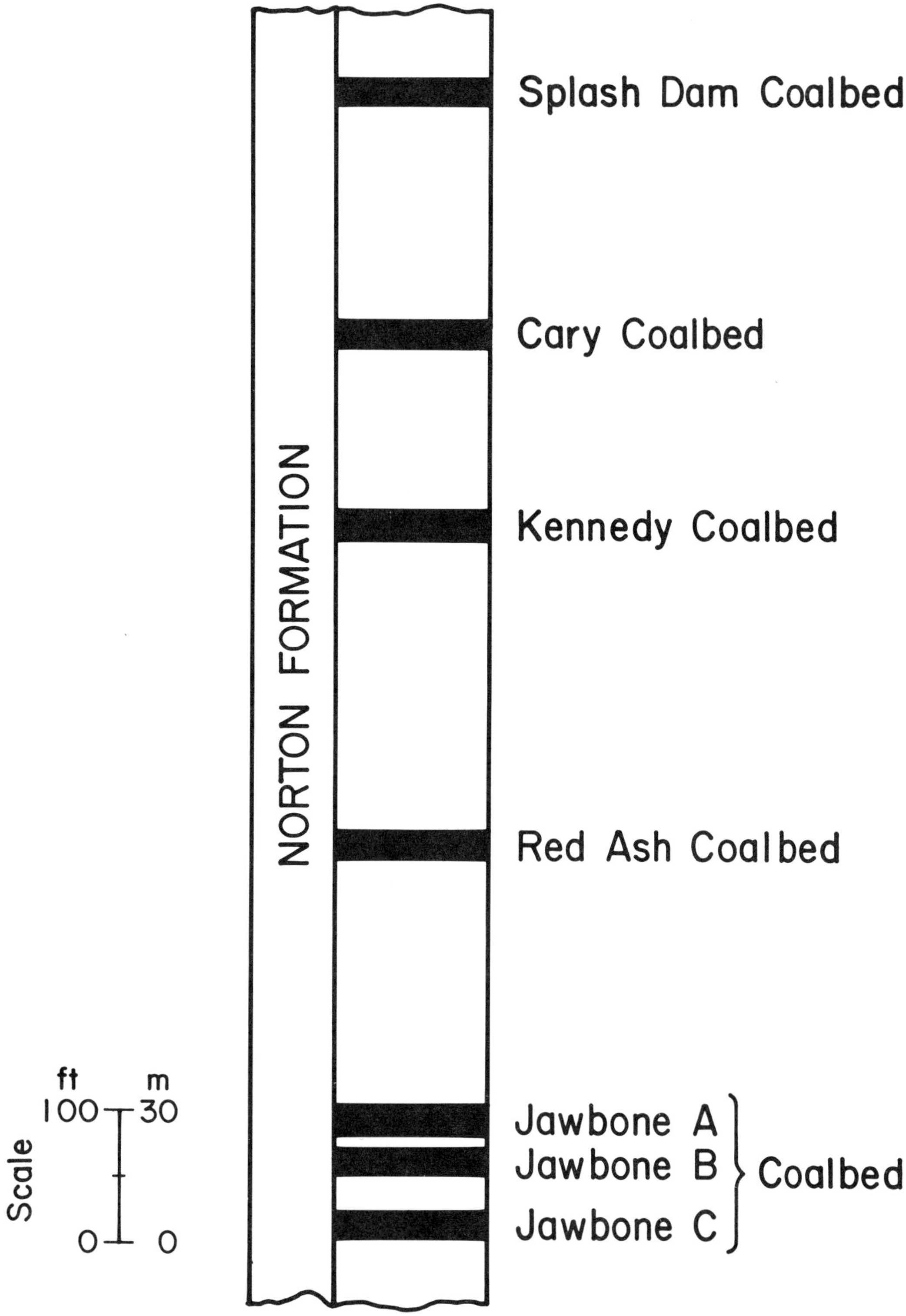

Figure 4. Coal seam stratigraphy in study area.

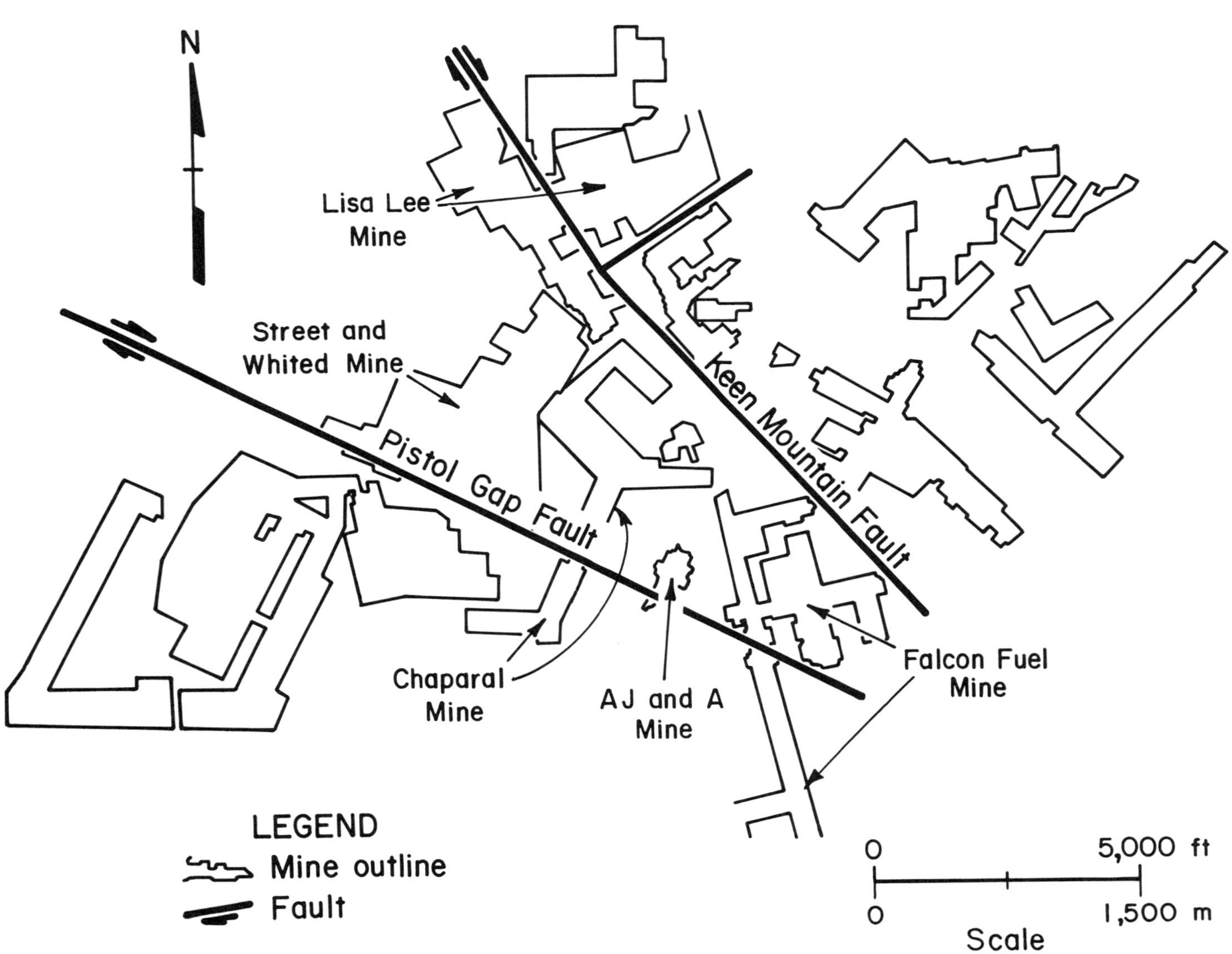

Figure 5. *Mines in study area affected by Keen Mountain and Pistol Gap faults.*

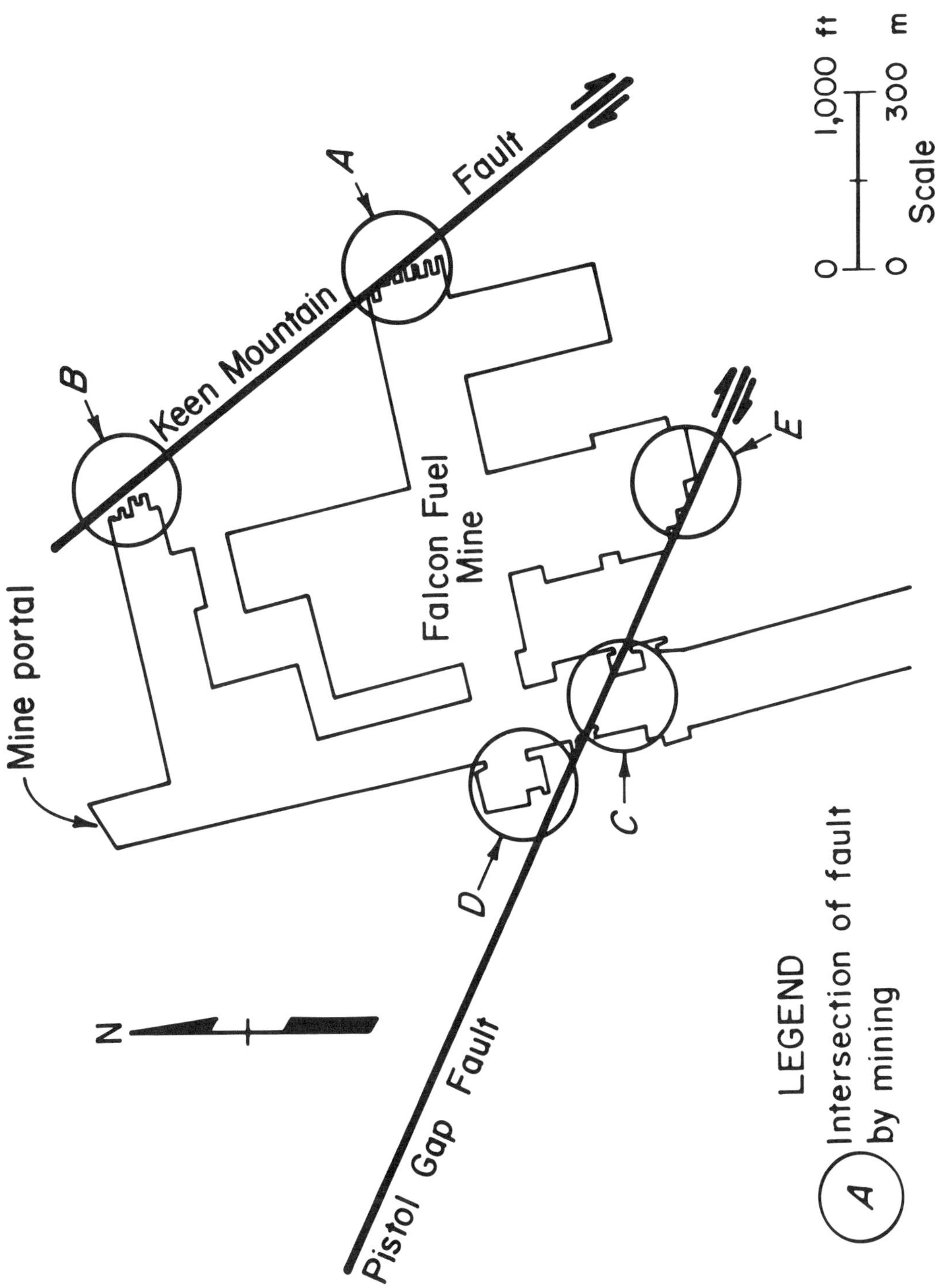

Figure 6. *Falcon Fuel Mine with developments (circled) affected by faults.*

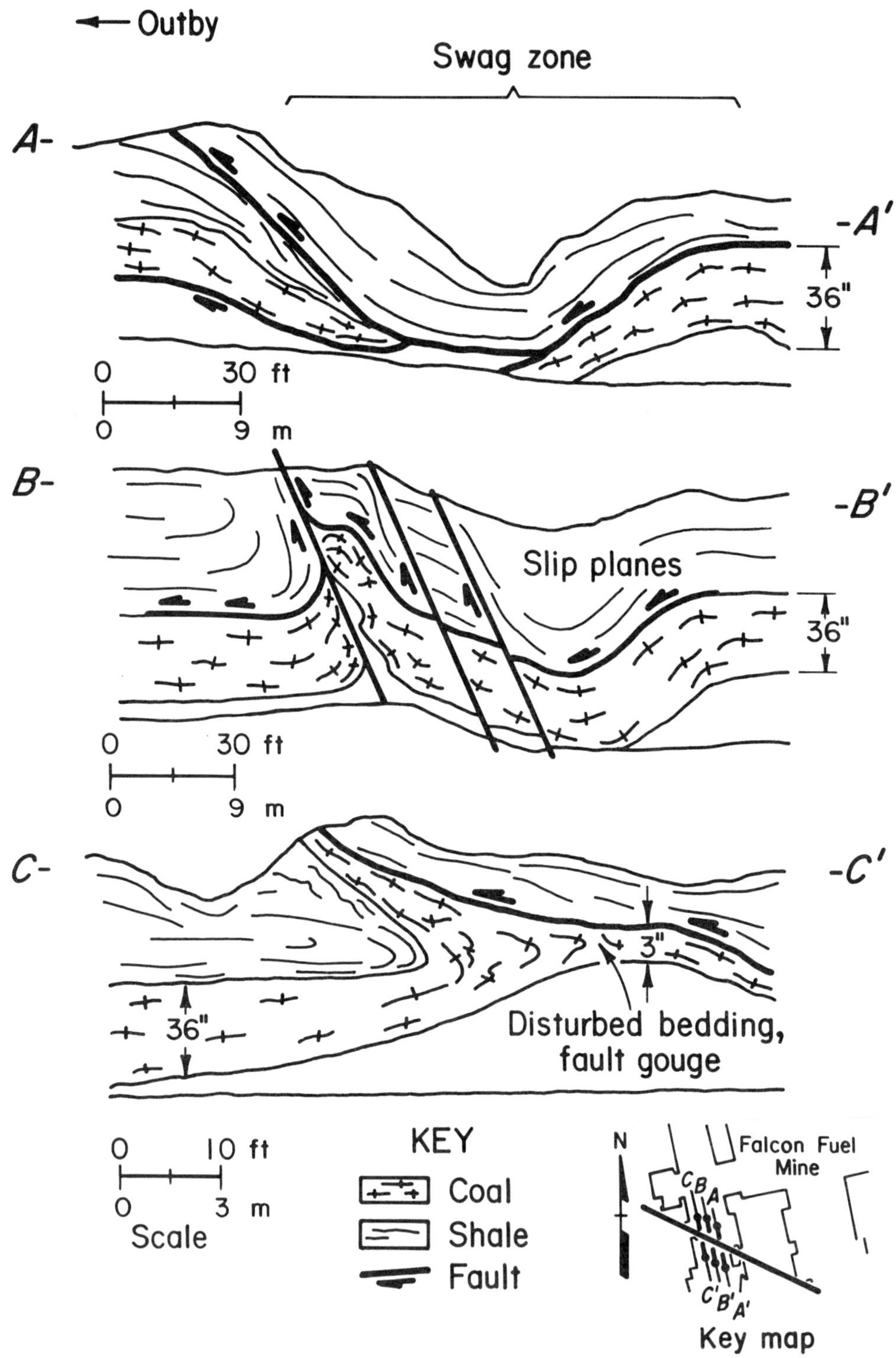

Figure 7. Series of cross sections of the Pistol Gap Fault exposed in the Falcon Fuel Mine.

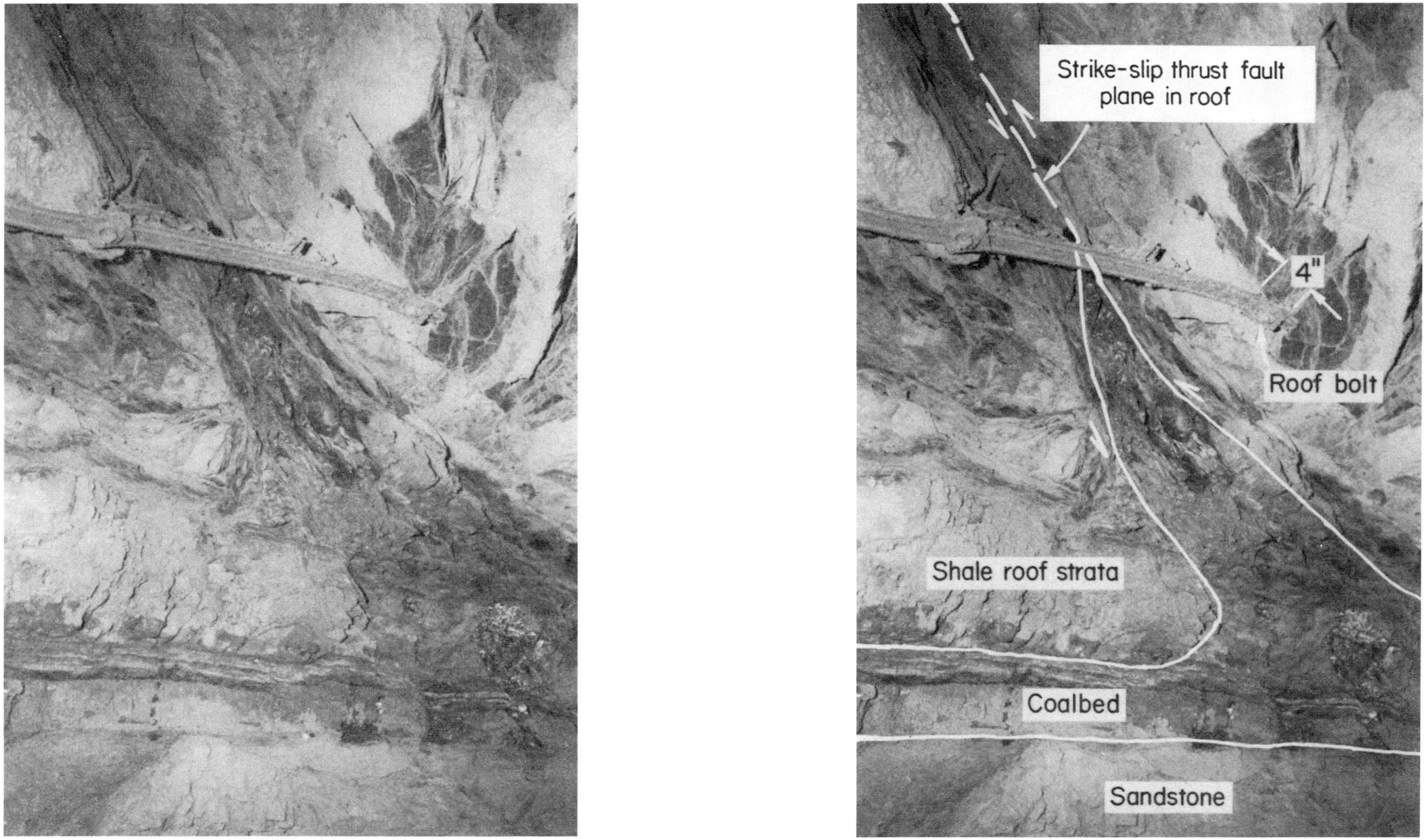

Figure 8. *Severe strata deformation and unstable roof caused by Pistol Gap Fault. The coal seam has been overturned and dragged upward.*

Figure 9. *Bedding-plane fault exposed at coal/roof interface. Vertical fractures isolate hazardous roof blocks.*

Figure 10. *Rehabilitation of a high roof fall in entry crossing the Pistol Gap fault zone.*

Figure 11 shows an interpretation of events in the formation of bedding-plane faults and coal seam swags at the site of the Pistol Gap Fault. Section A is the coal seam with the pre-existing strike-slip fault. Sections B-D are subjected to increasing lateral compression until complete failure occurs in section E. Section B shows the development of the swag, which precedes bedding plane failure in section C. This buckling of strata proceeds to failure in some places but in other instances does not. Because both of these events occur in close proximity (within several entries), the indication is that the compression is uneven, resulting in torsion and uneven disruption of strata. In sections C and D, the severed coal seam is telescoping past itself as the thrust fault shears upward into the roof. The strike-slip fault plane also is deformed. Continued compression also warps the coal seam in the vicinity of the vertical failure.

The Falcon Fuel Mine intersects the Pistol Gap Fault in several other locations (Figure 6). At location E, an 11-entry section intersected the fault, which trends N36°W at this location. Roof conditions were too poor to continue and the section was stopped. When these faults are intersected, the operator usually has no idea how far the fault zone might extend. Generally, the fault is probed laterally with a continuous miner for a break to see if the fault can be easily crossed. If not, the section is stopped and development continues elsewhere if an acceptable crossing place exists. At location C, it was necessary to drive through the fault zone to reach other reserves (Figure 6). This procedure was both time consuming and costly.

Chaparal Mine

The Chaparal Mine, which has been developed in the Jawbone A&B coal seam, also intersected the Pistol Gap Fault while driving a development to the southwest (Figure 5). To minimize production loss and hazardous roof conditions, the operators were forced to narrow the development in order to cross the fault. In addition, fewer crosscuts were turned, which made pillars more than twice as long as normal. Roof conditions were poor with falls over 4.6 m (15 ft) high which had to be supported by elaborate terraced cribbing (Figure 12).

Figure 13 shows two cross sections through the fault zone. Cross section A-A' shows a large swag zone (approximately 70 m (230 ft) wide) inby the fault zone. The coal seam has been buckled down 1-1.5 m (4-5 ft) before returning to normal elevation outby the fault zone. The coal seam swag, which is a precursor to strata rupture, occurs on only one side of the fault zone. In addition, small-scale anticlines are caused by upward buckling of the seam. Bedding plane failure is evident because of a well-defined fault gouge zone between several of the fault planes. Failure also is indicated by the coal seam being smeared out to only several inches thick. In both sections, the fault ramps up into the roof, causing a roof fall upon mining of the area.

Cross section B-B' shows the same failure reaction to compression. In this instance the coal nearest B' is split in the middle of the seam by compression from the direction of B, causing upward thrusting of the coal seam into the roof and an overturning of the fault. A normal fault occurs near B. This is due to tension near the edge of the fault zone. The coal seam and the roof strata are highly disturbed due to shortening from the compression. Characteristically, the coal seam becomes unusually thick near the main failure. This false thickening is due to compressional forces which deform and push the coal up against a fault plane.

Street And Whited Mine

The Street and Whited Mine, which has been developed in the Jawbone A&B coal seam, has been reduced to a small retreat operation, mainly because of the problems it had in crossing the Pistol Gap Fault (Figure 5). The mine intersects the fault in two places (Figure 14, location A and cross section A-A'). A seven-entry section driven to the south intersected the fault at cross section A-A'. The operators tried to cross the fault with two entries, but the coal seam pitched upward into the roof at a 30° angle The mining machine was

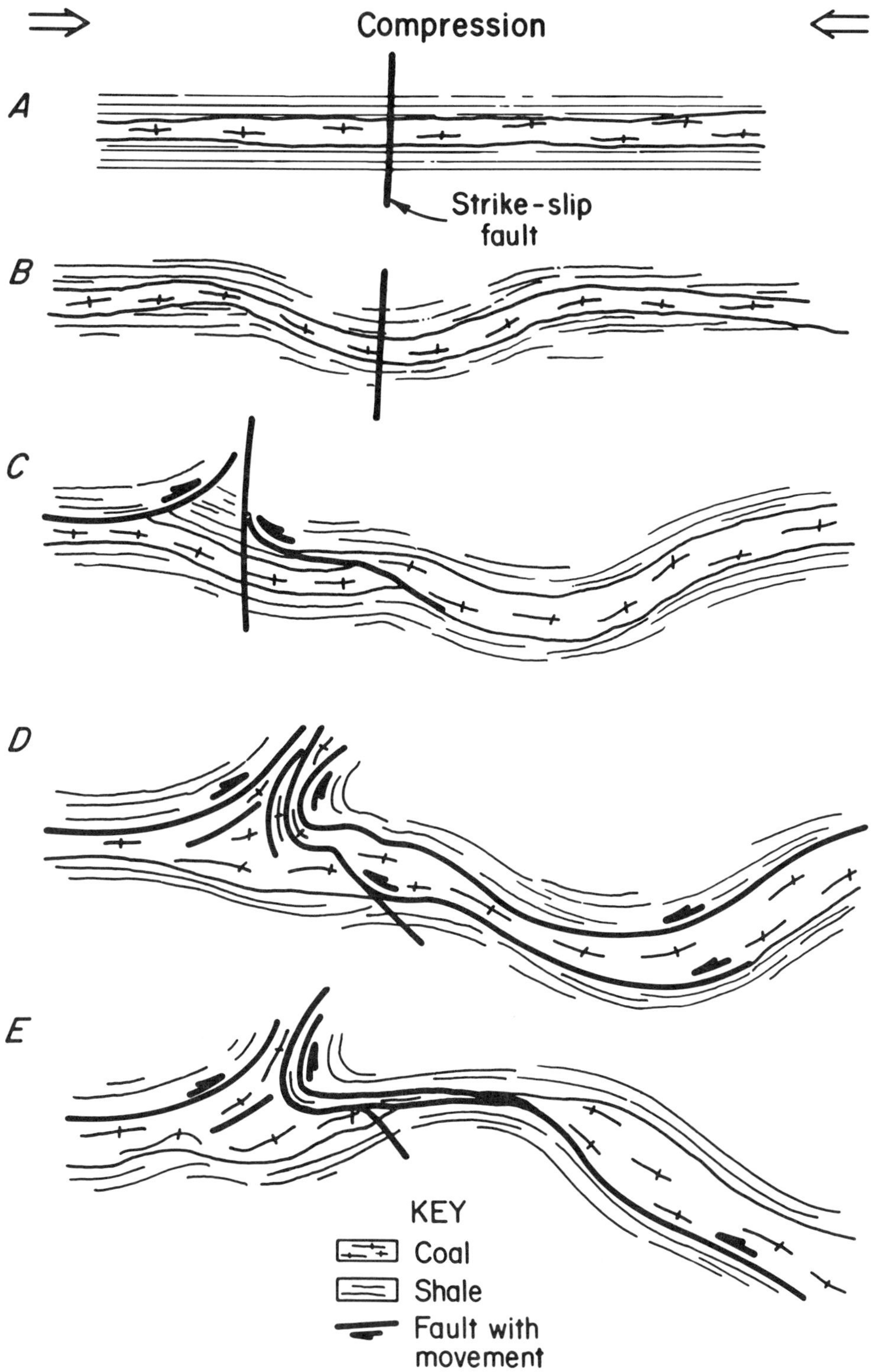

Figure 11. *Interpretation of events in the formation of bedding-plane faults due to lateral compression. Section A is the coal seam with the strike-slip fault present. The subsequent sections (B-D) are subject to increasing lateral compression until complete failure in section E.*

Figure 12. *Two-tiered cribbing platform built to support high roof fall in the Chaparal Mine.*

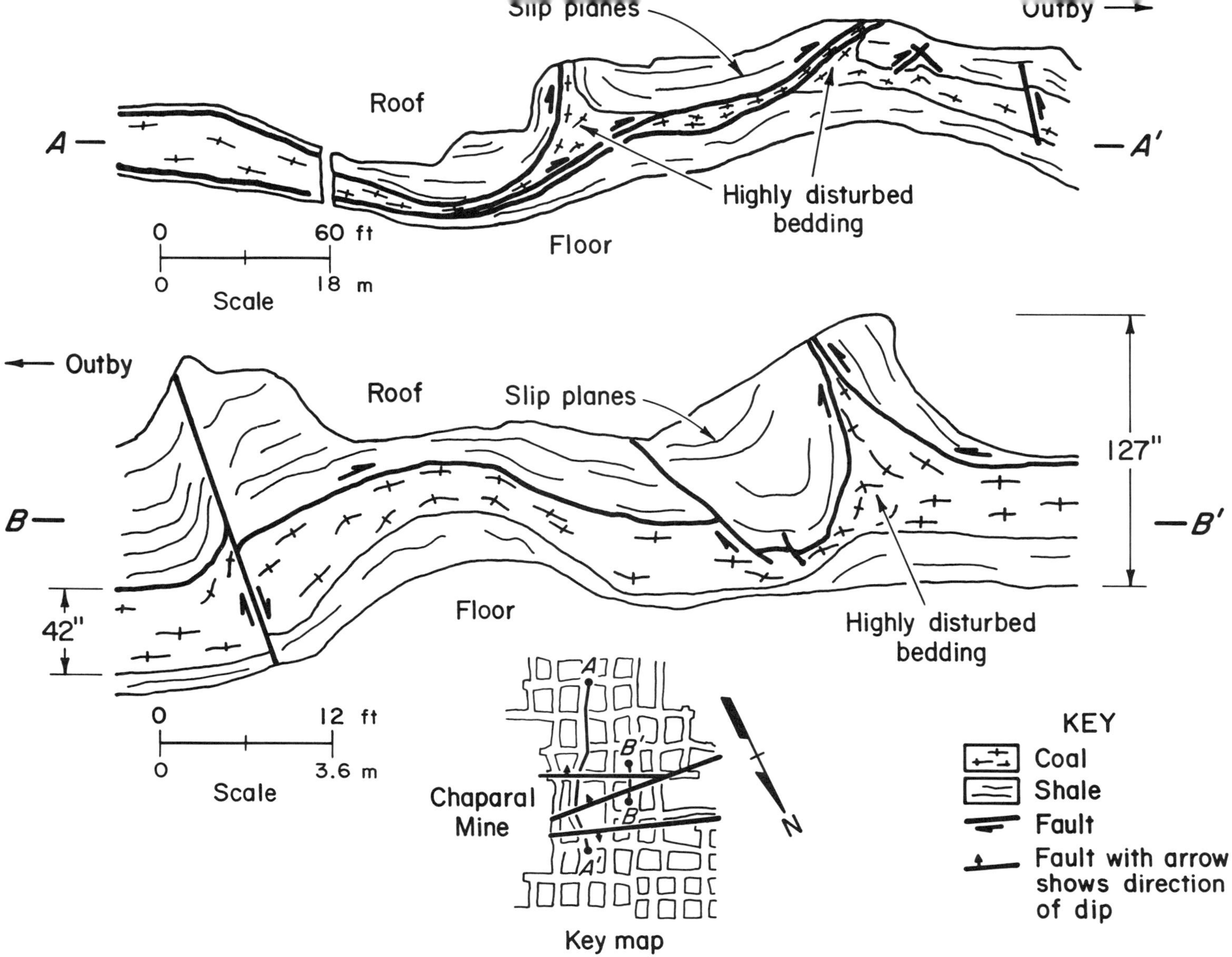

Figure 13. *Cross sections of Pistol Gap Fault exposure in the Chaparal Mine.*

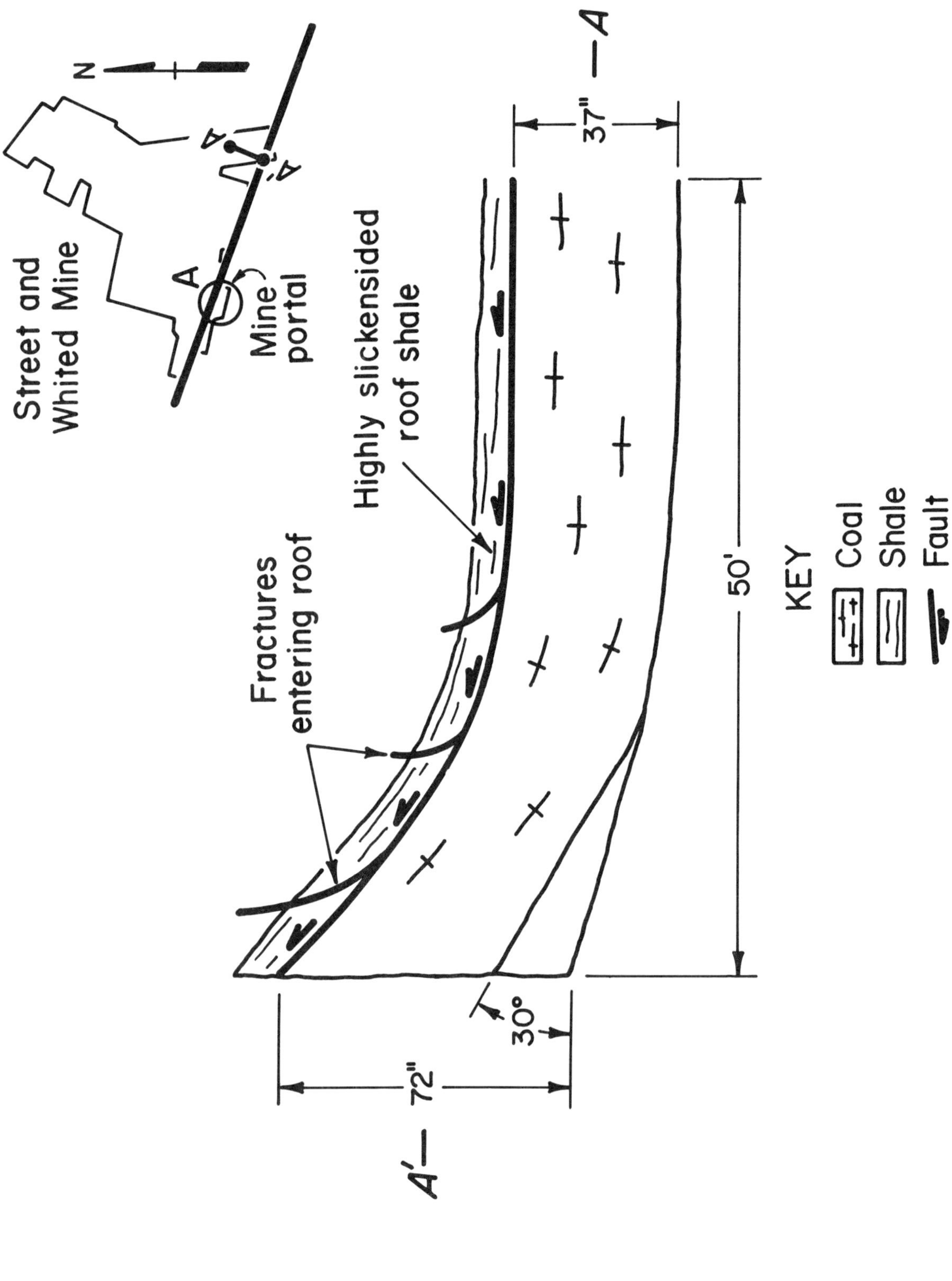

Figure 14. *Cross section of a mine entry intersecting the Pistol Gap Fault in the Street and Whited Mine.*

unable to mine at such a steep angle, and production was halted. Roof conditions began to deteriorate in the two advancing entries, with large roof falls developing. A bedding-plane fault is evident at the roof-coal interface. A series of small thrust faults also splinter off the bedding-plane fault and enter the roof shales. These faults indicate compressional failure. Because of limited reserves on the southwestern side of the fault zone, the operators have decided not to attempt to cross the fault zone.

The other location where the mine intersected the fault is in the main entry just inby the mine portal (Figure 14, location A). Here the disturbance is limited to merely a 15-cm (6-in.) normal type of fault offset and a rippled coal seam. Again, the fault movement resembles hinge movement with the pivot somewhere northwest of A.

AJ And A Mine

The AJ and A Mine, which has been developed in the Red Ash coal seam, is located in a coal seam approximately 52 m (170 ft) above the Street and Whited Mine. At least seven different operators have attempted to work this mine. All the business failures have been directly related to the effects of the Pistol Gap Fault on roof conditions. The mine is located on the fault trace, between the Falcon Fuel and Chaparal mines (Figure 5).

Figure 15 is a cross section of the Pistol Gap Fault as it is exposed by mining. The total offset of the coal seam is 2.4 m (8 ft). This offset mainly is due to the coal seam being downbuckled in the inby direction because of compression. After downbuckling, the coal seam was sheared off by a bedding-plane fault running along the floor, and the fault then cut upward through the coal. The fault runs along the top of the coal seam, smearing the coal to 5 cm (2 in.) thick before cutting up into the roof and overturning (Figure 16). Another opposite-dipping thrust fault is seen on the outby side of the fault zone (Figure 15). This fault flattens to a bedding plane fault and runs along the coal/roof interface. The overlying roof shales are highly contorted and disturbed as though dragged along above the fault plane with slip movement resisted by the coal seam. The presence of the two opposite-dipping, bedding-plane thrust faults shows failure on both sides of the fault zone. After initial buckling, the faults propagating along bedding met resistance at roughly the same point and failed into the roof. The coal seam itself shows signs of slippage and disburbance and becomes overthickened as the fault enters the roof. The AJ and A Mine, which is stratigraphically higher than the other mines, intersects the fault directly in line with its linear trace as do the other mines (Figure 5). This indicates that the Pistol Gap Fault plane is vertical and planar.

Lisa Lee Mine

The Lisa Lee Mine, which has been developed in the Jawbone A&B coal seam, has been severely affected by the Keen Mountain Fault and a local spur fault possibly related to the Keen Mountain Fault (Figures 5 and 17). Four development sections have intersected fault traces, and a fifth set of entries is expected to intersect a fault in the near future.

At location A, a five-entry section being driven southeast was stopped when the coal was interrupted by a suspected fault (Figure 17). The coal seam thickened up to 2.5 m (8.2 ft) in one entry and 2.2 m (7.2 ft) in another entry adjacent to the disturbance. Coal height is normally 0.70-0.82 m (2.3-2.7 ft) at this mine. The coal seam then abruptly ended. Although the section is now flooded and inaccessible, it is suspected that the coal on the southeastern side of the fault was thrust up against and over the northwestern block, thereby causing the overthickening. This type of overthrusting is found on other low-angle bedding-plane faults in the area. Figure 17 shows that this inferred fault follows a corridor between the Lisa Lee Mine and the Terri Lynn

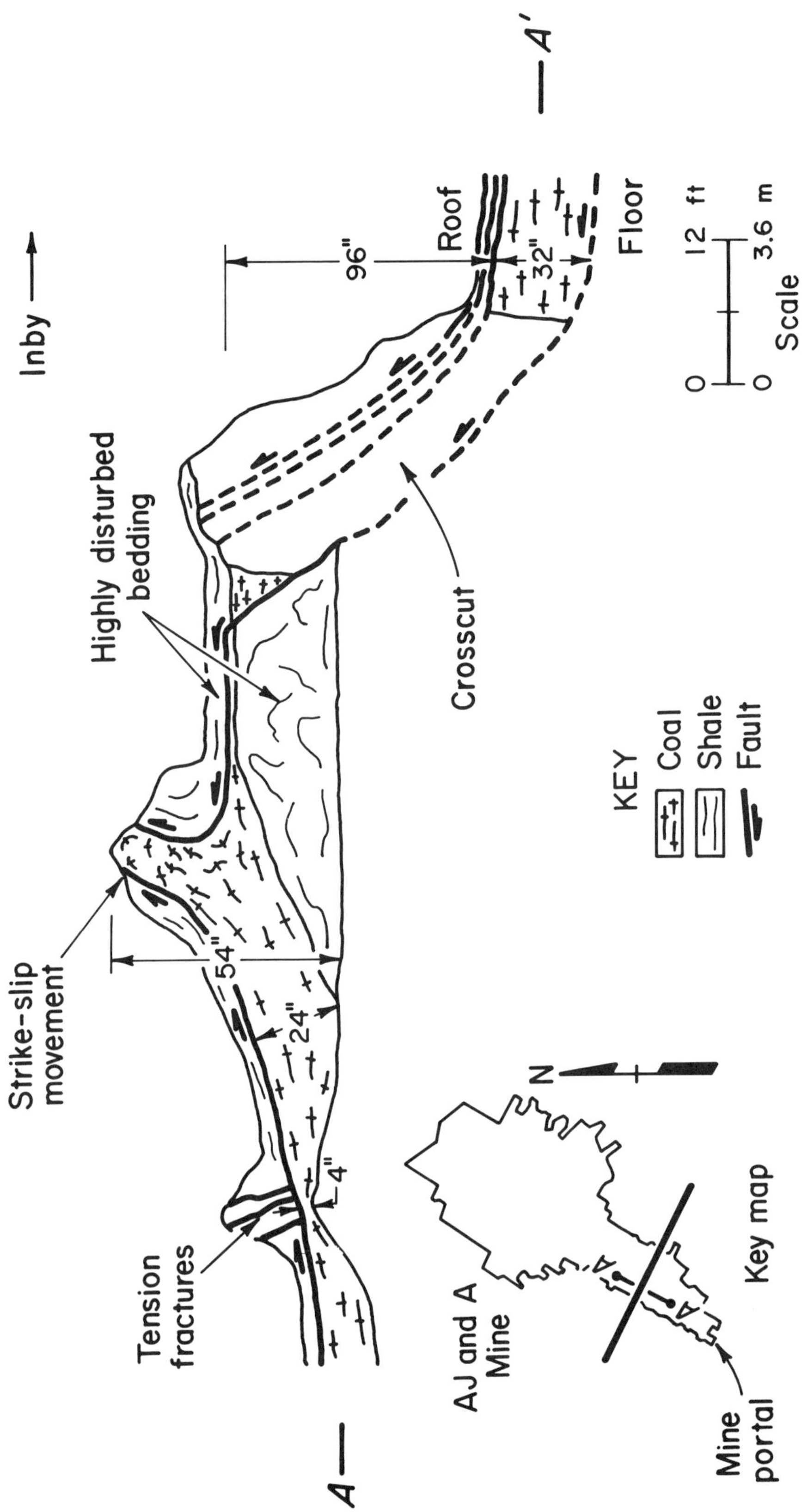

Figure 15. *Cross section of the Pistol Gap Fault at the AJ and A Mine.*

Figure 16. *Pistol Gap Fault exposed in rib at the AJ and A Mine.*

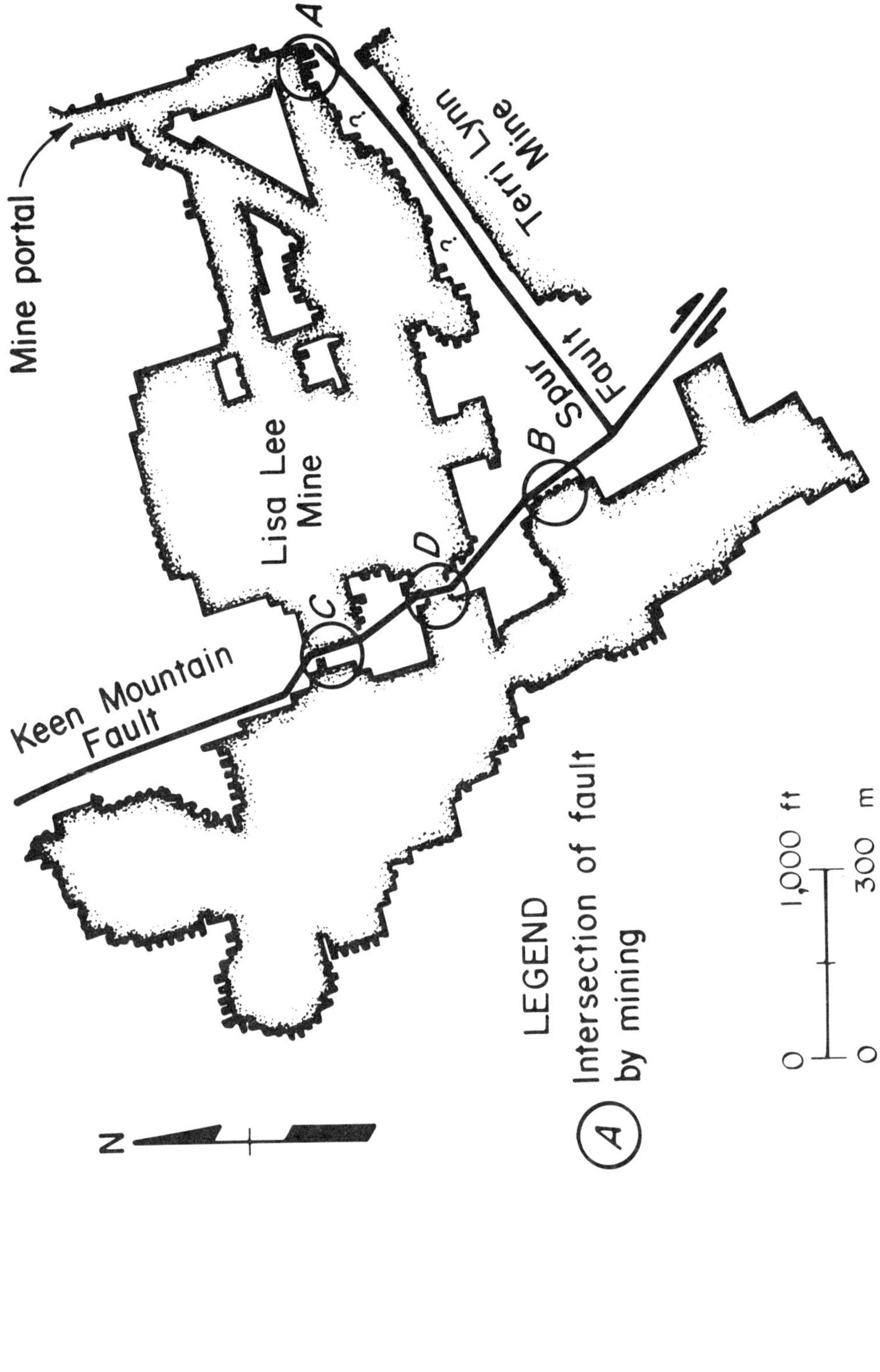

Figure 17. *Lisa Lee Mine outline and areas (circled) affected by the Keen Mountain Fault.*

Mine, which is located directly to the southeast, and the fault continues to adversely affect the Lisa Lee Mine along its southeastern border.

The Keen Mountain Fault is intersected by the Lisa Lee Mine in four places. The section that was being driven to the northeast at location B was completely stopped when it intersected the fault (Figure 17). The coal seam "ran out", and no attempt was made to drive through the zone and relocate the coal. At location C, a six-entry development was narrowed to one entry in crossing the fault. This single entry causes a severe restriction to ventilation. In addition, poor roof conditions occur at this fault intersection. The Keen Mountain Fault continues northwest through location C but is not exposed again because the coal seam splits are too thin to mine in this area.

At location D, the Keen Mountain Fault is intersected by a set of mains which provide access to the western side of the mine (Figure 17). Here, an eight-entry development was reduced to three entries to cross the fault zone. The effects of the fault were mapped in detail at this crossing. Figure 18 shows part of the fault zone in the cross section at location D. The zone is approximately 60 m (200 ft) wide and consists of a bedding-plane thrust fault, numerous roof fractures which do not offset the coal seam, and a large coal seam swag. The main coal seam offset is 1.2-1.3 m (3.9-4.3 ft), and the bed is downbuckled to the northeast. The main fault runs along bedding on the northeastern side of the offset for an undetermined distance until it abruptly enters the roof at a 60° angle, resulting in a 1.0-m (3.3-ft) high roof fall. Numerous horizontal slickensides and polished surfaces indicate extensive slippage between the coal and immediate roof shales. Compression in a northeast-southwest direction has caused the coal seam to be dragged up into the roof and locally thickened to over 1.8 m (72 in.). Slippage within the coal seam is obvious, and bedding is highly disturbed. The adjacent roof shales also are highly distorted. Failure occurred along the roof/coal interface and along the bedding planes within the coal seam.

The Keen Mountain Fault appears to have a much larger displacement at greater depth from the surface (450 m (1,500 ft) deep) than in the relatively shallow mines (60-120 m (200-400 ft) deep) and in the highwalls investigated. The Pocahontas #3 Seam at Island Creek's Beatrice-Pocahontas Mine (now closed) had been displaced up to 6 m (18 ft) by the fault at 450 m (1500 ft) of cover and only a few feet in the Jawbone A&B coal seam mines. At all exposures, strike-slip movement is indicated by slickensides and fault gouge.

USING GEOLOGIC INFORMATION TO RECOGNIZE STRUCTURAL FAULTS AND ASSOCIATED GROUND CONTROL HAZARDS

The recognition that the individual coal seam disruptions described above may be related to a single geologic structure is the first step necessary to allow hazard prediction. By documenting the occurrence of such composite deformation zones in local mines, a linear pattern of faulting may be established. By documenting the occurrence of these zones in stratigraphically separate mines, the vertical orientation of a fault may be described. Observation of the manner in which the coal seam is disturbed can indicate the mechanism of faulting which caused the disturbance. In the case of the Pistol Gap and the Keen Mountain faults, horizontal slickensides, overthrusting, fault gouge, and the coal seam swag can be used as criteria with which to recognize the fault.

It would be nearly impossible to delineate this type of faulting from drill hole information because the vertical displacement is so slight. Remotely sensed data, such as Landsat images and aerial photographs, may show the features as lineaments. Both the Keen Mountain and Pistol Gap faults are represented as lineaments on SLAR imagery. The Keen Mountain Fault was first mapped on SLAR imagery as an 23-km (14-mi.) long lineament. The Pistol Gap Fault can be identified as a lineament using Landsat imagery. Remote sensing

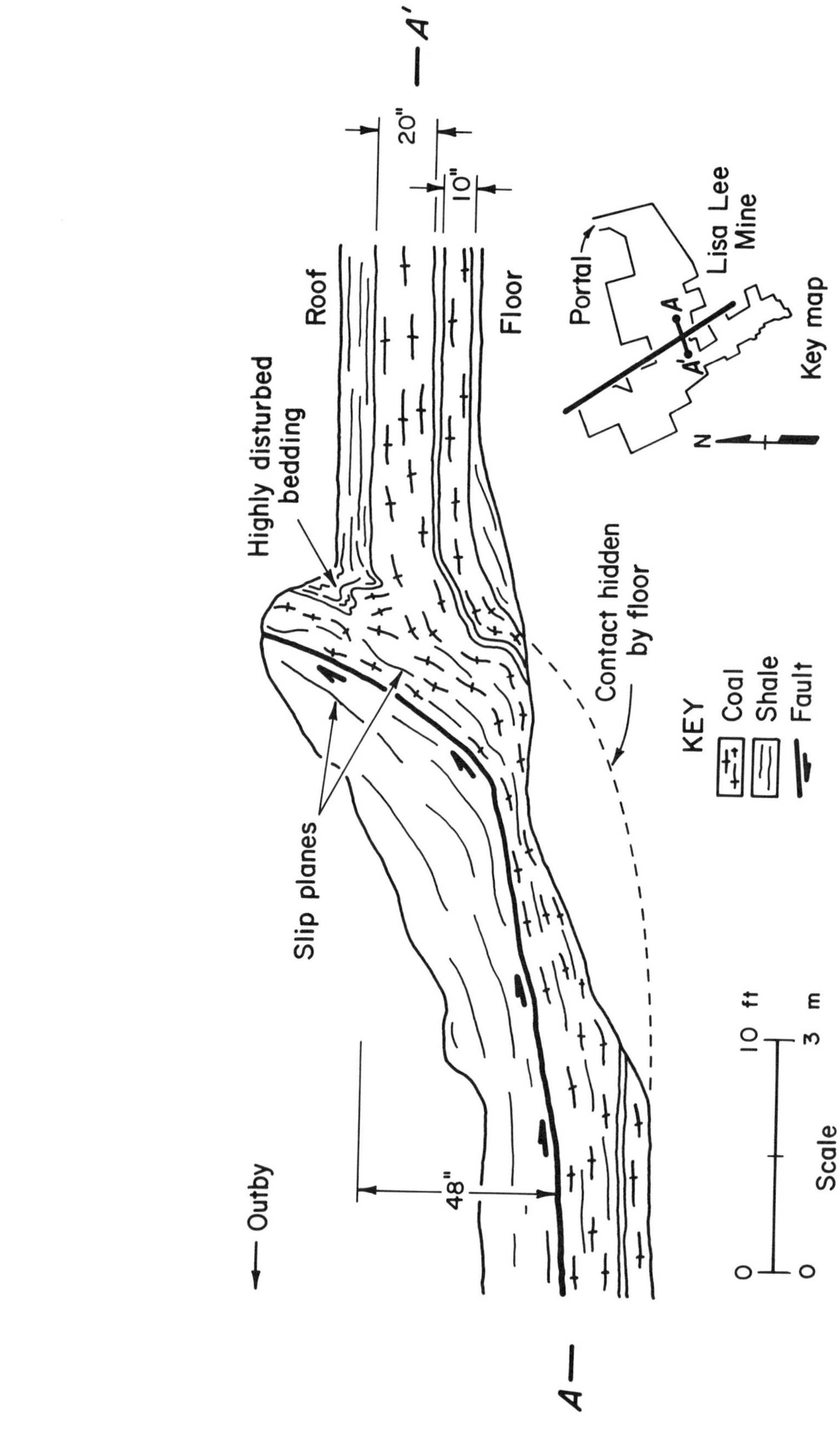

Figure 18. *Cross section of mining exposure of the Keen Mountain Fault at the Lisa Lee Mine.*

imagery may be used as a predictive tool to help demonstrate lateral continuity of faults and to highlight fault or fracture intersections, which may be particularly dangerous (Hylbert, 1977; Jansky and Valane, 1983). This information, along with underground mapping, still is necessary in order to determine the extent (fault zone width) and nature of the potential disturbance.

By consistent geologic mapping, both underground and from the air, it may be possible to project faults in advance of mining. The collection of available local mine information, including details as to methods found successful for supporting mine roof in these areas, would greatly improve safety in the mines. Such structures as these faults should not be a surprise to mine operators during mining if geologic concepts and interpretations are properly applied in advance and during development.

REFERENCES

Elder, C.H., P.W. Jeran, and D.A. Keck, 1974, Geologic structure analysis using radar imagery of the coal mining area of Buchanan County, Va.: U.S. Bureau of Mines Report of Investigations RI 7867, 29 p.

Hylbert, D.K., 1977, Developing geological structural criteria for predicting unstable mine roof rocks: U.S. Bureau of Mines, Contract Final Report #HO133018, 97 p.

Jansky, J.H., and R.F. Valane, 1983, Correlation of LANDSAT and air photo linears with roof control problems and geologic features: U.S. Bureau of Mines Report of Investigations RI 8777, 22 p.

McLoughlin, T.F., 1986, Explanation of the regional tectonic map of the southwestern coalfield of Virginia: Mining Safety and Health Administration Informational Report IR 1177, 17 p.

Miller, M.S., 1974, Stratigraphy and coal beds of Upper Mississippian and Lower Pennsylvanian rocks in Southwestern Virginia: Virginia Division of Mineral Resources, Bulletin 4, 211 p.

Miller, R.L., and J.O. Fuller, 1954, Geology and oil resources of the Rose Hill District--The fenster area of the Cumberland Overthrust Block--Lee County, VA.: Virginia Geological Survey Bulletin, v. 71, 383 p.

Wentworth, C.K., 1921, Russell Fork Fault of Southern Virginia: Journal of Geology, v. 29, p. 351-369.

STRESS IN UNDERGROUND COAL MINES: A PRACTICAL REVIEW OF ORIGIN, RECOGNITION, AND CONTROL

Bruce A. Collins
Mid-Continent Resources, Inc.
P.O. Box 500
Carbondale, CO 81623

ABSTRACT

Manifestations of stress in underground coal mines vary from most benign — barely perceptible "weight," or deformation of cap pieces on timbers — to catastrophic — massive roof falls, pillar failures, and outbursts. The prevention of injury to miners and damage to equipment as well as to the mine itself depends on the accurate recognition and interpretation of intensifying stress. Where possible, simple testing for stress effects, and ultimately, avoidance of stressed areas or application of stress-release techniques where avoidance is impossible, should be adopted. This becomes increasingly more important as "easier" deposits are exhausted and workings extend under deeper cover and into areas of geologic disruption.

INTRODUCTION

Stress, here used to describe any force which results in deformation or dilation of mine openings, has been an omnipresent factor in underground coal mining since the first primitive workings were extended more than a few feet into a coal bank centuries ago. The effects of stress release (strain) events, such as bottom heave, roof falls, coal and rock bursts, and so on, which can severely disrupt mining operations as well as threaten miners' lives, have been recognized since these earliest times. For most of the history of coal mining, indeed until the development of the roof bolt in the 1940s, areas of high stress, usually identified only after severe problems had developed, were avoided to the extent possible. However, as mines in the United States work ever deeper, and as the "easier" deposits are worked out, stress-related problems become harder to avoid. Few large-scale mining operations now are devoid of such phenomena, whether recognized as such or not.

It is the intent of this paper to discuss in rather broad terms the origins of the stresses found in coal mines, the recognition of the effects produced by these stresses, simple and practical methods for detecting stress, and practical methods for avoiding, minimizing, or controlling stress buildup and stress-release events. As such, complicated state-of-the-art electronic stress measurement methods, complex computer analyses of mountains of data (from satellite lineaments to microseismic profiles), and so on are not discussed. The author does not wish to minimize the importance of such work, only to emphasize practical approaches to the problem that can be initiated by any mine geologist or engineer. For a more theoretical approach, the reader is referred to such works as Stability in Coal Mining (Brawner and Dorling, 1979) and the many publications in this area issued by the U.S. Bureau of Mines. A great deal of foreign literature also is available, particularly from Germany, Great Britain, the Soviet Union, Poland, South Africa, and Australia (e.g., Airuni and Bolchinsky, 1981; Lu, 1983; Leighton, 1985; Salamon and Oravecz, 1976; Serata, 1987).

ORIGIN OF COAL MINE STRESSES

Gravity

The overriding source of stress in coal mines is, of course, gravity, and the simplest gravitational stress is overburden pressure. Ideally, a flat coal seam underlying a flat surface with homogeneous overburden would be subject to only the even lithostatic pressure at that depth. If the roof, floor, and coal ribs (sides) of a mine opening all exceed that pressure in strength, and are not significantly different among themselves, the opening will be stable. That is, there will be no convergence or stress-release events, rapid or slow, which would tend to close the opening.

However, such ideal conditions rarely, if ever, occur. Topographic variations result in uneven stress on the coal seam and adjacent rocks, leading to nonvertical stress components under slopes, ridge lines, and narrow valley bottoms (Moebs and Bauer, 1989, p. 43). Similar nonvertical stress components develop in dipping coal seams and are further complicated by topographic components. Add to this nonhomogeneous overburden rocks (which is usually the case), differing rock and coal strengths, and other common phenomena such as faults, rolls (tight folds caused by differential compaction or local structural anomalies), igneous and clastic intrusions, washouts, and so on, all of which result in nonvertical stress components, and it becomes clear that the gravitational stress environment of most coal seams is quite complex.

Deformational

In situ stresses that are not strictly gravitational are present in many coal basins in or adjacent to areas of regional tectonic activity. These stresses commonly are oriented over fairly large areas. More common, but usually much more restricted in lateral extent, are stresses adjacent to and within soft-sediment deformational structures such as channels, rolls, and clastic dikes. Formed in active delta-plain environments, many coal seams are adjacent to or cut by channels varying in size from a few feet to over a mile in width, and up to several hundred feet thick. Differential compaction and other penecontemporaneous and post-depositional soft-sediment phenomena can lead to significant disturbance of the coal and surrounding sediments, including faulting and folding (see Nelson, 1983), all of which phenomena disrupt the "normal" stress environment. In at least some instances, these features probably retain some differential stress from their time of formation.

Interstitial Fluids

The presence, abundance, and nature of interstitial fluids can impose stress on mine openings or intensify those stresses already acting. Water, and more rarely petroleum, can lubricate bedding planes and fractures, both usually serving to reduce catastrophic stress-release events by promoting more easy and therefore slower movements in the rock. The presence of gases, however, tends to increase stress and the intensity of stress-release events. This probably is due to compressibility of the gases and, relative to methane and carbon dioxide in coal at least, to the fact that considerable amounts of these gases are held loosely in the complex molecular structure of coal. In response to excess stress, these gases can be released as the coal is fractured, a process which the rapidly expanding gas then perpetuates as it exerts increased stress within breaking coal particles.

Lithology

Lithology can play a part in stress transmission and even origin. Some rock types, such as porous sandstone, may compress more under load than others, such as crystalline limestone. Relaxation, or expansion, of the compressed rock may occur on exposure, resulting in induced stress on pillars, artificial support systems, and even the intervening rocks. Under some circumstances, laminated sandstone, siltstone, and mudstone can form a beam which resists stress more than massive bodies of any of these rock types by themselves. In one western coal mine, a thick massive sandstone was so strongly and closely jointed that large blocks fell out shortly after mining, in spite of resin roof bolting, whereas an overlying sequence of laminated siltstone formed a competent roof. The tendency of any rock type to fracture will determine to some extent the abundance and nature of joints and, therefore, the rock's ability to withstand stress adjacent to an opening.

Exposure of rocks containing swelling clays in the mine environment, either in the floor by natural drainage or in the roof through wet drilling, can have serious consequences. In a rock slope being driven to access two coal seams, large slabs of mudstone around holes that were wet-drilled for roof bolts were literally forced to break and fall because confined bentonitic clay layers up to 15 cm (6 in.) thick, dampened by the drilling, swelled and pushed the rock into the mine opening.

Mining Induced

Stress is induced by mining in several ways. Clearly any opening in rock diverts or deflects the "normal" stress field present before the opening was created. This creates a nonvertical stress component around the margins of the opening and, in conjunction with the normal stress field, results in stress concentrations near the opening. The number, size, and shape of the openings all have a critical effect in the nature of the "stress envelope" that surrounds the mine opening (Serata and Lawshe, 1989, p. 158). Likewise, the distribution of openings results in a complex stress distribution that may have significant effects on newer workings adjacent to these openings in the same coal seam or in another seam above or below. Stress interactions between workings in different coal seams have caused considerable difficulties in mines separated by over 150 m (500 ft) of "competent" rock.

Rock Strength Differences

Finally, proximity to the mine opening of rock units of widely differing strengths, such as a very strong roof, a strong roof with a weak floor, strong immediate and main roof with an intervening weak unit, or a strong immediate and main floor with an intervening weak unit will alter stress transmission characteristics around a mine opening, increasing stress on some portions of the system while decreasing it on others. Very strong roof causes difficulties in room and pillar mining because the roof will bend instead of breaking and falling as pillars are removed. This causes loading up of remaining pillars or stumps, ultimately resulting in pillar bursts (in some locales termed "bumps"). Similarly, on longwall faces, a bending roof behind the face can result in face bursts, and in at least one case I know of, the virtual destruction of the self-advancing, hydraulic face supports.

A strong roof with a weak floor is an almost-automatic prescription for bottom heave, even under shallow cover. A weak unit between stronger immediate and main roof can separate from the main roof and break the lower immediate roof, sometimes with little or no warning over fairly large areas. Likewise, a weak unit between stronger immediate and main floor units can be squeezed out from under pillars, breaking the immediate floor and causing bottom heave, again occasionally abruptly enough to cause damage and injury. In one known case, a severe instantaneous bottom heave event (known locally and appropriately as a "bounce") literally bounced a shuttle car off the floor up to the mine roof in an 2.4-m (8-ft) thick coal seam, injuring the operator and another employee standing nearby.

RECOGNITION OF STRESS IN COAL MINES

The recognition of the results of stress (i.e., strain) in coal mines is easy. Bottom heave, rib sloughage, roof falls, pillar bursts, bumps, outbursts and rock bursts generally are hard to ignore (Bauer, 1985). Yet, interpreting these phenomena, defining the stresses that produced them, and recognizing pre-catastrophic signs of stress are at once more difficult and more important.

Detailed interpretations and descriptions of the phenomena associated with stress in coal mines are beyond the scope of this paper. Again, the reader is referred to the many publications of the U.S. Bureau of Mines, lists of which are available from the Publication Distribution Division (P.O. Box 18070, Pittsburgh, PA 15236; Telephone 412-892-4338).

Initial signs of stress can be subtle, often very much so. Slight bottom heave may not be recognized until a piece of equipment, or a miner for that matter, which formerly fit in an entry starts rubbing the roof. This may be true even in openings timbered along one or both sides, with heave only occurring in the center. Timbers and associated cap pieces and wedges often provide early clues to sagging top or heaving bottom, and only experience can differentiate between the two. Stress on timbers is first noticeable as the timber itself squeezes or punches through the cap piece and wedge, then progressively deforms and finally breaks. Numerous timbers deformed or broken in the same manner and direction are dead giveaways to oriented stress. Many observers automatically associate punched caps and wedges, deformed and broken timbers with sagging roof. However, these phenomena are often associated with bottom heave, even possibly more often than not considering the fact that coal seam underburden is commonly softer than overburden (fireclay and other swamp-bottom materials as opposed to washed-in clastics). Observe the roof carefully in any area where deformed wooden support components are found. If it generally is competent-appearing, without fracturing, obvious breakup areas, deformed roof bolts, slickensided surfaces, and so on, chances are that the deformation is being caused by bottom heave rather than roof-related problems.

Occasionally, in timbered entries, over time all timbers originally installed vertically (or at some other consistent angle) will be observed to lean in one direction, or timbers on one side of an opening will lean or deform while those on the other will not. In such cases, an oriented stress is clearly at work. It may be as simple as a directional movement of roof relative to floor, or it may be a complicated mixture of such movement coupled with resultant oriented bottom heave and pillar failure.

Deformation of roof bolt plates, most commonly "cupping" or apparent pulling of the bolt up through the plate, is another relatively common sign of stress that can vary from barely noticeable to obvious. With conventional (expansion shell) roof bolts, cupping results from sag anywhere between the anchoring shell and the head of the bolt so long as the resulting weight of the unsupported rock exceeds the deformation strength of the plate. With resin-grouted bolts, where such deformation was formerly thought to be impossible, such results can be more complicated. Deformation may represent partial or complete failure of the resin due to faulty resin, improper installation, or failure of the roof rock between the bolts. The latter case ultimately would leave the bolts themselves quite competently supporting a narrow column of rock around them while everything between the bolts to some horizon above their installed height has fallen. Commonly, when initial support has been lost in this fashion, falls will continue, even to above the bolted height, until a self-supporting layer is reached. Such failures are usually easy to differentiate from those resulting from using bolts that are too short. In cases of the latter, massive falls to some horizon above the bolts usually occur first, with little or no plate deformation.

A common form of roof bolt plate deformation occurs in roof containing frequent "pots", "kettlebottoms", or similar features, most frequently with circular to hemispherical slickensided outlines of generally unknown origin. The feature separates from the main roof, with its weight settling on the bolts within its margins. Plate deformation or even failure in such cases is common, but unless the unconfined weight of the separated rock exceeds the support capacity of the bolts, actual catastrophic falls are rare. In one case I know of, explosive plate failures — plates literally blasted off the bolt by the weight of the separating rock — accompanied by opening slickensided fractures around the margins of "pots" often as large as an intersection (6 m (20 ft) in diameter) occurred frequently, with no significant falls. Unless the volume of separated rock is known, however, it must be assumed in such cases that the bolts are being stretched and eventually will fail. Supplemental roof support should be installed if the opening is to be maintained for any length of time.

Shearing of roof bolts and deformation or failure of other metal, cement, cinder block, or heavy wood roof support systems, such as stoppings, cribs, square sets, and arches, are more serious signs of stress and usually demand more immediate attention. Whereas supplemental support is of utmost concern in such cases, it remains necessary to determine the cause of the deformation or failure, and specifically if it is oriented or not, in order to prevent or at least minimize future occurrences to the extent possible. Simple squeezing, collapse, shattering, or vertical deformation of wooden cribs, block cribs, stoppings, or metal entry-support systems generally indicates vertical (gravity) roof failure. However, even here the cause may vary from simple incompetent roof unit failure to cutting of competent roof due to strong coal and floor combined with regional jointing. Pushing over of cribs, failure of stoppings in one direction but not in another, and deformation of arch legs on one side of an entry but not the other are all signs of directional stress.

PRACTICAL METHODS OF MEASURING STRESS EFFECTS

The "measurement" of the effects of stress in underground coal mines can range from the simplest -- "Hey, Joe, every damn timber in 5 Right is leaning east!" -- to complete mine-wide computer monitoring of load cells, extensometers, and even hydraulic pressure in individual legs of supports on longwall faces. (An experimental system has been installed by the U.S. Bureau of Mines Denver Research Center (Conover et al., 1990).) There are, however, a few relatively simple tests for certain types of stress effects.

Probably the simplest method of determining whether or not a roof fracture, which might be the result of something as local as the sag of a thin layer of incompetent shale or as regional as basinal tectonic tension, is worsening is the driving of a nail or survey spad into small cracks, or a timber wedge into larger ones. If, over time, the nail, spad or wedge falls out, the fracture is enlarging.

Once they are recognized, effects of directional forces over time can be measured by the rate at which the tilt angle increases on timbers originally installed vertically. For example, the angle of the timbers can be measured with a Brunton or similar compass or graduated level on an hourly, daily, weekly, or monthly basis, depending on how rapidly deformation is occurring.

Convergence, or the rate of roof sag, bottom heave or both, can be accurately measured using a "pogo stick." The pogo stick is a spring-loaded device used to erect brattice curtains for ventilation. With the device fully extended, the inner tube is marked in quarter-inch (or centimeter) divisions with an indelible marker pen and placed vertically in the area to be measured. Readings, again taken at intervals determined by the gross rate of the mining activity, determine the rapidity and magnitude of the effect being measured. Change of the pogo stick from vertical, again measured by a simple level device, also can be determined.

Pillar or barrier stresses, which can lead to outbursts, catastrophic pillar failures, or massive bounces which usually are manifested by rapid episodes of bottom heave, are generally much more difficult to evaluate, even in mines or areas of mines where they previously have occurred. In general, in an area where minor bumps, bounces, or breaking noises of coal and/or rock have been increasing in frequency and/or magnitude, or where such noises, even if only moderate, suddenly stop, a significant stress-release event may be expected. The problem is that, unless there is at least some previous experience, it is impossible to determine what direction such release may take. In one episode I know of, a buildup of fracture noise in an area of soft coal under deep cover, followed by near-total silence, produced a moderate bounce hours later, accompanied by bottom heave measured in the inches range. A second incident occurred in a pillar area where slabbing was practiced under low cover (less than 92 m (300 ft)). A massive pillar failure, over more than 40 hectares (100 acres), accompanied by significant surface subsidence, occurred with minimal warning: only a few loud bumps minutes before the onset of the failure.

REMEDIATION MEASURES

Once concentrated stress buildup has been recognized, either by observation or unfortunately by the occurrence of catastrophic failure, there are a few relatively simple methods for future detection and stress release. Directional stresses manifested by leaning timbers, stoppings, sheared roof bolts, and so on identify themselves. These stresses can be partially to wholly relieved by changing mining direction in the case of roof problems, or changing direction or cutting or shooting the floor if the problem lies there. The principal idea is to shift the stress buildup away from the mine opening.

Areas of sagging roof, detected by falls, spreading pans, pogo sticks, etc., can be stabilized by grouting, truss bolting, taking down of thin incompetent material, or changing the direction of mining. Bottom heave can be alleviated by changing mining direction, cutting, or shooting, all of which shift the stress accumulation either deeper, below incompetent materials which heave, or to the adjacent pillars. Simple grading of areas of sagging top or heaving bottom may maintain required height for operating purposes, but rarely provides actual stress relief.

Excess stress on entry or longwall faces or barrier or chain pillars can be detected by test drilling. Auger drilling into the coal seam obviously will produce a known volume of cuttings per unit of depth, dependent on the diameter of the hole. If the coal is under excessive stress, this stress will attempt to close the hole as it is drilled, producing excessive cuttings. In addition, such drilling may itself result in minor stress release bumps or bounces. In areas where outbursts are at least partially related to the release of large amounts of coal seam gases (usually methane and/or carbon dioxide), drilling can be accompanied by the emission of large amounts of gasses. Semi-quantitative criteria for stress release techniques can be developed on a site-specific basis using such data (Haramy et al., 1988). These techniques are relatively sophisticated, involving such methods as water infusion, which fractures and lubricates the coal or rock, or large-diameter hole drilling and volley firing, where explosives are used to fracture the coal or rock. Both techniques result in the shifting of the stress back into the coal or rock away from the working face. The reader is referred to Haramy et al. (1988) and Varley (1986) for more detailed discussions of these methods.

CONCLUSIONS

Whereas the effects of stress from a multitude of origins have been evident in underground coal mines since the first primitive workings extended only a few feet beyond surface "bank" openings, these effects, their detection, and the control of the stresses themselves are becoming more significant as modern mines are developed into deeper and more geologically complex environments. The recognition of stress effects, from the crushing of cap pieces on timbers to the shearing of roof bolts, the occurrence of directional rib sloughage, bottom heave, roof falls, pillar failures, and outbursts of coal or rock, is important in mine planning in order to prevent injuries to miners and damage to mine openings and equipment.

Whereas sophisticated methods for the measurement of these effects are available, several relatively qualitative techniques are available to assess the existence and severity of the problem. These include simple angle and convergence measurements and drilling techniques. Stress control can be accomplished by changes in mine opening orientation and by cutting, drilling, and explosives techniques to shift stress away from mine openings.

REFERENCES

Airuni, A., and M. Bolchinsky, 1981, New methods of predicting and preventative measures for regular and instant methane emissions in the USSR mines: Annales des Mines de Belgique, no. 6, p. 503-531.

Brawner, C.O., and I.P.F. Dorling, eds., 1979, Stability in coal mining: San Francisco, Miller Freeman, 496 p.

Bauer, E.R., 1985, Recognizing signs of excessive loading in underground coal mines: U.S. Bureau of Mines Informational Circular IC 9017, 30 p.

Conover, D., K. Hanna, and T. Muldoon, 1990, Mine-wide monitoring applications in ground control research, in S.S. Peng, ed., Proceedings of the 9th International Conference on Ground Control in Mining, Morgantown, West Virginia, June 4-6, 1990: Morgantown, West Virginia University, p. 135-141.

Haramy, K.Y., J.P. McDonnell, and L.A. Beckett, 1988, Control of coal mine bursts: Mining Engineering, v. 40, no. 4, p. 263-267.

Leighton, F., 1985, Microseismic techniques to provide warning of coal, gas and rock outbursts, in Proceedings of the 21st International Conference on Safety in Mines, Sydney, Australia, October 21-25, 1985: Rotterdam, A.A. Balkema, p. 75-80.

Lu, P.H., 1983, Stability evaluation of retreating longwall chain pillars with regressive integrity factors, in Proceedings of the 5th Congress of the International Society for Rock Mechanics, Melbourne, Australia, April 10-15, 1983: Rotterdam, A.A. Balkema, p. E37-E40.

Moebs, N.N., and E.R. Bauer, 1989, Appalachian roof instability: Coal, v. 26, no. 2, p. 43-45.

Nelson, W.J., 1983, Geologic disturbances in Illinois coal seams: Illinois State Geological Survey, Circular 530, 47 p.

Salamon, M.D.G., and K.I. Oravecz, 1976, Rock mechanics in coal mining: Chamber of Mines of South Africa, PRD Series No. 198, 119 p.

Serata, S., 1987, Integration of field instrumentation and computer simulation: Development and application of SPDR technology, in Proceedings of the 2nd International Symposium on Field Measurement in Geomechanics, Kobe, Japan, April, 1987: p. 613-624.

Serata, S., and M. Lawshe, 1989, Stress control methods help to optimize underground mine design: Mining Engineering, v. 41, no. 3, p. 158-160.

Varley, F.D., 1986, Outburst control in underground coal mines, in A.W. Khair and S.S. Peng, eds., Proceedings of the 5th Conference on Ground Control in Mining, Morgantown, West Virginia, June 11-13, 1986: Morgantown, West Virginia University, p. 249-256.

COPING WITH TECTONIC STRESS IN THE ILLINOIS BASIN COAL FIELD

W. John Nelson and Robert A. Bauer
Illinois State Geological Survey
615 E. Peabody Drive
Champaign, IL 68120

ABSTRACT

The Illinois Basin coal field is subject to tectonic stress, with the major stress axis oriented east-west to east-northeast-west-southwest. This stress causes increased incidence of roof failure in north-south headings of underground mines. Failures commonly begin as post-mining compressive "kink zones" in laminated roof rocks. Jointing in the coal field is dominantly parallel with the major stress axis, as are local extensional fracture zones and open fissures. Also, some pre-mining faults, including small north-south thrust faults and an east-west strike-slip fault, may be products of contemporary stress. Joints and faults weaken the roof and admit water, and occasionally gas, into mine workings.

The most effective method to minimize stress-related problems in the Illinois Basin coal field is to drive mine headings 45° to the major stress axis.

INTRODUCTION

Tectonic stress originates deep within the earth and is the type of stress responsible for mountain building, continental drift, and most earthquakes. Miners and geologists long have understood the effects of past tectonic stress as exemplified by folds and faults in coal-bearing strata. Likewise, engineers long have known about vertical or overburden stress and its effect on mine openings. Yet, until the last decade, the fact that present-day in situ tectonic stress is responsible for ground failure in underground mines in the United States has not been recognized.

The stress field at any point within the earth can be resolved into three mutually perpendicular stress components, of which two are horizontal and the third is vertical. Under a condition of no tectonic stress, the two horizontal components are equal and the vertical component increases proportionally with depth. Where there is tectonic stress, the horizontal components are unequal. Depending on rock properties and mine design, the horizontal stress differential may be sufficient to cause directional ground failures in man-made excavations. Because tectonic stresses are regional in scope, they produce regional, directional patterns of ground failure. In the Illinois Basin, the maximum stress is oriented east-west to east-northeast-west-southwest and tends to produce failure in north-south headings.

This paper outlines the evidence for tectonic stress in the Illinois Basin coal field and describes how stress-related ground failure manifests itself. Also discussed are methods that operators can apply to characterize the stress field during the exploration and development phases of mining and ways to minimize the effects of stress.

Study Area

The Illinois Basin coal field includes Central and Southern Illinois, Southwestern Indiana, and Western Kentucky in the east-central United States. Bituminous coal seams of Pennsylvanian age, ranging from about 1.2 m to 3 m (4.0 ft to 10 ft) thick, are mined. The coal seams tend to be consistent in thickness across large areas and generally are horizontal or gently dipping, although several ancient tectonic fault zones occur in the southern part of the basin. Most underground mining is by continuous mining machines in panel room-and-pillar mines that are among the largest in the United States. Longwall mining, introduced during the 1970s, is playing an increasing role in coal recovery.

Because property boundaries generally run north-south and east-west (especially in the Illinois part of the basin), most mines are laid out with main headings north-south and east-west. This mining direction exacerbates stress-related ground failure.

MEASUREMENTS OF STRESS FIELD

Through analysis of seismograph records, the orientation of the focal (fault) plane and stresses responsible for an earthquake sometimes can be determined. Focal-plane analysis of four earthquakes in Southern Illinois shows that the maximum compressive stress was horizontal and east-west to east-northeast-west-southwest in each case (Figure 1).

The orientation of stress axes in a borehole can be determined by a technique known as hydrofracturing. A portion of the wellbore is isolated by means of rubber packers, and fluid is introduced at high pressure into the sealed portion, thereby fracturing the rock. The induced fractures are preferentially parallel to the major stress axis. The orientation of the fractures may be obtained from impression packers which inflate against the wellbore and obtain impressions of the fractures, or they may be photographed with a downhole camera. Haimson (1974) reported results of hydrofracturing tests in five shallow (about 100 m (330 ft)) wells in Southern Illinois (Figure 1). The indicated maximum compressive stress was horizontal and oriented N49°E to N72°E (average N62°E).

Another method for determining stress in wells is by observing borehole breakouts. Where the stress differential is high, the side walls of a well may break out, or fracture naturally, with a preferred orientation. Orientations of breakouts are determined by impression packer or downhole camera, as in hydrofracturing. Dart (1985) reported consistent orientation of side-wall failures in 17 wells, as deep as 2,100 m (6,900 ft), in the Illinois Basin (Figure 1). The maximum stress direction ranged from N35°E to S78°E.

Several methods are used to characterize stress within a mine. A typical method involves overcoring as employed by the U.S. Bureau of Mines. A vertical borehole is drilled in the mine roof, and a borehole deformation gauge is inserted. Then a hole of larger diameter is cored around the original hole containing the gauge. As overcoring proceeds, in situ stress is relieved in the cylindrical shell of rock. The cylindrical shell deforms, and the gauge measures the deformation. Overcoring or similar techniques have been applied at six coal mines in Illinois (Figure 1).

Blevins (1982) documented the major stress axis as N87°E in Hamilton County, Y. Chugh (1984, personal communication) documented the major stress axis as S67°E with in-mine stress measurements in Franklin County. Hanna et al. (1985) found the major stress axis to be N41°E in Christian County. Ingram and Molinda (1988) documented the major stress axis as being from N78°E to N88°E in Wabash County and from N75°E to N85°E in Saline County. Strain-relaxation measurements of core from Franklin County show the major stress axis as N76°E (Y. Chugh, 1984, personal communication).

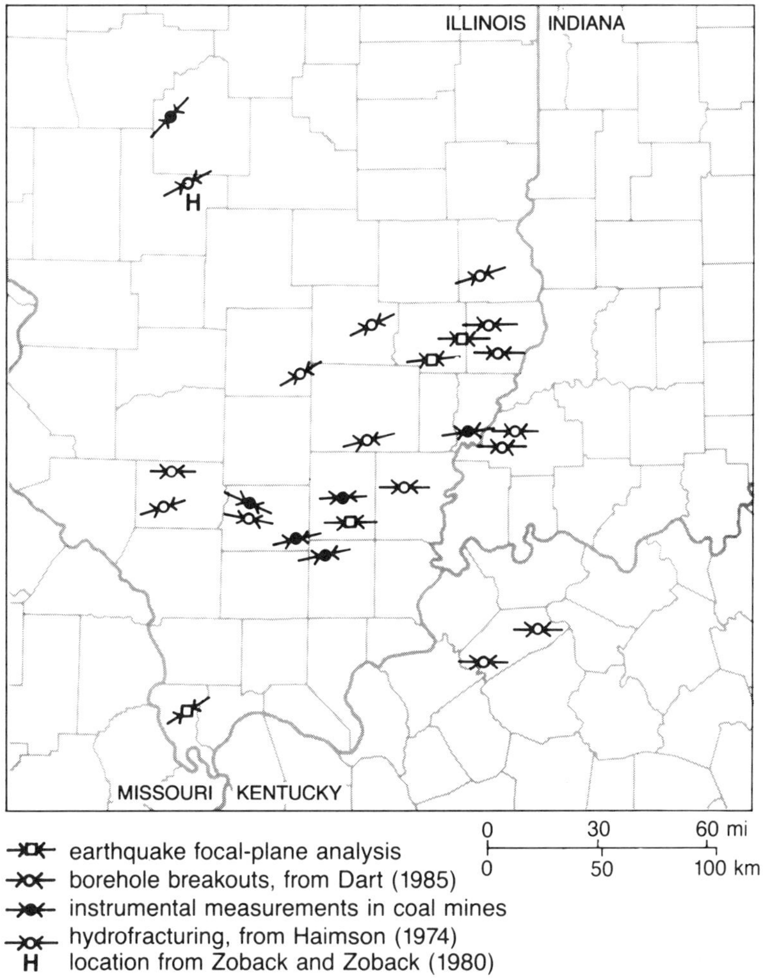

Figure 1. *Measured orientation of major stress axes in Illinois Basin coal field.*

The various in situ stress measurements in Illinois show a fairly consistent direction and magnitude. These stresses can produce mappable fractures and roof-failure patterns in underground coal mines.

STRESS FEATURES IN MINES

Oriented Roof Failures And "Kink Zones"

A tendency for roof failures to occur preferentially in north-south headings at some mines in Illinois has been known for more than 60 yrs (Illinois State Geological Survey (ISGS), unpublished open-filed field notes). In modern times, this tendency was documented first by Krausse et al. (1979), after detailed mapping of roof falls in mines in the Herrin Coal. Oriented failure has been observed at many mines in Southern Illinois and Western Kentucky.

Krausse et al. (1979, p. 21) coined the term "kink zone" for "narrow zones of compressional cracking and sagging which develop in the immediate roof after mining." Kink zones most commonly develop near the centerline of a mine opening. In some mines, they develop preferentially on one side of the entries. They are slightly sinuous in plan view, and may propagate for 100 m (330 ft) or more along the mine opening. A kink zone resembles a simple sag caused by leaving too broad a span of unsupported roof. However, a kink zone differs from a sag in that, on close inspection, a true kink zone displays unequivocal signs of lateral compression, such as pulverized rock along the center line and shearing (overlapping) of the ends of the broken rock layers (Figure 2). Lateral stress buckles the rock layers downward and jams their ends together.

A kink zone causes little trouble initially, aside from littering the floor with rock fragments. However, if left untended, the kink is likely to grow and may extend above the roof-bolt anchors and cause a major fall. The likelihood of a fall, of course, is greater if any other weaknesses, such as shear fractures, are present in the roof.

Figure 2. *Kink zone in north-south entry of a mine in Southern Illinois. Layers of top coal and gray shale in the roof sag noticeably. The roof was re-bolted after the kink zone developed.*

In the mines in Illinois and Western Kentucky where we have observed kink zones, kink zones are abundant in the north-south headings and virtually absent in east-west headings (Figure 3). This observation is consistent with the measured east-west to east-northeast-west-southwest major compressive stress.

Some lithologies are more susceptible than others to development of kink zones. Kink zones develop readily in brittle, thinly laminated or weak shale, siltstone, or sandstone. They rarely form in massive silty mudstone or siltstone, and have not been observed in thick-bedded to massive sandstone or limestone (Figure 3). The number and thickness of laminations and the inter-layer tensile and shear strength probably are more critical than the strength of the rock between bedding planes. Roof type appears to be far more critical than depth of mining in determining whether oriented roof falls or kink zones will appear.

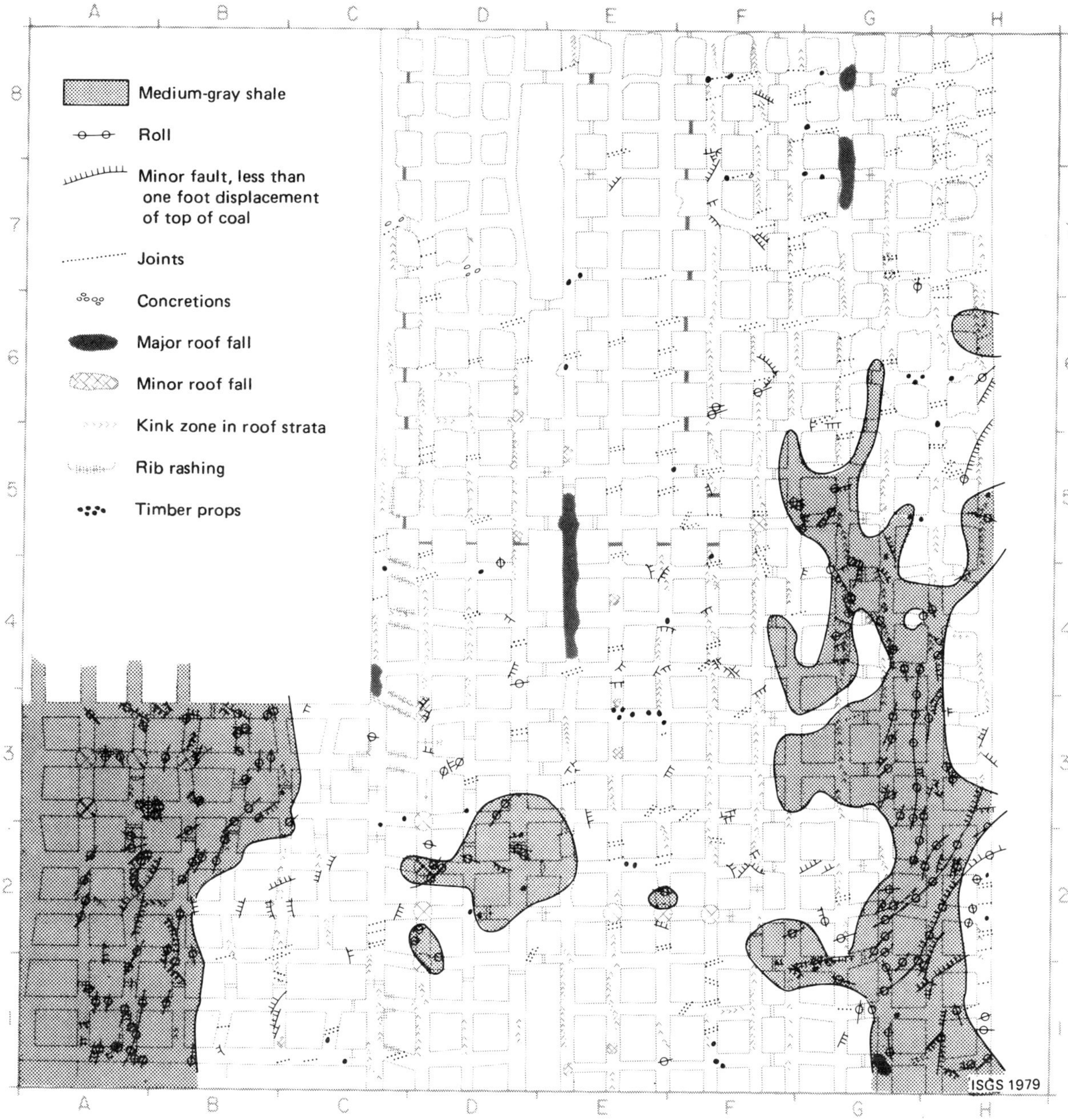

Figure 3. *Map of a portion of an underground mine in Jefferson County, Illinois (from Krausse et al., 1979). Major roof falls and kink zones developed only in N-S headings. Kink zones develop preferentially in dark gray, well-laminated shale (unshaded), and seldom appear in medium gray, poorly laminated shale. Study area is located about 100 m (330 ft)east of the Rend Lake Fault Zone, a pre-Tertiary extensional structure.*

Local structural features, especially faults, influence the development of kinks and directional roof falls. In Wabash County, Illinois, kink zones were observed only on the downthrown (west) side of a fault that has about 40 m (130 ft) of throw (Ingram and Molinda, 1988). Numerous kink zones and directional roof falls were mapped in a mine in Jefferson County, Illinois, a short distance east of the Rend Lake Fault Zone, on the downthrown side of the fault zone (Figure 3). Another study area in the same mine about 3 km (2 mi.) east of the fault zone contained few kinks (Krausse et al., 1979). A local stress concentration seemed to exist on the downthrown (east) side of the Rend Lake Fault Zone at Old Ben Mine No. 24 in Franklin County, Illinois. North-trending roof falls, severe kink zones, and "bumps" are expressions of that local stress (Bauer and DeMaris, 1982). In all three cases, the faults involved are north-trending, high-angle normal faults that developed in an ancient stress field substantially different from the present one. The field observations seem to suggest stress buildups on the downthrown sides of faults; however, overcoring at the Wabash Mine indicates similar magnitude of stress on both sides of the fault (Ingram and Molinda, 1988). Further study is needed on the relationship of stress to geologic structure.

Several mines in Southern Illinois have severe north-south roof failure, but kink zones have not developed. In these mines the coal is overlain by 10 m (33 ft) or more of homogeneous, light to medium gray, silty mudstone or slightly fissile laminated shale. The rock has low compressive strength (20-30 MPa (2,900-4,400 psi)) and deteriorates when exposed to moisture. The high lateral stresses sometimes produce breakout in north-south entries (Figure 3). More commonly, roof falls develop in north-south entries, have a high, arched profile, and extend 1-3 m (3-10 ft) above the top of the coal. The rock may simply crumble under compression, rather than forming kink zones.

Joints

Systematic planar jointing is developed in the roof strata of mined coal seams throughout the Illinois Basin. Intensity and spacing of joints vary greatly depending upon many factors, chiefly upon lithology. Hard, brittle, well-laminated rocks best exhibit jointing. The black fissile shales that directly overlie commercial coals in much of the basin are particularly well-jointed. Joint spacing in these rocks commonly ranges from 5 cm to 30 cm (2 in. to 12 in.). Joints are prevalent in most fissile gray shales and siltstones, but they are rare or absent in mudstones. Massive sandstones and limestones contain large joints spaced several meters to tens of meters apart.

Throughout the basin, the predominant direction of jointing in roof shales is parallel with the current major stress axis (Nelson and Bauer, 1987). This direction is N60-70°E in most of Illinois, N80°E to east-west in southeastern Illinois and Western Kentucky, and about N80°E in most of Indiana. A secondary joint set, perpendicular to the primary joints and major stress axis, developed at some sites. Primary joints do not parallel the regional major stress axis in a few places near pre-Tertiary fault zones in Southern Illinois and in Kentucky. Joints near fault zones generally are oriented parallel with the stress fields that caused the faulting. For example, north-south joints along the Rend Lake Fault Zone in Southern Illinois probably developed under the stress regime of east-west extension that produced the faults. Another area where joints do not fit the current stress field is Southern Indiana. According to Ault et al. (1985), primary joints in Southern Indiana dominantly trend N30-35°E. Noting that joints elsewhere in Indiana are parallel to the contemporary major stress axis, Ault et al. (1985) suggested that Southern Indiana lies in a different stress province than the rest of the state. No in situ stress measurements have been made in Southern Indiana to test that theory. High-angle normal faults which trend north-northeast are numerous in Southern Indiana; joints and faults might have formed in the same extensional, pre-Tertiary stress field.

Closely spaced joints occur in some mines that have high in situ stress. A vertical fracture zone 7-30-cm (3-12-in.) wide, penetrating the entire coal seam and the roof strata, was observed at the Inland No. 2 Mine (Figure 4). The zone strikes slightly north of east, parallel to ordinary joints and the major stress axis in this

Figure 4. *Zone of very closely spaced vertical fractures in the mine roof parallel to major stress axis as measured in this mine.*

mine as measured by means of strain gauges within the zone (Blevins, 1982). Individual fractures are spaced a few millimeters apart. Some fractures in the zone are open, others are coated or filled with calcite and pyrite. In places, small fragments of roof shale have shifted several centimeters downward into the coal along open fractures; however, the zone as a whole has no detectable offset. In the black shale and limestone roof of another deep mine in Southern Illinois, miners reported that fractures, which trend N80-85°E, widen after mining, producing loud pops and cracks. This effect was reported to be most pronounced at intersections where joints cross the longest open span (entries at this mine are driven northeast and northwest, oblique to the joints). Although ominous-sounding, these noisy joints seldom caused ground failure.

Joints parallel with the contemporary major stress axis have been observed elsewhere in the Northeastern U.S. (Engelder, 1982; Foote, 1983). Such joints are release fractures — fractures that develop under anisotropic horizontal stress as vertical (overburden) stress is removed. Not all east-west to east-northeast-west-southwest joints in the Illinois Basin are necessarily recent phenomena. Nevertheless, the present stress would widen and intensify any pre-existing joints having such a favorable orientation.

Joints are not a major cause of roof failure at most mines in the Illinois Basin. In many cases, joints penetrate only a short distance above the coal. Such fractures promote slabbing of the immediate roof, but rarely induce large falls. Closely spaced joints present a greater hazard, which can be alleviated by installing additional roof support and by leaving larger pillars.

Faults

Small faults, which have orientations consistent with the contemporary stress field, are widespread in coal-bearing rocks of the Illinois Basin coal field (Figure 5). Types of faults observed include thrust faults, a strike-slip fault, and vertical faults that resemble unusually intense extensional joints. These faults appear to be unrelated to any ancient tectonic structures in the region (Nelson and Bauer, 1987).

Small thrust faults in coal mines of Southern Illinois and Western Kentucky were described by Nelson and Bauer (1987). Faults range from tens of meters to 2.5 km (up to 1.6 mi.) long, have a few cm to 3 m (up to 10 ft) of displacement, and dip mostly at 15-30° from horizontal. They penetrate coal seams and adjacent strata; full vertical extent is unknown. All these thrust faults strike north-south or nearly so (Figure 5), and thus are perpendicular to the primary joints and the current major stress axis.

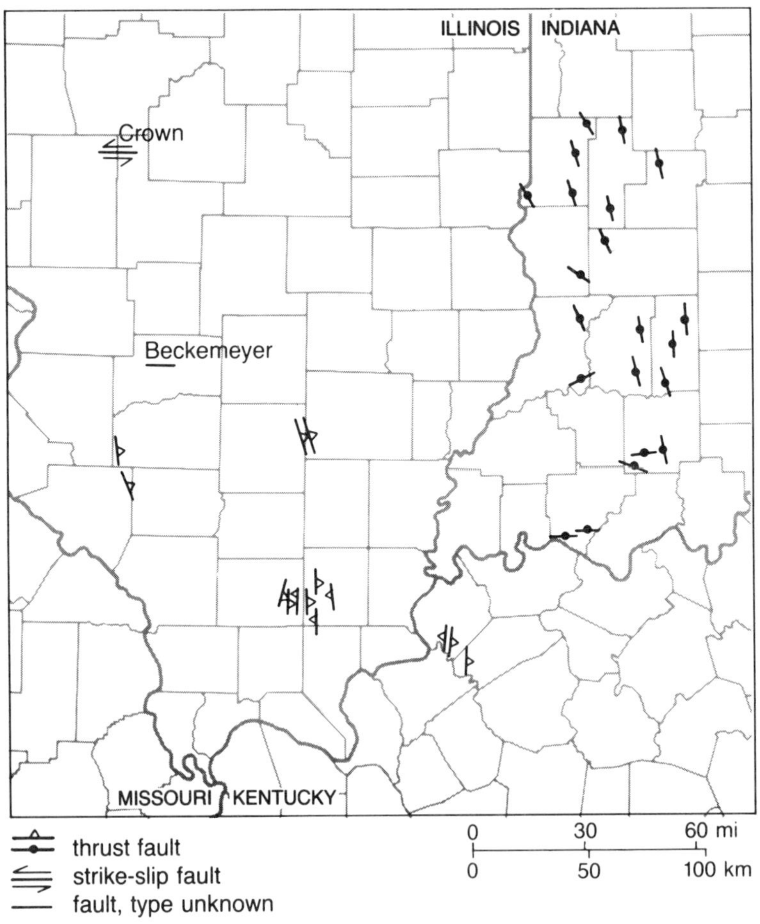

***Figure 5.** Faults believed to be products of the contemporary stress field in Illinois Basin coal field. Faults shown in Indiana are taken from Ault et al. (1985). Faults are drawn larger than they actually are for clarity.*

Similar thrust faults in surface coal mines in southwestern Indiana were reported by Ault et al. (1985) and attributed to contemporary stress. Most of the thrusts in Indiana strike N15-45°W (Figure 5). A few faults in southernmost Indiana strike nearly east-west, suggesting (along with the data on jointing) a different stress regime there (Ault et al., 1985).

An east-trending, left-lateral strike-slip fault, the Crown Fault, was mapped in the workings of the Crown II Mine, Macoupin County, Illinois (Nelson and Nance, 1980; Nelson, 1981; Ledvina et al., in press). To date, this fault has been mapped more than 5 km (3 mi.) along strike without either end being found. Vertical offsets are less than 50 cm (20 in.) in most places but reach 120 cm (47 in.). Horizontal displacement may be as great as 20 m (66 ft). Open vertical fissures occur along the main east-west fault and along innumerable northeast- to east-northeast-trending extensional fractures on either side of the main fault. These fissures are filled with a loose breccia of angular pieces of coal and rock (Figure 6), and lightly coated with fine crystals of calcite and other minerals. The thick mineral fillings that characterize pre-Tertiary faults of Southern Illinois are conspicuously absent. Large volumes of water, and seeps of gas and petroleum, commonly issue from

Figure 6. *Crevice in limestone roof associated with the Crown Fault. The crevice is lined with finely crystalline calcite and contains loose fragments of limestone. Ruler graduated in tenths of feet (0.1 ft=30 mm).*

the fissures along the Crown Fault when the fault is first intersected in mining. Similar features were described, sketched, and photographed at the now-abandoned Crown I Mine which is east of the Crown II Mine (J. Simon and J.A. Harrison, unpublished open-filed field notes, ISGS). If faults at the Crown I and Crown II mines connect, the fault is at least 11-km (6.8-mi.) long.

The lack of mineral fillings in open fissures along the Crown Fault suggests a young fault zone. Known ancient faults in the Illinois Basin generally are thoroughly mineralized. An east-west left-lateral fault is consistent with a major stress axis trending N60°E and a vertical secondary stress axis. The nearest measured in situ major stress axis is N41°E, and the least stress is vertical (Hanna, et al., 1985).

A fracture possibly similar to the Crown Fault was intersected in 1937 in workings of the Beckemeyer Coal Company in Clinton County, Illinois. The structure was described as an east-west-trending, vertical zone of pulverized coal and shale approximately 30-cm (12-in.) wide. It was observed only in two rooms of the mine. A field sketch shows slight offset of a shale parting in the coal seam along the fracture zone. Gas, oil, and saltwater entered the mine along the fracture zone. Two gas explosions, in which six miners were killed and several others severely burned, resulted when open cap lamps ignited gas from the fault (L.C. McCabe and G.V. Cohee, unpublished 1938 field notes, ISGS).

Several open fissures were observed in the roof of two coal mines in southwestern Illinois. The open fissures strike east-northeast and are in close proximity to small thrusts that strike north-northwest. These fissures are 10-20-cm (4-8-in.) wide and extend only a few meters or tens of meters along strike. Although the roof rock is severely shattered and loose breccia is present, the coal does not appear to be affected (Figure 7). Water flowed from the thrust faults and fissures when coal was first mined.

Some of the dangers these faults pose have already been mentioned. Gas outbursts along the Crown Fault reportedly have triggered the automatic power shutoffs on face equipment. Large quantities of gas, either from coal seams or from petroleum reservoirs, can accumulate in the wide fissures along some of these faults. Initial water seepage from the Crown Fault and some thrust faults has hampered production. Large sections of roof may fall without warning along faults. The thrust faults, which penetrate the roof at a low angle, isolate huge wedges of rock which are ready to fall when the underlying coal is taken out. Because displacements of the coal are slight, miners do not always recognize these faults sufficiently to take appropriate precautions.

Regional Extent and Origin of Stress

The stress regime of the Illinois Basin appears to prevail across most of Eastern North America. Sbar and Sykes (1973) and Zoback and Zoback (1980) found the major stress axis to be horizontal and northeast-southwest across most of the United States between the Rocky Mountains and Appalachian Mountains and north of the Gulf Coastal Plain. Gough (1984), Bell and Babcock (1986), and Gough and Gough (1987) showed that all of Canada east of the Rocky Mountains lies within the same stress province. Zoback and Zoback (1980) regarded the in situ stress as the result of viscous drag from the underlying asthenosphere as the North American Plate drifts southwestward from the Mid-Atlantic Ridge. Gough et al. (1983) proposed alternatively that a northeastward flow in the mantle transmits shear by viscous drag through the asthenosphere to the base of the lithosphere.

Figure 7. An open fracture zone, trending east-northeast, in the roof of an underground coal mine in southwestern Illinois. Note that the coal is not disturbed.

IDENTIFYING IN SITU STRESS

Pre-Mining Assessment

A northeast-southwest to east-west major stress should be assumed to exist throughout the Illinois Basin coal field, and stress should be investigated and considered in the planning of all new mines. Because changing mine layouts after mining commences is costly, the stress field should be characterized before the mine is designed. Stress requires concern, particularly when the coal seam to be mined is overlain by more than a few meters of mudstone, shale, siltstone, or laminated sandstone. Stress is less likely to cause roof failures when competent beds of limestone or massive, well-indurated sandstone lie within 3-5 m (10-16 ft) of the coal. Stress-related faults and joints, however, may weaken any kind of roof.

The first step in evaluating stress conditions at a potential mine site is to gather existing data. Instrumental data, of course, are of prime interest. So are observations on faults, joints, kink zones, directional roof failure, and related conditions at nearby active and abandoned mines. Such information is publically available from state and federal agencies such as the ISGS and the U.S. Bureau of Mines. Mine maps may provide clues as to the preferred direction to drive entries. For example, in the Clyde Mine, which operated from 1882 to 1906 in Macoupin County, west-central Illinois, main entries were driven northeast and northwest. Although the roof was a thick, relatively weak gray shale, some rooms at the Clyde Mine were driven as wide as 9 m (30 ft). A nearby mine with the same type of shale roof had severe instability in north-south headings only 6 m (20 ft) wide (Nelson, 1987).

If preliminary assessment indicates that stress is likely to pose a problem, or if assessment fails to characterize the stress regime accurately, instrumental measurements are in order. Such measurements can be made in conjunction with routine test drilling.

Hydrofracturing (Haimson, 1974) determines the stress directions and magnitudes, and borehole breakout studies (Dart, 1985; Bell and Babcock, 1986) show the orientation of the major stress direction. During an exploration program the magnitude and directions of the principal stresses can be best determined by hydrofracturing, strain-relaxation measurements, or oriented core. A new method of measurement is performed with a stress meter, which is an inflatable packer that has strain measuring points protruding through the side walls of the packer. This method potentially can measure stresses in fissile shales and weaker, fractured shales and claystones where hydrofracturing may have difficulty producing a vertical fracture.

Assessment During Mining

The appearance during mining of any of the features and conditions described in the section on "Stress Features in Mines" should prompt a thorough investigation to determine whether stress is truly involved and, if so, to characterize the stress regime. Accurate mapping is essential to identify effects of in situ stress. Many operators today routinely map roof falls and geologic conditions in their mines. Roof-fall maps should contain or be accompanied by as much information as possible on size, shape, and timing of roof failures. Troublesome areas should be revisited from time to time and carefully examined for kink zones, systematic fractures, and small faults. Most stress-related fractures are subtle and difficult to spot in freshly mined areas, particularly where they are behind a thick coating of rock dust. Kink zones and open fissures also may take some time to develop before they become noticeable and initiate roof failures. These features become much more obvious in old workings, particularly little-used areas where rock dust has scaled off the coal and roof.

Stress suspected from mapping and visual observations should be confirmed instrumentally. Measurements made within the mine are probably best performed by means of overcoring, strain relaxation, or stressmeter. The sampling locations for these tests should be in boreholes that extend well away from the influences of the mine entryway, mains, submains, or panels. Stress measurements in tunneling operations typically are performed at distances away from the tunnel of about three times the radius of the tunnel. If a pressure arch is developing or has developed over a panel, submains, or mains, then the operator may need to drill a greater distance to reach a point where mining no longer influences the in situ stress.

COPING WITH TECTONIC STRESS

In an effort to reduce stress-induced ground failures, at least four mine operators in the Illinois Basin have changed the direction of mining. At the Inland Steel Company No. 2 Mine in Hamilton County, Illinois, mining directions were changed from north-south and east-west to N45°E and N45°W after the major stress axis was found to trend nearly east-west. A marked reduction in roof failure and in percentage of waste rock loaded with the coal resulted from the change (Blevins, 1982). At the Kerr-McGee Coal Company Galatia No. 5 Mine in Saline County, Illinois, kink zones and roof failures were reduced by driving panel entries N30°E and S30°E instead of north-south (Ingram and Molinda, 1988). Also in Saline County, roof problems were reduced by turning entries northeast and northwest at the Kenellis Energies, Inc. Brushy Creek Mine (Cecil Newcom, personal communication, 1984). A similar change was made recently at Peabody Coal Company's Martwick Mine in Muhlenberg County, Kentucky; the results there have not yet been evaluated fully. Also, at several longwall mines in Illinois, where the companies are committed to north-south and east-west orientations for their workings, the longwall panels are oriented east-west to maximize stability in headgate and tailgate entries.

The best direction to mine probably is 45° to the major stress axis. This direction minimizes the length of mine opening along which kink zones can propagate. It also minimizes the span affected by joints, thrust faults, and extensional fissures. To ensure optimum orientation of entries, the major stress axis must be measured as was done in at least three of the cases cited above.

When other considerations preclude angling the main headings, less radical steps such as angling or staggering crosscuts may yield benefits. Other strategies for reducing stress-induced failure have been less successful. In an attempt to relieve stress at the Inland No. 2 Mine, miners cut vertical slots in the coal and roof at the faces of north-south headings. After one week the slots had converged as much as 8 cm (3 in.). After 10 days, the roof sagged noticeably. Although slot-cutting seemed to reduce spalling of the roof, the practice was discontinued (Blevins, 1982). Other strategies at this mine included driving sacrificial entries (entries driven ahead of adjacent critical entries to relieve stress) and alternating box-cuts left and right. These efforts were judged to be failures (Blevins, 1982). Sacrificial entries at a mine in east-central Illinois also failed to relieve stress in adjacent critical headings.

Other destressing techniques have been applied in hardrock mining and in deep, burst- or bump-prone coal mines in mountainous regions. Such techniques probably are not cost-effective in shallow, high-sulfur coal deposits of the Illinois Basin. The simplest, low-cost solution is to turn the mine headings 45° to the major stress axis (at least in room-and-pillar mining). Jeremic (1981) showed the same solution for tectonic stress problems in coal mines in Western Canada.

SUMMARY

The Illinois Basin coal field is subject to tectonic stress in which the major stress axis is oriented east-west to east-northeast-west-southwest. This stress creates compressional roof failure in entries perpendicular to the major stress axis, and several types of small faults. Characterizing stress conditions should be part of any coal exploration program. The most effective method to minimize stress-related ground failure in this coal field is to orient entries 45° to the major stress axis.

REFERENCES

Ault, C.H., D. Harper, C.R. Smith, and M.A. Wright, 1975, Faulting and jointing in and near surface mines of southwestern Indiana: U.S. Nuclear Regulatory Commission, Report NUREG/CR-4117, 27 p.

Bauer, R.A., and P.J. DeMaris, 1982, Geologic investigation of roof and floor strata, longwall demonstration, Old Ben Mine No. 24: Illinois State Geological Survey, Contract/Grant Report 1982-2, 49 p.

Bell, J.S., and E.A. Babock, 1986, The stress regime of the Western Canadian Basin and implications for hydrocarbon production: Bulletin of Canadian Petroleum Geology, v. 34, no. 3, p. 364-378.

Blevins, C.T., 1982, Coping with high lateral stresses in an underground coal mine, in S.S. Peng and J.H. Kelley, eds., Proceedings of the 2nd Conference on Ground Control in Mining, Morgantown, West Virginia, July 19-22, 1982: Morgantown, West Virginia University, p. 137-141.

Dart, R., 1985, Horizontal-stress directions in the Denver and Illinois basins from the orientation of borehole breakouts: U.S. Geological Survey Open-File Report 85-733, 41 p.

Engelder, T., 1982, Is there a genetic relationship between selected regional joints and contemporary stress within the lithosphere of North America?: Tectonics, v. 1, no. 2, p. 161-177.

Foote, G.R., 1982, Fracture analysis in northeastern Illinois and Northern Indiana: Urbana, University of Illinois, Unpublished Ph.D. Thesis, 193 p.

Gough, D.I., 1984, Mantle upflow under North America and plate dynamics: Nature, v. 311, p. 428-433.

Gough, D.I., C.K. Fordjor, and J.S. Bell, 1983, A stress province boundary and tractions on the North American Plate: Nature, v. 305, p. 619-621.

Gough, D.I., and W.I. Gough, 1987, Stress near the surface of the earth: Annual Review of Earth and Planetary Science, v. 15, p. 545-566.

Haimson, B.C. 1974, A simple method for estimating in situ stresses at great depths, in Field testing and instrumentation of rock: American Society for Testing and Materials, Special Technical Publication 554, p. 156-182.

Hanna, K., K. Haramy, and D. Conover, 1985, Field investigations of roof and pillar stability in coal mine intersections, in Y.P. Chugh, ed., Proceedings of 2nd Conference on Ground Control Problems in Illinois Coal Basin, Carbondale, Illinois, May 29-31, 1985: Carbondale, Southern Illinois University, p. 76-83.

Ingram, D.K., and G.M. Molinda, 1988, Relationship between horizontal stresses and geologic anomalies in two coal mines in Southern Illinois: U.S. Bureau of Mines Report of Investigations 9189, 18 p.

Jeremic, M.L., 1981, Coal mine roadway stability in relation to lateral tectonic stress--Western Canada: Mining Engineering, v. 33, no. 6, p. 704-709.

Krausse, H.-F., H.H. Damberger, W.J. Nelson, S.R. Hunt, C.T. Ledvina, C.G. Treworgy, and W.A. White, 1979, Roof strata of the Herrin (No. 6) Coal and associated rock in Illinois--A summary report: Illinois State Geological Survey, Illinois Minerals Note 72, 54 p.

Ledvina, C.T., S. Sroka and C. Shabica, in press, The principles of coal mining geology: Examples from the Herrin (No. 6) and Springfield (No. 5) coals of Illinois: Chicago, Northeastern Illinois University, Discrete Features, v. 1.

Nelson, W. J., 1981, Faults and their effect on coal mining in Illinois: Illinois State Geological Survey, Circular 523, 38 p.

Nelson, W. J., with contribution by P.J. DeMaris and R.A. Bauer, 1987, The Hornsby district of low-sulfur Herrin Coal in Central Illinois (Christian, Macoupin, Montgomery, and Sangamon Counties): Illinois State Geological Survey, Circular 540, 40 p.

Nelson, W.J., and R.A Bauer, 1987, Thrust faults in southern Illinois Basin--Result of contemporary stress?: Geological Society of America Bulletin, v. 98, no. 3, p. 302-307.

Nelson, W.J., and R.B. Nance, 1980, Geological mapping of roof conditions, Crown II Mine, Macoupin County, Illinois: Illinois State Geological Survey, Reprint 1981A, 9 p.

Sbar, M.L., and L.R. Sykes, 1973, Contemporary compressive stress and seismicity in Eastern North America: An example of intra-plate tectonics: Geological Society of America Bulletin, v. 84, p. 1861-1882.

Zoback, M.L., and M. Zoback, 1980, State of stress in conterminous United States: Journal of Geophysical Research, v. 85, p. 6113-6165.

THE EFFECTS OF GEOLOGY ON COAL MINE-INDUCED SUBSIDENCE IN THE UNITED STATES

William F. Kane
Department of Civil Engineering
The University of Tennessee
Knoxville, TN 37996-2010

Larry R. Powell
U.S. Bureau of Mines
5629 Minnehaha Ave., South
Minneapolis, MN 55417

ABSTRACT

Consideration of the consequences of subsidence is becoming increasingly important as the need for more energy and minerals conflicts with the land-use needs of an expanding population. European empirical methods for predicting coal mine-induced subsidence are based on geometry, and do not work without modification in the United States because geology plays a major role in controlling the manifestation and amount of surface subsidence. The United States has several major bituminous coal regions, all formed under varying depositional conditions at different geologic times. As a result, the stratigraphy and structural geology of each of these areas is quite different. Mining methods include room-and-pillar, high-extraction room-and-pillar, longwall, and shortwall approaches. Each of these methods, when combined with the existing geology, produces a subsidence style characteristic of a given area.

INTRODUCTION

The geology of coal-forming environments is a complex melange resulting from shifting shorelines, rivers, and swamps. Because of the rapidly changing depositional environments present in a region where peat accumulation may occur, geological variations abound. These variations result in various ground control problems underground such as roof falls, floor heave, and pillar punching.

In all undisturbed strata, the processes of deposition and tectonic change have achieved an equilibrium over geologic time. Overburden weight is the main cause of the natural stress field within the strata, although residual tectonic stresses impose horizontal forces which may be important in some geological settings.

Mining disturbs the original stress field, causing the strata to deform. When a sufficiently large space is created, the immediate roof breaks and falls into the cavity. The collapsing material increases in volume, occupying a greater space than in its undisturbed state. The geometry of the collapse is dependent on the mechanism of failure of the individual roof beds, which is controlled by the macro-structure of the strata and their tensile and shear strengths. At some height above the cavity, caving stops and the collapsed material begins to accept the superincumbent weight of the overburden, causing recompaction of the caved material. Strata above the caved zone sag rather than break, causing bed separations.

The results of this process are a combination of bending, fracturing, shearing, separating, caving, and heaving, all of which are dependent on the overburden lithology (Figure 1). In addition, position and thickness of individual beds, especially strong competent beds, influence the migration of the caving mechanism to the surface. This migration is complicated by the presence of joints, bedding planes, and other discontinuities whose character changes with rock lithology. At the surface, layers of unconsolidated material alter the transfer of the stresses produced in the rock layers and can result in some form of surface subsidence.

The surface effects of coal mine-induced subsidence can be minimized by implementing proper strata-control measures developed from knowledge gained during the precalculation of ground movements. Such prediction methodology has been developed and widely used in Europe. However, many of these empirical observations of subsidence are largely geometric and ignore the effects of local geological variations. This raises questions concerning the validity and accuracy of these methods if they are extended for predictions throughout the United States. To develop a prediction method to fit general conditions, Allen (1978) suggested that the effect of geological differences should be determined by collecting data at selected mines in different geological settings, then characterizing the subsidence mechanisms from the geological and surface movement data. Even with a very general model, however, geological differences between coal fields in Europe and the United States are considerable. Kratzsch (1983) and Whittaker and Reddish (1989) have described the causes, prediction, and mitigation techniques related to subsidence in general. This paper attempts to focus on the geological aspects of subsidence in the United States so that a better understanding of the effects of geology might lead to improved prediction and mitigation.

FACTORS AFFECTING SUBSIDENCE

The same geological factors that control the stability of underground openings also control the caving characteristics of overburden strata. When combined with a particular mining method, a characteristic style of surface subsidence will occur. This has been found to vary from one coal field to another as variations in the depositional environment and strata are reflected in surface expressions of subsidence. The United States has several major bituminous coal regions (Figure 2), all formed under varying depositional conditions at different geologic times.

Mine conditions and methods interact with the geologic constraints to produce different subsidence styles at the surface. Shadbolt (1977) has classified the propagation of ground movements from the underground excavation to the overlying surface into three processes: mining, structural, and geologic characteristics. Structural factors relate to the configuration and construction of surface structures and how they interact with ground movements. In addition, the time-dependent properties of the overburden which affect the development of ground movements should be considered. Time is the link between mining and geological conditions which affect the level of stress and creep in the rock mass. Only mining and geological factors will be discussed in more detail, along with site characteristics.

Mining Factors

The principal mining factors that determine the surface expression of subsidence are the width and depth of the excavated panel, the excavated thickness of the coal seam, the mining methods used, and stowing. These factors have a profound influence on the stress reorganization and caving processes generated by excavation. Consequently, subsidence predictions for longwall mining traditionally have been developed

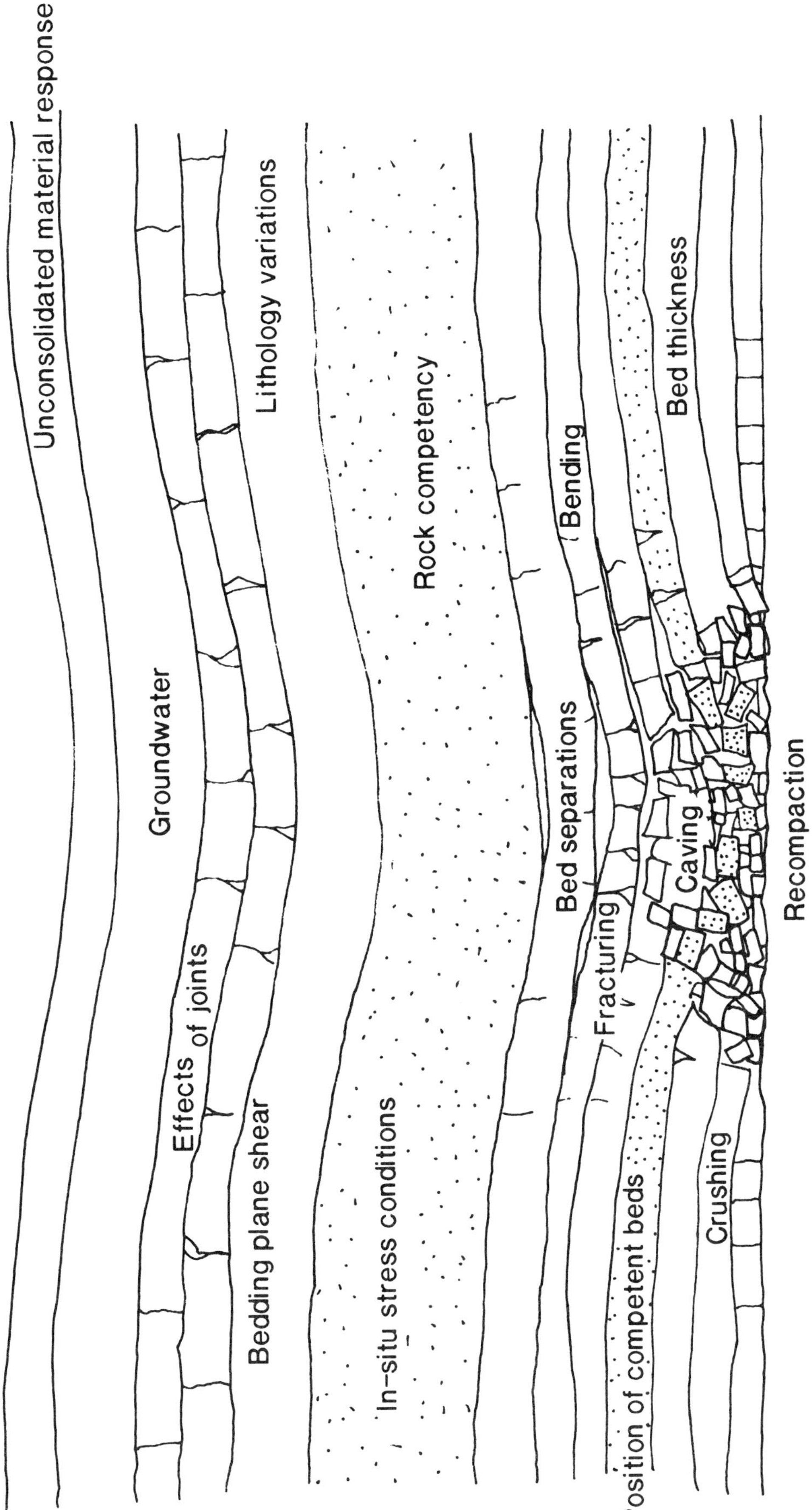

Figure 1. *Interactive processes involved in the manifestation of surface subsidence.*

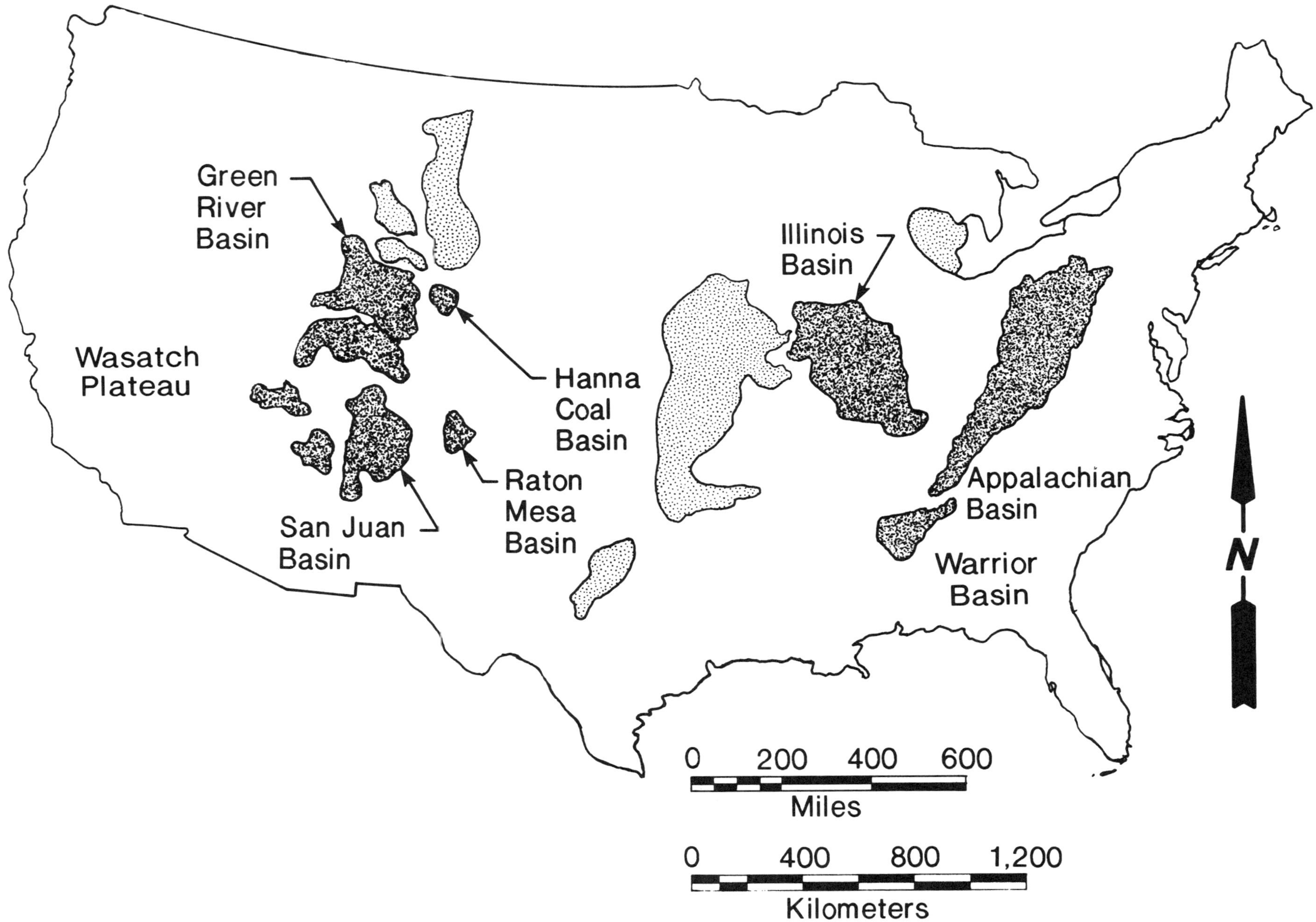

Figure 2. *Bituminous coal fields of the United States. Dark stippling indicates basins discussed in this paper.*

around simple parameters that describe the panel dimensions, the panel depth, and the application of stowing.

Because ground movement is spatial in nature, points on the surface subside downward and are displaced inward towards the center of the excavation. Points over the center of the excavation will subside vertically. The inward differential displacement is compressive. The area between the limits of movement and the edge of the excavation are subject to extension. From these displacements, movements that can develop at the surface are subsidence, tilt, curvature, horizontal displacement, and strain. A simplified section through a longwall panel is illustrated in Figure 3. The extent of the area of ground surface affected will be determined by the angle of draw, projected from the edge of the extracted area. The strain curves indicate the zones of positive strain (tension) at the edges of the subsidence trough, and negative strain (compression) in the middle, with the transition points lying approximately above the solid coal ribs.

Displacements generated at mine level are transmitted to the surface along lines that radiate from sources of deep movement. The movement of one point on the surface is affected by the movements within a circular area at depth (Figure 4). If the width of the working is such that on projecting the lines of draw inward from the edge of the working they meet exactly at the surface, then the critical width has been extracted for that depth (Figure 5b). When this occurs, the subsidence near the middle of the trough is the maximum possible subsidence. The angle of draw is directly related to the critical width of extraction, being the inverse tangent of one-half the critical width-to-depth ratio. As shown in Figure 5, two other mining conditions affect the shape of the settlement profile. If the panel width is less than the area of influence, then subsidence at the centerline is less than the maximum possible and the panel is at a subcritical width (Figure 5a). However, if the panel width is greater than the area of influence, then maximum subsidence extends over some area at the panel center (Figure 5c).

The amount of subsidence depends heavily on the degree of extraction achieved. Total extraction, resulting in collapse of the coal seam roof and overlying strata, would be expected to produce the greatest subsidence. Decreasing the extraction increases the amount of support, thus subsidence should decrease. However, the mining method used to achieve near-total extraction also may affect the amount of subsidence. Room-and-pillar mining followed by pillar extraction produces less subsidence than longwall mining (Hunt, 1980) because the former method removes support more gradually than longwall methods. This allows the immediate roof to bulk more upon caving, or because it is not a total extraction process, a certain percentage of stumps and fenders are left in place, thereby providing some support. Dunrud (1976) attributes reduced subsidence from retreat pillar mining to the development of a compression arch in the overburden which bridges the mining panel. Aynsley and Hewitt (1960) suggest that arching develops better in room-and-pillar mining because there is less ground in motion at any one time.

Site Characteristics

The conditions at a given site are related to both the coal geology and man-made alterations of the ground. Terrain characteristics also may be important. Terrain features have been shown by measurement and observation to have an influence upon the development of mine subsidence. Greater subsidence was observed over crests of ridges than in the intervening valleys (Gentry and Abel, 1978). Horizontal strain is greater in the down-slope direction of the pre-mining surface contours (Draper, 1964; Perz, 1958). In Pennsylvania, slopes of 12% to 15% were associated with damage to structures using mine design techniques shown to be adequate for level terrain (Cortis, 1969). However, other studies of coal mines beneath rugged terrain (Dahl and Choi, 1973; Kohli et al., 1980) suggest that topography plays only a minor role in subsidence characteristics.

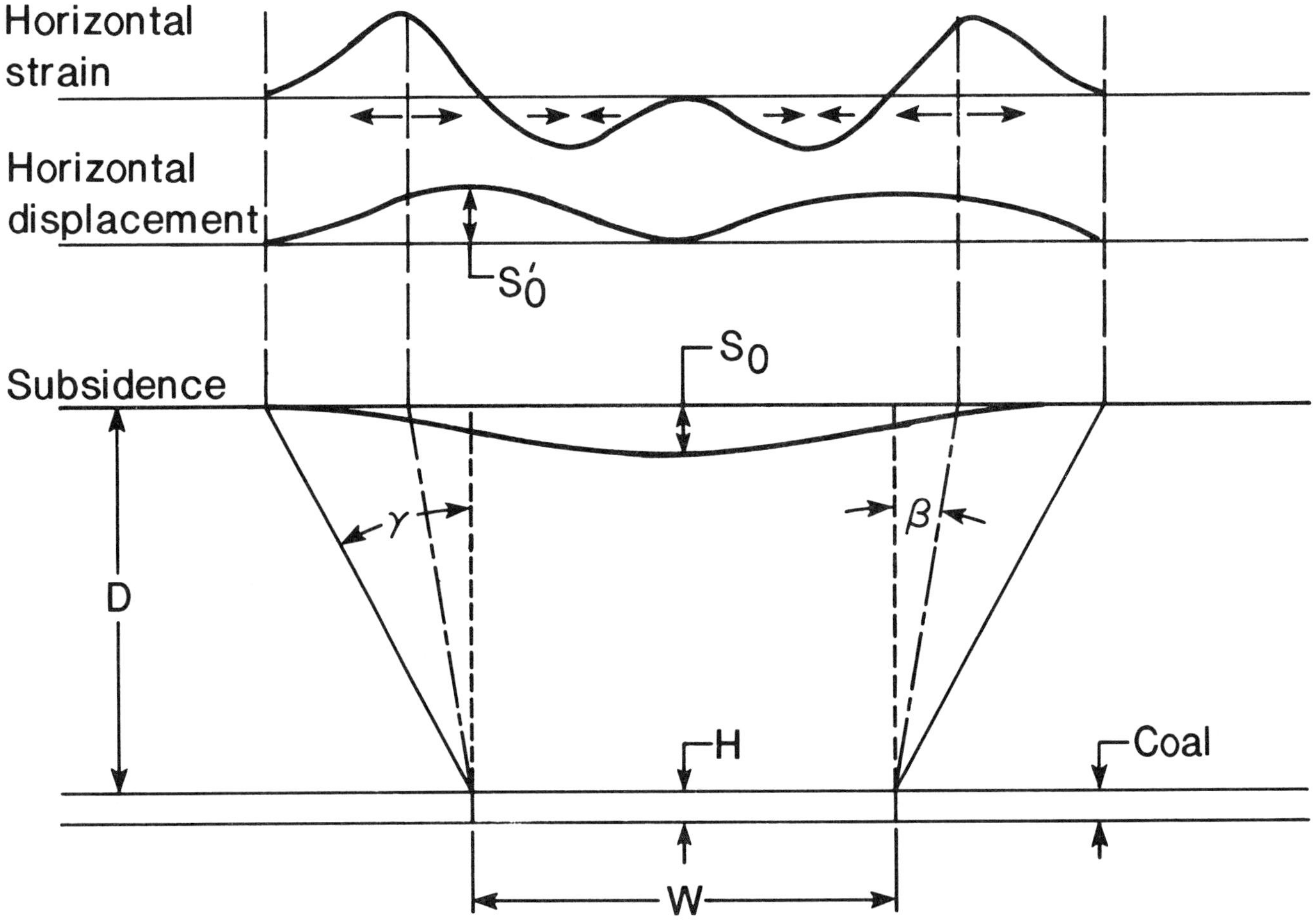

KEY

β = Angle of break
γ = Angle of draw
→← = Compression
D = Depth of coal seam
H = Extraction height
S_0' = Maximum possible slope for given conditions
S_0 = Maximum possible vertical subsidence given thickness of extracted seam
←→ = Tension
W = Width of extraction

Figure 3. *A simplified section through a longwall panel.*

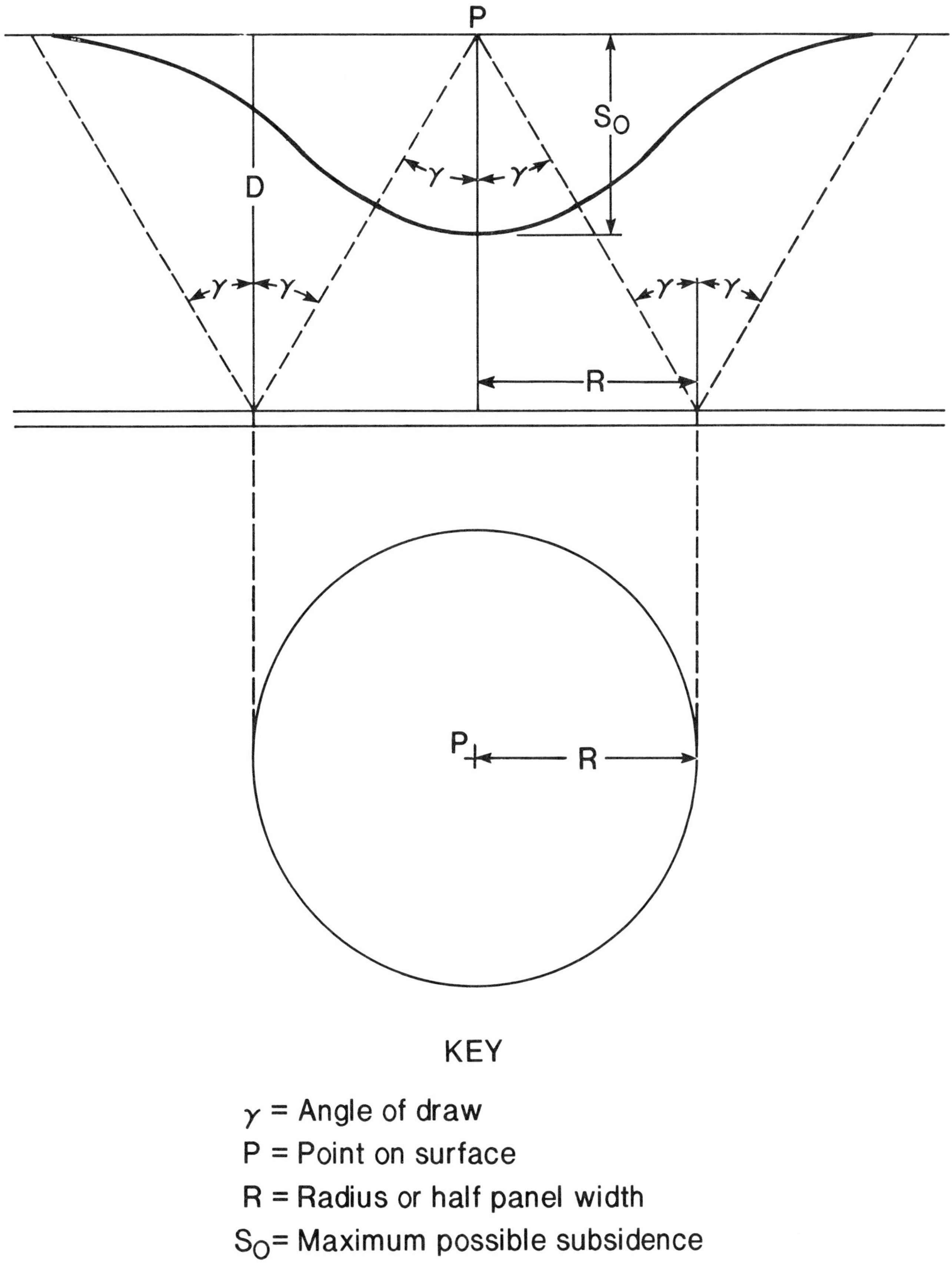

Figure 4. The movement of one point on the surface is determined by the movements within a circular area at depth.

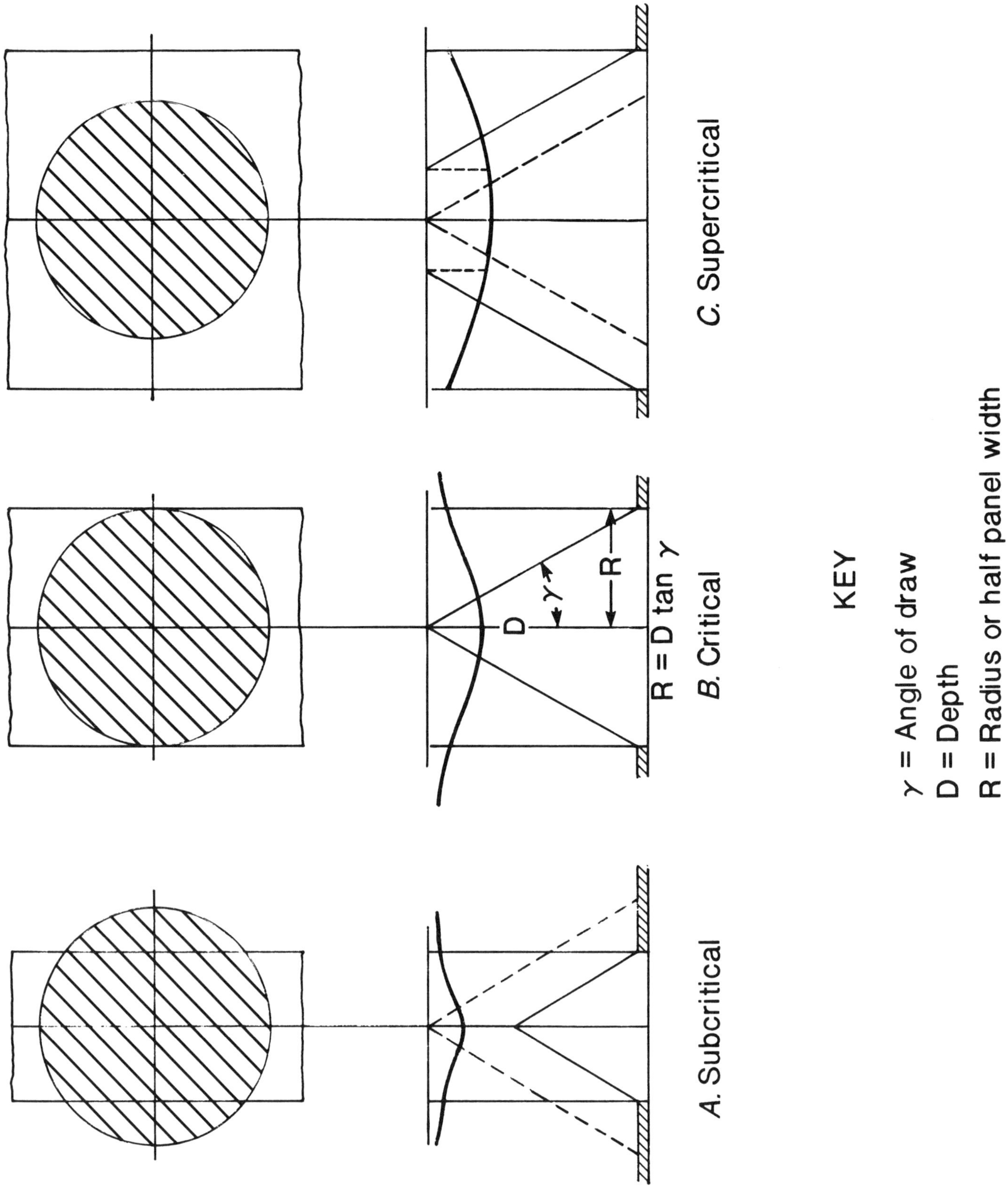

Figure 5. *Mining conditions that affect the shape of the settlement profile.*

Other site characteristics which influence subsidence are soils, groundwater, and previous mining. The nature of soils has been shown to influence the angle of draw (Crane, 1929; Pottgens, 1978) and soil thickness influences the amount of strain transmitted to the surface (Whittaker and Breeds, 1977). The presence of several mines can cause locally large settlements and affect the size of the area over which movements occur (Briggs and Ferguson, 1933; O'Rourke and Turner, 1979). Groundwater may activate instability and has an impact on consolidation of the overburden. In the presence of water, the weight of the strata increases while developing pore pressure reduces the shear strength of the rock mass.

Geologic Influences

Stratigraphic variation in the coal measure rocks of the U.S. is the result of the depositional environments within the ancient coal-forming swamps. Constantly changing conditions in the ancient fluvial-deltaic system, in response to tectonic subsidence, depositional processes, and differential compaction of sediments, cause abrupt vertical and lateral facies changes. Rock sequences are cyclic, but rarely are entire cyclothems preserved because erosional and depositional conditions change rapidly. Variability is reflected by the lack of vertical and lateral continuity of rock units, changes in grain size, lithologic variation, and thickness of individual beds.

Superimposed on this ever-changing depositional setting are tectonic controls. Deep-seated movements along basement structures can cause slow downwarping or uplift in areas, temporarily stabilizing a depositional environment in geologic time and position. This controls the distribution of lithologies by stacking or building up units of similar lithologic composition and causing them to thin or thicken laterally. The boundary between such areas is often a zone of abrupt lithologic change occurring over a relatively short distance. Features created in this manner include paleochannels, scours, pinchouts, crevasse splays, and clay veins. These features also will control the caving, bulking, and recompaction characteristics of the roof strata, all of which affect the amount of subsidence reaching the surface.

All the factors contributing to rapid changes in lithology affect the overall strength of the rock unit. Sandstone and limestone generally are the stronger lithologies found in coal measure rocks. A depositional setting affects bed thickness. Massive units are considered more competent than laminated units; therefore, massive sandstone and siltstone are considered stronger than laminated varieties. Massive mudstones are considered stronger and more resistant to bending than fissile shales. Gradational facies contacts are stronger than abrupt erosional contacts or changes in lithology. Bedding planes and facies contacts are not always horizontal, but can be inclined and lenticular.

Depositional and structural conditions influence the mechanical characteristics of the overburden which dictate the characteristics of the subsidence trough developed on the surface. Important geotechnical properties of rocks controlling subsidence are density (compressive, tensile, cohesive, and shear strengths), bulk, elastic, and shear moduli, Poisson's ratio, and slake durability (Dunrud, 1976). These properties are controlled by the composition, texture, cement, and structure of the bedrock and surficial material. Rock masses derive much of their strength from confining stresses and attendant increases in the frictional resistance across bedding, stratification, joints, and faults (Dunrud and Osterwald, 1978). These rock and rock mass properties affect the amount of strain experienced on the surface above mining panels in the U.S. (Dahl and Choi, 1973; Kohli et al., 1980; Triplett, 1983).

Paleodepositonal Controls

Many depositionally controlled factors have been identified as influencing cavability and the resulting subsidence and associated ground movements (Bell, 1975; Hunt, 1980; Shadbolt, 1977). Horne et al. (1978) examined the depositional environments of Appalachian coal seams and found that the type of roof falls that occur was predictable. They attempted to relate specific roof problems to particular paleoenvironments (Figure 6). For example, graywacke sandstone roof is common where migrating channels existed. This usually is in an upper delta-plain fluvial or transitional lower delta-plain environment. Lag deposits of shale and coal pebbles, usually formed near the bottom of channels, may weaken the roof. Beach barrier deposits contain clear, well-cemented orthoquartzitic sandstones, but as a result of their brittle nature, are often jointed and can result in massive failures. Where sandstones are flat-bedded or interbedded with shales, roof traits depend on bed thickness. Thin beds, as well as thick beds accompanied by slickensided fractures, are prone to breaking. Beds 0.6-3.0 m (2-10 ft) thick are considered to be less prone to breaking because of the bridging effect the thick rock beam provides. These beds are characteristic of the flanks of any depositional environments prone to periodic flooding, such as the sides of distributary mouth bars and splay deposits. They are found in either lower delta-plain or transitional lower delta-plain areas.

Coarsening-upward sequences, lithologies grading from shale upwards through sandstones, are characteristic of bay fills and are found in the transitional lower delta plain. Carbonaceous black shales also are typical of the transitional lower delta plain or the lower delta plain. They may be bay-fill or lagoonal deposits, which are indicative of low-energy transgression across the surface of the coal swamp. Rock formed in this environment tends to be brittle and jointed.

Animal burrows within shales, siltstones, and sandy shales may appreciably lower the strength of the rock. Depositional environments where a large degree of animal activity might occur are those open to influxes of marine or brackish water. These would be back-barrier, lower delta-plain, and transitional lower delta-plain environments.

"Seat earths" are silty clays which served as the soil for growing plants. As such, they were subject to a great deal of disturbance by root activity. Slickensided surfaces are very common in these rocks and the combination of root penetrations, fine grain sizes, and slickensides makes for cavable roof. The areas most likely to contain seat earth are the fluvial upper delta plain and the transitional lower delta plain.

Differential compaction with consequent slickensided sandstone-shale interface is conducive to easy caving. Channel boundaries in lower delta-plain or the back-barrier environments commonly exhibit differential compaction. Slumps occur on the cutbank sides of meandering stream channels. Resulting blocks commonly are large and heavily slickensided. Tidal channels in back-barrier environments and channels of the fluvial upper delta-plain and transitional lower delta-plain environments are most likely to contain such structures.

The competency of the overburden has been reported to influence the rate and duration of subsidence movements (Brauner, 1973; Flaschentrager, 1958; Grond, 1951; Orchard and Allen, 1977; Potts, 1949; Wardell, 1954). Correlations between stratigraphy and subsidence have been investigated (Dunrud, 1976; Kapp, 1978; Shadbolt, 1977; Wardell, 1971). Statistical studies indicate that subsidence is reduced with increasing amounts of strong, competent lithologies in the overburden (Abel and Lee, 1980; Peng and Cheng, 1980; Tanandand and Powell, 1981 and 1984; Department of Mining and Minerals Engineering, 1987).

Several studies have indicated that the type of strata in the overburden influences the angle of draw (Adamak and Jeran, 1980; Crane, 1929; Drent, 1957; Dunrud, 1976; Knothe, 1957; Louis, 1922; Marr, 1959; Wardell, 1959), with smaller angles of draw being more common in stronger overburden. The bending of competent strata into mine workings can cause surface heave beyond the edge of the panel of as much as 12 mm (0.47 in.) (Auchmuty, 1931; Draper, 1964; Maize et al., 1941; Montz, 1930).

At the mine level, the amount of caving that occurs within the overburden largely determines the style of surface subsidence. This caving results from the interaction of various strata and structure in the near and

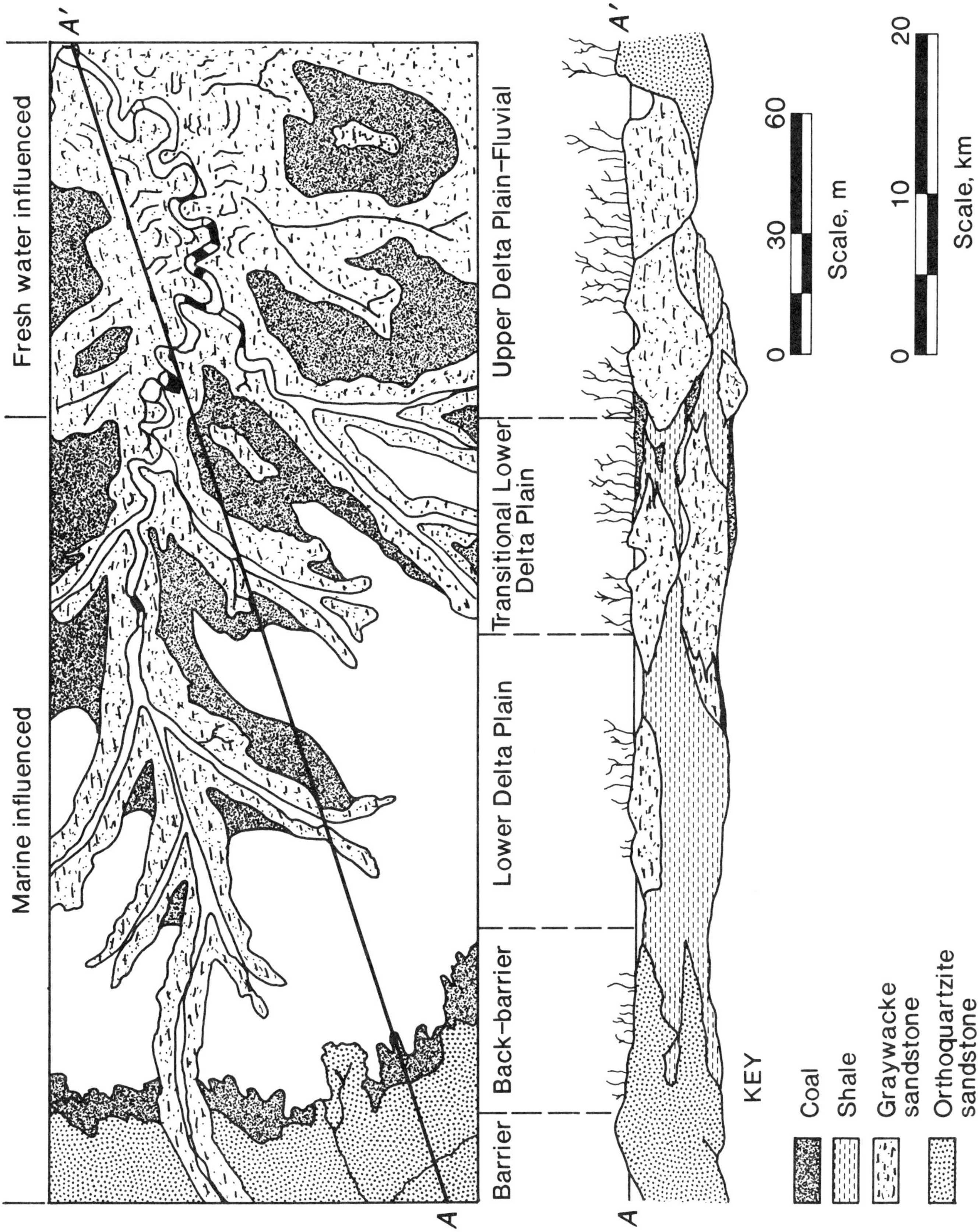

Figure 6. *Depositional environments of coal (after Horne et al., 1978).*

immediate roof. These features include paleochannels, scours, pinchouts, slickensides, crevasse splays, clay veins, joints, bedding planes, and faults.

Paleochannels are trough-like masses of cross-bedded sandstone often referred to as "sandstone rolls" (Figure 7). These features consist of sedimentary rocks which were deposited in ancient stream beds and tidal channels. In places, the channels cut down into the peat beds and subsequently were filled with sand as they were abandoned. Caving associated with these bodies results from the occurrence of weaker rocks (e.g., coal or shale poorly bonded to the adjacent sandstone channel fills). When mining occurs beneath the channel boundary, rocks along the weak interface are likely to pull away easily.

Pinch-outs are the narrowing of beds or rock in a horizontal direction until they disappear. The lateral thinning of a sedimentary stratum during deposition is a possible mode of formation of such features. In contrast, Moebs and Ellenberger (1982), note that pinch-outs often are found along the boundaries of channels, where they may be the result of the same cutting action that formed the channels. Mining beneath a pinch-out results in a cantilever rock beam which develops tensile stresses and subsequently fails.

Crevasse splays form where streams breach their natural levees, depositing sand on top of mud flats bordering the channel. Repeated episodes of flooding result in interbedded sequences of relatively thin layers of micaceous sandstones and shales. These units can be up to 9 m (30 ft) thick and have a sheet-like or lenticular appearance (Moebs and Ellenberger, 1982). Their inherent weakness is attributable to the lack of cohesion between the sandstone and shale layers and can be exacerbated by plant remains or thin carbonaceous layers on the bedding planes. This results in a sequence of isolated rock beams which are too thin to span a wide opening.

Structural Features

Structural features such as joints, faults, bedding planes, and folds are important in controlling subsidence (Alder et al., 1943; Crane, 1931; Shadbolt, 1977). Where strong, massive, jointed strata occur near the surface, localized strain concentrations are found (Whittaker and Breeds, 1977). Bedding planes influence relative movements (Crane, 1929; Robertson, 1949), especially between rock types of diverse properties, by accenting the strain across the boundary. The effect is increased when mining in inclined strata.

Lithologic distribution in the overburden controls bedding and jointing. Several studies show that joint frequency (spacing) is dependent on lithology (Babcock, 1976; McQuillan, 1973; Nickelsen, 1976; Priest and Hudson, 1976; Sterns and Freedman, 1972). Joint frequencies also have been found to decrease with increasing bed thickness (Harris et al., 1960; Hodgson, 1961). Situations that are particularly conducive to large caving volumes occur where joints are closely spaced or where joint sets parallel the mine entries of abandoned room-and-pillar mines.

Bedding planes are more frequent in sequences associated with rapid changes in rock types (Diessel and Moelle, 1965). Bedding planes often include large numbers of plant fossils which contribute to weak bonding and increased cavability.

Slickensides are one of the most common of all coal strata features. A slickensided surface is one that is polished and smoothly striated due to movement along a surface within the rock. Because of this polishing, the slickensided region forms a surface of low shear strength in the rock along which additional movement easily can take place. Slickensides may result from a number of causes. They can be found anywhere the sediments have slid past each other because of differential compaction or faulting. Milici et al. (1982) determined that a large proportion of slickensides in coal mine roof strata in Virginia resulted from tectonic

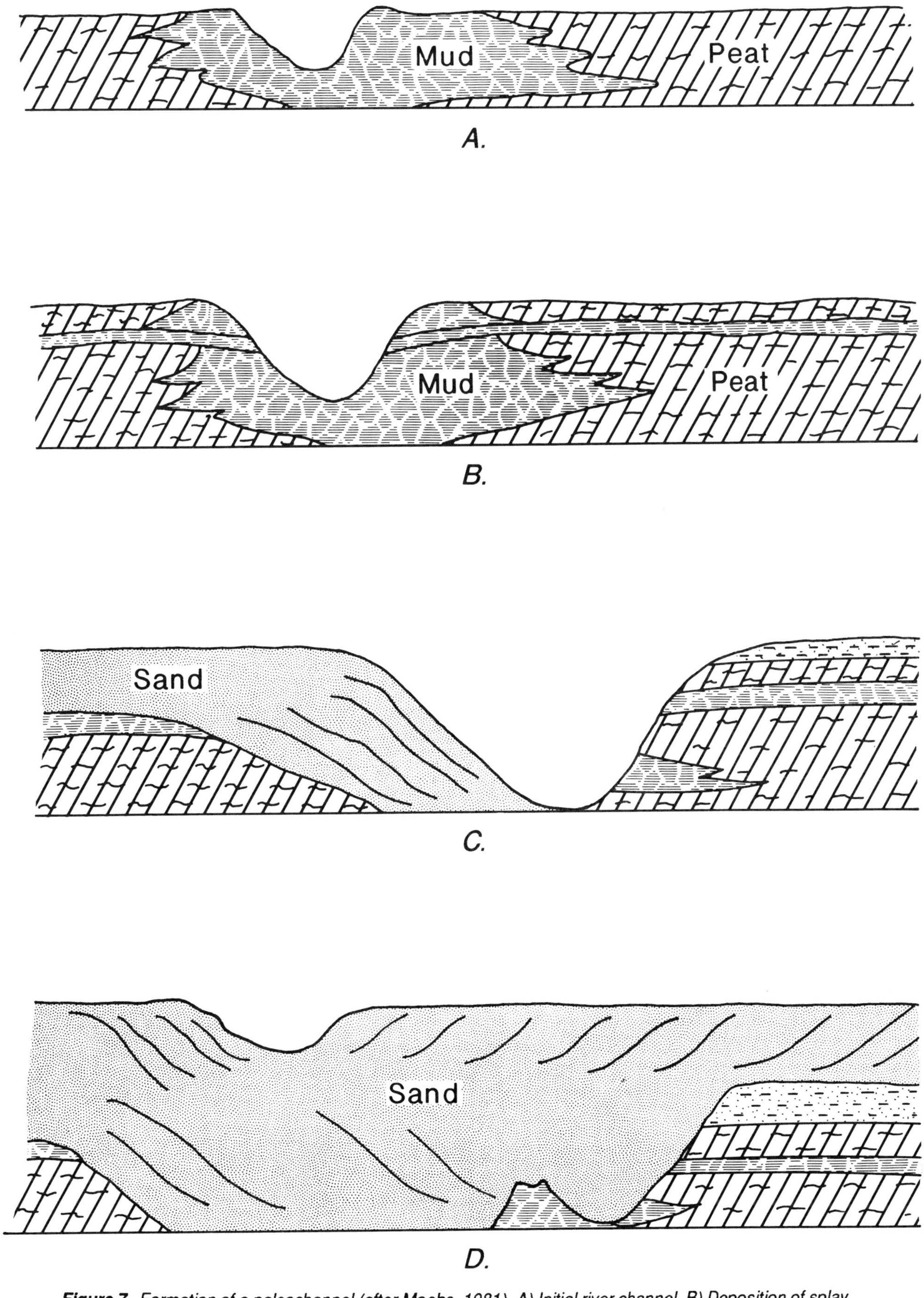

__Figure 7.__ Formation of a paleochannel (after Moebs, 1981). A) Initial river channel. B) Deposition of splay deposit due to flooding. C) Downcutting and scouring due to changes in flow regime. D) Replacement of original fine-grained material with sand.

movements. Similar features are associated with decollements elsewhere in the Appalachian Plateau (Harris and Milici, 1977).

Faults within the subsidence area commonly affect the limit of subsidence by providing planes of weakness that can localize subsidence (Briggs, 1929). In addition, they can cause the formation of surface steps in the subsidence profile with a subsequent effect on the draw angle (Kratzsch, 1983). Nelson (1981) divided faults into two groups: tectonic faults and non-tectonic faults. Tectonic faults arise from forces acting deep within the earth, whereas non-tectonic faults are local in nature and are formed by differential compaction or slumping of unlithified sediments.

Lithology

Caving height is influenced by the bulking factor, which is controlled by the lithology of the strata (Dunrud and Osterwald, 1978; Haycocks and Breeds, 1979; Louis, 1922; Piggott and Eynon, 1977; Price et al., 1969; Wardell and Wood, 1966). Abel and Lee (1980) investigated the different effects that sandstone, shale, and limestone in the overburden have on subsidence. They concluded that increased amounts of sandstone and limestone decreased the angle of draw, limestone having three times the effect of sandstone. Increased amounts of sandstone in the overlying strata were found to increase maximum subsidence. This relationship between the competency of the material and the angle of draw is due to the lower bulking factors of limestone and sandstone as compared to shale (or other weak rocks). The lower the bulking factor, the less the subsidence cavity fills in with broken rock, thereby increasing the maximum amount of subsidence, up to the limit of the size of the original underground opening. Limestone tends to bulk less than sandstone, which accounts for the difference noted above in their relative effects on subsidence.

The presence of, and distance to, rider coals are important parameters. Rider coals develop when natural stream levees are breached and sediment covers sections of adjoining swamps. When the flood waters abate, swamps re-expand and supply material for continued peat deposition. These seams, which may vary from fractions of a centimeter to several meters thick, lack tensile strength and, if located in close proximity above mine workings, can extend caving upward.

Other geologic factors which control caving include gas and water pressure in the immediate roof (Headlee, 1944). Wier (1970) also identified the presence of expandable clays, such as illite and montmorillonite, as a factor.

COAL-PRODUCING REGIONS OF THE UNITED STATES AND THEIR SUBSIDENCE CHARACTERISTICS

Major coal-producing basins in the United States can be divided on an east-west basis. The Eastern coal basins include the Appalachian Basin and the Illinois Basin. In the west, the major producing region is in the Rocky Mountain area. These regions are shown in Figure 2.

Appalachian Basin

The Appalachian Basin extends from Northeastern Pennsylvania southwest to Alabama. Coal rank ranges from high-volatile bituminous in the west and south to low-volatile bituminous to the east near the Allegheny Structural Front (Lyons, 1986). Close to the Front, rocks are folded and faulted and become separated into subbasins. The coal is of Pennsylvanian-Permian age and appears to be formed primarily by a barrier-bar type of coastal plain with deltaic sedimentation and consequent peat swamps (Englund et al., 1986; Horne et al., 1978).

Subsidence characteristics vary by subbasin. The geologic structure of the Pennsylvania coal fields tends to be relatively flat-lying beds with dips of 2°-8° (Keystone Coal Industry Manual, 1987). Slight folding has resulted in a series of relatively flat northeast-to-southwest trending synclines and anticlines. Surface topography is marked by deeply incised dendritic stream valleys. Angle of draw is approximately 15° for the limit of major surface deformation and 25° for the limit of measurable surface deformation (Moebs and Barton, 1985; Trevits and King, 1986). Sometimes large-scale downslope surface movements occur on the flanks of valleys. Tanandand and Powell (1981; 1984) found that a local increase in sandstone or limestone in the rock units would result in a decrease in maximum subsidence and also delays the rate at which the subsidence wave propagates. Similar characteristics can be found in Northern and Central West Virginia and Eastern Kentucky which have similar geology. A large number of sinkhole-type subsidence events have been recorded in Pennsylvania (Bruhn et al., 1978). It was found that this style of subsidence occurred with overburden thicknesses of less than 50 m (160 ft) and generally within 150 m (490 ft) of the outcrop.

Along the Allegheny Front in southeastern West Virginia and Southwestern Virginia, folding and faulting play a major role in the caving characteristics of the roof strata. Here angles of draw range from 12° to 34° with the relative percentages of hard rock, sandstone and limestone, controlling the amount of surface subsidence that occurs (Karmis et al., 1981). Hasenfus (1984) was able to account for the effects of remnant pillars in abandoned workings and demonstrated the difference in subsidence for these conditions when compared to longwall mines. The roof failure mechanism in this region was shown by Kane (1985) and Kane and Karmis (1986) to be significantly controlled by both stratigraphic and structural features. These include the presence of paleochannels and decollement faults with the concomitant presence of large areas of weak slickensided material. Large, massive barrier-bar sandstones, although clean, were found to be heavily jointed and prone to collapse.

In the coal fields of Tennessee, subsidence generally is very small because coal seams tend to be thin, and thick sandstone layers are common in the overburden. There are no working longwall mines in the region, and the thin seams tend to lend themselves to strip mining. Retreat room-and-pillar mining also is practiced, but what subsidence has occurred is unknown because surface lands above the coal fields are relatively unpopulated.

En echelon normal faulting, trending north-northwest and associated with graben structures, has been associated indirectly with highly fractured zones of roof rock in the Warrior Basin of Alabama (G.L. Owen, 1986, verbal communication). These normal faults commonly are connected by east-west trending zones in which the rock is densely fractured and contains water or gas under high pressure. Subsidence characteristics are variable and, as yet, not many data are available. Present longwall mining is relatively deep at 460 m to 600 m (1,500 ft to 1,970 ft) in depth (Strata Control/Rock Mechanics Dept., 1987), and subsidence ratios should be small.

Shallower, less than 300-m (984-ft) deep, workings in the Warrior Basin present a problem. Subsidence prediction in the western part of the basin is complicated by the presence of an overlap of thick Cretaceous units. Soils to the east are thinner, residual sands and clays from the underlying Pottsville Formation which simplify conditions for subsidence prediction. Eastward along the structural front, in the Cahaba coal field near Birmingham, Alabama, shallow abandoned mines in steeply dipping seams pose a significant threat to surface structures as the city expands along the neighboring ridges.

Illinois Basin

The Illinois Basin also contains coal of Pennsylvanian age. The coal is associated with marine strata, and deposition was controlled by tectonic events (Lyons, 1986). A broad coastal plain with deltas containing extensive peat swamps has been envisioned for the region (Trask and Palmer, 1986).

Surface subsidence and consequent damage are significant in the Illinois Basin due to the fact that relatively thick coal deposits and longwall mining techniques are coupled with a thick mantle of highly plastic loess soil. This basin is characterized by the presence of limestone beds and a very small angle of draw compared to other coal-producing areas.

Bauer and Hunt (1981) concluded that smaller percentages of subsidence and smaller angles of draw in Illinois, as compared to other areas, indicated that the rock overburden in Illinois is stiffer and more resistant to subsidence. Variations in subsidence factors were found to correspond to variations in regional geology. Geologic factors contributing to these variations included coal thickness, coal strength, presence or absence of thick limestones in the roof, and proximity to major structural features. Another study in Illinois (Hunt, 1980), using data compiled by Young (1916), concluded that subsidence in the Illinois Basin decreased with increasing depth and that regional geology affected the magnitude of subsidence by controlling the mechanical properties of the overburden.

Bauer and Hunt (1981) compared draw angles with percentages of shale, sandstone, and limestone in the overburden; no relationships were apparent. Although the range in the draw angles was significant (11° to 26°), lithologic variation was extremely small. Variations in sandstone and limestone characteristically were less than 20%. The draw angle, however, was found to increase in proportion to the number of geologic structures in a given area. For those areas containing such structural features as monoclines and fault systems, the average draw angle was approximately 20°. In areas not influenced by regional structural geology, the draw angle averaged about 13°. They concluded that any influence lithology may have on draw angle is concealed by the influence of geologic structures.

Rocky Mountain Region

The Rocky Mountain coal region is characterized by a number of coal fields. Coals generally are Cretaceous in age, and the major bituminous coal fields are described here.

The San Juan Basin is a northwest-trending asymmetric feature in the southern part of the Southern Rocky Mountains Province. It includes parts of northwestern New Mexico and southwestern Colorado. Dips are generally very small, less than 1°, although Fassett (1986) has recorded dips as high as 40° and older coal bearing strata may be vertical to overturned in the eastern part of the basin. Coals are of Upper Cretaceous age. Fassett and Hinds (1971) suggested that the depositional environment was a landward development of peat deposits behind a thick stack of strandline sandstone during a regressive-marine depositional regime.

The Green River Basin contains coals of primarily subbituminous rank and range up to 13 m (43 ft) in thickness (Lyons, 1986). This basin extends from southwestern Wyoming into Utah and Colorado. Strata tend to dip gently except around the Rock Springs Anticline. Dips as steep as 45° have been recorded there by Roehler (1986). Roehler (1986) used the coastal plain of South Carolina as an analogous depositional model. The basin is believed to originally have been part of a narrow lagoon between two plains. The center of the lagoon contained a forest swamp which enlarged and formed the thick peat deposits.

Subsidence in the New Mexico and Utah areas does not present as many problems as other regions simply because of a sparse population in the coal mining regions. Dunrud (1976) has described some of the effects of local geology on subsidence. In the Geneva Mine area of the Sunnyside mining district of Central Utah, he relates the failure of an outcrop barrier pillar which was designed as too small for the final stress abutment. Failure of the barrier resulted in a large cantilevered rock beam above the mined area which produced tension cracks on the surface. Tension cracks as wide as 1.5 m (5.0 ft) were observed in the massive sandstones which outcrop at the surface. In the compression zone of the beam, fractured bulges and small anticlines were observed in the sandstone where overburden depths were approximately 275 m (902 ft). In the Somerset mining district in Colorado, Dunrud (1976) suggests that burned coal already has caused surface subsidence and tension cracks prior to mining. For the two areas examined, Dunrud (1976) suggests that draw angles of 15° to 21° and a subsidence ratio of 0.31 are applicable. The width-to-depth ratio for critical extraction is 0.66. The reason for the low subsidence ratio is not clear and may be due to a temporary compression arch bridging the mined-out panels. Gentry and Abel (1978), in a study of the York Canyon Mine near Raton, New Mexico, determined a draw angle of 15.5° and a subsidence ratio of 0.7.

In Utah, in the Wasatch Plateau Field, many subsidence problems are attributed to the presence of burned-out coal seams and the resulting low-strength clinker (Mining Division, 1984). The placement of a stress abutment on barrier pillars containing a substantial portion of clinker results in crushing and surface subsidence. Up to 8 m (30 ft) of subsidence has been recorded with overburden thicknesses of 370 m (1210 ft). Subsidence ratios of 0.7 to 0.8 generally are the case, but Fry (1989, verbal communication) points out that it may take as long as five years for the subsidence process to be completed. In addition, it is necessary to mine approximately five successive 1,200-m (3,900-ft) long by 200-m (660-ft) wide longwall panels before achieving maximum subsidence. It appears that the presence of a thick (75 m (250 ft)) sandstone layer bridges any smaller caved region. This sandstone layer also may be responsible for relatively small angles of draw, often less than 20°. Exact draw angles are extremely hard to determine due to other parameters such as the clinker beds and steps in the subsidence profile caused by faults within the mines. Surface cracks exist where the sandstone outcrops. In other areas, a mudstone is present and ductile enough to prevent manifestation of cracks.

The Hanna Basin of south-central Wyoming is one of the more complex coal basins in the state. Coal-bearing formations are the Upper Cretaceous-Paleocene Ferris and the Paleocene-Lower Eocene Hanna formations. Besides coal, cross-bedded conglomeratic sandstones, siltstones, claystones, and shales make up the stratigraphic package. The depositional model is that of a fluvial environment replete with meandering streams (Glass and Roberts, 1980). The basin is the northwest part of a structural trough which is divided by a northeast-southwest trending anticline (Keystone, 1987). Faults are common.

Subsidence in Wyoming generally is confined to old, abandoned, relatively shallow (30-60 m (100-200 ft) in depth) underground workings. Active mines are all surface operations. Most of the subsidence problems stem from collapse of roof strata or pillar crushing. Pillar punching generally is not a problem. Two styles of subsidence have been observed: 1) troughing, a sagging of the surface, and 2) chimney subsidence which is expressed as conical depressions or open sinkholes at the surface (Karfakis et al., 1988). The type of subsidence was found to be a function of overburden depth. Sinkholes were much more prone to form when overburden was less than 30 m (100 ft) thick (Murphy, 1986). Karfakis (1986) analyzed chimney subsidence in the area above abandoned room-and-pillar mines in the Hanna Basin and determined that such subsidence occurred with overburden thicknesses up to 73 m (240 ft), with most occurring where the overburden was less than 30 m (100 ft) thick. Karfakis (1986) suggested bulking of caved material limited chimney development in thicker overburden, although burning coal seams may have an influence on how high caving will extend.

SUMMARY

The coal fields of the United States are characterized by different styles of subsidence. The major factors governing subsidence are geology and mining method. Geologic influences are dependent on the interrelationship of the depositional and tectonic history of a basin. This determines the overburden characteristics of the stratigraphic package. Thickness and depth of the coal, as well as rock mass properties, significantly affect the caving characteristics of the overburden. Mining methods and surface topography also play major roles.

ACKNOWLEDGMENTS

We wish to acknowledge the contributions of D. L. Boreck, U.S. Bureau of Mines Denver Research Center, and R.C. Fry, Utah Power Company, Salt Lake City, Utah, who provided many useful suggestions on improving the manuscript.

REFERENCES

Abel, J.F., Jr., and F.T. Lee, 1980, Lithologic controls on subsidence: New York, Society of Mining Engineers-AIME, Preprint 80-314, 16 p.

Adamak, V. and P.W. Jeran, 1981, Evaluation of existing predictive methods for mine subsidence, in S.S. Peng and M. Harthill, eds., Proceedings of the Workshop on Surface Subsidence Due to Underground Mining, Morgantown, West Virginia, November 30-December 2, 1981: Morgantown, West Virginia University, p. 88-97.

Alder, H., A. Walker, and L. Walker, 1942, Subsidence and its bearing on mining methods: Institution of Mining Engineers, Transactions, v. 102, n. 2, p. 302-326 and 421-424.

Allen, A.S., 1978, Basic questions concerning mine subsidence in the United States: American Association of Engineering Geologists, Bulletin, v. 15, n. 2, p. 147-161.

Auchmuty, R.L., 1931, Subsidence and ground movement in a limestone mine caused by longwall mining in a coal bed below: Transactions, American Institute of Mining, Mettalurgical, and Industrial Engineers, v. 94, p. 27-50.

Aynsley, W.J., and G. Hewitt, 1960, Subsidence observations over shallow workings, including pneumatic stowing and rapidly-advancing faces: The Mining Engineer, v. 120, p. 552-569.

Babcock, E.A., 1976, Bedrock jointing on the Alberta Plains, in R.A. Hodgson, S.P. Gay, Jr., and J.Y. Benjamins, eds., Proceedings of the First International Conference on New Basement Tectonics, Salt Lake City, Utah, June 3-7, 1974: Utah Geological Association Publication No. 5, p. 142-152.

Bauer, R., and S.R. Hunt, 1981, Strain and time characteristics of subsidence from coal mining in Illinois, in S.S. Peng and M. Harthill, eds., Proceedings of the Workshop on Surface Subsidence Due to Underground Mining, Morgantown, West Virginia, November 30-December 2, 1981: Morgantown, West Virginia University, p. 207-218.

Bell, F.G., 1975, Site investigations in areas of mining subsidence: London, Newnes-Butterworths, 168 p.

Brauner, G., 1973, Subsidence due to underground mining (in two parts). 1. Theory and practices in predicting surface deformation: U.S. Bureau of Mines Information Circular IC 8571, 56 p.

Briggs, H. and W. Ferguson, 1933, Investigation of mining subsidence at Barbauchlaw Mine, West Lothian: Institute of Mining Engineers, Transactions, v. 85, p. 303-334.

Bruhn, R.W., M.O. Magnuson, and R.E. Gray, 1978, Subsidence over the mined-out Pittsburgh Coal: New York, American Society of Civil Engineers, Preprint 3253.

Cortis, S.E., 1969, Coal mining and protection of surface structures are compatible: Mining Congress Journal, v. 55, n. 6, p. 84-89.

Crane, W.R., 1929, Subsidence and ground movement in the copper and iron mines of the Upper Peninsula, Michigan: U.S. Bureau of Mines Bulletin 295, 65 p.

Crane, W.R., 1931, Essential factors influencing subsidence and ground movement: U.S. Bureau of Mines Information Circular IC 6501, 14 p.

Dahl, H.D., and D.S. Choi, 1973, Some case studies of mine subsidence and its mathematical modeling, in E.R. Hoskins, Jr., ed., Proceedings of the 15th U. S. Symposium on Rock Mechanics, Custer State Park, South Dakota, September 17-19, 1973: New York, American Society of Civil Engineers, p. 1-22.

Department of Mining and Minerals Engineering, 1987, Prediction of ground movements due to underground mining in the Eastern United States coalfields: Volume I, Development of prediction methods: Blacksburg, Virginia Polytechnic Institute and State University, 205 p.

Diessel, C.F.K., and K.H.R. Moelle, 1965, The application of analysis of the sedimentary and structural features of a coal seam and its surrounding strata to the operations of mining, in General Proceedings of the 8th Commonwealth Mining and Metallurgical Congress: London, Institute of Mining and Metallurgy, p. 837-859.

Draper, J.C., 1964, Surface movement in the vicinity of pillars left in gob areas: Society of Mining Engineers-AIME, Preprint 64-FM3, 27 p.

Drent, S., 1957, Time-curves and the thickness of overlying strata: Colliery Engineering, v. 34, p. 271-278.

Dunrud, C.R., 1976, Some engineering geologic factors controlling coal mine subsidence in Utah and Colorado: U.S. Geological Survey Professional Paper 969, 39 p.

Dunrud, C.R., and F.W. Osterwald, 1978, Coal mine subsidence near Sheridan, Wyoming: Association of Engineering Geologists, Bulletin, v. 15, n. 2, p. 175-190.

Englund, K.J., J.F. Windolph, Jr., and R.E. Thomas, 1986, Origin of thick, low-sulphur coal in the Lower Pennsylvanian Pocahontas Formation, Virginia and West Virginia, in P.C. Lyons and C.L. Rice, eds., Paleoenvironmental and tectonic controls in coal-forming basins of the United States: Geological Society of America, Special Paper 210, p. 49-60.

Fassett, J.E., 1986, The non-transferability of a Cretaceous coal model in the San Juan Basin of New Mexico and Colorado, in P.C. Lyons and C.L. Rice, eds., Paleoenvironmental and Tectonic Controls in Coal-Forming Basins of the United States: Geological Society of America, Special Paper 210, p. 155-171.

Fassett, J.E., and J. S. Hinds, 1971, Geology and fuel resources of the Fruitland Formation and Kirtland Shale of the San Juan Basin, New Mexico and Colorado: U.S. Geological Survey Professional Paper 676, 76 p.

Flaschentrager, H., 1958, Considerations of ground movement phenomena: Colliery Engineering, v. 35, n. 8, p. 391-397.

Gentry, D.W., and J.F. Abel, Jr., 1978, Surface response to longwall coal mining in mountainous terrain: Association of Engineering Geologists, Bulletin, v. 15, n. 2, p. 191-220.

Glass, G.B., and J.T. Roberts, 1980, Coal and coal-bearing rocks of the Hanna Coalfield: Wyoming Geological Survey, Report of Investigations No. 22, 43 p.

Grond, G.J.A., 1951, The precise topographical measurements in coal mine underground works, in Proceedings of the International Conference About Rock Pressure and Support in the Workings, Liege, Belgium, April 24-28, 1951: Paper A3, 15 p.

Harris, L.D., and R.C. Milici, 1977, Characteristics of thin-skinned style of deformation in the Southern Appalachians and potential hydrocarbon traps: U.S. Geological Survey, Professional Paper 1018, 40 p.

Harris, J.F., G.L. Taylor, and J.L. Walper, 1960, Relation of deformational fractures in sedimentary rocks to regional and local structure: American Association of Petroleum Geologists Bulletin, v. 44, n. 12, p. 1853-1873.

Hasenfus, G.J., 1984, The prediction of surface subsidence due to room and pillar mining in the Appalachian Coalfield: Blacksburg, Virginia Polytechnic Institute and State University, Unpublished M.S. Thesis, 327 p.

Haycocks, C. and C.D. Breeds, 1979, Ground control simulation during longwall mining, in T.J. O'Neil, ed., Proceedings of the 16th Symposium on Application of Computers and Operations Research in the Mineral Industry, Tucson, Arizona, October 17-19, 1979: New York, Society of Mining Engineers--AIME, p. 528-536.

Headlee, A.J.W., 1944, Fracture zones in mine strata: Mining Congress Journal, v. 30, p. 57-60.

Hodgson, R.A., 1961, Regional study of jointing in Comb Ridge-Navajo Mountain Area, Arizona and Utah: American Association of Petroleum Geologists Bulletin, v. 45, n. 1, p. 1-38.

Horne, J.C., J.C. Ferm, F.T. Caruccio, and B.P. Baganz, 1978, Depositional models in coal exploration and mine planning in Appalachian Region: American Association of Petroleum Geologists Bulletin, v. 62, n. 12, p. 2379-2411.

Hunt, S.R., 1980, Surface subsidence due to coal mining in Illinois: Urbana, University of Illinois, Unpublished Ph.D. Thesis, 129 p.

Jeran, P.W., and J.H. Jansky,, 1983, A guide to geologic features in coal mines in the northern Appalachian coal basin: U.S. Bureau of Mines Information Circular IC 8918, 16 p.

Kane, W.F., 1985, Geological and geotechnical controls on the stability of coal mine entries: Blacksburg, Virginia Polytechnic Institute and State University, Unpublished Ph.D. Dissertation, 304 p.

Kane, W.F., and M. Karmis, 1986, Geologic and geotechnical controls on the stability of coal mine entries, in M. Karmis, ed., Proceedings of the International Symposium on Application of Rock Characterization Techniques in Mine Design, New Orleans, Louisiana, March, 1986: Littleton, Colorado, Society of Mining Engineers, p. 124-132.

Kapp, W.A., 1978, Subsidence investigations in the Northern Coalfield, New South Wales, and their application to the design of mine layouts in residential areas, in Proceedings of the 11th Commonwealth Mining and Metallurgical Congress, Hong Kong, May, 1978: London, Institute of Mining and Metallurgy, Paper 32, 11 p.

Karmis, M., C. Haycocks, B. Webb, and T. Triplett, 1981, The potential of the zone area method for mining subsidence prediction in the Appalachian Coalfield, in S.S. Peng and M. Harthill, eds., Proceedings of the Workshop on Surface Subsidence Due to Underground Mining, Morgantown, West Virginia, November 30-December 2, 1981: Morgantown, West Virginia University, p. 48-62.

Karfakis, M.G., 1986, Chimney subsidence--A case study, in Proceedings of the 27th U.S. Symposium on Rock Mechanics, Tuscaloosa, Alabama: Tuscaloosa, University of Alabama, p. 275-282.

Karfakis, M.G., S. Barnard, and J. Murphy, 1988, Subsidence abatement projects in Wyoming--An overview: New York, American Society of Civil Engineers, Geotechnical Special Publication No. 19, p. 33-52.

Keystone Coal Industry Manual, 1987: New York, McGraw-Hill, Inc., 1244 p.

Knothe, S., 1957, Observations of surface movements under influence of mining and their theoretical interpretation, in Proceedings of the European Congress on Ground Movements, Leeds, England, April 9-12, 1957: p. 210-218.

Kohli, K.K., Peng, S.S., and R.E. Thill, 1980, Surface subsidence due to underground longwall mining in the northern Appalachian coalfields: Society of Mining Engineers-AIME, Preprint 80-53, 8 p.

Kratzsch, H., 1983, Mining subsidence engineering: New York, Springer-Verlag, 543 p.

Louis, H., 1922, A contribution to the theory of subsidence: Institution of Mining Engineers, Transactions, v. 64, p. 257-273.

Lyons, P.C., 1986, Introduction and summary, in P.C. Lyons and C.L. Rice, eds., Paleoenvironmental and tectonic controls in coal-forming basins of the United States: Geological Society of America, Special Paper 210, p. 1-8.

Maize, E.R., E. Thomas, and H.P. Greenwald, 1941, Studies of roof movement in coal mines, 4. Study of subsidence of a highway caused by mining coal beneath: U.S. Bureau of Mines Report of Investigations RI 3562, 11 p.

Marr, J.E., 1959, The estimation of mining subsidence: Colliery Guardian, v. 198, p. 345-352.

McQuillan, H., 1973, Small-scale fracture density in Asmari Formation of Southwest Iran and its relation to bed thickness and structural setting: American Association of Petroleum Geologists Bulletin, v. 57, n. 12, p. 2367-2385.

Milici, R.C., T.M. Gathright II, B.W. Miller, and M.R. Gwin,, 1982, Geologic factors related to coal mine roof falls in Wise County, Virginia: Appalachian Regional Commission, Contract Report No. CO-7232-80-I302-0206, 103 p.

Mining Division, 1984, 1983 Subsidence report: Huntington, Utah, Utah Power & Light Company, Unpublished Internal Report, 13 p.

Moebs, N.N., 1981, The geologic character of some coal wants at the Westland Mine in Southwestern Pennsylvania: U.S. Bureau of Mines Report of Investigations RI 8555, 25 p.

Moebs, N.N., and T.M. Barton, 1985, Short-term effects of longwall mining on shallow water sources, in Mine Subsidence Control, Proceedings of Bureau of Mines Technology Transfer Seminar, Pittsburgh, Pennsylvania, September 19, 1985: U.S. Bureau of Mines Information Circular 9042, p. 13-24.

Moebs, N.N., and J.E. Ellenberger, 1982, Geologic structures in coal mine roof: U.S. Bureau of Mines Report of Investigations RI 8620, 16 p.

Montz, H.W., 1930, Subsidence from anthracite mining: Transactions, American Institute of Mining Engineers, v. 88, p. 98-134.

Murphy, J., 1986, Abandoned mine lands subsidence mitigation in Wyoming, a land use approach, in M.G. Karfakis, ed., Proceedings of the Governor's Workshop on Mine Subsidence, Laramie, Wyoming, October 31, 1986: State of Wyoming, Department of Environmental Quality, p. 6-1 to 6-35.

Nelson, R.A., and R.B. Nance, 1980, Geologic mapping of roof conditions, Crown II mine, Macoupin County, Illinois: Society of Mining Engineers--AIME, Preprint No. 80-308, 9 p.

Nickelsen, R.P., 1976, Early jointing and cumulative fracture patterns, in R.A. Hodgson, S.P. Gay, Jr., and J.Y. Benjamins, eds., Proceedings of the First International Conference on New Basement Tectonics, Salt Lake City, Utah, June 3-7, 1974: Utah Geological Association Publication No. 5, p. 193-199.

Orchard, R.J., and W.S. Allen, 1975, Time dependence in mining subsidence, in Proceedings of the Symposium on Minerals and the Environment, London, England, June 4-7, 1974: London, Institute of Mining and Metallurgy, p. 643-659.

O'Rourke, T.D., and S.M. Turner, 1979, Longwall subsidence patterns: A review of observed movements, controlling parameters, and empirical relationships: Ithaca, New York, Cornell University, Geotechnical Engineering Report 79-6, 82 p.

Peng, S.S., and S.L. Cheng, 1980, Prediction of surface subsidence profile due to underground coal mining: Morgantown, West Virginia University, Department of Mining Engineering, Technical Report 80-5, 27 pp.

Perz, W., 1958, Subsidence observations in Austria: Colliery Engineering, v. 35, n. 11, p. 479-482 and 533-535.

Piggott, R.J., and P. Eynon, 1977, Ground movements arising from the presence of shallow abandoned mine workings, in J.G. Geddes, ed., Proceedings of the Conference on Large Ground Movements and Structures, Cardiff, Wales, July 4-7, 1977: London, Halsted Press, p. 749-780.

Pottgens, J.J.L., 1978, Ground movements by coal mining in the Netherlands, in S.K Saxena, ed., Proceedings of the International Conference on Evaluation and Prediction of Subsidence, Pensacola Beach, Florida, January 15-20, 1978: New York, American Society of Civil Engineers, p. 267-282.

Potts, E.L.J., 1949, Underground and surface strata movements in mining areas: Colliery Guardian, v. 179, p. 425-430.

Price, D.G., A.B. Malkin, and J.L. Knill, 1969, Foundations of multi-story blocks on the coal measures with special reference to old mine workings: Quarterly Journal of Engineering Geology, v. 1, p. 271-322.

Priest, S.D., and J.A. Hudson, 1976, Discontinuity spacings in rock: International Journal Rock Mechanics, Mining Sciences, and Geomechanics Abstracts, v. 13, p. 135-148.

Powell, L.R., 1989, Assessment of geological controls on longwall mine subsidence in the northern Appalachian Basin: U.S. Bureau of Mines, Twin Cities Research Center, Unpublished Report, 31 p.

Robertson, T., 1949, The geological aspects and their relationship to town planning in County Durham: Colliery Guardian, v. 179, p. 575-578.

Roehler, H.W., 1986, McCourt Sandstone Tongue and Glades Coal Bed of the Rock Springs Formation, Wyoming and Utah, in P.C. Lyons and C.L. Rice, eds., Paleoenvironmental and tectonic controls in coal-forming basins of the United States: Geological Society of America, Special Paper 210, p. 141-154.

Shadbolt, C.H., 1977, Mining subsidence--Historical review and state of the art, in J.G. Geddes, ed., Proceedings of the Conference on Large Ground Movements and Structures, Cardiff, Wales, July 4-7, 1977: London, Halsted Press, p. 705-748.

Sterns, D.W., and M. Friedman, 1972, Reservoirs in fractured rock: American Association of Petroleum Geologists, Memoir 16, p. 82-106.

Strata Control/Rock Mechanics Dept., 1987, Strata control profile: Brookwood, Alabama, Jim Walter Resources, Inc., Unpublished Internal Report, 404 p.

Tanandand, S., and L.R. Powell, 1981, Consideration of overburden lithology for subsidence prediction, in S.S. Peng and M. Harthill, eds., Proceedings of the Workshop on Subsidence Due to Underground Mining, Morgantown, West Virginia, November 30-December 2, 1981: Morgantown, West Virginia University, p. 17-29.

Tanandand, S., and L.R. Powell, 1984, Influence of lithology on longwall mining subsidence: Mining Engineering, v. 36, p. 1666-1671.

Trask, C.B., and J.E. Palmer, 1986, Structural and depositional history of the Pennsylvanian System in Illinois, in P.C. Lyons and C.L. Rice, eds., Paleoenvironmental and tectonic controls in coal-forming basins of the United States: Geological Society of America, Special Paper 210, p. 63-77.

Trevits, M.A., R.L. King, and B.V. Johnson, 1986, Overview of the Bureau of Mines subsidence research program, Preprint for the 88th Annual General Meeting of C.I.M., Montreal, Quebec, May 11-15, 1986: Montreal, Canadian Institute of Mining and Metallurgy, 14 p.

Triplett, T.L., 1983, Ground movements and deformation above mined panels in Appalachia: Blacksburg, Virginia Polytechnic Institute and State University, Unpublished M.S. Thesis, 216 p.

Wardell, K., 1954, Some observations on the relationship between time and subsidence: Institution of Mining Engineers, Transactions, v. 113, p. 471-483.

Wardell, K., 1959, The problems of analyzing and interpreting observed ground movement: Colliery Engineering, v. 36, p. 529-538.

Wardell, K., 1971, The effects of mineral and other underground excavations on the overlying ground surface, in Proceedings of the Conference on Geological and Geographical Problems of Areas of High Population Density, Washington, District of Columbia, 1970: Association of Engineering Geologists, p. 201-217.

Wardell, K., and J.C. Wood, 1966, Ground instability problems arising from the presence of old shallow mine workings: Proceedings of the Midland Soil Mechanics and Foundation Engineering Society, v. 7, p. 5-30.

Wier, C.E., 1970, Factors affecting coal mine roof rock in Sullivan County, Indiana: Proceedings of the Indiana Academy of Science, v. 79, p. 263-269.

Whittaker, B.N., and C.D. Breeds, 1977, The influence of surface geology on the character of mining subsidence, in Proceedings of the Conference on Geotechnics of Structurally Complex Formations, Capri, Italy: Milan, Italy, Associasione Geotechnica, p. 459-468.

Whittaker, B.N., and D.J. Reddish, 1989, Subsidence, prediction and control: New York, Elsevier, 528 p.

Young, L.E., 1916, Surface subsidence resulting from coal mining: Illinois State Geological Survey, Mining Investigation Bulletin 17, 112 p.

Chapter 6

Coal Quality Concerns

PREFACE

Coal quality has a direct impact upon the economics and, therefore, the mineability of a coal deposit. Rank of the coal is of concern for heating value, metallurgical properties, and gas content among other variables. Elements and compounds, such as sulfur and trace metals, contained in the coal and liberated during combustion and other processes, have become of increasing concern for air and water quality, especially downstream of power plants and abandoned mines. Revised clean air legislation in particular may make some of the higher sulfur-content coal seams uneconomic to use or at least require blending with low-sulfur coal or use of more efficient stack scrubbing equipment to meet emissions standards. Other elements in coal can cause other problems, such as sodium fouling of boilers and leaching of toxic metals into the groundwater from ash and slag after coal use.

The overview paper by Deul reviews the geological bases for variations in quality parameters in coal and limitations on beneficiation ("cleaning") of coal. The basic theme of geological origins of such variations is built upon by the other four papers.

Winston considers the causes of variations in several parameters over much of the Warrior coal basin, with emphasis upon the effect of those variations on mining and coalbed methane resources. Likewise, Goodarzi and Van der Flier-Keller look at the big picture for quality variations in Western Canadian coals. More areally restricted analyses are presented in the other two papers. Hower et al. consider the effects of geologic structure and deformation contemporaneous with coal deposition on coal quality in Eastern Kentucky. The finer details of coal structure and petrology as they affect quality variations in West Virginia are reported by Martin and Cavaroc.

Cross-Reference

The gross coal quality (i.e., rank) is a matter of concern for coalbed methane exploration and development as well, so some related material will be found in that chapter. As noted, coal quality affects reserves and mineability. Therefore, the Reserve Estimation papers, especially the one by Treworgy, will be of interest in connection with this chapter.

The environmental impact of coal utilization is intimately related to the coal quality and the compositions of spoil material and other nearby sediments. Finkelman and Brown view the elements in coal as both a hazard and a potential bonus in some cases. Püttmann et al. consider the potential impact of hydrocarbons from coal and its mined overburden on air and water quality. Wildeman reviews how compositional variations in coal impact production of acid drainage, and Klusman and Machemer discuss how to clean up what has been mobilized in the water. Finally, both Moran et al. and Beaver et al. consider how coal wastes and their components affect groundwater quality and open pit mine reclamation and reuse.

COAL QUALITY AND GEOLOGY

Maurice Deul
Department of Geology and Planetary Science
University of Pittsburgh
Pittsburgh, PA 15260

ABSTRACT

Coal is a rock with a wide range of compositions. Quality is more a matter of utilization specifications than of geological considerations. The principal concerns of utilization are the quantity and composition of the ash-forming ingredients and the content and modes of occurrence of sulfur in coal. Some minerals are common constituents of coal, and their occurrence depends upon the environments of deposition and on the geochemical transformations that proceed during diagenesis, deep burial, and subsequent geologic history. If the minerals in coal are present as large discrete particles, they can be removed from broken coal during beneficiation; this becomes difficult to impossible if the minerals are finely dispersed. Organically combined sulfur cannot be removed. This paper will attempt to review the geologic factors that generally influence the quality of coals.

INTRODUCTION

Coal is a rock and as such is subject to all the factors that form and transform rocks. This often is overlooked when coal is viewed mainly as a fuel and as an object of commerce. Because coal has a wide range of chemical compositions and of physical properties, there may be some confusion about the descriptive terms "rank" and "quality". Coal rank indicates the degree of metamorphism to which coal, as a rock, has been subjected. Therefore, a low-rank coal, like lignite, has not been subject to much thermal maturation, and a high-rank coal, like anthracite, has had a high degree of thermal maturation. However, high quality depends upon many other factors, most of which relate to utilization of the coal. There are high-quality, low-rank coals such as lignites which have a very low mineral matter content. There also are low-quality, low-rank coals which, because of weathering may have a high mineral matter content and excessively high moisture content. Similarly, there are low-quality, high-rank coals which have a very high carbon content, but also have a high content of organically combined sulfur (S) or a high content of fine-grained, disseminated mineral matter.

Coal is essentially a near end-member of a homologous series with 100% carbonaceous matter at one end of the series and with 100% mineral matter at the other end. This is tacitly recognized by chemists in their common usage of the mineral-matter-free or moisture-ash-free basis for treatment of coal analyses in experimental work and in process industries.

Whenever classification of coal has been a subject for discussion or study, the mineral matter content of coal is mathematically removed from consideration -- and rightly so for such a purpose. However, mathematical elimination of mineral matter does not eliminate the fundamental problem of mineral matter in coal.

Minerals are an integral part of a coal seam. Once this fact gains general acceptance, the genetic relations of the minerals found in coal can be properly appreciated. Many papers on the kinds of minerals, the distribution of minerals, and the mode of occurrence of minerals in coal have been published. The biochemical and geochemical origins of ash-forming ingredients in coal, the fate of mineral elements leached from degraded plants, and the origin of minor elements in coal are the subject of an earlier report (Deul, 1958).

Coal is a rock which is of great economic value and its quality is largely determined by the percentage of minerals present. Most persons concerned with coal are familiar with the mineral matter content of coal only through the residue remaining after combustion. However, geologists are concerned with coal as a rock as well as its value as a commodity, and in mapping coal formations throughout the world, they are faced with the problem of whether what appears to be coal should be called coal or something else. In such a situation, Schopf's (1956) definition of coal becomes very important. Here is his definition:

> "Coal is a readily combustible rock containing more than 50 percent by weight and more than 70 percent by volume of carbonaceous material formed from compaction or induration of variously altered plant remains similar to those in peaty deposits. Differences in the kinds of plant materials (type), in degree of metamorphism (rank) and range of impurity (grade) are characteristic of varieties of coal."

On a weight basis, Schopf (1956) extends the definition of coal to include one half of the homologous series discussed at the beginning of this paper (Figure 1). The lateral gradation of a coal seam with only a relatively small percentage of mineral matter into a carbonaceous shale is not uncommon in Cretaceous and later coal seams of the Western United States. Many thinly bedded deposits are not considered to be coal seams nor would they be included as such according to Schopf's definition. Yet these kinds of deposits probably contain a far greater geologic reserve of carbonaceous material than do the coal seams now included in reserve data compiled by the U.S. Geological Survey.

Geologic factors of sedimentation are the primary controls of the mineral content of coal seams. Purely chemical control of mineral deposition probably occurs only during the peat stage and during the early diagenetic stages of the coalification process. During the subsequent history of a coal seam, certain physical factors become important in controlling mineral deposition, but these operate directly in conjunction with chemical processes.

Coal is a product of geological as well as biological agencies. All too often the geological aspect of coal is ignored. In general, coal technologists and fuel chemists dealing with the whole coal as a substance consider ash-free or mineral matter-free coal as a starting point for processing or for calculation. This is a snare and a delusion unless one always keeps in mind the fact that mineral-matter-free or ash-free coal is only a concept and that the ideal is approached on rare occasions only by careful separation of selected vitrain.

It is entirely improper and quite inaccurate to consider minerals deposited with plant materials in a coal swamp as adventitious matter. Every naturally occurring body of water, from a droplet on the end of a stalactite to an ocean basin, is a suitable environment for mineral deposition. The very existence of coal is dependent upon a rare sequence of geologic events. Each stage contributing to the transformation of living plant material into coal also influences the migration of what commonly are termed mineral elements.

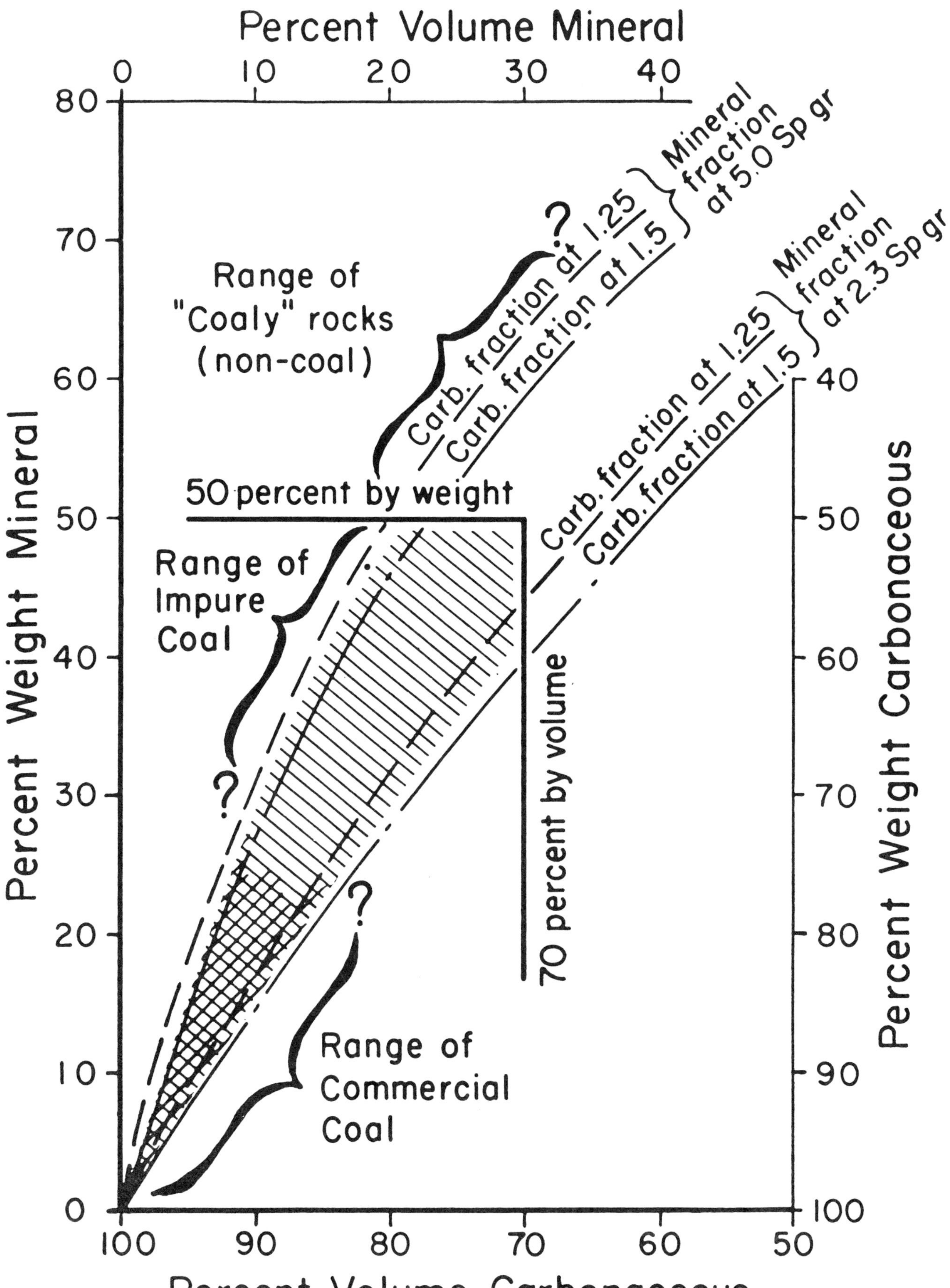

Figure 1. Relation of volume and weight of carbonaceous and mineral mixtures (from Schopf, 1956).

GENERAL CONSIDERATIONS OF ASH-FORMING INGREDIENTS

Total mineral matter in coal is taken to be the summation of all elements present which are not constituents of the coal substance — the organic matter itself — which consists of carbon (C), hydrogen (H), oxygen (O), nitrogen (N), and S.

There is a difference between "minerals" in coal and "mineral matter" in coal. The distinction is an actual one and is not only theoretical. "Minerals", in the geological sense, have distinct properties and, in large part, can be physically separated from the organic coal substance, the limitation of separation being physical distribution. "Mineral matter" in coal includes the minerals in coal and, in addition, should include those chemically bound elements other than C, H, O, N, and S. Some of these elements, like fluorine (F) and chlorine (Cl), are largely removed during combustion or ashing. Other elements, like boron (B) and beryllium (Be), make a large contribution to the total ash content because of the oxygen combined with them. In contrast, some C may be present as carbonates in minerals. Certainly H and O are present as water or hydroxide in clays. Additional O is present in the silica and alumina and other oxides in minerals. Nitrogen is rare in minerals in coal. Sulfur as pyrite and as sulfates usually exceeds organic S.

An ultimate analysis for C, H, S, N, and O accurately represents the composition of the organic matter in coal only when corrections are applied to account for those elements present in the minerals in the coal. Unless a complete mineral analysis is made to determine the composition and the amounts of minerals present in the coal, the corrections can be made only from formulas applied to the ash content of the coal. In the crudest of estimations, the percent ash is equated to mineral matter. Parr, among others, recognized this as only a gross approximation and made refinements by applying corrections based on the S content of the coal and sulfate in the ash together with other factors.

King et al. (1936) added factors which were varied depending upon experience with specific British coals. Millot (1958) has further modified the King et al. (1936) formula to specifically compensate for mineral matter in carbonaceous shales associated with British coals. These corrections have been given much attention by the fuel scientist and fuel engineer concerned with the inherent fuel value of coal as well as by the research chemist concerned with an accurate analysis of the reactive organic materials.

A thorough investigation of the nature of ash-forming ingredients in coal must consider the fact that the environment of deposition of plant debris — the coal swamp or peat bog — is in essence a medium in which chemical reactions are occurring. Materials are being dissolved and removed from the system; other substances are being added. Some of the materials added are relatively inert whereas others are highly reactive. The reaction vessel is large: for a small peat bog, on the order of acres in area, and for a Carboniferous coal swamp, thousands of square miles. Not all the reactions are proceeding simultaneously, nor will all of the reactions necessarily take place within a given limited area in the coal swamp. However, the summation of these reactions should yield a picture which can be logically interpreted.

Thus, the possible origins of ash-forming ingredients in coal are as follows:

1) From elements originally present in coal-forming plants and not lost during degradation and coalification.

2) From elements combined with organic matter during the peat stage.

3) From elements combined with organic matter during compaction and burial.

4) From elements combined with organic matter during weathering and oxidation.

5) From finely divided minerals deposited in the coal swamp.

6) From minerals formed in the coal during the peat state and during subsequent geologic history.

The products of these six sources of ash-forming ingredients can be related to terms commonly used by coal technologists. It previously has been suggested by Deul (1955) that the ash-forming ingredients consist of mechanically separable minerals, inseparable minerals, and non-volatile elements organically combined, complexed, or base-exchanged. The use of such terms is better than use of terms like "inherent ash" or "fixed ash", which to many persons implies that the ash is that derived from the original plant substance.

Separable ash and inseparable ash are terms essentially incorrect. One does not separate ash from unburned coal; one only can separate ash-forming ingredients. Anyone who has worked with high-ash coals and attempted to make appropriate corrections in an ultimate analysis will readily understand the inherent inaccuracy in using such a concept.

All ash-forming ingredients are not necessarily minerals. A certain percentage of them are bound with the organic matter as compounds or as complexes.

GEOLOGICAL FACTORS

The accumulation of peat in sufficient quantities to permit the ultimate formation of a coal seam of commercial importance requires a degree of stability short of equilibrium. Organic material must accumulate at a rate somewhat greater than that at which complete degradation occurs. However, minerals are continuously being supplied to the swamp, and with only a relatively small percentage of organic matter accumulating a carbonaceous shale forms, rather than a coal seam. A coal swamp, throughout its life, is a basin of deposition into which minerals are being brought by water and wind. These are mainly hydromica and kaolin minerals with minor amounts of quartz and traces of resistant accessory detrital minerals such as zircon, garnet, tourmaline, and other minerals, as reported by Gauger et al. (1934).

The clay minerals that occur as persistent partings in so many coal seams are typical of the geological control of sedimentation operating on a very large scale. The so-called "coal balls", which are large masses of limestone or dolomite found in great abundance within certain coal seams, owe their presence in coal to geological controls.

There are many references to occurrences of ore minerals, uranium (U), and other rare elements in coal seams, but except for pyrite and marcasite, most of these can be related to epigenetic processes of enrichment and deposition. Because coal is a rock, all the geologic processes that can modify or alter rocks also can act on coal seams. For example, there is no reason why zinc (Zn) minerals or lead (Pb) minerals may not be found in coal seams where fractures have been conduits for mineralizing fluids. Epigenetic U has been found in Western U.S. coal seams where surface waters have leached U from overlying welded tuffs. High concentrations of U also have been found in coal fragments in the Colorado Plateau type sandstone deposits of U ore. These are special circumstances of mineralization in a host rock that have little relationship to the genesis of coal.

Once again, it must be emphasized that geological conditions and geological processes are the primary controls in establishing the quality of an incipient coal seam. However, even beyond this first stage of development of coal, beyond the purely sedimentary stage, geological factors have an influence on the chemical and physical controls which will be discussed below.

Climate certainly influences the conditions of sedimentation, plant growth, and peat formation. A detailed study of the paleoclimate controls in the central Appalachian Basin is provided by Cecil et al. (1985). The geochemical factors controlling the quality of some Pennsylvanian-age coals are discussed by them in some detail.

CHEMICAL FACTORS

The chemical environment of the coal swamp can be effective in modifying the composition of minerals. The acidity of swamp waters is certainly an important factor in influencing the ultimate disposition of the suspended and detrital mineral matter brought into the peat bog. The acidity of 64 samples of peat taken 15 cm (0.5 ft) below the surface in 15 Minnesota peat bogs had an average pH of 4.2. The pH data are given in Table 1 and are taken from Passer's (1957) report.

These data probably are typical for most peat bogs not located in limestone or dolomite terranes. The high acidity of these peat bogs probably is surprising, and this is why these data are presented here. Coastal swamps, which may be brackish or which may alternate between fresh-water and marine environments, have a different range of acidity.

Generally, the pH of samples from any given peat profile will decrease from a minimum value near the top to a maximum at or near the base of the peat. One example is a peat profile from Rice Lake Bog in St. Louis County, Minnesota, where the pH varies from 3.6 to 5.5 as indicated in Table 2.

A relatively low pH may be an important factor in altering the character of the clay minerals deposited in a coal swamp. Grim (1953) believes that the leaching of alkalies and alkaline earths under such circumstances may account for the dominance of kaolinite in fresh-water lake sediments. The intense acid leaching of clays associated with coal seams may well account for the origin of the high-grade fire clays so commonly found with coal seams. Carroll (1958) experimented with precisely this type of system in the laboratory and found that, under anaerobic conditions in fresh water, as much as 62% of the ferric oxide present on the clay minerals tested was removed in only 23 days. By the action of anaerobic bacteria, the pH generally was reduced 2 units or more. This demonstrated that, by chemical activity the character of the sedimentary minerals may be changed and iron made available for further chemical reaction.

Bacteria are factors in the chemical control of mineral matter in coal. Sulfide minerals, and especially pyrite and marcasite, are among the most important minerals to be found in coal. The roles of these minerals in affecting the utilization of coal are beyond the scope of this paper. For further information, see Ensminger (1977) and Neavel (1981). Bacteria are active in two ways here: 1) one group of organisms will attack the protein in the plants and, as a consequence of that activity, will ultimately release hydrogen sulfide; 2) another group of bacteria, the Desulfovibrios of Zobell and Rittenberg (1948), reduce sulfate ions to sulfide ions. Quantitatively, the contribution of sulfide ion from reduced sulfate is far more important. Even insofar as the composition of surface waters entering a peat swamp is concerned, there are geological overtones. In basins of limited extent, geological controls are effective determinants of the chemical nature of the surface waters in a drainage area. An outstanding example is given by Davis (1958) in a discussion of the geologic control of the mineral composition of surface waters of the Southern Coast Ranges of California. Davis shows that the ratio of bicarbonate to sulfate concentrations is related to the lithology of the rocks exposed in the tributary drainage area and that the ratio may vary from 0.8 to 6.0 (dominantly bicarbonate) to a ratio of 0.02 to 0.7 (dominantly sulfate).

Peat Bog	County	No. Samples	pH
Rice Lake	St. Louis	17	3.8, 3.6, 3.3, 3.9, 3.4, 3.3, 3.4, 3.2, 3.2, 3.1, 3.3, 4.0, 3.9, 3.34, 3.5, 5.3
Floodwood State	St. Louis	8	4.1, 4.4, 4.6, 5.1, 4.3, 5.1, 4.0, 4.2
Corona	Carlton	5	3.0, 4.1, 4.1, 3.4, 3.4
Cook	St. Louis	3	3.9, 3.8, 3.9
Cusson	St. Louis	1	4.0
Big Falls	Koochiching	3	5.3, 5.3, 5.5
Red Lake 1	Beltrami	7	5.2, 5.2, 5.8, 4.7, 4.8, 5.5
Red Lake 2	Beltrami	2	5.5, 3.5
Rapid River	Lake of the Woods	3	4.5, 4.9, 5.2
Canyon	St. Louis	3	5.6, 5.6, 5.4
Elmer	St. Louis	3	4.6, 3.8, 3.9
Zim	St. Louis	3	4.4, 5.4, 4.8
Cotton Shaw	St. Louis	3	3.3, 3.1, 3.0
Payne	St. Louis	3	4.7, 5.8, 5.1
Prosit	St. Louis	1	4.8

Table 1. *pH of 64 samples taken at 15 cm (0.5 ft) beneath the surface from 15 Minnesota peat bogs located in five counties (from Passer, 1957).*

Depth (ft)	pH
0.5	3.6
1.5	3.5
2.5	3.5
3.5	4.1
4.5	4.0
5.5	3.9
6.5	4.1
7.5	4.4
8.5	4.5
9.5	4.4
10.5	5.1
11.5	5.0
12.5	5.5

Table 2. *Profile of pH at position CV, Rice Lake Bog, St. Louis County, Minnesota (from Passer, 1957, p. 4).*

The nature of the minerals deposited in the peat swamp and the chemical modification of these sedimentary minerals have been discussed briefly. It now is necessary to consider the interaction of physical and chemical forces in further transformations which may take place subsequent to the peat stage. When equilibrium is interrupted and a peat is buried under an increasing load of sediment, water is squeezed out of the peat and from the underlying sediments. The expelled water migrates either laterally or vertically upward under the compressive forces. During this stage of burial and compaction, the early diagenetic stage, chemical conditions are constantly changing. The expelled waters probably are enriched in iron, carbonate, silica, and sulfide ions as compared with water in a peat swamp. Replacement of organic structures by pyrite, carbonate minerals, and quartz may well take place at this stage. Replacement phenomena are far too complex to discuss in the light of present knowledge, but the fact of replacement as a process cannot be denied.

PHYSICAL FACTORS

Once a coal seam has been formed and the seam moisture is reduced to below 20%, as in a subbituminous rank coal, cleating of coal becomes a prominent physical feature. Cleats are closely spaced, parallel, vertical joints; their origin is attributed to pressure phenomena, to tectonic forces, and to forces of contraction following devolatilization (Stutzer and Noe, 1940; McCulloch et al., 1974)). These cleat surfaces often are coated with thin layers of mineral matter. Regardless of the mode of origin suggested for the cleats in coal, all observers agree that cleating occurs late in the history of a coal seam. Cleating is a physical phenomenon, and only after cleats have formed can minerals be deposited in the voids formed by such fracturing. In these voids are deposited minerals of a high degree of purity and which can be easily identified. Pyrite, kaolinite, calcite, gypsum, and halite are minerals which have been found on cleat surfaces of coal. Their origins obviously are similar despite the great difference in composition of the minerals. Pyrite and kaolinite were formed by recrystallization after solution of minerals already present in the coal seams. Calcite, gypsum, and halite were in part formed from ions migrating through the coal from outside the coal seam.

Erosion, a geological factor, may expose a coal seam to the atmosphere, and under such conditions, cleat fractures in coal may be enlarged by physical forces. Weathering processes now impose an entirely new set of chemical conditions upon the coal seam. Atmospheric oxidation and the action of ground waters effect changes in the mineral matter of coal which are even more startling than the changes in the organic matter. Whole new suites of minerals are formed whereas pyrite may be completely destroyed. Minerals like gypsum, jarosite, melanterite, copiapite, and many other sulfate minerals are formed from the oxidation products of pyrite. Allophane, an amorphous clay mineral which is chiefly a hydrous mixture of alumina and silica, may be an extreme alteration product of clay minerals within the coal and in the overlying sediments (Breger et al., 1955).

PRACTICAL CONSIDERATIONS IN COAL BENEFICIATION

Coal beneficiation, the processes by which undesirable components of as-mined coal are removed, is an engineering specialty. Usually, coal is marketed for a specific end use, and the specifications for the delivered coal are agreed to by both the buyer and the seller. Because of the very large tonnages involved, except for some very special markets, the quality parameters usually are limited to calorific value (BTU/lb.), percent ash residue, total S, and moisture. In addition, and especially for Western U.S., low-rank coals, ash-fusion and ash-softening temperatures and sodium (Na) or Cl contents may be specified because of boiler fouling problems.

During exploration and mine planning, core samples of coal are taken for analysis and to perform washability tests. From these cores, large mineral partings and thinly interbedded sedimentary rock are removed because they can be separated in practice by use of a rotary breaker before further processing. Depending on the mining method, further crushing may or may not be required. Ideally, only minimal size reduction should be done because the coarser the product the less costly is the entire process in terms of capital investment, water and separating agents, waste water treatment, fines disposal, and labor. Rarely is the ideal encountered. Usually, there is not a clear-cut separation between pure coal substance, free of minerals, and rock, free of combustible materials. What is achieved in practice is a separation, usually by gravity in a fluid medium, of pieces of coal substance with some disseminated minerals from pieces of rock with some coal substance. If the minerals are very finely disseminated, the rejected material may be largely coal substance but with enough minerals to cause the "coal" to sink in the separation medium. In such instances, further crushing may be required to "liberate" more "coal" from the "minerals". Here, unless we clearly understand that the minerals are a part of the rock we call coal, we can fall into the trap of calling the contained minerals "impurities".

The problem of designing a coal beneficiation plant is further complicated by not understanding fully that what will be mined will not always be identical to the coal defined by the core sampling program. Out-of-seam dilution occurs when rock is mined accidentally, as from roof falls, or purposely, as during extraction of roof or floor to maintain a uniform height when the coal seam thins, when overcasts are required, or when rock-filled fractures are encountered. Once a beneficiation circuit is overloaded with rock, the efficiency of the system is reduced, and coal may be produced from the plant that does not meet specifications.

Some coals, as mined, may have a high content of sodium chloride. It may be possible to clean such coals by water washing, but if the salt content is too high, the washing circuits may become highly concentrated in salt, and if water disposal is an environmental problem, the coal may not be marketable unless transported elsewhere for washing, a highly unlikely scenario.

Finally, there may be some coals which have so high an organically combined S content that such coals cannot be used unless a system is designed to extract the S from the flue gases in a marketable form.

SUMMARY AND CONCLUSIONS

Geological controls are the primary factors in determining the grade of coal insofar as the mineral content is concerned. These primary controls, operative in the geologic past, have influenced current practices in coal utilization. The chemical environment of the coal swamp certainly has played a role in modifying the minerals deposited in the swamp, which has acted as a reaction vessel within which pyrite and marcasite could form. Under changing physical conditions during burial and compaction, and later, if subjected to erosion and exposure to circulating groundwater and air, physical factors become dominant in controlling the chemical transformations that occur.

Perhaps it would have been more precise to discuss the factors which control the mineral content of coal all under only one classification — geochemical factors. In a broad sense, geochemistry is concerned with all the changes by which the chemical composition of earth systems are modified. Sedimentation, a geological process, is itself controlled by diastrophism, initiated by major geophysical and geochemical processes. Geochemical transformations occur during every stage of coalification, from the first accumulation of plant debris to the ultimate exposure and oxidation of a coal seam.

However, such broad concepts, unless one is accustomed to thinking in such terms, do not convey to the fuel chemist the idea that the variable mineral content of coal is dependent upon a relatively few fundamental processes. Therefore, an attempt has been made to weave into the discussion of chemical and physical factors the continuous thread of dependence upon geological control — and this, in essence, makes the controls geochemical.

The mineral content of coal is a natural consequence of its mode of deposition and its subsequent geological history and is not due to capricious causes. It is common knowledge that the mineral content of coal influences its utilization, often in a deleterious manner. Coal preparation as now practiced, can effectively reduce the mineral content of almost any coal. With a better understanding of how minerals in coal are formed, and how they occur, the technology of coal preparation and utilization can be further improved.

REFERENCES

Breger, M., M. Deul, and S. Rubinstein, 1955, Geochemistry and mineralogy of a uraniferous lignite: Economic Geology, v. 50, p. 206-226.

Carroll, D., 1958, Role of clay minerals in the transportation of iron: Geochimica et Cosmochimica Acta, 14, p. 1-28.

Cecil, C.B., R.W. Stanton, S.G. Neuzil, F.T. Dulong, L.F. Ruppert, and B.S. Pierce, 1985, Paleoclimate controls on Late Paleozoic sedimentation and peat formation in the central Appalachian Basin (U.S.A.), in T.L. Phillips and C.B. Cecil, eds., Paleoclimate controls on coal resources of the Pennsylvanian System of North America: International Journal of Coal Geology, v. 5, p. 195-230.

Davis, G.H., 1958, Geologic control of mineral composition of surface waters of the Southern Coast Ranges, California (abstr.): Geologic Society of America Bulletin, v. 69, p. 1680.

Deul, M., 1955, The origin of ash-forming ingredients in coal (abstr.): Economic Geology, v. 50, p. 103.

Deul, M., 1958, Biochemical and geochemical origins of ash-forming ingredients in coal, in Preprints of Papers Presented at Chicago, Illinois: American Chemical Society, Division of Gas and Fuel Chemistry, p. 169-182.

Ensminger, J.T., 1977, Coal: Origin classification and physical and chemical properties, in H.M. Braunstein, E.D. Copenhaver, and H.A. Pfuderer, eds., Environmental, health and control aspects of coal conversion: An overview, Volume 1: Oak Ridge National Laboratory, Prepared for Energy Research and Development Administration, Report ORNL.EIS-94, p. 2-1 to 2-104.

Gauger, A.W., E.P. Barrett, and F.J. Williams, 1934, Mineral matter in coal--A preliminary report: American Institute of Mining, Metallurgical, and Industrial Engineers Transactions, v. 108, p. 226-238.

Grim, R.E., 1953, Clay mineralogy: New York, McGraw-Hill Publishing Co., 384 p.

King, J.G., M.B. Maries, and H.E. Crossley, 1936, Formulae for the calculations of coal analyses to a basis of coal substance free of mineral matter: Journal of the Society of Chemical Industry, v. 55, p. 271-281.

McCulloch, C.M., M. Deul, and P.W. Jeran, 1974, Cleat in bituminous coalbeds: U.S. Bureau of Mines Report of Investigation RI 7910, 25 p.

Millot, J., O'N., 1958, The mineral matter in coal I--The water of constitution of the silicate minerals: Fuel, v. 37, p. 71-80.

Neavel, Richard C., 1981, Origin, petrography and classification of coal, in Martin A. Elliot, ed., Chemistry of coal utilization, 2nd Supplementary Volume: New York, John Wiley & Sons, p. 91-158.

Passer, M., 1957, Peat Samples from Minnesota Bogs: Duluth, University of Minnesota, Department of Chemistry, Report No. 16, 127 p.

Schopf, J.M., 1956, A definition of coal: Economic Geology, v. 51, p. 521-527.

Stutzer, O., and A.C. Noe, 1940, Geology of coal: Chicago, University of Chicago Press, 461 p.

Zobell, E.E., and S.C. Rittenberg, 1948, Sulfate-reducing bacteria in marine sediments: Sears Foundation Journal of Marine Research, v. 7, p. 602-617.

COAL QUALITY OF THE MARY LEE COAL GROUP AND ITS RELATIONSHIP TO COAL RESOURCE DEVELOPMENT, COALBED METHANE, AND GEOLOGIC HISTORY IN THE WARRIOR COAL FIELD, ALABAMA

Richard B. Winston[1]
Geological Survey of Alabama
P.O. Box O
Tuscaloosa, AL 35486-9780

ABSTRACT

Medium- and low-volatile bituminous coal is present in southwestern Jefferson and northeastern Tuscaloosa counties, Alabama. In this high-rank area, final coalification largely occurred after folding of the Blue Creek Anticline as indicated by near-horizontal isorank lines cutting across structural contours. Outside the high-rank area, coal rank is relatively uniform and shows no relationship to present-day depth of burial, indicating that coalification outside the high-rank area was completed prior to the deposition of the Cretaceous cover.

Low-ash (<12%), low-sulfur (<1%) coal occurs in a belt 15-25 km (10-15 mi.) wide on the southeastern edge of the Warrior coal field. Northwest of the low-ash, low-sulfur belt is a belt 25 km (15 mi.) wide where ash and sulfur content commonly are high (ash >18%, sulfur >2%) but variable.

Economic factors allow underground mines deeper than 300 m (1,000 ft) to exist only in the high-rank area. Coalbed methane production has occurred in areas with less than 37% volatile matter. Based on this criterion, southern Walker County is a potential area of future coalbed methane production.

INTRODUCTION

Understanding coal quality is important in coalbed methane exploration and in coal resource development. Rank and grade also influence the suitability of coal for power generation and coking applications. Rank and grade determine how much methane may have been generated by coal. A significant amount of thermogenic methane is not generated until coal reaches bituminous rank (Jüntgen and Klein, 1975), and high ash content may have an adverse effect on methane retention in coal (Close and Erwin, 1989).

The purpose of this paper is 1) to describe the regional variation in coal quality of the Mary Lee coal group in the Black Warrior coal basin of Alabama, 2) to show how variation in coal quality is related to the geologic history of Alabama and 3) to point out the implications of coal quality trends in the Mary Lee coal group for

1. Present Address: 18 Wilelinor Drive, Edgewater, MD 21037.

coal resource development and coalbed methane exploration both in the Mary Lee coal group and in other coal groups for which little information is available.

Geologic Setting

The Pennsylvanian-age Mary Lee coal group is in the Pottsville Formation in the Black Warrior Basin of Alabama. The Black Warrior Basin is a triangular region in northwestern Alabama and northeastern Mississippi (Figure 1a). The Ozark Uplift and the Nashville Dome form its boundary on the north. On the southeast, it is bordered by the Appalachian Valley and Ridge Province. On the southwest, it is bordered by the Central Mississippi Uplift and the Ouachita Mountains (Kidd, 1982). The portion of the Black Warrior Basin containing rocks of the Pottsville Formation is the Warrior coal field (Figure 1b). The Warrior coal field comprises the Plateau coal region to the north and the Warrior coal basin to the south (Figure 1b). The northern outcrop limit of the base of the Black Creek coal seam is the traditional boundary between the Black Warrior coal basin and the Plateau coal region (McCalley, 1899).

The southeastern border of the Black Warrior Basin is defined by the northwest limb of the Birmingham Anticlinorium, and by the Opossum Valley Fault (Figure 2) (Pashin et al., 1990). To the northwest of these structures is the Blue Creek Syncline, the tightly folded Blue Creek Anticline, the broad, flat-bottomed Coalburg Syncline and the gently folded Sequatchie Anticline. Structural relief in the Mary Lee coal group from the Coalburg Syncline to the outcrops on the Blue Creek Anticline locally exceeds 600 m (2,000 ft). The entire Paleozoic sequence dips to the southwest where it is covered by Cretaceous and Tertiary rocks of the Mississippi Embayment (Kidd, 1982; Thomas, 1988). The Bessemer cross-strike structural discontinuity (CSD; Figure 2) (Thomas and Bearce, 1986) occurs immediately to the southeast of the Warrior coal field, but has not been mapped within it.

Five coal seams make up the informal Mary Lee coal group in Alabama (Figure 3a). In ascending order, these are the Ream, Jagger, Blue Creek, Mary Lee, and Newcastle coal seams (Culbertson, 1964). The four upper coal seams join and split in a complex fashion. The Mary Lee coal group and its associated coarsening-upward rock sequence informally are called the "Mary Lee cycle" (Figure 3b; Pashin et al., 1990). For this paper, the Ream coal seam was not studied because few data are available and because it may represent a separate cycle (J.C. Pashin, personal communication, 1989).

The Mary Lee coal group in Alabama is an important economic resource. The original demonstrated reserve base of the Mary Lee coal group as of 1978 was 2,061,000,000 tonnes (2,267,000,000 tons) (36% of total demonstrated reserves in Alabama; Ward, 1984). Total production in 1987 from the Mary Lee coal group was 13,200,000 tonnes (14,500,000 tons) (58% of Alabama coal production), with 12,000,000 tonnes (13,000,000 tons) produced from underground mines and 1,400,000 tonnes (1,500,000 tons) produced from surface mines (Fanning and Moore, 1989; Tolson, 1988). The Mary Lee coal group is also an important drilling target for coalbed methane.

Previous Work

The coal quality in the Warrior coal field is incompletely understood. Most previous studies deal only with rank, and these either lack detail or deal only with a small area (Culbertson, 1964; Epsman et al., 1989; McFall et al., 1986; Murrie et al., 1976; Semmes, 1920, 1929). Only Epsman et al. (1989) mapped ash or sulfur (S) content, and they only deal with the Brookwood and Oak Grove degasification fields. Vitrinite reflectance data is summarized by Pashin et al. (1990) and Winston (1990). Pashin et al. (1990) and Winston (1990, 1991) also report some of the same results reported here.

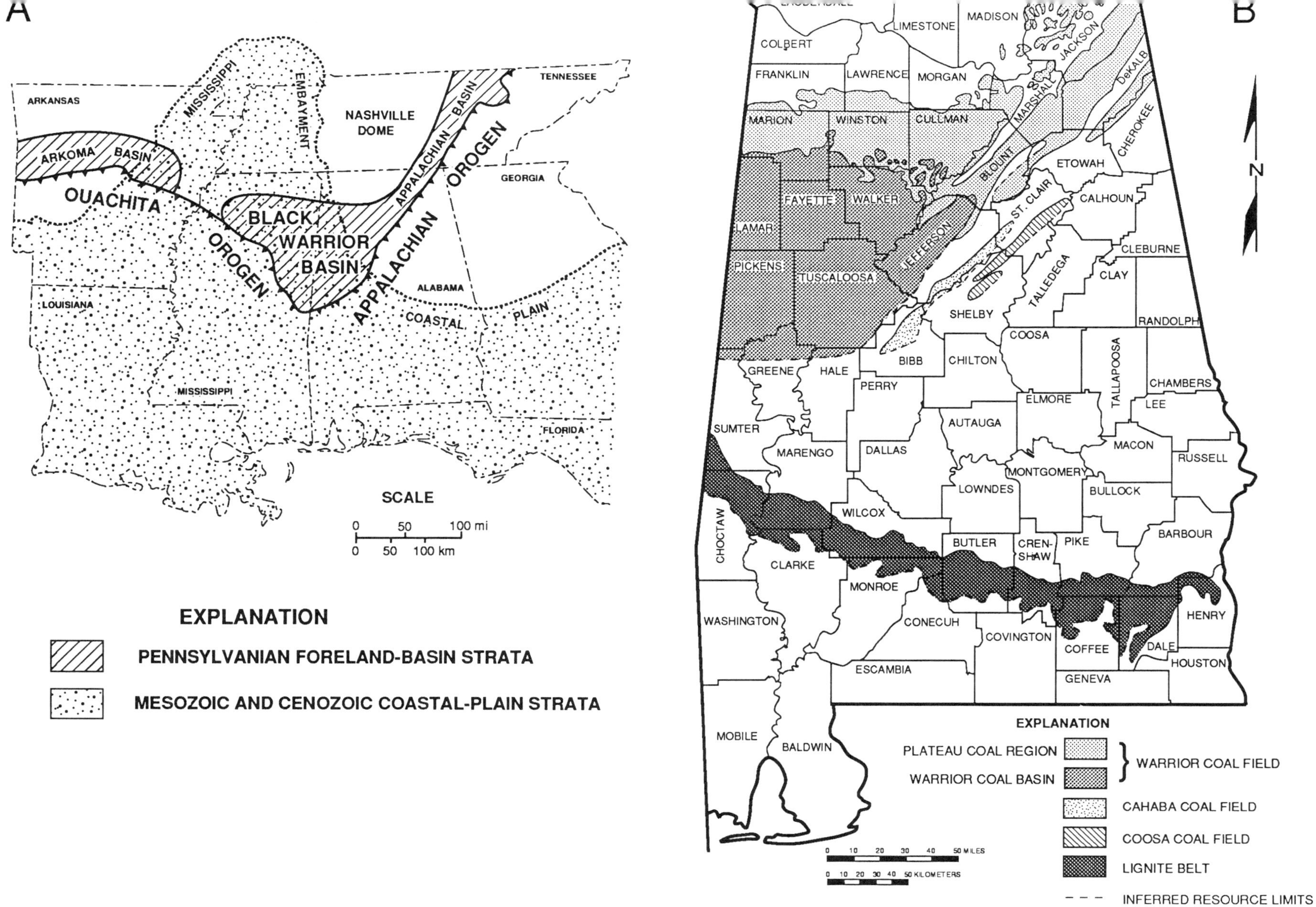

Figure 1. *A) Location and geologic setting of the Black Warrior Basin (modified from Pashin et al., 1990). B) Alabama coal fields (modified from Tolson, 1988).*

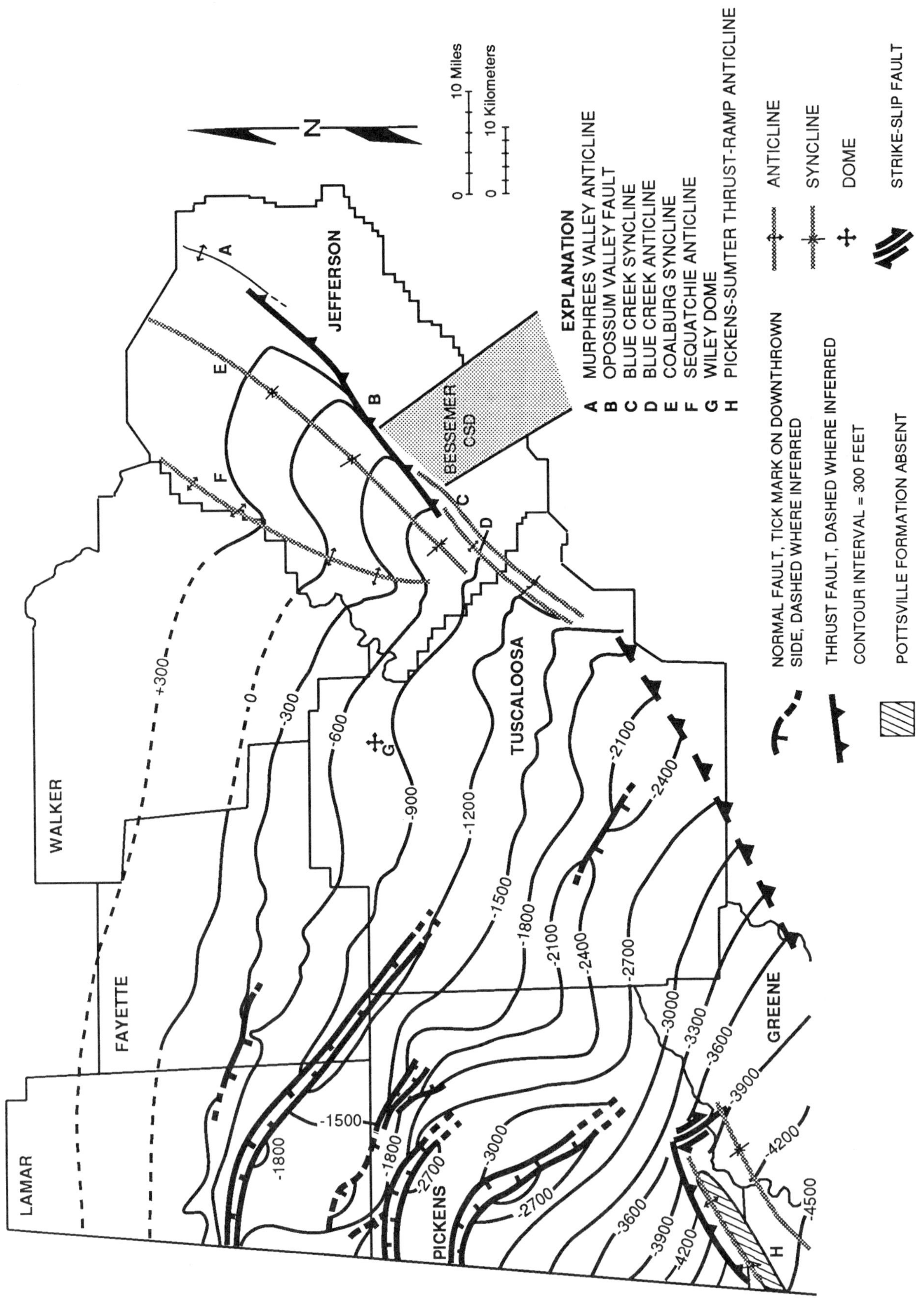

Figure 2. *Structural geology of the Black Warrior Basin in Alabama (modified from Pashin et al., 1990).*

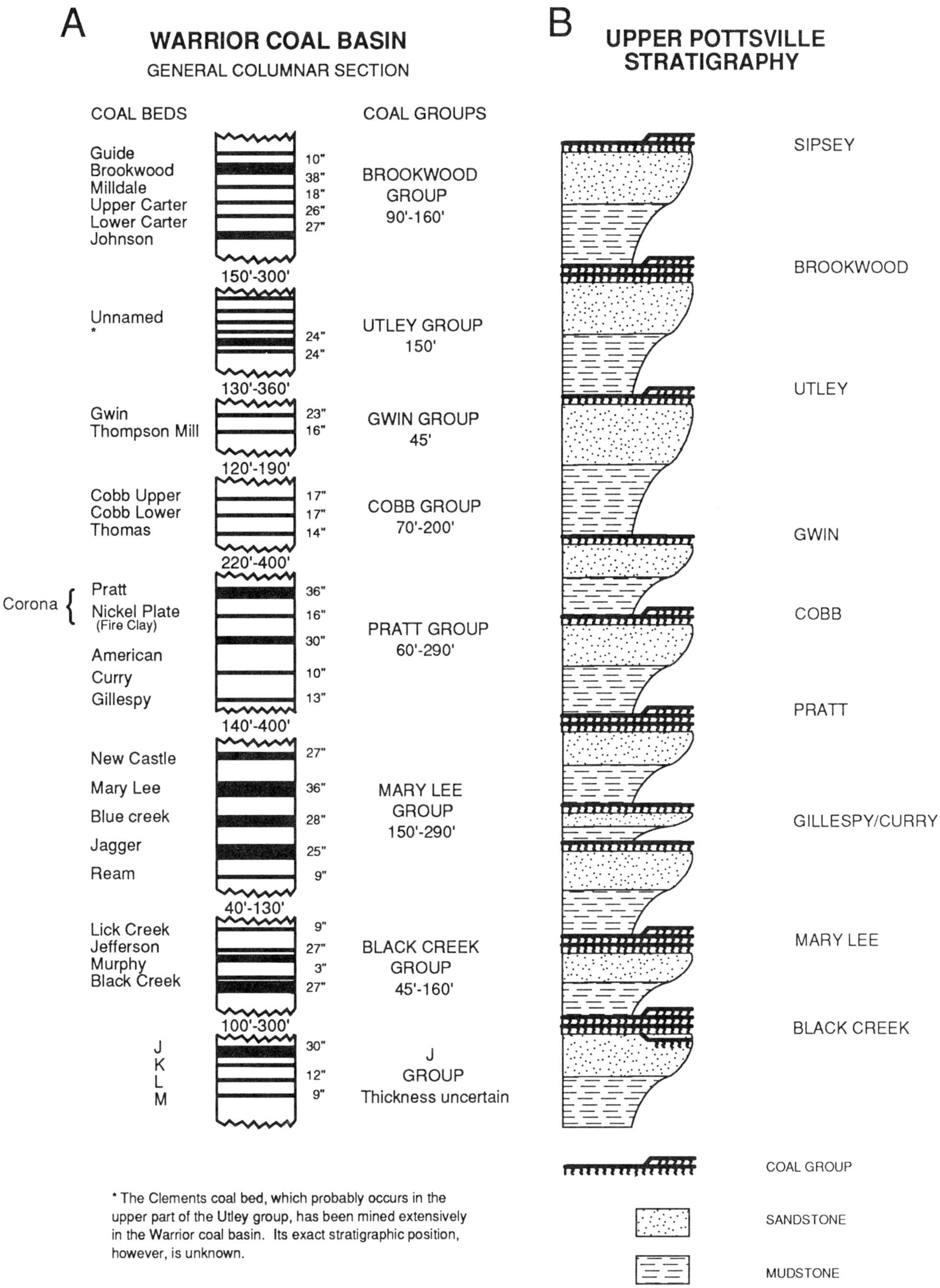

Figure 3. *A) Alabama coal groups (after Ward, 1986). B) Alabama sedimentary cycles (modified from Pashin et al., 1990).*

METHODS

Maps of volatile matter (dry, mineral-matter-free (dmmf) basis), calorific value in British thermal units (Btu) per pound (moist, mineral-matter-free (mmmf) basis), ash (dry basis), S (dry, ash-free (daf) basis), and vitrinite reflectance of the coal seams in the Mary Lee coal group were prepared using data from more than 400 localities (Figure 4). Data for this report were obtained from the files of the Geological Survey of Alabama and calculated to dmmf, mmmf, dry, or daf bases using the standard formulas in American Society for Testing Materials (ASTM) standards D388-88 and D3180-84 (American Society for Testing Materials, 1988a and c). Coal samples for chemical analyses are from localities throughout the Pottsville Formation outcrop area (Figure 4). Coal was sampled from cores, test pits, mines, and outcrops. Volatile matter, calorific value, ash, and S data are available for nearly all samples. Where analyses from more than one coal seam in the Mary Lee coal group were available, the average value was mapped for volatile-matter content, calorific value, and vitrinite reflectance. In areas where vitrinite reflectances from the Mary Lee coal group were not available, data from the Black Creek, Pratt, and Cobb coal groups were used to estimate the reflectance at the Mary Lee horizon. If measurements both above and below the Mary Lee horizon were available, the reflectance at the Mary Lee horizon was assumed to be between them. If measurements were available either only above or only below the Mary Lee coal group, the reflectance at the Mary Lee horizon was assumed to be the same as that of the measurement closest to the Mary Lee horizon. These data could be used because the difference in rank among the coal groups is slight at any one location as indicated by studies of cores containing coal from several coal groups. Use of these data allows a vitrinite reflectance map to be constructed for the portion of the Black Warrior coal basin which is beneath Cretaceous cover. For ash and S content, only data from the Blue Creek and Mary Lee coal seams were used. In preparing maps of ash content, no contours depicting more than 18% ash were drawn.

Coal analyses used in this report were provided by coal-mining companies to the Geological Survey of Alabama or were obtained from the USCHEM computer file of the National Coal Resource Data System (NCRDS) operated by the U.S. Geological Survey. Published analyses by Fieldner et al. (1925), Shotts (1956, 1960), and Fanning and Moore (1989) were used for areas where other data are scarce. Locality information for most of the published analyses is inadequate, so most of these analyses could not be used. Those analyses that were used could be assigned at least to a specific mine.

The method by which coal is treated before analysis may greatly influence analytical results. Some investigators use heavy-liquid separation to divide the coal into fractions of differing density before analysis. Density of the cleaned coal typically ranges from <1.50 to <1.65 g/cm^3. Many of the NCRDS analyses represent uncleaned coal. However, mineral partings thicker than 1 cm (0.38 in.) typically are removed from the samples that are not washed (American Society for Testing Materials, 1988d, ASTM standard D4596-86). This is done to obtain a sample equivalent to washed coal. As will be discussed later, the ash map reveals geologically significant patterns indicating that differences in sample handling do not seriously compromise the results. Use of analyses of only washed or only unwashed samples would have resulted in inadequate geographic coverage, so analyses of both types of samples had to be used.

Mean-maximum vitrinite reflectance was measured using standard procedures (Stach et al., 1982; American Society for Testing Materials, 1988b, ASTM standard D2798-85). Other vitrinite-reflectance measurements were compiled from published and unpublished sources (Geochem Laboratories, 1982; Hines, 1988; Levine et al., 1989; McFall et al., 1986; Robertson Research (U.S.) Inc., 1985). In many places, the coal group identifications given by the original authors were incorrect. The correct stratigraphic positions were determined based on correlation of geophysical logs on file at the Alabama Oil and Gas Board.

A vitrinite-reflectance profile was made in the area of high-rank coal using coal from a corehole drilled by Jim Walter Resources, Inc. Other vitrinite-reflectance measurements were made using channel and column

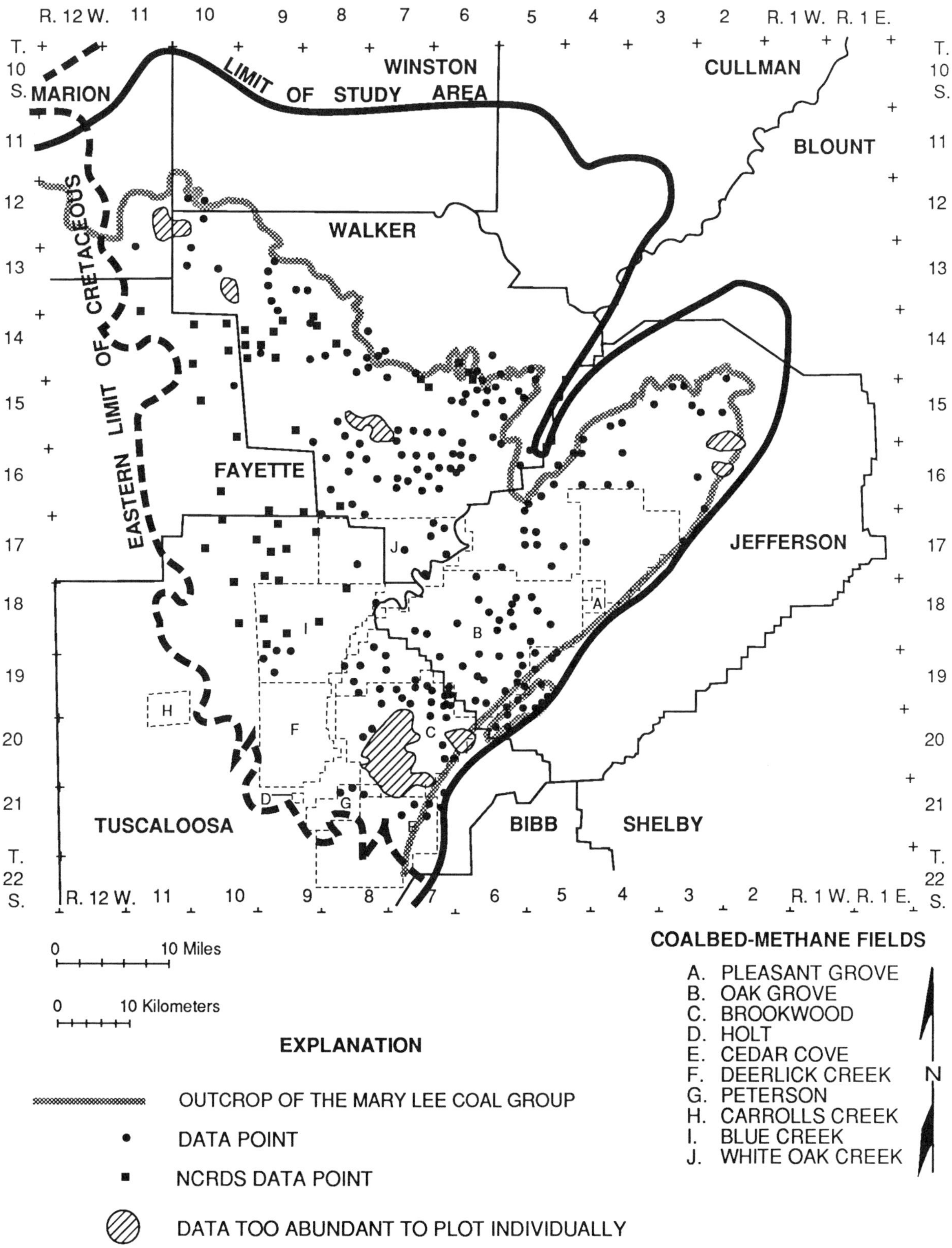

Figure 4. Sample locations for the Mary Lee coal group.

samples from outcrops and mines and cuttings from oil and gas wells. Samples from well cuttings were picked at depths where geophysical logs indicated coal was present. Such samples only were collected if examination of samples above and below the sampled horizon lacked abundant coal. This insured that little of the coal in the sample represented reworked material. Cuttings were not ground prior to mounting and polishing, and only one vitrinite reflectance measurement was made on each particle. Grinding was avoided to minimize the importance of particles derived by caving from higher coal seams. Weathered coal was avoided when making vitrinite reflectance measurements. Although some samples still may have experienced weathering which went undetected, this probably has not affected the results because weathering has little influence on vitrinite reflectance (Stach et al., 1982, p. 199-202).

RESULTS

Coal rank in the Mary Lee coal group, as determined by the ASTM classification, ranges from high-volatile B bituminous to low-volatile bituminous (Figures 5 and 6, Table 1). The rank of most of the coal is high-volatile A bituminous as defined by ASTM standard D388-88 (American Society for Testing Materials, 1988a). A small amount of high-volatile B bituminous coal occurs in northwestern Walker County. The highest rank coal is found in a "high-rank area" in northeastern Tuscaloosa and southwestern Jefferson Counties in the Coalburg Syncline immediately northwest of the Blue Creek Anticline. Within the high-rank area, volatile matter increases sharply to the southeast by about 6% across the northwest limb of the Blue Creek Anticline. The structural cross section shows that isorank lines are subhorizontal and are oblique to bedding in the northwest limb of the Blue Creek Anticline (Figure 7). The isorank lines cannot parallel the coal seams because where the Mary Lee coal group is present in outcrop, its volatile matter content is much higher (rank is lower) than it is at depth (Figure 7).

The high-rank area is elongated to the northwest. Its boundaries have been chosen using the maps of volatile matter content in the Mary Lee and Pratt coal groups. The northeast and southwest boundaries of the high-rank area can be defined by two straight lines extending into Walker County (Figure 5). Volatile matter is about 3-8% lower and calorific value is higher by about 400 Btu/kg (880 Btu/lb) of coal in this area than in

Rank	**Btu/lb (mmmf)**	**% Volatile Matter (dmmf)**[1]	**Approximate % Vitrinite Reflectance**[2]
Low volatile	—	14-22	1.5-2.0
Medium volatile	—	22-31	1.0-1.7
High-volatile A	>14000	>31	0.6-1.2
High-volatile B	13000-14000	>31	0.5-0.8
High-volatile C	11000-13000	>31	0.4-0.7

[1] from ASTM standard D388-88

[2] from Damberger et al. (1984)

Table 1. *Abbreviated rank classification of bituminous coal*

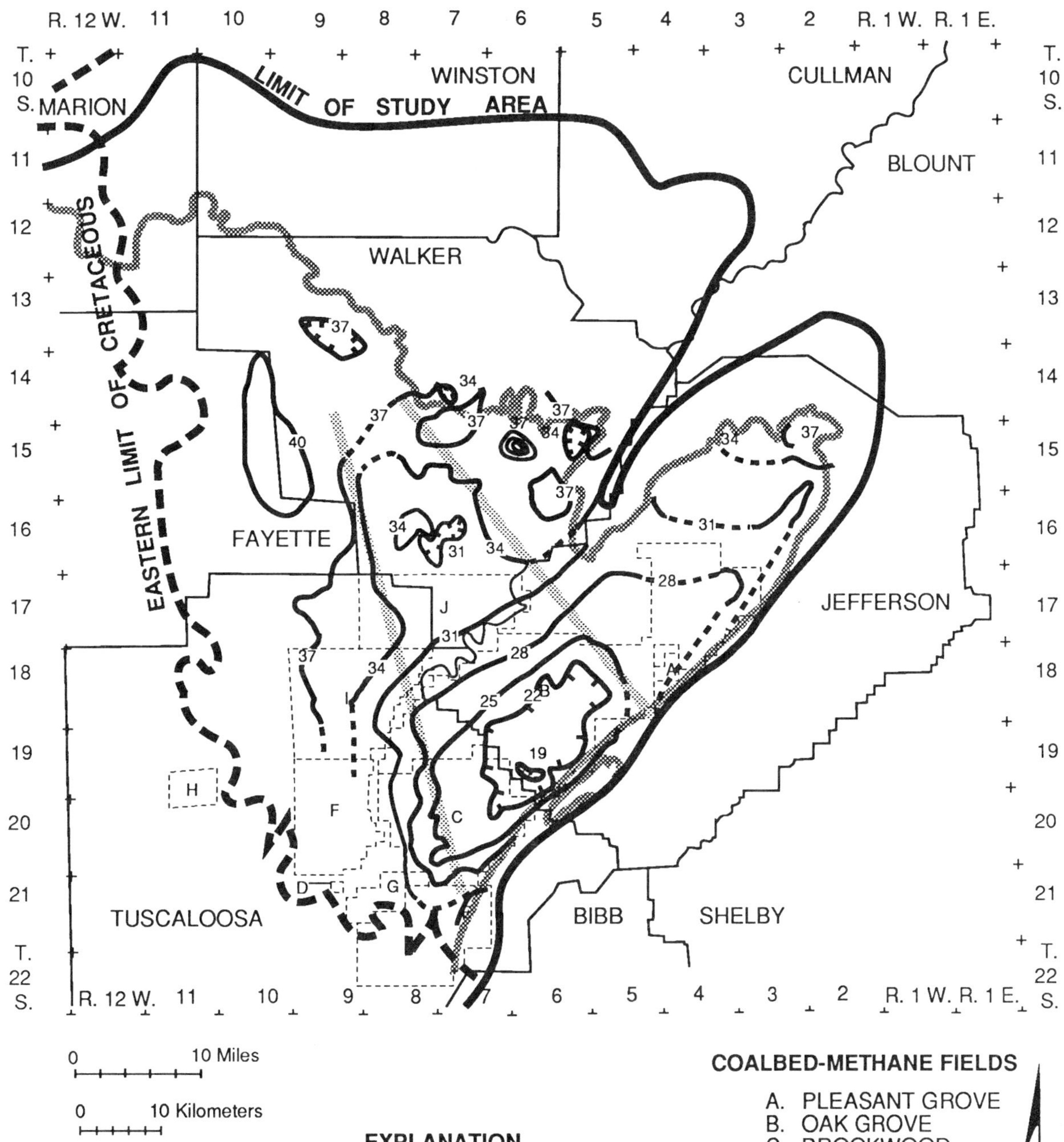

COALBED-METHANE FIELDS

A. PLEASANT GROVE
B. OAK GROVE
C. BROOKWOOD
D. HOLT
E. CEDAR COVE
F. DEERLICK CREEK
G. PETERSON
H. CARROLLS CREEK
I. BLUE CREEK
J. WHITE OAK CREEK

EXPLANATION

OUTCROP OF THE MARY LEE COAL GROUP

VOLATILE-MATTER CONTENT, DASHED WHERE INFERRED

CONTOUR INTERVAL = 3 PERCENT

LIMIT OF HIGH-RANK AREA

NOTE: SEE FIGURE 4 FOR SAMPLE LOCATION

Figure 5. *Volatile-matter content (dry, mineral-matter-free basis) of the Mary Lee coal group.*

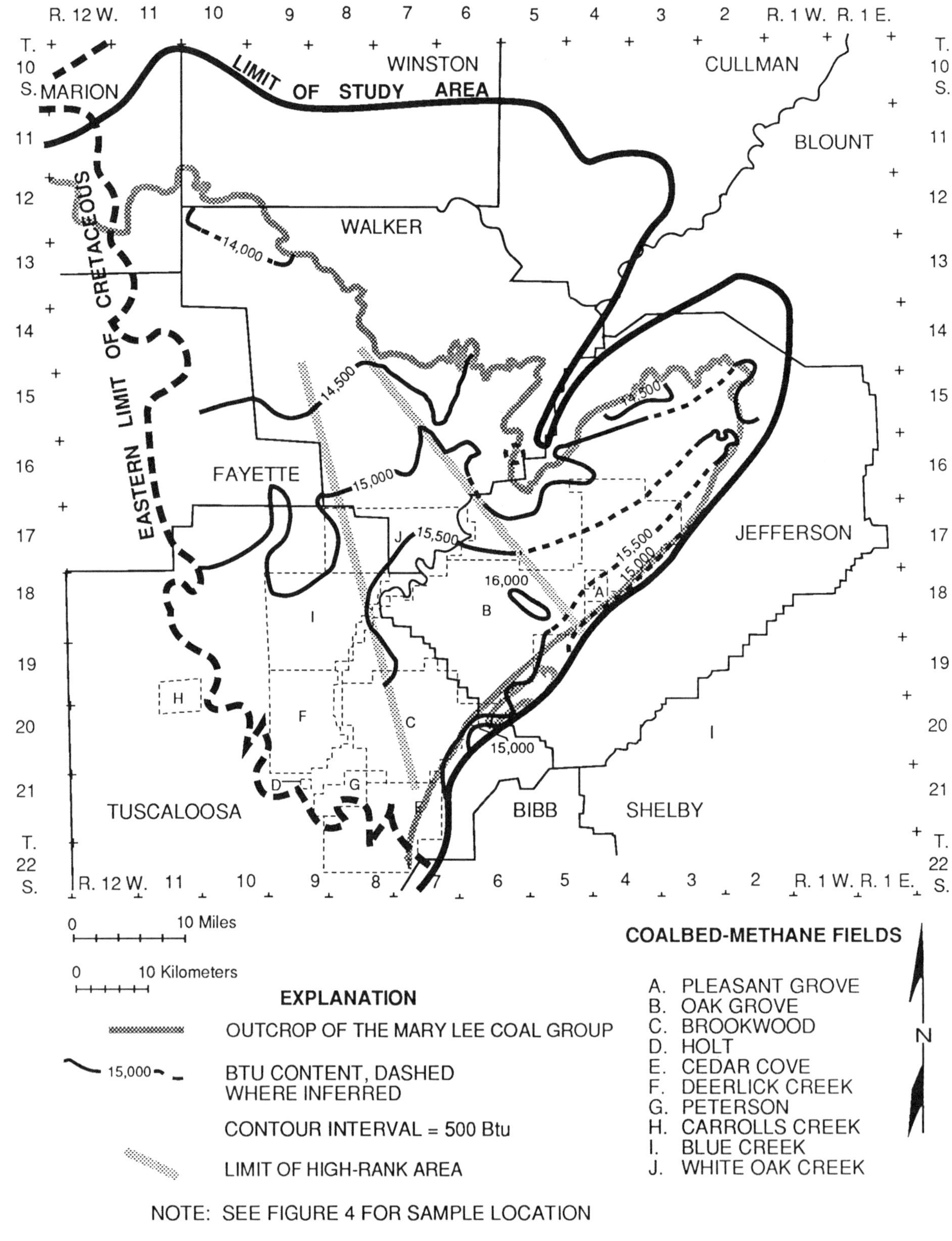

Figure 6. *Calorific value in Btu/lb (moist, mineral-matter-free basis) of the Mary Lee coal group.*

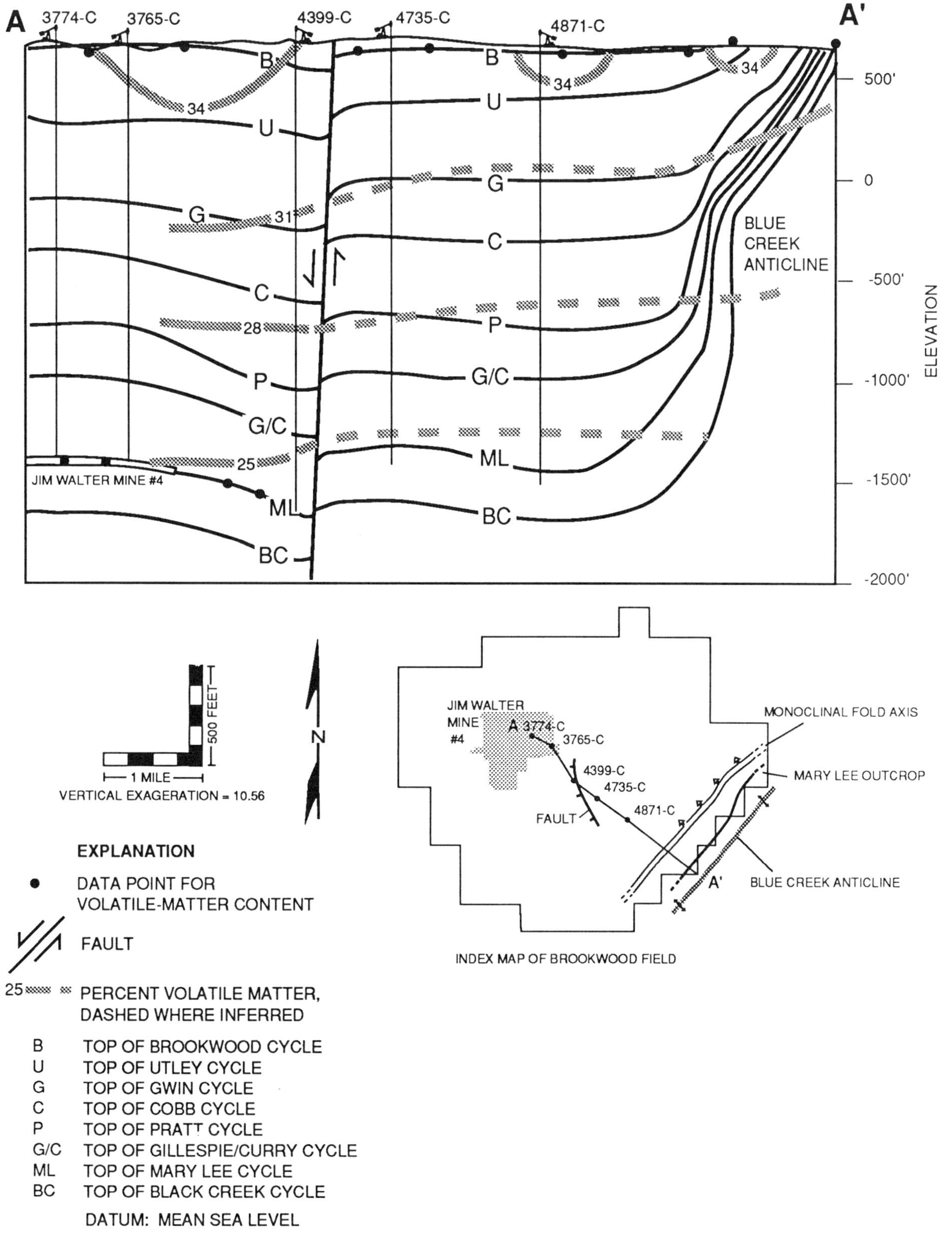

***Figure 7.** Structural cross section of upper Pottsville strata in the Brookwood field showing approximately horizontal isovols. Blue Creek Anticline axis from Epsman et al. (1988).*

areas immediately outside the northeast and southwest boundaries. The northeastern line extending from Jefferson County to Walker County is aligned with the southwest margin of the Bessemer CSD of Thomas and Bearce (1986). The southwestern line stretching from Tuscaloosa County to Walker County does not coincide with any known structural feature. As will be discussed later, these differences in rank are important in understanding the origin of the high-rank area.

Beneath the Cretaceous-age cover, virtually no chemical analyses of coal are available, so the best rank information comes from vitrinite reflectance. One vitrinite-reflectance value indicates that a small amount of medium-volatile bituminous coal is present in northern Sumter County (Figure 8 and Table 1). The rank of most other Mary Lee coal is apparently high-volatile A bituminous with rank possibly as low as high-volatile C bituminous occurring in Fayette County and northeastern Pickens County.

Coal with low ash content (<12%) is present in a belt 15-25 km (10-15 mi.) wide in the southeastern part of the Warrior coal field (Figure 9). To the northwest of the low-ash belt, there is another belt about 25 km (15 mi.) wide where ash content commonly is high (>18%), but is quite variable. The high-ash belt is wider in the west than in the eastern part of the Warrior coal basin. To the northwest of the high-ash belt, ash content generally is moderate (12-18%).

The area of Mary Lee and Blue Creek coals thicker than 1.2 m (4 ft) (Figure 9) (Murrie et al., 1976) corresponds closely with the low-ash belt of the Mary Lee and Blue Creek coal seams. (The Mary Lee coal of Murrie et al. (1976) is the Mary Lee and Blue Creek coal seams of the present study.) This indicates that differences in sample treatment have not obscured geologically significant trends in coal quality.

The regional pattern of S content can be divided into three belts which generally coincide with the ash-content belts (Figure 10). Most of the area in the low-ash belt also has a low S content (<1%). In the belt with high but variable ash content, S content commonly is high (>2%), but it also is quite variable. Sulfur content is especially high on the southwest end of this belt. To the northwest of the high-ash, high-sulfur belt is a belt where S content is low to moderate (from <1% to 2%).

DISCUSSION

Geologic Significance

On a basin-wide scale, coal rank is difficult to relate to regional structural trends. However, the decrease in rank to the southeast across the northwest limb of the Blue Creek Anticline in the high-rank area can be related directly to structure. Because the isorank lines are subhorizontal and do not parallel bedding on the anticline (Figure 7), coalification evidently postdated or occurred during the late stages of folding in the high-rank area. Outside the high-rank area, the timing of coalification in the Pottsville Formation relative to folding is not clear. However, the occurrence of lignite in the flat-lying Cretaceous-age Tuscaloosa Formation indicates that most coalification in the Pottsville Formation outside the high-rank area preceded deposition of Cretaceous-age sediments of the Mississippi Embayment.

The presence of the high-rank area is not due to a deeper depth of burial than areas to the northeast and southwest. Structural relief equivalent to the northwest limb of the Blue Creek Anticline would be required to explain the decrease in coal rank to the northeast and southwest of the high-rank area. Because no such structure exists (Figure 2), the high-rank area must be due to either greater heating (due to a higher paleogeothermal gradient) or a longer duration of heating than surrounding areas, possibly due to a later episode of coalification in the high-rank area that did not affect other areas. Gravity and magnetic data show no evidence of igneous intrusives which could explain the rank anomaly.

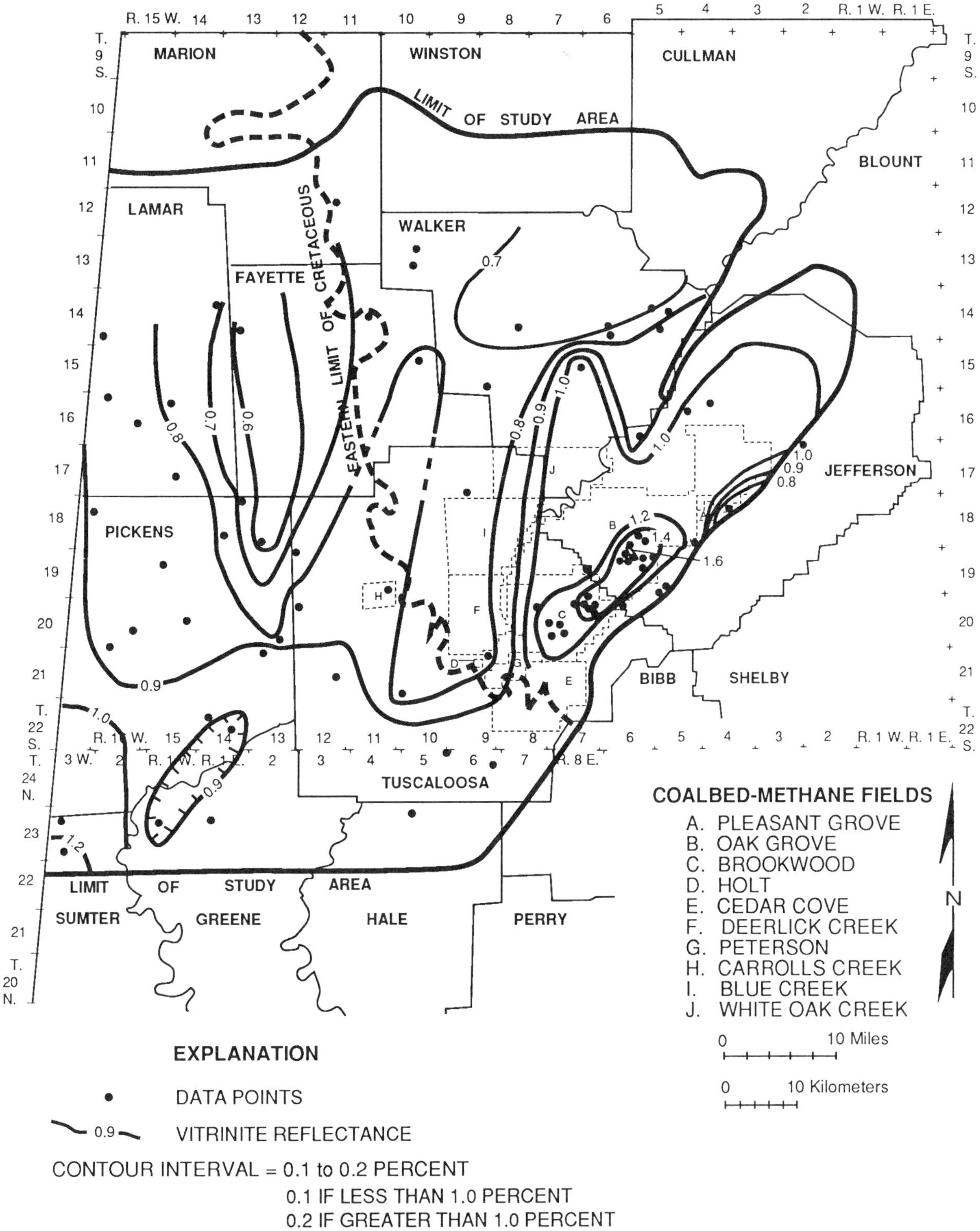

Figure 8. *Vitrinite reflectance map of the Mary Lee coal group in Alabama.*

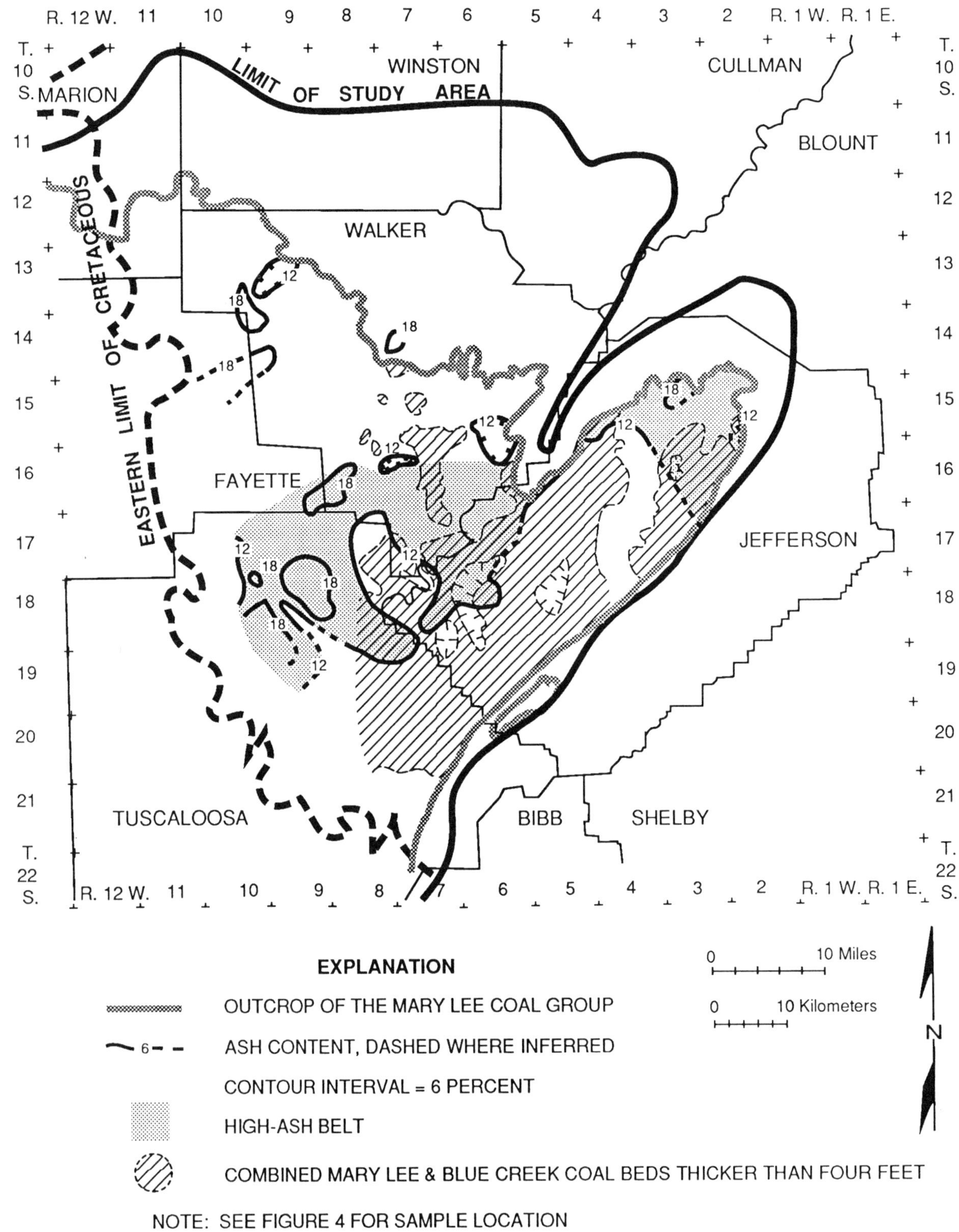

Figure 9. *Ash content (dry basis) of the Mary Lee and Blue Creek coal seams, Mary Lee coal group, with the 1.2-m (4-ft) thickness limits from Murrie et al. (1976).*

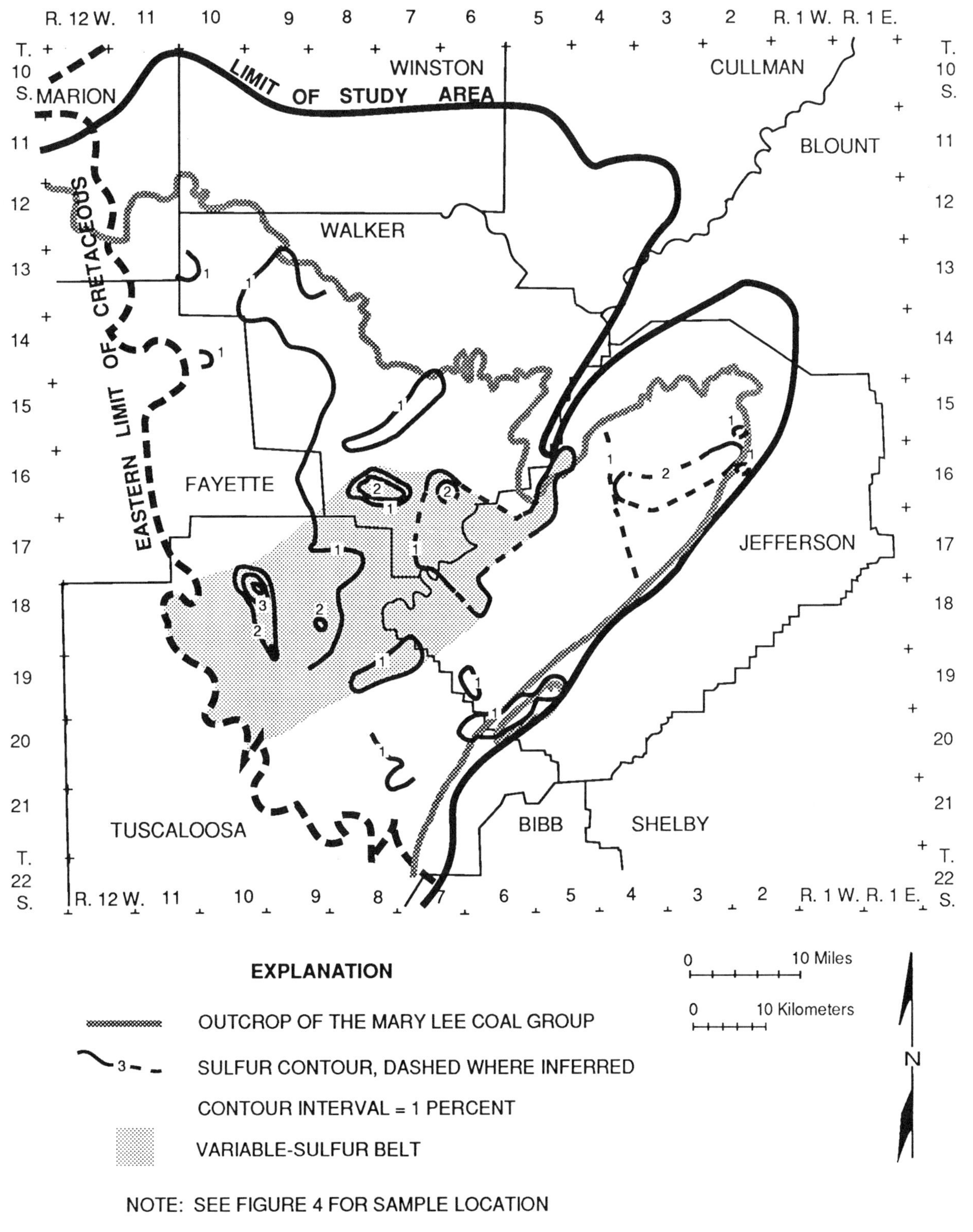

Figure 10. *Sulfur content (dry, ash-free basis) of the Mary Lee and Blue Creek coal seams, Mary Lee coal group.*

Vitrinite reflectance of the coal in Sumter County indicates that the rank is probably medium-volatile bituminous (Figure 8). The geology of the Pottsville Formation in the deeper, southwest part of the Alabama Black Warrior Basin is understood too poorly to interpret the higher rank at this time.

Trends in ash and S content in the Mary Lee Group are similar to those found in several other coal groups in Alabama (Pashin et al., 1990; Winston, 1991). The high-ash, high-sulfur belt may represent a low-lying area that was subject to more frequent flooding during peat accumulation and was influenced more by marine water after peat accumulation. Ongoing geological investigations of the surrounding rocks may aid in interpreting the trends in ash and S content.

Economic Significance

Coalbed Methane

The most productive coalbed methane field in Alabama so far has been the Brookwood Field (see Figure 4 for coalbed methane field locations); over two-thirds of the coalbed methane produced in 1987 came from that field (Tolson, 1988). In terms of gas production per well, the most productive fields in 1987 were the Brookwood and Cedar Cove fields (Tolson, 1988). Coal rank has an influence on coalbed methane production because the amount of methane generated during coalification is determined primarily by coal rank (Rightmire, 1984). Other factors, such as present depth, determine methane retention and producibility. In Alabama, the maximum volatile-matter content of the Mary Lee coal group in areas where coalbed methane has been produced has been less than 37%.

Target coal seams for coalbed methane production include the Black Creek through Cobb coal groups. In southern Walker County, volatile-matter content of the Mary Lee coal group is less than 34% in the northern end of the high-rank area (Figure 5). The Mary Lee coal group in this area is too shallow to be a good target for coalbed methane production. Nevertheless, this rank information is significant for coalbed methane exploration because the underlying Black Creek coal group is deep enough for a significant amount of methane still to be retained in the coal. Relatively little information is available about the rank of the Black Creek coal group in this region, but because the Black Creek coal group is deeper than the Mary Lee coal group, its rank should be equal to or higher than that of the Mary Lee coal group. (During preparation of this report, the White Oak Creek coal degasification field was established in this area and several coalbed methane wells were drilled. Two wells were productive. Most of the permitted wells in the field were cancelled by the operator without being drilled because of expected low productivity. No further drilling is expected in the near future.)

If the high-rank area of the Warrior coal field extends to the southeast into the Cahaba and Coosa coal fields, it may be significant for future coalbed methane production in those fields.

One area of present interest for coalbed methane development is along the trend of the Coalburg Syncline beneath the Cretaceous-age cover. Little information is available about coal rank in this area. Three vitrinite-reflectance measurements (0.92-0.97%) (Hines, 1988; Levine et al., 1989; Winston, 1990) indicate that the coal rank of the Mary Lee coal group is high-volatile A bituminous (Stach, 1982, p. 45), and that the coal probably has a volatile-matter content of about 35% (Stach, 1982, p. 45). A vitrinite-reflectance measurement of 0.86 (Winston, 1990) from the Carrolls Creek Field would correspond to a volatile-matter content of about 37% (Stach, 1982, p. 45). Two of the three coalbed methane wells in this field which reached the Mary Lee coal group reported a small amount of gas production in initial tests.

Thus far, all coalbed methane production in Alabama has been from areas with vitrinite reflectance of 0.8% or higher. Thus, vitrinite reflectance of more than 0.8% is favorable for coalbed methane exploration. In the San Juan Basin, however, areas with vitrinite reflectances as low as 0.5% have coalbed methane production (Rice et al., 1989). The area with vitrinite reflectance between 0.5% and 0.8% in Alabama has not been drilled yet for coalbed methane. Significant new reserves may occur in this area.

Coal Resource Development

Coal quality has a major influence on the location of mines in Alabama. All the deep underground mines (deeper than 300 m (1,000 ft)) in the Mary Lee coal group are in the area where coal rank is medium-volatile bituminous or higher (Figure 11). Mining of high-volatile A bituminous coal and high-volatile B bituminous coal occurs in underground mines less than 150 m (500 ft) deep and in surface mines. Mining thus is restricted to the vicinity of the outcrop belt and to the high-rank area due to economics. Areas of low ash and S content are large enough to support mining. Mining can occur even in the high-ash, high-sulfur belt because ash and S contents locally are low within the belt and may be blended to conform to market demands. The average sale price of coal in Alabama exceeds $44/tonne ($40/ton) due to high mining costs, but high rank and quality support the expense, thereby keeping Alabama coal competitive especially on a cost per million Btu basis.

CONCLUSION

The Mary Lee coal group is a major economic resource in Alabama. Over half of all coal production in the state comes from the group which also is a major target for coalbed methane production. The rank of the group ranges from high-volatile B bituminous to low-volatile bituminous. Some high-volatile C bituminous coal may be present beneath the Cretaceous-age cover, but information for this area is scarce. Coal of medium-volatile bituminous rank or higher occurs along the trend of the Coalburg Syncline in southwest Jefferson County, northeast Tuscaloosa County, and in Sumter County (Figures 5 and 8). The area of high-rank coal in Jefferson and Tuscaloosa counties is due either to a higher temperature of coalification than other areas (due to a higher paleogeothermal gradient) or to a longer period of high temperature than other areas. The high-rank area can economically support underground mines deeper than 300 m (1,000 ft). Where the rank is lower, surface mines and underground mines shallower than 150 m (500 ft) are possible.

The Mary Lee coal group is a major target for coalbed methane production. Coal rank is one factor that may limit production. Coalbed methane has been produced in Alabama from areas where the Mary Lee coal seam has a volatile-matter content less than 37%. Based on this criterion, some untapped areas of potential coalbed methane production are in southern Walker County and southern Tuscaloosa County.

ACKNOWLEDGEMENTS

I would like to thank R.L. Barnett, R.V. Chandler, R.M. Mink, J.C. Pashin, L.J. Rheams, E.G. Rogers, B.H. Tew, W.A. Thomas, and W.E. Ward for their assistance during this project and Jim Walter Resources, Inc. for permission to collect coal samples. This project was supported in part by Gas Research Institute contract 5087-214-1544 to the Texas Bureau of Economic Geology.

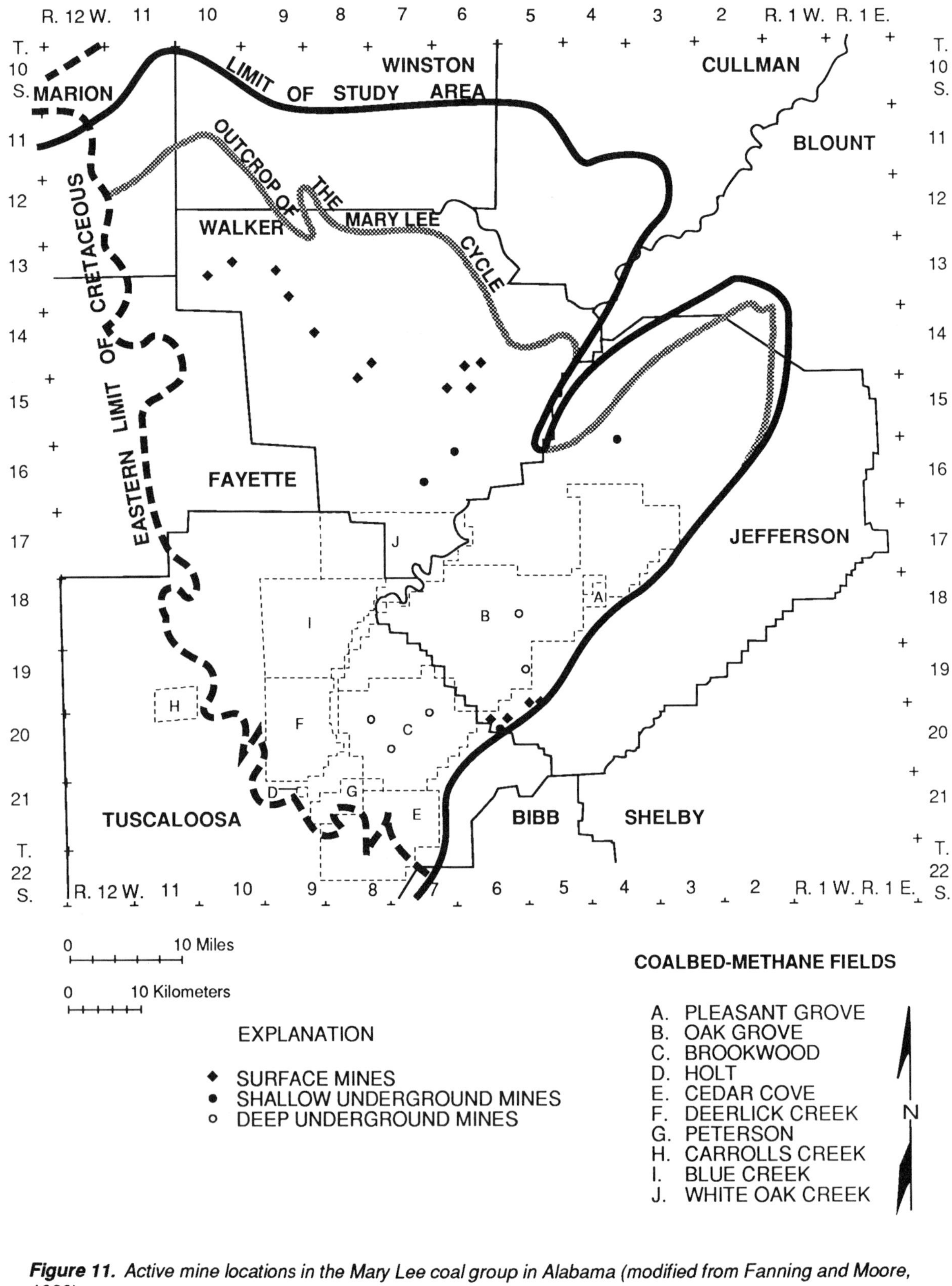

Figure 11. *Active mine locations in the Mary Lee coal group in Alabama (modified from Fanning and Moore, 1989).*

REFERENCES

American Society for Testing Materials, 1988a, D-388, Standard classification of coals by rank: Philadelphia, Annual Book of Standards, Vol. 05.05, p. 182-185.

American Society for Testing Materials, 1988b, D-2798, Standard method for microscopical determination of the reflectance of the organic components in a polished specimen of coal: Philadelphia, Annual Book of Standards, Vol. 05.05, p. 283-285.

American Society for Testing Materials, 1988c, D-3180, Standard method for calculating coal and coke analyses from as-determined to different bases: Philadelphia, Annual Book of Standards, Vol. 05.05, p. 319-321.

American Society for Testing Materials, 1988d, D-4596, Standard practice for collection of channel samples of coal in the mine: Philadelphia, Annual Book of Standards, Vol. 05.05, p. 383-384.

Close, J.C., and T.M. Erwin, 1989, Significance and determination of gas content data as related to coalbed methane reservoir evaluation and production implications, in Proceedings of the 1989 Coalbed Methane Symposium, April 17-20, 1989: Tuscaloosa, University of Alabama, Eastern Region Coalbed Methane Resource Center, p. 37-56.

Culbertson, W.C., 1964, Geology and coal resources of the coal-bearing rocks of Alabama: U.S. Geological Survey Bulletin 1182-B, 79 p.

Damberger, H.H., R.D. Harvey, R.R. Ruch, and J. Thomas, 1984, Coal characterization, in B.R. Cooper and W.A. Ellingson, eds., The science and technology of coal utilization: New York, Plenum Press, p. 7-45.

Epsman, M.L., G.V. Wilson, J.C. Pashin, J.S. Tolson, W.E. Ward, R.V. Chandler, R.B. Winston, K.E. Richter, R.P. Hamilton, and L.J. Rheams, 1989, Geologic evaluation of critical production parameters for coalbed methane resources in the Black Warrior Basin of Alabama. Annual report, Part II, Brookwood and Oak Grove Degasification Fields: Tuscaloosa, Geological Survey of Alabama, 178 p.

Fanning, B.J., and R. Moore, 1989. Surface and underground mines producing Alabama coal: Montgomery, Alabama Department of Economic and Community Affairs, 43 p.

Fieldner, A.C., H.M. Cooper, and F.D. Osgood, 1925, Analyses of mine samples: U.S. Bureau of Mines Technical Paper 347, p. 12-99.

Geochem Laboratories, 1982, Southern overthrust regional studies area II Alabama data shipment: Houston, unpublished proprietary report.

Hines, R.A., 1988, Carboniferous evolution of the Black Warrior foreland basin, Alabama and Mississippi: Tuscaloosa, University of Alabama, Unpublished Ph.D. Dissertation, 231 p.

Jüntgen, H., and J. Klein, 1975, Entstehung von Erdgas aus kohligen Sedimenten: Erdöl und Kohle, Erdgas Petrochemistrie, Ergängsband 1, p. 52-69.

Kidd, J.T., 1982, Structural geology of the Black Warrior Basin in Alabama, in L.J. Rheams and D.J. Benson, eds., Depositional setting of the Pottsville Formation in the Black Warrior Basin: Tuscaloosa, Alabama Geological Society, p. 27-33.

Levine, J.R., D.A. Thompson, W.R. Telle, and J.N. Thomas, 1989, A coalbed methane resource evaluation in southern Tuscaloosa County, Alabama: Tuscaloosa, University of Alabama, School of Mines and Energy Development, 90 p.

McCalley, H., 1899, Report on the Warrior Coal Basin: Tuscaloosa, Geological Survey of Alabama, 327 p.

McFall, K.S., D.E. Wicks, and V.A. Kuuskraa, 1986, A geologic assessment of natural gas from coal seams in the Warrior Basin, Alabama: Washington, Lewin and Associates, 84 p.

Murrie, G.W., W.P. Diamond, and S.W. Lambert, 1976, Geology of the Mary Lee group of coalbeds, Black Warrior coal basin, Alabama: U.S. Bureau of Mines Report of Investigations RI 8189, 49 p.

Pashin, J.C., W.E. Ward, R.B. Winston, R.V. Chandler, D.E. Bolin, R.P. Hamilton, and R.M. Mink, 1990, Geologic evaluation of critical production parameters for coalbed methane resources, Part II, Black Warrior Basin, Annual Report (August 1988-July 1989): Chicago, Gas Research Institute, 177 p.

Rightmire, C.T., 1984, Coalbed methane resource, in C.T. Rightmire, G.E. Eddy, and J.N. Kirr, eds., Coalbed Methane Resources of the United States: American Association of Petroleum Geologists, Studies in Geology Series #17, p. 1-13.

Rice, D.D., J.L. Clayton, and M.J. Pāwlewicz, 1989, Characterization of coal derived hydrocarbons and source-rock potential of coal beds, San Juan basin, New Mexico and Colorado, U.S.A.: International Journal of Coal Geology, v. 13, p. 597-626.

Robertson Research (U.S.) Inc., 1985, Oil generation in Black Warrior Basin, Alabama, a geochemical study: Kingwood, Texas, 328 p.

Semmes, D.R., 1920, Petroleum possibilities of Alabama, Part I, Northern Alabama: Geological Survey of Alabama Bulletin 22, p. 46-120.

Semmes, D.R., 1929, Oil and gas in Alabama: Geological Survey of Alabama Special Report 15, 408 p.

Shotts, R.Q., 1956, A compilation of complete analyses of Alabama coals published since 1925, Warrior and Plateau fields: Alabama State Mine Experiment Station Bulletin 6, 31 p.

Shotts, R.Q., 1960, Coal analyses made at the Alabama State Mine Experiment Station, 1944-60 and some other unpublished analyses: Alabama State Mine Experiment Station Bulletin 7, 39 p.

Stach, E., M.-Th. Mackowsky, M. Teichmüller, G.H. Taylor, D. Chandra, and, R. Teichmüller, 1982, Stach's Textbook of Coal Petrology (3rd Ed.): Berlin, Gebrüder Borntraeger, 535 p.

Thomas, W.A., 1988, The Black Warrior Basin, in L.L. Sloss, ed., Sedimentary cover--North American craton: Geological Society of America, The Geology of North America, Vol. D-2, p. 471-492.

Thomas, W.A., and D.N. Bearce, 1986, Birmingham anticlinorium in the Appalachian fold-thrust belt, basement fault system, synsedimentary structure, and thrust ramp, in T.L. Neathery, ed., Southeastern Section of the Geological Society of America: Geological Society of America, Centennial Field Guide Volume 6, p. 191-200.

Tolson, J.S., 1988, Alabama coal data for 1987: Geological Survey of Alabama, Information Series 58I, 131 p.

Ward, W.E., 1984, Reserve base of bituminous coal and lignite in Alabama: Geological Survey of Alabama, Circular 118, 102 p.

Ward, W.E., 1986, Coal occurrence and general geology in the Carbon Hill 7.5-minute quadrangle, Walker and Winston counties, Alabama: Geological Survey of Alabama, Map 150, 52 p.

Winston, R.B., 1990, Vitrinite reflectance of Alabama's bituminous coal: Geological Survey of Alabama, Circular 139, 54 p.

Winston, R.B., 1991, Preliminary report on coal quality trends in upper Pottsville Formation coal groups and their relationships to coal resource development, coalbed methane occurrence, and geologic history in the Warrior coal basin, Alabama: Alabama Geological Survey, Circular 152, 53 p.

GEOLOGICAL CONTROLS AND CONSTRAINTS ON THE CONCENTRATION OF ELEMENTS IN WESTERN CANADIAN COALS

F. Goodarzi[1]
Institute of Sedimentary and Petroleum Geology
Geological Survey of Canada
3303 - 33rd Street N.W.
Calgary, Alberta T2L 2A7
Canada

E. Van der Flier-Keller
Department of Geography
University of Victoria
Victoria, British Columbia V8W 2Y2
Canada

ABSTRACT

The distribution and concentration of major and minor elements in Western Canadian coals are controlled by the following parameters:

1) The Environment of Deposition: This is the most important factor in controlling the elemental concentration in coals, particularly the organically bound elements (elements associated with coal macerals; i.e., boron and sulfur). The environment of deposition of the coals studied is divided into freshwater, slightly brackish, and brackish based on sedimentological studies.

2) Weathering: Weathering has an important control on the concentration or depletion of elements. Weathered coals are depleted in organically bound elements. In contrast, they are enriched in inorganically bound elements (elements associated with coal minerals). These relationships were determined by examining weathered and fresh samples of the same seam.

3) Rank: The rank of coal controls the elemental composition and migration behavior of the elements with respect to groundwater, possibly by controlling the number of organic bonding sites. The concentration of As and alkaline group elements such as barium and sodium generally often are rank related and decrease with an increase in coal rank in Western Canada.

4) Geological Setting, Country Rock, and Groundwater: These three factors also are important in the concentration of elements in coal. The coal in an area rich in arsenic or uranium often is enriched in these two elements. The uranium oxide (U_3O_8) content in such a setting can be as high as 4,400 ppm in coal

---

1. Geological Survey of Canada Contribution No. 39889.

ash, as exemplified by the Saskatchewan lignites. Faulting and folding of a coal seam also may result in elemental enrichment (e.g., enrichment of the platinum group elements due to movement of platinum-rich solutions along fault planes). This appears to have occurred in the Blakeburn coals of south-central British Columbia. The activity of groundwater is important in the mobilization of elements, particularly in low-rank coals (lignite to subbituminous). Low-rank coals often act as aquifers and concentrate elements such as arsenic and boron. Examples would be the subbituminous coals of British Columbia and the lignites of Southern Saskatchewan.

INTRODUCTION

The nature and distribution of major, minor, and trace elements in coal and the degree of concentration of certain environmentally hazardous elements (i.e., arsenic (As), boron (B), bromine (Br), cadmium (Cd), chlorine (Cl), chromium (Cr), fluorine (F), molybdenum (Mo), lead (Pb), sulfur (S), and selenium (Se)) or industrially harmful elements (i.e., B, sodium (Na), phosphorus (P), and S) are important in coal utilization. The strict environmental standards in regard to utilization of coal set by governments and coal users dictate the production of an environmentally clean coal. Selective mining of coal reduces the input of hazardous elements such as iron (Fe), S, and others by eliminating the part of a seam enriched in these elements (Figure 1). Furthermore, mechanical cleaning and washing may reduce the amount of hazardous elements to an acceptable level.

Geochemical variation in coal is well documented (e.g., see Gluskoter et al., 1977, and Valkovic, 1982). This geochemical variation occurs worldwide and can be studied on various scales: 1) in single coal seams (Gluskoter et al., 1977; Goodarzi, 1988; Goodarzi and Van der Flier-Keller, 1988; Rimmer and Davis, 1986), 2) within coal deposits containing a number of seams (Goodarzi, 1987a and b; Harvey et al., 1983; Kraner et al., 1986; Van der Flier-Keller and Fyfe, 1987), and 3) between coal deposits in the same or different coal basins (Glick and Davis, 1984; Goodarzi, 1988; Goodarzi and Van der Flier-Keller, 1989; Harvey and Ruch, 1986; Van der Flier-Keller and Fyfe, 1987). The variation of elements in coal has been attributed to a number of geological factors: 1) the nature of the depositional environment (Chou, 1984; Finkelman, 1983; Gluskoter et al., 1977; Goodarzi, 1987a and b, 1988; Lindahl and Finkelman, 1986; Styan and Bustin, 1983; Swaine, 1962, 1983); 2) the original plant communities (Cecil et al., 1983; Styan and Bustin, 1983); 3) the coal rank (Goodarzi et al., 1985; Lindahl and Finkelman, 1986; Van der Flier-Keller and Fyfe, 1987); 4) the nature of the country rock, especially as it may affect detrital material supplied to the coal basin (Denson and Gill, 1956; Goodarzi and Van der Flier-Keller, 1988); 5) the hydrologic conditions (Goodarzi and Van der Flier-Keller, 1988; Spears, 1987; Van der Flier-Keller and Fyfe, 1988); and 6) the basin characteristics and tectonic setting (Fielding, 1987; Hobday 1987).

Western Canadian Coals

Major Canadian coal resources (about 246,000,000,000 tonnes (271,000,000,000 tons) out of a total of 248,000,000,000 tonnes (273,000,000,000 tons)) which are classed as resources of immediate interest are present in the three western provinces of British Columbia, Alberta, and Saskatchewan (Figure 2) (Darragh and Goodarzi, 1985; Smith, 1989). These resources can be divided geographically into the following regions (see Figure 2): (a) Insular belt, (b) Intermontane coals of British Columbia (various formations); (c) Mountain and Inner Foothills (Kootenay Group, Luscar Group, and equivalents); (d) Outer Foothills and Alberta Plains (Saunders, Belly River, and Edmonton groups); (e) Saskatchewan coals (Ravenscrag Formation) (Darragh and Goodarzi, 1985; Smith 1989). Production of coal in Western Canada is about 70,000,000 tonnes (77,000,000 tons) per year (Smith, 1989).

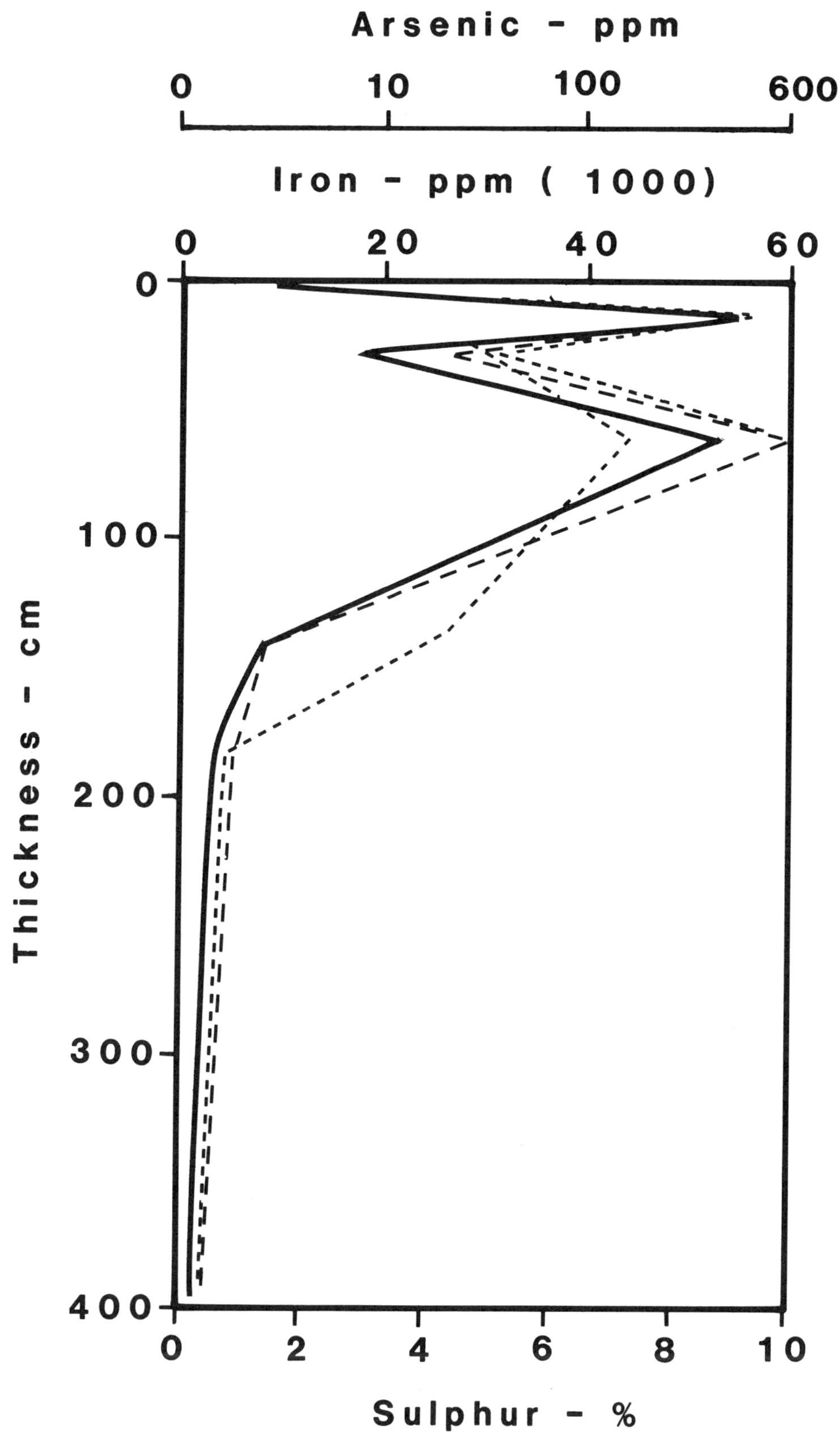

Figure 1. *Distribution of As, Fe, and S in a coal seam from Alberta showing the concentration of arsenopyrite (FeAsS) near the top of the coal seam.*

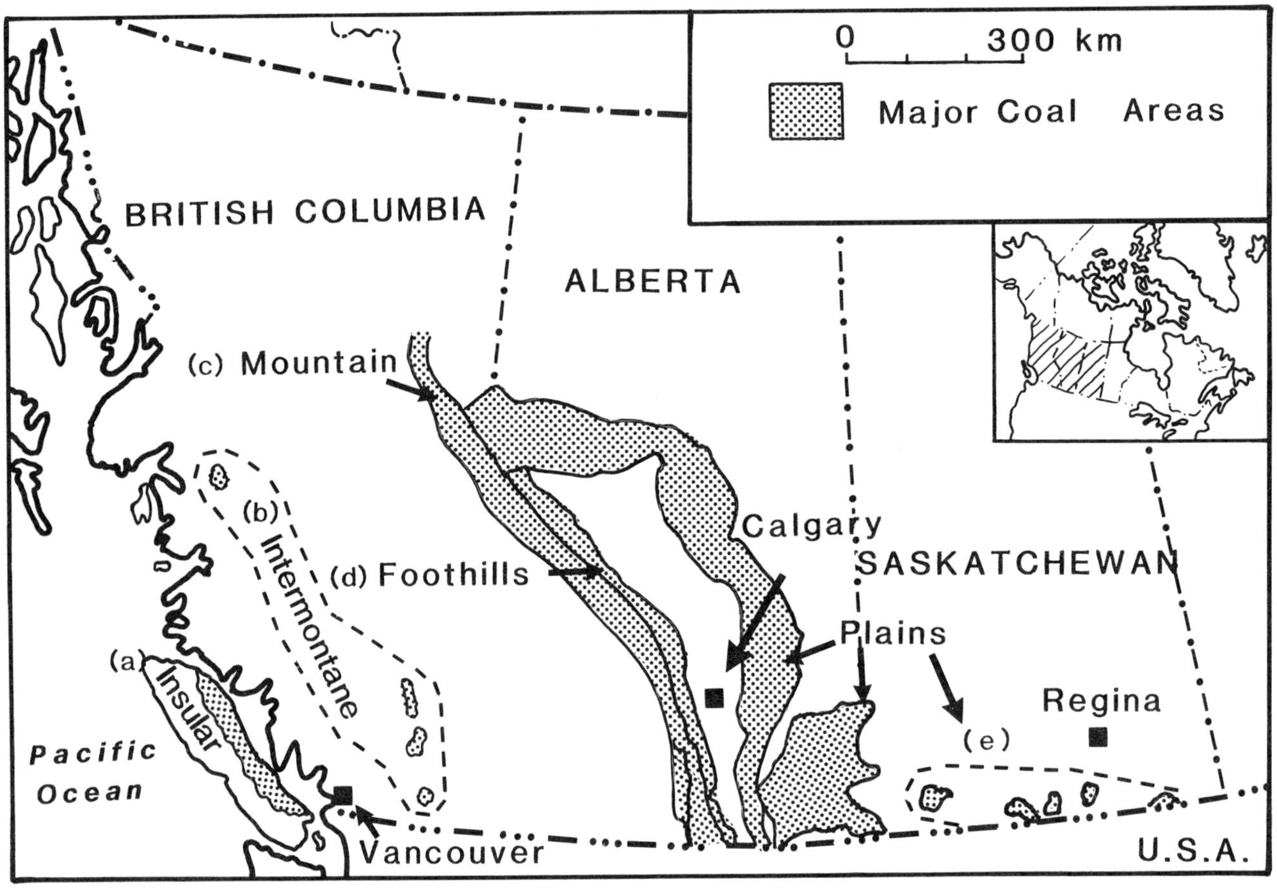

Figure 2. *Geographical distribution of Western Canadian coals.*

The coal deposits range in age from Jurassic/Cretaceous to Oligocene. Coal rank increases from lignite-subbituminous in Saskatchewan to anthracite in the Groundhog Field of northwest British Columbia (Bustin, 1984; Cameron, in press; Hughes and Cameron, 1985; Kalkreuth and McMechan, 1984). However, the lignite to subbituminous coals generally occur in the Plains (Alberta and Saskatchewan) and southern Intermontane regions (British Columbia), whereas bituminous to anthracitic coals occur in the Mountain, Foothills and Cordillera regions (Vancouver Island and northern Intermontane, such as the Groundhog Field).

A number of coal deposits from each geographic and geologic setting were chosen for this study on the basis of availability of geochemical data. The deposits include a) Insular Belt coals (Nanaimo and Comox deposits), b) Intermontane coals (Hat Creek No. 1 and No. 2, Tulameen, Merritt, and Telkwa deposits), c) Mountain and Inner Foothills (Sparwood Ridge (Westar Coal Co.), Eagle Mountain, Greenhills (Fording Mine), Byron Creek Mine, Crowsnest Resources, and Mount Allan and Tent Mountain deposits), d) Outer Foothills and Alberta Plains (Obed Marsh, Highvale, and Roselyn mines), and e) Saskatchewan (Willowbunch and Estevan fields).

Geochemistry Of Western Canadian Coals

Studies on geochemical characteristics of individual coal deposits in Western Canada have been carried out on the Canadian Cordillera coals (Vancouver Island coals)(Van der Flier-Keller and Dumais, 1988),

Intermontane coals (Dilles and Hill, 1984; Goodarzi, 1987; Goodarzi et al., 1985; Goodarzi and Van der Flier-Keller, 1988, 1989; Hill, 1988; Landheer et al., 1982; Nichols and D'Auria, 1981), Mountain and Inner Foothills coals (Goodarzi, 1987c, 1988; Goodarzi and Cameron, 1987; Goodarzi et al., 1985; Kronberg et al., 1981; Landheer et al., 1982; Newmarch, 1953; Van der Flier-Keller and Fyfe, 1987), Outer Foothills and Alberta Plains coals (Landheer et al., 1982), and Saskatchewan coals (Beaton and Goodarzi, in press; Landheer et al., 1982). Goodarzi et al. (1985) concluded that local conditions of deposition, sediment provenance, rank, and age of coals have a definite control on elemental concentrations in coals. The concentration range of elements in coal samples from a single formation is much smaller than the range in elemental concentration in coal samples from different formations.

The objectives of this paper are to discuss the geological controls and constraints on the concentration of elements in selected Western Canadian coal deposits, particularly in relation to a) nature of depositional environment, b) weathering, c) rank, and d) geological setting, nature of the country rock, and groundwater activity.

Analysis Methods

Coal was collected as channel samples for the whole of each seam. The elemental concentration for each seam is based on an average of 1-2 whole coal seam samples.

Trace element (except B) determinations were made by induced neutron activation analysis carried out on coal samples. Boron was determined using sodium hydroxide fusion followed by inductively coupled plasma (ICP) emission spectrometry (Pollock, 1975).

The platinum-group elements (PGE) and gold (Au) were determined using lead fire assay and direct-current plasma spectroscopy. The results are reported as "whole coal" analyses.

Analytical results that are reported as "less than" values represent the presence of an element in a particular sample at a concentration below the limit of quantitative accuracy for a particular analysis method. These low values, along with values falling outside of a standard deviation of ±20% of measured values, are discarded when performing statistical evaluations.

Statistics

Cluster analyses of trace elements in the coal samples were performed using the method described by Labonte and Goodarzi (1985). There are some elements, such as silver (Ag), Au, Cd, iodine (I), indium (In), iridium (Ir), niobium (Nb), nitrogen (N), Pr, tin (Sn), and zirconium (Zr), that were reported as being "less than" or as having standard deviations of ±20% of measured values which were not included in the statistical analyses.

Dendrograms, which are tree-like, two-dimensional graphs for depicting the results of a clustering process, were used to evaluate element-element and sample-sample comparisons. The clustering process is based on the magnitude of similarity of occurrences of elements between samples or elements within a sample, starting with the highest order of similarity and ending with the lowest. The object names (variables or samples) are spaced evenly along the x-axis. The fusion (clustering) of objects occurs at levels that are shown on the y-axis, which constitutes a similarity index and which has annotations that the user can use as a reference level to ascertain how many clusters are contained in the dendrogram. The x-axis represents the linear arrangement of objects, permuted so as to achieve the different fusions from which the tree-like structure is derived.

The dendrogram also depicts similarities within groups as well as between groups of objects. Therefore, the dendrogram method is capable of identifying similarities between coal samples.

DISCUSSION OF CONTROLS ON CONCENTRATIONS OF ELEMENTS

The Nature Of The Depositional Environment

The distributions of some elements (i.e., B, Cl, Cr, Na, and S) often are controlled by the environment of deposition, particularly during the early stages of coal accumulation. The environment of coal accumulation controls the Eh-pH of the system, and the above listed elements are particularly sensitive to Eh-pH changes. For example, coals deposited in brackish-water environments are formed under high-pH conditions which are caused by the presence of carbonates. This is in contrast to freshwater coals which are formed under low pH conditions. Marine-influenced coals often contain high concentrations of B, Na, Cl, and pyrite (Cohen et al., 1984; Goodarzi, 1987c) and of elements associated with pyrite, such as S, Fe, As, and Mo.

Coal forms in a) freshwater (inland, limnic-telmatic), b) slightly brackish (coastal, paralic), and c) marine-influenced (brackish) environments (Teichmüller and Teichmüller, 1982). There are differences in the mineral matter and elemental concentrations of coals deposited in freshwater, slightly brackish, and brackish environments. Freshwater coals contain low concentrations of pyrite (Cecil et al., 1982; Cohen and Spackman, 1977; Cohen et al., 1984), total S (<0.6%) (Cohen et al., 1984), and B (average = 20 ppm) (Goodarzi, 1987a; 1987; Goodarzi and Van der Flier-Keller, 1988; Swaine, 1983). Slightly brackish-water coals contain intermediate amounts of S (1.35%) (Cohen et al., 1984) and B (average = 50-90 ppm) (Goodarzi, 1988; Swaine, 1983). Brackish-water coals contain high amounts of pyrite and S (>1.8%) (Cohen et al., 1984) and B (average = 160 ppm) (Goodarzi, 1987; Swaine, 1983).

The concentration of B in coal has been used to determine the paleosalinities of depositional environments for Australian, Canadian, and U.S. coals (Bohor and Gluskoter, 1973; Chou, 1984; Gluskoter et al., 1977; Goodarzi, 1987a and b, 1988; Lindhal and Finkelman, 1986; Swaine, 1962, 1983). Swaine (1983) has grouped some Australian coals according to their B content, with freshwater coals containing up to 40 ppm, slightly brackish-water coals with 40-120 ppm, and marine-influenced coals with >120 ppm. Bohor and Gluskoter (1973) noted that some Illinois coals with <125 ppm B content were deposited in freshwater environments. Chou (1984) stated that coal overlain by marine rocks has higher B, S, Mo, and uranium (U) than coal overlain by fluvial and lacustrine rocks.

Goodarzi (1987a and b, 1988) and Goodarzi and Van der Flier-Keller (in press) have examined the B concentration of Western Canadian coals in five basins (Figure 3), and the B contents were compared in terms of the coal depositional environments. Insular Region coals from Vancouver Island (Figure 2) were deposited under brackish conditions (B = 109-499 ppm; Van der Flier-Keller and Goodarzi, in press). The brackish environment also is indicated by the presence of perhydrous vitrinite and by sedimentological and paleontological evidence.

In the Intermontane coal deposits (Figure 2) from British Columbia, B content is in the range of a freshwater environment (<50 ppm; Goodarzi, 1987a; Goodarzi and Van der Flier-Keller, 1988, 1989). Sedimentological evidence indicates these coals were accumulated in a freshwater lacustrine-fluvial environment (Camsell, 1913; Shaw, 1952; Williams and Ross, 1979; Wilson, 1979).

Seam No. 1 at Byron Creek, British Columbia (Figure 4), occurs near the base of the Mist Mountain Formation (Kootenay Group). It has higher B and Na, and lower thorium (Th) than seam No. 3, which was deposited above seam No. 1 (Figure 4), suggesting that seam No. 1 was deposited in brackish conditions and seam

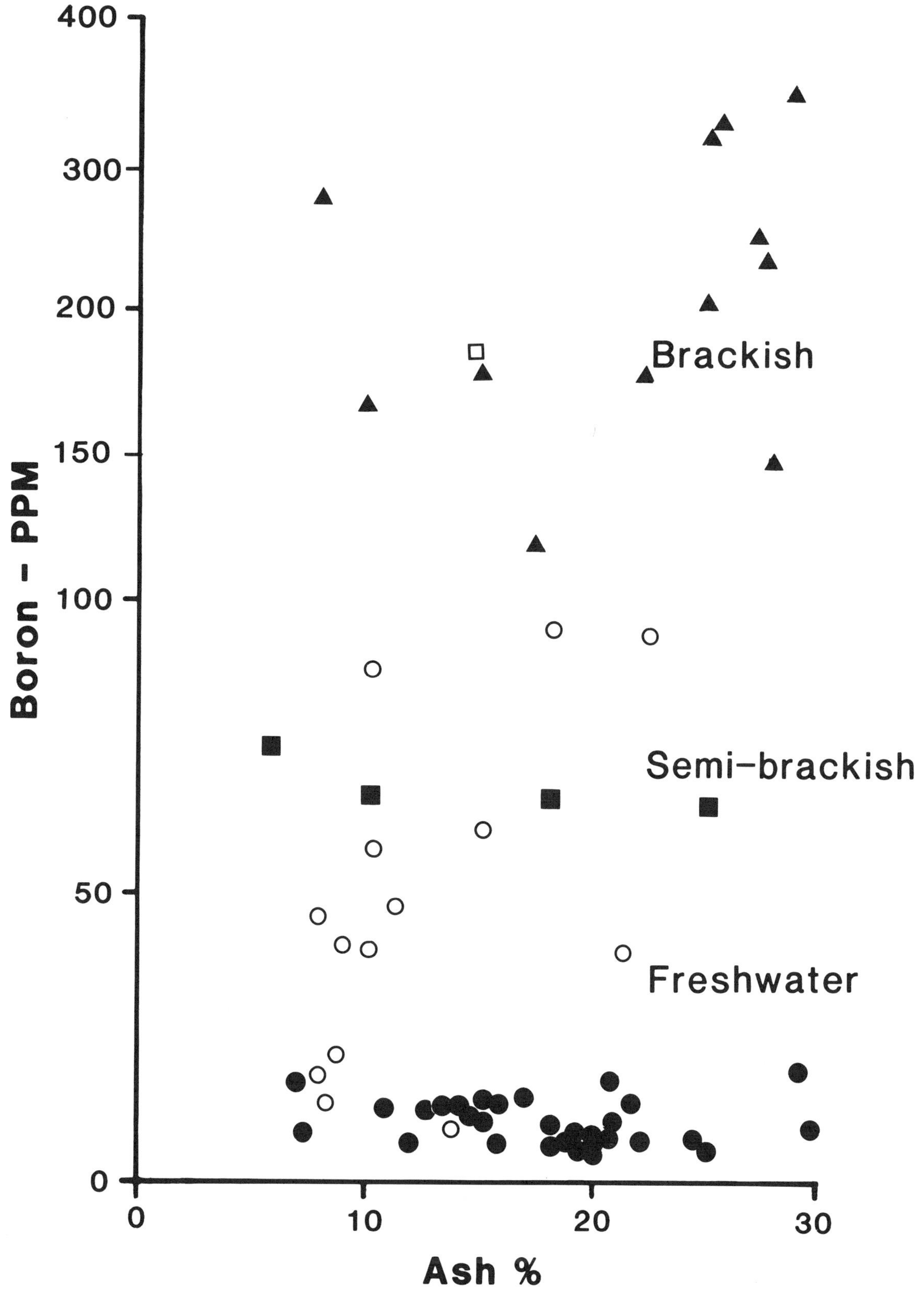

Figure 3. Concentration of B versus ash content in coals from Hat Creek (●) Obed Marsh (■), Fording (o), Byron Creek (□), and Insular coals (Δ). Each data point is representative of a coal seam.

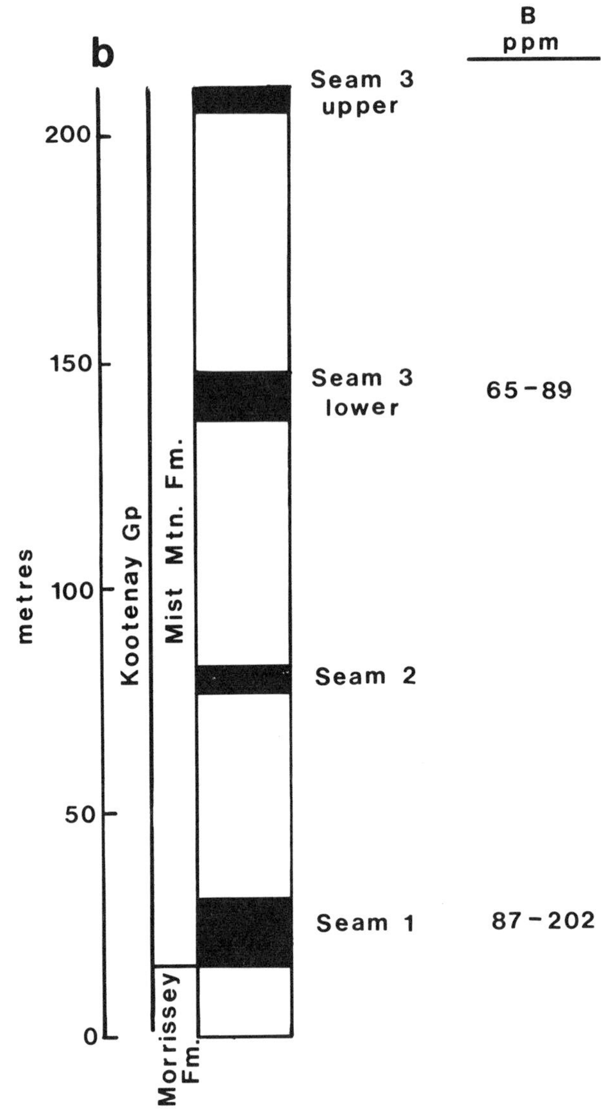

Figure 4. *Generalized section of coal-bearing strata from the Byron Creek Collieries in the Mist Mountain and Morrissey formations of the Kootenay Group.*

No. 3 in a slightly brackish environment. Additional evidence regarding the brackish-water character of seam No. 1 is the frequent occurrence of coprolites in this coal (Goodarzi, 1986) and the sedimentological evidence provided by Gibson (1985) for the Kootenay Group. Gibson (1985) suggested a prograding deltaic environment with the basal sandstone representing a beach ridge or coastal dune environment. Succeeding units are portrayed as deposited in fluvial-deltaic environments largely or totally uninfluenced by marine conditions. The seam at Byron Creek, which is most likely to be influenced by marine conditions using the sedimentological framework of Gibson (1985), is seam No. 1, which stratigraphically occurs close to the basal sandstone. However, seam No. 3, which occurs stratigraphically higher (150 m (490 ft)) than seam No. 1 at Byron Creek, was influenced by slightly brackish conditions. This is confirmed by the B content of seam No. 3 which ranges from 65 to 89 ppm (Figure 4).

Fording coals also belong to the Kootenay Group (Figure 5) and are located in the same basin as the Byron Creek coals. Dunlop and Bustin (1988) stated that the coal seams from the Eagle Mountain section of the

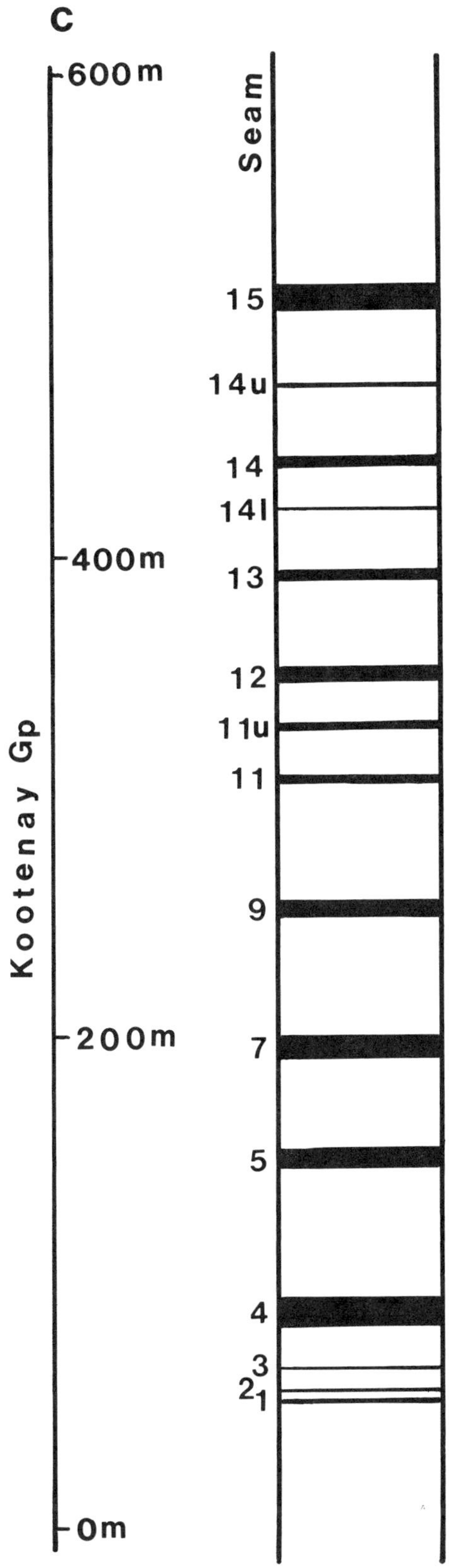

Figure 5. Generalized section of coal-bearing strata from Fording Coal Mine, British Columbia. Numbers refer to seam designations.

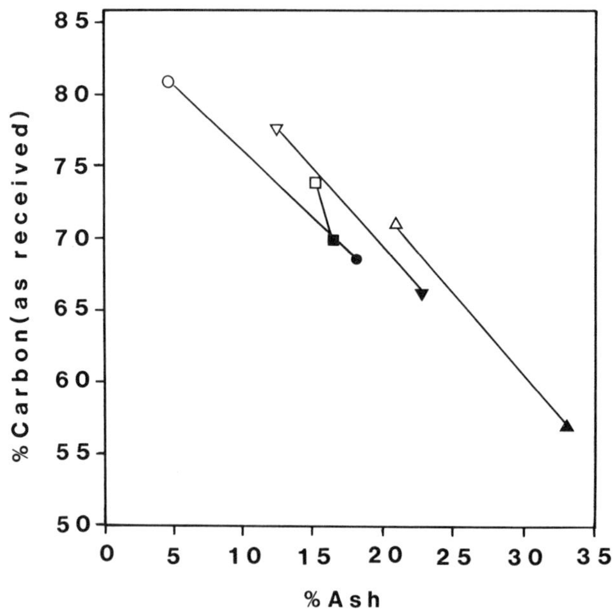

Figure 6. *Concentration of C (as received) versus ash content of coal. The open symbols represent fresh coal from the mine. The closed symbols represent weathered coal from outcrop. Connecting lines between symbols indicate samples from the same coal seam.*

Fording Mine accumulated in a nonmarine environment. Coal seams No. 1 and 2 were deposited in a coastal peat swamp behind dune ridge deposits. The stratigraphically higher coal seams No. 4 through No. 8 were deposited in a fluvial floodplain environment. In contrast, the upper coal seams No. 4 through No. 15 were formed in a floodplain-lacustrine environment. Goodarzi (1988) examined the B content of Fording coals (Eagle Mountain section) and his data suggest that the depositional environment for the coals is freshwater to slightly brackish (B = 9-85 ppm).

Obed Marsh coals belong to the Outer Foothills and Alberta Plains Region (Figure 2). The B content of these coals is 52 ppm, which is in agreement with the sedimentological interpretation of Jerzykiewicz and McLean (1980) who suggest that the coals were formed inland in a meandering river/lake environment.

All these studies indicate that paleosalinity, as determined by B content, should be supplemented by other lines of evidence such as sedimentology, paleontology, and organic petrology. Gluskoter et al. (1977) stated that B should be used with caution in the determination of paleosalinity. High B content of a coal simply could be due to the activity of brine groundwater (Beaton and Goodarzi, in press). Goodarzi (1987a and b, 1988) and Goodarzi and Van der Flier-Keller (1989) have used the concentration of elements such as B, Cr, and S combined with available sedimentological data to determine the environment of deposition of Western Canadian coals. The Byron Creek, Hat Creek, and Fording coals (Figure 3) were used as standards for determination of depositional environments. The data suggest that the coals from the Saskatchewan Plains Region (i.e., Estevan, Willow Bunch, and Cypress coal fields), most coals in the Alberta Plains and Outer Foothills Region (Obed Marsh and Highvale mines), Mountain and Inner Foothills Region (most of Fernie and Elk Valley coal fields), and Intermontane coals are formed under freshwater to freshwater-slightly brackish environments (Figure 3). In contrast, the coals from the Insular Basin, British Columbia (Figure 3), and some Mountain and Outer Foothills coals (seam No. 1, Byron Creek Collieries) (Figure 3) were formed in brackish environments.

Weathering

Weathering is the natural oxidation of coal which occurs at low temperature (<30°C) over a long period of time (years). Weathering relatively increases the oxygen (O), moisture, and ash contents, but decreases the fixed carbon and calorific value of coal (Ettinger et al., 1966; Goodarzi, 1987b; Van Krevelen, 1961). Figure 6 shows the variation of C versus percent ash in fresh coals and their weathered outcrops in Western Canadian coals.

Goodarzi (1987b) examined the elemental concentrations of fresh Kootenay coals and their weathered outcrops. He found that weathering depletes the organically bound elements, especially the halogens (i.e., Br and Cl) and organic S (Figures 7 and 8), while concentrating the inorganically bound elements (i.e., aluminum (Al) and inorganic S) (Figures 8 and 9). Results of Goodarzi and Van der Flier-Keller (1988) indicate that the concentration of halogens (Cl and Br) is depleted in a suite of 12-yr old core samples of subbituminous coal from Zone A of the Hat Creek coal deposit No. 2 when compared to relatively fresh (1-yr old) coal samples from the same zone of the Hat Creek coal deposit No. 1 (Figure 10).

The above results indicate that element concentrations from weathered outcrops or samples do not reflect the true element concentrations in fresh coal.

Rank

Variation of organically bound elements (i.e., carbon (C), hydrogen (H), N, and O) with rank is well documented (Van Krevelen, 1961). In contrast to H, N, and O, the concentration of C increases with the rank of coal. There is little reference to the variation of elements other than C, H, N, and O with increasing rank in Western Canadian coals.

Goodarzi et al. (1985) compared the elemental distribution in both coal and coal ash with increasing rank (%R_o = 0.20-2.14) from Western Canada using statistical analyses (Labonte and Goodarzi, 1985). The coals fall into two groups, one group made up of lignites and subbituminous coals, though from different formations and basins, whereas bituminous and semi-anthracite coals from the Kootenay Group fall into the second group (Figure 11).

In contrast to this study, Goodarzi and Cameron (1987) examined the concentration of elements in medium-volatile bituminous to semi-anthracite coal (%R_o = 1.30-2.45) from the Mount Allan section, Alberta (Figures 12 and 13). Concentrations of some elements (i.e., U, hafnium (Hf), and rare earths) shows a uniform

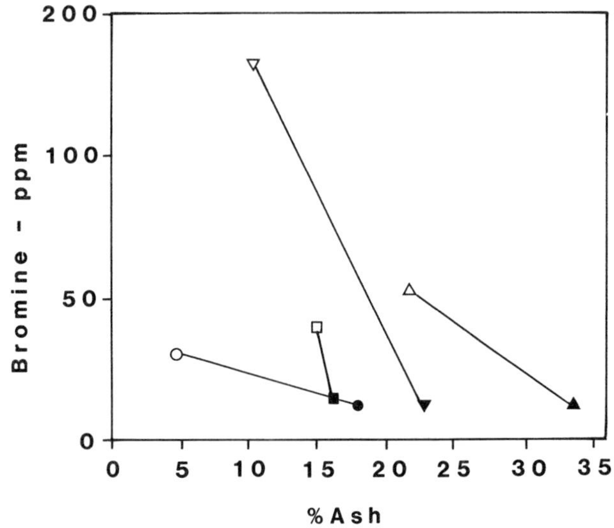

Figure 7. Concentration of Br versus ash content of coal. Legend same as for Figure 6.

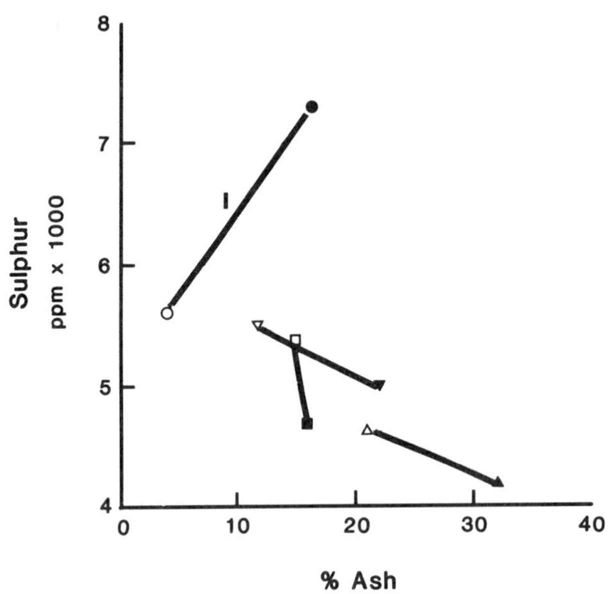

Figure 8. Concentration of total S (as received) versus ash content of coal. Legend same as for Figure 6 (I = inorganically bound S).

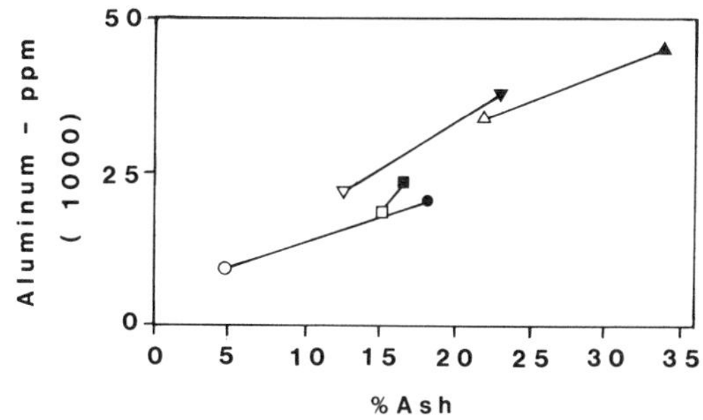

Figure 9. *Concentration of Al versus ash content of coal. Legend same as for Figure 6.*

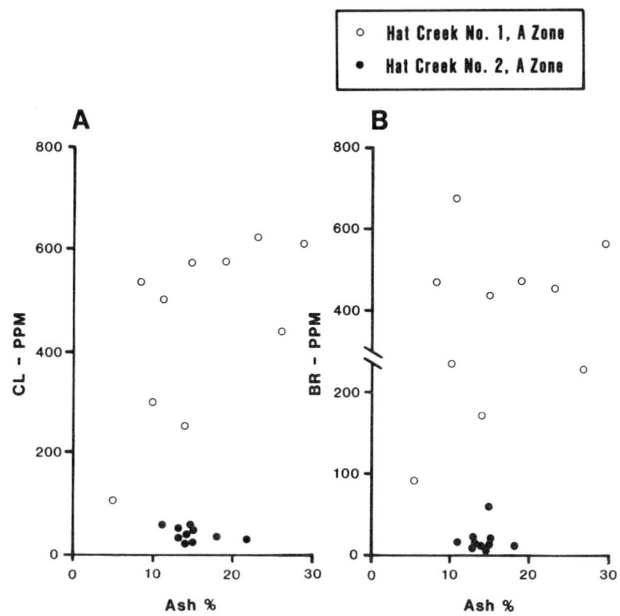

Figure 10. *Variation of Br and Cl versus ash content for coals in Zone A of Hat Creek deposits No. 1 (o) and No. 2 (•) (modified from Goodarzi and Van der Flier-Keller, 1988). Each data point represents a coal seam.*

decrease with increasing rank, whereas concentrations of other elements, such as Br, decrease with rank (Figure 14). The Mount Allan samples fall into two populations, one consisting of seams of semi-anthracitic rank and the upper end of the low-volatile bituminous range, whereas the other population is comprised of the remainder of the low-volatile coals along with those of medium-volatile bituminous rank (Figure 15).

The concentration and distribution of As, Na, and Ba in Western Canadian coals of different ages (Oligocene-Cretaceous) and basins can be correlated partially to the coal rank. The concentration of As decreases with increasing rank from lignite-subbituminous to bituminous-semi-anthracitic coals (Figure 16a). However, there are coals which do not follow this trend (i.e., the high-volatile bituminous coals from the Comox Basin) which have a high As content (155 ppm, Van der Flier-Keller and Dumais, 1987). Both Ba and Na decrease with increasing rank from lignite to high-volatile bituminous coal (Figures 16b and 16c), in agreement with the results of Lindahl and Finkelman (1986) who stated that concentrations of Ba and Na

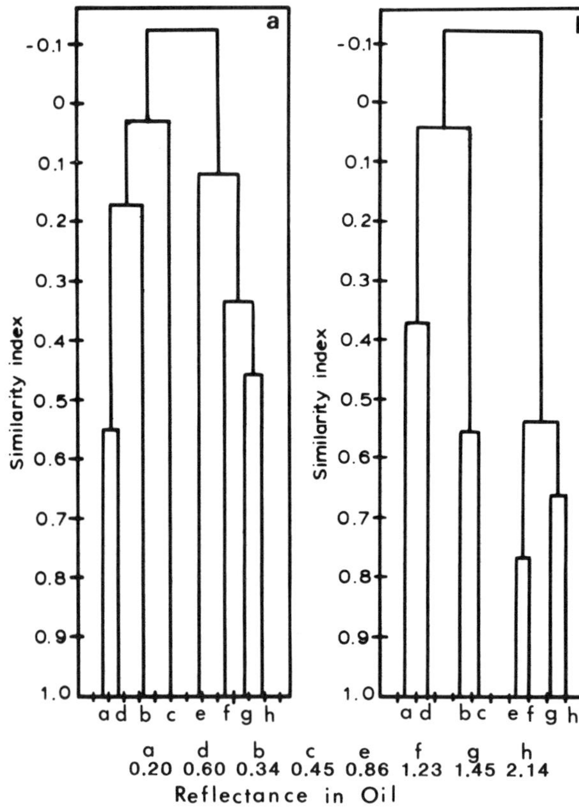

Figure 11. *Dendrogram showing the similarity index between coals A) based on trace elements in coal and B) based on trace elements in coal ash (after Goodarzi et al., 1985).*

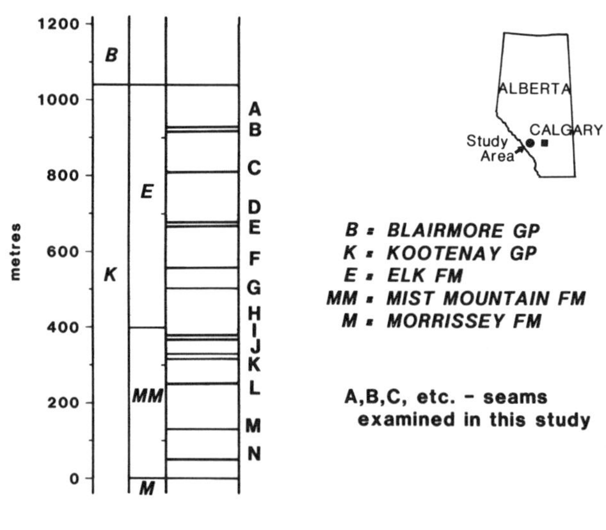

Figure 12. *Locality map and schematic representation of the Mount Allan section (after Goodarzi and Cameron, 1987).*

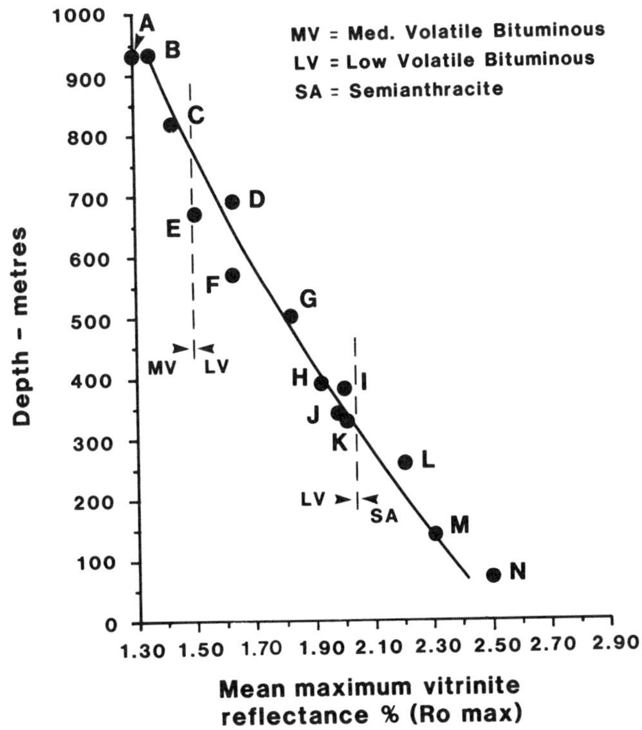

Figure 13. *Depth-reflectance profile for the Mount Allan section (redrawn from Hughes and Cameron, 1985).*

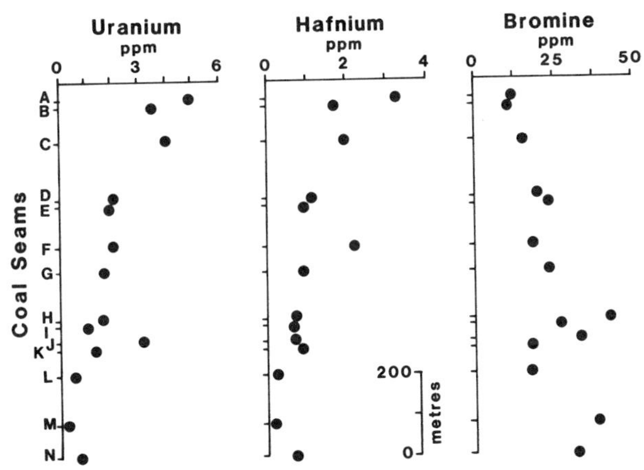

Figure 14. *Variation in concentrations of Br, Hf, and U with increasing depth. Each data point represents a coal seam (after Goodarzi and Cameron, 1987).*

decrease with increasing rank for U.S. coals. This relationship between the elements As, Na, and Ba with rank is more noticeable in lower-rank coals (lignite to subbituminous), perhaps due to the presence of functional groups, such as -COOH, which provide sites for absorption of these elements. The amount of functional groups decreases with increasing rank, and this is evident by the small changes in the concentrations of As, Na, and Ba above the first coalification jump (>0.50 $\%R_o$) (Figures 16a-c).

The concentration of vanadium (V) in Western Canadian subbituminous to semi-anthracite coals decreases with increasing rank (Figure 16d). However, lignites from the Alberta Plains and Saskatchewan regions do

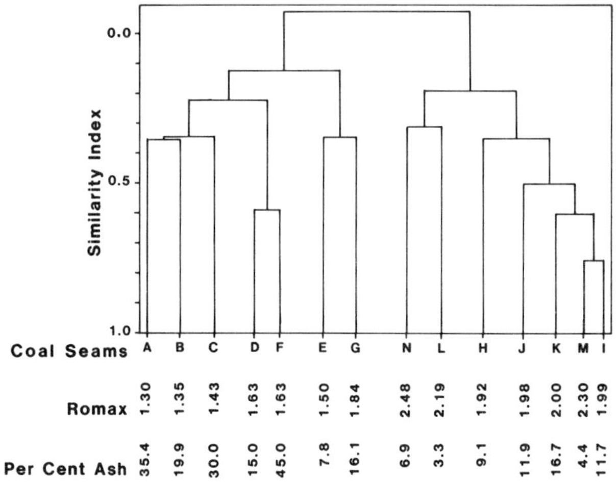

Figure 15. Similarity index of coal samples from Mount Allan section (redrawn from Goodarzi and Cameron, 1987).

not follow the rank relationship, and their V content often is low (Figure 16d). The low V content in the Plains coals possibly is due to the source rocks being depleted in V. Chlorine is an element with organic affinity in coal; therefore, its concentration should decrease with increasing coal rank. In fact, the concentration of Cl shows an increase with increasing rank (Figure 17). In addition, the coals appear to fall into two groups based on their depositional environments: a) freshwater to slightly brackish-water coals with low Cl content, and b) brackish-water coals with high Cl content (Figure 17). However, this variation in Cl content may not reflect the true geochemistry of Cl with increase in rank due to the sensitivity of Cl to weathering and its depletion in weathered outcrops (Goodarzi, 1987b). Possible inputs of Cl by secondary NaCl-bearing groundwater (Beaton, 1989) also should be considered.

Geological Setting, Nature Of Country Rocks, And Activity Of Groundwater

The geochemical characteristics of coals also are controlled by the nature of the country rocks, particularly through the activities of associated groundwater (Freeze and Cherry, 1979). The influence of country rocks may be limited to only coal-bearing strata or extended throughout a coal deposit and coal field. The following examples from Western Canada show the influence of country rock and groundwater on the geochemistry of coals.

Blakeburn Open Cast Mine, Tulameen, South-central British Columbia

The Tulameen coals are high-volatile bituminous in rank and are of Eocene-Oligocene age (Goodarzi and Van der Flier-Keller, 1988). The petrology and geochemistry of a 200-m (660-ft) long section of this coal in the Blakeburn Mine are reported by Goodarzi and Van der Flier-Keller (1988, 1989). The coal section is disrupted by a major fault and some minor faults which offset the entire coal-bearing sequence by up to 30 m (100 ft) (Van der Flier-Keller and Goodarzi, 1989). The geochemical study of these coals indicates that the coals are enriched in platinum group elements, particularly in coals associated with minor faults (up to 10,000 ppb platinum (Pt) and 6,100 ppb paladium (Pd)) (Van der Flier-Keller and Goodarzi, 1989). The enrichment of platinum group elements (PGE) in the faulted Blakeburn coals is due to increased surface area of the coal associated with deformation and resulting increased flow of groundwater which has flowed through the PGE-rich ultramafic complex overlying and/or surrounding the coal-bearing strata.

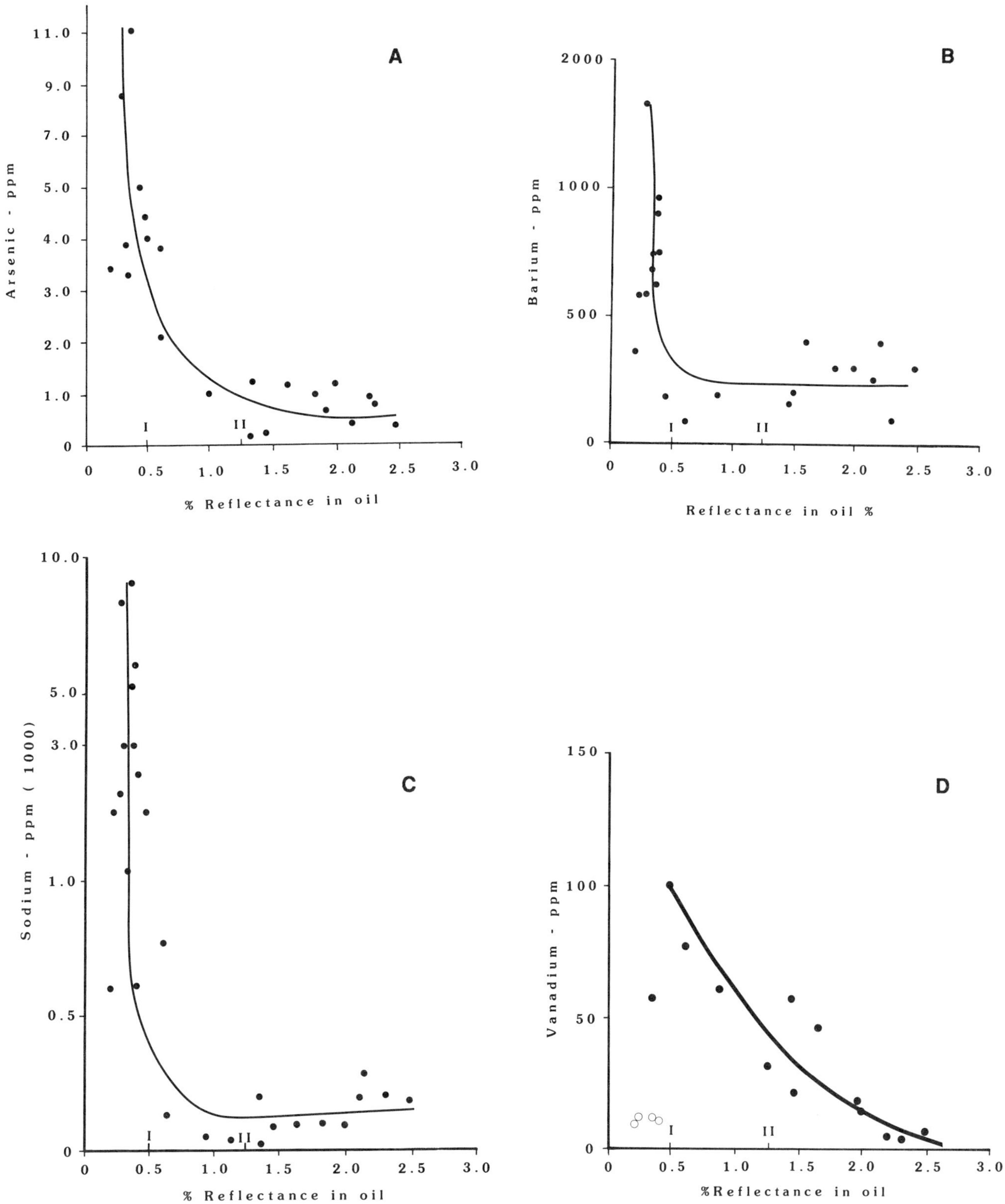

Figure 16. *Variation of elements versus rank in Western Canadian coals. Each data point represents a coal seam. A) As, B) Ba, C) Na, and D) V. Coals from the Plains are shown by open symbols. I and II indicate the coalification stages.*

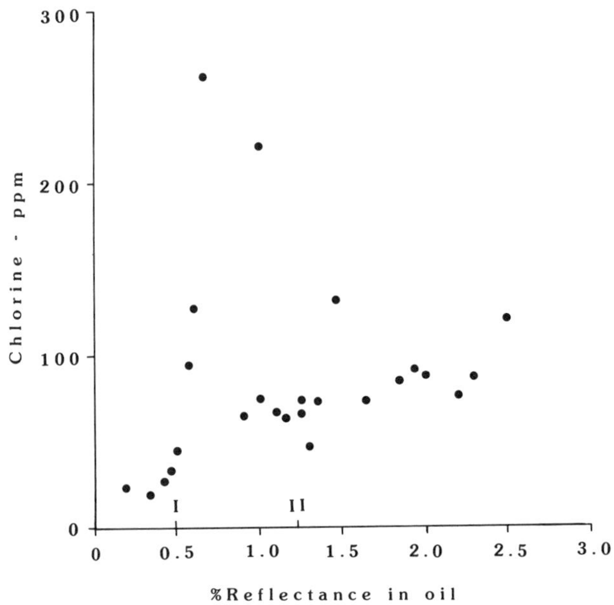

Figure 17. *Variation of Cl versus rank in Western Canadian coals. Each data point represents a coal seam.*

British Columbia and Saskatchewan Coal Fields

High concentrations of As and S in the Hat Creek coals and of Th and U in coals from Cypress Hills is attributed to the activity of groundwater. The Hat Creek coal deposits are located in Intermontane British Columbia (Figure 2). Coals from these deposits are lignite to subbituminous in rank and Eocene to Oligocene in age (Goodarzi, 1985; Goodarzi and Gentzis, 1987). The geochemistry of these coal deposits is reported by Goodarzi (1987a), Goodarzi and Hill (1989), and Goodarzi and Van der Flier-Keller (1988). Coal deposit No. 2 at Hat Creek, British Columbia is 510 m (1,700 ft) thick and contains four coal zones from top to bottom (A, B, C, and D, respectively) (Figure 18). Each coal zone contains a number of coal seams (Goodarzi and Gentzis, 1987). The rank of coal increases with depth ranging from lignite (%R_o = 0.35) at the top of the section to subbituminous B (%R_o = 0.48) at the base (zone D).

The higher concentrations of As and S for zones A and B (Figure 19) are related to rank (by way of porosity) and activity of the groundwater. The porosity of coal is related to the molecular structure of coal and is rank related, particularly for coal of bituminous rank (Van Krevelen, 1961). In general, the porosity of coal increases with decreasing rank (Cartz and Hirsch, 1960). Because of this relationship between rank and porosity, low-rank coals (lignite to subbituminous) are good aquifers (Van der Flier-Keller and Fyfe, 1987). Therefore, they may concentrate or lose elements as a result of water percolating through them. Because As and S are organically bound elements, these elements may have been leached out of volcanic rocks occurring in the area (Campbell et al., 1977) and brought into the two upper zones (A and B) by groundwater. The decreases in the concentrations of As and S from zones A to D also is related to a rank increase in the same direction (Figure 14).

The Cypress Hills coal deposit is located in southwestern Saskatchewan. Coals are lignite in rank and Paleocene in age (Beaton, 1989; Cameron, in press). Cameron and Birmingham (1970) examined the coals from the Cypress Hills area. Sporadic occurrences in the coals near the town of Eastend reveal high uranium oxide (U_3O_8) contents (825 ppm in coal, 4,400 ppm in ash). The highest value for uranium oxide occurs in

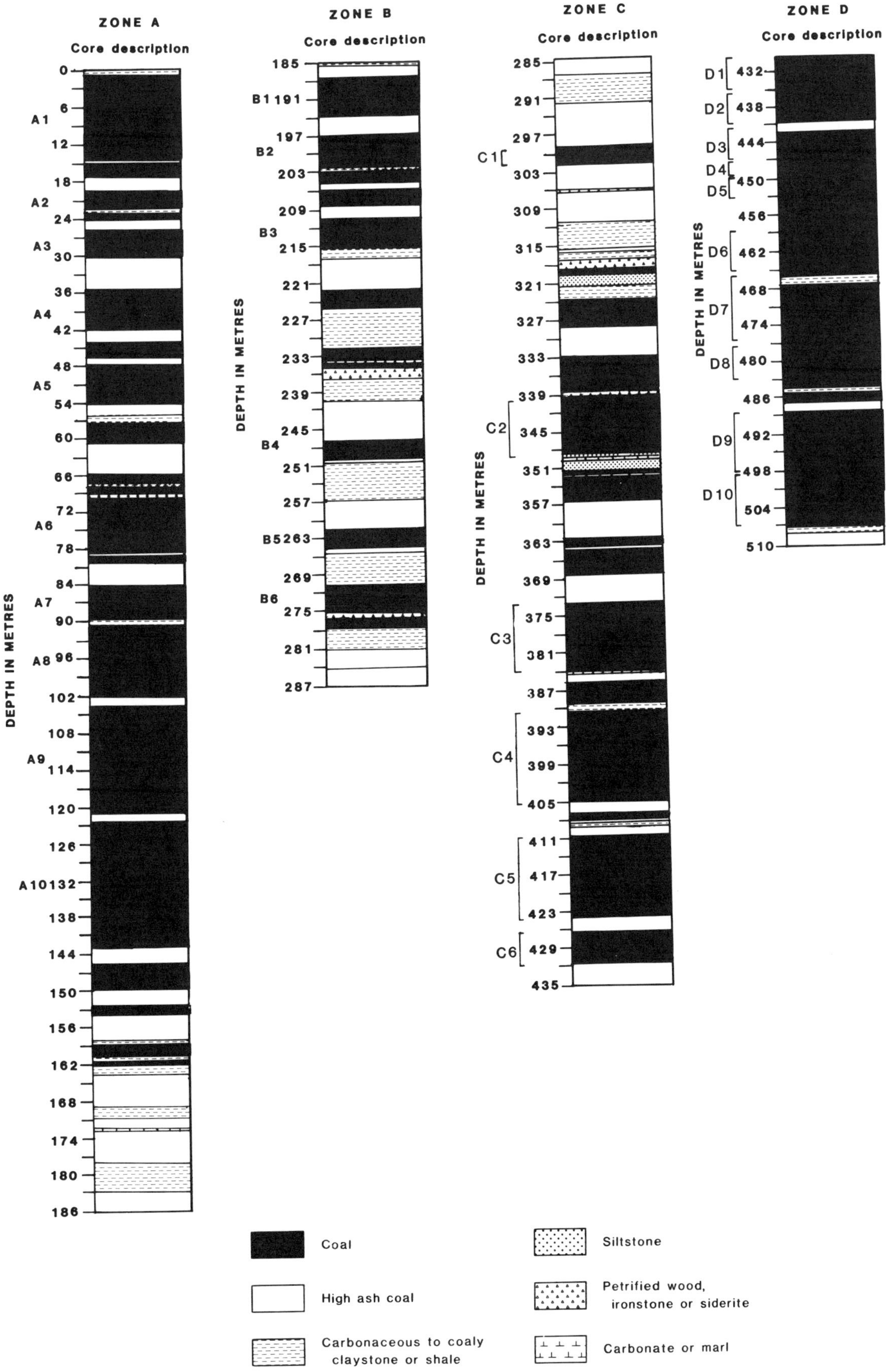

Figure 18. *Generalized section of coal-bearing strata from Hat Creek coal deposit No. 2.*

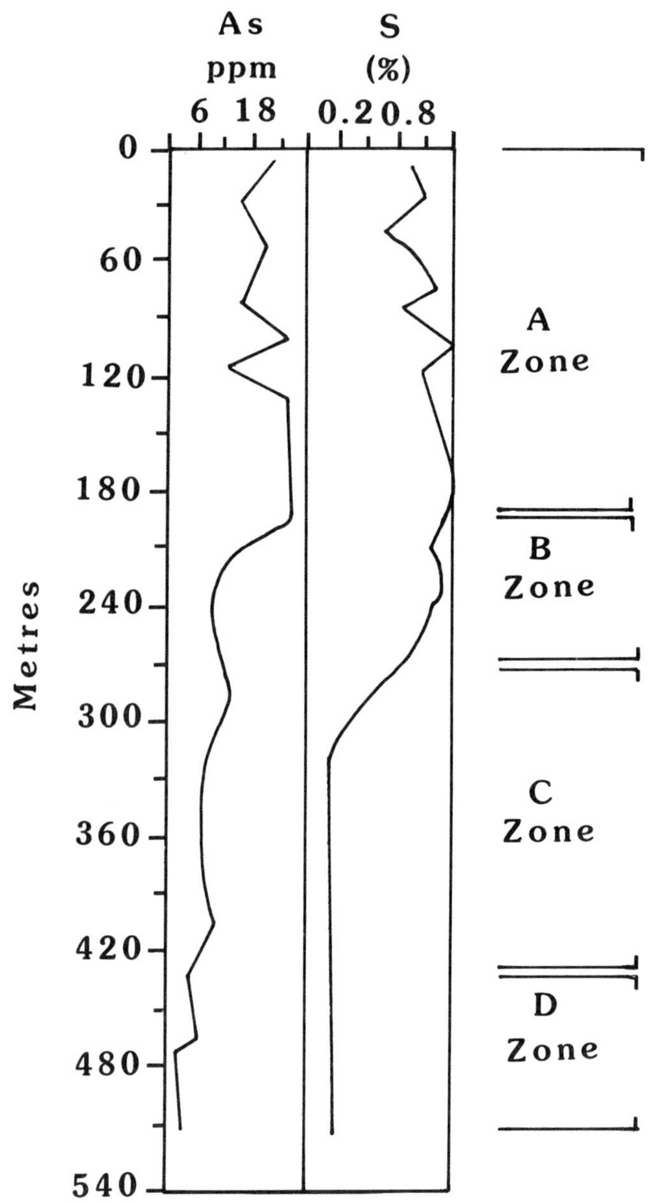

Figure 19. *Variation of As (ppm) and total S (%) in coal zones from Hat Creek coal deposit No. 2. Each data point represents a coal seam.*

the coal samples taken from the upper part of the coal seams. This is similar to observations by Ting (1972) which indicate that high concentrations of U and Mo occur in coals underlying volcanic rocks in North Dakota. Uranium in North Dakota is concentrated mostly in the tops of coal seams. The distribution of U in coal seams is attributed by Ting (1972) to downward movement of groundwater. This is similar to results from Cameron and Birmingham (1970) indicating an enrichment of uranium oxide due to leaching of U from overburden rocks, which include volcanoclastics (Frazer et al., 1935), and its deposition in the coal.

The sedimentological evidence indicates that the lignite deposits of Southern Saskatchewan were deposited in freshwater environments (Blackburn, 1982; Nambudiri, 1987). However, the environment of deposition as determined by B (109-234 ppm) is indicative of a slightly brackish to brackish environment of deposition for some of these lignites (Beaton and Goodarzi, in press). This is contradictory to the palynological, paleobotanical, and sedimentological evidence (Blackburn, 1982; Nambudiri, 1987). Broughton (1979, 1985) has stated that the coal fields of Southern Saskatchewan were formed in depression areas resulting from deep, underlying salt dissolution collapse structures. Up to 180 m (590 ft) of salt has been dissolved from deep, underlying Devonian evaporites in this area.

The detailed geochemical study of Beaton (1989) indicates that the Saskatchewan lignites containing high amounts of B and Na are in part influenced by the upwelling of brine. In contrast, lignites which are not in direct contact with the groundwater have low B content (10-27 ppm) indicating the true level of B for coals formed in freshwater (Goodarzi, 1987a). Clearly the high B and Na contents in some of the lignites in Saskatchewan is an overprint from the country rocks through groundwater activity.

SUMMARY

There are a number of geological parameters that influence the distribution and concentration of elements in Western Canadian coals. These are 1) depositional environment and rank of coal, 2) weathering, 3) nature of the country rock, and 4) activity of groundwater. The above parameters are the most important geological controls on the elemental composition of coals. Coals deposited in freshwater environments have low concentrations of B. In contrast, coals deposited in brackish or marine-influenced environments have relatively high B contents. Brackish-water coals, in general, contain higher amounts of sulfides and halogens. Rank of coal also controls the elemental composition of coal, particularly As and alkaline groups such as Na and Ba.

Weathering has a significant effect on organically bound elements (i.e., C, S, and particularly halogens). These elements are depleted in outcrop coal samples when compared to fresh samples of the same coals. The reverse is evident for inorganically bound elements.

One of the most important geological factors controlling the influx of some elements into coals is the nature of country rock and activity of groundwater. Low-rank coals particularly are affected by these parameters. High As, B, Na, U, and PGE contents in coals can be due to leaching of country rock which contains these elements and their subsequent absorption by coal from the same groundwater. The absorption of elements is greater the lower the rank of the coal.

ACKNOWLEDGMENTS

The manuscript benefitted from critical review by Dr. A. R. Cameron of the Geological Survey of Canada and Dr. T. Gentzis of the Alberta Research Council. We thank both of them.

REFERENCES

American Society for Testing Materials, 1978, Annual Book of ASTM Standards, Part 26: Proximate analysis of coals and coke, D3172-73, and Ultimate analysis of coal and coke, D3176-74: Washington, p. 380 and p. 390-391 (respectively).

Beaton, A.P., 1989 The organic petrology and geochemistry of lignite from the Paleocene Ravenscrag Formation, Southern Saskatchewan: University of Regina, Energy Research Unit, Open-File Report No. 203, 200 p.

Beaton, A.P., and F. Goodarzi, in press, The petrology, mineralogy and elemental composition of a lignite seam from the Paleocene Ravenscrag Formation, Southern Saskatchewan: International Journal of Coal Geology.

Blackburn, D.T., 1982, Palaeobotany of coal-bearing strata from the Cretaceous and Tertiary of Western Canada and Ellesmere Island: Geological Survey of Canada, Open File Report 3-128, 128 p.

Bohor, B.F., and H.J. Gluskoter, 1973, Boron in illite as a paleosalinity indicator of Illinois coals: Journal of Sedimentary Petrology, v. 43, no. 4, p. 945-956.

Broughton, P.L., 1985, Geology and resources of the Saskatchewan coalfields, in T.H. Patching, ed., Coal in Canada: Canadian Institute of Mining and Metallurgy, Special Volume 31, p. 87-98.

Broughton, P.L., 1979, Origin of coal basins by salt solution tectonics in Western Canada: University of Cambridge, Unpublished Ph.D. Thesis, 289 p.

Bustin, R.M., 1984, Coalification, level and their significance in the Groundhog coalfield, north-central British Columbia: International Journal of Coal Geology, v. 4, p. 21-44.

Cameron, A.R., in press, Regional patterns of reflectance distribution in lignites of the Ravenscrag Formation, Saskatchewan, Canada: International Journal of Coal Geology.

Cameron, A.R., and T.P. Birmingham, 1970, Radioactivity in Western Canadian coals: Geological Survey of Canada, Paper 70-52, 35 p.

Campbell, D.D., L.T. Jory, and C.R. Saunders, 1977, Geology of Hat Creek coal deposits: Canadian Institute of Mining and Metallurgy Bulletin, v. 70, p. 99-108.

Camsell, C., 1913, Geology and mineral deposits Tulameen District, British Columbia: Geological Survey of Canada, Memoir 26, 188 p.

Cartz, L., and P.N. Hirsch, 1960, A contribution to the structure of coals from X-ray diffraction studies: Transactions of the Royal Society of London, Series A, v. 252, p. 557-604.

Cecil, C.E., R.W. Stanton, F.T. Dulong, and J.J. Renton, 1982, Geologic factors that control mineral matter in coal, in R.H. Filby, ed., Atomic and nuclear methods in fossil energy research: New York, Plenum Press, p. 323-335.

Cecil, C.E., R.W. Stanton, F.T. Dulong, and J.J. Renton, 1983, Recent developments on the origin of mineral matter in coal, in Proceedings of the International Conference on Coal Science, August, 1983, Pittsburgh, Pennsylvania: IEA, p. 381.

Chou, C.L., 1984, Relationship between geochemistry of coal and nature of strata overlying in Herrin coal in the Illinois Basin, U.S.A.: Memoirs of the Geological Society of China, v. 6, p. 269-280.

Church, B.N., 1977, Geology of Hat Creek Basin, summary of field activities: British Columbia Department of Mines, p. 104-108.

Cohen, A.D., and W. Spackman, 1977, Phytogenic organic sediments and sedimentary environments in Everglades-mangrove complex: Palaeontographica, v. 172, p. 125-149.

Cohen, A.D., W. Spackman, and P. Dolsen, 1984, Occurrence and distribution of sulphur in peat forming environment of Southern Florida: International Journal of Coal Geology, v. 4, p. 73-96.

Darragh, A., and F. Goodarzi, 1985, Canadian coals--Their distribution and potential: Journal of Coal Quality, v. 5, no. 1, p. 35-39.

Denson, N.N., and J.R. Gill, 1956, Uranium-bearing lignite and its relation to volcanic tuffs in Eastern Montana and North and South Dakota: U.S. Geological Survey Professional Paper 300, p. 413-418.

Dilles, S.J., and P.A. Hill, 1984, Vertical distribution of copper and zinc in coal zones A, B, C, D, Hat Creek, British Columbia: Economic Geology, v. 79, p. 1936-1940.

Dunlop, R.L., and R.M. Bustin, 1988, Sedimentology and depositional environment of Mist Mountain Formation, at Eagle Mountain, Southeast Canadian Rocky Mountains: Bulletin of Canadian Petroleum Geology, v. 35, no. 4, p. 389-415.

Ettinger, E., E. Lidin, and I. Eremin, 1966, Natural factors influencing coal sorption properties, II: Gas capacity of coals found in weathering zone of coal deposit: Fuel, v. 45, p. 277-282.

Falini, F., 1965, On the formation of coal deposits of lacustrine origin: Geological Society of America Bulletin, v. 76, p. 1317-1346.

Fraser, F.J., F.H. McLearn, L.S. Russell, P.S. Warren, and R.D.T. Wickenden, 1935, Geology of Southern Saskatchewan: Geological Survey of Canada, Memoir 176, 137 p.

Freeze, R.A., and J.A. Cherry, 1979, Groundwater: Englewood Cliffs, New Jersey, Prentice-Hall Inc., 604 p.

Gibson, D.W., 1985, Stratigraphy, sedimentology and depositional environments of the coal-bearing Jurassic-Cretaceous Kootenay Group, Alberta and British Columbia: Geological Survey of Canada Bulletin 357, 108 p.

Glick, D.C., and A. Davis, 1984, Variability in the inorganic content of United States' coals--A multivariate statistical study, Final Report--Part 10: Pennsylvania State University, DOE Contract Report DE-AC22-80PC30013, 404 p.

Gluskoter, H.J., R.R. Ruch, W.G. Miller, R.A. Cahill, G.B. Dreher, and J.K. Kuhn, 1977, Trace elements in coal: Occurrence and distribution: Illinois State Geological Survey, Circular 499, 154 p.

Goodarzi, F., 1985, Organic petrology of Hat Creek Deposit No. 2, British Columbia: International Journal of Coal Geology, v. 5, p. 377-396.

Goodarzi, F. 1986, Comparison of morphology and reflectance of macerals from a tectonically thickened coal seam: Geological Survey of Canada, Paper 86-1B, p. 671-678.

Goodarzi, F., 1987a, Concentration of elements in lacustrine coals from Zone A, Hat Creek Deposit No. 1, British Columbia, Canada: International Journal of Coal Geology, v. 8, p. 247-268.

Goodarzi, F., 1987b, Comparison of elemental distribution in fresh and weathered samples of selected coals in the Jurassic-Cretaceous Kootenay Group, British Columbia, Canada: Chemical Geology, v. 63, p. 21-28.

Goodarzi, F., 1987c, Elemental concentrations in Canadian coals 2: Byron Creek collieries, British Columbia: Fuel, v. 66, p. 250-254.

Goodarzi, F., 1988, Elemental distribution in coal seams at the Fording Coal Mine, British Columbia, Canada: Chemical Geology, v. 68, p. 129-154.

Goodarzi, F., and A.R. Cameron, 1987, Distribution of major, minor and trace elements in coals of the Kootenay Group, Mount Allan, Alberta: Canadian Mineralogist, v. 25, p. 555-565.

Goodarzi, F., A.E. Foscolos, and A.R. Cameron, 1985, Mineral matter and elemental concentrations in Western Canadian coals: Fuel, v. 64, p. 1599-1605.

Goodarzi, F., and T. Gentzis, 1987, Depositional setting, determined by organic petrography, of the Middle Eocene Hat Creek No. 2 Coal Deposit, British Columbia, Canada: Bulletin of Canadian Petroleum Geology, v. 35, p. 197-211.

Goodarzi, F., and P.A. Hill, 1989, Intermontane coal at Hat Creek, British Columbia: Review of macerals, minerals, tonsteins, and elements, in T. Thanasuthipitak and P. Qunchanum, eds., Proceedings of the International Symposium on Intermontane Basins, Geology and Resources, Chiang Mai, Thailand: [s.n.], p. 89-98.

Goodarzi, F., and E. Van der Flier-Keller, 1988, Distribution of major, minor and trace elements in Hat Creek Deposit No. 2., British Columbia, Canada: Chemical Geology, v. 70, p. 313-333.

Goodarzi, F., and E. Van der Flier-Keller, 1989, Organic petrology and geochemistry of Intermontane coals from British Columbia, III--The Blakeburn Opencast Mine in Tulameen, British Columbia, Canada: Chemical Geology, v. 75, p. 227-247.

Goodarzi, F., and E. Van der Flier-Keller, 1990, Variation of elements in the southern Intermontane coals of British Columbia, Canada: Journal of Coal Quality, v. 9, no. 2, p. 62-65.

Harris, L.A., H.E. Barrett, and O.C. Kopp, 1981, Elemental concentrations and their distribution in two bituminous coals of different paleoenvironments: International Journal of Coal Geology, v. 1, p. 175-193.

Harvey, R.D., R.A. Cahill, C.L. Chou, and J.D. Steele, 1983, Mineral matter and trace elements in the Herrin and Springfield Coals, Illinois Basin Coal Field: Final Report: Illinois State Geological Survey, Report #EPA-CR806654 (EPA-600/7-84-036), 162 p.

Harvey, R.D., and R.R. Ruch, 1986, Mineral matter in Illinois and other U.S. coals, in K.S. Vorres, ed., Mineral matter and ash in coal: American Chemical Society, Symposium Series 301, p. 10-40.

Hill, P.A., 1988, Vertical distribution of minerals in coal zones A, B, C, D, Hat Creek, British Columbia: International Journal of Coal Geology, v. 10, p. 141-153.

Hobday, D.K., 1987, Gondwana coal basins of Australia and Africa: Tectonic setting, depositional systems and resources, in A.C. Scott, ed., Coal and coal-bearing strata: Recent advances: Geological Society, Special Publication No. 32, p. 219-233.

Hughes, J.D., and A.R. Cameron, 1985, Lithology, depositional setting and coal rank-depth relationships in the Jurassic-Cretaceous Kootenay Group at Mount Allan, Cascade coal basin, Alberta: Geological Survey of Canada, Paper 81-11, 41 p.

Jerzykiewicz, T., and J.R. McLean, 1980, Lithostratigraphical and sedimentological framework of coal-bearing Upper Cretaceous and Lower Tertiary strata, Coal Valley area, central Alberta Foothills: Geological Survey of Canada, Paper 79-12, 47 p.

Kalkreuth, W., and M.E. McMechan, 1984, Regional pattern of thermal maturation as determined from coal-rank studies, Rocky Mountain Foothills and front ranges north of Grande Cache, Alberta--Implications for petroleum exploration: Bulletin of Canadian Petroleum Geology, v. 32, p. 249-271.

Karner, F.R., H.H. Schobert, S.K. Falcone, and S.A. Benson, 1986, Elemental distribution and association with inorganic components in North Dakota lignites, in K.S. Vorres, ed., Mineral matter and ash in coal: American Chemical Society, Symposium Series 301, p. 70-89.

Kim, H., 1979, Depositional environment and stratigraphic subdivision, Hat Creek No. 1 deposit, British Columbia, in Proceedings of the Fourth Annual Meeting of the Canadian Institute of Mining and Metallurgy, Vancouver, British Columbia, October 25-27, 1979: Vancouver, British Columbia Hydro and Power Authority, p. 1-20.

Kronberg, B., J.R. Brown, W.S. Fyfe, M. Pierce, and G.C. Winder, 1981, Distributions of trace elements in Western Canadian coal ashes: Fuel, v. 60, p. 59-63.

Labonte, M., and F. Goodarzi, 1985, Use of the dendograph for data processing in fuel science: Fuel, v. 64, p. 1177-1178.

Landheer, F., H. Dibbs, and J. Labuda, 1982, Trace elements in Canadian coals: Ottawa, Ontario, Environment Canada, Environmental Protection Service Report No. 3-AP-82-6, 41 p.

Lindahl, P.C., and R.B. Finkelman, 1986, Factors influencing major, minor and trace element variation in U.S. coals, in K.S. Vorres, ed., Mineral matter and ash in coal: American Chemical Society, Symposium Series 301, p. 61-69.

Nambudiri, V., 1987, Peat forming flora in the Paleocene Ravenscrag Formation, Southern Saskatchewan: University of Regina, Energy Research Unit Report No. 175, p. 28.

Newmarch, C.B., 1953, Geology of the Crowsnest coal basin: British Columbia Department of Mines, Bulletin No. 33, 107 p.

Nichols, C.L., and J.M. D'Auria, 1981, Seam and location differentiation of coal specimens using trace element concentrations: Analyst, v. 106, p. 874-882.

Pollock, E.N., 1975, Trace impurities in coal by wet chemical methods, in S.P. Babu, ed., Trace elements in fuel: American Chemical Society, Special Paper 141, p. 23-24.

Rimmer, S.M., and A. Davis, 1986, Geological controls on the inorganic composition of Lower Kittanning Coal, in K.S. Vorres, ed., Mineral matter and ash in coal: American Chemical Society, Symposium Series 301, p. 41-52.

Shaw, W.S., 1952, The Tulameen coalfield, British Columbia: Geological Survey of Canada, Paper S2-19, 13 p.

Smith, G.G., 1989, Coal resources of Canada: Geological Survey of Canada, Paper 89-4, 146 p.

Spears, D.A., 1987, Mineral matter in coals, with special reference to the Pennine Coalfields, in A.C. Scott, ed., Coal and coal-bearing strata: Recent advances: Geological Society, Special Publication No. 32, p. 171-185.

Styan, W.B., and R.M. Bustin, 1983, Petrography of some Fraser River Delta peat deposits: coal maceral and microlithotype precursors in temperate-climate peats: International Journal of Coal Geology, v. 2, p. 321-370.

Swaine, D.J., 1962, Boron in New South Wales Permian coals: Australian Journal of Earth Sciences, v. 26, no. 6, p. 265-266.

Swaine, D.J., 1983, Geological aspects of trace elements in coal, in S.S. Augustithis, ed., The significance of trace elements in solving petrogenetic problems and controversies: Athens, Greece, Theophrastus Publication, p. 521-532.

Teichmüller, M. and R. Teichmüller, 1982, Fundamentals of coal petrography, in E. Stach et al., eds., Coal petrology: Berlin-Stuttgart, Gebrüder Borntraeger, p. 5-82.

Ting, F.C., 1972, Petrographic and chemical properties of selected North Dakota lignites, in F.T.C. Ting, ed., Depositional environments of the lignite-bearing strata in western North Dakota: North Dakota Geological Society, Miscellaneous Series No. 50, p. 63-68.

Valkovic, V., 1983, Trace elements in coal, Vol. 1: Boca Raton, Florida, CRC Press, 210 p.

Van der Flier-Keller, E., and S. Dumais, 1988, Inorganic matter content and specialized element potential of the Nanaimo and Comox coalfields, Vancouver Island: British Columbia Ministry of Energy, Mines and Petroleum Resources, Paper 1988-1, p. 435-440.

Van der Flier-Keller, E., and W.S. Fyfe, 1985, Uranium-thorium systematics of two Canadian coals: International Journal of Coal Geology, v. 4, p. 335-353.

Van der Flier-Keller, E., and W.S. Fyfe, 1987, Geochemistry of two Cretaceous coal-bearing sequences: James Bay Lowlands, N. Ontario and Peace River Basin, N.E. British Columbia: Canadian Journal of Earth Sciences, v. 5, p. 1038-1052.

Van der Flier-Keller, E., and W.S. Fyfe, 1988, Mineralogy of Lower Cretaceous coals from the Moose River Basin, Ontario and Monkman, British Columbia: Canadian Mineralogist, v. 26, p. 343-353.

Van der Flier-Keller, E., and F. Goodarzi, 1988, Regional variations in coal quality in the Canadian Cordillera: Geological Society of America Abstracts with Program, v. 20, no. 7, p. A89-A90.

Van der Flier-Keller, E., and F. Goodarzi, 1989, Effect of structure and tectonics on coal geochemistry--Some examples from British Columbia, in W. Langenberg, ed., Proceedings of the Western Canada Coal Geoscience Forum, Edmonton, Alberta, April 24-25, 1989: Alberta Research Council, Information Series No. 103, p. 233-240.

Van der Flier-Keller, E., and F. Goodarzi, in press, Regional variations in coal quality in the Canadian Cordillera, in P. McCabe, ed., Distribution and Quality of Cretaceous Coals: Geological Society of America.

Van Krevelen, D.W., 1961, Coal: Amsterdam, Elsevier, 512 p.

Williams, V.E., and C.A. Ross, 1979, Depositional setting and coal petrology of the Tulameen Coalfield, south-central British Columbia: American Association of Petroleum Geologists Bulletin, v. 63, p. 2058-2069.

Wilson, M.V.H., 1979, Middle Eocene freshwater fishes from British Columbia: Royal Ontario Museum of Life Sciences Contributions, v. 113, p. 388-399.

STRUCTURAL CONTROLS ON THE PETROLOGY AND GEOCHEMISTRY OF THE POND CREEK COAL SEAM, PIKE AND MARTIN COUNTIES, EASTERN KENTUCKY

James C. Hower and James D. Pollock[1]
University of Kentucky
Center for Applied Energy Research
3572 Iron Works Pike
Lexington, Kentucky 40511

Thomas B. Griswold[2]
Kentucky Energy Cabinet
Lexington, Kentucky 40578

ABSTRACT

The Pond Creek coal seam provides examples of the use of a broad spectrum of coal characterization techniques to delineate coal quality trends. Petrology, geochemistry, and palynology all point to a syndepositional influence on peat characteristics by the northeastward-plunging Belfry Anticline. The Pond Creek coal seam and the split between the coal seam and the rider coal thin over the anticline. To the northwest of the anticline, the Pond Creek coal seam has a high pyrite-content zone at the top. The coal to the southeast generally has a low sulfur content with a distinctive lithologic zonation. The thickening of clastic lithotypes and the thinning of the bright lithotypes to the east (southeast of the anticline) provides further evidence for the contemporaneous development of the anticline. The recognition of structural control on the coal quality allows for further definition of coal quality trends.

INTRODUCTION

The integrated study of a wide variety of coal quality parameters in conjunction with the study of the depositional setting, including the tectonic regime, of a coal deposit is a powerful tool in assessing its mineability. An understanding of the reasons for changes in coal quality provides a sounder basis for resource exploitation than does the simple mapping of a few quality parameters. Long-term mining plans must be thought out not only in terms of contracts but also in terms of how the coal fits in the context of both the contract specifications and environmental standards for emissions and waste disposal.

---

1. Present Address: Kentucky Revenue Cabinet, Frankfort, KY 40601.

2. Present Address: Rhodes & Associates, Inc., Lexington, KY 40504.

The study of the Pond Creek coal seam in Eastern Kentucky provides an example of the use of a number of analytical tools in the study of a coal deposit. One of the most fundamental tools to be used is the relationship of the coal deposit to the regional structure. In particular, if the coal shows signs of a penecontemporaneous influence from growing geologic structures, then the changes in the characteristics of the coal relative to the structure can possibly be extrapolated beyond the immediate study area. The Pond Creek coal seam is a mature resource, providing us the advantage of having numerous mine exposures at which to study the coal. However, the approach of studying both the lateral and vertical variability of coal is even more important in the exploration stage.

POND CREEK COAL SEAM

The Pond Creek (also known as the Lower Elkhorn and No. 2 Gas in West Virginia) coal seam ranks among the most important energy resources in the Appalachian Basin. As seen on Table 1, the Pond Creek Coal annually provides 35-40% of the mine production in Pike County, Kentucky. Pike County is second nationally to Campbell County, Wyoming, in coal production and leads the U.S. in underground production. The Pond Creek Coal also is mined in Martin County, Kentucky. The correlative Blue Gem, Path Fork, and Imboden coals are mined to the southwest of Pike County.

Much of the discussion that follows centers on the variations in coal quality across Pike and Martin Counties. Analyses from our coal quality averages, based on the Kentucky Geological Survey (Currens et al., 1987) and all performed according to ASTM specifications, for four 7.5-min. topographic quadrangles are given in Table 2. Only whole-channel analyses were used in the compilation of the table. We also collected and analyzed bench/lithotype samples at the same sites as many of the whole-channel samples (reported in Hower and Bland, 1989, and Hower and Pollock, 1988).

Stratigraphic And Structural Setting

The Pond Creek coal seam lies within the Middle Pennsylvanian Breathitt Formation (Figure 1), which is up to 760 m (2,500 ft) thick and consists of siltstones, sandstones, and shales with minor amounts of marine shales, limestones, and most of the major coals of Eastern Kentucky. In west-central Pike County, the Pond Creek Coal generally is overlain by shale and silty shale containing nodules of the marine Campbell Creek

Quadrangle	Pond Creek		All Coals	
	Mines	10^3 tons	Mines	10^3 tons
Jamboree	6	1750.6	22	2664.5
Lick Creek	14	3139.2	31	3852.2
Meta	6	1674.7	33	3608.2
Thomas[1]	2	1873.8	26	6249.1
Others	59	4900.6	---	---
Pike County	85	11465.0	425	31384.5
Martin County	2	1873.8	83	13338.1

[1] Pond Creek Coal production is in Martin County; other mines are in Pike, Martin, and Floyd Counties.

Table 1. *Production of Pond Creek Coal and other coals in Pike and Martin Counties, Kentucky (Kentucky Department of Mines and Minerals, 1987).*

	Jamboree	Lick Creek	Meta	Thomas
Moisture (%, as determined)	2.07	2.10	3.06	3.41
st. dev.	0.73	1.26	0.95	0.47
range	1.22-4.30	1.06-7.20	2.03-5.00	2.15-3.99
no. samples	26	31	10	14
Ash (%, dry)	8.88	9.98	9.53	9.09
st. dev.	2.00	2.59	2.37	1.58
range	4.69-12.24	3.90-17.51	5.47-13.73	7.36-12.53
no. samples	26	31	10	14
Sulfur (%, dry)	0.62	0.69	0.72	1.30
st. dev.	0.20	0.24	0.14	0.30
range	0.48-1.46	0.49-1.37	0.47-1.04	0.56-1.83
no. samples	26	31	10	14
Heating value (MJ/kg, daf)[1]	35.5	35.4	34.8	34.3
st. dev.	0.4	0.5	0.6	0.2
range	34.6-36.0	33.9-36.5	34.2-35.7	34.0-34.6
no. samples	26	31	10	14
Free swelling index	7.1	7.12	3.2	2.1
st. dev.	0.7	0.8	0.9	1.0
range	6-8.5	5-8	2-4.5	1-5
no. samples	21	26	8	14
Hardgrove grindability index	52.8	51.4	43.5	43.3
st. dev.	6.3	3.6	2.1	2.5
range	45-62	48-57	42-45	41-46
no. samples	10	7	2	3
Ash softening temperature (°C, reducing atmosphere)	1454	1519	1535	1321
st. dev.	123	88	81	152
range	1241-1589	1247-1601	1421-1599	1154-1482
no. samples	10	16	4	5
Chlorine (%, dry)	0.09	0.09	0.15	0.31
st. dev.	0.02	0.01	0.06	0.04
range	0.06-0.12	0.07-0.10	0.10-0.19	0.26-0.35
no. samples	9	8	2	4
Fe_2O_3 (%, ash basis)	6.81	6.58	5.43	13.35
st. dev.	4.49	3.76	0.95	3.61
range	3.37-21.81	3.00-17.16	4.02-6.64	9.46-23.01
no. samples	20	23	8	13

[1] Btu/lb = 429.92 MJ/kg

Table 2. *Average quality of Pond Creek coal seam based on analyses in Center for Applied Energy Research database for four 7.5-min. topographic quadrangles in study area.*

	Jamboree	Lick Creek	Meta	Thomas
CaO (%, ash basis)	0.67	1.66	4.50	4.81
st. dev.	0.48	1.23	3.30	2.21
range	0-1.35	0-4.50	1.49-11.49	1.58-8.28
no. samples	20	23	8	13
SiO_2 (%, ash basis)	54.10	52.81	47.76	42.39
st. dev.	3.99	4.73	4.19	4.40
range	45.71-59.20	44.31-59.91	38.87-51.77	36.11-50.39
no. samples	20	23	8	13
Al_2O_3 (%, ash basis)	30.35	30.29	30.80	27.18
st. dev.	2.75	2.48	3.12	2.43
range	22.39-33.61	26.05-35.20	27.82-36.07	22.86-31.00
no. samples	20	23	8	13
Vitrinite (%, mmf)	74.0	80.2	76.3	75.3
st. dev.	3.2	3.2	2.5	2.0
range	68.1-79.4	75.1-87.0	73.0-80.3	71.1-78.1
no. samples	20	25	8	14
Fusinite+macrinite (%, mmf)	6.7	7.6	9.2	6.7
st. dev.	1.4	1.7	2.6	1.6
range	4.5-9.5	4.4-9.9	6.5-14.5	3.8-9.1
no. samples	20	25	8	14
Seminfusinite (%, mmf)	7.8	4.0	5.0	5.7
st. dev.	2.2	1.5	1.4	1.2
range	4.7-12.7	1.9-7.6	2.3-7.0	3.9-7.5
no. samples	20	25	8	14
Micrinite (%, mmf)	3.8	2.3	1.5	3.0
st. dev.	0.9	1.2	0.6	1.1
range	2.5-5.7	0-4.5	0.9-2.5	1.6-5.4
no. samples	20	25	8	14
Liptinite (%, mmf)	7.6	5.8	7.9	9.2
st. dev.	1.9	0.8	1.0	1.5
range	5.5-11.9	4.4-7.6	6.1-8.9	7.2-11.5
no. samples	20	25	8	14
Vitrinite (oil, 546 mm)	1.05	0.99	0.89	0.85
st. dev.	0.05	0.05	0.05	0.02
range	0.99-1.15	0.87-1.06	0.84-0.95	0.83-0.89
no. samples	19	25	8	14

Table 2 (continued). *Average quality of Pond Creek coal seam based on analyses in Center for Applied Energy Research database for four 7.5-min. topographic quadrangles in study area.*

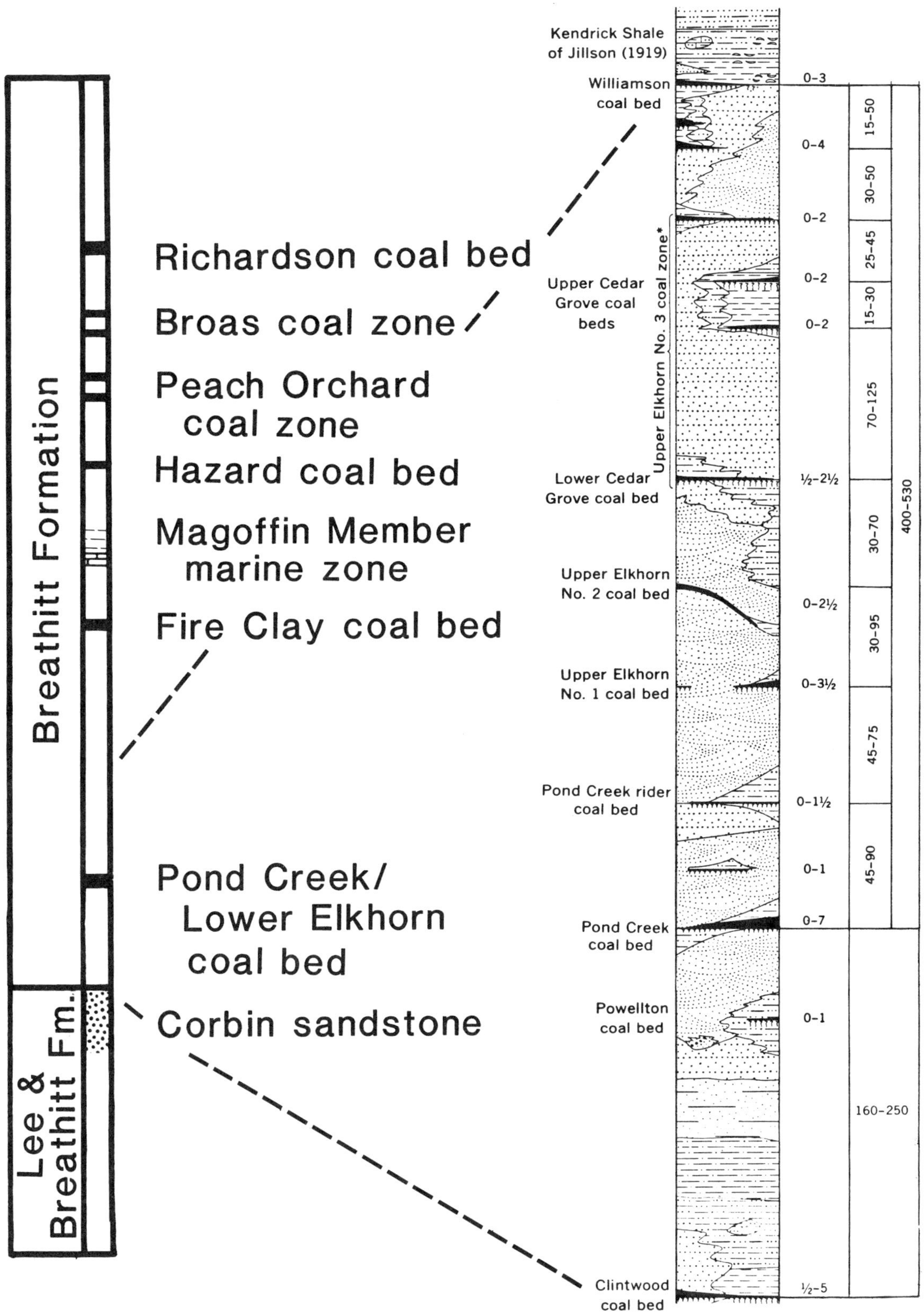

Figure 1. Relative position of the Pond Creek coal seam in the Middle Pennsylvanian Breathitt Formation. Detailed section from Jamboree 7.5-min. Quadrangle (Outerbridge and Van Vloten, 1968) (thicknesses in feet).

Limestone. The interval above the Pond Creek coal seam often contains a rider coal. In eastern Pike County, channel sandstones commonly overlie the coal. Further details of the stratigraphic setting can be found on the geologic maps of the region, such as for the maps of the 7.5-min. topographic quadrangles discussed in this study (McKay and Alvord, 1969; Outerbridge and Van Vloten, 1968; Rice, 1963; Wolcott and Jenkins, 1966), and in summaries in McDowell (1986) and Chesnut (1988).

The clastic wedge of coal-bearing Pennsylvanian-age rocks thickens (Rice, 1981) and dips to the southeast, towards the axis of the Appalachian Basin (Coskren and Rice, 1979). The Rome Trough, an east-west basement structure which may have caused thickening of the late Paleozoic section (Ammerman and Keller, 1979), is located in northern Pike and southern Martin counties.

The Pond Creek horizon in Pike County, Kentucky, is part of a northeastward-dipping monocline, at an elevation of about 550 m (1,800 ft) along the Virginia border falling to about an elevation of 60 m (200 ft) along the border of Martin County (Hunt et al., 1937). Elevation of the coal seam increases to over 200 m (660 ft) to the northwest in central Johnson County (Figure 2). The monoclinal dip is interrupted in central Pike County by the northwestward-plunging, N20°E-trending Belfry Anticline. The anticline, with prominent development near the town of Belfry in northeastern Pike County (see map in Rice et al., 1977), has slopes of as much as 30 m/km (160 ft/mi.) normal to the anticlinal axis. In central Pike County, the structure is asymmetric with a greater dip on the northwestern side.

COAL QUALITY AND THE BELFRY ANTICLINE

Regional variation in coal quality and, consequently, the mineability of the Pond Creek coal seam are linked to the development of the regional structure. Coal rank increases from high-volatile B bituminous in central Floyd County to high-volatile A/medium-volatile bituminous along the Virginia/Pike County, Kentucky, border. The regional trends in rank are more a function of basin-wide influences than of local influences (Hower and Rimmer, 1991). Rank-influenced coal properties, such as free-swelling index and Hardgrove grindability index, therefore, should vary independently of structure-induced variations to the extent that they are dependent on coal rank. Coal properties, however, display a complex interdependence on a number of variables. Hardgrove grindability, for example, is a function of both rank and maceral/microlithotype composition as well as of the amount and type of mineral matter (Hower and Wild, 1988).

The influence of the development of the Belfry Anticline on the quality of the Pond Creek coal seam was determined through the study of the petrology (Hower and Pollock, 1988), ash geochemistry (Hower and Bland, 1989), and palynology (Helfrich and Hower, 1991) of the coal. Palynology is available only on samples from a limited portion of the study area. The following discussion is drawn, without further citation, from those studies. Generalized megascopic descriptions are shown on Figure 3, and a summary of the coal character is shown on Figure 4.

Region To Northwest Of Belfry Anticline

To the west and northwest of the Belfry Anticline, the coal typically has a zone of dull coal (durain, dull clarain, and bone) in the lower portion of the coal seam and again at the top of the coal seam. The lower and upper thirds of the coal seam have relatively high K_2O/Al_2O_3 ratios (up to 0.10 compared to 0.01 in the middle benches), suggesting a significant clastic influence on the ash geochemistry. The basal lithotypes generally are enriched in zinc (Zn), copper (Cu), nickel (Ni), cobalt (Co), chromium (Cr), manganese (Mn), and vanadium (V) relative to the upper lithotypes. A dull clarain/durain in the lower half of the coal seam has high TiO_2 concentrations (as much as 5.75%, ash basis (8.35% ash, dry), at site 4 in the Thomas 7.5-min. Quadrangle), accompanied by significant zirconium (Zr) enrichment. The TiO_2-Zr association suggests a

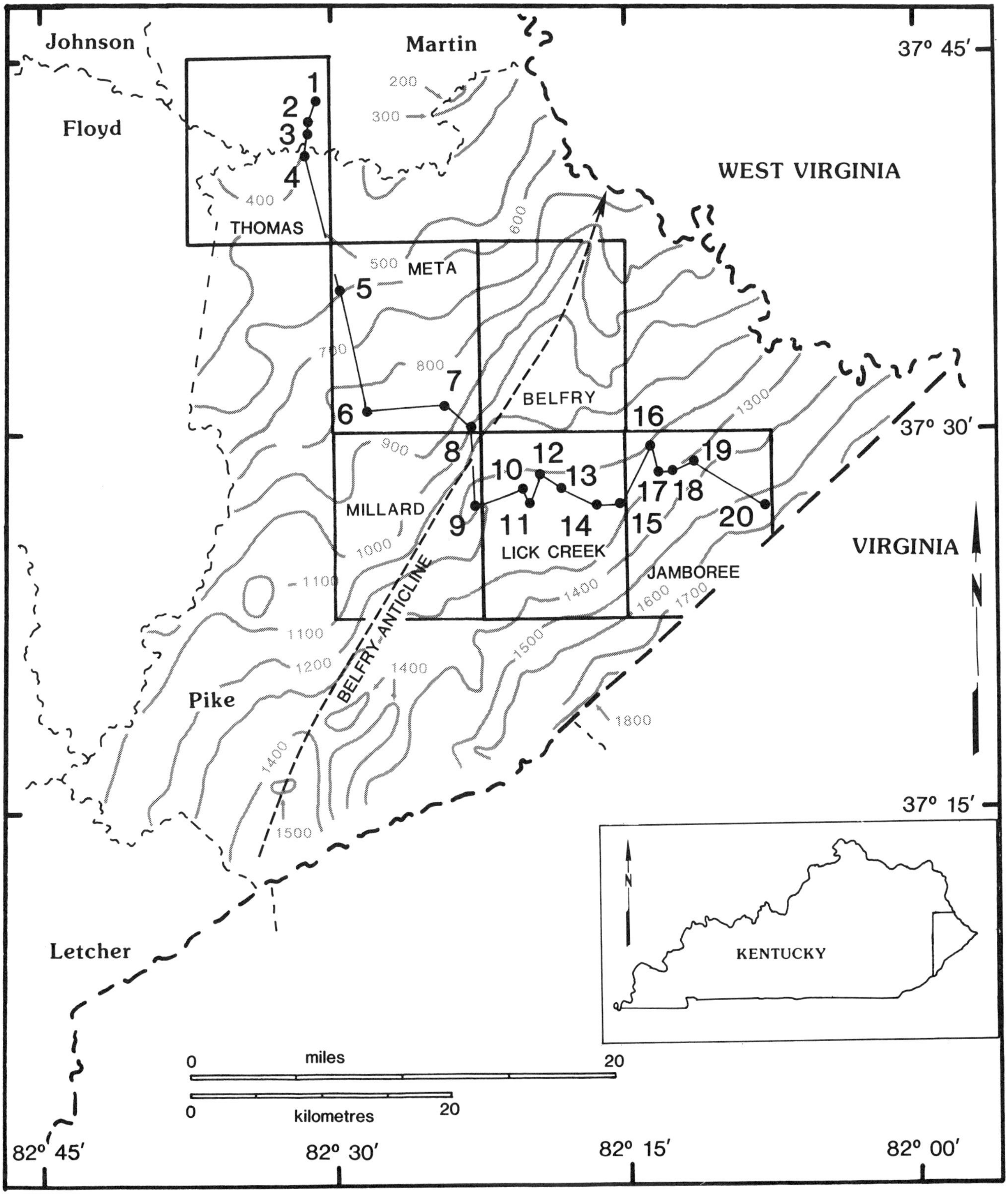

Figure 2. Pond Creek sample sites, 7.5-min. quadrangles, and elevation (in feet) of base of Pond Creek coal seam (after Hunt et al., 1937).

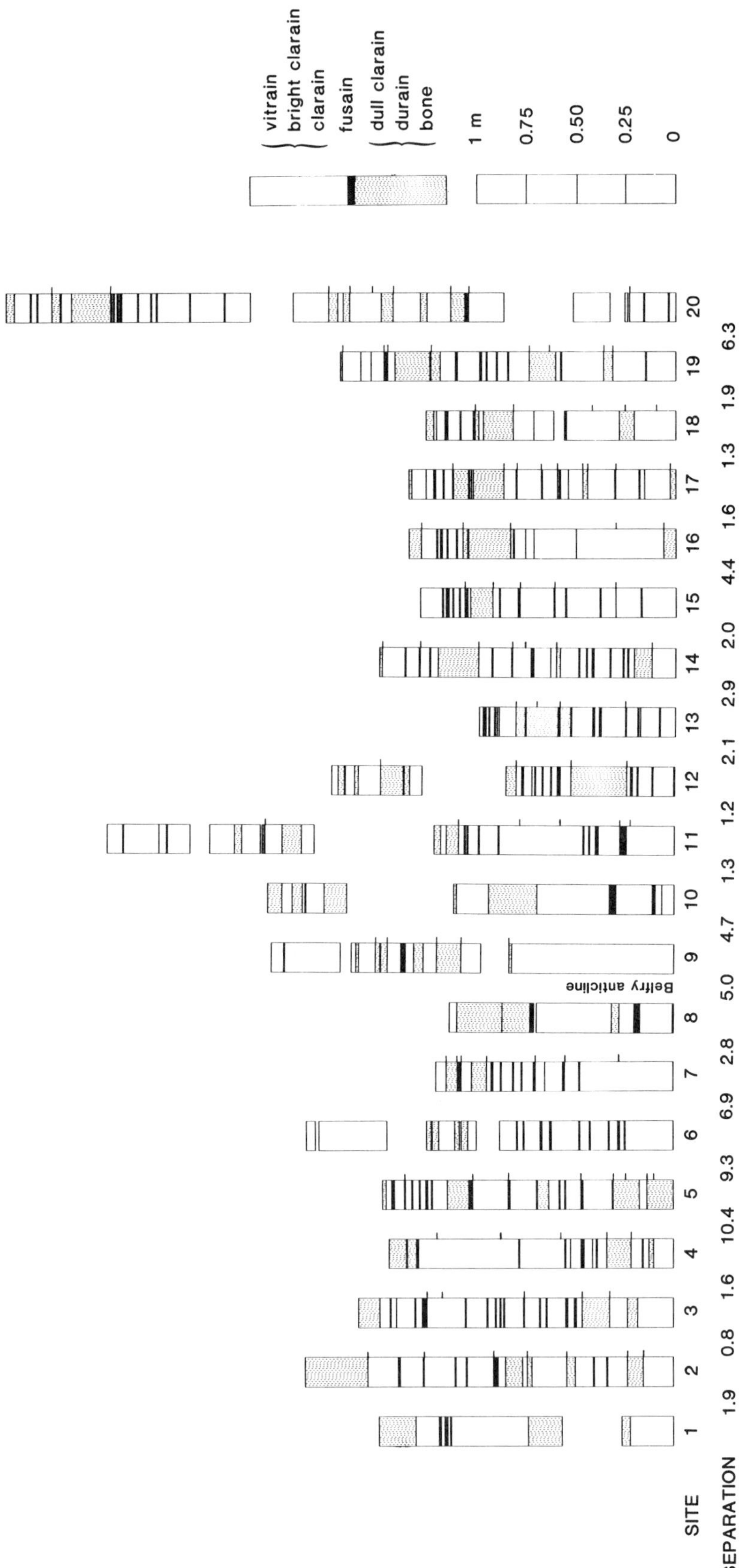

Figure 3. *Generalized megascopic lithology along cross section from Thomas Quadrangle to Jamboree Quadrangle.*

Zone	NORTHWEST OF ANTICLINE	PROXIMAL TO ANTICLINE	Zone	SOUTHEAST OF ANTICLINE
	• High K_2O/Al_2O_3	• Thinning of main seam	7	• Durain • High V, Cr, Cu
Top 1/3	• Dull lithotype at top of coal	• Thinning of split between main seam and rider	6	• Bright lithotypes • Low trace metals
	• High S_T and Fe_2O_3 at some sites		5	• Bone • 20-30% ash • 71.6-77% vitrinite • Low trace metals
Middle 1/3	• High CaO with enrichment in Ba and Sr and, to a lesser extent, Mn, Cu, and Ni		4	• Bright lithotypes with abundant fusain • High Mn, Fe_2O_3, CaO zones • Upward decrease in Mn, Zn, Cu, Ni, Co, Ba, Sr, Fe_2O_3, and CaO
Bottom 1/3	• Basal enrichment in Zn, Cu, Ni, Co, Cr, Mn, and V		3	• Dull lithotype • Increasing Sr, Ba, Mn • Lowest bench with lateral consistency in palynology
	• High TiO_2 and Zr in durain/dull clarain		2	• Bright lithotype • Increasing Zn, Cu, Ni, Co, and Sr
			1	• Durain (not present at all sites) • Low Zn, Cu, Ni, Co, and Sr • High K_2O_3/Al_2O_3

Figure 4. *Generalized geochemical trends in three regions discussed in text.*

strong clastic influence. CaO content rises through the middle of the coal seam, as much as 26.86% (ash basis; 3.99% ash, dry) in the site 5 samples, with accompanying enrichment in Mn, Cu, and Ni, and particularly barium (Ba) and strontium (Sr), through portions of the zone. Total sulfur (S) and Fe_2O_3 are highest at the top of the coal seam, but both can exhibit marked short-distance variation. At site 4, the whole-coal sample has 1.83% S_T (dry) with a top bench containing 5.73% S_T (dry) and 49.05% Fe_2O_3 (ash basis; 13.97% ash, dry), whereas a whole-channel sample 110 m (360 ft) to the west has 0.56% S_T (no bench samples collected). The lower two-thirds of the coal seam at the high-sulfur site has less than 0.60% S_T. Short-distance variations, from less than 1% S_T to over 1.5% S_T, are known to exist in mines in the Meta Quadrangle.

The Rome Trough sites in the Thomas and northern Meta quadrangles (sites 1-5) have higher chlorine (Cl) contents than sites to the east of the Belfry Anticline (0.3% Cl vs. 0.1% Cl, respectively). The higher Fe_2O_3

content associated with the higher-sulfur zones leads to a lower ash-softening temperature for the northwestern sites, particularly the Thomas Quadrangle sites, as compared to the southeastern sites.

Exposures Proximal To Belfry Anticline

The main bed of the Pond Creek Coal thins across the trace of the Belfry Anticline from the sites in the Meta Quadrangle through to the sites in the Lick Creek Quadrangle. Thinning of the coal also has been observed to the northeast in the Belfry Quadrangle where the anticline is prominent. The split between the main seam and the rider coal thins across the anticline. The rider coal is mined along with the main seam in portions of the Millard and Lick Creek quadrangles, thereby yielding a thicker mined section in those mines. The rider coal, or the upper split in the latter cases, generally is higher in S than the main bed of the Pond Creek Coal. The main seam has diagnostic lithologic features, in particular the location of durains, which permit its recognition in the expanded sections where the rider coal and main seam are mined together.

Region To Southeast Of Belfry Anticline

To the east and southeast of the Belfry Anticline, the Pond Creek Coal typically has seven lithologic zones or benches. Extensive studies of the coal seam at sites 14 through 19 (see Figure 5) have revealed a number of petrographic, geochemical, and palynologic trends.

The basal zones, the zone 1 durain and the zone 2 bright lithotypes (as well as the zone "O" bright lithotype which underlies zone 1 at site 19), exhibit diverse palynomorph assemblages that have relatively little lateral continuity. Zone 1 does not appear until sites 16 and 17 and is not present at site 18. Zone 2 is the basal

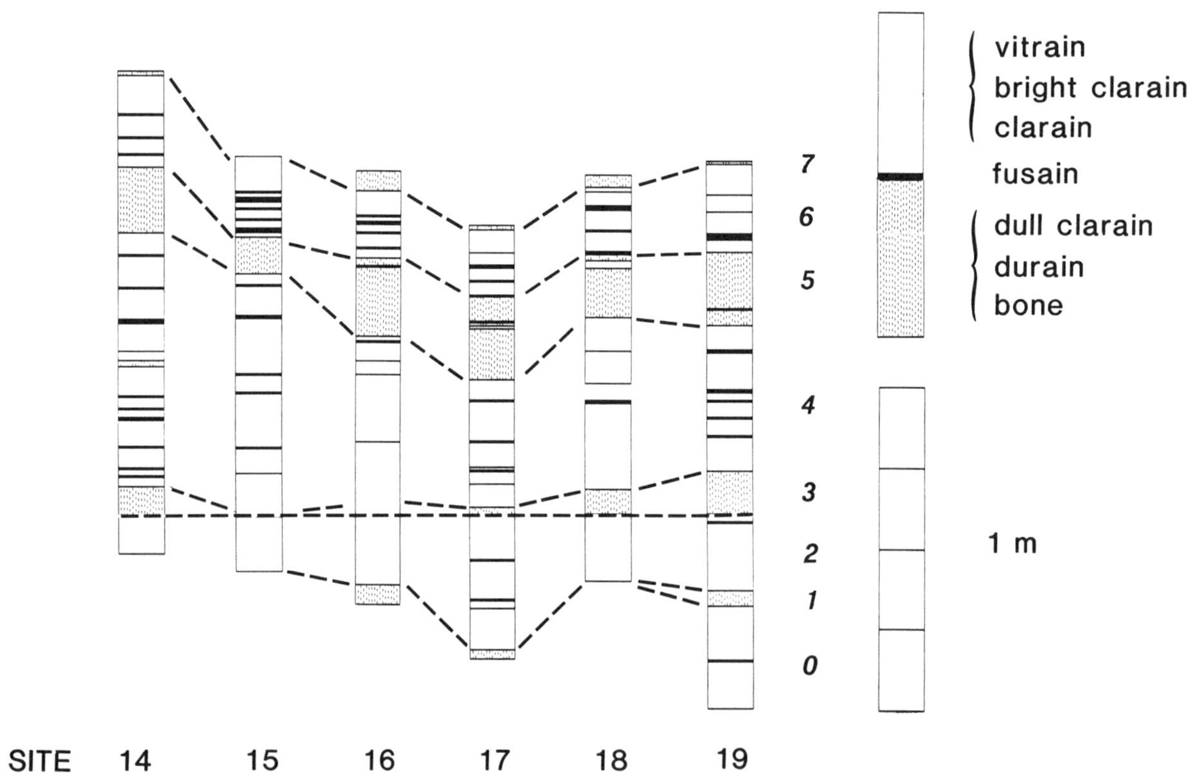

Figure 5. *Generalized megascopic lithology of sites 13 to 19 emphasizing thickening of basal lithotypes to the east and the thinning of the zone 4 bright lithotypes to the east. Horizontal datum at base of zone 3 lithotype. Lithologic zones designated in numerals to right of site 19.*

lithotype at western sites 14 and 15 and at site 18. The basal durain, with maceral group composition $V_{22.2}$ $I_{55.7}$ $L_{22.1}$ to $V_{38.2}$ $I_{49.9}$ $L_{11.9}$, has a relatively high K_2O/Al_2O_3 ratio and low concentrations of Zn, Cu, Ni, Co, and Sr as compared to the overlying benches. Manganese remains low in zone 2, in places below the limits of detection, whereas Zn, Cu, Ni, Co, and Sr increase.

Zone 3, a thin dull zone ($V_{62.4}$ $I_{30.0}$ $L_{7.6}$ to $V_{73.9}$ $I_{19.2}$ $L_{6.9}$) increases in thickness to the east. It is absent to the west of site 14 (Figure 5). This bench is the first with some lateral consistency in palynology, generally being dominated by herbaceous lycopod forms. Strontium, Ba, and Mn increase through zone 3 to zone 4.

Zone 4, composed of bright lithotypes (with abundant fusains) containing over 80% vitrinite (mineral-free basis), is thickest at western sites 14 and 15. Zone 4 does thin to the west of site 14 (Figure 5). The zone is dominated by arborescent lycopods but secondary trends in sphenopsids, ferns, tree ferns, and herbaceous lycopods point to a slightly greater elevation, with a relatively "dry land" flora, at the western sites. Manganese concentration reaches its highest level, over 6,400 ppm (ash basis; 3.16% ash, dry) in the bench immediately below the 5.2-cm (2.0-in.) thick pyrite layer at site 18, the only megascopic pyrite observed east of the anticline. The high-Mn content bench also has the highest Fe_2O_3 (60% with only 0.56% S_T) and highest CaO contents of any of the benches at site 18. Decreases in Mn, Zn, Cu, Ni, Co, Ba, and Sr occur in the upper portions of zone 4. Concentrations of Fe_2O_3 and CaO also decrease, and K_2O increases, at the top of the zone.

Zone 5, a bone coal with 20-30% ash (dry) and 71.6-77.0% vitrinite, and the zone 6 bright lithotypes both have high K_2O/Al_2O_3 ratios and low concentrations of trace metals as compared to the underlying benches. The bone coal tends to be thicker in the east whereas the bright lithotypes show little lateral change in thickness. A thin durain, as little as 0.3 cm (0.1 in.) thick, caps the Pond Creek coal seam. The zone 7 durain has a group maceral concentration of $V_{13.0}$ $I_{47.5}$ $L_{39.5}$ to $V_{29.7}$ $I_{40.3}$ $L_{30.0}$ and high V, Cr, and Cu concentrations with low Ba, Sr, and CaO compared to the underlying benches.

Summary Of Evidence For Penecontemporaneous Anticline/Peat Development

Overall, several lines of evidence in the sites to the east of the anticline point to the development of the Belfry Anticline, and related structures, as being contemporaneous with the development of the Pond Creek swamp (illustrated on Figure 5 for sites 13 to 19). The basal lithotypes (zones "O", 1, 2) thicken as a unit to the east. The clastic lithotypes, zones 3 and 5, also increase in thickness to the east. The thick zone 4 bright lithotypes are thickest in the west, at site 14, where a "dry-land" flora was noted. The brackish-to-marine influence seen in the high Fe_2O_3-CaO-Mn lithotype capped by a pyrite band within zone 4 at site 18 suggests that the eastern sites may have been periodically submerged during zone 4 development. The relative elevation of the sites near the anticline apparently provided a favorable site for peat development, leading to the greater thickness of the bright zones. Together with the overall thinning of the coal and the thinning of the split between the rider and the main coal seam across the trace of the anticline, the evidence strongly suggests penecontemporaneous tectonism.

Was the northwest-dipping monocline also developing along with the Pond Creek swamp? The contrast in coal quality across the anticline, the higher-S, higher-Cl sites to the northwest versus the low-sulfur sites to the southeast, suggests that this indeed was the case. The anticline thus was a barrier between the brackish-to-marine influenced area to the northwest and the region of low-sulfur coals to the southeast.

IMPLICATIONS FOR REGIONAL ASSESSMENTS OF COAL QUALITY

Baseline studies such as this provide a view of the regional variation in coal quality and of the depositional controls which produced the observed variations. Study of the Pond Creek coal seam suggests that the scale of coal quality variation varies across the region. The area to the northwest of the anticline has significant differences in quality, most notably total S, on the mine-unit scale (about 100 m (330 ft)). Variations to the southeast of the anticline generally appear to occur on a quadrangle scale.

If the penecontemporaneous development of the Belfry Anticline did influence the deposition of the Pond Creek coal seam, a further study in the investigation of regional coal quality would be to determine its influence on other coals. The Upper Elkhorn No. 2 coal seam is thicker, with lower S and ash contents, at sites near the anticline than at sites farther to the west in Floyd County (Raione and Hower, 1984). The Taylor coal seam at the base of the Magoffin Member, the Hazard coal seam, the Hazard No. 7 coal seam, and the Hazard No. 8 coal seam (the latter two are correlative to the Peach Orchard coal zone; see Figure 1) all are thin or split and have high S and ash contents through the same general region that the Upper Elkhorn No. 2 has diminished quality. The Taylor through Hazard No. 8 coals exhibit greater thickness and better quality farther to the west in Perry County (Hower et al., 1991).

This regional view of the Pond Creek coal seam evolved through the succession of studies of coal petrology, geochemistry, and palynology. None of the tools alone could have provided as thorough an understanding of the coal as did all three together. The recognition of the structural control on the coal quality allows for further definition of coal quality trends. Not all coal quality assessments necessarily will have all of the tools available here (for example, palynology can be a time-consuming step). However, the coal geologist should call upon as many diagnostic parameters as possible, starting with the lithologic description of the coal seam and progressing to the laboratory investigations, to best assess coal quality in the region of interest.

ACKNOWLEDGMENTS

We wish to thank Garry D. Wild, Gerald A. Thomas, Charles E. Jones, and Alan E. Bland, all presently (G.D.W., G.A.T.) or formerly (C.E.J., A.E.B.) with the Center for Applied Energy Research (CAER), and Charles T. Helfrich, Eastern Kentucky University, for their assistance in the studies of the Pond Creek coal seam. Research at the CAER was funded by the Kentucky Energy Cabinet, Commonwealth of Kentucky.

REFERENCES

Ammerman, M.L., and G.R. Keller, 1979, Delineation of Rome Trough in Eastern Kentucky by gravity and deep drilling data: American Association of Petroleum Geologists Bulletin, v. 63, p. 341-353.

Chesnut, D.R., Jr., 1988, Stratigraphic analysis of the Carboniferous rocks of the Central Appalachian Basin: Lexington, The University of Kentucky, Unpublished Ph.D. Thesis, 297 p.

Coskren, T.D., and C.L. Rice, 1979, Contour map of the base of the Pennsylvanian System, Eastern Kentucky: U.S. Geological Survey, Map MF-1100.

Currens, J.C., L.J. Bragg, and J.C. Hower, 1987, Analysis of coal samples from the Big Sandy District, Kentucky: Kentucky Geological Survey, Series XI, Information Circular 20, 421 p.

Helfrich, C.T., and J.C. Hower, 1991, Palynologic and petrographic variation in the Pond Creek coal bed, Pike County, Kentucky: Organic Geochemistry, v. 17, in press.

Hower, J.C., and A.E. Bland, 1989, Geochemistry of the Pond Creek coal bed, Eastern Kentucky coal field: International Journal of Coal Geology, v. 11, p. 205-226.

Hower, J.C., and J.D. Pollock, 1988, Petrology of the Pond Creek coal bed in Eastern Kentucky: Organic Geochemistry, v. 12, p. 297-302.

Hower, J.C., and S.M. Rimmer, 1991, Coal rank trends in the Central Appalachian Coalfield: Virginia, West Virginia, and Kentucky: Organic Geochemistry, v. 17, in press.

Hower, J.C., E.J. Trinkle, J.D. Pollock, and C.T. Helfrich, 1991, Influence on penecontemporaneous tectonism on thickness and quality of Breathitt Formation coals, eastern Kentucky, in S. Suboleski, ed., One point two--New geologic perspectives on Central Appalachian low-sulfur coal supply: Coal Decisions Forum, in press.

Hower, J.C., and G.D. Wild, 1988, Relationship between Hardgrove grindability index and petrographic composition for high-volatile bituminous coals from Kentucky: Journal of Coal Quality, v. 7, p. 122-126.

Hunt, C.B., G.H. Briggs, Jr., A.C. Munyan, and G.R. Wesley, 1937, Coal deposits of Pike County, Kentucky: U.S. Geological Survey Bulletin 876, 95 p.

Kentucky Department of Mines and Minerals, 1987, Annual report: Commonwealth of Kentucky, 146 p.

McDowell, R.C., 1986, The geology of Kentucky--A text to accompany the geologic map of Kentucky: U.S. Geological Survey Professional Paper 1151-H, 76 p.

McKay, E.J., and D.C. Alvord, 1969, Geologic map of the Lick Creek Quadrangle, Pike County, Kentucky: U.S. Geological Survey Geologic Quadrangle Map GQ-716.

Outerbridge, W.F., and R. Van Vloten, 1968, Geologic map of part of the Jamboree Quadrangle, Pike County, Kentucky: U.S. Geological Survey Geologic Quadrangle Map GQ-775.

Raione, R.P., and J.C. Hower, 1984, Petrographic characterization of the Upper Elkhorn No. 2 coal zone of Eastern Kentucky: University of Kentucky Institute for Mining and Minerals Research, Report No. DOE/PC/30223-11, 167 p.

Rice, C.L., 1963, Geology of the Thomas Quadrangle, Kentucky: U.S. Geological Survey Geologic Quadrangle Map GQ-227.

Rice, C.L., 1981, Introduction: The stratigraphic framework of the Pennsylvanian rocks in Eastern Kentucky, in J.C. Cobb et al., eds., Coal and coal-bearing rocks of Eastern Kentucky: Kentucky Geological Survey, Guidebook for Geological Society of America Coal Division Field Trip, p. 2-5.

Rice, C.L., R.G. Ping, and J.L. Barr, 1977, Geologic map of the Belfry Quadrangle, Pike County, Kentucky: U.S. Geological Survey Geologic Quadrangle Map GQ-1369.

Wolcott, D.E., and E.C. Jenkins, 1966, Geologic map of the Meta Quadrangle, Pike County, Kentucky: U.S. Geological Survey Geologic Quadrangle Map GQ-497.

LITHOTYPE CONTROL OF TRACE ELEMENT CONCENTRATIONS IN SOME SOUTHEASTERN WEST VIRGINIA COALS

Jeffrey W. Martin[1] and Victor V. Cavaroc
Department of Marine, Earth and Atmospheric Sciences
North Carolina State University
P.O. Box 8208
Raleigh, NC 27695-8208

ABSTRACT

An ability to relate local concentrations of trace elements to easily recognizable and mappable coal lithotypes would provide a useful source of coal quality information in mine site planning. Forty-three Carboniferous-age bituminous coal samples from southeastern West Virginia were analyzed for 30 trace elements to test if such relationships could be identified. Results of chemical (INAA, XRF, HTA) and mineralogical (XRD) analyses show that medium- and high-ash (8.6 to 36.2% HTA) dull-luster coal lithotypes (durains and durain-rich clarains) contain significantly higher concentrations of quartz and clay minerals and the elements Al, Ce, Cs, Cr, Eu, Sc, Se, Si, Th, Ti, U, V, and Zr than low- and medium-ash (1.9 to 12.3% HTA) bright-luster lithotypes (vitrains and vitrain-rich clarains). Bright coals contain more pyrite (and possibly calcite) and are relatively enriched in Sb, Ca, Cl, Co, Fe, Sr, S, and Zn. The elements As, Br, Mg, and K exhibit no detectable associations with coal lithotypes. Significant amounts of elements associated with dull coals probably are contained in detrital mineral matter incorporated during the accumulation of coal-forming peats. Elements concentrated in bright coals likely reflect higher proportions of authigenic and/or diagenetic minerals.

INTRODUCTION

Plans to expand mining operations must consider the quality of the coal and reject material likely to be mined. The concentrations of various trace elements have become an important consideration of mine product quality due to increased concern about the environmental impact of coal usage and reject disposal (e.g., Wallace and Berry, 1978). Also, certain elements can cause technological difficulties in the use of coal and coal products (e.g., Renton, 1982). The ability to predict trace element concentrations during mine site evaluations and mine planning could result in more efficient, cost-effective operations.

This paper will demonstrate that the variations in concentrations of many critical trace elements within the coals of a local area correlate with the megascopic coal lithotypes. The latter are defined on the basis of physical appearance and can be identified easily in outcrops or during core logging without using more time-consuming and costly laboratory petrographic analyses. This study also indicates that a relatively small number of analyses from samples selected on the basis of coal lithotype are adequate to characterize trace element distributions associated with coals in the local mining area. We emphasize that the results reported in this paper apply to only this local study area. If similar relationships occur in coals elsewhere, the specific trace element concentration values can differ in keeping with differences in geological settings.

1. Present Address: Ground Water Associates, 771 Brooksedge Plaza Drive, Westerville, OH 43081.

The two important lithotype components of bituminous coal seams are "bright-" and "dull-luster" coals, which usually are tabulated in drill logs during exploratory or developmental core drilling. The thickness distributions of these coal lithotypes within seams often are mapped using core data during site evaluations. A close association between trace element content and bright- or dull-coal components readily can be converted into estimations of the quality of mine product using different mining plans.

COAL LITHOTYPES

Banded bituminous coals primarily are composed of two types of altered plant matter (McCabe, 1984; Parks and O'Donnell, 1956; Stach, 1975a; Teichmüller, 1975; Teichmüller and Teichmüller, 1975; Thiessen and Sprunk, 1935). Bright coal (also termed vitrain or anthraxylon) has a bright, glassy luster. It is black in color, relatively soft, contains well-developed jointing (cleats), and breaks with a conchoidal fracture. Coal petrographic studies show that bright coals formed from peats that are rich in fragments of humic or "woody" plant tissues, such as stems, branches, bark, and roots. Dull coal (also termed durain, attritus, or splint) has a dull to slightly silky luster and ranges from black to steel gray in color. It is relatively hard and massive and breaks with a splintery to hackly fracture. Dull coal originated from herbaceous and subaquatic peats rich in partially oxidized and finely-divided plant remains, spores, pollen, cuticles, resins and wax-secreting tissues. Cameron and Schruben (1983) and Casagrande and Erchull (1976) have described arborescent and herbaceous peats as potential modern analogs for bright and dull coals, respectively.

The megascopic coal lithotype classification used by the U.S. Geological Survey (USGS) (Davis, 1978; Schopf, 1960) divides bituminous coals into banded and non-banded types. Banded (or attrital) coals consisting of interlaminated bright and dull materials are subdivided further on the basis of luster (bright, moderately bright, midlustrous, moderately dull, dull). Nonbanded coals, including cannel and boghead, are relatively rare (McCabe, 1984; Schopf, 1960; Sprunk et al., 1940). These coals have a massive structure and consist entirely of dull material derived mainly from spores and algae, respectively. Research by Cameron (1978), Davis (1978), and Esterle and Ferm (1986) showed that megascopic classifications using the terms bright coal and dull coal are useful in describing a gross picture of coal composition and seam stratigraphy and may be related in a general sense to coal compositional properties.

In this study, only banded coals are investigated because the relative volume of nonbanded coals is minute in the local study area. Coal samples selected for analysis also were restricted to bright or dull lithotypes, as these two types constituted 96% of the coal seams measured and sampled (77% bright, 19% dull, 4% midlustrous). Figure 1 shows the relationship of the study samples with the subdivisions proposed for pure and impure coals based on ash content (American Society for Testing Materials, 1988; Wood et al., 1983). The term "bone coal" commonly is associated with impure coals. However, this term often reflects variable local usage, sometimes including cannel coal and coaly shales and mudstones (Bates and Jackson, 1980; Schopf, 1960; Wood et al., 1983). By definition, bone coal does not specify the type or weight percentages of the impurities (American Society for Testing Materials, 1988; Wood et al., 1983). The USGS suggests that the use of this term be discontinued (Wood et al., 1983).

A simple bright/dull lithotype classification is appropriate for the purposes of this study for several reasons. First, this classification method represents the primary subdivision of banded coal lithotypes. Second, the bright/dull classification is a straightforward field technique based entirely on apparent coal physical properties and can be replicated easily in other mining areas. Third, bright and dull lithotypes represent the two major environments found in modern peat deposits, arborescent and herbaceous peats, respectively.

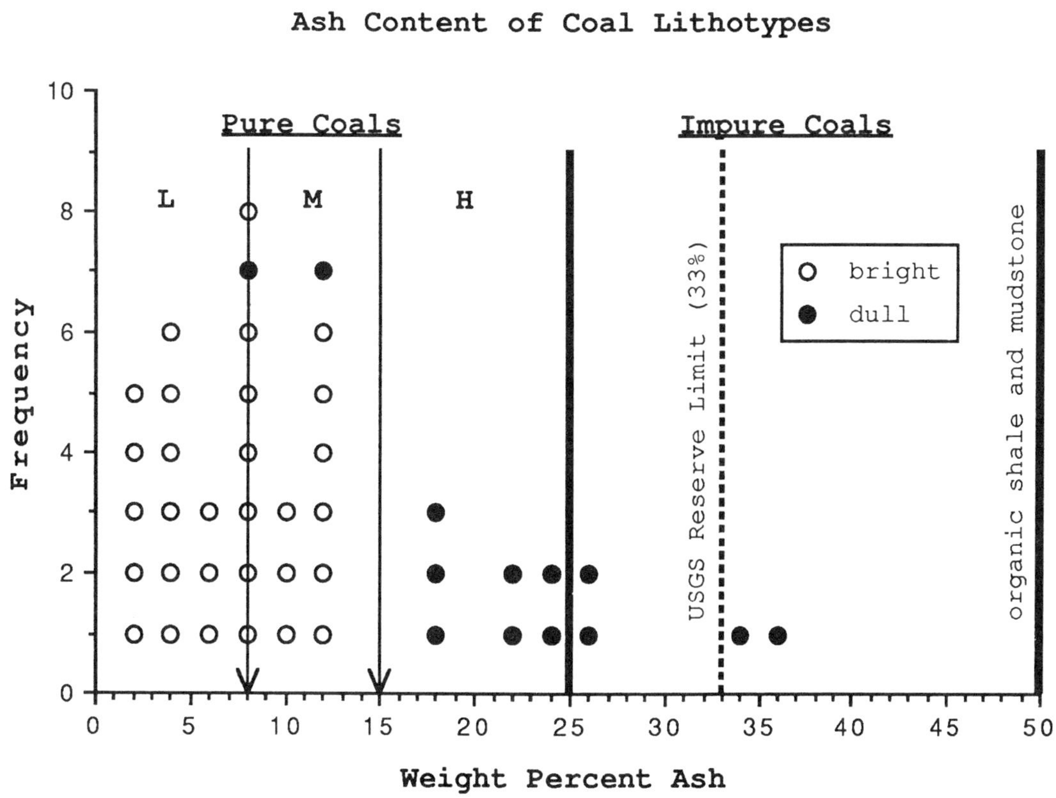

Figure 1. *This figure shows the relationships of the samples (solid and open circles) analyzed in this study to coal nomenclature based upon percent ash (American Society for Testing Materials, 1988; Wood et al., 1983). The subdivision of pure coal into L, M, and H reflects low-, medium-, and high-ash contents. The heavy dashed line dividing impure coal is the ash limit (33%) used by the USGS in reserve analyses (Wood et al., 1983).*

PREVIOUS STUDIES

Few studies of trace elements specifically have investigated relationships with coal lithotypes. Zubovic et al. (1966) compared concentrations of beryllium (Be), boron (B), titanium (Ti), vanadium (V), chromium (Cr), cobalt (Co), copper (Cu), gallium (Ga), germanium (Ge), lanthanum (La), molybdenum (Mo), nickel (Ni), tin (Sn), yttrium (Y), and zinc (Zn) in undifferentiated block coal samples with concentrations in constituent vitrain and fusain. Reidenour et al. (1967) found that dull coal lithotypes in the Lower Kittanning Seam of Western Pennsylvania contain higher sulfur (S) concentrations than bright lithotypes. Baumann (1982) showed that vitrain, bright clarain, and dull clarain of the Herrin (No. 6) Seam of Illinois can be statistically differentiated using concentrations of cadmium (Cd), calcium (Ca), Cu, iron (Fe), lead (Pb), and Ti. Miller (1984) used factorial and discriminant function analyses to classify bright, semibright, and dull lithotypes of the Hance Seam in southeastern Kentucky on the basis of antimony (Sb), aluminum (Al), arsenic (As), Co, Fe, Mo, Ni, scandium (Sc), S, thorium (Th), Ti, uranium (U), V, and zirconium (Zr) concentrations. Data presented by Karner et al. (1984) suggest relationships between Ca, magnesium (Mg), sodium (Na), and S concentrations in lithotypes of North Dakota lignites.

STUDY AREAS AND SAMPLING PROCEDURES

A total of 43 coal samples (30 bright, 13 dull) were collected from three areas in southeastern West Virginia (Figure 2). Thirty-three of these were taken from freshly cleaned coal seam outcrop faces in the central portion of Webster County. Seams that contained both well-developed bright and dull beds preferentially were sampled (Figure 3). In addition, four samples were collected from an underground mine in the Cedar Grove Seam in Boone County, and six were chosen from cores of the Pocahontas No. 3 Seam in Raleigh County. These ten additional samples were analyzed to assure that observed differences in element concentrations between outcrop samples were not the result of weathering processes. The Boone and Raleigh County samples also provide an assessment of the in-seam lateral distribution of trace element concentrations. The total sample set was designed to include a variety of stratigraphic units and fluvial-deltaic depositional facies (Figure 3). Details of the stratigraphic and sample relationships are presented in Martin (1990).

Samples consisted of an approximately 5-cm (2-in.) square column cut through each bright and dull bed. Bright coals were designated as those containing more than 60% visible bright (vitrain-rich) material, whereas dull coals contained more than 60% visible dull (durain-rich) material. Coal units of intermediate composition (<60% bright, <60% dull) constituted only about 4% of the measured coals and were not sampled. These three classes are equivalent to megascopic lithotypes described by Cameron (1978), Davis (1978), Hunt and Hobday (1984), Sprunk et al. (1940), and Thiessen and Sprunk (1935). Bright beds thicker than 24 cm (0.75 ft) were divided into upper, middle, and lower subsamples to examine the vertical variability of ash and trace element contents. Most dull beds are less than 24 cm (0.75 ft) thick and were not divided into vertical subsamples.

Seams were sampled upwards from the base to prevent contamination of the underlying beds by overlying material. No samples were collected within 1.3 cm (0.5 in.) of a boundary between vertical changes in lithotype. Sampling at exposures where coal seams were directly overlain by sandstone was avoided. Cheek and Donaldson (1969) suggest that the waters of the sands and coal-forming peats may have intermingled before lithification, possibly altering the minor element contents of the peat.

ANALYTICAL METHODS

Chemical And Mineralogical Analyses

Figure 4 indicates the elements investigated and the analytical technique (XRF = X-ray fluorescence; INAA = induced neutron activation analysis) used for each element. A variety of elements, with a wide range of geochemical properties, were selected, many of which are of potential economic and/or environmental importance. High-temperature ash (HTA) contents also were determined for the coal samples. X-ray diffraction (XRD) was used to investigate the mineral content of selected coal samples. INAA, XRF, and XRD analyses were performed on whole coal samples. Martin (1990) provides a detailed discussion of sample preparation and analytical procedures.

Statistical Analysis

Statistical techniques used to determine relationships between lithotypes and element concentration were analysis of variance (ANOVA), Duncan multiple range (DMR), correlation, and cluster analysis using the SPSS/PC+ software package (Norusis, 1986a and 1986b). Nonparametric statistical tests (Kolmogorov-Smirnov and Mann-Whitney) also were used to either evaluate data distribution normality or supplement ANOVA tests. Davis (1986) provides an overview of these techniques which are similar to those previously used in other coal trace-element investigations (Baumann, 1982; Harvey et al., 1983; Miller, 1984).

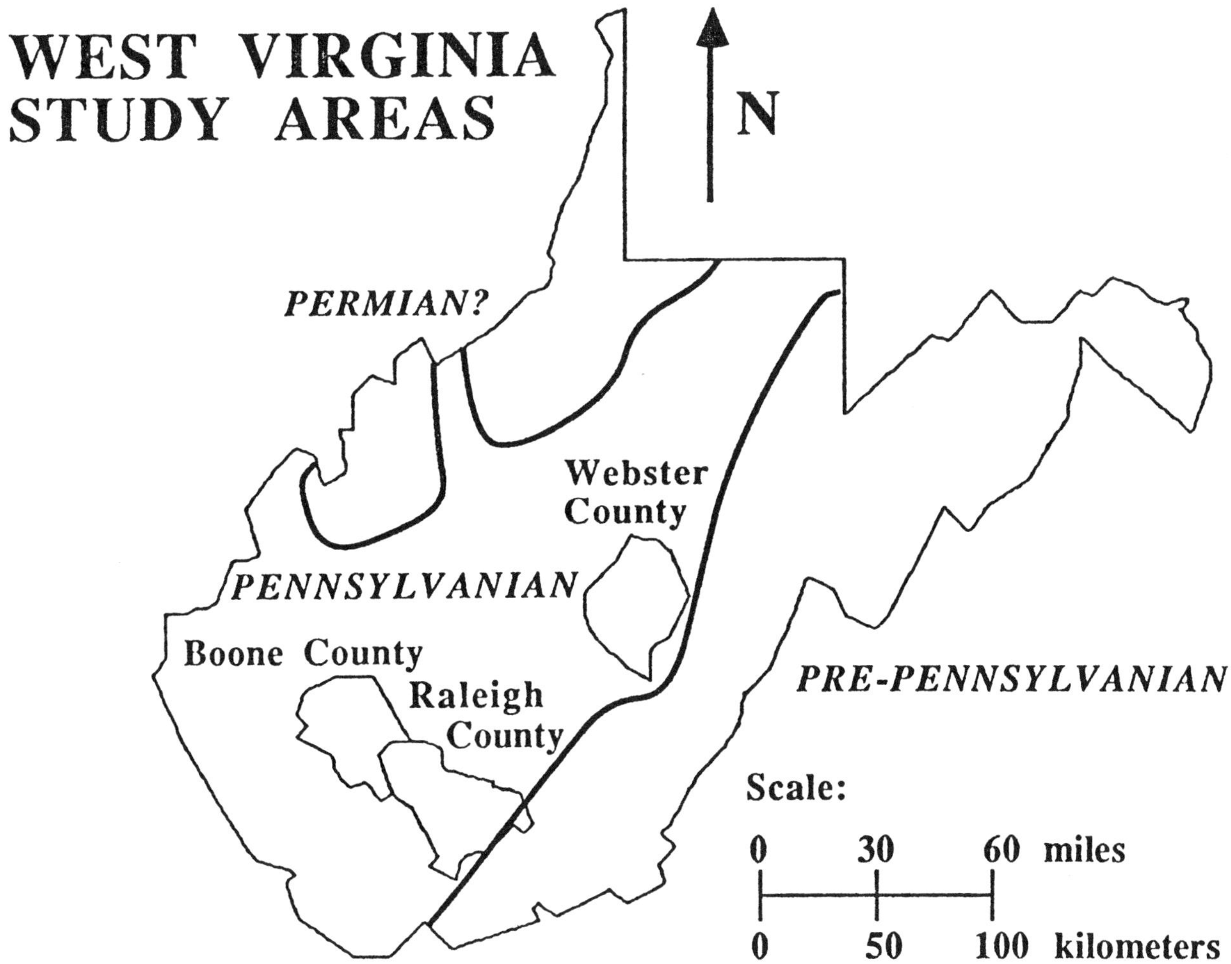

Figure 2. Locations of Webster, Boone, and Raleigh counties (study area for this report) in Southern West Virginia.

RESULTS

Chemical And Mineralogical Analyses

Table 1 contains the results of INAA and XRF analyses of trace element concentrations in the coal samples. High-temperature ash data also are shown in Table 1. Analytical errors for the INAA and XRF analyses are available in Martin (1990). Threshold (or "less than") values were obtained for barium (Ba), Mo, Ni, rubidium (Rb), and silver (Ag) in nearly all sample analyses. Table 2 contains the average threshold values for total, bright, and dull coal lithotypes. These five elements consequently were eliminated from subsequent statistical analyses and were treated only semi-quantitatively. The anomalously high Ca value (1.23%) of sample 28 also was eliminated from statistical analysis. Results of XRD analysis of eight selected coal samples (samples 1, 5, 11, 13, 18, 26, 31, and 39 in Table 1) are shown in Table 3.

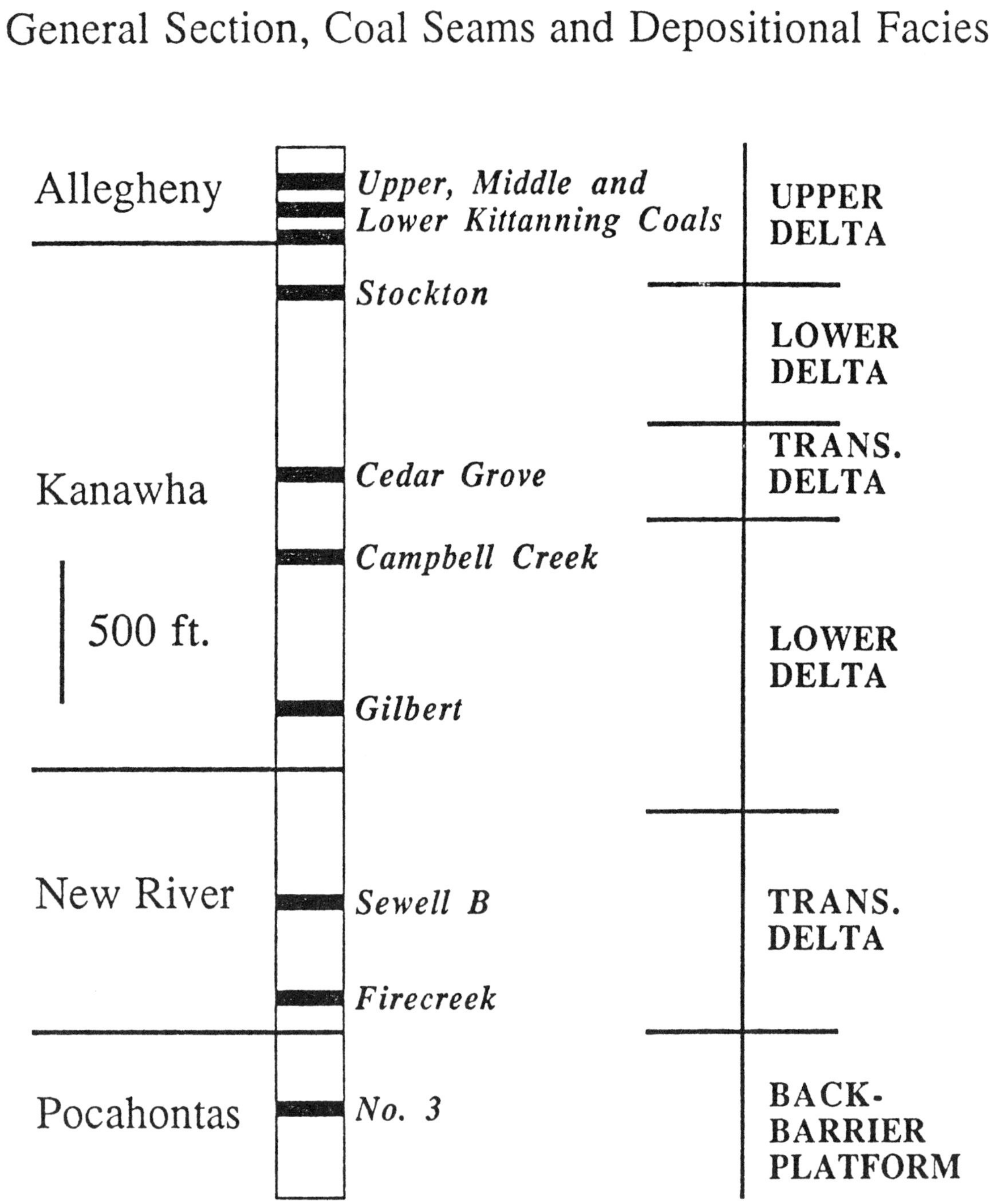

Figure 3. *General stratigraphic section of the study area. Sampled coal seams and inferred depositional environments are indicated.*

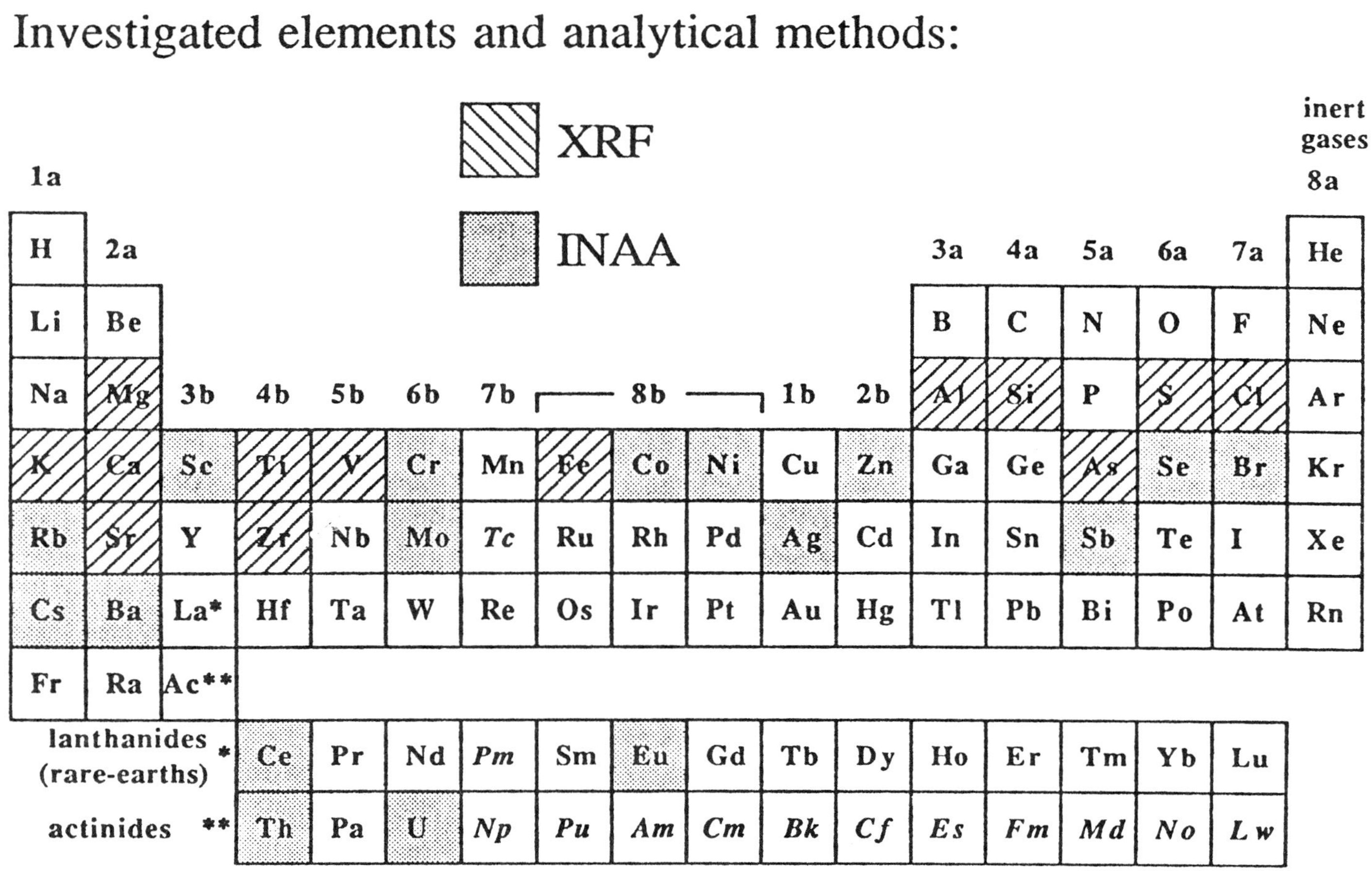

Figure 4. *Periodic chart showing investigated elements and the respective analytical techniques used for each element (XRF = X-ray fluorescence; INAA = induced neutron activation analysis).*

Statistical Characterization Of Trace Element Data

Averages and standard deviations for ash and trace element values of total, bright, and dull coal distributions are listed in Table 4. For the samples analyzed, dull coals contain significantly more ash than bright coals (22% versus 7.2% average HTA, respectively). This general relationship also was found in studies of coals of the Appalachian Plateau region by Sprunk et al. (1940; 7.6% versus 3.3% HTA) and U.S. subbituminous and bituminous coals by Parks and O'Donnell (1956; 8.3% versus 4.0% HTA). Both bright and dull coals collected from Webster County have higher ash contents than these regional averages, possibly because of the close proximity to an uplifted orogenic front. These relationships are consistent with ash data presented in the Webster County Geological Survey (Reger, 1920) which shows that ash contents of undifferentiated Webster County coals are higher on average (8.0% HTA) than those presented for undifferentiated coals in Sprunk et al. (1940; 5.5% HTA) and undifferentiated West Virginia coals in Parks and O'Donnell (1956; 6.4% HTA).

Bright coals contain higher mean concentrations of Sb, As, bromine (Br), Ca, chlorine (Cl), Co, Fe, strontium (Sr), S, and Zn than dull coals. Dull coals are characterized by higher mean concentrations of percent ash, Al, cerium (Ce), cesium (Cs), Cr, europium (Eu), Mg, potassium (K), Sc, selenium (Se), silicon (Si), Th, Ti, U, V, and Zr. Most bright- and dull-coal trace element distributions tend to be positively skewed, with kurtosis values that indicate both peaked and flattened curves. Kolmogorov-Smirnov tests for normality show that only two total coal (Br, Cl), 6 bright coal (Br, Cl, Se, Si, V, Zr), and 13 dull coal (Al, As, Br, Cl, Cr, Mg, Se, Si, Sr, S, Ti, V, Zr) element distributions approximate a normal distribution (0.75 probability; Martin, 1990). A simple logarithmic transformation markedly improves these proportions. Miller (1984) examined probability plots of untransformed versus logarithmic element data and likewise concluded that most trace element distributions tend to be log-normally distributed.

Comparable trace element distributions have been determined for Appalachian coals by Headlee and Hunter (1955) and Miller (1984). Rather than sample intervals based on coal lithotypes, Gluskoter et al. (1977) analyzed 23 coals that were collected as a mixture of seam channel, bench, run-of-mine, and plant-washed samples. Their (total coal) mean element values and ranges more closely approximate those determined for bright coals in this study. Comparative concentration values (Table 4) for total coal and dull coal in this study are considerably higher than those for Gluskoter et al. (1977). This discrepancy probably exists because dull coals in this study included high-ash samples which are not included in their study. Dull material tends to have a higher ash content and specific gravity than bright coal in general (Chandra and Taylor, 1975; Parks and O'Donnell, 1956; Sprunk et al., 1940), and is more likely to be removed during coal washing processes.

Tests Of "No Difference" In Mean Element Concentrations Between Coal Lithotypes

ANOVA and Mann-Whitney tests of "no difference" between logarithimic-transformed bright and dull lithotype means for ash and trace elements are shown in Table 4. An equivalent, randomized-block ANOVA design was used to remove the effects of variation due to differences between depositional environments of the coal seam (Martin, 1990). The results did not change the significant relationships in Table 4, indicating that depositional environments do not act as an overriding control of minor elements independent of coal lithotypes.

The Mann-Whitney test determines the probability that two independent samples belong to the same group or population. Unlike ANOVA, the underlying distribution does not need to be specified, and it works well with small data sets. Equivalent results for both the ANOVA and Mann-Whitney tests support the preceding assumption of log-normality. Significantly different mean values (0.10 level) support most of the differences between bright and dull coal lithotypes previously inferred from the mean concentration data. The exceptions are that no significant differences could be detected between the mean values of As, Br, Mg, and K

Sample No.	Seam	Litho-type	%Ash	%Al	Sb	As	Ba	Br	%Ca	Ce	Cs	%Cl	Cr	Co	Eu	%Fe	%Mg	Mo	Ni	%K
1	PC	br	2.5	0.51	0.2	13	<183	43	0.13	15	<0.1	0.14	4	3	0.3	0.39	<.001	<1	14	0.05
2	PC	br	6.1	1.59	1.3	6	<261	40	0.13	23	0.4	0.12	12	3	0.3	0.74	0.014	<2	<24	0.14
3	PC	br	3	0.74	0.5	1	<165	40	0.13	17	<0.1	0.18	9	4	0.2	0.34	0.001	<1	<14	0.04
4	PC	du	24.5	4.09	0.3	10	<518	24	0.11	55	<0.3	0.01	36	2	1.1	0.39	0.021	<6	<45	0.09
5	PC	du	25.6	4.39	0.4	11	<839	28	0.18	144	0.9	0.02	49	2	3	0.4	0.052	<8	<60	0.13
6	PC	du	23.2	4.57	<0.3	12	<975	17	0.12	157	0.8	0.02	50	2	3.3	0.4	0.046	<6	<70	0.15
7	GB	br	8.2	0.85	2.2	19	<389	17	0.18	9	0.5	0.08	34	10	0.2	3.19	0.058	<2	<41	0.27
8	GB	br	1.9	0.27	<0.1	4	<142	26	0.23	7	<0.1	0.12	4	10	0.1	0.63	0.03	<1	<12	0.05
9	GB	br	3	0.44	0.7	5	<365	12	0.23	4	<0.2	0.15	9	16	0.1	0.79	0.022	<1	<28	0.06
10	CC	br	11.8	0.89	0.8	61	<271	22	0.22	20	0.7	0.09	13	1	0.2	3.5	0.082	<2	<24	0.28
11	CC	br	11.8	0.7	1.8	64	<236	28	0.2	11	0.3	0.11	6	1	0.1	3.5	0.046	<2	<21	0.19
12	CC	br	12.3	1.51	0.4	26	<494	15	0.17	17	0.5	0.16	10	7	0.3	3.5	0.059	<2	<40	0.23
13	ST	br	3.9	0.36	0.4	32	<234	34	0.27	<1	0.3	0.24	3	2	0.2	3.34	0.037	2	<19	0.08
14	FC	br	10.8	2.52	0.9	24	<476	25	0.16	41	<0.2	0.15	28	12	0.8	0.88	0.016	<3	<39	0.06
15	FC	br	4.5	1.16	0.3	10	<305	29	0.16	25	<0.1	0.22	11	24	0.4	0.51	0.012	<2	28	0.06
16	FC	br	8.2	1.85	14.8	15	<927	29	0.16	30	0.5	0.16	19	59	0.6	0.52	0.032	<4	<74	0.21
17	SWb	br	7.2	1.4	2.2	9	<425	42	0.27	10	<0.2	0.09	10	12	0.7	1.35	0.098	<2	<38	0.22
18	SWb	br	8.7	1.28	<0.1	8	<239	20	0.39	18	0.3	0.11	12	10	0.4	0.87	0.116	<2	<23	0.18
19	SWb	br	8.5	1.72	1.1	8	<297	31	0.17	37	0.2	0.11	17	10	0.8	0.68	0.037	4	<25	0.14
20	SWb	br	2.3	0.39	2.6	2	<321	22	0.18	8	<0.2	0.22	6	16	0.3	0.45	0.011	<2	<34	0.09
21	SWb	br	3.8	0.69	0.1	8	<214	17	0.21	12	<0.1	0.18	5	13	0.3	1.11	0.029	3	<19	0.04
22	SWb	br	11.6	2.52	5	6	<732	37	0.15	40	2.5	0.16	38	44	0.7	0.97	0.078	<8	<61	0.29
23	CG	br	5.7	0.93	4.1	8	<486	37	0.21	15	0.6	0.23	13	20	0.6	0.71	0.02	<5	<39	0.15
24	CG	br	4.5	0.76	1.5	6	<306	18	0.38	14	<0.2	0.23	6	10	0.5	0.82	0.01	5	<31	<0.04
25	FC	du	33.2	6.48	<0.7	11	<1807	32	0.12	108	<0.7	0.02	87	1	2.3	0.61	0.035	<16	<125	0.09
26	SWb	du	26.3	5.03	4.4	19	<1926	34	0.13	148	1.1	0.05	74	11	3	0.68	0.112	<19	<131	0.37
27	CG	du	21.3	3.37	<0.2	10	<580	11	0.12	73	4.8	0.07	30	6	1.3	0.53	0.115	<4	<42	0.59
28	CG	du	8.6	1.2	0.2	2	<273	1	1.23	38	<0.1	0.12	9	5	0.7	0.53	0.016	<2	25	<0.04
29	LK	br	12.2	2.65	0.8	10	<348	37	0.13	<2	1.4	0.16	18	10	0.6	1.65	0.086	<2	<32	0.43
30	LK	br	7.2	1.54	0.6	7	<249	32	0.15	23	0.6	0.22	14	7	0.4	1.13	0.058	4	<23	0.28
31	LK	br	9.2	1.61	1	24	<326	40	0.15	21	0.5	0.18	19	5	0.4	1.98	0.029	6	<33	0.17
32	MK	br	12.1	1.62	1	9	<387	16	0.21	24	0.6	0.06	24	8	0.3	2.22	0.064	5	41	0.17
33	MK	br	4.9	1.01	<0.5	8	<311	50	0.27	24	<0.1	0.14	12	4	0.3	1.61	0.065	<8	<22	0.05
34	UK	br	6.2	0.96	1.5	11	<268	18	0.18	14	0.7	0.17	8	5	0.4	1.56	0.049	14	<24	0.26
35	UK	br	4.7	0.79	0.8	12	<238	35	0.2	19	0.2	0.18	10	3	0.4	1.45	0.022	6	<23	0.09
36	UK	br	7.4	1.71	1	11	<280	31	0.15	22	<0.1	0.14	17	4	0.4	0.99	0.019	6	<23	0.14
37	UK	br	10.3	2.24	0.3	5	<308	21	0.14	21	0.7	0.17	18	4	0.4	0.68	0.076	<1	<25	0.65
38	LK	du	17.9	3.31	0.7	9	<500	53	0.13	37	0.5	0.09	42	8	0.6	0.55	0.042	<5	<44	0.24
39	LK	du	12.6	2.95	0.4	11	<364	26	0.15	36	<0.2	0.13	31	7	0.5	0.52	0.034	7	<34	0.15
40	MK	du	21.3	2.25	<0.2	14	<405	10	0.19	35	<0.2	0.05	41	4	0.5	2.57	0.059	<3	<36	0.06
41	UK	du	36.2	4.44	0.3	8	<969	7	0.11	65	0.6	0.03	75	1	1.5	0.47	0.074	<5	<67	0.02
42	UK	du	17.4	2.17	0.8	6	<526	41	0.13	72	1.4	0.1	54	2	1.1	0.55	0.031	<4	<39	0.17
43	UK	du	17.3	1.56	0.3	5	<325	38	0.13	43	<0.1	0.08	31	2	1	0.4	0.01	<4	<24	0.04

Table 1. *Ash and trace element analyses (% = weight percent; all other elements in ppm). Lithotypes (br = bright; du = dull) are based on field classification.*

Sample No.	Seam	Lithotype	Rb	%S	Sc	Se	%Si	Ag	Sr	Th	%Ti	U	V	Zn	Zr
1	PC	br	<5	0.86	1	2	0.37	<0.4	53	0.6	0.015	<0.6	13	<3	12
2	PC	br	<7	1.19	3	5	1.98	<0.7	57	3.2	0.14	1.5	43	<7	40
3	PC	br	<4	0.88	2	3	0.72	<0.5	61	1.7	0.061	0.7	24	<3	22
4	PC	du	<12	0.68	6	7	7.76	1.5	42	11.6	0.403	2.5	110	<7	136
5	PC	du	<16	0.62	12	22	8.59	<2.0	60	14.9	0.375	2.5	91	<11	135
6	PC	du	<17	0.55	12	13	8.07	<2.3	58	17.8	0.458	1.7	109	<16	166
7	GB	br	<11	0.86	6	2	1.65	<1.1	73	1.2	0.069	<0.9	26	<10	29
8	GB	br	<5	0.7	1	<1	<0.16	<0.4	106	0.5	0.016	<0.3	8	8	11
9	GB	br	<9	0.77	5	<1	0.28	<0.9	111	0.5	0.021	<0.3	19	<7	22
10	CC	br	<10	1.01	3	5	1.44	<0.7	82	2.2	0.068	0.9	20	12	16
11	CC	br	<9	1.32	1	6	0.72	<0.6	73	0.8	0.031	0.8	11	<5	11
12	CC	br	<16	2.52	4	3	1.67	<1.1	106	1.5	0.059	<0.4	20	31	17
13	ST	br	<8	2.52	2	4	0.16	<0.6	124	0.4	0.009	<0.5	6	34	<11
14	FC	br	<11	0.68	8	6	2.83	<1.2	131	5.4	0.054	1.4	43	<9	27
15	FC	br	<9	0.59	3	3	1.26	<0.7	175	2.1	0.032	<0.5	17	<5	12
16	FC	br	<24	0.63	10	4	2.5	<2.1	254	3	0.173	1.7	56	64	30
17	SWb	br	<10	0.8	6	2	2	<1.0	107	1.7	0.072	1.4	29	60	20
18	SWb	br	<7	0.74	2	3	2.29	<0.6	154	1.9	0.117	0.7	30	25	37
19	SWb	br	<8	0.88	4	4	2.83	<0.7	78	3.5	0.109	1.6	35	5	29
20	SWb	br	<10	0.63	3	<1	0.44	<0.9	73	1	0.035	<0.7	17	<9	13
21	SWb	br	<7	0.59	1	2	0.52	<0.5	247	0.4	<0.009	<0.3	7	<5	<11
22	SWb	br	<20	0.75	14	<2	3.12	<1.8	93	5.2	0.174	<2.3	63	<14	46
23	CG	br	<12	1.05	6	8	1.56	<1.1	86	1.8	0.077	3.4	33	78	22
24	CG	br	<8	0.9	5	5	0.72	<0.9	134	1.3	0.026	1.6	14	32	11
25	FC	du	<30	0.24	37	8	8.54	<4.3	47	37.2	0.178	<4.6	43	<34	77
26	SWb	du	<39	0.65	32	18	7.08	<4.4	83	16.4	0.248	6.8	82	<21	79
27	CG	du	30	0.67	10	23	6.92	<1.4	56	8.5	0.338	<1.1	82	<11	79
28	CG	du	<7	0.5	3	6	1.98	<0.7	174	2.8	0.102	0.9	28	15	22
29	LK	br	18	1.71	4	7	3.1	<0.8	118	3.4	0.131	1.2	37	22	29
30	LK	br	11	1.37	3	7	1.94	<0.7	148	2.8	0.072	1.5	27	23	24
31	LK	br	<9	1.8	4	7	2.02	<0.9	123	3.7	0.104	2.7	35	32	25
32	MK	br	<11	1.66	6	10	3.21	<1.0	82	4	0.171	2.4	49	34	41
33	MK	br	<7	1.42	2	4	1.01	<0.7	284	1.9	0.099	<3.0	26	<6	15
34	UK	br	<10	1.64	2	7	1.42	<0.8	70	1.4	0.049	<0.6	27	18	18
35	UK	br	<7	1.34	2	9	0.98	<0.7	105	1.6	0.037	<0.7	20	17	14
36	UK	br	<7	0.92	4	4	2.11	<0.7	89	2.9	0.101	1.3	42	<4	26
37	UK	br	12	0.89	4	4	2.79	<0.8	105	3.7	0.186	0.7	55	<7	38
38	LK	du	<12	0.94	8	12	5.9	<1.2	95	9.8	0.439	2.2	125	<9	125
39	LK	du	<9	0.93	6	11	3.5	<1.1	105	5.9	0.253	1.8	82	<10	61
40	MK	du	<10	1.11	8	9	6.26	<1.2	123	10.7	0.434	1.4	123	<11	106
41	UK	du	<19	0.6	15	15	11.52	<2.3	86	18.9	0.527	1.7	154	<17	150
42	UK	du	15	0.96	8	9	7.03	<1.3	83	11.5	0.511	1.5	146	<10	121
43	UK	du	7	0.79	4	8	6.82	<0.8	88	10.1	0.686	3.3	197	<6	179

Table 1 (continued). *Ash and trace element analyses (% = weight percent; all other elements in ppm). Lithotypes (br = bright; du = dull) are based on field classification.*

Element	Total Coal	Bright Coal	Dull Coal
Ba	<469	<339	<770
Mo	<5	<4	<7
Ni	<38	<30	<57
Rb	<12	<10	<17
Ag	<1.2	<0.9	<1.9

Table 2. *Mean threshold concentrations (in ppm) for Ba, Mo, Ni, Rb, and Ag in coal samples.*

Mineral	Bright Coal (vitrain-rich)	Dull Coal (durain-rich)
Quartz	x	X
Kaolinite-Type Clay Mineral	x	X
Illite-Type Clay Mineral	x	X
Pyrite	x	ND
Calcite	x ?	ND

Table 3. *Minerals identified through XRD analyses of whole coal samples (x = mineral present; X = more intense diffraction peaks; ND = not detected; ? = fewer than three diffraction peaks identified).*

concentrations for bright and dull coals. ANOVA and Mann-Whitney tests were not performed for Zn as only one non-threshold value was available for the dull coal group. However, comparison of the Zn threshold values for dull coal samples with values obtained for bright samples (Table 1) suggests that bright coals contain notably higher Zn concentrations. Miller (1984) also found that bright coals contain significantly higher amounts of Sb, Ca, Co, Fe, Sr, and S, whereas dull coals were relatively enriched in Cr, Sc, Th, Ti, and Zr.

The cluster dendrogram (Figure 5) illustrates the effectiveness of the significantly different trace elements in classifying bright and dull coal samples on the basis of overall trace element content. In the cluster analysis computation, Cs, Sb, and U data sets contained too many threshold values to be included. Squared euclidean distance was used as the distance measure between standardized values; the clustering method was average linkage between groups (Norusis, 1986b). Eight of the 43 samples (six bright and two dull) were rejected by the program because of missing values for one or more elements. The remaining coal samples cluster into bright and dull groups with a correct classification rate of 98%. One medium-ash dull coal (sample 28) was classified with the low- to medium-ash, bright coals (Figure 1).

Element	Total Coal: Mean	Total Coal: Std. Dev.	Bright Coal: Mean	Bright Coal: Std. Dev.	Dull Coal: Mean	Dull Coal: Std. Dev.	Lithotype With Larger Mean	ANOVA Probability**	Mann-Whitney Probability**
% Al	1.93	1.44	1.24	0.67	3.52	1.50	dull	<0.01	<0.01
Sb	1.5	2.57	1.8	2.85	0.9	1.36	bright	0.08	0.04
As	13	12.76	14	14.91	10	4.10	bright	0.62	0.94
Br	28	11.23	29	9.98	26	13.92	bright	0.18	0.35
% Ca	0.18	0.06	0.20	0.07	0.13	0.03	bright	<0.01	<0.01
Ce	38	38.10	19	9.35	78	45.90	dull	<0.01	<0.01
Cs	0.9	0.96	0.6	0.54	1.5	1.52	dull	0.02	0.02
% Cl	0.13	0.06	0.16	0.05	0.06	0.04	bright	<0.01	<0.01
Cr	24	20.50	14	8.61	47	21.52	dull	<0.01	<0.01
Co	9	10.91	11	12.36	4	3.06	bright	<0.01	<0.01
Eu	0.7	0.77	0.4	0.18	1.5	1.01	dull	<0.01	<0.01
% Fe	1.18	0.97	1.40	1.02	0.66	0.58	bright	<0.01	<0.01
% Mg	0.05	0.03	0.04	0.03	0.05	0.03	dull	0.47	0.63
% K	0.18	0.14	0.17	0.13	0.19	0.15	dull	0.75	0.91
Sc	7	7.22	4	2.88	13	10.33	dull	<0.01	<0.01
Se	7	5.04	5	2.19	12	5.43	dull	<0.01	<0.01
% Si	3.28	2.88	1.64	0.94	6.92	2.35	dull	<0.01	<0.01
Sr	107	53.50	117	57.37	85	35.88	bright	0.03	0.05
% S	1.00	0.49	1.12	0.52	0.71	0.23	bright	<0.01	<0.01
Th	5.6	7.05	2.2	1.38	13.6	8.46	dull	<0.01	<0.01
% Ti	0.17	0.17	0.08	0.05	0.38	0.16	dull	<0.01	<0.01
U	1.9	1.22	1.5	0.76	2.4	1.60	dull	0.04	0.03
V	52	45.17	28	15.09	106	45.51	dull	<0.01	<0.01
Zn	32	19.81	33	19.96	14.5	-	dull	-	-
Zr	51	48.51	23	10.08	110	45.00	bright	<0.01	<0.01
% ash	11.6	8.47	7.2	3.38	22.0	7.61	dull	<0.01	<0.01

** probability of no difference between bright and dull means

Table 4. *Mean element concentrations in coal lithotypes with ANOVA and Mann-Whitney tests of no difference (% = weight percent; all other elements in ppm).*

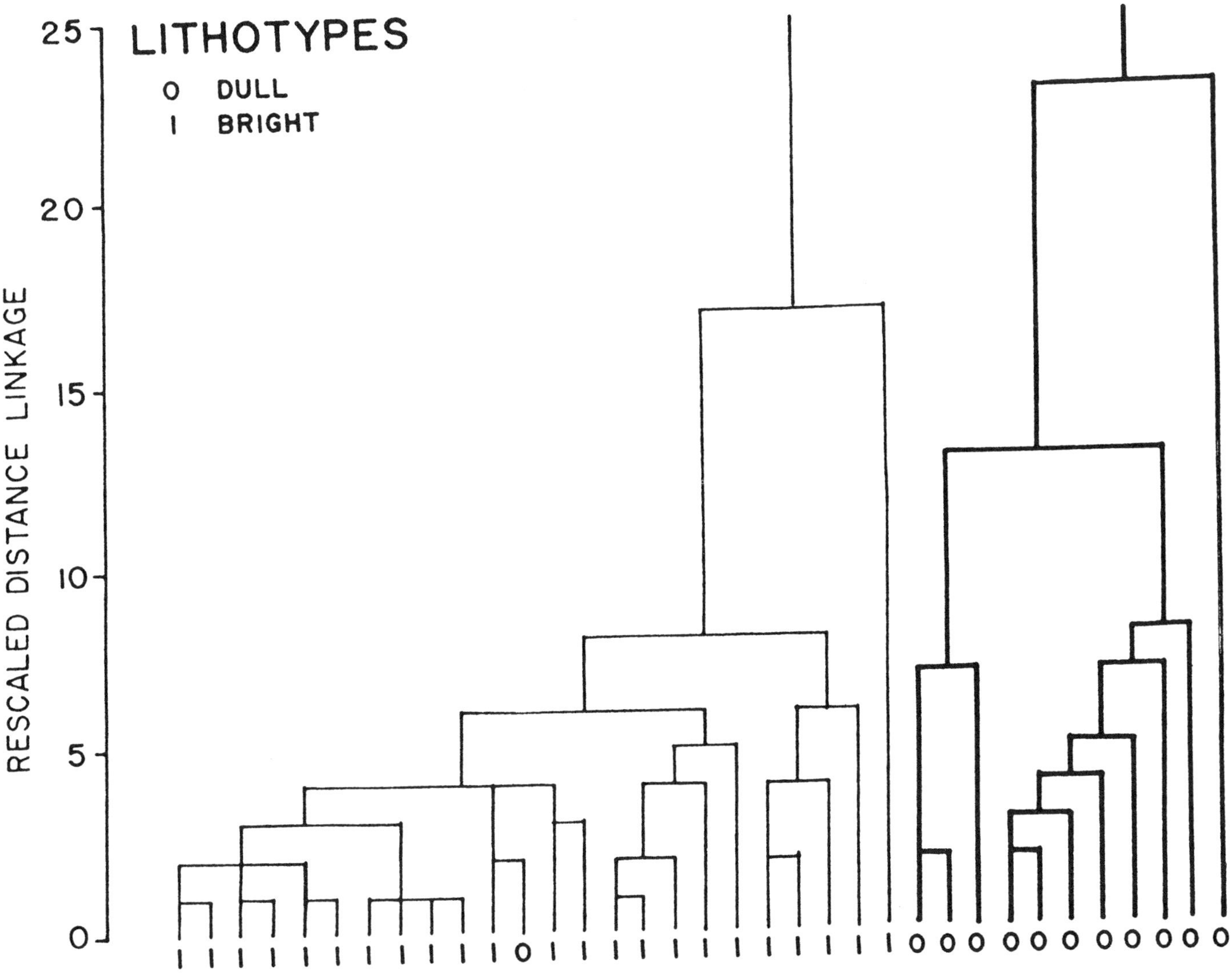

Figure 5. Cluster analysis dendrogram illustrating the effectiveness of grouping bright and dull coal lithotypes based on trace element contents. The misclassified sample is a medium-ash dull coal.

Relation Of Coal Lithotype Mineralogy To Trace Element Distributions

XRD analyses of coal samples indicate that both bright and dull lithotypes contain quartz, a kaolinite-type mineral, and an illite-type mineral. Pyrite and calcite were identified exclusively in some of the bright lithotype samples. Pyrite also could be visually observed in some bright coal samples. Diffraction peaks of quartz and clay minerals in dull coals (Figure 6) are more intense than those obtained for bright coals, indicating a higher concentration of these minerals in dull coal. This is consistent with the higher mean ash content determined for dull coal (22.0%) versus that of bright coal (7.2%; Figure 1).

Previous research by Finkelman and Stanton (1978), Kuhn et al. (1980), Finkelman (1980, 1981), Harvey et al. (1983), and Miller (1984) indicates that the distributions of most trace elements are related primarily to

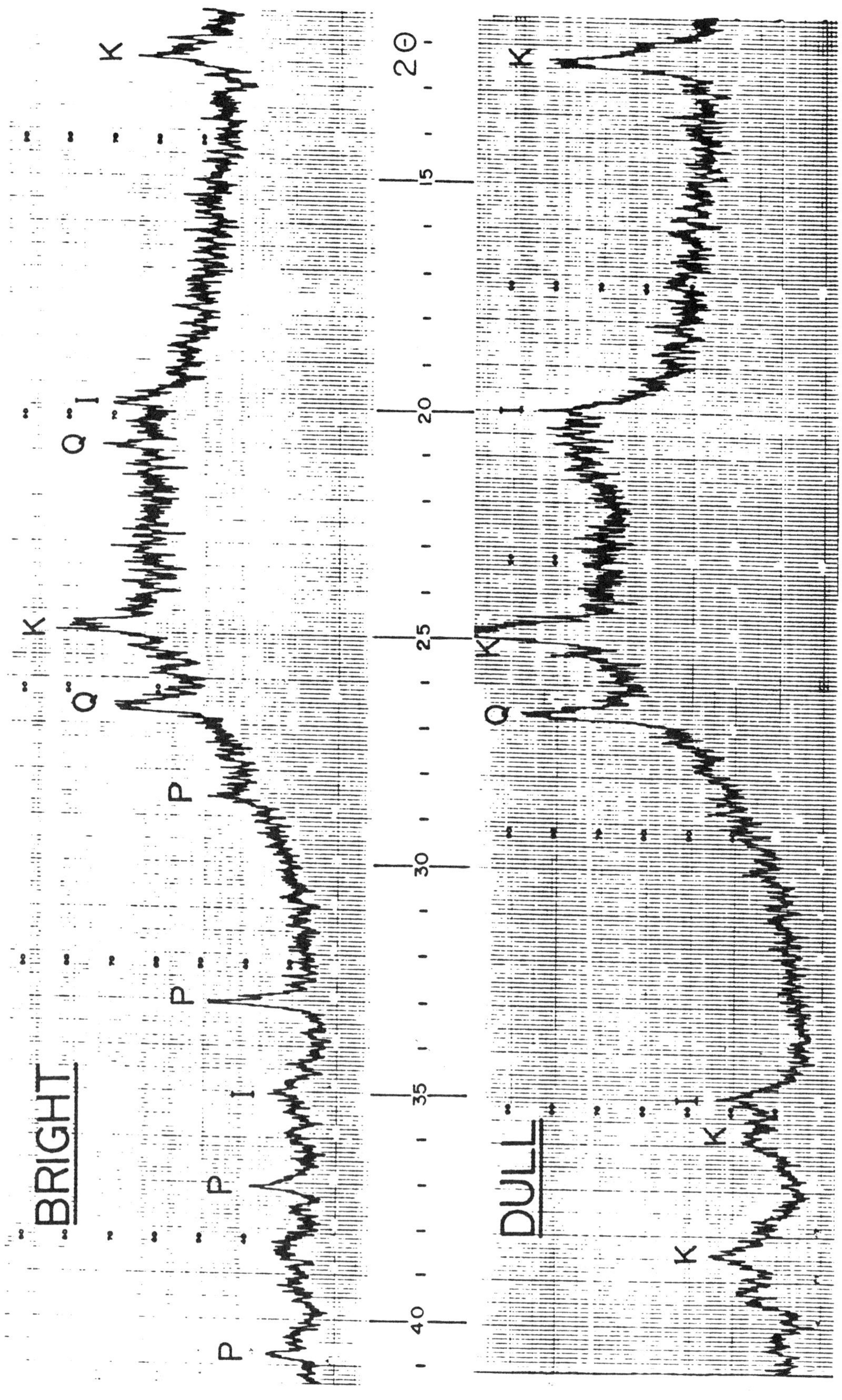

Figure 6. *X-ray diffraction patterns for typical bright and dull coals. Note the presence of more intense quartz (Q), kaolinite-type (K), and illite-type (I) mineral peaks in the dull sample, as well as its lack of pyrite (P) peaks. High proportions of quartz and clay minerals are indicative of the incorporation of detritus into the coal-forming environment.*

the distribution of mineral matter in coal. Similarly, differences in the overall trace element contents of bright and dull coals examined in this study appear to be related to differences in lithotype mineral content. Higher pyrite and calcite contents in bright coals are consistent with high mean Ca, Fe, and S concentrations. High mean values for Sb, Co, and Zn may be the result of the incorporation of these elements as trace constituents of pyrite; Sr is likely present as a trace component in calcite. The greater proportion of quartz and clay minerals in dull coals probably accounts for high mean Al and Si values. Clay minerals may include Cr, Cs, Th, Ti, U, and V as trace constituents. These and many other elements with high mean concentrations in dull coal (Ce, Eu, Sc, and Zr) also could be contained in trace silicate, oxide, or phosphate minerals such as feldspars, zircon, tourmaline, rutile, ilmenite, magnetite, xenotime, and monazite.

Four elements, Sb, Br, Cl, and S, are present in higher mean concentrations in the total group of coal samples (Table 4) than in a group of 13 associated shale and mudstone samples analyzed by Martin (1990). The latter include roof shale, shale parting, and seat earth samples. On average, the shale and mudstone samples contained 1.4 ppm Sb, <8 ppm Br, and 0.60% S, as compared to 1.5 ppm Sb, 28 ppm Br, and 1.00% S in the coal. Shales were not analyzed for Cl, but mean shale values (160-180 ppm) provided by Clark (1989) are significantly less than the mean total coal value of 1,300 ppm. Previous researchers (Finkelman, 1980; Kuhn et al., 1980; Harvey et al., 1983) have proposed that significant proportions of these elements are contained in coal organic compounds.

Vertical Variation Of Trace Element Concentrations In Bright Coals

Table 5 contains ANOVA test results of no difference between mean values of upper, middle, and lower subsamples from six bright coal units (Table 1: samples 7-12, 14-22, and 29-31). Visual examination of mean subsample data suggests that most elements are present in higher concentrations toward the top and bottom of the bed. However, only the vertical differences for Cr, Sc, U, and V are sufficiently pronounced to be detected (0.10 level) by ANOVA. Overall, the vertical variation of element concentrations within the bright coals appears to be markedly less than that between bright and dull coal lithotypes. DMR tests (0.10 level) in Table 5 indicate that the central part of the beds contains markedly lower concentrations of Cr, Sc, and V, whereas U is concentrated at the base. In addition, the DMR test recognizes significant increases in Sb and Zr in the basal subsamples and Sr depletion in the upper subsamples. Similar distribution patterns have been observed in both lignites (Karner et al., 1986) and bituminous coals (Baldwin, 1975; Cecil et al., 1979; Harvey et al., 1983).

Lateral Variation Of Trace Element Concentrations In Bright And Dull Coals

Five sets of paired correlative lithotype samples from the Cedar Grove and Pocahontas No. 3 seams were used to investigate the lateral distribution of elements. Two pairs of samples (bright/bright, dull/dull) were selected from sites about 600 m (2,000 ft) apart in a deep mine in the Cedar Grove Seam. Three pairs of samples (bright/bright, dull/dull, bright/dull) were chosen from cores of the Pocahontas No. 3 Seam located 4.1 km (2.6 mi.) apart. In both cases, the sampling sites are directly overlain by shale partings which thicken toward the seam margins as illustrated in Figure 7. These partings represent periods of flooding and detrital influx into the coal swamps. The partings in the Cedar Grove Seam are laterally transitional with fluvial channel sandstones (Mathew et al., 1978).

Because the number of samples is too small for systematic statistical tests, comparisons of the coal trace element concentrations are illustrated in Figure 7. The relative lateral changes in element concentrations are the same for both seams. Changes of more than 10% in element concentrations between sampling sites are associated with trends "A", "B", "C", and/or "D" in Figure 7. In general, elements preferentially concentrated in dull coals tend to increase as the ash (mineral matter) content of the coal increases (A), which is usually in the direction of thickening shale partings (C). On the other hand, most of the elements which characterize

Element	Upper (U)	Subsample: Middle (M)	Lower (L)	ANOVA Probability**	Duncan Multiple Range Test***
% Al	1.45	0.94	1.61	0.35	
Sb	1.6	0.7	3.8	0.15	M < L
As	21	17	14	0.88	
Br	27	25	27	0.96	
% Ca	0.19	0.22	0.17	0.35	
Ce	18	16	25	0.65	
Cs	0.9	0.4	0.8	0.62	
% Cl	0.13	0.16	0.15	0.45	
Cr	18	9	19	0.08	M < U, L
Co	10	11	23	0.46	
Eu	0.5	0.3	0.5	0.37	
% Fe	1.84	1.29	1.41	0.68	
% Mg	0.06	0.05	0.04	0.95	
% K	0.22	0.13	0.18	0.35	
Sc	5	2	7	<0.0	M < U, L
Se	5	4	4	0.95	
% Si	1.91	1.34	2.07	0.69	
Sr	97	151	127	0.21	U < M
S	0.95	0.89	1.22	0.63	
Th	2.5	1.4	2.9	0.23	
% Ti	0.07	0.05	0.11	0.25	
U	1.2	1	2	0.07	U, M < L
V	28	17	38	0.03	M < U, L
Zn	32	19	42	0.31	
Zr	22	19	28	0.19	M < L
% ash	8.8	6.3	8.8	0.46	

** probablity of no difference between subsample means
*** significance level for DMR test is 0.10

Table 5. *Mean element concentrations in bright coal subsamples with results from ANOVA and Duncan Multiple Range tests (% = weight percent; all other elements in ppm).*

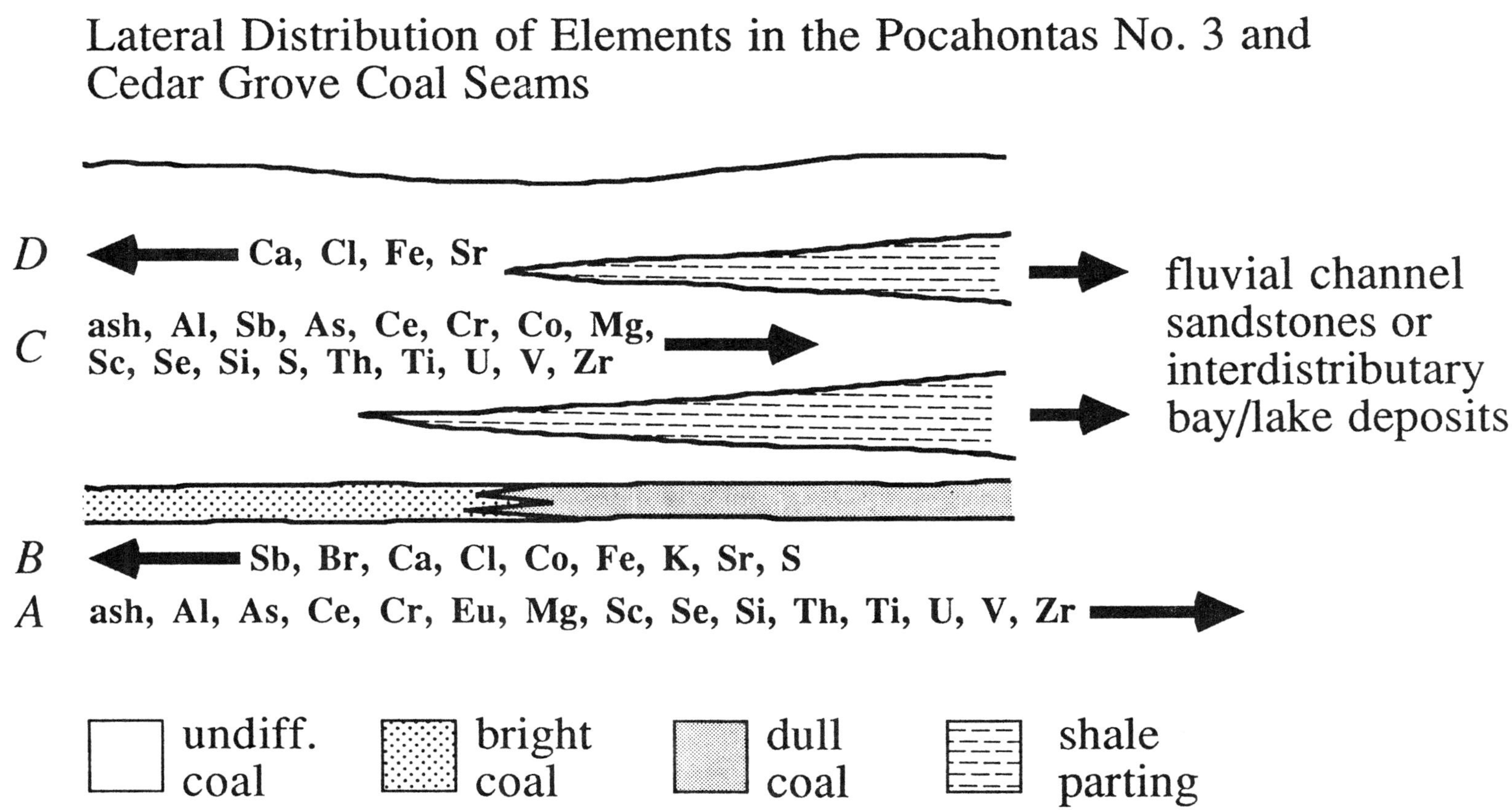

Figure 7. *Schematic diagram that summarizes the lateral distribution of investigated elements within correlative coal seams. Four trends are illustrated. "A" and "B" are changes in trace element concentrations associated with lateral coal lithotype transitions. "C" and "D" are the lateral variation of trace element concentrations associated with thickening shale partings in undifferentiated coal containing both bright and dull lithotypes.*

bright coal increase toward areas of the seams containing fewer shale partings and lower amounts of ash/mineral matter (B and D).

The spatial distribution of ash content and most elements within the Cedar Grove and Pocahontas No. 3 seams appears to be controlled mainly by lithotype distribution and the proximity of shale partings. These observations are important because they lend support to the previously discussed statistical analyses of trace element concentrations in bright and dull lithotypes.

IMPLICATIONS FOR THE ORIGINS OF TRACE ELEMENTS IN COAL

This study indicates that the in-seam distributions of all investigated elements except As, Br, Mg, and K are primarily a function of the local area's coal lithotypes. Differences between bright and dull coal lithotypes account for more than half of the total variation measured for 19 of 24 elements (Figure 8). Nine (69%) of the 13 elements associated with dull coals exhibit a significant positive correlation (at the 0.10 level) of greater than 0.55 with percent ash in bright and/or dull coal groups (Table 6). In contrast, all eight elements preferentially concentrated in bright coals exhibit either low or negative correlation coefficients with ash content (Martin, 1990). The in-seam lateral distribution of percent ash, as well as most of the elements, appears to be related to the positions of shale partings and dull coal lithotype units. These associations suggest potential genetic relationships between the formation of bright and dull coal lithotypes and the incorporation of many trace elements in coal seams.

Different types of precursor plant materials provide a plausible model for explaining the variation in trace element content between coal lithotypes. Some researchers (Cecil et al., 1979; Ruppert et al., 1985; Schofield and Haskin, 1964) have proposed that the concentrations of many coal trace elements are related primarily to biological accumulation by coal-forming flora. Bright coals are derived mainly from woody material, which should contain significantly lower quantities of ash and most trace elements than the herbaceous vegetation, partially oxidized plant matter, and non-woody plant tissue from which dull coals originated (Conner and Shacklette, 1975; Renton, 1982). Casagrande and Erchull (1976), Cameron and Schruben (1983), and Sarnecki (1983) suggest relationships between vegetation types and ash and trace element contents of modern peats.

However, conclusions that relate element distributions in coal to modern peat-forming vegetation largely are speculative. Plant-soil-atmosphere interactions are extremely complex (Omrod, 1984; Robb and Pierpoint, 1983; Shkolnik, 1984). Utilization of trace elements by plants has changed during the course of evolution (Boichenko, 1976), and most of the plant species which flourished during the Carboniferous are extinct. Also, floral assemblages appear to be related to depositional settings (DiMichele et al., 1984; Marley et al., 1978; Staub and Cohen, 1979) which introduced trace elements into the coal-forming peats from external water and detrital sources (Bailey, 1981; Chou 1984). Depletion of Al, Eu, Ce, Co, Fe, Zn, S, K, and Mg from rooted seat earths underlying coal seams relative to roof shales and shale partings (Martin, 1990) may indicate possible element accumulation by flora. This effect, however, would be complicated by direct chemical leaching of these paleosols.

Detrital minerals likely were incorporated into the investigated coal seams as fluvial detritus and airborne dust. Finkelman and Stanton (1978), Bailey (1979), Finkelman (1980), Davis et al. (1984), Miller (1984) and Otte (1984) are among those who presented data suggesting that significant proportions of many elements present in coal are related to detrital matter. Additional evidence is provided by the lateral increase in percent ash and element concentrations with increased proximity of fluvial detritus observed in the Cedar Grove and

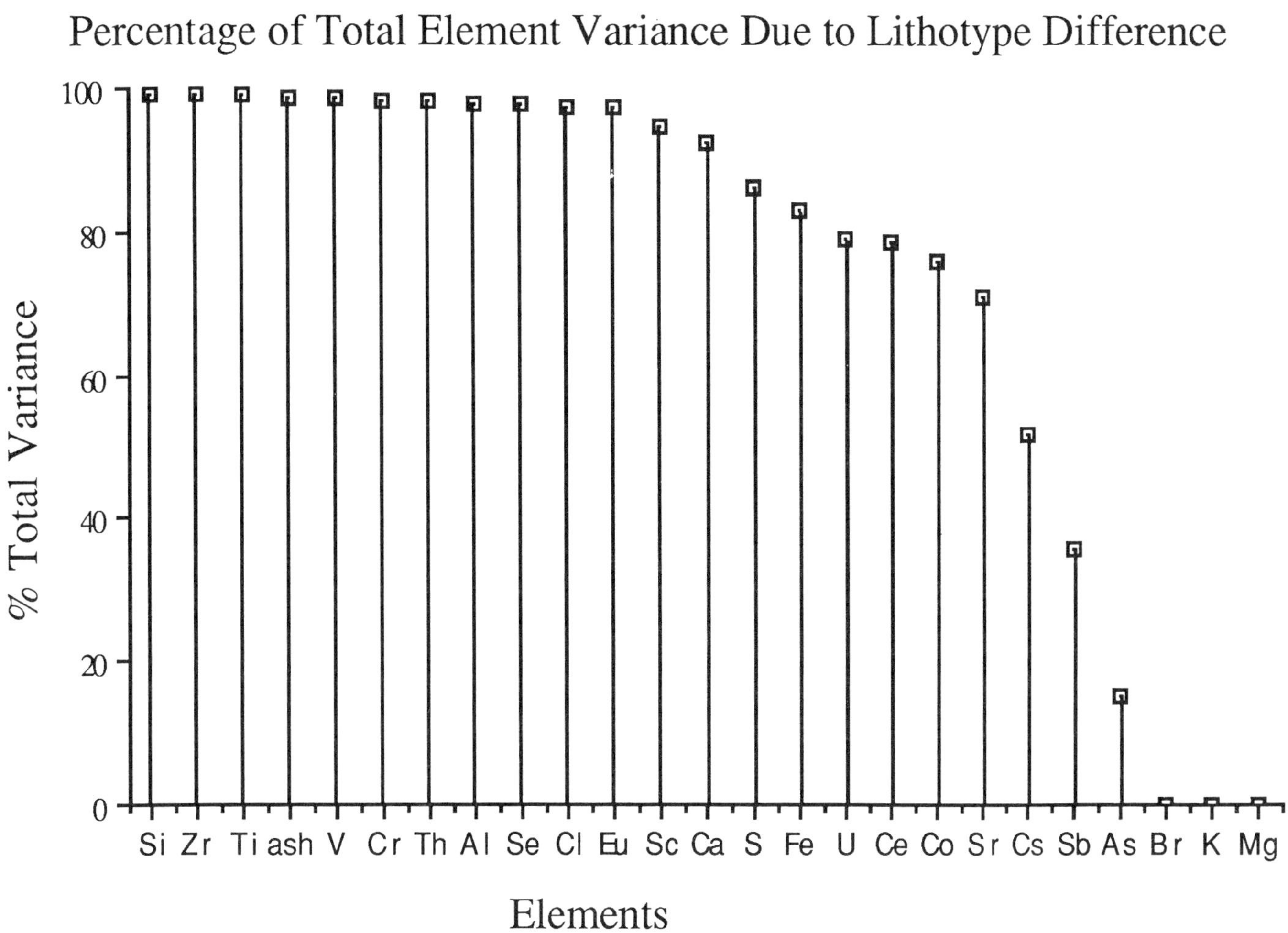

Figure 8. Percentage of the total variation in concentrations that is accounted for by differences between lithotypes for each element. The percentage is calculated from ANOVA data tables using the following formula: [(lithotype variance) - (residual variance)]/ (lithotype variance).

Pocahontas seams (Figure 7). Nearly all of the elements preferentially associated with dull coal (plus Sb, As, Co, and Mg) show such an increase.

Previous researchers (Sprunk et al., 1940; Hacquebard, 1951; Smith 1962; Reidenour et al., 1967; Stach, 1975b; Teichmüller, 1975) have interpreted some durain-rich dull coals as subaquatic organic accumulations which readily could have received fluvially transported detrital matter. Reidenour et al. (1967) specifically proposed that dull coals of the Allegheny Formation in Western Pennsylvania are of a subaqueous origin. The high proportions of quartz and clay minerals, together with high concentrations of ash and constituent trace elements such as Al, Cs, Cr, Si, Th, Ti, and V observed in dull coals in this study support this hypothesis. Also, several dull coal units examined in this study grade laterally into shale partings which are transitional with fluvial sandstone deposits at seam margins (Figure 7 and Martin, 1990). Low- and medium-ash dull coals commonly grade into high-ash dull coals which are transitional with shale partings (Sprunk and O'Donnell, 1942; Stach, 1975b).

Elements associated with bright coals include the carbonate- and sulfide-related elements Sb, Ca, Co, Fe, Sr, S, and Zn, as well as Cl. Calcium, Fe, and Sr tend to increase laterally toward the more central,

Element	Lithotype with Higher Conc.	Ash Correlation: Bright	Dull	In-Seam Lateral Conc. Increase Towards: Shale Partings	Bright Coal	Dull Coal
Al	dull	0.73	0.82	X		X
Cr	dull	0.62	0.85	X		X
Eu	dull	0.66	0.55			X
Sc	dull	0.59	0.68	X		X
Si	dull	0.75	0.92	X		X
Th	dull	0.65	0.80	X		X
Ti	dull	0.59		X		X
V	dull	0.56		X		X
Zr	dull	0.71		X		X
Ce	dull			X		X
Cs	dull			X	X	
Se	dull			X		X
U	dull			X		X
Ag	dull?					
Ba	dull?					
Mo	dull?	(Threshold Values Only)				
Ni	dull?					
Rb	dull?					
Ca	bright				X	
Cl	bright		-0.84		X	
Co	bright			X	X	
Fe	bright				X	
S	bright		0.61		X	
Sb	bright					
Sr	bright		-0.64		X	
Zn	bright					
As	no diff.			X		X
Br	no diff.				X	
K	no diff.	0.63			X	
Mg	no diff.	0.58		X		X

Table 6. *Summary of associations between coal lithotypes and trace elements.*

detrital-starved portions of the Cedar Grove and Pocahontas No. 3 seams; Ca, Cl, Co, Fe, Sr, and S tend to increase laterally toward areas containing bright coal (Figure 7). Accumulation of these elements as dissolved pore water constituents and subsequent precipitation as authigenic and/or diagenetic carbonates and sulfides in wood-rich peat or lignite beds may account for the significantly higher concentrations in bright coals. Similarly, groundwater precipitation of diagenetic carbonates and sulfides in the cleats of bituminous coals may account for the high concentrations of these elements in bright coals (Renton, 1982). Cleats in bright coals tend to be more numerous and well-developed than those in dull coals; therefore, bright coals should contain greater amounts of cleat mineralization. Pyrite mineralization was observed along the cleats within some bright coal units in the study areas, but not within dull units.

Marginal enrichment (and in some cases depletion) of bright coal beds in Sb, Cr, Sc, Sr, U, V, Zr, and the other investigated elements (Table 5) may reflect a greater availability of mineral matter and mineral-rich solutions during the beginning and end of wood-rich peat accumulation (Zubovic et al., 1964). Marginal enrichment could be the result of post-burial migration of elements within the peat deposit (Sarnecki, 1983) or preferential groundwater flow along contacts with coal seam clastic components (roof shales, shale partings, seat earths; Nicholls, 1968).

CONCLUSIONS

This study demonstrates that the concentrations of 20 trace elements in several bituminous coal seams in a portion of southeastern West Virginia are closely associated with easily field-identifiable, bright and dull coal lithotypes. Significantly higher concentrations of Al, Ce, Cs, Cr, Eu, Sc, Se, Si, Th, Ti, U, V, and Zr are characteristic of medium- to high-ash dull coals. Antimony, Ca, Cl, Co, Fe, Sr, S, and Zn are more highly concentrated in the low- to medium-ash bright coals. Arsenic, Br, Mg, and K are not concentrated preferentially in either lithotype.

The origins and diagenetic histories of coal trace elements are varied, complex, and difficult to document. However, correlations found to exist between element concentrations and coal lithotypes in the local study area have the potential to provide an excellent empirical tool for use by industry in mine planning and expansion. Other researchers such as Miller (1984) have found similar associations for selected elements. Wherever such predictable ranges of element concentrations can be obtained by selective sampling of the dominant coal lithotypes of an area, the results readily can be extrapolated to potential mine product based on existing maps of the seam's component lithotypes. Unfortunately, past methods of sample collection for chemical analyses used throughout most of the coal industry (i.e., seam channel, bench, and run-of-mine samples) generally disregard coal lithotypes, thereby excluding much of this data as a potential source of mine planning information.

ACKNOWLEDGEMENTS

Westmoreland Coal Company of Crab Orchard, West Virginia, Pardee and Curtin Company (Blackwood Land Company) and Peeker Run Coal Company of Webster Springs, West Virginia, and the American Association of Petroleum Geologists (1987 Grants-in-Aid Program) contributed financial and material assistance. Dr. R. V. Fodor, Department of Marine, Earth and Atmospheric Sciences, and Jack Weaver, Manager of Nuclear Services, North Carolina State University, provided valuable assistance with the coal analyses. Special thanks go to Phil McClung and Matthew Hall of Pardee and Curtin, and to Dr. Alan Dennis and John Bullock of Westmoreland Coal Company for their ideas and encouragement during the early stages of the study. Dr. Dennis was kind enough to critically review the manuscript in its draft stage.

REFERENCES

American Society for Testing Materials, 1988, Standard definitions of terms relating to megascopic description of coal and coal seams and microscopical description and analysis of coal: Philadelphia, Designation D-2796-88, p. D8-D10.

Bailey, A., 1981, Chemical and mineralogical differences between Kittanning coal from marine-influenced vs. fluvial sequences: Journal of Sedimentary Petrology, v. 51, no. 2, p. 383-395.

Baldwin, H.E., 1975, The vertical and lateral distribution of trace elements in the Fire Clay coal seam near Hazard, Kentucky: Columbia, University of South Carolina, Unpublished M.Sc. Thesis, 58 p.

Bates, R.L., and J.A. Jackson, 1980, Glossary of geology (2nd Ed.): Falls Church, Virginia, American Geological Institute, p. 75.

Baumann, D.R., 1982, The occurrence and distribution of mineral matter in coal lithotypes in the Herrin (No. 6) coal seam under marine and nonmarine influences: Carbondale, Southern Illinois University, Unpublished M.Sc. Thesis, 151 p.

Boichenko, E.A., 1976, Metal compounds in the evolution of plants: Izvestiya Akademii Nauk SSSR, Biological Series, v. 3, p. 378-385.

Cameron, A.R., 1978, Megascopic description of coal with particular reference to seams in southern Illinois, in R.R. Dutcher, ed., Field description of coal: American Society for Testing Materials, Special Technical Publication 661, p. 9-32.

Cameron, C.C., and P. Schruben, 1983, Variations in mineral matter content of a peat deposit in Maine resting on glacio-marine sediments, in R.R. Raymond and M.J. Andrejko, eds., Mineral matter in peat: Its occurrence, form and distribution: Los Alamos, New Mexico, Los Alamos National Laboratory, p. 63-76.

Casagrande, D.J., and L.D. Erchull, 1976, Metals in Okefenokee peat-forming environments: Relation to constituents found in coal: Geochimica et Cosmochimica Acta, v. 40, p. 387-393.

Cecil, C.B., R.W. Stanton, S.D. Allshouse, R.B. Finkelman, and L.P. Greenland, 1979, Geological controls on element concentrations in the Upper Freeport coal bed: Fuel, v. 24, no. 1, p. 230-235.

Chandra, D., and G.H. Taylor, 1975, Gondwana coals, in E. Stach et al., eds., Stach's textbook of coal petrology (2nd Ed.): Gebrüder Borntraeger, Berlin, p. 139-158.

Cheek, R., and A. Donaldson, 1969, Sulfur facies of the Upper Freeport Coal of northwestern Preston County, West Virginia, in A.C. Donaldson, ed., Some Appalachian coals and carbonates: Models of ancient shallow water deposition: West Virginia Geological and Economic Survey, Field Trip Guidebook for 1969 Geological Society of America Annual Meeting, p. 279-307.

Chou, C.-L., 1984, Relationship between geochemistry of coal and the nature of strata overlying the Herrin Coal in the Illinois Basin, U.S.A.: Memoirs of the Geological Society of China, no. 6, p. 269-280.

Clark, K.F., 1989, Mineral composition of rocks, in R.S. Carmichael, ed., Practical handbook of physical properties of rocks and minerals: Boca Raton, Florida, CRC Press, Inc., Vol. I, Section I, p. 1-138.

Conner, J.J., and H.T. Shacklette, 1975, Background geochemistry of some rocks, soils, plants and vegetables in the conterminous United States: U.S. Geological Survey Professional Paper 574-F, 168 p.

Davis, A., 1978, Compromise in coal seam description, in R.R. Dutcher, ed., Field Description of Coal: American Society for Testing Materials, Special Technical Publication 661, p. 33-40.

Davis, A., S.J. Russell, S.M. Rimmer, and J.D. Yeakel, 1984, Some genetic implications of silica and aluminosilicates in peat and coal: International Journal of Coal Geology, v. 3, p. 293-314.

Davis, J.C., 1986, Statistics and data analysis in geology (2nd Ed.): New York, John Wiley and Sons, 646 p.

DiMichele, W.A., T.L. Phillips, and R.A. Peppers, 1984, The influence of climate and depositional environment on the distribution and evolution of Pennsylvanian coal-swamp plants, in B.H. Tiffney, ed., Geological factors and the evolution of plants: New Haven, Connecticut, Yale University Press, p. 223-256.

Esterle, J.S., and J.C. Ferm, 1986, Relationship between petrographic and chemical properties and coal seam geometry, Hance Seam, Breathitt Formation, southeastern Kentucky: International Journal of Coal Geology, v. 6, p. 199-214.

Finkelman, R.B., 1980, Modes of occurrence of trace elements and minerals in coal: An analytical approach, in R.H. Filby, ed., Atomic and nuclear methods in fossil energy research: New York, Plenum Press, p. 141-149.

Finkelman, R.B., 1981, Recognition of authigenic and detrital minerals in coal (abstr.): Geological Society of America, Abstracts with Programs, v. 13, no. 7, p. 451.

Finkelman, R.B., and R.W. Stanton, 1978, Identification and significance of accessory minerals from a bituminous coal: Fuel, v. 57, no. 12, p. 763-768.

Gluskoter, H.J., R.R. Ruch, W.G. Miller, G.B. Cahill, G.B. Dreher, and J.K. Kuhn, 1977, Trace elements in coal: Occurrences and distributions: Illinois State Geological Survey, Circular 499, 154 p.

Hacquebard, P.A., 1951, The correlation by petrographic analysis of the No. 5 Seam of the St. Rose and Chimney Corner coalfields, Inverness County, Cape Breton Island, Nova Scotia: Canada Geological Survey Bulletin 19, 42 p.

Harvey, R.D., R.A. Cahill, C.-L. Chou, and J.D. Steel, 1983, Mineral matter and trace elements in the Herrin and Springfield Coals, Illinois Basin coal field: Illinois State Geological Survey, Contract/Grant Report 83/4, 162 p.

Headlee, A.J.W., and R.G. Hunter, 1955, Part V--The inorganic elements in the coals, in H.P. Price, ed., Characteristics of mineable coals of West Virginia: West Virginia Geological Survey Report, Volume 13(A), p. 37-122.

Hunt, J.W., and D.K. Hobday, 1984, Petrographic composition and sulphur content of coals associated with alluvial fans in the Permian Sydney and Gunnedah basins, Eastern Australia, in R.A. Rahmani and R.M. Flores, eds., Sedimentology of coal and coal-bearing sequences: Oxford, Blackwell Scientific Publications, Special Publication Number 7 of the International Association of Sedimentologists, p. 61-84.

Karner, F.R., D.R. Kleesattel, S.A. Benson, and H.H. Schobert, 1984, Optical and microbeam investigation of lithotypes of the Beulah-Zap Lignite, North Dakota (abstr.): Geological Society of America, Abstracts with Programs, v. 16, no. 6, p. 555.

Karner, F.R., H.H. Schobert, S.K. Falcone, and S.A. Benson, 1986, Elemental distribution and association with inorganic and organic components in North Dakota lignites, in K.S. Vorres, ed., Mineral matter and ash in coal: American Chemical Society, Symposium Series 301, p. 70-89.

Kuhn, J.K. , F.L. Fiene, R.A. Cahill, H.J. Gluskoter, and N.F. Shimp, 1980, Abundance of trace and minor elements in organic and mineral fractions of coal: Illinois State Geological Survey, Environmental Geology Notes 88, 67 p.

Marley, W.E., R.M. Flores, and V.V. Cavaroc. 1978, Coal accumulation in Upper Cretaceous marginal deltaic environments of the Blackhawk Formation and Star Point Sandstone, Emery, Utah: Utah Geology, v. 6, p. 25-40.

Martin, J.W., 1990, Trace element distribution within lithotypes and depositional settings of some south-central West Virginia coal seams: Raleigh, North Carolina State University, Unpublished M.Sc. Thesis, 313 p.

Mathew, D., J. Ferm, B. Bedford, K. Mathew, M. Williams, and C. Tipper, 1978, Reserves and mining potential of coal seams, Penn Virginia and Kelly Hatfield leases: Beckley, West Virginia, Hampton Division, Westmoreland Coal Company, 23 p.

McCabe, P.J., 1984, Depositional environments of coal and coal-bearing strata, in R.A. Rahmani and R.M. Flores, eds., Sedimentology of coal and coal-bearing sequences: Oxford, Blackwell Scientific Publications, Special Publication Number 7 of the International Association of Sedimentologists, p. 13-42.

Miller, T.R., 1984, The chemical variation in the Upper Hance coal seam, southeastern Kentucky: Lexington, University of Kentucky, Unpublished M.Sc. Thesis, 130 p.

Nicholls, G.D., 1968, The geochemistry of coal-bearing strata, in D. Murchison and T.S. Westoll, eds., Coal and coal-bearing strata: New York, American Elsevier Publishing Company, Inc., p. 269-308.

Norusis, M.J., 1986a, SPSS/PC+ for the IBM PC/XT/AT: Chicago, SPSS Inc., 643 p.

Norusis, M.J., 1986b, SPSS/PC+ for the IBM PC/XT/AT, Advanced Statistics: Chicago, SPSS Inc., 332 p.

Omrod, D.P., 1984, Impact of trace element pollution on plants, in M. Treshow, ed., Air pollution and plant life: New York, John Wiley and Sons Ltd., Environmental Monographs and Symposia, p. 291-319.

Otte, L.J. , 1984, Controls over the quantity and distribution of mineral sediment in North Carolina peat deposits (abstr.): Geological Society of America Abstracts with Programs, v. 16, no. 3, p. 185.

Parks, B.C., and H.J. O'Donnell, 1956, Petrography of American coals: U.S. Bureau of Mines Bulletin 550, 193 p.

Reger, D.B., 1920, Webster County and portion of Mingo District, Randolph County, south of Valley Fork of Elk River: West Virginia Geological Survey, County Report, 682 p.

Reidenouer, D., E.G. Williams, and R.R. Dutcher, 1967, The relationship between paleotopography and sulfur distribution in some coals of Western Pennsylvania: Economic Geology, v. 62, p. 632-647.

Renton, J.J., 1982, Mineral matter in coal, in R.A. Meyers, ed., Coal structure: New York, Academic Press, p. 283-326.

Robb, D.A., and W.S. Pierpoint, 1983, From metals and micronutrients: Uptake and utilization by plants: London, Academic Press Ltd., Annual Proceedings of the Phytochemical Society of Europe, No. 21, 341 p.

Ruppert, L.F., C.B. Cecil, R.W. Stanton, and R.P. Christian, 1985, Authigenic quartz in the Upper Freeport coal bed, west-central Pennsylvania: Journal of Sedimentary Petrology, v. 55, no. 3, p. 334-339.

Sarnecki, J.C., 1983, Selected trace element anomalies in a Front Range bog, Larimer County, Colorado, in R.R. Raymond and M.J. Andrejko, eds., Mineral matter in peat: Its occurrence, form and distribution: Los Alamos, New Mexico, Los Alamos National Laboratory, p. 179-188.

Schofield, A., and L. Haskin, 1964, Rare-earth distribution patterns in eight terrestrial materials: Geochimica et Cosmochimica Acta, v. 28, p. 437-446.

Schopf, J.M., 1960, Field descriptions and sampling of coal beds: U.S. Geological Survey Bulletin 1111-B, p. 25-70.

Shkolnik, M.Ya., 1984, Trace elements in plants: Amsterdam, Elsevier Science Publishers B.V., Developments in Crop Science, Vol. 6, 463 p.

Smith, A.H.V., 1962, The palaeoecology of Carboniferous peats based on miospores and petrography of bituminous coals: Proceedings of the Yorkshire Geological Society, v. 33, Part 4, no. 19, p. 423-474.

Sprunk, G.C., W.H. Ode, W.A. Selvig, and H.J. O'Donnell, 1940, Splint coals of the Appalachian Region: Their occurrence, petrography, and comparison of chemical and physical properties with associated bright coals: U.S. Bureau of Mines Technical Paper 615, 59 p.

Sprunk, G.C., and H.J. O'Donnell, 1942, Mineral matter in coal: U.S. Bureau of Mines Technical Paper 648, 67 p.

Stach, E., 1975a, The lithotypes of humic and sapropelic coals, in E. Stach, M.-Th. Mackowsky, M. Teichmüller, G.H. Taylor, D. Chandra, and R. Teichmüller, eds., Stach's textbook of coal petrology (2nd Ed.): Berlin, Gebrüder Borntraeger, p. 132-137.

Stach, E., 1975b, Seam identification, in E. Stach, M.-Th. Mackowsky, M. Teichmüller, G.H. Taylor, D. Chandra, and R. Teichmüller, eds., Stach's textbook of coal petrology (2nd Ed.): Berlin, Gebrüder Borntraeger, p. 310-315.

Staub, J.R., and A.D. Cohen, 1979, The Snuggedy Swamp of South Carolina: A back-barrier estuarine coal-forming environment: Journal of Sedimentary Petrology, v. 49, no. 1, p. 133-144.

Teichmüller, M., 1975, Origin of the petrographic constituents of coal, in E. Stach, M.-Th. Mackowsky, M. Teichmüler, G.H. Taylor, D. Chandra, and R. Teichmüller, eds., Stach's textbook of coal petrology (2nd Ed.): Berlin, Gebrüder Borntraeger, p. 176-239.

Teichmüller, M., and R. Teichmüller, 1975, The geological basis of coal formation, in E. Stach, M.-Th. Mackowsky, M. Teichmüller, G.H. Taylor, D. Chandra, and R. Teichmüller, eds., Stach's Textbook of Coal Petrology (2nd Ed.): Berlin, Gebrüder Borntraeger, p. 5-53.

Thiessen, R., and G.C. Sprunk, 1935, Microscopic and petrographic studies of certain American coals: U.S. Bureau of Mines Technical Paper 564, 71 p.

Wallace, A., and W.L. Berry, 1978, Trace elements in the environment — Effects and potential toxicity of those associated with coal, in H.K. Wali, ed., Ecology and coal resource development, Vol. 1: New York, Pergamon Press, p. 95-114.

Wood, G.H., Jr., T.M. Kehn, M.D. Carter, and W.C. Culbertson, 1983, Coal resource classification system of the U.S. Geological Survey: U.S. Geological Survey Circular 891, 65 p.

Zubovic, P., T. Stadnichenko, and N.B. Sheffey, 1964, Distribution of minor elements in coal beds of the Eastern Interior Region: U.S. Geological Survey Bulletin 1117-B, p. B1-B41.

Zubovic, P., T. Stadnichenko, and N.B. Sheffey, 1966, Distribution of minor elements in coals of the Appalachian Region: U.S. Geological Survey Bulletin 1117-C, p. C1-C37.

Chapter 7

Environmental Impacts

PREFACE

As noted in the prefaces for the Coal Quality Concerns and Underground Coal Gasification chapters, the impacts of coal extraction and utilization are a major area of concern to governments, the public, and the coal industry in general. Indeed, the primary contact most of the public has with the coal industry nowadays is in the form of end products (e.g., electricity) and information or first-hand contact concerning abandoned mines, acid drainage, acid rain, and other negative aspects of past and present coal use. Environmental activism and regulation are ever-present parts of any coal project and will remain so.

Nonetheless, the coal producing and using industries are cleaner today than ever before. Regulations force the matter upon those who won't comply voluntarily. However, in general, there is a serious concern among members of the industry for maintenance of good environmental quality.

The role of geology and geologists in this effort is discussed by Dutcher and Dutcher in the overview paper. As represented by the other seven papers in this chapter, there are many facets to the environmental impacts of coal utilization, especially in the areas of mining and coal use/refuse. The challenges are considerable, particularly as environmental regulations become more stringent over time. The opportunities for applications of geology to coal-related environmental problems are great in turn.

Moran et al. provide examples of problems and solutions involved in open pit mining in Alberta. Finkelman and Brown consider the possibility of turning environmental liabilities of coal composition into worthwhile byproducts, rather than trying to cure some of the later environmental impacts of trace elements. Püttmann et al. bring to light a new environmental wrinkle, the release of aromatic hydrocarbons (some with carcinogenic properties) from coal and mined overburden in coal mines and stockpiles. Wildeman and Klusman and Machemer present the opposite ends of the acid drainage problem, how it forms and how to remedy it through "natural" means (wetlands), respectively.

How do you handle the ash and other wastes from coal use (such as from power plants)? Beaver et al. address some of the disposal considerations, especially with respect to impact on groundwater supplies.

The paper by Goodarzi and Gentzis reviews the effect of natural burns on the coal and its surroundings. Although abandoned mine fires are not explicitly discussed, the principles are the same and the paper is applicable to this ongoing mined-land problem.

Cross-Reference

To some degree, the estimation of coal reserves nowadays is related to the ultimate environmental impact of the coal mining or utilization. Treworgy and Sites et al. point out some of these considerations in the Reserve Estimation chapter.

Coalbed methane development generally is a relatively benign process. However, water quality can be impacted by discharge of water from coal wells, and this is briefly touched upon by Mitchell and Pappajohn. Likewise, water quality (especially groundwater) is a major concern for underground coal gasification projects. This is the emphasis of the paper by Beaver et al. in that chapter.

The main paper of cross-interest in the Mining chapter is the one by Kane and Powell on subsidence. Subsidence remains the foremost impact of underground mines on surface properties, especially in populated coal mining areas. This paper will provide basic insights on what controls the subsidence process.

Of course, the Coal Quality Concerns papers have direct bearing on environmental impact of coal use. They all discuss the origin of the trace elements and other material in coal which have to be dealt with after mining and during and after coal use.

COAL UTILIZATION, THE ENVIRONMENT, AND THE ROLE OF GEOLOGISTS

L.A.F. Dutcher and R.R. Dutcher
P.O. Box 128
Carbondale, IL 62903

COAL AND THE ENVIRONMENT

The United States has demonstrated reserves (coal potentially mineable with current technology) sufficient to meet demand for at least 300 yrs (Illinois Coal Association, 1990, p. 5) at the 1990 production rate of 930,900,000 tonnes (1,024,000,000 tons) (Energy Information Administration, 1991, p. 6). These deposits, underlying about 13% of the country's land area, are the energy equivalent of about 2,000,000,000,000 bbl of crude oil — more than triple the world's known oil reserves. About 85% of coal used in the nation goes to the electric utility industry to generate 55% of the electricity produced in the United States. The remaining buyers include steel, other industrial facilities, and retail dealers to a limited extent. Future increases in demand will depend upon economic growth of domestic and foreign markets and the price of competing fuels. However, future demand especially will depend upon the scope of government regulation and the progress of research and technology development in reducing environmental impacts resulting from mining and utilization and in advancing clean coal-burning technologies.

More than half of the coal mined in the United States is produced in surface mines with the remainder from underground mines. When coal seams of sufficient thickness and quality are near the surface (generally less than 46-60 m (150-200 ft) deep), contour, area, or open pit mining methods are used. When the economic limit of an existing surface mine is reached, punch, auger, or conventional underground mining methods may be used to extend the life of the mine. When coal seams lie farther below, shafts or slopes must be constructed to reach and recover the coal and allow for room and pillar or longwall mining methods. "Room and pillar" refers to a system of extraction through regularly-sized openings (rooms) leaving large pillars of undisturbed coal which support the roof of the mine. Longwall mining is an advanced, high-productivity, capital-intensive technique in which self-advancing hydraulic chocks support the mine roof temporarily so that a rotating cutting drum or steel plow can be pulled across a coal face which is several hundred feet in width. When roof supports are advanced, the mine roof is allowed to collapse behind them. This usually results in almost immediate subsidence of the land surface in an amount less than the thickness of the seam. An important environmental problem, especially in agricultural areas, involves the correlation of geologic parameters with extraction method so that guidelines can be developed to maximize coal recovery while still mitigating effects of subsidence on the hydrology and soil productivity of farmland as well as on undermined structures.

Prior to the 1950s, most coal was used locally or regionally, depending upon transportation costs. At that time, the U.S. railroad industry responded to potential competition from the construction of an Ohio pipeline (which never was used) by developing the unit-train concept. Up to the late 1960s, prior to the effects of the Federal Clean Air Act of 1970, 90% of U.S. coal was produced from privately-owned seams in the eastern part of the country, western coal having been of little interest to users outside the western states because of a generally lower Btu value than coals found in the East and Midwest. The use of western coals soared in the Midwest during the 1970s in response to the Clean Air Act due to their much lower sulfur contents, on average (some of it 0.6-0.7%) than eastern coal. Today, unit trains carrying more than 10,000 tons (Illinois Coal Association, 1990, p. 67) are moving low-sulfur coal 1,600 km (1,000 mi.) or more from western states to the Midwest and South. About two-thirds of the cost of western coal delivered to the Midwest is attributable to transportation, raising the cost of electricity to homeowners and industry. Only one pipeline presently is transporting coal as a slurry: from Arizona to Nevada over a distance of 439 km (273 mi.). Others are planned but remain controversial for political and environmental reasons.

Coal can be processed in a variety of ways for chemicals, chars, and liquid and gaseous fuels. However, nearly all coal used in the U.S. is utilized directly in the unprocessed solid form or in its carbonized form as coke. One notable exception is the use of coal by Eastman Chemicals of Tennessee to produce acetic anhydride via the Lurgi Gasifier. The byproducts of the production of coke result in coal tar pitch which is the sole source of binder for carbon anodes used in aluminum smelting and for carbon electrodes used by the steel industry. Coal tar from coke manufacture is also a major source of the chemicals benzene, toluene, and xylene (BTX), and it is used in the manufacture of carbon-carbon composites and carbon fibers.

In coal basins such as the San Juan and Warrior basins where the seams contain large quantities of methane, production of the gas occurs directly through drill holes that often produce vast quantities of water prior to flow. This form of energy exploitation, in addition to low-Btu synthetic natural gas (SNG) from in situ combustion, may be applicable to coal seams that are too deep or too thin to be economically mined. Detailed geologic planning will be essential for these types of activity as it is in conventional mining operations.

The 1973 Middle East oil embargo led to a re-evaluation of the U.S. coal extraction and utilization industries. Our extensive reliability on coal had declined from a World War II peak level of 567,000,000 tonnes (624,000,000 tons) in 1947 to a new annual low of 345,000,000 tonnes (380,000,000 tons) in 1961 (Ward, 1984, p. 10) due to the availability of inexpensive oil and gas. The resulting government press for a dramatic expansion of coal production in the East and Midwest (Yancik, 1979, p. 20), and introduction of coal production on a large scale in the West, ran headlong into a "heightened sensitivity to environmental consequences of our actions" (Davenport, 1979, p. 3). This belief that environmental conditions and concerns should not be sacrificed had developed in the U.S. (National Environmental Policy Act) and other countries in the 1960s.

The job of resolving this energy-environment conflict fell in large part to the Carter Administration's National Energy Plan (NEP). In late 1976, Congress gave the Department of the Interior the necessary authority to plan comprehensively for public lands (Davenport, 1979, p. 4-5). The coal-leasing system was revised, with issues related to other minerals, water, timber, wildlife, and recreation to be considered together for the first time in issuing coal leases on public lands. Long-existing public debate centering around unreclaimed surface-mined lands (nearly 400,000 hectares (1,000,000 acres) by 1974 (Davenport, 1979, p. 6)) and mine water pollution resulted in several attempts at regulatory legislation. President Carter signed the Surface Mining Control and Reclamation Act in August 1977 giving the Office of Surface Mining (OSM) responsibility for its enforcement. Other new legislation impacting the coal industry at this time included the Mine Health and Safety Act of 1977 and the Clean Air Act Amendments of 1977.

The Clean Air Act Amendments passed by Congress in 1990 have given targeted utilities a further deadline of 1995 to cut SO_2 emissions to 1.1 kg (2.5 lb) of SO_2 per million British Thermal Units (Btu), with a further reduction to 0.54 kg (1.2 lb) SO_2 by the year 2000. A potential future issue emerging for energy planners is the greenhouse effect or so-called "global warming".

ROLE OF GEOLOGISTS AND GEOSCIENCE IN ENVIRONMENTALLY SOUND COAL UTILIZATION

Geologists and the geosciences have found a role in all stages of the determination and mitigation of environmental impacts of coal exploitation, reclamation, and utilization. These roles have included the communication of geologic principles and information to both federal and state lawmakers in order to formulate laws to attain the desired goals of society, data gathering and initial planning to assure surface mining operations are conducted in a manner to protect the environment, preparation of applications for mining to be submitted to the appropriate authorities, and enforcement of environmental regulations. Also included is the more traditional role of carrying out scientific investigations.

Assessment of environmental impacts at any specific site requires an enlightened evaluation of the interaction between the proposed activities of mining, transportation, and/or end-use operation and the pre-activity characteristics of the site. Model studies by the U.S. Geological Survey (Leopold et al., 1971) provide "a checklist of factors defining the existing environment and possible aspects of the mining operation that can interact with these" characteristics (Ward, 1984, p. 295). Such matrices are essential for preparation of Environmental Impact Statements (EISs) as well as in the planning of data gathering in the field.

The Permanent Regulatory Program of OSM (Federal Register, 1979, p. 15351-15376) established the following geologic areas for assessment and regulation:

1) coal exploration,

2) pre-existing site geology, hydrology, water quality, and land use,

3) alternative water supply,

4) cross sections, maps, and plans,

5) protection of the hydrologic balance during and after mining operations,

6) management of boreholes, and

7) extensive evaluation of alluvial valley floors in arid and semi-arid areas west of the 100th meridian.

Both sections 779.25 and 784.23(c) of the Federal Register (U.S. Government Printing Office, 1979, p. 15357 and 15370, respectively) require that specified maps, plans, and cross sections "shall be prepared by or under the direction of and certified by a qualified professional engineer or professional geologist" with the assistance of other appropriate experts.

Thus, the need for geologic studies and geologists in assessing and mitigating environmental impacts, long claimed by the practitioners of the science, came to be "legitimized" in the public eye by the inclusion of geologists in teams of specialists assembled by coal mining companies, government and regulatory agencies, and research institutions. The roles of geologists evolved to encompass interpretation and planning for compliance with the legal and scientific requirements of the Surface Mining Act, as well as participation in its enforcement and the advancement of research designed to solve associated fundamental problems.

History has demonstrated that principles developed by scientists and technicians only have an impact if they are utilized and contained in existing law (Dwyer, 1979, p. 216). Scientists have the knowledge which must be communicated in order to make the necessary changes to achieve the goals of adequate environmental controls. Papers contained in this section present a variety of geologic situations and demonstrate the need for an active role by geologists in all stages of the determination and mitigation of environmental impacts of coal exploitation, reclamation, and utilization.

REFERENCES

Davenport, J.M., 1979, U.S. coal development policy: Resolving the energy-environment conflict, in M.K. Wali, ed., Ecology and coal resource development: New York, Pergamon Press, Vol. 1, p. 3-8.

Dwyer, J.W., 1979, Enforcement of P.L. 95-87 (The Surface Mining Control and Reclamation Act of 1977) Who does what? And when? And where?, in M.K. Wali, ed., Ecology and coal resource development: New York, Pergamon Press, Vol. 1, p. 215-225.

Illinois Coal Association, 1990, 1990 Illinois Coal Facts: Springfield, Illinois, 68 p.

Leopold, L.B., F.E. Clark, B. Hanshaw, and J.R. Balsley, 1971, A procedure for evaluating environmental impact: U.S. Geological Survey Circular 645, 13 p.

U.S. Energy Information Administration, 1991, Coal distribution, January-December, 1990: Washington, Report DOE/EIA-0125 (90/4Q), p. 6.

U.S. Government Printing Office, 1979, Dept. of Interior, Office of Surface Mining Reclamation and Enforcement: Surface Coal Mining and Reclamation Operations Permanent Regulatory Program: Washington, Federal Register, Vol. 44, No. 50, Book 3, p. 15311-15463.

Ward, C.R., 1984, Coal geology and coal technology: Melbourne, Blackwell Scientific Publications, 345 p.

Yancik, J.J., 1979, Coal's response to the challenge in the National Energy Plan: Demand and supply projection through 1985, in M.K. Wali, ed., Ecology and coal resource development: New York, Pergamon Press, Vol. 1, p. 13-25.

ENVIRONMENTAL IMPACT OF SURFACE COAL MINING AND RECLAMATION IN THE PLAINS REGION OF ALBERTA, CANADA

S.R. Moran, M.R. Trudell, T.M. Macyk, and D.B. Cheel
Terrain Sciences Department
Alberta Research Council
P.O. Box 8330, Postal Station F
Edmonton, Alberta T6H 5X2
Canada

A.E. Howard
Alberta Agriculture
Research Centre
Jail House Rd.
Lethbridge, Alberta T1J 4C7
Canada

ABSTRACT

Between 1979 and 1988, the Plains Hydrology and Reclamation Project (PHRP) studied the impacts of surface coal mining in the plains of Alberta on the hydrology and agricultural capability of reclaimed landscapes. The most important adverse hydrologic impact of surface mining is removal of the shallow aquifers that provide the majority of agricultural water supplies. In some areas, replacement water supplies are available from beneath the base of mining. Chemical quality of groundwater in mine spoil is degraded significantly relative to that in pre-existing coal aquifers. Surface mining has essentially no impact on water quality outside of mined areas.

The agricultural capability of reconstructed soil landscapes generally is equivalent to, although less variable than, that of unmined landscapes. Physical and chemical properties of well-drained reconstructed soils are improved through downward leaching of sodium and through a decrease in bulk density. Surface wetness and soil salinity result in degradation of agricultural capability in small areas of reclaimed sites.

INTRODUCTION

Between 1979 and 1988, the Plains Hydrology and Reclamation Project (PHRP) investigated interactions of groundwater, soils, and geology and successful reclamation of surface coal mines in the plains of Alberta. The overall goal of PHRP was 1) to predict the long-term success and the hydrologic impacts of current reclamation practices and 2) to develop reclamation technology that will allow necessary modification of current practice to assure long-term success and mitigate deleterious environmental consequences.

The first phase of the study, which was completed in 1984, included characterization and instrumentation of two study areas, the Battle River study area, which included the Diplomat, Vesta, and Paintearth mines, and the Wabamun study area, which included the Highvale and Whitewood mines (Figure 1). Small-scale surface mining at both the Diplomat and Vesta mines in the Battle River mining area began in the 1950s or early 1960s. The scale of the operations at the Vesta Mine was increased in the late 1970s with the introduction of a large, modern dragline. Mining at the Diplomat Mine came to an end during the latter phases of our study, in the mid-1980s. At both the Diplomat and Vesta mines, the study sites were situated in areas that were mined during the transition to modern, larger-scale mining practices. At both mines, initial instrumentation, which was installed in 1979 and 1980, was situated in areas of older mining that were reclaimed to pre-modern standards. Later instrumentation, which was installed between 1985 and 1987, was situated in freshly reclaimed areas that had been mined using current practice. The Paintearth Mine was opened in the early 1980s and all instrumentation was installed in freshly reclaimed sites. At both the Highvale and Whitewood mines, the instrumented areas were located in pits that had been mined during the early to mid-1970s using modern mining and reclamation practices. Active mining continued in other pits of these mines throughout the project.

Research from the first phase of study resulted in focusing the second phase of the project on 1) potential salinization of reconstructed soils from shallow groundwater, 2) potential deterioration of capability for agriculture as a result of differential subsidence, and 3) potential changes in the chemical and physical characteristics of reconstructed soils.

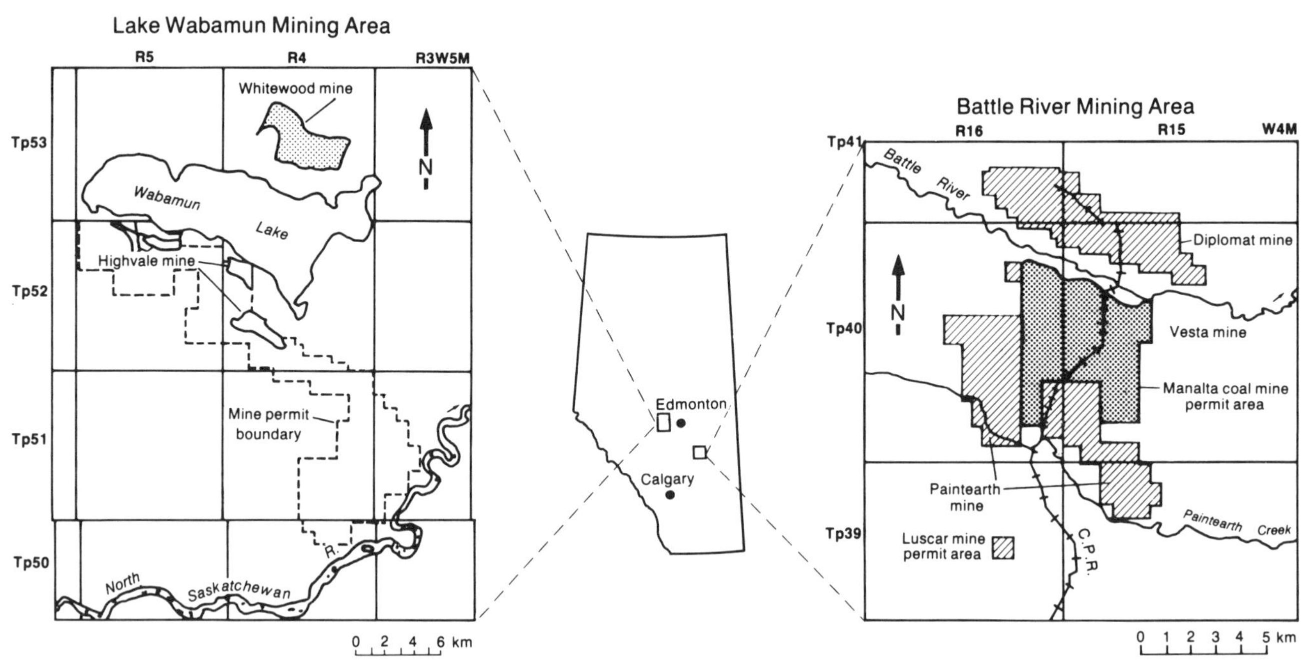

Figure 1. Location of the Battle River and Wabamun study areas of the Plains Hydrology and Reclamation Project.

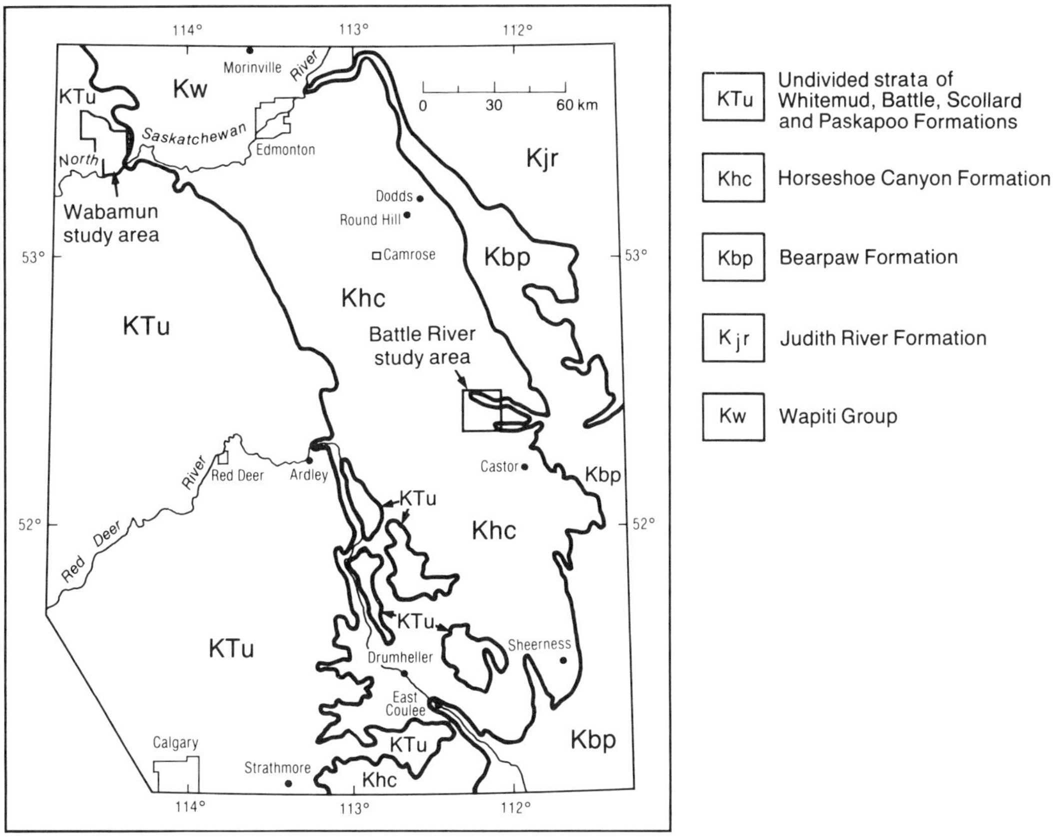

Figure 2. *Map showing geologic setting of the Plains Hydrology and Reclamation Project study areas.*

CHARACTERISTICS OF THE STUDY AREAS

Battle River Study Area

The Battle River study area is located in the Aspen Parkland Ecoregion (Strong and Leggat, 1981, Map 1), which is a transition zone between boreal forest and grassland environments. Annual precipitation is about 400 mm (16 in.), with about 67% occurring during the growing season. Potential evapotranspiration exceeds precipitation by about 200 mm/yr (8 in./yr). Mean summer temperatures are in the range of 12.5°C to 14.5°C. Natural vegetation in the region is characterized by fescue grassland with aspen groves commonly occurring along with willows in rings around the numerous ponded depressions, or sloughs, that dot the landscape (Strong and Leggat, 1981, p. 14-16). Soils in the Battle River study area are dominantly Orthic Black and Dark Brown Chernozems, Solonetzic Black, and Solonetzic soils (Macyk and Maclean, 1987). The land with the highest capability for agriculture generally is farmed in crop rotations that include cereal grains and canola, with forages being grown on land with somewhat lower capability. The poorest farm land generally is kept in pasture or is uncleared parkland or bush.

The Battle River study area is underlain by rocks of the Bearpaw and Horseshoe Canyon formations of Late Cretaceous age (Figure 2). The Bearpaw Formation, which in the study area is at least 75-m (250-ft) thick, consists dominantly of marine shale, but contains three persistent sandstone beds (Figure 3). The Battle River Bed, which rests on a marine marginal sandstone at the base of the Horseshoe Canyon Formation, generally consists of two coal seams separated by a thin parting (Maslowski Schutze et al., 1986). The

Paintearth Bed, which lies from 16 m (52 ft) to 20 m (66 ft) above the Battle River Bed, consists of two to three mineable seams separated by partings (Figure 3). The Battle River Bed is mined at the Diplomat, Vesta, and Paintearth mines. The Paintearth Bed is absent in initial cuts at both Vesta and Paintearth mines, but is mined in most of the workings of these two mines. The Horseshoe Canyon Formation is overlain by as much as 10 m (33 ft) of Pleistocene glacial drift, consisting dominantly of silty till.

The two coal seams that are being mined in the Battle River study area are the principal shallow aquifers in the area, accounting for at least 38% and 24%, respectively, of the 34 domestic wells in the area. These aquifers are overlain and separated by an aquitard sequence consisting of interlensing clay, shale, siltstone, and clayey sandstone. The sandstone unit that underlies the Battle River Bed is fine grained, contains abundant interstitial clay, and has too low a hydraulic conductivity to be an aquifer within the study area. The three sandstone beds within the Bearpaw Formation supply 24% of the domestic wells. The thick sequence of shale that overlies and separates these units is a major aquitard. Potentiometric gradients trend downward everywhere, except beneath the floors of the two deep valleys that cross the area. Groundwater flow in and beneath the Battle River Bed is toward the valley of the Battle River. Groundwater flow directions in the Paintearth Bed and the thin sand beds within the glacial drift are controlled by outcrops of the unit in the sides of smaller topographic features such as creeks and hill slopes.

The mining operations at the three mines in the Battle River area are similar, differing only in detail. Wheeled scrapers are used to remove topsoil and subsoil at all three mines. At the Diplomat Mine, a large stripping shovel was used to remove overburden in most of the mine. In the later stages of the operation, small tracked draglines were used in limited areas. At the Vesta and Paintearth mines, large draglines (45.9 m^3 (1,620 ft^3)) are used as the prime earthmover. At all three mines, coal is loaded using shovels and front-end loaders and trucked to the nearby powerplant. Spoil pile leveling and grading operations generally are separated from the active pit by only one or two ridges. Spoil is graded primarily using tracked dozers. At all three mines, topsoil and subsoil generally are replaced directly by the wheeled scrapers. In some parts of the mines, however, limited stockpiling of subsoil and topsoil has been required.

Wabamun Study Area

The Wabamun study area is located in the Boreal Mixedwood Ecoregion (Strong and Leggat, 1981, Map 1), which is a transition zone between mixed conifer-deciduous forest and parkland environments. Annual precipitation is about 440 mm (17 in.), with more than 70% occurring during the growing season. Potential evapotranspiration exceeds precipitation by about 200 mm/yr (8 in./yr). Mean summer temperatures are in the range of 10.5°C to 13.5°C. Natural vegetation in the region is dominated by woodlands comprised of aspen, balsam poplar, and white spruce (Strong and Leggat, 1981, p. 26-30). Soils are dominantly Grey Luvisols with significant areas of Solonetzic soils. The land with the highest capability for agriculture generally is farmed in crop rotations that include cereal grains and canola, with forages being grown on land with somewhat lower capability. The poorest farm land generally is kept in pasture or is uncleared.

The Ardley Coal Zone, which is mined in the Wabamun study area consists of six continuous coal seams, with the upper two being the thickest and most important (Figure 3). Overlying the coal zone, the Paskapoo Formation consists of a sequence of continental fluvial sediment (Maslowski Schutze, 1987). The basal, dark grey, flood-basin shale is overlain by a complex sequence of thick, apparently structureless sandstone that in places grades laterally into interbedded shale, siltstone, and sandstone. This overbank facies, which comprises most of the overburden in the majority of the study area is overlain by a sequence of interbedded shale, mudstone, and minor siltstone which represents a return to flood-basin conditions (Figure 3). Throughout the study area, the bedrock is folded, faulted, and crushed by intense glacial deformation (Fenton et al., 1986). A variable thickness of glacial drift, which contains abundant disturbed and transported blocks

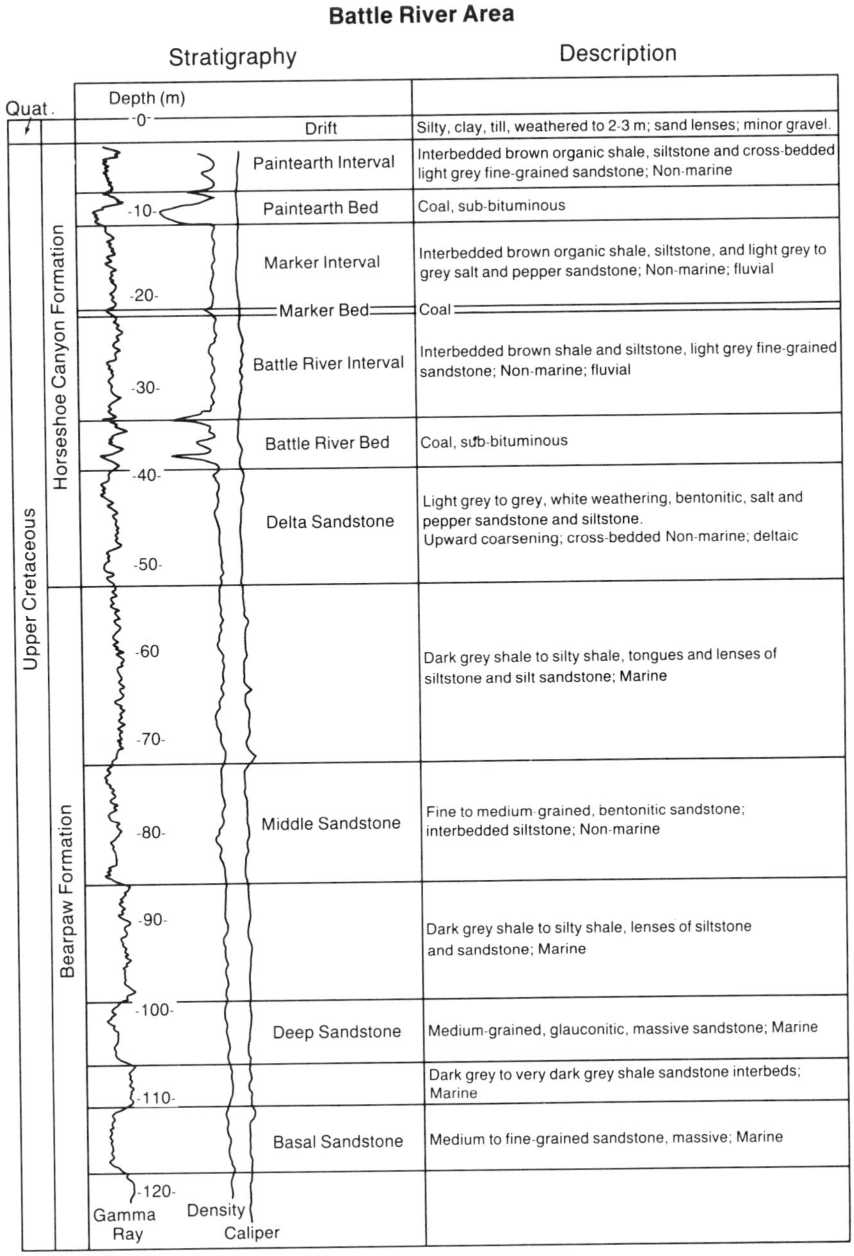

Figure 3a. *Generalized stratigraphic column for the Battle River study area (Lower Horseshoe Canyon Coal Zone).*

of locally derived bedrock, overlies the Paskapoo Formation. The drift consists dominantly of clayey to silty till.

The groundwater system at the Highvale Mine study area is characterized by downward flow beneath the uplands to the south and downward and lateral flow toward the northeast throughout the mine area. Discharge conditions generally are present only at the subcrop or outcrop of the sandstone and coal aquifers along the edge of Lake Wabamun. The sandstone unit that lies above the Ardley Coal Zone is the principal aquifer in the study area, supplying 62% of 196 domestic wells. Isolated sand bodies in the glacial drift sequence account for another 17% of the domestic wells. The coal zone, which is a minor aquifer, supplies 18% of the

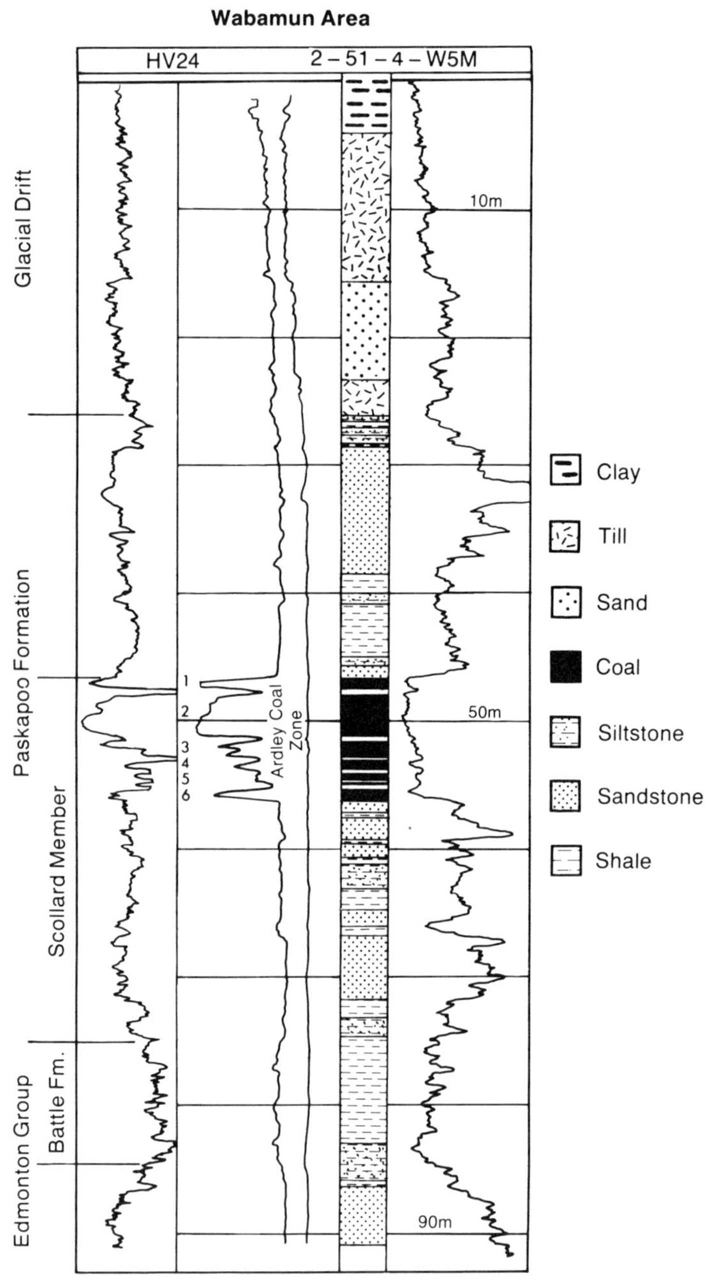

Figure 3b. *Generalized stratigraphic column for the Wabamun (Highvale) study area (Ardley Coal Zone).*

wells. Only 3% of the wells are completed in the scattered, discontinuous, generally slightly permeable sandstone beds that occur beneath the Ardley Coal Zone.

Mining operations at the mines in the Wabamun area are very similar to those described previously in the Battle River area.

ENVIRONMENTAL IMPACT OF MINING AND RECLAMATION

Decreased Groundwater Supply Capability Of Reclaimed Areas

The most important hydrological impact of surface mining of coal in the plains of Alberta is a reduction in groundwater supply capability within mined areas. Only limited impacts have been observed or are anticipated outside of the areas disrupted by mining.

In areas of potential surface mining of coal in the plains of Alberta, the groundwater supplies used by the rural population are derived almost entirely from either fractured coal seams or sandstone overlying the coal (Trudell, 1986, p. 9-10; Trudell et al., 1986, p. 12-13). Surface mining removes these aquifers and replaces them with mine spoil. Premining overburden, from which the spoil is derived, consists dominantly of silty to clayey bedrock or glacial drift which have very low hydraulic conductivities (Figure 4) (Trudell and Faught, 1987; Trudell et al., 1987). The hydraulic conductivity of the spoil overburden is less than that of the pre-mining coal aquifers but somewhat greater than that of the original overburden (Figure 5). At this hydraulic conductivity, the spoil is not capable of supplying water to wells (Trudell, 1986; Trudell and Moran, 1986, p. 14-15; Trudell et al., 1986).

Aquifers capable of replacing the shallow coal aquifers that were removed by mining are present beneath the base of disturbance only in limited areas. Laterally persistent, sheet-like sand bodies occur in the Bearpaw Formation beneath the lowermost coal zone in the Horseshoe Canyon Formation at sites such as the Battle River site (Maslowski Schutze et al., 1986). The geologic and hydrologic setting of these sand bodies at depths below actively flushed groundwater-flow systems, however, generally results in water that is too saline for human, and in places, animal consumption. At these depths, flow paths and residence times are long, flow velocities are low, and diffusion of sodium chloride from adjacent marine shale beds produces brackish water. At sites where the sand beds in the Bearpaw Formation are recharged through deeply incised glacial meltwater valleys, such as in the eastern part of the Battle River study area (Trudell et al., 1987; Trudell et al., 1986), the water quality in these aquifers can be acceptable for human consumption.

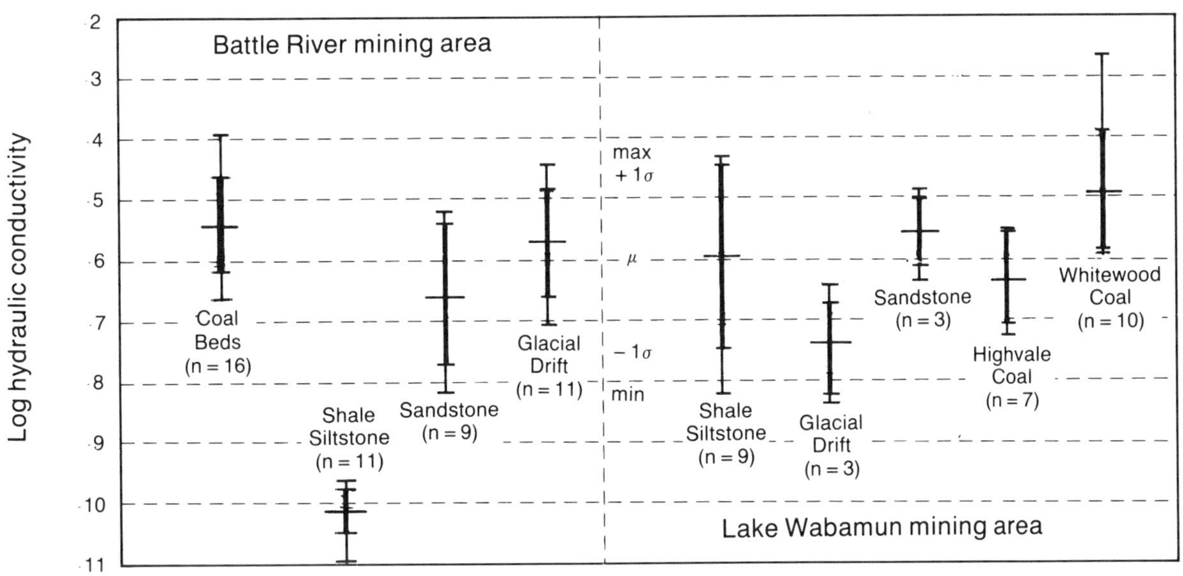

Figure 4. *Summary of hydraulic conductivity data for overburden in the plains of Alberta.*

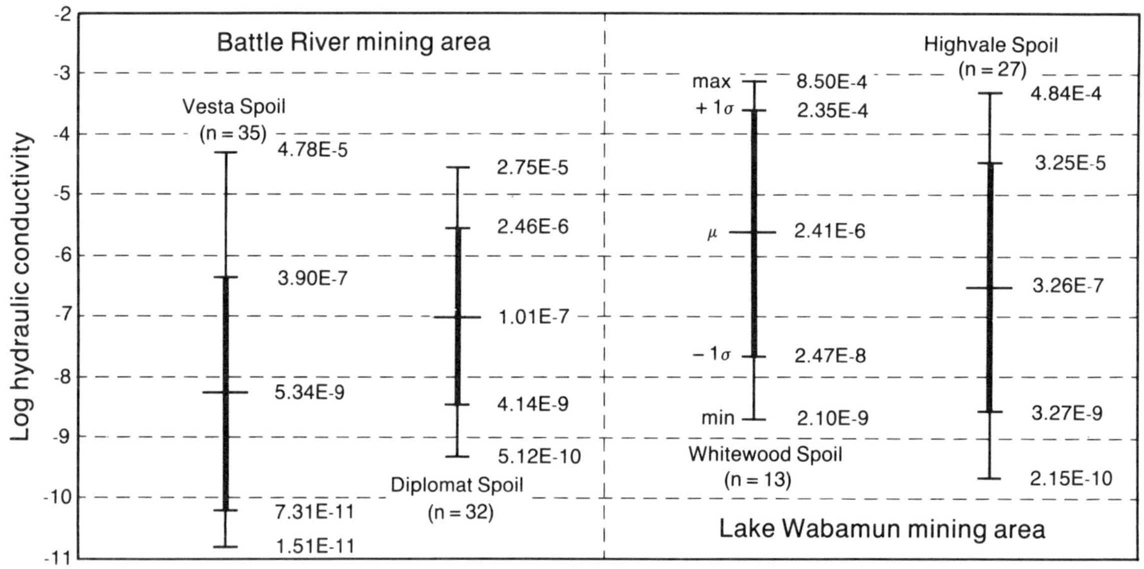

Figure 5. *Summary of hydraulic conductivity data for mine spoil in the plains of Alberta.*

At mine sites in the Ardley Coal Zone, such as at the Lake Wabamun mining area, the best potential post-mining groundwater supplies involve development of deeper, laterally persistent coal seams or discontinuous sand bodies. Inadequate well yield and chemical quality are potential limitations to development of replacement water supplies in deeper coal seams. Discontinuous sand bodies that range in thickness from a few meters to tens of meters and in width from a few hundred meters to several kilometers are widespread in coal-bearing sequences throughout the plains of Alberta. Location of these sand bodies, however, is difficult to nearly impossible without detailed geological investigations.

The groundwater in mine spoil in the plains of Alberta, in virtually all cases, is more saline and has a substantially different chemical composition than the water in premining aquifers (Trudell et al., 1988). Increases in total dissolved solids (TDS) ranged from 2.4 to 5.9 times the pre-mining levels (Figure 6). Sodium and sulfate account for the major increase in most areas, although calcium and magnesium are significantly increased in spoil of the lower Horseshoe Canyon Coal Zone at the Battle River mining area. Major ion concentration in spoil water derived from a mix of drift and bedrock (Vesta and Highvale mines) is greater than in spoil that contains dominantly glacial drift (Diplomat and Whitewood mines). It is not known whether the greater salinity of spoil groundwater at the Battle River mining area relative to that in the Wabamun mining area (Figure 6) is a function of intrinsic, geological differences in the overburden characteristics or results from differences in the climates in the two areas.

On the basis of observed groundwater chemistry and experimental weathering of overburden materials, we conclude that the principal source of salt in the spoil groundwater is solution of pre-existing salts. In the semi-arid environment of east-central Alberta, secondary carbonate and sulfate salts are ubiquitous in the soil zone over much of the landscape. Although oxidation of either organic or pyritic sulfur in previously unweathered overburden is occurring (Moran et al., 1984, p. 6-7), evidence suggests that it does not make a significant contribution to the dissolved salt load as has been postulated elsewhere in Western North America (Groenewold et al., 1983; Moran and Cherry, 1977; Moran et al., 1978b).

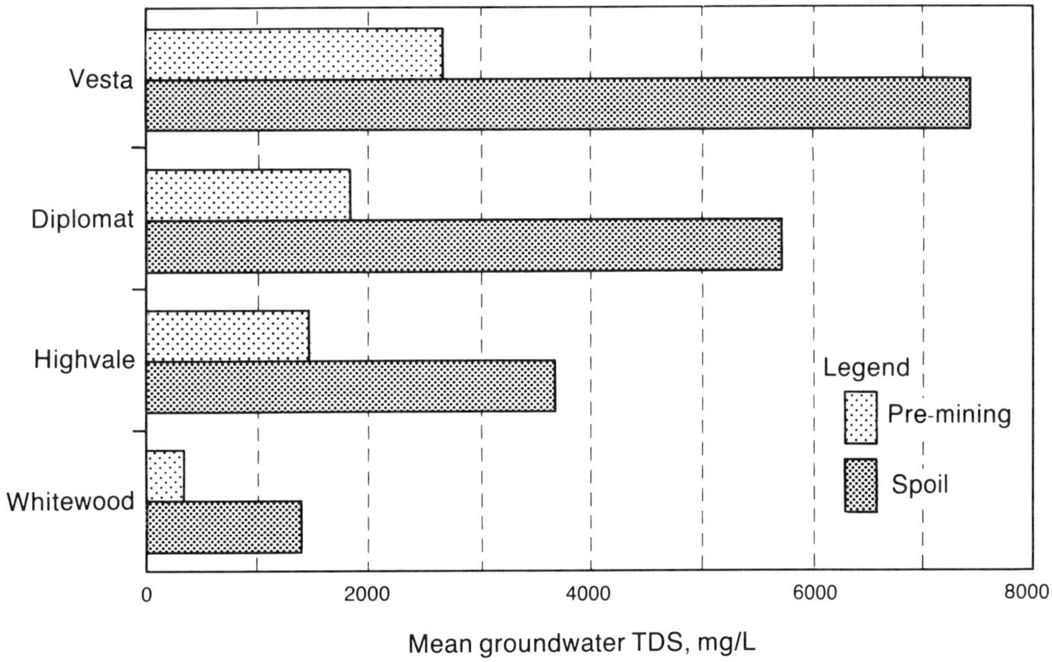

Figure 6. *Comparison of total dissolved solids (TDS) concentrations in groundwater in pre-mining aquifers and mine spoil in the plains of Alberta.*

At both the Battle River mining area (Trudell et al., 1986) and the Lake Wabamun mining area (Trudell and Moran, 1986), water levels have been lowered only in wells that are a few hundred meters from the mine. At distances greater than about 1 km (0.6 mi.) from the mine, no effect on water levels has been observed. These observations are consistent with data from mines throughout the plains of Western North America that, like the sites studied in Alberta, are located in areas of groundwater recharge (Moran et al., 1978a; Van Voast and Reiten, 1987).

Groundwater Contamination By Spoil-Derived Water

Although groundwater in spoil at the Battle River site has TDS concentrations from 5,000 to 12,000 mg/L (Trudell et al., 1988) and hydraulic head values that are higher than those in deeper potable aquifers (Trudell et al., 1986), contamination of aquifers beneath mined areas by downward migration of spoil-derived groundwater is not considered likely. The spoil is separated from the deeper sandstone aquifers by 30-50 m (100-160 ft) of sediment that consists of low-permeability shale and siltstone. The spoil generally is expected to drain laterally.

At one location adjacent to the Vesta Mine, spoil-derived groundwater has been detected in a shallow sand aquifer at distances as great as 1.5 km (0.93 mi.) from the mine and within the Battle River Bed at distances as great as 1.0 km (0.62 mi.) (Trudell et al., 1989). At this site, ponded depressions and shrub-covered terraces within the elevated reclaimed landscape at the boundary of the mine enhance groundwater recharge and result in water levels that, at least locally, are higher than the nearby unmined area. Migration of spoil-derived groundwater into adjacent aquifers can be minimized or prevented in most settings if care is taken in construction of berms and in landscaping reclaimed areas so that excessive trapping of snow by shrubs and ponding of surface water is avoided.

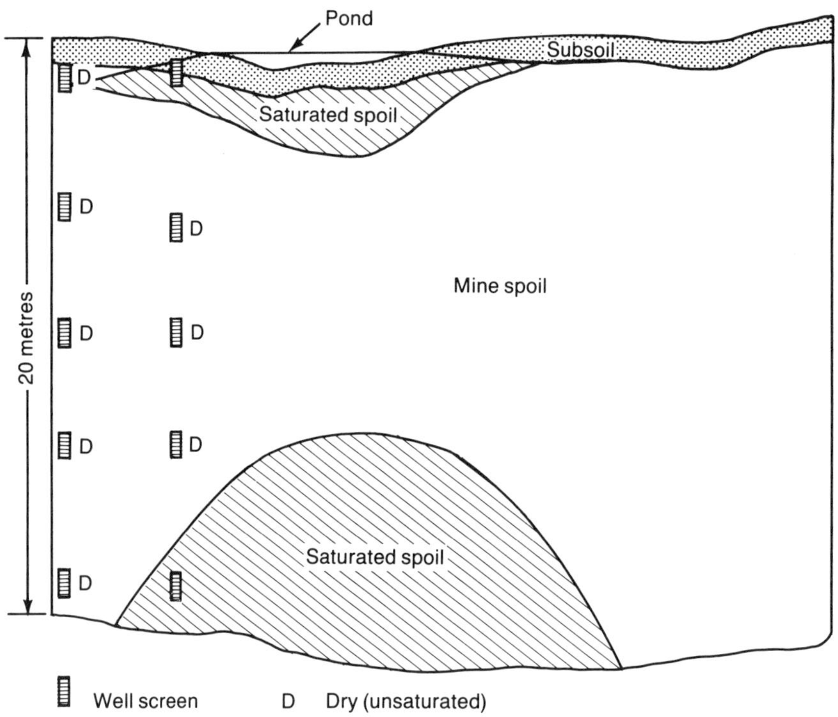

Figure 7. *Hydrologic setting of a pond perched above the water table in a subsidence depression at the Vesta Mine.*

Contamination Of Surface Water By Discharge Of Spoil-Derived Groundwater

Groundwater discharging from mine spoil does not appear to pose a significant threat to the chemical quality of surface water. Along the 10-km (6-mi.) reach of the Battle River Valley that traverses the Diplomat and Vesta mines, 25 springs and seeps were sampled (Trudell, 1988). Fourteen of these springs were spoil-derived, and the remainder were from unmined aquifers. The discharge from the spoil-derived springs and seeps constituted about twice that of the unmined aquifers (103 m^3/day (3,640 ft^3/day) versus 50 m^3/day (1,800 ft^3/day)), but the salt loading was more than four times as great (449 kg/day (990 lb/day) versus 106 kg/day (234 lb/day)). The impact of these discharges on the Battle River, however, is almost undetectable even at the minimum long-term average daily flow of the Battle River (Trudell, 1988, p. 126-133). The entire salt loading from the Battle River study area results in only a 2.29% increase in salt loading in the river. Discharge from only the spoil-derived springs represents an increase of 1.86% in the salt loading in the river.

Maintenance Of The Capability Of Reclaimed Landscapes For Agriculture

The regulations that govern surface mining of coal in Alberta require that reclaimed land be restored to an agricultural capability equivalent to that of the land prior to mining. Modern reclamation practice in Alberta generally involves grading spoil into long smooth surfaces that are flat or gently sloping. The overall topographic configuration of mined areas generally is slightly elevated relative to adjacent unmined areas because of the "swelling" or bulking of the replaced spoil. A layer of from 1.0 m to 1.5 m (3.2 ft to 4.9 ft) of "subsoil" is placed on the graded spoil using scrapers. About 15 cm (6 in.) of topsoil then is spread over the subsoil.

The premining landscape consists of a complex mosaic of soils, which reflect differences in parent material and landscape setting (Macyk and Maclean, 1987). Within a limited area, where the same parent material occurs in the same topographic setting, the soil is the same and the physical and chemical properties are quite uniform. Abrupt changes in chemical and physical properties occur within short distances as one passes laterally from one soil to another. Mining mixes the soil zone materials to produce a reconstructed soil mantle of uniform thickness with more homogeneous chemical and physical properties than in the premining landscape (Macyk, 1986, p. 40-41).

The capability of reconstructed soils for agriculture generally is a reflection of the premining capability (Macyk, 1987). The properties of the reconstructed soil are an average of those of the premining soils, and consequently, the agricultural capability of reconstructed soil reflects a mix of the premining capabilities. The capability of a reconstructed landscape is more uniform than, and generally is not as good as the best nor as poor as the lowest, capability in the premining landscape.

The salinity and sodium (Na) content of reconstructed soils in well-drained sites has been observed to decrease in upland sites studied in the Battle River area between 1984 and 1987 (Macyk et al., 1988). This trend was observed in the topsoil at seven of nine locations. The pattern was somewhat more complex in the subsoil, with salinity and Na concentration decreasing at some depths and increasing at others. The overall pattern is consistent with downward flushing of salts with no evidence that Na is moving upward from the spoil into the subsoil. Macyk et al. (1988) concluded that the capability of the reclaimed landscape is improving over time.

During grading of spoil and placement and grading of subsoil, discrete layers of more dense, compacted soil are formed by dozers and scrapers. These zones of more dense material form barriers to infiltrating water and to root penetration. Where these compacted layers or "pans" persist for extended periods, they can significantly decrease the capability of the post-mining landscape. In some areas of the United States, compaction is seen as a major problem for reclamation (Jansen and Hooks, 1988; McCormack, 1987, p. 23-25; Smout, 1987, p. 124-125). Our studies indicate that density is progressively decreasing over time at most sites, which is interpreted to reflect frost action and increasing root penetration by forage crops (Macyk et al., 1988). On this basis we conclude that the physical characteristics and, therefore, the capabilities of reconstructed soils are improving with time.

Ephemeral ponding of water in depressions that form by differential subsidence of spoil decreases the capability of a reclaimed landscape for agriculture in two ways: 1) ponding in the spring disrupts seeding and, during the summer, it drowns crops; 2) evaporation from the saturated soil around perched ponds in reclaimed landscapes has the potential to produce sodic, saline soils. Hydraulic conductivity of sodic, bedrock-derived spoil is at least two orders of magnitude lower than that of the drift-derived subsoil. As a result, water ponded in subsidence depressions is perched above the regional water table in the spoil. Figure 7 is a schematic cross section showing the setting of one such perched, ephemeral pond at the Vesta Mine. This site displays transient conditions at an early stage of spoil resaturation. The saturated zone at the base of the spoil is expected to expand over time. Data on recharge rates in this setting, however, suggest that it is unlikely that the water table will ever rise near enough to the surface to intersect the pond. Evaporation from the saturated soil surfaces and transpiration from plants around the edges of these depressions induces flow outward from the ponds and upward from the upper surface of the spoil.

Soil salinity has been observed to form within a few years in lowland settings in reclaimed landscapes where extensive ponding of surface water develops. This type of setting occurs where thin overburden results in the reclaimed surface being at or below the pre-mining grade. Deep depressions formed by final cuts in these settings become the site of permanent ponds, which result in rapid resaturation of the spoil. Groundwater levels rise to the vicinity of the land surface over extensive areas within one or two decades. Soil salinity develops through evaporation where the water table is within 1-2 m (3-6 ft) of the surface. Salts migrate both outward from the spoil beneath the ponds and downward from adjacent upland areas (Moran et al., 1986).

Potential negative effects on agricultural capability in lowland settings can be minimized by appropriate land use. Where attempts are made to grow crops in lowland settings in reclaimed areas, the agricultural capability is expected to be progressively degraded as the salinity becomes worse. Optimum productivity of lowland settings is maintained, however, by using these areas for forage or pasture.

ACKNOWLEDGEMENTS

This research was sponsored by the Reclamation Research Technical Advisory Committee (RRTAC) of the Alberta Land Conservation and Reclamation Council, with financial support from the Heritage Savings Trust Fund and from the Alberta Research Council. The assistance and support of Luscar Ltd., Manalta Coal Ltd., Alberta Power Limited, and TransAlta Utilities Corporation, without which the study would not have been possible, is gratefully acknowledged.

REFERENCES

Fenton, M.M., M.R. Trudell, J.G. Pawlowicz, C.E. Jones, S.R. Moran, and D.J. Nikols, 1986, Glaciotectonic deformation and geotechnical stability in open pit coal mining, in R.K. Singhal, ed., Geotechnical stability in surface mining: Boston, A.A. Balkema Publishers, p. 225-234.

Groenewold, G.H., R.D. Koob, G.J. McCarthy, B.W. Rehm, and W.M. Peterson, 1983, Geological and geochemical controls on the chemical evolution of subsurface water in undisturbed and surface-mined landscapes in Western North Dakota: North Dakota Geological Survey, Report of Investigation No. 79, 151 p.

Jansen, I.J., and C.L. Hooks, 1988, Excellent agricultural soils after surface mining: Mining Engineering, v. 40, no. 11, p. 1044-1047.

Macyk, T.M., 1986, Characterization and variability of soils reconstructed after surface mining in central Alberta: Alberta Land Conservation and Reclamation Council, Report #RRTAC 86-20, 146 p.

Macyk, T.M., 1987, An agricultural capability rating system for reconstructed soils: Alberta Land Conservation and Reclamation Council, Report #RRTAC 87-13, 27 p.

Macyk, T.M. and A.H. Maclean, 1987, Soil survey of the Plains Hydrology and Reclamation Project--Battle River project area: Alberta Land Conservation and Reclamation Council, Report #RRTAC 87-10, 33 p.

Macyk, T.M., Z.W. Widtman, and M.E. Pigot, 1988, Measurement of change in reconstructed soils at the Paintearth Mine (1983-1987): Edmonton, Terrain Sciences Department, Alberta Research Council, Report prepared for Luscar Ltd., 121 p.

Maslowski Schutze, A., 1987, Geology of the Highvale study site: Plains Hydrology and Reclamation Project: Alberta Land Conservation and Reclamation Council, Report #RRTAC 87-11, 65 p.

Maslowski Schutze, A., R. Li, M.M. Fenton, and S.R. Moran, 1986, Geology of the Battle River study site: Plains Hydrology and Reclamation Project: Alberta Land Conservation and Reclamation Council, Report #RRTAC 86-4, 70 p.

McCormack, D.E., 1987, Formulation of soil reconstruction and productivity standards, in C.L. Carlson and J.H. Swisher, eds., Innovative approaches to mined land reclamation: Carbondale, Southern Illinois University Press, p. 19-30.

Moran, S.R., and J.A. Cherry, 1977, Subsurface-water chemistry in mined-land reclamation: Key to development of a productive post-mining landscape, in Proceedings of the Second Annual General Meeting of the Canadian Land Reclamation Association, August 17-20, 1977, Edmonton, Alberta: Edmonton, Canadian Land Reclamation Association, 29 p.

Moran, S.R., J.A. Cherry, P. Fritz, W.M. Peterson, M.H. Sommerville, S.A. Stancel, and J.H. Ulmer, 1978a, Geology, groundwater hydrology, and hydrogeochemistry of a proposed surface mine and lignite gasification plant site near Dunn Center, North Dakota: North Dakota Geological Survey, Report of Investigation No. 61, 263 p.

Moran, S.R., G.H. Groenewold, and J.A. Cherry, 1978b, Geologic, hydrologic, and geochemical concepts and techniques in overburden characterization for mined-land reclamation: North Dakota Geological Survey, Report of Investigation No. 63, 152 p.

Moran, S.R., A.E. Howard, and M.R. Trudell, 1986, Groundwater-induced soil salinization around pothole sloughs in east-central Alberta, in G. van der Kamp and M. Madunicky, comps., Proceedings of the Third Canadian Hydrogeological Conference, International Association of Hydrogeologists, April 20-23, 1986, Saskatoon, Saskatchewan: Edmonton, International Association of Hydrogeologists, p. 302-312.

Moran, S.R., M.R. Trudell, A. Maslowski Schutze, A.E. Howard, T.M. Macyk, and E.I. Wallick, 1984, Plains Hydrology and Reclamation Project: Results of five-year's study, in Proceedings of the Ninth Annual Meeting, Canadian Land Reclamation Association, August 21-24, 1984, Calgary, Alberta: Edmonton, Canadian Land Reclamation Association, 21 p.

Smout, G.A., 1987, Field experience with soil reconstruction in the Midwest, in C.L. Carlson and J.H. Swisher, eds., Innovative approaches to mined land reclamation: Carbondale, Southern Illinois University Press, p.123-142.

Strong, W.L., and K.R. Leggat, 1981, Ecoregions of Alberta: Alberta Department of Energy and Natural Resources, ENR Technical Report Number T/4, 64 p.

Trudell, M.R., 1986, Post-mining groundwater supply potential of the Highvale Mine site: Alberta Land Conservation and Reclamation Council, Report #RRTAC 86-7, 25 p.

Trudell, M.R., 1988, Impacts of surface mining on chemical quality of streams in the Battle River mining area, in M.R. Trudell, comp., Plains Hydrology and Reclamation Project: Spoil groundwater chemistry and its impacts on surface water: Alberta Land Conservation and Reclamation Council, Report #RRTAC 88-2, p. 108-135.

Trudell, M.R., and R.L. Faught, 1987, Premining groundwater conditions at the Highvale Mine site: Alberta Land Conservation and Reclamation Council, Report #RRTAC 87-12, 83 p.

Trudell, M.R., R.L. Faught, and S.R. Moran, 1987, Hydrogeology and groundwater chemistry of the Battle River mining area: Plains Hydrology and Reclamation Project: Alberta Land Conservation and Reclamation Council, Report #RRTAC 87-9, 88 p.

Trudell, M.R., R.L. Faught, and S.R. Moran, 1988, Chemistry of groundwater in mine spoil, central Alberta, in M.R. Trudell, comp., Plains Hydrology and Reclamation Project: Spoil groundwater chemistry and its impacts on surface water: Alberta Land Conservation and Reclamation Council, Report #RRTAC 88-2, p. 1-107.

Trudell, M.R., and S.R. Moran, 1986, Generalized procedures for assessing post-mining groundwater supply potential in the plains of Alberta: Alberta Land Conservation and Reclamation Council, Report #RRTAC 86-3, 30 p.

Trudell, M.R., S.R. Moran, and T.M. Macyk, in press, Potential contamination of shallow aquifers by surface mining of coal: Alberta Land Conservation and Reclamation Council, 63 p.

Trudell, M.R., G.J. Sterenberg, and S.R. Moran, 1986, Post-mining groundwater supply at the Battle River site, east-central Alberta: Alberta Land Conservation and Reclamation Council, Report #RRTAC 86-6, 49 p.

Van Voast, W.A., and J.C. Reiten, 1987, Groundwater changes associated with surface coal mining in the northern Powder River Basin, in C.L. Carlson and J.H. Swisher, eds., Innovative approaches to mined land reclamation: Carbondale, Southern Illinois University Press, p. 331-356.

COAL AS A HOST AND AS AN INDICATOR OF MINERAL RESOURCES

Robert B. Finkelman
U.S. Geological Survey
956 National Center
Reston, VA 22092

Robert D. Brown, Jr.
U.S. Bureau of Mines
2401 E St. NW, M.S. 9702
Washington, DC 20241

ABSTRACT

The potential for using coal and coal waste as a mineral resource has been considered for more than 100 yrs. In theory, the coal currently being mined could provide at least half of the U.S. annual consumption of Be, Co, Ga, Ge, Hf, Nb, Se, Sr, Te, Tl, and Y, as well as significant contributions of many other elements. However, the potential has never been realized, and no significant byproducts have been produced from coal or coal wastes during the past 25 yrs.

Although extracting byproducts from coal waste may not be currently economical, the association of minerals with coal can be exploited in two other ways. Byproduct recovery from high-sulfur coal may prove to be economically feasible, especially if the resultant cleaned coal can be marketed or used as an energy source. The chalcophile elements, those elements having an affinity for sulfide minerals, offer the greatest potential for byproduct recovery. These elements can be relatively highly concentrated in coal, and they can occur as cleat-filling minerals that are amenable to physical or chemical separation and concentration. Also, several chalcophile elements can be extracted simultaneously, and their removal can have environmental benefits.

Perhaps the greatest use for the minerals in coal is as geochemical indicators of mineralization. The minerals in coal may not be economical to extract, but high concentrations may be indicative of nearby mineralization. For example, the occurrence of exceptionally high concentrations of zinc, silver, and lead in a coal northwest of the Llano Uplift in Texas may warrant detailed mineral exploration in that area.

INTRODUCTION

The potential for using coal and coal waste as mineral resources has intrigued both the coal and mineral industries for many years. According to Burnet (1982), literature references and patents indicate that considerable work has been done on resource recovery from coal since the 1930s. In addition, a recent workshop on resource recovery from coal (Waide, 1982) and an international conference on utilization of coal ash (Burnet, 1986) show that the topic is still current. Recently, Swaine (1988, p. 879) credited V.M. Goldschmidt (Goldschmidt, 1935; Goldschmidt and Peters, 1933) with "...being the first person to suggest that coal could be a commercial resource for some elements, in particular, germanium...". However, minerals associated with coal were exploited long before Goldschmidt's suggestions in the mid-1930s.

Jenney (1903) described pre-1900 mining operations in which minerals were extracted from coal and coal-bearing rocks. These mining operations included extraction of sphalerite and galena from coal in Missouri, recovery of vanadium (V) from a lignitic coal in Argentina, extraction of antimonial silver (Ag) ores associated with bituminous coal in Peru, and recovery of Ag from a lignitic sandstone in Utah. Jenney (1903, p. 462) noted that the Utah deposit reached its "height of production in 1877-79." Jenney (1903) also mentioned the occurrence of gold (Au) in coal from Utah and from two locations in Wyoming. One of the Wyoming occurrences is discussed in detail by Stone (1912).

These early, successful mining operations were not followed by other attempts to extract minerals from coal. Cobb et al. (1979) examined the potential of recovering cadmium-bearing sphalerite from coal mine waste in Illinois. However, to date there has been no effort to extract the sphalerite commercially. Perhaps the only example comparable in this century to the mining activities cited by Jenney (1903) was the limited production of uranium (U) from lignites in North and South Dakota in the mid-1960s (Noble, 1973; Schnable, 1975).

Despite the poor production record, we believe that there is reason for continued interest. We suggest that the potential exists for using coal and coal waste to lead to nearby mineral resources, provided that we apply our knowledge properly. We envision three aspects to this concept of mineral resources from coal: mineral byproduct recovery from coal and coal waste, coal as a byproduct of mineral extraction, and trace element concentrations in coal as a geochemical indicator of nearby mineralization.

MINERAL RESOURCES FROM COAL

Byproduct Recovery from Coal or Coal Wastes

Almost 816,000,000 tonnes (900,000,000 tons) of coal are produced annually in the United States from which about 91,000,000 tonnes (100,000,000 tons) of waste products are produced (Staff, 1989). This yield places coal waste second only to sand and gravel in volume of material produced in the United States (Staff, 1989). Only about 20% of coal waste products are being used, principally in the construction industry (Burnet, 1986). The remainder is disposed of at considerable cost in mine backfills, spoil and refuse piles, and in solid-waste landfills.

This vast untapped resource may contain valuable contributions of strategic or critical elements. Using nationwide means (from the U.S. Geological Survey's National Coal Resources Data System) for ash yield and element concentrations, calculations indicate that coal and coal ash produced in the United States contain at least half of the U.S. annual consumption of aluminium (Al), beryllium (Be), cobalt (Co), gallium (Ga), germanium (Ge), hafnium (Hf), niobium (Nb), selenium (Se), strontium (Sr), tellurium (Te), thallium (Tl), yttrium (Y), and the rare earth elements, as well as significant contributions of many other elements (Table 1 and Figure 1). Technology is available to extract many of these elements from coal ash. Detailed studies have been conducted on the extraction of Al, cadmium (Cd), Ga, Ge, iron (Fe), molybdenum (Mo), titanium (Ti), U, zinc (Zn), and other elements (Burnet, 1986). However, extraction of these elements has not been economic, except for the limited production of U during the mid-1960s and some production of sulfuric acid from sulfide minerals in coal.

Even though technology exists for some byproduct recovery, there are several reasons for the absence of successful operations and the lack of attempts at byproduct recovery:

Element	Current import dependence of U.S. (in %)[1]	U.S. annual consumption[1,3]	Avg. concentration in U.S. coal (ppm)[2]	Potential amount available annually from coal[3]
Al	97	5.4 million mt	14,000	15 million mt
As	100	24,000 mt	6.4	7,000 mt
Be	13	320 st	2	2,000 st
Bi	>75	3.7 million lb	0.7	1.4 million lb
Co	84	9,600 st	6	6,000 st
Hf	Not Available	55 st	0.7	700 st
Li	0	3,100 st	16	16,000 st
Nb	100	7.5 million lb	3	6 million lb
Se	58	660 mt	3	3,300 mt
Th	Not Available	30 mt	3	3,300 mt
Ti	8	23,000 st	800	800 million st
Tl	100	2,800 lb	1	2 million lb
V	54	10 million lb	22	44 million lb
Y	100	14 mt	8	8,800 mt

1 From Staff (1989).
2 From the National Coal Resources Data System.
3 mt = metric tons; st = short tons

Table 1. *Maximum annual amount of element potentially available from U.S. coal.*

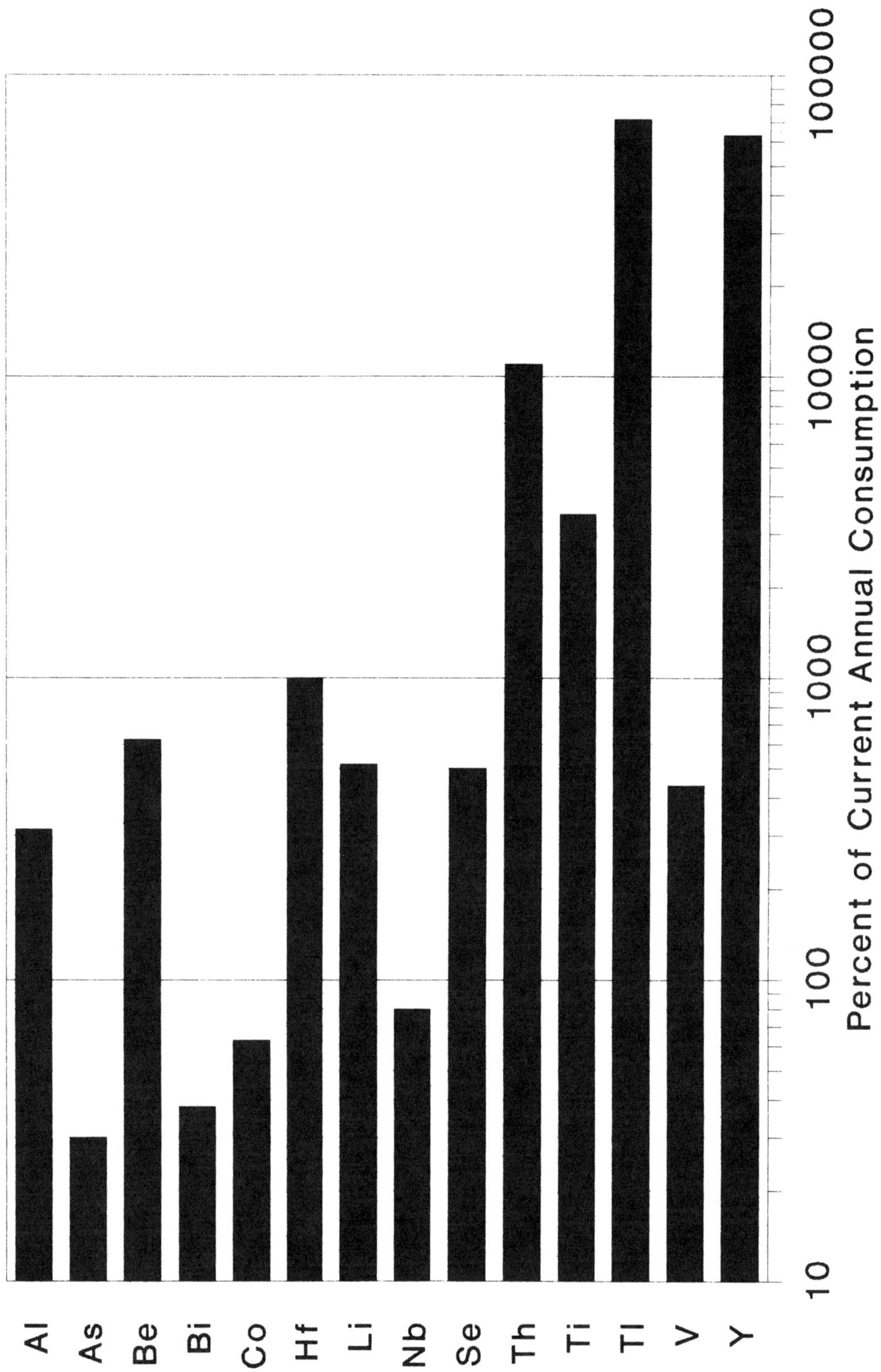

Figure 1. *Potential annual byproduct recovery from coal.*

1) The sources of coal and coal waste are widely dispersed. There are more than 4,000 coal mines in 27 states, about 1,000 coal-burning powerplants in every state, and more than 500 coal preparation plants and more than 500 industrial coal users in 30 states (Keystone, 1987). With only one exception, no U.S. mine produced more than 11,800,000 tonnes (13,000,000 tons) of coal in 1986. Fewer than a dozen power plants burned more than 4,500,000 tonnes (5,000,000 tons) of coal annually; none of the power plants burned more than 9,000,000 tonnes (10,000,000 tons) (Keystone, 1987).

2) Much of the U.S. ash production is from low-ash and low-sulfur coals having considerably lower trace element concentrations than the national average.

3) There is little information on the chemical forms of elements in coal, making it difficult to devise widely applicable, efficient extraction procedures.

4) Metals markets are depressed. During the first half of the 1980s, prices of most major metal commodities fell steadily. By 1986, prices of several major metals reached lows not encountered since the 1930s (Secretary of the Interior, 1988). These price declines followed a period where the U. S. mineral industry had grown at a slower pace than the rest of the economy during business expansions and had declined faster during recessions. A declining demand for minerals and increased competition from low-cost foreign sources forced mineral industry cutbacks (U.S. Bureau of Mines, 1986).

Even so, we believe there is reason for hope for selected elements of strategic importance. Recent increases in demand for minerals and the outlook for new and expanding markets, particularly for advanced materials, may give rise to economic hope where only gloom existed a few years ago. Among the more attractive possibilities are bismuth (Bi), Se, antimony (Sb), Cd, Ag, and Tl, as well as Au, platinum (Pt) group elements, Te, and indium (In).

These elements offer several advantages:

1) The elements generally have chalcophile tendencies. That is, they have an affinity for sulfur (S), and therefore, all these elements have been concentrated in coal by the same processes. Many of these elements occur in cleat and fracture-filling minerals. As such, they are in relatively large grains that can be concentrated readily by existing coal cleaning procedures.

2) Most are highly concentrated in coal with respect to crustal abundances. For example, according to Valkovic (1983), Se (on an ash basis) is concentrated at about 100 times its crustal abundance. Arsenic (As) is the next most highly enriched element; on an ash basis, it is enriched about 10 times its crustal abundance. Tellurium, Cd, Sb, mercury (Hg), Mo, and lead (Pb) also exhibit enrichment with respect to crustal abundance. These enrichment factors are derived from average concentrations of these elements in coal. However, these elements can occur in coal in extraordinarily high concentrations, many times their average concentration in coal. Some of these remarkable occurrences are noted in the next section.

3) Most chalcophile elements are volatile and, therefore, may be collected from flue gas or from flue gas deposits.

4) They all readily form soluble compounds that can be extracted by a single chemical process.

5) Most are environmentally deleterious; therefore, extraction from coal or coal waste is advantageous.

Recovery of minerals from coal could increase the domestic supply of metals needed for the manufacture of advanced materials. For example, five of the candidate elements for recovery from coal — Bi, Nb, Tl, Ti, and

Y — are important in the area of superconductivity. For new high-temperature superconductors, the amount of Y potentially available is exciting, but the feasibility of recovery of Bi and Tl is much greater because of differences in their modes of occurrence. The larger grain size and higher specific gravity of the sulfide hosts for Bi and Tl likely will allow easier physical separation and chemical extraction. Yttrium will be more difficult to recover from coal because of the fine-grained texture of the clay and phosphate minerals in which it resides. Whereas there appears to be an adequate supply of Nb and Ti for low-temperature superconductors, even for the Superconducting Super-Collider, any increase in the supply of materials for high-temperature superconductors is important because their uses will be limited to small-scale applications unless supplies can be increased (Sorrell, 1988).

Coal as a Byproduct of an Extraction Process

Some coals are extraordinarily enriched in certain trace elements. For example, a coal in Alabama has almost 2,200 ppm arsenic (i.e., about 12,500 ppm As in the ash). A coal in Texas has 300-600 ppm Ag in the ash. The ash of a coal in Illinois contains almost 2,000 ppm Cd. Coals containing almost 8,000 ppm Ge in the ash are known from several locations. Lignites in Mississippi average more than 10 ppm Se. Coals having several thousand ppm U have been found in several western states. These and other exceptional concentrations of elements in coal are on file in the U.S. Geological Survey's National Coal Resources Data System (NCRDS). Figure 2 shows the major U.S. coal basins and the locations of selected coal samples having anomalously high concentrations of various elements.

Coals containing such high element concentrations generally are not mined because of environmental and coal-quality (high ash yield) concerns. However, mining of these coals for their mineral resources and subsequent use of the cleaned coal as a byproduct to generate heat may make some of these deposits economically viable. An example of this synergistic relationship can be seen in the mining of clays and associated lignites in Texas (Ayres, 1989).

Coal as a Geochemical Indicator of Mineralization

Using plants for biogeochemical prospecting for mineral resources has been successfully applied for a number of chemical elements (Dunn, 1986, Carlisle et al., 1988). Because coal is primarily derived from plant matter, coal may be used in a similar way to perform paleo-biogeochemical prospecting. Moreover, the complexing of chalcophile elements enhances coal's potential as an indicator of mineralization. Although many of the extremely high elemental concentrations in coal may not be economic to extract, they may be indicative of nearby mineralization. Many of the 13,000 coal analyses in the NCRDS are from sedimentary basins having virtually no other chemical analyses. The data for coal offer a unique opportunity to assess the mineral resource potential of these basins. The coal may provide more than a window to allow us to detect mineralization. The ability of coal to complex and concentrate many elements may magnify our view through the window and make detecting mineralization easier.

An example of how this concept can be applied can be seen in an exceptionally high concentration of Ag detected in a Pennsylvanian bituminous coal sample from the Fort Worth Basin in north-central Texas (Figure 3). One of the four coal samples analyzed from the basin had 300-500 ppm Ag in the ash. At $6/oz for Ag, this concentration represents approximately $22 of Ag per tonne ($20 of Ag per ton) of coal. The sample also contained about 5,000 ppm Zn and 500 ppm Pb in the ash. The Zn occurred in large (1 mm) fracture-filling sphalerite grains (Figure 4). Preliminary data indicate that the Ag probably is associated with the sphalerite. Even if the coal does not contain economic concentrations of these elements, their presence in the coal may be an indication of nearby mineralization.

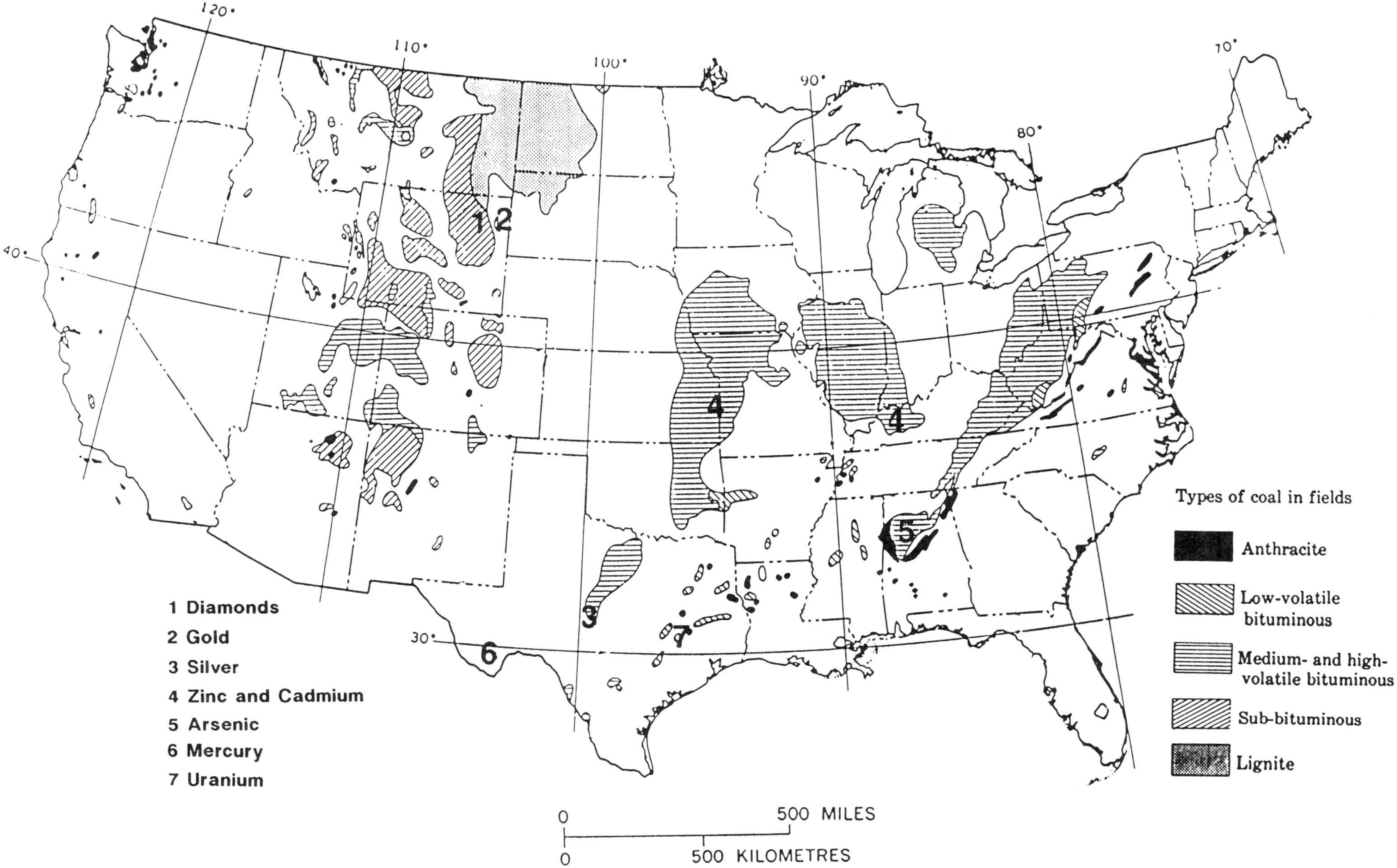

Figure 2. *Map of the United States depicting coal basins and the locations of selected coal samples having anomalous metal concentrations.*

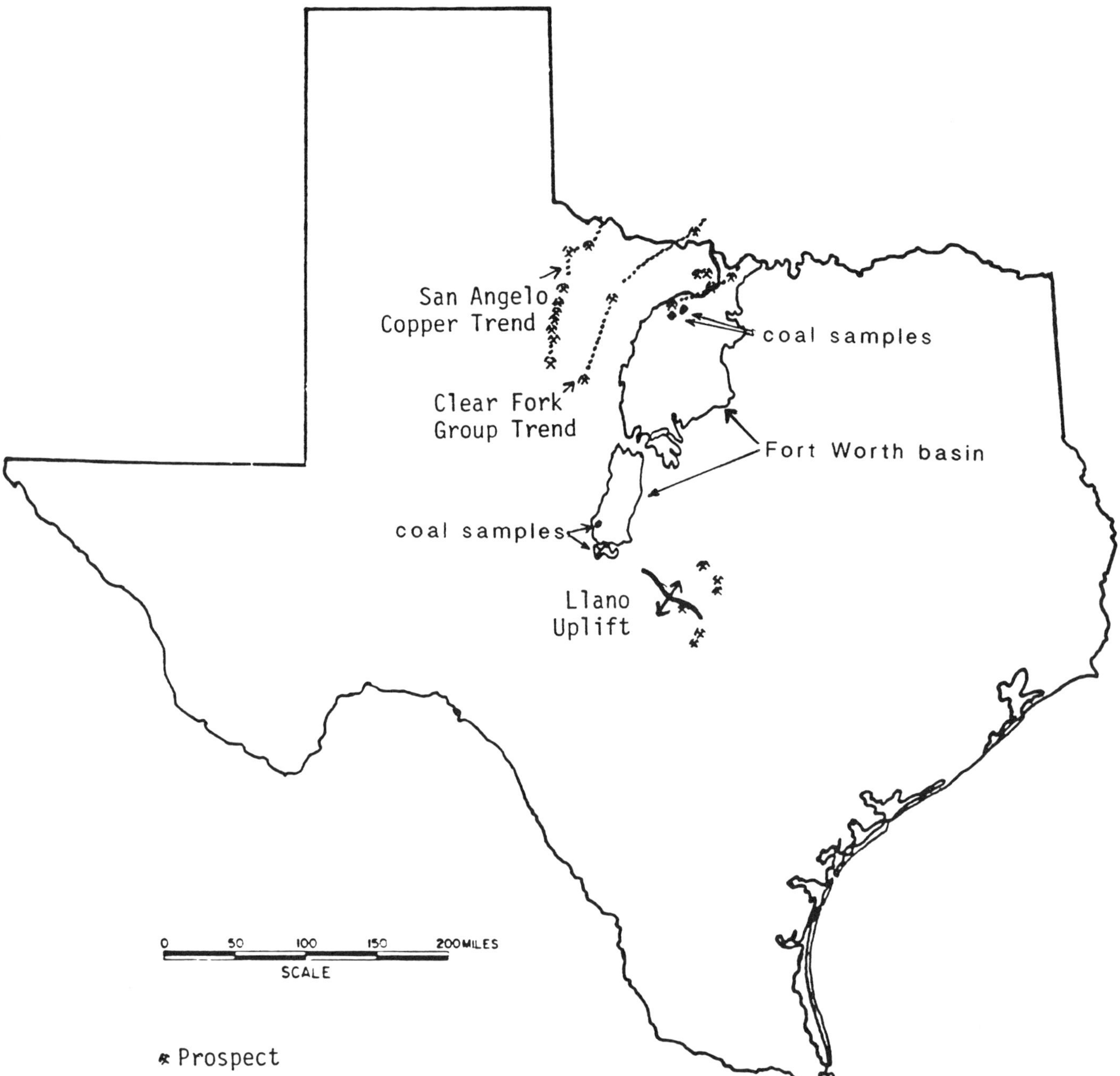

Figure 3. Map of Texas showing locations of mineral prospects and coal samples in the vicinity of the Fort Worth Basin.

The sample having a high Ag concentration and the sample having a high Zn concentration both occur in the southernmost part of the basin about 80 km (50 mi.) northwest of the Llano Uplift. The other two coal samples from the northern part of the Fort Worth Basin have average values for all the chalcophile elements. Zinc, Pb, and Ag have been mined in an area east and southeast of the Llano Uplift. Barnes (1956) speculated that the Zn, Pb, and Ag in the vicinity of the Llano Uplift were deposited from upward-moving magmatic solutions, perhaps related to Middle Pennsylvanian igneous activity in the Llano area. Perhaps the distribution of Ag and Zn may indicate the source of the mineralizing solutions.

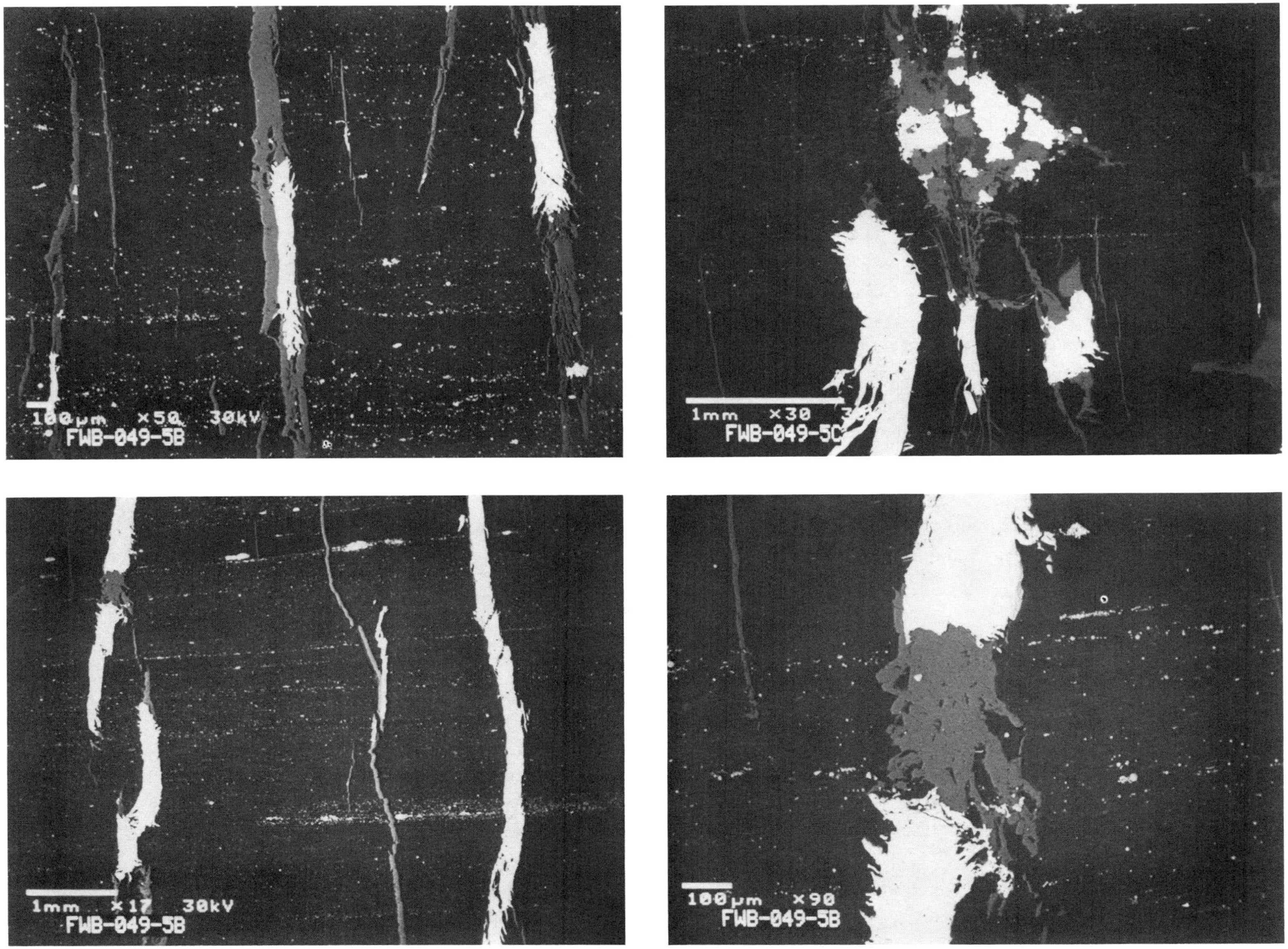

Figure 4. *Scanning electron photomicrographs showing sphalerite (white) and calcite (grey) vein fillings in polished blocks of coal from the Fort Worth Basin.*

The analysis of the coals of the Fort Worth Basin shows that mineralization came from the south; this direction is consistent with the source of the igneous activity in the Llano area. The occurrence of extraordinarily high concentrations of Ag, Zn, and Pb in coal from the southern part of the basin indicates a high degree of mineralization in the area west-northwest of the Llano Uplift and may warrant a detailed exploration of that area for economic mineral deposits.

CONCLUSIONS

Mineral byproducts were extracted economically from U.S. coal in the late 19th and early 20th centuries. More recently, researchers have concluded that extraction of minerals from coal ash is economically marginal at best.

Nevertheless, two other aspects to the association of minerals and coal should be given careful consideration. Byproduct recovery from low-grade (high-sulfur, high-ash) coal may prove to be economically feasible, especially if the resultant cleaned coal can be marketed or used as an energy resource. Low-grade coal generally has substantially higher concentrations of chalcophile elements than does most coal being mined. The potential of using coal as a biogeochemical prospecting tool to help locate hidden mineral resources should be further evaluated.

REFERENCES

Ayres, W. B., Jr., 1989, Geology of the Elgin-Butler clay pits, in W. B. Ayers, Jr., J.A. Breyer, and R.B. Finkelman, eds., Depositional Settings of Texas Lignites: American Geophysical Union, 28th International Geological Congress, Field Trip Guidebook T173, p. 31-32.

Barnes, V. E., 1956, Lead deposits in the Upper Cambrian of Central Texas: Texas Bureau of Economic Geology, Report of Investigations No. 26, 68 p.

Burnet, G., 1982, An overview: Resources from coal wastes and ash, in C.H. Waide, ed., Proceedings of the Resources from Coal, Coal Wastes and Ash Workshop, Upton, New York: Brookhaven National Laboratory, BNL 51679, p. 17-23.

Burnet, G., 1986, Newer technologies for resource recovery from coal combustion solid wastes: Energy, v. 11, no. 11/12, p. 1363-1375.

Carlisle, D., W.L. Berry, and I.A. Kaplan, eds., 1988, Organic matter, biological systems and mineral exploration: New York, Prentice Hall, Ruby Volume, 465 p.

Cobb, J. C., J.M. Masters, C.G. Treworgy, and R.J. Helfinstine, 1979, Abundance and recovery of sphalerite and fine coal from mine waste in Illinois: Illinois State Geological Survey, Mineral Note 71, 11 p.

Dunn, C. E., 1986, Biogeochemistry as an aid to exploration for gold, platinum and palladium in the northern forests of Saskatchewan, Canada: Journal of Geochemical Exploration, v. 25, p. 21-40.

Goldschmidt, V. M., 1935, Rare elements in coal ashes: Industrial and Engineering Chemistry, v. 27, no. 9, p. 1100-1102.

Goldschmidt, V. M., and C.L. Peters, 1933, Über die anreicherung seltener Elemente in Steinkohlen: Nachrichten von der Gesellschaft der Wissenshaften zu Gotteingen, Math.-Physik. Klasse, Heft 4, p. 371-386.

Jenney, W. P., 1903, The chemistry of ore-deposition: American Institue of Mining, Metallurgical, and Industrial Engineers Transactions, v. 33, p. 445-498.

Keystone Coal Industry Manual, 1987: New York, McGraw-Hill Inc., 1,244 p.

Noble, E. A., 1973, Uranium in coal, in E.R. Landis, ed., Mineral and water resources of North Dakota: North Dakota Geological Survey, Bulletin 63, p. 80-83.

Schnable, R. W., 1975, Uranium, in Mineral and water resources of South Dakota: Report prepared by the U.S. Geological Survey for the Committee on Interior and Insular Affairs, United States Senate, p. 172-176.

Secretary of the Interior, 1988, The mineral position of the United States--1988: Washington, U.S. Department of the Interior, 122 p.

Sorrell, C. A., 1988, The new superconductors - An overview: Minerals and Materials, U.S. Bureau of Mines, June-July, p. 7-15.

Staff, 1989, Mineral commodity summaries: Washington, U.S. Bureau of Mines, 191 p.

Stone, R. W., 1912, Coal near the Black Hills, Wyoming-South Dakota: U.S. Geological Survey Bulletin 499, 66 p.

Swaine, D. J., 1988, Victor Moritz Goldschmidt's contributions to coal science: Fuel, v. 67, p. 877-879.

U. S. Bureau of Mines, 1986, The mineral position of the United States - the past 15 Years (1971-1985): Washington, U.S. Department of the Interior, 45 p.

Valkovic, V., 1983, Trace elements in coal: Boca Raton, Florida, Chemical Rubber Co. Press., v., 1, 210 p.

Waide, C. H., ed., 1982, Proceedings of the Resources from Coal, Coal Wastes and Ash Workshop, Upton, New York: Brookhaven National Laboratory, BNL 51679, 171 p.

ANALYSIS OF AROMATIC HYDROCARBONS IN OVERBURDEN FROM COAL MINES: ASSESSMENT OF THE ENVIRONMENTAL IMPACT

Wilhelm Püttmann, Kai Steffens, and Wolfgang Goßel
Lehrstuhl für Geologie, Geochemie und Lagerstätten des Erdöls und der Kohle
RWTH Aachen
Lochnerstrasse 4-20
51 Aachen
Federal Republic of Germany

ABSTRACT

Bituminous coals are known to contain significant amounts of petroleum-like hydrocarbons trapped in the pore system of the macromolecular network. With increasing thermal stress acting upon coals under the particular conditions in their deposits, the relative amount of volatile aromatic hydrocarbons within the trapped bitumen increases until the coal reaches the rank of medium-volatile bituminous coals. Coking coals particularly are rich in compounds such as benzene, biphenyl, naphthalene, and their alkylated derivatives. Potential hazardous environmental impact of these hydrocarbons has to be considered when mining or reclaiming overburden because approximately 10% overburden material related to coal mining is coal. Exposure to long-term weathering destroys the pore system of coals, which might result in a release of highly volatile bituminous coal constituents into the atmosphere. This view is supported by analyses of coals present in overburden material.

In the present study, the on-line combination of thermodesorption coupled to gas chromatography and mass spectrometry is shown to be an appropriate tool for the detection and quantitation of hydrocarbons of a wide boiling temperature range present in solid sample material. The method is preferentially suitable for the analysis of highly volatile constituents such as benzene, toluene, and xylenes, which are not quantitatively analyzed using conventional solvent-extraction methods.

INTRODUCTION

Analytical Aspects

Investigations concerning light hydrocarbons (lighter than C_{10}) in sedimentary organic matter have been carried out extensively in order to study the origin, migration and accumulation of crude oils in the geosphere (Schaefer et al., 1978a). A suitable method for the determination of light hydrocarbons was found in the hydrogen stripping of the rock samples in a flow system prior to gas chromatography analysis (Leythaeuser et al., 1979; Schaefer et al., 1978a and b; Schaefer and Pooch, 1982) . Results have shown that, in sediment samples, the light hydrocarbons primarily are composed of saturated constituents. However, with increasing depth of the samples analyzed, the amount of aromatic hydrocarbons such as benzene and toluene increased significantly (Schaefer et al., 1978), probably due to the increasing thermal stress acting on the sedimentary organic matter with depth. The application of the hydrogen stripping method to the study of a high-volatile bituminous coal and of a roof rock of a coal seam, both originating from the Ruhr District, revealed the presence of benzene and toluene in relatively low amounts (Leythaeuser et al., 1979).

Apart from the hydrogen stripping method, thermal mobilization of hydrocarbons from rock samples also has been applied in petroleum geochemistry for the quantitative analysis of hydrocarbons in the molecular range from C_1 to C_{15} (Schaefer and Pooch, 1982). It is well established that, up to a temperature of 320°C, there is thermal mobilization of the hydrocarbons trapped in the macromolecular structure (kerogen).

At higher temperatures, pyrolysis effects have to be considered (Bjorøy et al., 1985). "Flash thermal desorption" of oil shale samples at a temperature of 358°C has shown that the occurrence of pyrolysis effects is indicated by the generation of pristenes in the desorbed material (Crisp et al., 1986).

Based on these findings "thermodesorption" of coals at 300°C has been applied in order to analyze the composition of hydrocarbons present in coals of a wide coalification range using gas chromatography (Schaefer and Püttmann, 1987) and gas chromatography/mass spectrometry methods (Püttmann et al., 1988). The advantage of this method is in the simultaneous detection of light hydrocarbons (heavier than C_5) and heavy hydrocarbons (lighter than C_{35}).

Because of the very small sample amount used for analysis (~2 mg), it might be difficult to extrapolate from the extracted amount of the hydrocarbons in the sample analyzed to the total amount in an pile of overburden material. However, previous investigations carried out with a bituminous coal have shown that the scatter of hydrocarbon yields due to inhomogeneities of the samples amounts only to approximately 20% (Schaefer and Püttmann, 1987). This is explainable by the well known effect of homogenization of hydrocarbons in bituminous coals due to the process of spreading the mobile organic compounds throughout the whole seam (Allan and Larter, 1983).

In particular, the analysis of light hydrocarbons trapped in solid sample material is impossible when solvent-extraction methods are applied. During evaporation of the solvent, volatile constituents with boiling points below 170°C are either completely removed or masked in quantitative analysis. Therefore, constituents such as benzene, toluene, and xylenes (BTX) are not recovered when extract analysis is applied.

Overburden From Coal Mines--General Remarks

In the past, the environmental impact of the mined overburden or spoil from coal production has been seen mainly in the mobilization of inorganic trace elements from the overburden dumping areas (Van Berk, 1987). Analyses concerning the breakdown of organic material under the influence of weathering effects were restricted to the measurement of CO and CO_2 as the proposed dominant released products (Beier, 1985). However, it has been known for more than 20 yrs that coals and adjacent shales contain crude-oil-like bituminous constituents physically trapped in their pore systems (Brooks and Smith, 1967; Ouchi and Imuta, 1963). The abundant aromatic hydrocarbons present in the extractable bitumen have been analyzed in detail by Radke et al. (1982). There is no doubt about the toxicity of the aromatic hydrocarbons detected, because most of the compounds (i.e., alkylnaphthalenes and alkylphenanthrenes) are of significant ecotoxic and partly carcinogenic potential (Jensen and Hites, 1983; Ott et al., 1978). Moreover, the environmental impact of BTX aromatics in overburden material from coal mines has to be considered because these compounds are present abundantly in medium- to low-volatile bituminous coals (Schaefer and Püttmann, 1987) and to a minor extent in shales associated with coals (Leythaeuser et al., 1979).

A complete conversion of the aromatic hydrocarbons present in coal-containing overburden material to CO and CO_2 under the process of weathering can be excluded. Instead, one has to think about a chemical transformation of the aromatic hydrocarbons into polar oxygen-containing compounds. This view is supported by the detection of aldehydes and ketones in extracts from Canadian coal samples which had been exposed to long-term weathering conditions (Püttmann et al., 1987). The polar compounds are much more water-soluble compared to the precursor hydrocarbons. Therefore, the well known reduction of the organic carbon content during weathering might be a result of water washing of polar compounds generated in the shales and coals. Additionally, a significant change in the adsorption potential of coal under the influence of weathering might result in an easier release of trapped hydrocarbons.

The present study is focused on the analysis of light aromatic hydrocarbons in coals and related overburden from the major mining districts in the Federal Republic of Germany. Figure 1 indicates the extents of the coal fields and, in particular, the mines which provided samples for this study.

EXPERIMENTAL METHOD

The sample material investigated was ground to a grain size smaller than 0.2 mm (0.008 in.) and transferred into a glass crucible (10 mm by 1 mm (0.4 in. by 0.04 in.) inner dimensions). The sample weight was determined using a "micro-scale autobalance" (a product of Perkin Elmer). The weight of the analyzed sample material was approximately 2 mg. Subsequently, the glass crucible was placed into the autosampler, a device commonly used with elemental analyzers, which was mounted on top of the gas chromatography injector system. The autosampler enables the introduction of the sample mechanically into the heated gas chromatography injector under an atmosphere of helium used as carrier gas for the subsequent gas chromatography analysis. The injector was operated in the splitless mode at a temperature of 300°C. Thermodesorption is achieved under the helium flow with a desorption time of 10 min. The desorbed material is trapped in the initial part of the gas chromatography column by maintaining the gas chromatography oven temperature at -50°C (with a cryogenic unit from Varian). In order to avoid cross contamination, the glass crucible is removed from the injector-liner after each run.

The gas chromatography analyses were carried out with a Varian 3700 gas chromatograph using a 25-m (82-ft) long, fused-silica column (0.25 mm (0.01 in.) inner diameter) coated with chemically bonded SE-54 as a stationary phase. After completion of the desorption process the oven temperature was raised from -50°C to 300°C at a rate of 4°C/min. and kept at maximum temperature for 15 min.

The gas chromatography capillary column was passed directly through a heated tunnel into the ion source of a Finnigan Mat 8200 mass spectrometer which was operated in EI mode using an electron energy of 70 eV, an emission current of 1.0 mA, and an ion source temperature of 240°C. Mass spectra were recorded by scanning cyclically from 50 to 700 mass units per 1.1 sec. For the data acquisition, an INCOS data system was used. The detection of individual compounds is carried out by mass chromatography of related molecular ions and by comparison of the retention behavior with published data. The quantitation of compounds is determined using the method described in Püttmann and Goßel (1988). In order to be sure that pyrolysis did not affect the coal samples analyzed during thermal treatment, mass chromatography for compounds with a mass-to-charge ratio (m/z) of 83 was carried out for the recognition of n-alkenes and pristenes, respectively. None of the analyzed samples revealed detectable amounts of alkenes.

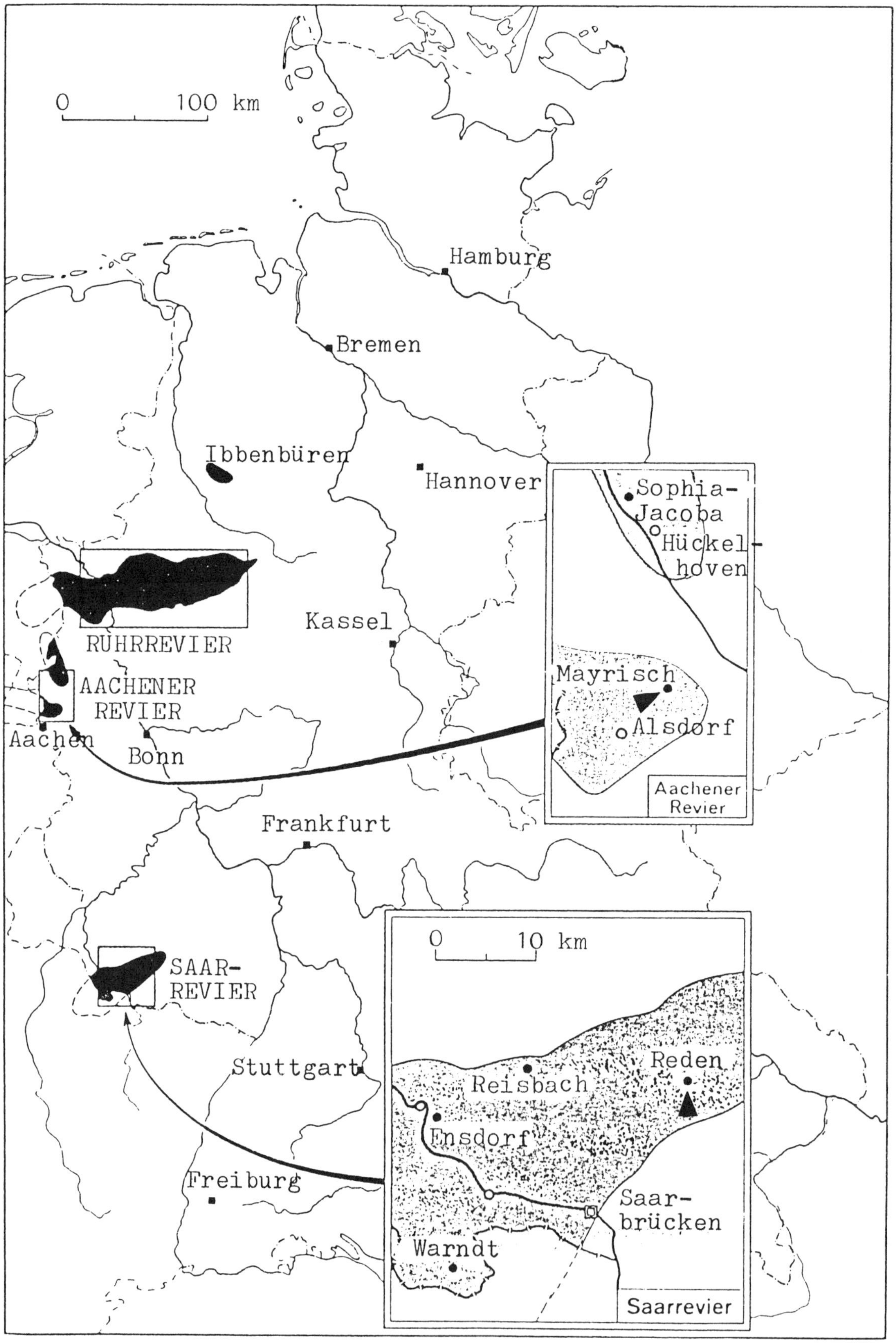

Figure 1. *Location map (modified from Brecht et al., 1988) showing the major coal mining districts (Revier) in the Republic of Germany from which the analyzed coal and overburden samples originate. Two mines from which investigated samples originate are marked by triangles.*

RESULTS AND DISCUSSION

Analysis Of Bituminous Coals

Results of gas chromatography/mass spectrometry analyses of coal samples originating from three major coal mining districts in the Federal Republic of Germany are shown in Figure 2. Coal sample 1 represents a freshly produced sample of low-volatile bituminous coal from the mining district near Aachen. Within the desorbed bitumen (Figure 2a), seven compounds are predominant (indicated by their ring structure notation on Figure 2a). These compounds are benzene (1), toluene (2), xylenes (3), naphthalene (5), 2-methylnaphthalene (6), biphenyl (9), and phenanthrene (13) (numbering system for compounds is shown in Figure 2c and Table 1). The amounts of these compounds are listed in Table 1. Benzene (1) and toluene (2) are the most abundant constituents. The unsubstituted, higher molecular weight aromatic hydrocarbons (5, 9, and 13) dominate significantly over their alkylated counterparts. It can be assumed that this particular relative composition is a result of desalkylation reactions primarily affecting the alkylaromatic hydrocarbons at higher coalification stages because of the enhanced temperature influence (Radke et al., 1982). Aromatic hydrocarbon compositions dominated by unsubstituted constituents also are known from extracts of buried charcoal (Raymond et al., 1989). Coals of lower rank, in general, are characterized by a more complex mixture of trapped hydrocarbons, including saturated hydrocarbons and alkylated aromatic hydrocarbons.

A typical chromatogram of hydrocarbons, desorbed from a high-volatile bituminous coal from the Saar District is shown in Figure 2b. The complexity of the composition of compounds in the higher boiling temperature range is caused by the abundance of alkylated naphthalenes, biphenyls, dibenzufurans, and fluorenes. These compounds are generated in coals of a very narrow rank range at the end of the high-volatile bituminous coal stage (Püttmann and Kalkreuth, 1989). The generation of these compounds partly can be explained by the thermal destruction of buried lignin. Potential precursor structures, proposed by Freudenberg (1964), are abundantly present in coniferous lignin which is thought to have made up the bulk of the ultimately generated coal (Van Krevelen, 1961).

An explanation for the presence of the aromatic hydrocarbons detected in coal samples 1 and 2 (Table 1) can be given by the thermal degradation of resistant biopolymers buried in the coal seams. It previously has been shown that vitrinites obtained from bituminous coals yield abundant phenols upon pyrolysis (Senftle et al., 1986). With increasing rank of the coals the amount of phenols in the pyrolysis products decrease (Larter, 1989) to approximately zero at the end of the medium-volatile bituminous coal stage. It also is known that, during coalification in high-volatile bituminous coals, a significant amount of water still is generated (Jüntgen and Karweil 1966). This water presumably originates from the condensation of phenols which in turn make up the bulk of lignin-like polymers present in coals (Hayatsu et al., 1987). Based on the proposed lignin structure (Freudenberg, 1964), such condensation products should consist mainly of biphenyl, dibenzofuran, fluorene, and their alkylation products. The identification of some of these compounds in coal sample 2 has been carried out by mass chromatography of related molecular ions (Figure 3). The compounds identified are listed in Table 3. Thus, the onset of the generation of these compounds at the end of the high-volatile bituminous coal stage is not surprising. However, at higher rank, the amount of these compounds decreases again, as shown in the case of coal sample 1. The hydrocarbons desorbed from this coal are dominated by BTX aromatics. A possible explanation for the compositional changes with increasing rank is in the degradation of the biphenyls, dibenzofurans, and fluorenes due to cleavage reactions. Alternatively, a direct source of BTX from the degradation of lignin has to be considered.

During coalification, the naphthalenes are generated at a much lower temperature than the one which is required for the lignin degradation. Naphthalenes begin to appear in coal extracts together with the coalification jump from subbituminous to bituminous coals (Püttmann et al., 1986). Naphthalenes were shown

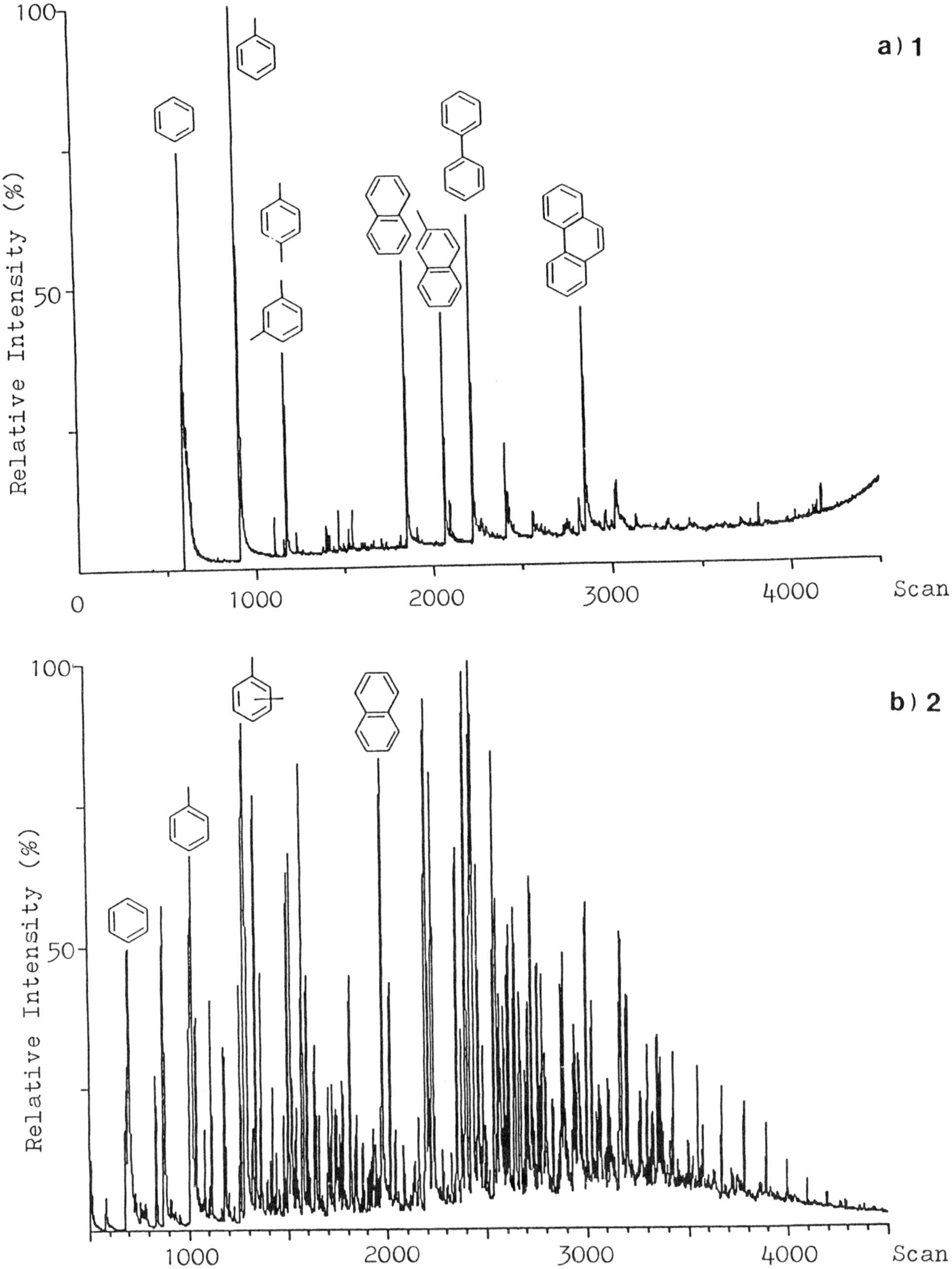

Figure 2. *Reconstructed ion current chromatograms (gas chromatography/mass spectrometry analysis) of hydrocarbons desorbed at 300°C from unweathered coal samples 1, 2, and 3. For peak identification and results of quantitation, see Table 1.*

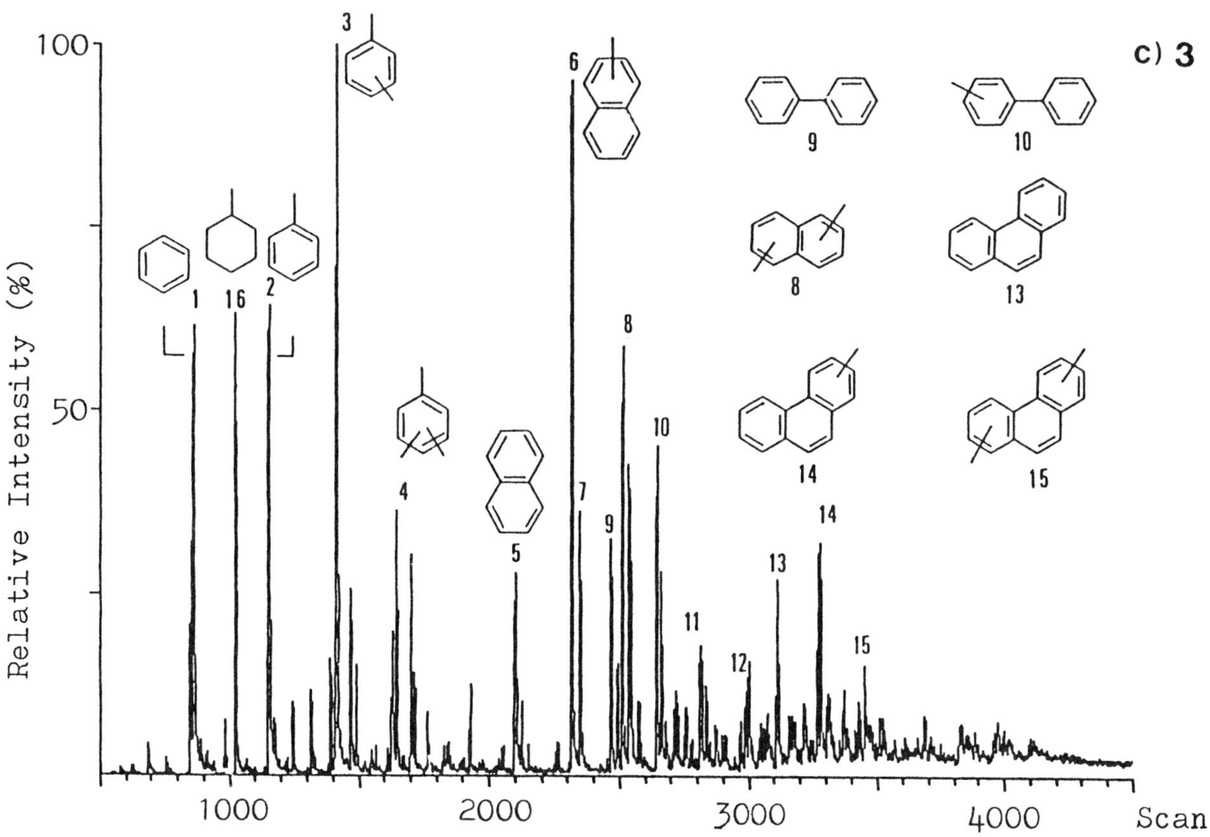

Figure 2 (continued). *Reconstructed ion current chromatograms (gas chromatography/mass spectrometry analysis) of hydrocarbons desorbed at 300°C from unweathered coal samples 1, 2, and 3. For peak identification and results of quantitation, see Table 1.*

often to be associated with a sporinite-derived kerogen (Horsfield, 1989). Therefore, biological precursor molecules for the naphthalenes presumably are enriched in the spores.

Additionally, a series of n-alkanes is detectable in the desorbed hydrocarbons from coal sample 2. N-alkanes are present in large amounts in coals at the beginning of the high-volatile bituminous coal stage (Püttmann et al., 1987) and decrease constantly with increasing coalification probably due to cleavage reactions induced by the thermal stress. In the lower boiling temperature range of the desorbed bitumen (Figure 2b), benzene and its alkylated derivatives are detected in relatively large amounts (Table 1), which is atypical for high-volatile bituminous coals. However, based on previous studies it is known that Saar District coals show technological properties (i.e., coking capabilities) that are not predictable from commonly used coal classification parameters such as vitrinite reflectance (Püttmann et al., 1986). In total, the bitumen composition shows similarities with crude oils except for a higher content of aromatic hydrocarbons. This might explain why, in the Saar District, a high-wax type of crude oil, the so-called "Frankenholz oil", is thought to originate from coal measures (Tissot and Welte, 1984). The close relationship between crude oils and the hydrocarbon composition of solvent extracts from particular coals sheds light on the environmental hazard which might be connected with the surface storage of coals as part of the mined overburden material. It has been speculated by Teichmüller (1974) that the well-known loss of coking properties, particularly of Saar District and Illinois Basin coals during storage, might be a result of evaporation of volatile constituents. This assumption is supported by the present study because the Saar District coal investigated here is rich in BTX aromatics (Table 1), which might be released from the coal during storage.

Compound	Sample 1 (mg/kg)	Sample 2 (mg/kg)	Sample 3 (mg/kg)
1. benzene	198.8	119.5	122.3
2. toluene	91.5	114.3	143.0
3. xylenes	29.3	320.9	313.1
4. C_3-benzene	5.9	141.7	54.3
5. naphthalene	56.4	72.8	72.9
6. 2-methylnaphthalene	29.7	87.4	91.9
7. 1-methylnaphthalene	4.9	60.0	46.7
8. biphenyl	61.4	48.7	48.4
9. dimethylnaphthalenes	3.5	221.8	159.2
10. methylbiphenyls	28.9	47.6	81.5
11. trimethylnaphthalenes	0.9	170.0	36.8
12. C_4-naphthalenes	-	60.7	5.9
13. phenanthrene	55.3	44.1	62.6
14. methylphenanthrenes	11.6	106.6	89.7
15. dimethylphenanthrenes	0.6	95.0	44.9

Table 1. *Results of the quantitation of aromatic hydrocarbons in three selected, unweathered coal samples (values in mg compound/kg coal). Sample 1 is from a low-volatile bituminous coal from the mining district near Aachen. Sample 2 is from a high-volatile bituminous coal from the Saar District. Sample 3 is from a medium-volatile bituminous coal from the Ruhr District.*

Coal sample 3 originates from overburden material coming from the Ruhr coal mining district. The material was used in 1987 for road construction material (Anonymous, 1987). The reconstructed ion current from gas chromatography/mass spectrometry analysis of the desorbed bitumen is shown in Figure 2c. The rank of the coal is in the medium-volatile bituminous coal range, which is indicated by a methylphenanthrene index of 1.78 determined from the relative intensities of phenanthrene and methylphenanthrenes (Radke et al., 1982). The total composition of the bituminous constituents is similar to the one obtained from a recently investigated Ruhr District coal of high coking potential (Püttmann et al., 1988) . The intensity of alkylated aromatic hydrocarbons in relation to the unsubstituted counterparts is much higher as compared to coal sample 1 (Table 1). Thus, in this coal the desalkylation process is still at the beginning stage. However, compared to the Saar District coal, a smaller amount of alkylated naphthalenes is detected (Table 1). Additionally, n-alkanes are of minor intensity. According to the quality of the total bitumen composition, the coal can be placed between the other two samples investigated.

The compound distribution pattern allows one to consider the coal as coking coal of high quality. A typical feature of this coal as well as of other coking coals is the occurrence of methylcyclohexane as a dominant compound in the low boiling temperature range. In higher-rank coals, methylcyclohexane is missing in the desorbate as shown in the case of sample 1 (Figure 2a).

Compound	Sample 4 (mg/kg)	Sample 5 (mg/kg)
1. benzene	-	-
2. toluene	-	-
3. xylenes	2.3	-
4. C_3-benzene	5.7	0.8
5. naphthalene	8.8	2.2
6. 2-methylnaphthalene	18.3	6.6
7. 1-methylnaphthalene	10.7	2.4
8. biphenyl	19.2	6.8
9. dimethylnaphthalenes	45.7	9.3
10. methylbiphenyls	36.6	10.5
11. trimethylnaphthalenes	9.8	2.5
12. C_4-naphthalenes	1.2	0.6
13. phenanthrene	36.7	17.8
14. methylphenanthrenes	55.0	22.1
15. dimethylphenanthrenes	26.8	10.5

Table 2. *Results of the quantitation of aromatic hydrocarbons in coal present in overburden of the coal mining district near Aachen (values in mg compound/kg coal). Sample 4 is medium-volatile bituminous coal particles from mined overburden material. Sample 5 is low-volatile bituminous coal particles from mined overburden material.*

Analysis Of Weathered Overburden Material

In order to compare the bitumen compositions of freshly produced coals with those of coals exposed to weathering conditions at the surface, several samples of older overburden have been investigated. Results obtained from two analyses are shown in Figure 4. Figure 4a shows the reconstructed ion current (gas chromatography/mass spectrometry analysis) of the hydrocarbons desorbed from a medium-volatile bituminous coal, from the mining district near Aachen, which has been mixed with overburden for more than 10 yrs. Results show that highly volatile constituents are completely depleted. C_2- and C_3-benzenes are present in small amounts (Table 2) as compared to the amounts detected in a freshly produced coal sample of a similar rank (coal sample 3, Table 1). Additionally, the intensities of naphthalenes are reduced significantly. Biphenyls and phenanthrenes are the dominant constituents of this severely weathered sample. The amount of phenanthrenes is approximately in the order of magnitude commonly observed in related unweathered coal samples (Tables 1 and 2). The most striking observation is the almost total removal of the monoaromatic compounds and the depletion of diaromatic systems. A similar rapid removal of smaller aromatic molecules has been observed in the study of a sediment contaminated with crude oil (Teal et al., 1978).

The second sample investigated (sample 5) shows a similar profile of the desorbable hydrocarbons in the gas chromatography/mass spectrometry analysis (Figure 4). However, the amount of the hydrocarbons is much lower compared to sample 4. This might be due to the higher rank of sample 5. With increasing rank of recently produced coals, the absolute amount of higher boiling temperature constituents decreases, which leads to lower intensities of residual hydrocarbons after weathering occurs. Up until now, knowledge about

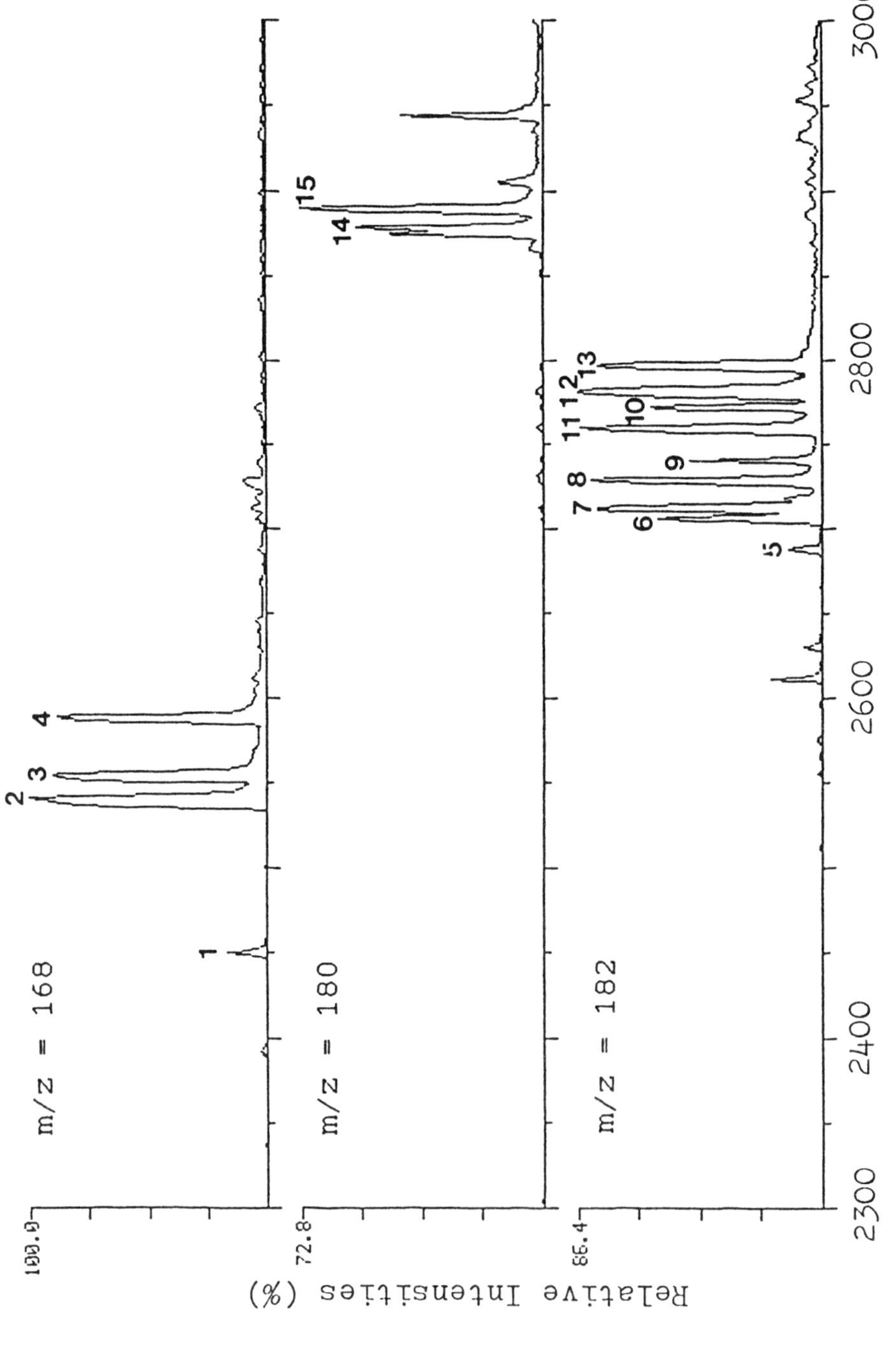

Figure 3. *Mass chromatograms of m/z = 168, m/z = 180, and m/z = 182 in the scan range from 2,300 to 3,000 for the recognition of selected biphenyls, fluorenes, and dibenzofurans present in the desorbate of sample no. 3. The compounds identified are listed in Table 3.*

Compound
1. 2-methylbiphenyl
2. 3-methylbiphenyl
3. 4-methylbiphenyl
4. dibenzofuran
5. 3-ethylbiphenyl
6. 3,5-dimethylbiphenyl
7. 3,3'-dimethylbiphenyl
8. 3,4'-dimethylbiphenyl
9. 4,4'-dimethylbiphenyl
10. 3,4-dimethylbiphenyl
11. 4-methyldibenzofuran
12. 2- + 3-methyldibenzofuran
13. 1-methyldibenzofuran
14. 2-methylfluorene
15. 1-methylfluorene

Table 3. *Compounds identified in mass chromatograms of m/z = 168, m/z = 180, and m/z = 182 obtained by gas chromatography/mass spectrometry analysis of sample 2 (see Figure 3).*

the pathways for the removal of volatile constituents has been poor. In the case of contaminated sediments, a combination of solubilization, evaporation, and (possibly) microbial degradation has been discussed. Based on the study of biodegradation of crude oils, it is well established that microbial activities result in a selective removal of individual alkylated naphthalenes (Rowland et al., 1986). A preliminary investigation of the alkylnaphthalene patterns of weathered coal samples could not provide indications for a selective removal of single naphthalenes, which argues against the occurrence of major microbial activities using aromatic hydrocarbons as the carbon source.

Thus, solubilization and/or evaporation appear to be more important effects contributing to the primary depletion of BTX aromatics. Within the series of aromatic hydrocarbons trapped in bituminous coal, the smaller molecules (i.e., benzene and toluene) are characterized by a high degree of water solubility. This might explain the rapid removal of these compounds from coals exposed to weathering conditions. However, due to the low boiling point of benzene (80°C), possible evaporation of this compound has to be considered.

In order to investigate whether benzene is released from coal in overburden material during slight heating, a headspace experiment using a bath temperature of 80°C was undertaken. Crushed fragments of coal sample 1 were investigated. The result of the gas chromatography/mass spectrometry analysis of the headspace gas after heating the sample for 10 min. is shown in Figure 5. The only compound detectable was benzene, which indicates that, for example, under solar irradiation and the accompanying heating of the coal, evaporative release of benzene has to be considered as a likely process in the weathering environment.

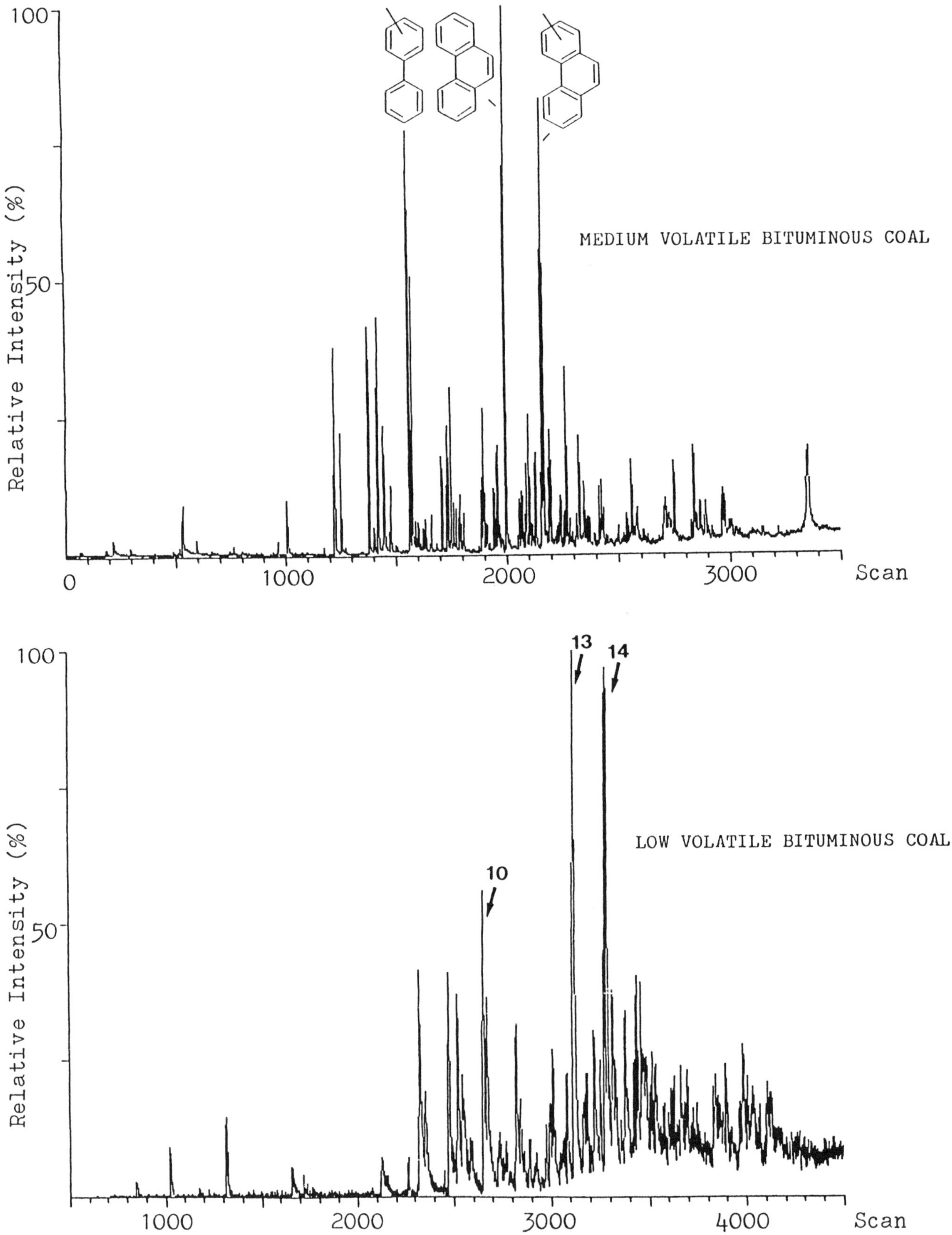

Figure 4. *Reconstructed ion current chromatograms (gas chromatography/mass spectrometry analysis) of hydrocarbons thermally desorbed (300°C) from weathered coal samples originating from the mining district near Aachen. Results of quantitation are shown in Table 2.*

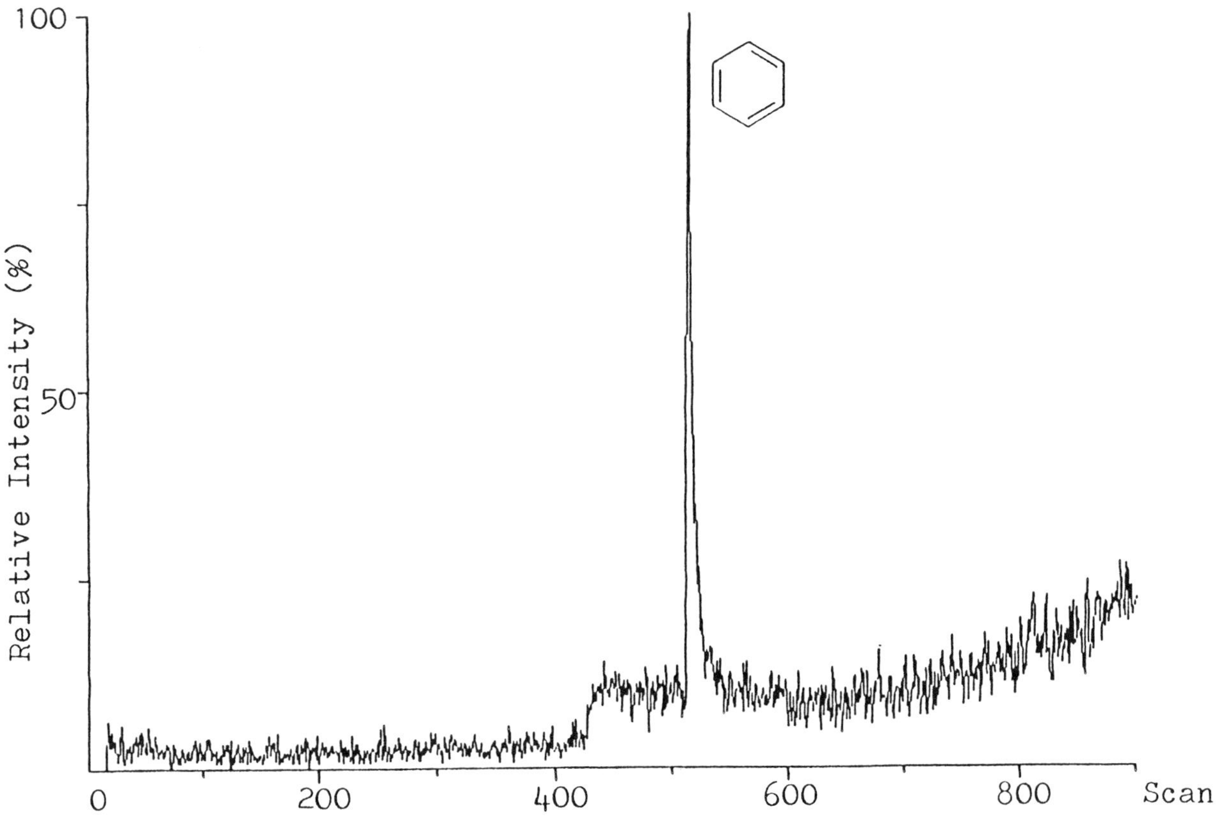

Figure 5. *Gas chromatography/mass spectrometry analysis of headspace gas desorbed from crushed coal particles (sample 1) in a septum-sealed glass tube at 80°C. Analytical conditions were similar to those applied for thermodesorption-gas chromatography/mass spectrometry analysis.*

CONCLUSIONS

Freshly produced coals and coals exposed to weathering conditions in spoil piles have been comparatively studied using the method of thermodesorption combined with gas chromatography/mass spectrometry analysis. The most striking difference observed when comparing the compositions of desorbed hydrocarbons is in the relative contents of highly volatile aromatic constituents such as benzene, toluene, and xylenes (BTX) and biphenyls. These compounds are present in abundance in coals which have reached the end of the high-volatile bituminous coal stage and in medium-volatile bituminous coals. In weathered coal of this rank range present in weathered spoil material, these compounds have disappeared.

Based on the approximation that about 10% of the material in spoil piles is coal, the amount of BTX transferred from the geosphere to the atmosphere/hydrosphere can be estimated to about 45 g/tonne (2 oz/ton) overburden for coals from the rank range mentioned above. In addition, about 11,000,000 tonnes (12,000,000 tons) of cleaned coal are stored at present at the surface as a stockpile. Based on the assumption that only 30% of the coals used as reserve are in the critical rank range, about 1,800 tonnes (2,000 tons) of BTX are transferred together with this coal to the earth's surface and might be released during weathering.

The potential pathways for release of these compounds are as follows:

1) direct evaporation into the atmosphere followed by photo-oxidation,

2) solubilization in water followed by evaporation, and

3) microbial/chemical alteration to polar, water-soluble metabolites.

First analyses (headspace gases) indicate that at least the removal of benzene during weathering might be a result of evaporation. However, microbial or chemical alteration of BTX into polar metabolites such as phenols, aldehydes, and carboxylic acids also have to be considered. These compounds are able to leave the spoil piles through surface and subsurface water movement.

More research is necessary to clarify and quantify the different pathways by which BTX aromatics have left the overburden material. Nonetheless, past research and the results presented here indicate the potential for the introduction to the environment of these organic compounds which are of concern for their impacts on air and water quality in the vicinity of coal mines and power plants.

REFERENCES

Allan, J., and Larter, S.R., 1983, Aromatic structures in coal maceral extracts and kerogens, in M. Bjorøy et al., eds., Advances in organic geochemistry--1981: Chischester, England, Wiley, p. 534-545.

Anonymous, 1987, "Friesenspieß" nimmt Formen an: Ruhrkohle, Heft 5, p 7.

Beier, E., 1985, Gasaustausch von Steinkohlen und anderen Stoffen bei jahrzehntelanger Lagerung an Luft: Erdöl und Kohle, v. 38, p. 127-129.

Bjorøy, M., H. Solli, K. Hall, and P. Leplat, 1985, Analysis of source rocks, reservoir rocks and cap rocks by combined thermal extraction and pyrolysis-gas chromatography, in B. Thomas et al., eds., Petroleum geochemistry in exploration of the Norwegian Shelf: London, Graham and Trotman, p. 327-337.

Brecht, C., H.-G. Goethe, H. Krämer, H. Reintges, and H.-G. Willing, 1988, Jahrbuch 88/89 Bergbau, Ol und Gas, Elektrizität, Chemie: Essen, Verlag Glückauf GmbH, p. 28.

Brooks, J. D., and J.W. Smith, 1967, The diagenesis of plant lipids during the formation of coal, petroleum and natural gas: Geochimica et Cosmochimica Acta, v. 31, p. 2389-2397.

Crisp, P.T., J. Ellis, J.W. De Leeuw, and P.A. Schenck, 1986, Flash thermal desorption as an alternative to solvent extraction for the determination of C_8-C_{35} hydrocarbons in oil shales: Analytical Chemistry, v. 58, p. 258-261.

Freundenberg, K., 1964, Entwurf eines Konstitutionsschemas für das Lignin der Fichte: Holzforschung, v. 18, p. 3-9.

Hayatsu, R., R.E. Winans, R.L. McBeth, R.G. Scott, L.P. Moore, and M.H. Studier, 1987, Lignin-like polymers in coal: Nature, v. 278, p. 41-43.

Horsfield, B., 1989, Practical criteria for classifying kerogens: Some observations from pyrolysis-gas chromatography: Geochimica et Cosmochimica Acta, v. 53, p. 891-901.

Jensen, T.E., and R.A. Hites, 1983, Aromatic diesel emissions as a function of engine conditions: Analytical Chemistry, v. 55, p. 594-599.

Jüntgen, H., and J. Karweil, 1966, Gasbildung und Gasspeicherung in Steinkohlenflözen: Erdöl und Kohle, v. 19, p. 251-258 and 339-334.

Larter, S., 1989, Chemical models of vitrinite reflectance evolution: Geologische Rundschau, v. 78, p. 349-359.

Leythaeuser, D., R.G. Schaefer, and B. Weiner, 1979, Generation of low molecular weight hydrocarbons from organic matter in source beds as a function of temperature and facies: Chemical Geology, v. 25, p. 95-108.

Ott, F.S., R.P. Harris, and C.M. O'Hara, 1978, Acute and sublethal toxicity of naphthalene and three methylated derivatives to the estuarian copepod, Eurytemora Affinis: Marine Environmental Research, v. 1, p. 48-58.

Ouchi, K., and K. Imuta, 1963, The analysis of benzene extracts of Youbari Coal II--Analysis by gas chromatography: Fuel, v. 42, p. 445-456.

Püttmann, W., C.B. Eckardt, and R.G. Schaefer, 1988, Analysis of hydrocarbons in coal and rock samples by on-line combination of thermodesorption, gas chromatography and mass spectrometry: Chromatographia, v. 25, p. 279-288.

Püttmann, W., and W. Goßel, 1988, Analytik von aromatischen Kohlenwasserstoffen in Abraummaterial aus dem Steinkohlen-Bergbau: Wissenschaft und Umwelt, v. 3, p. 123-130.

Püttmann, W., and W. Kalkreuth, 1989, Comparison of hydrocarbon compositions in a sequence of humic coals, cannel coals and oil shales from the Pictou Coalfield, Nova Scotia: Atlantic Geology, v. 25, p. 93-103.

Püttmann, W., K. Steffens, and W. Kalkreuth, 1987, Chemical characterization of natural weathering in coal, in J.A. Moulijn et al., eds, Proceedings of the International Conference on Coal Science 1987: Amsterdam, Elsevier, p. 411-414.

Püttmann, W., M. Wolf, and E.M. Wolff-Fischer, 1986, Chemical characteristics of liptinite macerals in coal, in D. Leythaeuser and J. Rullkötter, eds., Advances in organic geochemistry--1985: Organic Geochemistry, v. 10, p. 625-632.

Radke, M., H. Willsch, D. Leythaeuser, and M. Teichmüller, 1982, Aromatic components in coal: Relation of distribution pattern to rank: Geochimica et Cosmochimica Acta, v. 46, p. 1832-1848.

Raymond, A.C., S.Y. Liu, D.G. Murchison, and G.H. Taylor, 1989, The influence of microbial degradation and volcanic activity on a Carboniferous wood: Fuel, v. 68, p. 66-73.

Rowland, S.J., R. Alexander, R.I. Kagi, D.M. Jones, and A.G. Douglas, 1986, Microbial degradation of aromatic components of crude oils: A comparison of laboratory and field observations: Organic Geochemistry, v. 9, p. 153-161.

Schaefer, R.G., D. Leythaeuser, and B. Weiner, 1978a, Single-step capillary column gas chromatographic method for extraction and analysis of sub-parts per billion (10^9) amounts of hydrocarbons (C_2-C_8) from rock and crude oil samples and its application in petroleum geochemistry: Journal of Chromatography, v. 167, p. 355-363.

Schaefer, R.G., and H. Pooch, 1982, Thermal mobilization of hydrocarbons from small-size rock samples. Application in petroleum geochemistry: Chromatographia, v. 16, p. 257-260.

Schaefer, R.G., and W. Püttmann, 1987, Analysis of hydrocarbons in coal by means of thermodesorption-gas chromatography: Journal of Chromatography, v. 395, p. 203-215.

Schaefer, R.G., B. Weiner, and D. Leythaeuser, 1978b, Determination of sub-nanogram per gram quantities of light hydrocarbons (C_2-C_9) in rock samples by hydrogen stripping in the flow system of a capillary gas chromatograph: Analytical Chemistry, v. 50, p. 1848-1854.

Senftle, J.T., S.R. Larter, B.W. Bromley, and J.H. Brown, 1986, Quantitatic chemical characterization of vitrinite concentrates using pyrolysis-gas chromatography: Rank variation of pyrolysis products: Organic Geochemistry, v. 9, p. 345-350.

Teal, J.M., K. Burns, and J. Farrington, 1978, Analyses of aromatic hydrocarbons in intertidal sediments resulting from two oil spills of No. 2 fuel oil in Buzzard Bay, Massachusetts: Journal of Fish Research Board of Canada, v. 35, p. 510-520.

Teichmüller, M., 1974, Entstehung und Veränderung bituminöser Substanzen in Kohlen in Beziehung zur Entstehung und Umwandlung des Erdöls: Fortschritte der Geologie Rheinland und Westfalen, v. 24, p. 65-112.

Tissot, B.P., and D.H. Welte, 1984, Petroleum formation and occurrences (2nd Ed.): Berlin, Springer-Verlag, p. 252.

Van Berk, W., 1987, Hydrochemische Stoffumsetzungen in einem Grundwasserleiter--beeinflußt durch eine Steinkohlen-Bergehalde: Landesamt für Wasser und Abfall des Landes NRW, Besondere Mitteilungen zum Deutschen Gewässer kundlichen Jahrbuch Nr. 49, p. 1-175.

Van Krevelen, D.W., 1961, Coal: Amsterdam, Elsevier, 514 p.

DRAINAGE FROM COAL MINES: CHEMISTRY AND ENVIRONMENTAL PROBLEMS

Thomas Wildeman
Department of Chemistry and Geochemistry
Colorado School of Mines
Golden, CO 80401

ABSTRACT

Much of the research on coal-mine drainage chemistry was conducted a decade ago, and now increased environmental awareness has brought about renewed interest in the findings. Consideration of the trace minerals and elements in coal points to the possible generation of acidic waters upon weathering, especially when pyrite is present. When pyrite weathers, it produces H^+ and Fe^{3+} which catalyze the incongruent weathering of other carbonates and sulfides. In this weathering mechanism, catalysis by bacteria is important. Of the environmental problems in coal mine drainage, the mineral acidity of the water is the most serious. This is caused not only by the H^+, but also by Mn^{4+}, Fe^{3+}, and Al^{3+} that are found or generated within the drainage. Case studies in Kentucky, Pennsylvania, Illinois, and Colorado show that the abundance and form of pyrite in the deposit and in the overburden determines the level of acidity and the concentration of heavy metal pollutants in the drainage. Recent trends in environmental enforcement that emphasize integrated stream water standards and biotoxicity assays point to the possibility that the concentrations of heavy metals in coal mine drainages may cause environmental concern.

INTRODUCTION

Acid mine waters are not new. Their production was noted in Roman times, and their possible toxicity was reported by Agricola in De Re Mettallica (D.K. Nordstrom, personal communication, 1989). Research on the refinement of the causes for acid mine drainage production also is not new. Most of the important ideas on the mechanism of production were generated in the 1960s and 1970s. The primary reactants are pyrite (FeS_2), water, and ultimately oxygen; and important catalysts are bacteria, particularly *Thiobacillus ferrooxidans*. Because many of the ideas on the cause of acid mine drainage were established about a decade ago, they can be found in texts and monographs that often are easier to locate than the primary literature sources. This review will draw extensively on these secondary sources so that the reader can more readily augment this paper. For each section, the useful monographs will be cited.

To define the subject, Table 1 shows the concentrations of constituents that are routinely determined in coal mine drainages, the constituents in a comparable metal mine drainage, the abundance ranges of these elements in coals, and the maximum allowable contaminant levels for public drinking water. The references to the data are noted at the bottom of Table 1. For the coal information, the monographs by Bouska (1983) and Valkovic (1983) are useful. Manahan (1984) gives a good explanation of the environmental effects of each constituent.

Coal Mine Drainage						
Substance	United States	Illinois	Kentucky	Big Five Tunnel	Drinking Water Standards	Coal
Al	—	37	—	18	—	14000
Fe	0.6 - 220	57	50 - 500	50	0.3	16000
Mn	0.3 -12	6.4	—	32	0.05	100
Cu	0.01 - 0.17	—	—	1.6	1.0	19
Zn	0.03 - 2.2	—	—	10	5.0	39
Cd	0.01 - 0.10	—	—	0.03	0.01	1.3
Pb	0.01 - 0.40	—	—	0.01	0.05	16
As	0.002 - 0.20	—	—	0.02	0.05	15
pH	3.2 - 7.9	3.0	1.8 - 3.5	2.6	6.5 - 8.5	—
SO_4^{2-}	—	1300	500 - 12000	2100	250	—
Ref.	1	2	3	4	5	6

1. The 10-90% concentration range of 23 acid drainages from coal mines throughout the United States taken from the EPA effluent limitations document (U.S. Environmental Protection Agency, 1982).
2. Median of 110 drainages from coal refuse disposal sites in Southern Illinois as compiled by Proudan et al. (1982).
3. Regional estimates from Caruccio (1979).
4. A typical metal mine drainage from the Front Range Mineral Belt of Colorado collected by Wildeman and Laudon (1989).
5. Compiled from U.S. Government Printing Office (1988). For mine drainages, effluent limits in milligrams/Liter are: Fe, 7.0 daily maximum, and 3.5 monthly average; Mn, 4.0 daily maximum, and 2.0 monthly average; pH between 6.0 and 9.0 at all times. For the other substances in the table, there are no written restrictions (U.S. Environmental Protection Agency, 1982).
6. Average of U.S. coal compiled by Valkovic (1983).

Table 1. *Concentrations of environmentally important constituents in acid mine drainages and in coal. For waters, the concentrations are in milligrams/Liter; for coal, in parts per million.*

For the coal mine drainages in the Illinois and Kentucky coal regions, the concentrations of the major contaminants are quite similar, and these compare well with the values for the whole U.S. In the U.S. Environmental Protection Agency (EPA) document on effluent limitations for coal mining (U.S. Environmental Protection Agency, 1982), tests were reported on that were done to determine whether drainages from Western U.S. coal mines and Eastern U.S. anthracite mines should be separate categories. No case could be made for subcategories. The Big Five Tunnel (Wildeman and Laudon, 1989) is a metal-mine drainage, and the concentration of most of the constituents fall well within the ranges for U.S. coal-mine drainages. There appears to be reasonable cause to group all acid mine drainages together rather than split the waters into a number of categories. If this is done, then differences from the usual chemistry can be more successfully investigated. Prodan et al. (1982) give means, standard deviations, minimum values, and maximum values for 110 effluents from abandoned coal-refuse sites in Illinois, and the numbers give a good indication of how far waters vary from the median values and from the ranges reported in Table 1.

For the metal and coal mine drainages, the concentrations of iron (Fe), manganese (Mn), aluminum (Al), and sulfate (SO_4^{2-}) are in the same range. For a coal seam, the possibility of large abundances of copper (Cu), zinc (Zn), cadmium (Cd), lead (Pb), and arsenic (As) exists. However, other than the U.S. Environmental Protection Agency (1982) document, data on concentrations of these constituents in effluents are difficult to find. These heavy metals do exist in metal-mine drainages and are well documented (Wildeman, 1983;

Wildeman et al., 1974; Wildeman and Laudon, 1989). In this section, the geochemistry of the weathering of pyrite will be developed and this will be related to the weathering reactions of the minerals responsible for the other contaminants in mine drainage. The chemistry of mine drainages from the Central City mining district in Colorado will be used as an example of how the weathering reactions are interrelated. Finally, the role of hydrology in the production of acid mine drainage will be reviewed.

Pyrite Oxidation

In coal mining situations, pyrite is the mineral that is responsible for acid drainage problems. This same mineral is also the cause of the problems in metal mining situations. Understanding how pyrite weathers is essential to understanding the causes of the problem and the relationships between coal- and metal-mining pollution problems. Stumm and Morgan (1981) review the chemistry of pyrite weathering and the following description is summarized from their text. The overall stoichiometric reactions are as follows:

$$FeS_2\,(s) + 7/2\,O_2 + H_2O \rightarrow Fe^{2+} + 2\,SO_4^{2-} + 2\,H^+ \qquad [1]$$

$$Fe^{2+} + 1/4\,O_2 + H^+ \rightarrow Fe^{3+} + 1/2\,H_2O \qquad [2]$$

$$Fe^{3+} + 3\,H_2O \rightarrow Fe(OH)_3 + 3\,H^+ \qquad [3]$$

$$FeS_2 + 14\,Fe^{3+} + 8\,H_2O \rightarrow 15\,Fe^{2+} + 2\,SO_4^{2-} + 16\,H^+ \qquad [4]$$

The accepted reaction path for the dissolution of pyrite is:

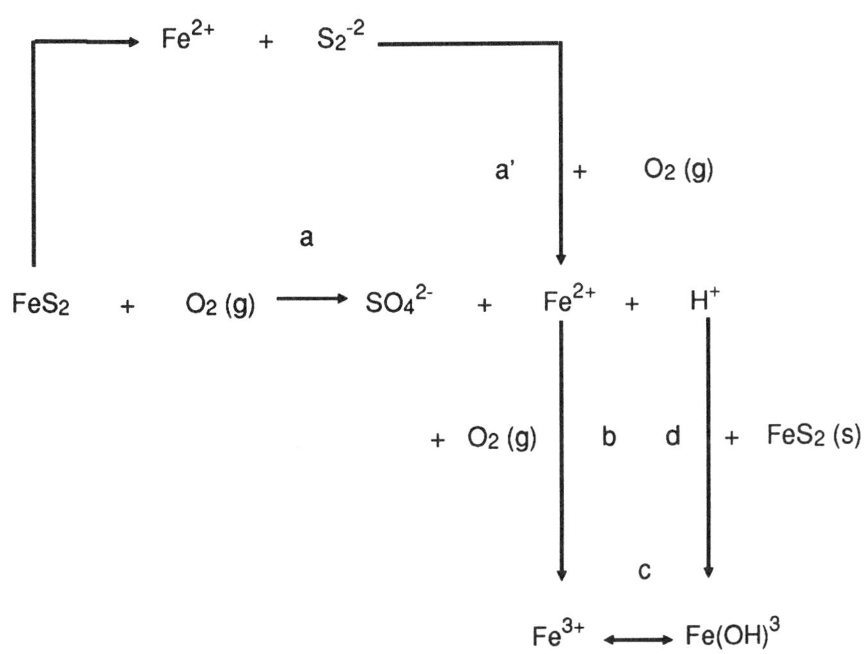

Key features of the stoichiometry and reaction path are as follows:

1) Weathering is by oxidation. Because pyrite formation only occurs in a reducing environment, oxygen gas from outside the deposit is the ultimate oxidant.

2) Hydrogen ions (H^+) are produced by the oxidation. For every mole of pyrite oxidized, two moles of H^+ are produced by the oxidation to sulfate (reaction 1), and two moles of H^+ are produced upon the precipitation of ferric hydroxide (reactions 2 and 3).

3) Because ferric hydroxide is so insoluble, pyrite oxidation is among the most acid producing of all weathering reactions.

4) The slow step in the reaction path is oxidation in solution of Fe^{2+} to Fe^{3+}. Sulfur (S) oxidation is relatively rapid.

5) Once the weathering has produced Fe^{3+}, this species can rapidly oxidize pyrite as shown in reaction 4 and step d of the reaction path. Therefore, Fe^{3+} cannot persist in the presence of pyritic minerals.

6) Steps a and d can be separated in time and space to enable the production of acid drainage from different environments.

Microorganisms can significantly catalyze the rate of steps a, b, and d in the mechanism. The monograph by Erlich (1981) is a good review of how weathering reactions can be mediated by bacteria. Thiobacillus ferrooxidans can accelerate the rate of steps b and d by orders of magnitude. Thiobacillus thiooxidans can catalyze step a. Bacteria are necessary to increase the rate of pyrite weathering to the point that pollution problems will occur (Taylor et al., 1984; van Everdingen and Krouse, 1988).

Recent studies on the stable isotope geochemistry of the sulfate in acid mine drainage have added some refinements to the pyrite weathering mechanism (Taylor et al., 1984; van Everdingen and Krouse, 1988). Reaction 4 is found to be a major cause of sulfide oxidation, and this reaction does not directly use molecular oxygen. Therefore, flooding mine workings to eliminate air-pyrite contact may not necessarily stop pyrite weathering. Weathering could continue by bacterial mediation of reactions 1 and 2 in the unsaturated zone in the soil and then by reaction 4 in the flooded workings. In addition, where pyrrhotite is present along with pyrite, production of acid drainage apparently is more widespread (M. Kalin, personal communication, 1988).

Incongruent Weathering

The concept of congruent and incongruent reactions is important to pyrite weathering and to reactions that form other constituents in acid mine drainage (Garrels and McKenzie, 1971). To demonstrate incongruence, consider manganese in coal which exists as rhodochrosite ($Mn(CO_3)$) (Valkovic, 1983). Below pH = 4, the $Mn(CO_3)$will react accordingly:

$$Mn(CO_3) + 2\,H^+ \rightarrow Mn^{2+} + H_2O + CO_2\,(g) \qquad [5]$$

CO_2 gas can escape because it is slightly soluble in water, and if this occurs, $MnCO_3$ cannot be reprecipitated in an acidic solution. This is an example of incongruent weathering. Some other reaction or severe altering of solution conditions is necessary to cause reprecipitation of the reaction products. Reaction 5 is the basis for how Mn exists in coal mine drainages as Mn^{2+}. Reactions 1-4 show that Fe^{3+} and SO_4^{2-} in mine drainage cannot be changed back to pyrite through the reversal of a simple reaction.

Other sulfide minerals can weather by congruent reactions. A possible reaction for the weathering of sphalerite (ZnS) is

$$ZnS\ (s) + H_2S \leftrightarrow Zn^{2+} + HS^- + OH^- \qquad [6]$$

However, if ferric ion is present, it can oxidize the bisulfide ion in the same way as in the dissolution of pyrite:

$$8\ Fe^{3+} + HS^- + 4\ H_2O \rightarrow 8\ Fe^{2+} + SO_4^{2-} + 9\ H^+ \qquad [7]$$

In contact with an acid mine drainage solution, ZnS also will be weathered in a manner that cannot be reversed easily.

Tables 1 and 2 list the chemistry of some acid mine drainages. Iron, Mn, and SO_4^{2-} dominate the constituents in coal mine drainages; reactions 1-5 explain their presence. In drainages from metal and coal mines, Cu, Zn, Cd, Pb, and As often are present in amounts detrimental to the environment. Reaction 6 explains the presence in solution of these base metal cations. The presence of Al in mine drainages is explained best by acidic solutions causing the dissolution of clays (Wildeman, 1983). Groundwater hydrology, fluctuations in rainfall, and the manner of ore deposition also can affect mine drainage chemistry (Wildeman, 1981). The model for the chemistry of the Argo Tunnel is an example of what these other factors can do (Wildeman, 1983). However, reactions 1-7 are basic to the system and the other factors cause secondary changes in the rate and extent of these reactions.

Mineral Acidity

Of all the environmental problems related to mine drainage, the low pH is the most troublesome. Not only do the pHs of the drainages shown in Table 1 fall far out of bounds from the drinking water standards, but increasing the pH to within the drinking water standards also is necessary for long-term removal of all the

Mine Drainages			
Constituent	**Central Zone**	**Intermediate Zone**	**Peripheral Zone**
Al	25 - 100	n.d.	n.d.
Fe	200 - 700	2 - 170	0.5 - 4
Mn	90 - 120	20 - 40	1.0 - 5.0
Cu	6 - 60	0.1 - 5	< 0.01 - 0.11
Zn	60 - 400	7 - 100	0.3 - 8.0
Cd	0.2 - 2.9	< 0.01 - 0.20	< 0.01 - 0.04
Pb	0.1 - 0.5	< 0.01 - 0.20	< 0.01 - 0.06
As	0.2 - 2.7	< 0 .001 - 0.01	< 0.001
SO_4^{2-}	2300 - 4000	900 - 1300	240 - 800
pH	2.1 - 2.7	4.0 - 6.0	5.4 - 6.9

Table 2. *Dissolved constituents in mine drainages of the Central City mining district from Wildeman (1981) and Wildeman et al. (1974). All concentrations are in milligrams/Liter except pH; n.d. means not detected.*

other pollutants. Consequently, most every pollutant removal method relies on raising the pH (Wildeman and Laudon, 1989; Klusman and Machemer, this volume). In addition, for acid-base stability of most natural water, buffering by the carbonic acid-bicarbonate-carbonate system is the most likely method (Stumm and Morgan, 1981). This begins to occur at a pH of between 5 and 6. Any effluent that is released into natural surface waters should be at a pH above 6 to insure that it will not harm the existing ecosystem.

However, the low pH is not caused just by the presence of H^+ ions. Examination of reaction 3 shows that Fe^{3+} hydrolyzes forming $Fe(OH)_3$ precipitate and H^+. Fe^{3+} should be considered an acid the same as H^+. Al^{3+} and Mn^{4+} also will strongly hydrolyze, thereby forming H^+. Because the pH of an acid drainage depends on all these chemical factors, the term "mineral acidity" or just "acidity" is given to the situation. Acidity is defined operationally by how the analysis is conducted (American Public Health Association, 1988). In the analysis, hydrogen peroxide is added, the solution is boiled, and then titrated with standard sodium hydroxide to a pH of 8.2. If the water contains appreciable concentrations of Fe, Mn, and Al, the solubility products of these metal hydroxides will determine which constituents will hydrolyze by a pH of 8.2 and contribute to the acidity.

Using solubility product data from Lindsay (1979), Al^{3+}, Fe^{3+}, and Mn^{4+} will completely hydrolyze by this pH of 8.2, but Fe^{2+} and Mn^{2+} at the concentrations encountered in mine drainages still will be in solution. However, Fe^{2+} can oxidize according to reaction 2 and Mn^{2+} can do likewise. These constituents should be considered potential contributors to the acidity. Because the acidity analysis calls for addition of hydrogen peroxide and boiling, it is certain that Fe^{2+} and Mn^{2+} are oxidized to some extent and counted in the measure of mineral acidity.

Step b in the reaction path shows that the oxidation of Fe^{2+} to Fe^{3+} is slow, and experience with the treatment of mine drainages shows the oxidation of Mn^{2+} to be even slower (Watzlaf, 1988; Klusman and Machemer, this volume). This slow oxidation implies a long-term release of mineral acidity which can cause the reversal of some treatment methods that rely on hydroxide precipitation (Watzlaf, 1988). Slow oxidation also is responsible for the persistence of mine drainage conditions long after the water has breached the surface. For example, the red and roily nature of surface waters associated with mining is caused by the slow oxidation of Fe^{2+} and its subsequent precipitation as $Fe(OH)_3$.

CASE STUDIES OF MINE DRAINAGE PROBLEMS

In analyzing studies of acid mine drainage, there is one common cause: the presence of pyrite. The studies investigate how the availability of the other reactants, such as water or oxygen, affect the water quality (Prodan et al., 1982; Wildeman, 1981), how the nature of the pyrite might affect the water quality (Caruccio, 1979; Caruccio et al., 1977; Wildeman, 1983), and how the nature of the overburden might be used to predict the occurence of acid drainage problems (Caruccio, 1979; Erickson and Hedin, 1988; Hedin and Erickson, 1988).

Coal Mine Drainage Studies

In Illinois, Prodan et al. (1982) studied runoff water quality from 110 coal refuse disposal sites. As seen in Table 1, these waters are typical of coal mine drainage problems. Two sites were reclaimed by recontouring the waste piles, adding limestone and lime to the surface of recontoured piles, and adding a 30-cm (12-in.) thick topping of natural soil material. It was hoped that these steps would reduce the water and oxygen contact with the pyrite and that the alkaline materials would neutralize any acid which was produced previously. At one site, there was success; however at the other, water quality turned progressively poorer after initial good conditions. The turn to poorer quality was attributed to the inability of reclamation to keep the necessary reactants of oxygen and water away from the pyrite. In particular, it was suggested that the stored potential of Fe^{3+} salts in the pile kept the weathering of pyrite continuing through reaction 4.

In the 1960s, various people were investigating why some beds in the Appalachian coal field had acid drainage problems and others did not. The accumulated hypotheses then were put to test in Eastern Kentucky by Caruccio (Caruccio et al., 1977; Caruccio, 1979). A major cause for the production of acid waters was suggested to be the type of pyrite in the deposit and not the amount, and the type of pyrite could be related to the depositional environment (paleoenvironment). In particular, coals deposited in marine or brackish-water environments contained appreciable amounts of framboidal pyrite and this form was highly reactive to oxidative weathering. Tests of the hypothesis did indeed show that some coal seams that contained framboidal pyrite did produce highly acidic water; however, others did not. There were some seams that contained framboidal pyrite and whose drainages contained appreciable sulfate, and yet, the drainages were not acidic. The presence of sulfate in the water showed that, indeed, the pyrite was reacting. Further studies revealed that some paleoenvironments also generated calcareous material in the surrounding sediments in amounts sufficient to neutralize the acidity produced.

These findings led to a model for the prediction of the chemistry of water draining from a coal seam, with the model being subdivided into four hydrogeochemical environments. The four environments roughly can be divided into the matrix shown in Figure 1. In the matrix, the amount of pyrite refers to reactive, framboidal pyrite and not coarse-grained, stable mineral, and "carbonate" refers to reactive minerals that will produce alkalinity in the water. If the environment generated a situation where the two constituents in the matrix have opposite abundances, then the water chemistry is quite predictable. When the abundances of the pyrite and

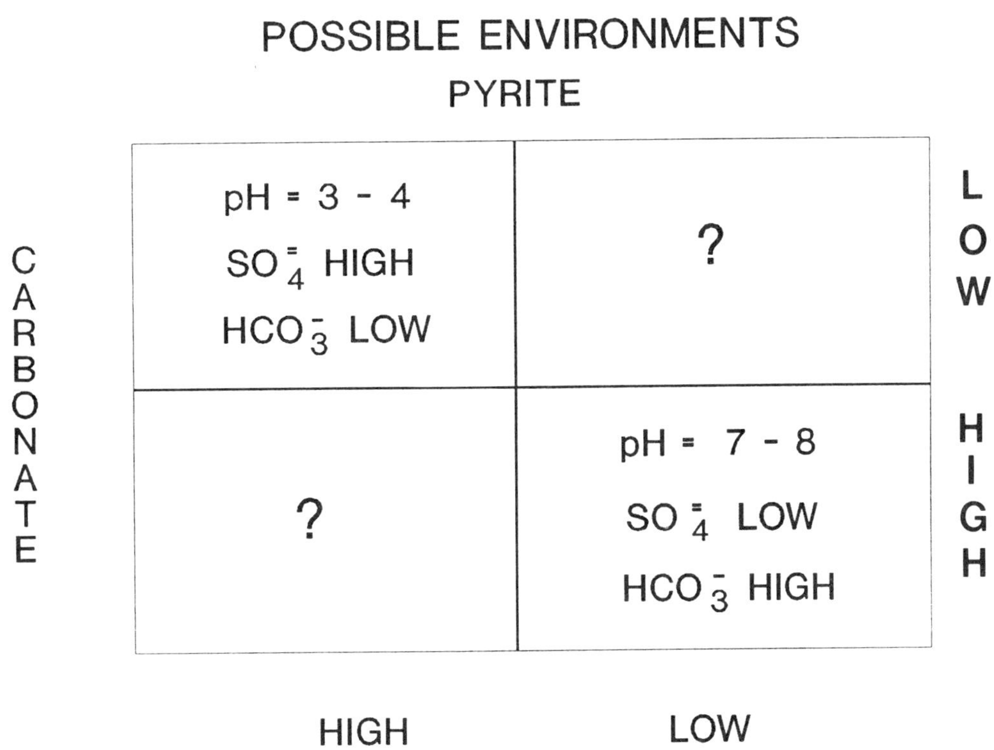

Figure 1. *A matrix adapted from Caruccio (1979) showing how abundances of calcareous materials as carbonates and framboidal pyrite in the overburden can affect mine drainage chemistry.*

carbonate are both high or low, then results are uncertain. One key factor in uncertain environments is the placement of the calcareous material above and hydrologically antecedent to the acid-producing strata. This situation typically will produce a neutral water with high sulfate concentrations. Application of the model to prediction of coal mine drainage problems requires a study of the paleoenvironments in a coal field, a petrographic analysis of the forms of pyrite in a coal seam, analysis of the chemistry of regional groundwaters to determine the pH, sulfate concentration, and alkalinity, and analysis of the strata over the coal seam for carbonate and pyrite.

Overburden Analysis

The Caruccio study (Carrucio et al., 1977; Carrucio, 1979) and others related to it no doubt influenced the requirement that potentially acid-producing strata be identified and assessed as part of the mine permitting process. This has led to a number of studies on how the composition of the overburden is related to acid production and what analyses of the overburden best predict acid drainage potential (diPretoro and Rauch, 1988; Erickson and Hedin, 1988; Hedin and Erickson, 1988). diPretoro and Rauch (1988) found that the alkalinity of the overburden as determined by net neutralization potential is the best predictor of acid drainage. Total S content bears little relation to mine drainage quality. However, the ratio of net neutralization to total S content is significantly related to net alkalinity of a mine drainage. Erickson and Hedin (1988) studied how the acid-base account method for overburden analysis works at 32 sites where prediction is difficult. This would correspond to the questionable areas in the Figure 1 matrix. They were able to draw boundaries for prediction but were not capable of making quantitative predictions for these doubtful sites. One of the primary problems they encountered was the heterogeneity of the overburden and the variability of spoils seeps. Methods for aggregating the data were devised to minimize these problems.

In an attempt to find better methods of prediction, Hedin and Erickson (1988) used a laboratory simulation of the weathering process to assess acid production potential. The level of sulfate in the drainage was strongly correlated to the sulfate produced by weathering and this was strongly correlated to the amount of total S in the overburden samples. However, there was still no strong correlation between any of the parameters measured in the weathering tests and the alkalinity of the mine drainage. Presently, limits can be placed on when coal-mine drainage problems will occur, but good predictive tools in questionable situations still are not available.

Heavy-Metal Studies

The levels of heavy metals in coal mine drainages are little reported, and these metals may prove to be causing problems when bioassay techniques are used for environmental assessment (Horning and Weber, 1985; Peltier and Weber, 1985). The effluent guideline study by the U.S. Environmental Protection Agency (1982) did report concentration ranges of heavy metals in coal mine drainages and these are given in Table 1. Note that the concentrations of Cd, As, and Pb can range above drinking water standards. Table 1 also shows that the average abundances of Cd, As, and Pb in coal are at levels appreciably above average crustal abundances (Krauskopf, 1979). How do these heavy metals originate in coal mine drainages, and is there a problem with concentrations in effluents being higher than drinking water standards?

The studies by Wildeman and co-workers (Wildeman, 1981, 1983; Wildeman et al., 1974) on the Central City mining district in Colorado give some perspective on how pyrite affects the concentrations of trace heavy metals in mine drainages. This is a typical example of a zoned hydrothermal deposit of gold and base metals (Sims et al., 1963). The distribution of minerals from the high-temperature Central Zone to the lower-temperature Peripheral Zone is shown in Figure 2. The chemistry of drainages emanating from mines in the various zones is summarized in Table 2. The striking feature about the chemistry of these waters is that Cd, Zn, and Pb are in lowest concentration in the Peripheral Zone even though the ore minerals for these metals are in

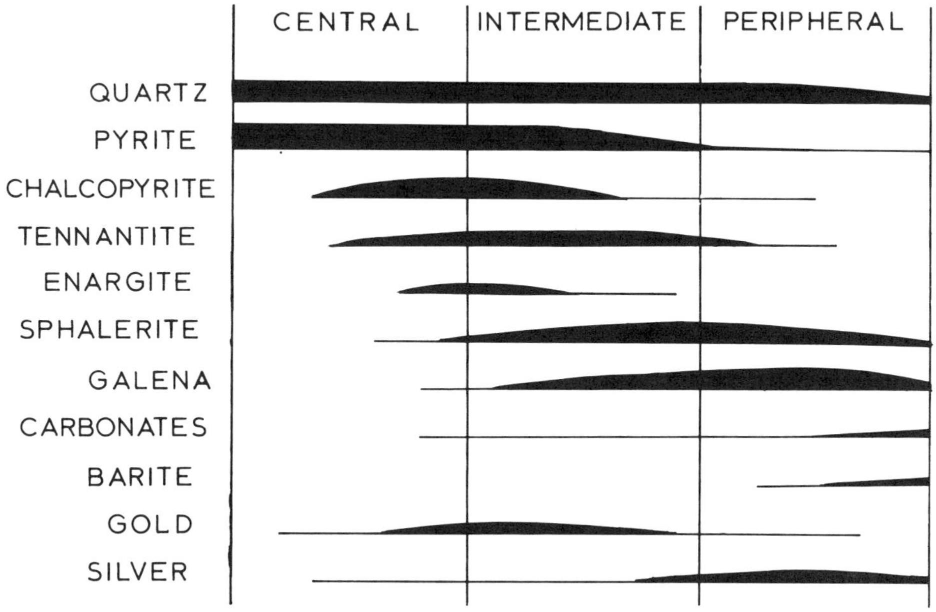

Figure 2. *Zonal distribution of vein-forming minerals in the Central City mining district taken from Sims et al. (1963).*

highest abundance in that zone. The concentration of all the contaminant metals in the drainages correlates with the abundance of pyrite in the ore. Fe^{3+} and H^+ in the groundwater catalyze the dissolution of the other sulfides to such an extent that they become important constituents in the drainage from a metal mine even though the base metals may be in low abundance in the deposit.

When coal deposits are considered, all the heavy metals listed in Table 1 are associated with the pyrite and other sulfide minerals in coal and its associated overburden (Bouska, 1981; Valkovic, 1983). Consequently, when the pyrite weathers, the products of that weathering are highly likely to release trace heavy metals from the coal. Beyond the U.S. Environmental Protection Agency (1982) study, it is difficult to find information on concentrations of heavy metals in coal drainages. However, Watzlaf (1988) gives some insight into why this is the case. In the treatment of acid drainages, Mn is the most difficult metal to remove. As shown in Table 1, investigations by the U.S. Environmental Protection Agency (1982) did find significant concentrations of heavy metals in untreated coal mine drainages. However, it was found that if Mn was reduced to 2 mg/L in the effluent, the heavy metals also were reduced to acceptable levels. Therefore, limitations on these metals were not promulgated, and the limitation on Mn of 2 mg/L was established. Watzlaf (1988) has determined that this guideline is reasonable. However, the sludge produced is quite unstable and subject to resolubilization.

HYDROLOGY RELATED TO ACID MINE DRAINAGE

Although the presence of pyrite is definitely the key factor that determines mine drainage quality issuing from underground adits, there have been some studies that show how groundwater hydrology is involved. In a long-term study of the Argo Tunnel drainage in Idaho Springs, Colorado, Wildeman (1983) found that the chemistry of the water varied little with the seasons and precipitation events. To explain the findings he used

DIFFUSE	CONDUIT
1. No response to climatological change.	1. Responds to climatological changes.
2. Little fluctuation in flow.	2. Obvious fluctuations in flow.
3. No suspended solids in the water.	3. Carries suspended solids at times of high runoff.
4. Water temperature may not change throughout the year.	4. Water temperature changes with the seasons.
5. Parameters indicative of concentration, such as conductivity (μmho/cm) and hardness, do not change with the climate.	5. Parameters indicative of concentration, such as conductivity (μmho/cm) and hardness, show obvious changes with storms and runoff.
6. Specific concentrations of ions show little change with the climate.	6. Specific concentrations of ions show obvious changes with storms and runoff.
7. Residence time of months for the water in the aquifer.	7. Residence time of days for the water in the aquifer.

Table 3. *Differences between diffuse and conduit aquifers (from Wildeman, 1983).*

aquifer models developed to explain the chemistry of carbonate springs in Pennsylvania (Jacobsen and Langmuir, 1974; Shuster and White, 1971). Two simple models for groundwater systems are generated: the conduit flow system and the diffuse flow system. These two models can be treated as the end members of all recharge systems. The properties that distinguish the two systems are given in Table 3. In both aquifers, recharge is from the surface, through the soil vadose zone, and down to the groundwater table (Wildeman, 1983). The Argo Tunnel drainage is an example of a primarily diffuse aquifer.

A surprising characteristic of a diffuse aquifer is that when annual recharge occurs in spring and the flow of water does rise slightly, some constituents in the water will increase in concentration. Wildeman (1983) found that all the metals associated with pyrite dissolution increased in concentration during the spring recharge. He suggested that pyrite weathering is a slower reaction and that if the weathering products are retained in microfaults in the vadose zone above the water table, then the reaction is most favored. This water then is released from the faults during spring recharge.

Many of the adits in Colorado that Wildeman studied are regional systems that serve to lower the groundwater and expose deeper deposits. Some adits in the Eastern U.S. are of this type, but the greater concern is with overburden exposed during strip mining and with coal refuse piles. Caruccio and coworkers have done extensive research related to this problem (Caruccio, 1979; Caruccio et al., 1977; Snyder and Caruccio, 1988).

During studies on how overburden was related to acid mine drainage, Caruccio (1979) noted that the weathering reactions that produced acidity appeared to be much slower than those that produced alkalinity. Because of the slow rate, this suggests that pyrite oxidation is kinetically controlled, whereas because of the faster rate, carbonate dissolution apparently is controlled through equilibrium processes. He suggested that situations that produced frequent flushing intervals of the overburden should reduce acid mine drainage.

Recently, Snyder and Caruccio (1988) tested a comparable hypothesis on two surface coal mine backfills. Through careful monitoring of the water budget, they were able to separate the shallow subsurface flow that is associated with rapid recharge from the slow, deep groundwater recharge. They found that the base flow associated with deep groundwater carries the acidity. This water is associated with the spring recharge which sustains the acid mine seeps for the rest of the water year. The results of the hydrology studies in Colorado and the Eastern U.S. correlate quite well.

MEASURES FOR UNDERSTANDING AND CONTROLLING PROBLEMS

Handling the environmental aspects of a coal mining operation can be just as costly as the actual coal extraction. Furthermore, abatement of the problem is far more troublesome than planning to avoid the problem. Consequently, prior analysis and a complete environmental design are essential to a successful operation. Within this prior analysis should be the following:

1) A study conducted of the physical groundwater hydrology of the region that estimates how the region accepts rapid changes due to precipitation events, where the water table is located, and how base flow below the water table is recharged.

2) A study conducted of the chemical hydrology that includes a complete chemical analysis of the surface waters, springs and seeps, and groundwaters. This survey definitely should include analyses of acidity, alkalinity, Fe, Mn, sulfate, and bicarbonate. For some constituents, samples should be collected and analyzed monthly so that recharge characteristics of aquifers can be assessed.

3) A study conducted of the overburden that includes the identification and estimate of the abundances of minerals. Measures of the net neutralization potential and the maximum potential acidity should be made (diPretoro and Rauch, 1988; Erickson and Hedin, 1988). If overburden analysis and the chemistry of seeps and springs point to the questionable areas in the Figure 1 matrix, laboratory simulations of weathering should be attempted. A thorough environmental plan dealing with questionable overburden is necessary.

4) A study conducted of the coal seam to determine the paleoenvironment and the form of pyrite present. Other minor minerals in the coal also should be determined.

5) A survey conducted of the regional coal mining operations. This survey should include their analysis of the overburden, paleoenvironment, forms of pyrite, chemical characteristics of seeps and springs, and environmental plans.

If acid mine drainage problems already exist, abatement is likely to require a continuous treatment process. Any measures that can be taken to reduce the volume of water requiring treatment should be taken. In this case, assessment of whether the drainage is a diffuse or conduit aquifer is useful. Surface water infiltration into conduit aquifers should be eliminated. Reducing recharge into groundwater base flow also should be attempted. After the water flowing through the deposits or spoils has been reduced as much as possible, then soil covers and alkaline amendments can be used. If drainage flow has been reduced to less than 0.1 m^3/min. (3.5 ft^3/min.) and is not subject to large fluctuations in chemistry or discharge rate, then treatment methods using wetlands (Klusman and Machemer, this volume; Wildeman and Laudon, 1989) have a much better chance of success.

REFERENCES

American Public Health Association, 1988, Standard methods for the examination of water and wastewater (16th Ed.): Washington, Method 402, p. 265-269.

Bouska, V., 1981, Geochemistry of coal: New York, Elsevier Publishing Co., 284 p.

Caruccio, F.T., 1979, The nature of acid mine drainage reactions and their relation to overburden analysis, in M.K. Wali, ed., Ecology and coal resource development: New York, Pergamon Press, p. 775-781.

Caruccio, F.T., J.C. Ferm, J. Horne, G. Geidel, and B. Baganz, 1977, Paleoenvironment of coal and its relation to drainage quality: U.S. Environmental Protection Agency, Document No. EPA-600/7-77-067, 107 p.

diPretoro, R.S., and H.W. Rauch, 1988, Use of acid-base accounts in premining prediction of acid drainage potential: A new approach from northern West Virginia, in Mine drainage and surface mine reclamation, Volume I: U.S. Bureau of Mines Information Circular IC 9183, p. 2-10.

Ehrlich, H.L., 1981, Geomicrobiology: New York, Marcel Dekker, Inc., 393 p.

Erickson, P.M., and R.S. Hedin, 1988, Evaluation of overburden analytical methods as means to predict post-mining coal mine drainage quality, in Mine drainage and surface mine reclamation, Volume I: U.S. Bureau of Mines Information Circular IC 9183, p. 11-20.

Garrels, R.M., and F.T. Mackenzie, 1971, The evolution of sedimentary rocks: New York, W.W. Norton, 397 p.

Hedin, R.S., and P.M. Erickson, 1988, Relationships between the initial geochemistry and leachate chemistry of weathering overburden samples, in Mine drainage and surface mine reclamation, Volume I: U.S. Bureau of Mines Information Circular IC 9183, p. 21-28.

Hedin, R.S., D.M. Hyman, and R.W. Hamack, 1988, Implications of sulfate-reduction and pyrite formation processes for water quality in a constructed wetland: Preliminary observation, in Mine drainage and surface mine reclamation, Volume I: U.S. Bureau of Mines Information Circular IC 9183, p. 382-388.

Horning, W.B., and C.I. Weber, 1985, Short-term methods for estimating the chronic toxicity of effluents and receiving waters to freshwater organisms: U.S. Environmental Protection Agency, Document No. EPA/600/4-85/014, 162 p.

Jacobsen, R.L., and D. Langmuir, 1974, Controls on the quality variations of some carbonate spring waters: Journal of Hydrology, v. 23, p. 247-265.

Krauskopf, K.B., 1979, Introduction to geochemistry (2nd Ed.): New York, McGraw-Hill Co., 617 p.

Lindsay, W.L., 1979, Chemical equilibria in soils: New York, John Wiley & Sons, 449 p.

Manahan, S.E., 1984, Environmental chemistry (4th Ed.): Boston, Willard Grant Press, 612 p.

Peltier, W.H., and C.I. Weber, 1985, Methods for measuring the acute toxicity of effluents to freshwater and marine organisms (3rd Ed.): U.S. Environmental Protection Agency, Document No. EPA/600/4-85/013, 216 p.

Prodan, P.F., L.M. Mele, and J.P. Schubert, 1982, Runoff water quality and hydrology at coal refuse disposal sites in Southern Illinois, in D.H. Graves, ed., Proceedings of the Symposium on Surface Mining, Hydrology, Sedimentology, and Reclamation, Louisville, Kentucky, December, 1982: Louisville, University of Kentucky, p. 57-63.

Shuster, E.T., and W.B. White, 1971, Seasonal fluctuations in the chemistry of limestone springs: A possible means for characterizing carbonate aquifers: Journal of Hydrology, v. 14, p. 93-128.

Sims, P.K., M.A. Drake, and E.W. Tooker, 1963, Geology and ore deposits of the Central City mining district, Gilpin County, Colorado: U.S. Geological Survey Professional Paper 359, 231 p.

Snyder, D.T., and F.T. Caruccio, 1988, The partitioning of flow components of acidic seeps from surface coal mines and the identification of acid producing horizons within the backfill, in Mine drainage and surface mine reclamation, Volume I: U.S. Bureau of Mines Information Circular IC 9183, p. 59-66.

Stumm, W., and J.J. Morgan, 1981, Aquatic chemistry (2nd Ed.): New York, John Wiley & Sons, 780 p.

Taylor, B.E., M.C. Wheeler, and D.K. Nordstrom, 1984, Stable isotope geochemistry of acid mine drainage: Experimental oxidation of pyrite: Geochimica et Cosmochimica Acta, v. 48, p. 269-678.

U.S. Environmental Protection Agency, 1982, Development document for effluent limitations guidelines and standards for coal mining: Washington, Document No. EPA 440/1-82/057, 660 p.

U.S. Government Printing Office, 1988, Code of federal regulations, Title 40, Protection of the environment: Washington, Parts 100-149, p. 530-531 and p. 608.

Valkovic, V., 1983, Trace elements in coal, Vol. 1: Boca Raton, Florida, CRC Press Inc., 210 p.

van Everdingen, R.O., and H.R. Krouse, 1988, Interpretation of isotopic compositions of dissolved sulfates in acid mine drainage, in Mine drainage and surface mine reclamation, Volume I: U.S. Bureau of Mines Information Circular IC 9183, p. 147-156

Watzlaf, G.R., 1988, Chemical stability of manganese and other metals in acid mine drainage sludge, in Mine drainage and surface mine reclamation, Volume I: U.S. Bureau of Mines Information Circular IC 9183, p. 83-90.

Wildeman T.R., 1981, A water handbook of metal mining operations: Fort Collins, Colorado Water Resources Research Institute, Report 113, 84 p.

Wildeman, T.R., 1983, Chemistry of the Argo Tunnel: Quarterly of the Colorado School of Mines, v. 78, no. 4, p. 31-37.

Wildeman, T.R., D.L. Cain, and A.J. Ramirez, 1974, The relation between water chemistry and mineral zonation in the Central City mining district, Colorado, in Proceedings of the National Symposium on Water Resources Problems Related to Mining, Golden, Colorado, June, 1974: American Water Resources Association, Proceedings No. 18, p. 219-229.

Wildeman, T.R., and L.S. Laudon, 1989, The use of wetlands for treatment of environmental problems in mining: Non-coal mining applications, in D.H. Hammer, ed., Proceedings of the International Conference on Constructed Wetlands for Wastewater Treatment, Chattanooga, Tennessee, June, 1988: Ann Arbor, Michigan, Lewis Publishing, p. 221-231.

REVIEW OF PROCESSES OPERATING IN CONSTRUCTED WETLANDS TO ATTENUATE ACID MINE DRAINAGE

Ronald W. Klusman and Steven D. Machemer
Department of Chemistry and Geochemistry
Colorado School of Mines
Golden, CO 80401

ABSTRACT

Acid drainage from coal mines occurs where oxygenated ground water contacts materials containing pyrite, which in turn pollutes streams with acidic water and dissolved metals. There are a number of natural processes which operate to attenuate acid mine drainage. This paper reviews these processes, including exchange of metals by organic-rich substrates occurring in constructed wetlands, sulfate reduction with precipitation of iron and other sulfides, precipitation of ferric and manganese oxyhydroxides, adsorption of metals by ferric and manganese oxyhydroxides, and metal uptake by living plants.

These processes can be enhanced by the use of constructed wetlands to treat acid drainage. Several case histories are presented which illustrate both successes and failures in the use of wetlands to treat acid coal mine drainage. These examples are taken from locations in both the Eastern and Western United States.

INTRODUCTION

Acid mine drainage is produced when there is increased access of oxygen and water to materials containing pyrite. This access usually is provided by the underground and surface mining of coal and other sulfide-containing materials. The formation of pyrite in a peat swamp is due to the diagenetic conversion of organic sulfur (S) to sulfide (S^{2-}) ion and the reduction of sulfate (SO_4^{2-}) in connate water of the peat to S^{2-} ion by microorganisms. Any iron (Fe) which is present in the peat as detritus or adsorbed by the peat is reduced to the ferrous form (Fe^{2+}), reacting with S^{2-} to form iron sulfides which are diagenetically converted to pyrite. Sparling and Hennick (1974) and Altschuler et al. (1983) review the process of pyrite formation in a peat swamp. The coalification of the peat is a longer-term diagenetic process, extending beyond the formation of iron sulfides and other sulfides.

The oxidation of pyrite in coal is a complex process which frequently is simplified to two steps. The oxidation of the S^{2-} to SO_4^{2-} is followed by the oxidation of Fe from Fe^{2+} to the ferric form (Fe^{3+}). The second step is accelerated greatly by the apparent ubiquitous presence of particular genera of microorganisms, but still can be a relatively slow process. The result is acidic drainage from open adits, gob piles, and other locations where there are increased rates of oxidation of pyritic materials. Wildeman (this volume) reviews in detail the processes involved in the oxidation of pyrite.

There are a number of natural processes which operate to attenuate the negative impacts of acid mine drainage. This paper examines these processes and how they are leading to the development of new technology which enhances the natural removal of acidity and metals from acid mine drainage. The new technology which is being investigated is the use of constructed wetlands to provide passive treatment of acid mine drainage. Some engineering considerations are necessary in order to modify the hydrologic characteristics of the substrate, which in turn enhance and optimize one or more of the processes described in subsequent sections.

These processes can be enhanced by the construction of an engineered artificial wetland with the objective of providing relatively long-term, low-cost treatment of mine drainages and seeps. The wetlands constructed to date have not been operating sufficiently long to know what the expected life might be. The natural processes operating in a wetland include the following:

1) Exchange of metals by an organic-rich substrate, which is typically a blend of natural and composted materials,

2) Sulfate reduction with precipitation of iron sulfides and other sulfides,

3) Precipitation of ferric and manganese (Mn) hydroxides,

4) Adsorption of metals by ferric oxyhydroxides,

5) Metal uptake by living plants.

These processes are reviewed in some detail, according to their relative importance for the improvement of water quality from mine drainages. The final section reviews data from selected case histories of constructed and natural wetland treatment of acid mine drainage.

PROCESSES INVOLVED IN ACIDITY REDUCTION AND METAL REMOVAL

Exchange Of Metals By Organic Matter

Exchange of dissolved metals by a substrate such as peat or organic-rich blends is likely a mechanism whereby the metals are temporarily retained in a wetland, thereby allowing other processes to operate. The organic matter in a wetland is high in humic and fulvic acids which remove metals by an acid-base reaction. Humic and fulvic acids are poorly characterized natural materials which currently are separated from peats and soils by extraction with a base. The separation and characterization of these complex materials are beyond the scope of this discussion. Aiken et al. (1985) discuss these substances in detail. For purposes of discussion, both groups of compounds can be considered to have several carboxyl and phenolic functional groups attached to a larger organic molecule. The exchange process primarily is controlled by the carboxylic functional groups and can be represented by the following reactions:

$$RCOOH \leftrightarrow RCOO^- + H^+ \qquad [1]$$

$$RCOO^- + M^+ \leftrightarrow RCOOM \qquad [2]$$

The R- represents the larger part of the humic or fulvic acid which does not actively participate in the acid-base reaction, and the exact composition is not important. The carboxylic acid functional group is represented by -COOH and dissociates to the carboxyl group, $-COO^-$, and free protons (reaction 1). The amount of this dissociation is a function of pH, with the carboxylic acid favored at low pH (i.e., conditions of higher acidity). This dissociation is quantified by an experimentally determined equilibrium constant as

$$K_{\alpha} = \frac{[RCOO^-][H^+]}{[RCOOH]} \qquad [3]$$

where the terms in brackets represent the activity or, approximately, the concentration of the respective chemical species. The term negative logarithm of $[H^+]$ is the pH of the aqueous solution containing the humic or fulvic acid. The dissociated acid or conjugate base represented by $RCOO^-$ can bind to dissolved metals as in reaction 2. The strength of this metal-organic complex is represented by a stability constant or binding constant (K_{β}). Morel (1983) summarizes some data on the binding of metal cations by humic materials.

As an acid mine drainage encounters aqueous species such as bicarbonate, which provide alkalinity, the pH rises, favoring dissociation of the carboxylic acid group and shifting reaction 1 to the right. The $RCOO^-$ group then binds the metal M, forming the carboxylic acid complex as represented by reaction 2. The pH of water associated with natural wetlands tends to be low, indicating the dominance of reaction 1 in controlling the acid-base chemistry of this environment. This is in contrast with the acid-base chemistry of most natural waters being controlled by bicarbonate.

The amount of dissociation for reaction 1 can be estimated using the Henderson-Hasselbach equation (Thurman, 1985):

$$pH = pK_{\alpha} + \log\frac{[A^-]}{[HA]} \qquad [4]$$

The term $[A^-]$ represents the activity or, approximately, the concentration of the dissociated carboxylic acid group, and [HA] represents activity of the protonated carboxylic acid group. The pK_{α} is the negative logarithm of the K_{α} value and, for humic materials, averages approximately 4.2. For a wetland with a pH of 4.0, equation 4 becomes

$$4.0 = 4.2 + \log\frac{[A^-]}{[HA]} \qquad [5]$$

Rearranging and clearing the logarithm gives:

$$0.63 = \frac{[A^-]}{[HA]} \qquad [6]$$

The proportion of [HA] is 1-$[A^-]$, and solving for $[A^-]$ gives 0.39. This means 39% of the carboxylic acid groups are dissociated and available to complex metals from the aqueous phase. This also means that simple dilution of an acidic water in a wetland does not have a major effect on the pH. Dilution causes reaction 1 to shift to the right, producing more free protons or acidity.

Some natural wetlands such as ombrotrophic fens receive virtually all of their mineral matter from precipitation. Wetlands which receive dissolved cations primarily from the surrounding drainage basin are characterized as minerotrophic. Even in minerotrophic wetlands, the dissolved cations tend to be relatively low in concentration. If drainage from a coal mine enters a wetland, there will be a tendency for exchange of protons from the humic and fulvic acids for metal cations from the drainage, as represented by reactions 1 and 2. The mine drainage nearly always contains higher concentrations of calcium (Ca), magnesium (Mg), Fe, and Mn than the wetland, thereby driving reaction 2 to the right. The magnitude of the exchange is a function of the activity, or effective concentration, of the dissolved metals and the pH of the wetland.

Wieder and Lang (1986) measured various fractions, including he exchangeable Fe, aluminum (Al), and Mn in four wetlands, using a sodium pyrophosphate extraction. The wetlands ranged from ombrotrophic (Red Lake) to minerotrophic (Big Run Bog), and Buckles Bog was intermediate in composition. Tub Run Bog was minerotrophic and impacted by mine drainage for a period of 40 yrs. Table 1 indicates the pyrophosphate-extractable portion was the largest component of the metals retained by the wetlands.

The retention of Fe by ion exchange with the peat can be characterized by experimentally adding Fe to peat samples and using a Langmuir-type expression to describe the results. The peat samples were rinsed with distilled water, followed by sequential addition of increasing concentrations of Fe^{2+}. The form of the Langmuir expression used by Wieder and Lang (1986) is given by

$$Fe\ adsorption = \frac{YMAX * X}{K + X} \qquad [7]$$

where Fe adsorption is the cumulative Fe^{2+} adsorbed by the dry peat sample, YMAX is the maximum Fe^{2+} adsorbed, K is the Fe^{2+} concentration at which 50% of YMAX is achieved, and X is the Fe^{2+} concentration of the filtrate at each step in the sequential iron-loading process on a given peat sample.

Figure 1 from Wieder and Lang (1986) is the isotherm for the Fe^{2+} adsorption by peat samples from three of the wetlands. Table 2 lists the K values and YMAX for the experiments. It must be emphasized that ion

	Red Lake, MN	**Buckle's Bog, MD**	**Big Run Bog, WV**	**Tub Run Bog, WV**
Sample size	5	5	10	38
Fe (μmol/g dry mass)				
Total (HCL extractable)	45±8	101±17	300±46	869±96
Pyrophosphate extractable	23±6	47±10	269±44	484±44
Oxalate extractable	8.4±1.1	17±2	45±9	299±42
Dithionite extractable	8.7±1.8	24±4	17±2	186±29
FeS_2	2.4±0.2	2.1±0.4	6.1±0.5	7.6±0.8
FeS	1.0±0.3	0.3±0.04	3.2±1.0	3.1±0.3
Al (μmol/g dry mass)				
Total (HCL extractable)	65±11	109±16	159±26	243±12
Pyrophosphate extractable	35±5	61±8	140±24	109±12
Oxalate extractable	6.2±1.5	11.2±0.9	12.4±2.2	31.9±1.9
Dithionite extractable	2.9±1.3	7.4±2.0	9.3±0.7	20.4±1.6
Mn (μmol/g dry mass)				
Total (HCL extractable)	1.5±0.3	5.3±0.8	2.5±0.4	4.9±2.0
Pyrophosphate extractable	1.3±0.2	4.7±0.7	2.2±0.3	2.5±0.8
Oxalate extractable	0.4±0.04	0.7±0.2	0.2±0.05	1.2±0.6
Dithionite extractable	0.06±0.03	0.6±0.03	0.2±0.02	1.1±0.3
S (μmol/g dry mass)				
Total (HCL extractable)	48.6±2.5	87.2±5.4	134±8	108±6
Organic	41.2±2.1	77.2±5.4	114±8	83.9±4.7
FeS_2	4.9±0.5	4.3±0.8	12.1±1.0	15.1±1.5
FeS_{2-}	1.0±0.3	0.3±0.04	3.2±1.0	3.1±0.3
SO_4^{2-}	1.5±0.3	5.5±0.6	5.2±0.5	5.8±0.5
Organic matter (%)	84.1±1.4	86.3±2.3	89.6±1.4	55.9±2.8

Table 1. *Iron, Al, Mn, and S chemistry and organic matter concentration in surface (0-20 cm (0-8 in.) deep) peat; means ± standard errors (modified from Wieder and Lang, 1986).*

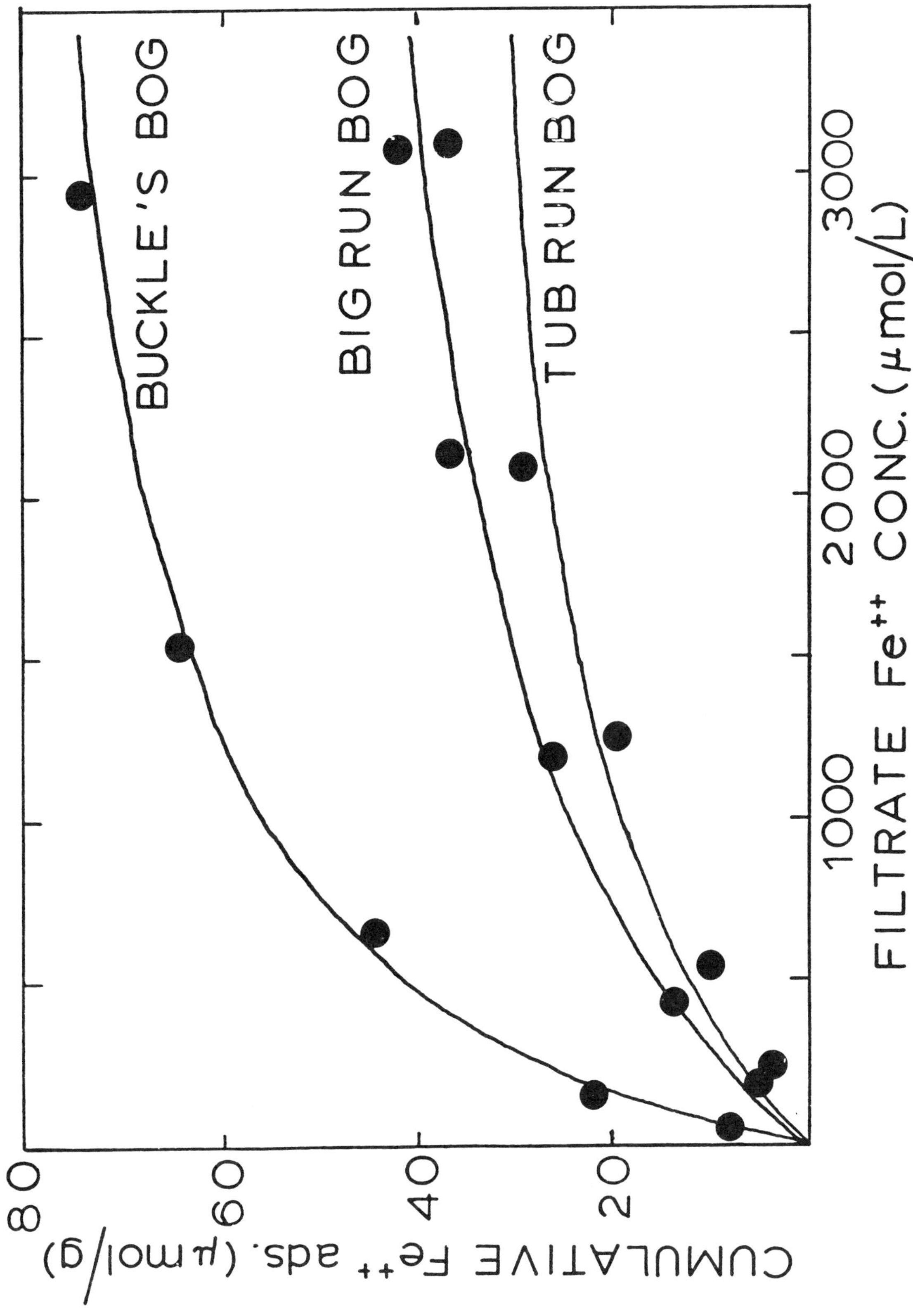

Figure 1. *Ferrous Iron adsorption by peat samples from three bogs. The plotted points are means ± standard errors.*

Site	K	YMAX	R^2
Buckle's Bog	564 ± 152^a	88.1 ± 7.3^a	96
Big Run Bog	1257 ± 563^a	57.1 ± 10.4^b	85
Tub Run Bog	1115 ± 543^a	42.3 ± 8.1^b	60

Table 2. *Estimated Langmuir parameters for Fe^{2+} adsorption by Buckle's Bog (n=5), Big Run Bog (n=10), and Tub Run Bog (n=32) peat. K (μmol Fe^{2+}/L) and YMAX (μmol Fe^{2+}/g dry mass) values are means ± asymptotic standard errors (modified from Wieder and Lang, 1986). K and YMAX values with the same letter superscripted do not differ significantly (pairwise t-tests using asymptotic standard errors of the parameter estimates; $p \leq 0.05$). R^2 values indicate the percentage of the uncorrected sum of squares explained by the regression.*

exchange is a dynamic process which is a function of the activity or effective concentration of the individual cations. A change in the composition of the mine drainage will shift equilibrium within the wetland and cause adsorption of ions of higher activity and desorption of ions of lower activity. Wieder et al. (1987b) demonstrated that adsorption of Fe^{2+} by a peat was matched by the desorption of potassium (K), sodium (Na), Ca, Mg, and protons (Figure 2). Wieder et al. (1988) demonstrated a displacement of Mn and Ca from organic matter by an increase in dissolved Al in a drainage water. Laboratory studies of model wetlands indicate the substrate can be a source of Al, Mn, Ca, and Mg, but a sink for Fe^{2+} (Wieder et al., 1990). There is also a direct relationship between the magnitude of the exchange capacity and the percentage of organic matter in the wetland (Wieder, 1985).

The important function of exchange reactions from the standpoint of treatment of acid mine drainage is the temporary retention of the metal cations on the substrate of the wetland. This increase in residence time allows other microbially mediated metal-removal processes to operate. The attainment of equilibrium is limited by time-dependent kinetic factors determining the rates of reaction. The retention of metals by exchange increases the degree or extent of a reaction.

The organic-rich substrate also produces the reducing conditions necessary for another potential removal process, that of SO_4^{2-} reduction. Without additional metal removal and acidity reduction processes beyond that of cation exchange, the capacity of a wetland soon will be saturated and passive treatment will no longer be effective.

Microbial Sulfate Reduction

Sulfate reduction is a natural process which can occur under reducing conditions present below the surface of a natural or constructed wetland. This process has not been taken advantage of in the first generation of constructed wetlands. The process offers considerable potential for metal removal from acid drainage and is being incorporated into some newer constructed wetlands (Hedin et al., 1988; Hedin and Nairn, 1990; McIntire and Edenborn, 1990; Wildeman and Laudon, 1989; Wildeman et al., 1990).

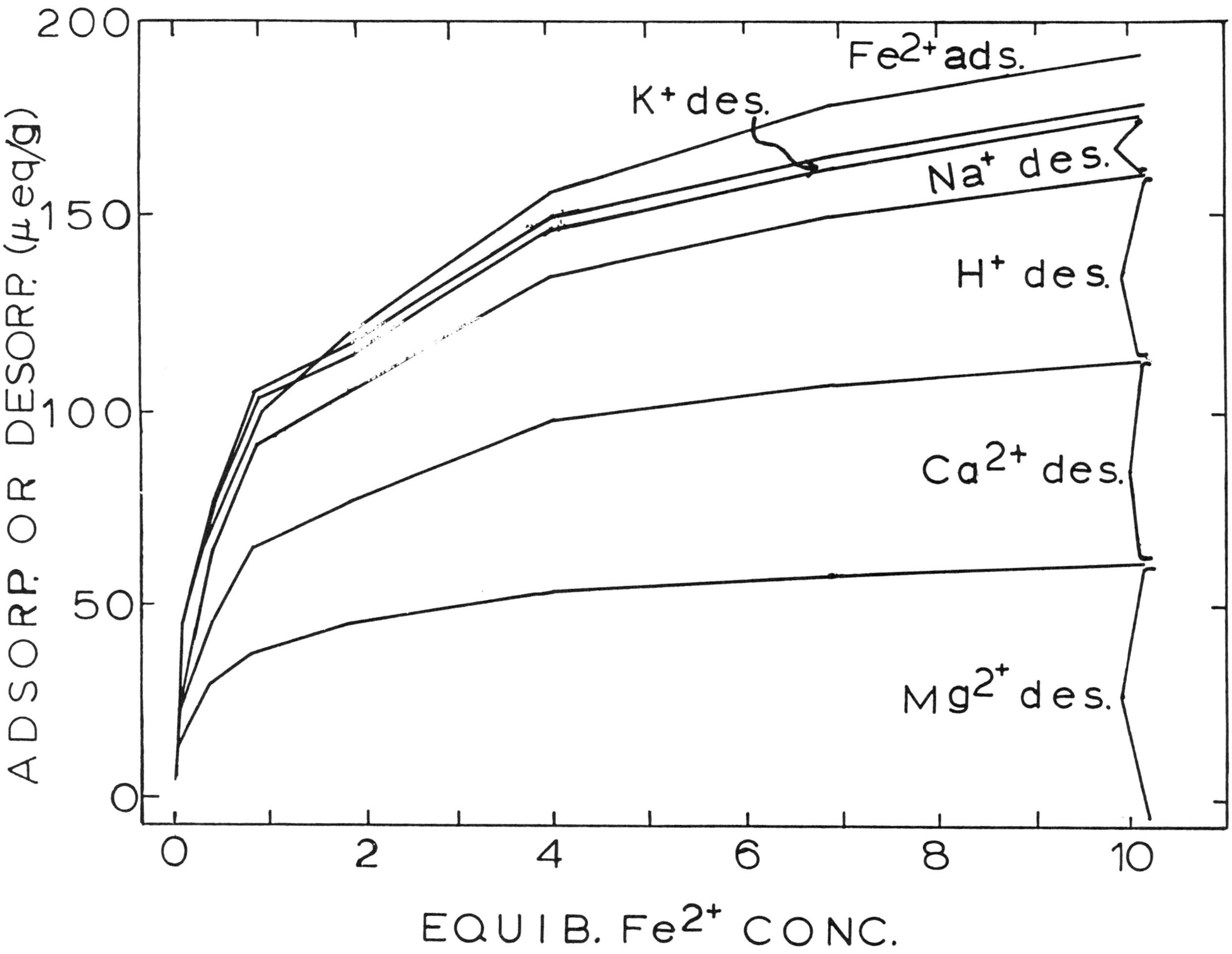

Figure 2. *Adsorption of Fe^{2+} onto Sphagnum peat by cation exchange and the corresponding desorption of Mg^{2+}, Ca^{2+}, H^{+}, Na^{+}, and K^{+} (modified from Wieder and Lang, 1986).*

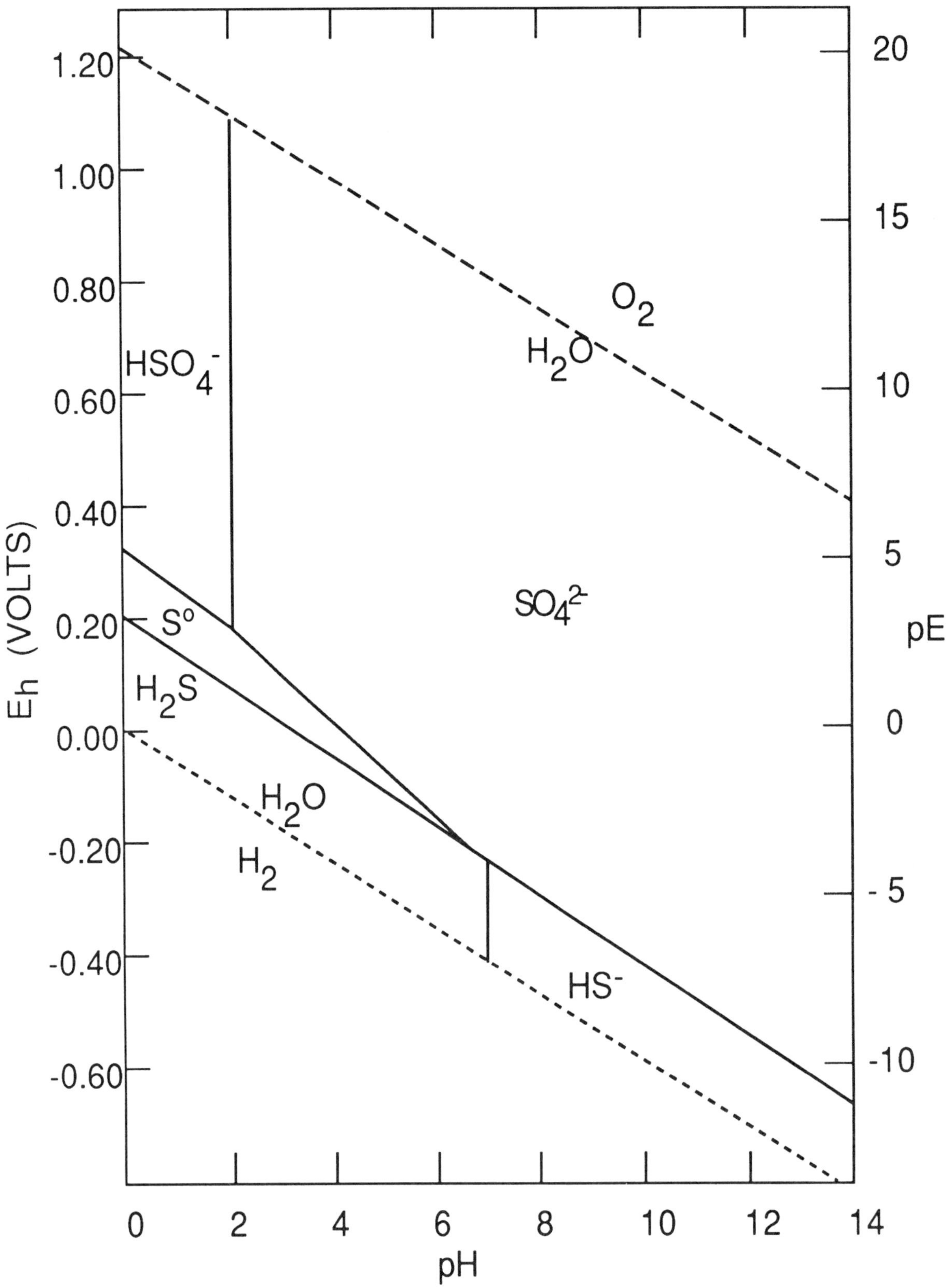

Figure 3. *Aqueous S speciation, as a function of Eh and pH, at 960 mg/L (10^{-2} M) as SO_4^{2-}, 25°C, and 1 atm. pressure (from Decker et al., 1987).*

The distribution of aqueous S species with redox potential, or Eh(pE), and pH can be calculated from thermodynamic data (Stumm and Morgan, 1981). Figure 3 is the Eh-pH diagram for aqueous S species with a total dissolved S concentration equal to 10^{-2}M. This corresponds to a SO_4^{2-} concentration which is typical for an acid mine drainage. Under oxidizing conditions over a wide range of pH, SO_4^{2-} is the dominant species. The presence of decaying organic matter in the organic substrate of a wetland rapidly depletes oxygen, creating reducing conditions or a low pE favorable for the production of hydrogen sulfide. When molecular oxygen is not available, microbes use a series of available sources in turn and, in the process, create more reducing conditions.

The more important microbial reactions under reducing conditions in a wetland are dissimilatory SO_4^{2-} reduction and methanation. Organic matter is a complex material but can be represented for chemical reaction purposes as "CH_2O". Using this simplified representation, SO_4^{2-} reduction is coupled with organic matter oxidation:

$$SO_4^{2-} + 2\text{"}CH_2O\text{"} + 2H^+ \rightarrow H_2S + 2H_2O + 2CO_2 \qquad [8]$$

This reaction occurs at a pH<7.0. At a pH>7.0, the reaction can be written as

$$SO_4^{2-} + 2\text{"}CH_2O\text{"} + H^+ \rightarrow HS^- + 2H_2O + 2CO_2 \qquad [9]$$

An alternative way of writing reaction 9 is

$$SO_4^{2-} + 2\text{"}CH_2O\text{"} \rightarrow HS^- + 2HCO_3^- + H^+ \qquad [10]$$

Methanation can be represented by

$$2\text{"}CH_2O\text{"} \rightarrow CH_4 + CO_2 \qquad [11]$$

An alternative form is

$$2\text{"}CH_2O\text{"} + H_2O \rightarrow CH_4 + HCO_3^- + H^+ \qquad [12]$$

The net result of microbial dissimilatory SO_4^{2-} reduction is a decrease in SO_4^{2-} and an increase in bicarbonate in the interstitial waters. This produces hydrogen sulfide and the methanation reaction produces methane which can be lost to the atmosphere. The rotten egg odor observed when the bottom sediments in a wetland are disturbed is evidence of the sulfate-reduction reaction. The loss of hydrogen sulfide to the atmosphere also represents a loss of acidity from the water and the production of bicarbonate, which in turn can neutralize additional acidity. The methanation reaction occurs in similar biogeochemical environments, but does not consume acidity.

Sulfate reduction by microorganisms has the potential of being a significant metal-removal and acidity-reduction process operating in a wetland receiving water high in SO_4^{2-}. Anaerobic conditions are required for the reduction of SO_4^{2-} by microorganisms, which are readily attained at depths of a few inches and below in both natural and constructed wetlands. The process consumes organic matter and SO_4^{2-}, and in turn produces aqueous carbon dioxide and hydrogen sulfide which partially can be lost to the atmosphere. It must be emphasized that in order for significant SO_4^{2-} reduction to occur, the high-sulfate acid mine drainage must flow through the anaerobic portion of the wetland. The typical natural wetland and most constructed wetlands do not have sufficient permeability to fully take advantage of this process.

The hydrogen sulfide also can precipitate iron sulfides, which gradually convert to pyrite. Berner (1971) suggests that pyrite does not form directly, but rather that metastable forms such as greigite (Fe_3S_4) and mackinawite (FeS) are initially formed. The reaction of Fe^{2+} with bisulfide ion in the reduced zone of a wetland can be represented by

$$Fe^{2+} + HS^{-} \rightarrow FeS + H^{+} \quad [13]$$

The conversion to pyrite can be represented by the reaction

$$FeS + S^{o} \rightarrow FeS_2 \quad [14]$$

This sequence has been demonstrated experimentally and produces the framboidal pyrite found in coals (Altschuler et al., 1983). Berner (1971) addresses the rate of conversion of soluble Fe^{2+} to pyrite in a marine environment. Tuttle (1988) and Laudon (1988) discuss the forms of sulfides found in anaerobic environments in more detail. These iron sulfides can coprecipitate other metal sulfides of significant environmental concern. Sulfate reduction, in the simplest aspects, is a reversal of the process which created the initial acidity.

Figure 4 is an Eh-pH diagram for Fe, modified from Hem (1970) to reflect the composition of a moderately acidic mine drainage. The diagram demonstrates that, under the reducing or anaerobic conditions found in the substrate of a wetland, pyrite will be the thermodynamically stable phase expected to form.

Microbial SO_4^{2-} reduction has been studied for a substantial period of time in the marine environment, which is high in SO_4^{2-}, and is summarized by Goldhaber and Kaplan (1975) and Jorgenson (1977, 1982, 1983). Until recently, freshwater environments have not been considered high enough in dissolved SO_4^{2-} to attract interest in studying SO_4^{2-} reduction. The introduction of acid mine drainage into wetlands by chance, and interest in the effect of wetlands on acid deposition from the atmosphere, has spurred research into the process. Just as acid mine drainage has been modified by wetlands, the effects of acid rain can be partially ameliorated by flowing over wetlands.

An early study of this process was by Tuttle et al. (1969) who investigated unusual changes in water chemistry in an acidic stream as it passed through a dam made by wood wastes from a sawmill. They found significant decreases in SO_4^{2-} and Fe, and increases in pH as the water seeped through the sawdust dam. Microorganisms of the genera Desulfovibrio and Desulfotomaculum were isolated from the sawdust and from the drainage below the dam.

Smith and Klug (1981) studied SO_4^{2-} reduction in an eutrophic lake that was not impacted by acid mine drainage. Wieder and Lang (1986) and Wieder et al. (1987a, 1987b) have studied the process of dissimilatory SO_4^{2-} reduction in non-impacted and acid-impacted wetlands of the Appalachian region. Herlihy and Mills (1985) and Herlihy et al. (1987) have studied the SO_4^{2-} reduction process in a Virginia lake which is receiving acid mine drainage.

Precipitation Of Ferric And Manganese Oxyhydroxides

The oxidation of Fe is the second, slower step in the production of acid mine drainage. In many cases, it is not complete at the time the water emerges from the source, but continues and reaches completion in the stream receiving the drainage. The result is the production of ferric hydroxide, or "yellow boy", with undesirable effects on the biota of the stream. This process is catalyzed by bacteria of the genera Thiobacillus, Leptothrix, and Metallogenium. This process occurs in a thin aerobic zone which can exist in both natural and constructed wetlands. The microbial growth is enhanced at solid-liquid interfaces such as rocks or plant parts below the water surface. Taylor et al. (1984) studied in great detail the formation of iron oxyhydroxides

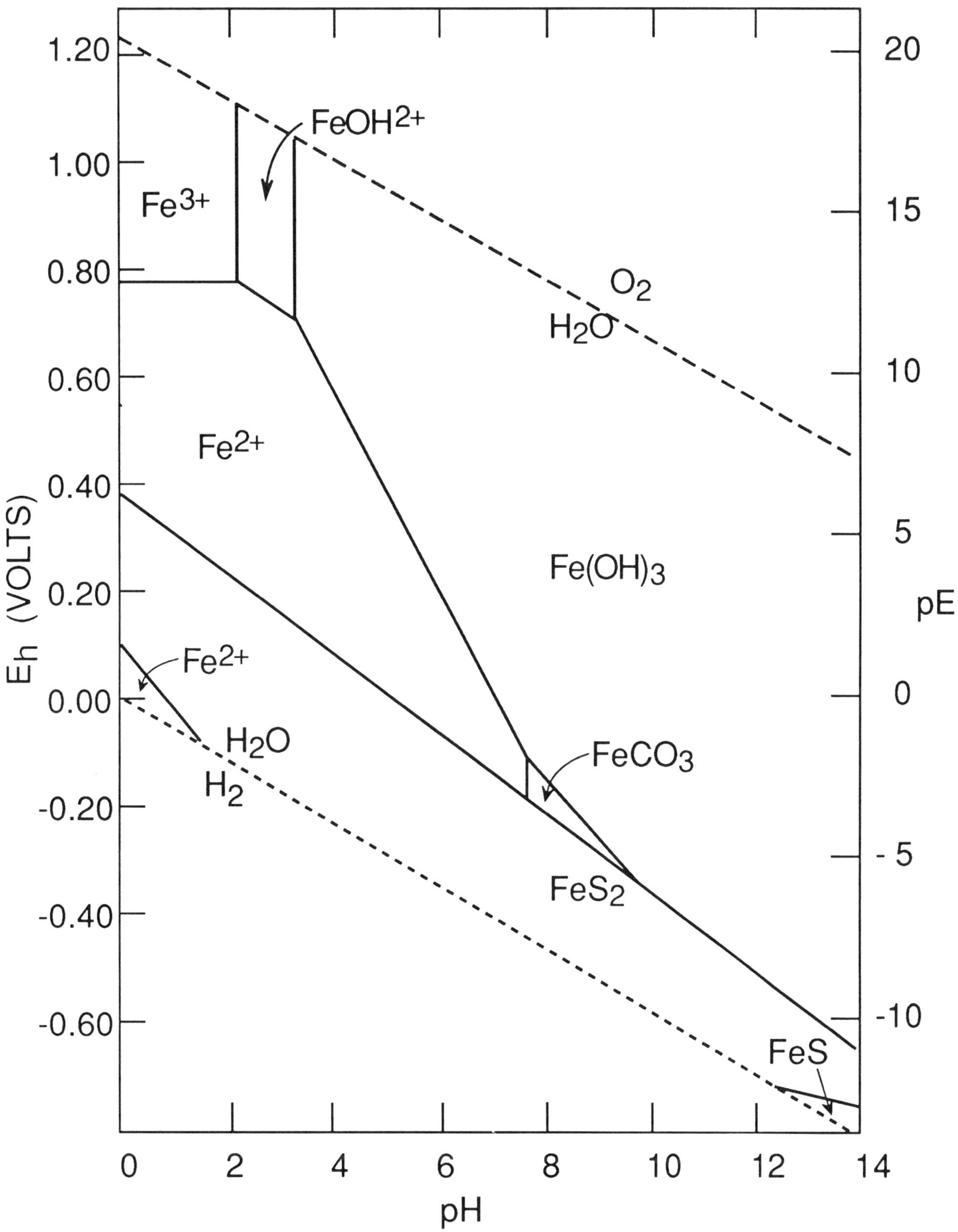

Figure 4. *Eh and pH diagram for solid and dissolved forms of Fe at 25°C and 1 atm pressure. Activity of S species = 960 mg/L (10^{-2} M) as SO_4^{2-}, carbon dioxide species = 6.1 mg/L (10^{-4} M) as bicarbonate, and dissolved Fe = 5.6 mg/L (10^{-4} M).*

or Fe plaque on the roots of Typha latifolia (cattail) in acid mine drainage. By slowing the water flow in a wetland, the bacteria have time to complete the oxidation of Fe and its subsequent removal. The biogeochemistry of Fe and Mn has been reviewed by Singer and Stumm (1970), Lundgren and Dean (1979), Marshall (1979), Nealson (1983), Gwynfryn-Jones (1986), and Ehrlich (1990).

Manganese removal by oxidation and precipitation is more difficult. Figure 5 is an Eh-pH diagram for Mn which indicates the larger field of stability for the soluble divalent Mn than for the Fe^{2+} on Figure 4. Coal mine operators using active chemical treatment to raise the pH and reduce the Fe and Mn commonly use lime or sodium hydroxide, followed by settling in a pond. Large amounts of alkali are needed to raise the pH to excessively high levels in order to accelerate the oxidation and precipitation of Mn. The reactions for Mn can be represented by (from Watzlaf, 1985)

$$2Mn^{2+} + O_2 + 4H^+ \rightarrow 2Mn^{4+} + 2H_2O \qquad [15]$$

$$Mn^{4+} + 2H_2O \rightarrow MnO_2 + 4H^+ \qquad [16]$$

The regulatory limits of 2.0 mg Mn/L for an average of 30 consecutive days usually can be met only by addition of large excesses of alkali. This has the unfortunate consequence of raising the pH too high, requiring an additional expense of subsequent acid addition to bring the effluent into the pH range of 6 to 9.

The rate of oxidation of Mn^{2+} is pH-dependent. Marshall (1979) reports that oxygenation of Mn requires several hours even at a pH greater than 8.5. Figure 6 illustrates the strong pH dependence of oxidation rate for both Fe and Mn. Iron oxidation occurs at an acceptable rate at pH values near neutrality, making the regulatory limit relatively easy to reach in a settling pond. The rate of oxidation for Mn is considerably slower, and cannot be brought into an acceptable range without raising the pH above 9.0. The net result frequently is coprecipitation of partially oxidized Mn onto iron oxyhydroxide.

It should be noted that SO_4^{2-} reduction and Fe and Mn oxidation are processes that are mutually incompatible. Sulfate reduction requires anaerobic conditions, whereas Fe and Mn oxidations require aerobic conditions. In a natural wetland, the zone of oxidation is normally a relatively thin layer on the surface. This implies there is a boundary zone in which the mutually exclusive processes may be counteracting one another. Ehrlich (1987) even reports the reduction and solubilization of manganese oxide by anaerobic respiration of microbes. The engineering aspects of designing a staged wetland which takes advantage of both reduction and oxidation processes have not been worked out yet. Most researchers feel staging is necessary, and a single wetland will not be as effective in reducing the acidity and removing both Fe and Mn to the degree necessary.

Adsorption Of Metals By Ferric Oxyhydroxides

The adsorption of metals by solids is controlled by the nature of the surface of the solids involved. Electrical double-layer theory initially was proposed by Helmholtz in the late 1800s and has undergone extensive modification which continues today. Figure 7 is a simplified view of a solid surface which indicates a negative charge on the surface of the solid which in turn attracts cations. The distribution of ions within the electrical double layer is not the same as that of the free solution. The tendency for binding of the ion to the surface is a function of the charge and size of the hydrated ion. Ions of higher positive charge and smaller size tend to be bound more tightly to a negatively charged surface. Overall, the solid-water system must be electrically neutral, and a statistical increase of anions at slightly greater distances is expected, in order to balance the charge of the cations adsorbed against a negative surface.

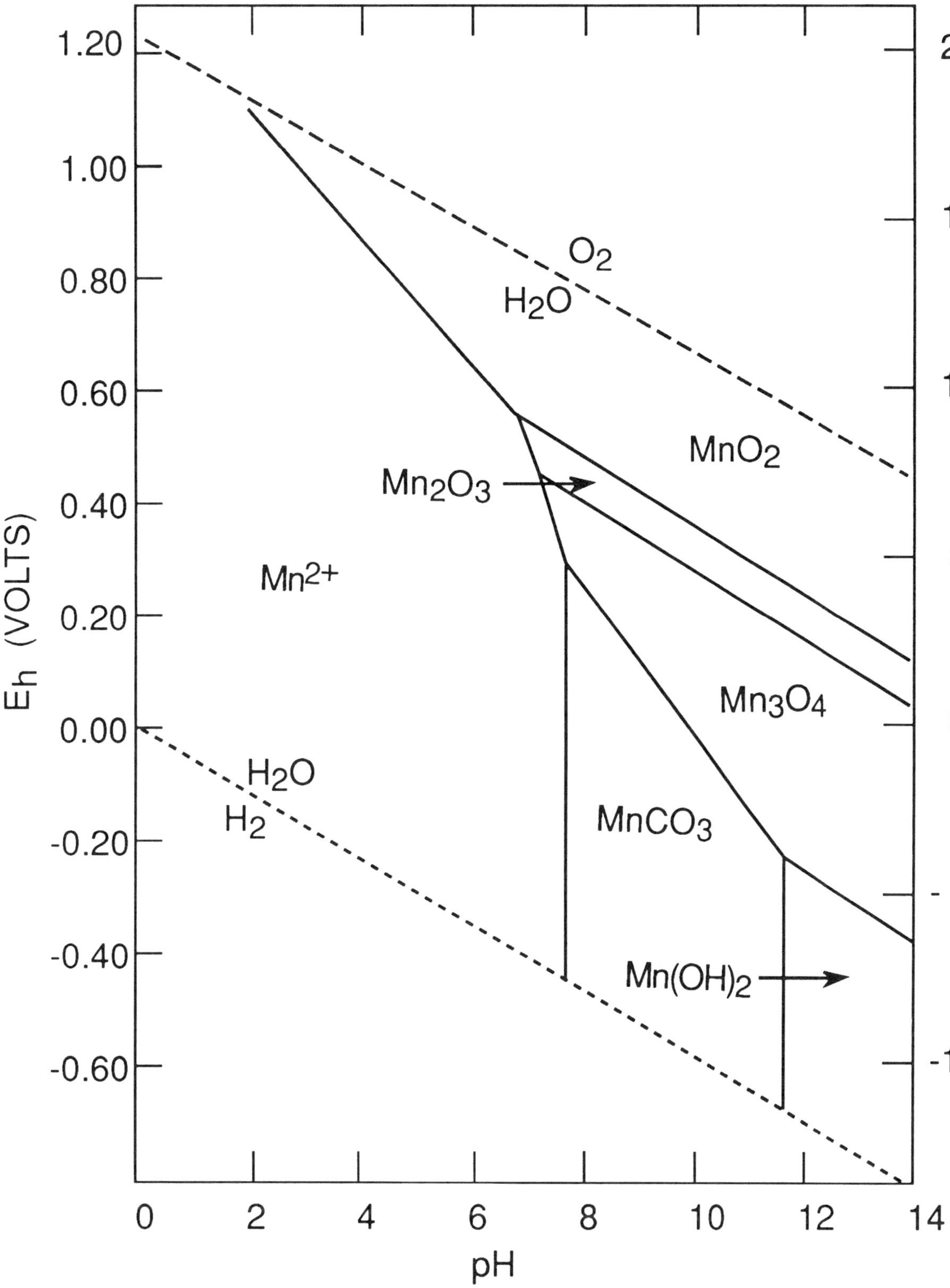

Figure 5. *Eh-pH diagram for Mn = 5.5 mg/L (10^{-4} M) and carbon dioxide species = 6.1 mg/L (10^{-4} M) as bicarbonate (modified from Hem, 1970).*

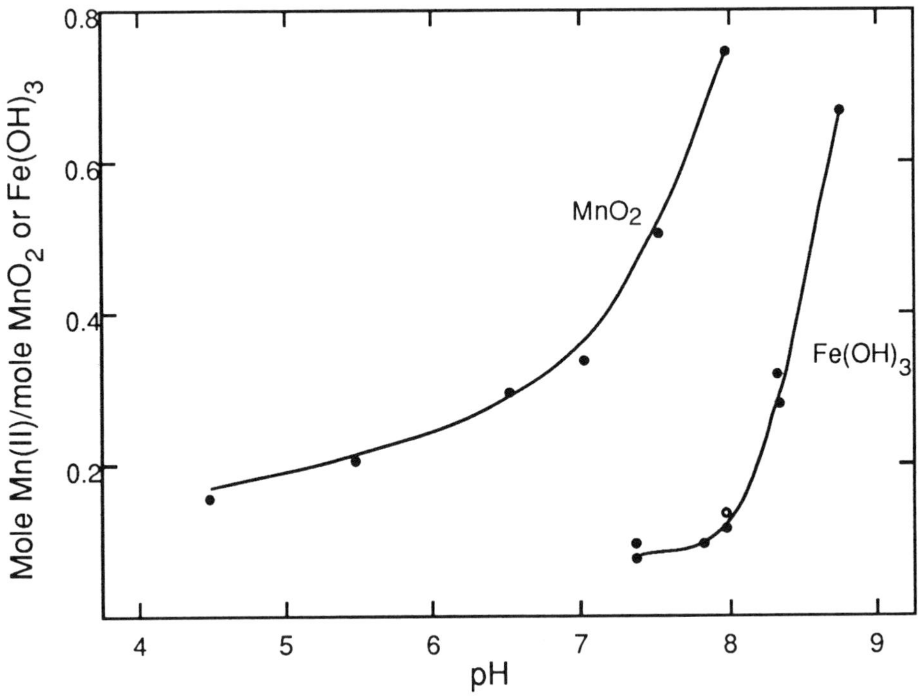

Figure 6. *Removal of Mn^{2+} and Fe^{3+} by oxygenation at different pH values in bicarbonate solutions (from Marshall, 1979).*

The sign and the magnitude of the surface charge is a function of pH. At high pH, a solid tends to appear more negative, attracting cations. At lower pH, the surface can become positive and will attract anions. The pH where the surface appears neutral is called the "zero point of charge" (ZPC) or "isoelectric point". For silicates, the ZPC is in a pH range of 2-4, but for ferric oxyhydroxides the ZPC is in the pH range of 6-8. As a result, ferric oxyhydroxides do not adsorb cations significantly until the pH of the system approaches neutrality. Stumm and Morgan (1981) discuss electrical double-layer theory in detail.

The adsorption process is dynamic to some extent. Freshly precipitated ferric oxyhydroxides readily adsorb cations such as Ca or Mn, but displacement is possible if the chemistry of the aqueous phase changes significantly. The adsorption process and retention of cations is particularly sensitive to a decrease in pH. If the flow of the mine drainage changes significantly with season, the ability for effective removal of cations by a wetland also will fluctuate. Aging of a precipitate of ferric oxyhydroxide generally improves the retention of adsorbed cations. Ferric oxyhydroxides are produced continually in the oxidizing zone of a wetland which continually receives mine drainage, thereby providing new surfaces for the continued adsorption of other cations.

Marshall (1979) reports that the precipitation of Mn^{4+} oxide frequently is non-stoichiometric. The initial precipitation is slow, but once initiated, additional Mn^{2+} will be adsorbed from solution. The adsorbed Mn is then slowly converted into Mn^{4+} oxide. X-ray diffraction indicates a poor degree of crystallinity and a range of composition from $MnO_{1.3}$ (Mn_3O_4) to $MnO_{1.95}$. Figure 8 illustrates the pH dependence of adsorption of Mn by both ferric hydroxide and Mn^{4+} oxide.

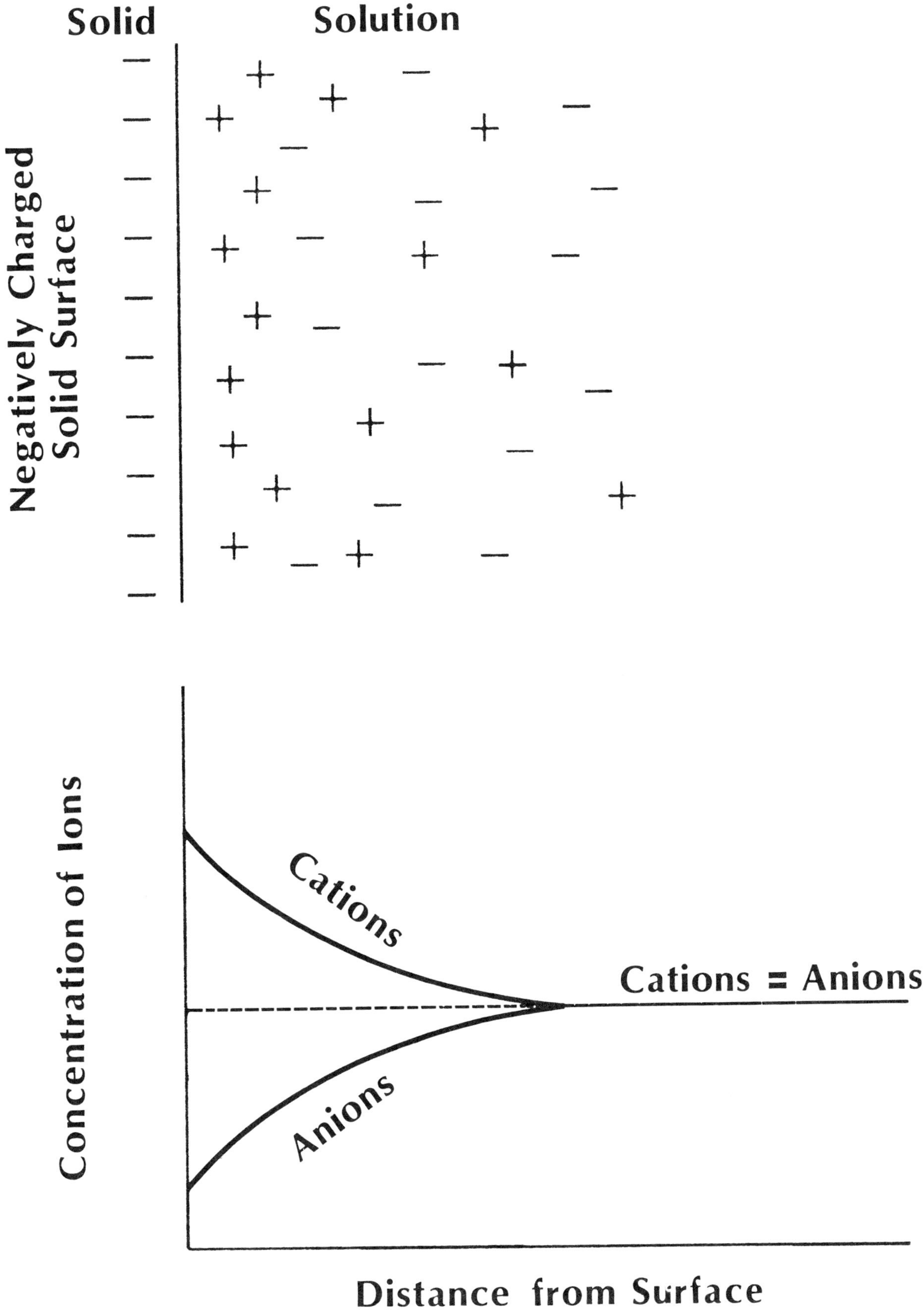

Figure 7. *Schematic drawing of the electrical double layers in which excess negative charge on a solid surface is balanced by an excess concentration of cations in solution near the solid-solution interface (from Drever, 1982).*

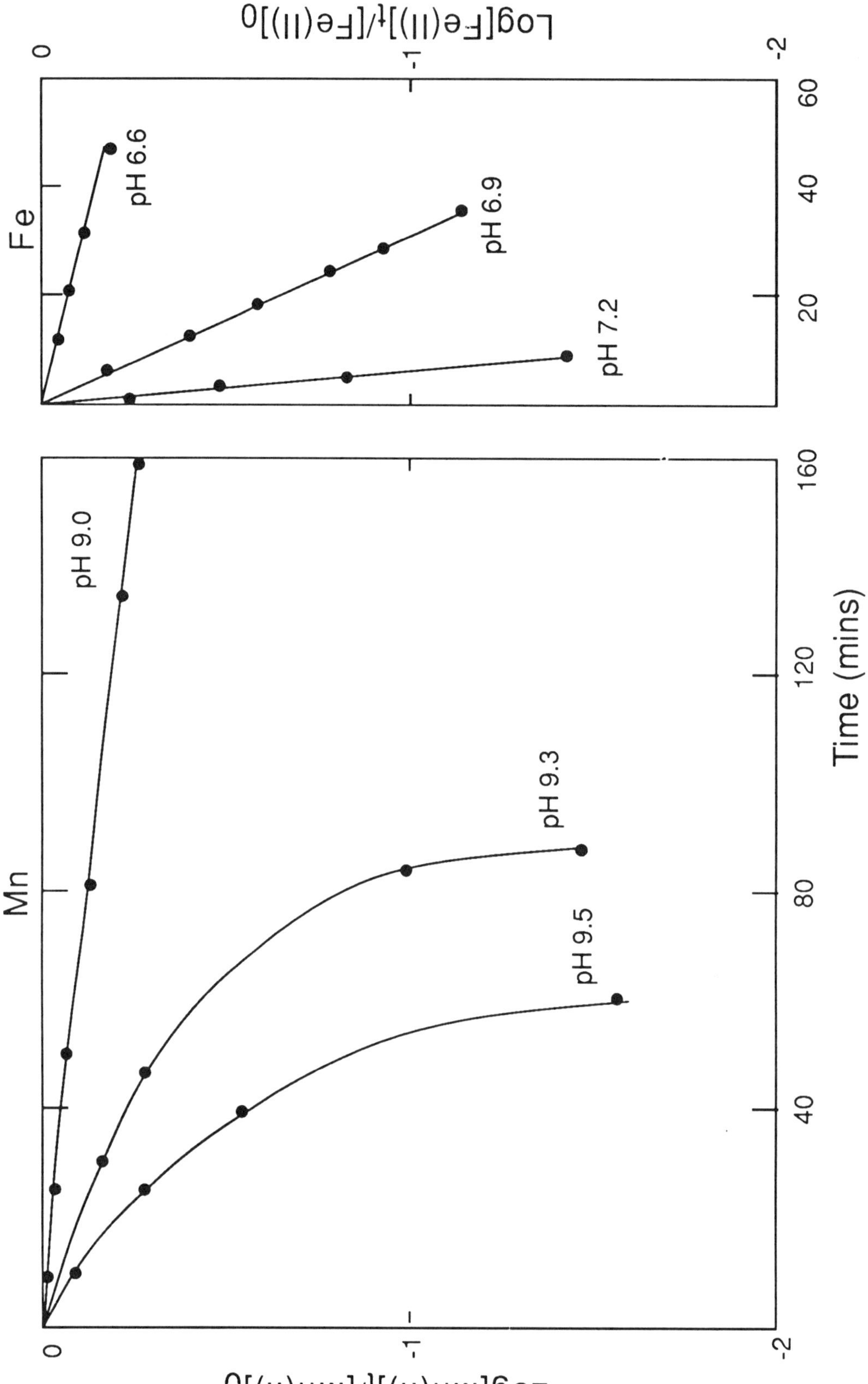

Figure 8. *Mn^{2+} adsorption by MnO_2 and $Fe(OH)_3$ as a function of pH at 25°C (from Marshall, 1979).*

Metal Uptake By Living Plants

Metal uptake by living plants generally is considered minor in the improvement of the quality of coal mine drainage (Emerick et al., 1988; J. Kriessl, personal communication, 1988). Plants growing in seepage from mines frequently show elevated concentrations of metals relative to control areas (McHerron, 1986). The thesis by McHerron (1986) suggested that Sphagnum dominated wetlands may concentrate Fe and the vascular plants may concentrate Mn. Mass balance calculations typically indicate 1-2% metal removal by this mechanism. The death of the plant will recycle the material back into the wetland and, therefore, the removal is not permanent. Sencindiver and Bhumbla (1988) found less than 1% of the Fe transported into the wetland by the acid mine drainage was retained in the tissue of the living plants. Hiel and Kerins (1988) also report metal uptake by the plants as a minor contribution. A few specialized plants can hyperaccumulate some toxic elements.

A more significant contribution of the plant is to provide organic carbon to the substrate which is necessary for two of the other processes, those of cation exchange by organic matter and subsequent SO_4^{2-} reduction. Plants of the genera Typha and Carex are tolerant of the acidity produced by mine drainage and are widely used (Girts and Kleinmann, 1986; Brodie et al. 1988). The long-term effectiveness of the wetlands-based treatment of acid mine drainage is dependent upon the establishment and maintenance of a vegetative cover over the substrate which is providing the treatment.

CASE HISTORIES OF WETLANDS-BASED TREATMENT OF ACID MINE DRAINAGE

Natural wetlands have peat as the primary substrate. Peat has a high water content, but a low permeability which does not allow transmission of water through the substrate. The flow in a natural wetland is primarily over the surface. Peat also is composed primarily of refractory organic materials left over after natural decay processes have consumed the more useful components. Peat is not capable of providing sufficient nutrients for effective biogeochemical treatment of mine drainage.

Constructed wetlands are an engineered approach to using the natural processes to treat mine drainage. Typically, constructed wetlands use a high-organic content substrate such as manure, composted manure, sewage sludge, mushroom compost, and combinations of these materials. Peat may not be used at all, or is a minor component. Constructed wetlands offer promise of low capital investment and low maintenance costs for decreasing acidity and trace metals in mine effluents. These primary trace metals of environmental interest in acid drainage from coal mines are Fe and Mn. The effectiveness of the artificial wetland approach to acid mine drainage treatment over time periods of tens of years is not yet known.

A relatively large number of wetlands have been constructed in various regions of the U.S., but primarily have been tested in the Appalachian coal province. Girts and Kleinmann (1986) have summarized the characteristics, engineering parameters, and operation of approximately 25 constructed wetlands. Most are effective, at least temporarily, in the removal of Fe and increasing the pH. The removal of Mn and SO_4^{2-} is uniformly less effective.

Wieder (1989) surveyed design parameters and treatment effectiveness of 137 constructed wetlands. Most provided some improvement in water quality, particularly for pH, and reductions in acidity and Fe. Only 15 provided sufficient improvement in water quality to meet required water quality standards. Wieder (1989) also concluded that a lack of significant correlations between design parameters and treatment efficiency preclude the identification of critical characteristics of a performing wetland.

Hedin et al. (1988) suggest that more effective utilization of dissimilatory SO_4^{2-} reduction in constructed wetlands will improve the overall effectiveness in water treatment. The SO_4^{2-} reduction will require new designs in constructed wetlands which allow significant flow of acid mine drainage through an anaerobic zone of wetlands which are deeper than most which have been constructed to date.

The three case histories to be described were selected from the recent literature and illustrate some of the designs which have been used. All use oxidation in the surface layer of constructed wetlands as the primary mechanism for the removal of Fe. The next generation of constructed wetlands likely will combine reduction and oxidation processes in series for more effective removal of metals and decreasing of acidity. The research of Hedin and Nairn (1990), McIntire and Edenborn (1990), and Wildeman et al. (1990) are demonstrating the effectiveness of anaerobic (reducing) wetlands in removal of Fe and trace toxic metals. The coupling of a reducing wetland and an oxidizing wetland in series will be necessary to remove Fe, trace metals, and Mn. Computer modeling of the biogeochemical processes using an equilibrium computer code is supportive of the use of an anaerobic wetland, followed by an aerobic wetland to effectively remove Fe, trace metals, and Mn (Klusman, 1990).

TVA-Constructed Wetlands In The Tennessee Valley

Brodie et al. (1988) reported on a series of wetlands constructed at Tennessee Valley Authority (TVA) facilities to treat seepage from mines, coal storage piles, ash storage, and coarse mine waste piles. Eleven wetlands ranging in size from 3.5 m^2 (38 ft^2) to 113 m^2 (1,216 ft^2) per average flowing liter-minute were constructed. The primary species transplanted from local sites into the wetlands was cattail (Typha latifolia) with lesser amounts of Scirpus, Eleocharis, and Carex. The water depth in the constructed wetlands ranged from 15-30 cm (9.3-19 in.) with a few small pools up to 1 m (3.3 ft) in depth. The wetlands were fertilized with phosphorous (P) and K. Specific substrates are not given for most of this group of constructed wetlands.

The wetlands discharge was planned to meet state effluent guidelines of total Fe<3.0 mg/L, total Mn<2.0 mg/L, pH of 6.0-9.0 s.u., and nonfilterable residues (NFR) <35.0 mg/L. Table 3 summarizes the physical characteristics of the 11 wetlands, the chemical character of the influent water, and the discharged water from the wetland. The water quality generally is improved in all cases, and meets regulatory limits in most cases. Those effluents which did not meet the regulatory limits underwent further chemical treatment in order to meet the guidelines. Even in those sites where chemical treatment was necessary, the cost of chemical treatment was reduced compared to what would have been necessary without wetlands treatment.

Examination of Table 3 indicates the wetlands are more effective in decreasing the Fe content of the effluent than in decreasing the Mn. This is consistent with the data discussed in the section on "Precipitation of Ferric and Manganese Oxyhydroxides." The relatively slow precipitation of manganese oxyhydroxides results in less effective treatment in these wetlands which are shallow and emphasize treatment by oxidation and precipitation.

The pH actually decreased in the effluent of three of the wetlands. This probably reflects the relatively slow hydrolysis of Fe^{3+} during the second step in the formation of acid mine drainage (Wildeman, this volume). The oxidation of the Fe^{2+} to Fe^{3+} and subsequent hydrolysis reached completion in the wetland, producing the additional acidity. This is not entirely detrimental, as the oxidation would have reached completion in the downstream receiving waters if it had not occurred in the wetland.

Total costs for construction of nine of the wetlands treatment systems are shown in Table 4. These wetland construction costs were amortized within a period of one year by the reduction in purchases of chemicals and in cost of treatment. Brodie et al. (1988) also provide guidelines for estimating the size of wetland necessary for treatment of acid mine drainage. They suggest an area of 2 m^2/ppm (21 ft^2/ppm) Fe and 7

Wetlands System	Date Initiated Operation	Area (m^2)	Number Cells	Influent Water Parameters (mg/L)				Effluent Water Parameters (mg/L or L/min.)						Treatment area (m^2/mg/min.)	
				pH	Fe	Mn	NFR	pH	Fe	Mn	NFR	Flow		Fe	Mn
												Ave	Max		
WC 018	6-86	4800	3	5.6	150.0	6.8		3.9	6.4	6.2		70	1495	0.2	4.2
King 006	10-87	9300	3	4.2	153.0	4.9	40.0					379	2271	0.2	5.0
Imp 4	11-85	2000	3	4.9	135.0	24.0	42.0	4.6	3.0	4.0	6.0	42	49	0.4	2.0
950 NE	9-87	2500	2	6.0	11.0	9.0	19.0	6.6	0.5	0.2	49.0[1]	348	1673	0.7	0.8
RT - 2	9-87	7300	3	5.7	45.2	13.4		6.7	0.8	0.2	2.0	238	681	0.7	2.3
Imp 2	6-86	11000	5	3.1	40.0	13.0	9.0[2]	3.1	3.4	14.0	0.8[2]	400	2200	0.7	2.1
Imp 3	10-86	1200	3	6.3	13.0	5.0	28.0	6.8	0.8	1.9	4.7	87	379	1.1	2.8
WC019	6-86	25000	3	5.6	17.9	6.9		4.3	3.3	5.9		492	6360	2.8	7.4
950 - 1 & 2	1976	3400	3	5.7	12.0	8.0	20.0	6.5	1.1	1.6	5.4	83	341	3.4	5.1
Imp 1	5-85	5700	4	6.3	30.0	9.1	57.0[3]	6.5	0.9	2.1	2.8	53	227	3.6	11.8
Col 013	10-87	9200	5	5.7	0.7	5.3		6.7	0.7	13.5		288	408	45.6	6.0

[1] one effluent sample to date
[2] one sample, July 1987
[3] from preconstruction in-stream sample

Table 3. *TVA acid drainage wetlands treatment summary (modified from Brodie et al., 1988).*

Wetlands System	Area ha	% Equipment	% Labor	% Overhead	Total ($000s)	$/m^2	$/ft^2
RT 2	0.7	32.6	53.4	14.0	26.4	3.58	0.33
Imp 2	1.1	70.2	12.3	17.5	57.0	5.25	0.49
WC 018 & 019	2.9	—	—	—	209.0	6.98	0.65
Imp 1	0.5	19.0	49.6	31.4	42.9	7.46	0.69
950 NE	0.3	38.1	46.0	15.9	35.2	13.80	1.28
Imp 4	0.2	28.2	26.8	45.0	28.0	14.12	1.32
King 006	0.9	28.6	43.3	28.1	131.7	14.21	1.32
Imp 6	0.1	34.8	43.5	21.7	40.2	32.03	2.98

average cost = $12.18/m^2 = $1.13/ft^2

Table 4. *Wetland system construction costs (modified from Brodie et al., 1988).*

Location	**Pre-wetland (1983-1985)**	**Winter 1985-6**	**Spring 1986**	**Summer 1986**	**Fall 1986**	**Winter 1986-7**	**Spring 1987**	**Summer 1987**	**Fall 1987***
Seep	6.16	6.45	6.40	6.40	6.46	6.34	6.25	6.36	6.39
Cell 1	—	6.07	6.52	6.40	6.16	6.48	6.45	6.33	6.29
Cell 2	—	5.92	6.46	6.22	6.16	6.48	6.40	6.31	6.27
Cell 3	—	6.03	6.40	6.22	6.47	6.36	6.27	6.19	5.99
Ditch End	5.88	6.07	6.50	6.24	6.16	6.50	6.43	6.26	6.18

* after site modifications of August 31, 1987

Winter = December-February
Spring = March-May
Summer = June-August
Fall = September-November

Table 5. *Average pH values by season for each wetland cell and the ditch outfall (modified from Stark et al., 1988). For pre-wetland means, n=5; n=4-7 for post-wetland means.*

m^2/ppm (75 ft^2/ppm) Mn for drainages with pH<5.5. For drainages with pH>5.5, an area of 0.7 m^2/ppm (8 ft^2/ppm) Fe and 2 m^2/ppm (21 ft^2/ppm) Mn is recommended.

Simco #4 Constructed Wetland, Coshocton, Ohio

Stark et al. (1988) described a wetland constructed to treat seepage from an abandoned, but plugged, coal mine near Coshocton, Ohio. The discharge prior to construction of the wetland averaged 454 L/min. (120 gal./min.), pH of 5.7-6.3 s.u., total Fe of 145-241 mg/L, total Mn of 3.1-6.0 mg/L, and acidity of 140-389 mg/L.

The wetland was constructed with three segments or cells in series with a total area of 3,000 m^2 (32,000 ft^2) (Figure 9a). The bottom of the wetland was constructed with a 15-cm (6-in.) thick layer of crushed limestone. The limestone was covered with a 45-cm (18-in.) thick layer of compost, into which lime was incorporated. Cattail (Typha latifolia) rhizomes were planted at a density of 3-4/m^2 (0.3-0.4/ft^2). Periodic liming with agricultural limestone and fertilization with P were used to increase algal growth.

During early stages of operation, there was little change in the pH of the effluent after passing through the wetland, but substantial Fe was removed. Manganese concentrations in the influent and effluent remained essentially constant during the first 2 yrs of operation.

After 2 yrs of operation, the effluent began to violate the standard for Fe and the wetland was modified to reduce the flow through cell 1 (Figure 9b). Hay bales were added as flow baffles and to increase the retention time and promote microbial activity. The flow also decreased due to drought conditions at the time of the wetland modification which complicates the interpretation of the data on water quality improvement.

Table 5 summarizes the pH measurements at various points in the drainage-wetland system. Improvement in the pH is marginal, although the drainage generally is above the regulatory value of 6 s.u. Table 6 contains the data for Fe in the drainage-wetland system, and Table 7 summarizes the percent decrease in Fe at various points in the system. The data suggest a greater efficiency in the summer than in the winter, although this interpretation is confounded somewhat by the system modification and the change in flow rate. Tables 6 and 7 are shown in graphical form in Figure 10. The seasonal effect is apparent in the mine drainage and the wetland effluent during the first year, but is not as apparent in the second year. The concentration of Fe in the effluent continues to decrease during the fall of the second year while the concentration in the mine drainage is beginning to rise.

Manganese concentrations in the mine drainage ranged from 1.66-602 mg/L, and averaged 2.12 mg/L, since construction of the wetland. Manganese concentration at the outlet averaged 2.28 mg/L, indicating a slight increase between the drainage and the wetland outlet. This was attributed to an additional, small, high Mn seep entering the wetland from the side, but the difference may not be statistically significant. This also demonstrates little effectiveness in the ability of the wetland to remove Mn. Continuing research by Stark et al. (1990) confirms the increase in efficiency of Fe removal with time through 1989 and the superior summer efficiency over winter efficiency.

Tracy Wetlands Near Great Falls, Montana

Hiel and Kerins (1988) constructed two wetlands in the Sand Coulee drainage near the town of Tracy, Montana. The larger wetland has an area of 420 m^2 (4,500 ft^2) and treats a mine drainage with a flow ranging from 30-57 L/min. (7.9-15 gal./min.). The drainage has a pH of 1.7 s.u., total Fe of 284 mg/L, total Al of 178 mg/L, total Mn of 1.51 mg/L, and SO_4^{2-} of 2,618 mg/L. The smaller wetland has an area of 108 m^2 (1,160 ft^2) and treats a drainage with a flow ranging from 23-30 L/min. (6.0-7.9 gal./min.). This mine drainage has a pH of 3.1 s.u., total Fe of 148 mg/L, total Al of 46.7 mg/L, total Mn of 1.2 mg/L, and SO_4^{2-} of 1,560 mg/L.

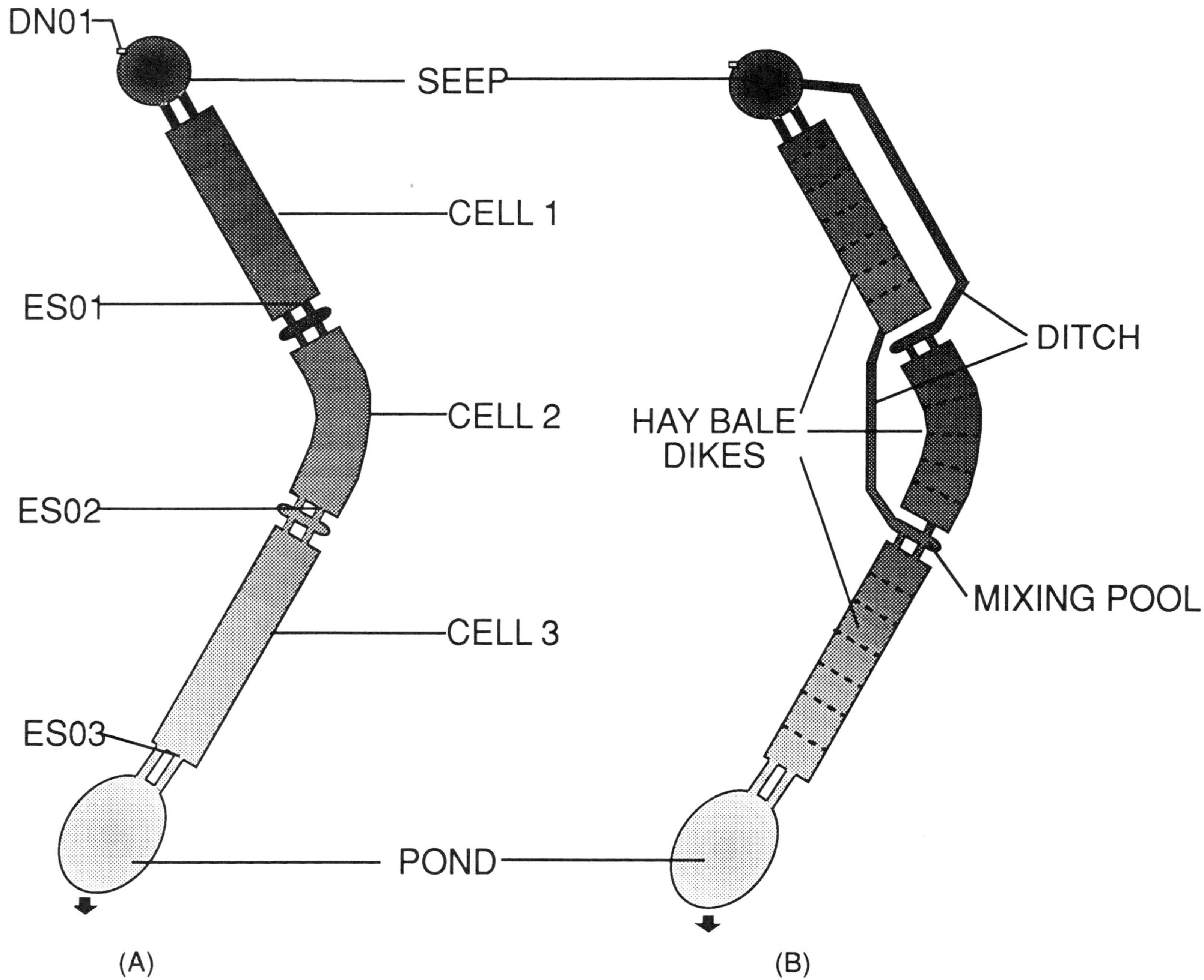

Figure 9. *Diagrams of Simco #4 wetland (modified from Stark et al., 1988). A) Water sampling points shown at right. Water sampling point DNO2 is not shown, but is approximately 100 m (330 ft) downstream from the pond outlet. B) Diagram after site modifications of August 31, 1987. Hay bale dikes were spaced at approximately 11-m (35-ft) intervals.*

Location	Pre-wetland (1983-1985)	Winter 1985-6	Spring 1986	Summer 1986	Fall 1986	Winter 1986-7	Spring 1987	Summer 1987	Fall 1987*
Seep	180 (range: 144-241)	116	106	127	144	113	100	87	100
Cell 1	—	99	93	98	123	100	78	68	62
Cell 2	—	93	81	77	105	86	63	59	59
Cell 3	—	84	65	56	74	60	42	34	26
Ditch End	122	75	55	32	58	52	38	26	19

* after site modifications of August 31, 1987

Winter = December-February
Spring = March-May
Summer = June-August
Fall = September-November

Table 6. *Average total Fe (mg/L) by season for each wetland cell and the ditch outfall (modified from Stark et al., 1988). For pre-wetland means, n=5; n=4-7 for post-wetland means.*

Component	Winter 1985-6	Spring 1986	Summer 1986	Fall 1986	Winter 1986-7	Spring 1987	Summer 1987	Fall 1987*
Cell 1	14	12	22	17	17	21	22	38
Cell 2	6	12	20	14	15	17	17	41
Cell 3	12	18	27	29	30	33	42	56
Ditch	10	16	42	22	8	13	22	26
Cum. Wetland	28	37	56	53	51	55	62	73
Cum. End of Ditch	35	47	75	59	53	62	70	81

* after site modifications of August 31, 1987

Winter = December-February
Spring = March-May
Summer = June-August
Fall = September-November

Table 7. *Average percent reduction in total Fe concentration by season for each wetland cell, the ditch, cumulative (cum.) for the wetland proper, and end of ditch (modified from Stark et al., 1988). For each season, n=4-7.*

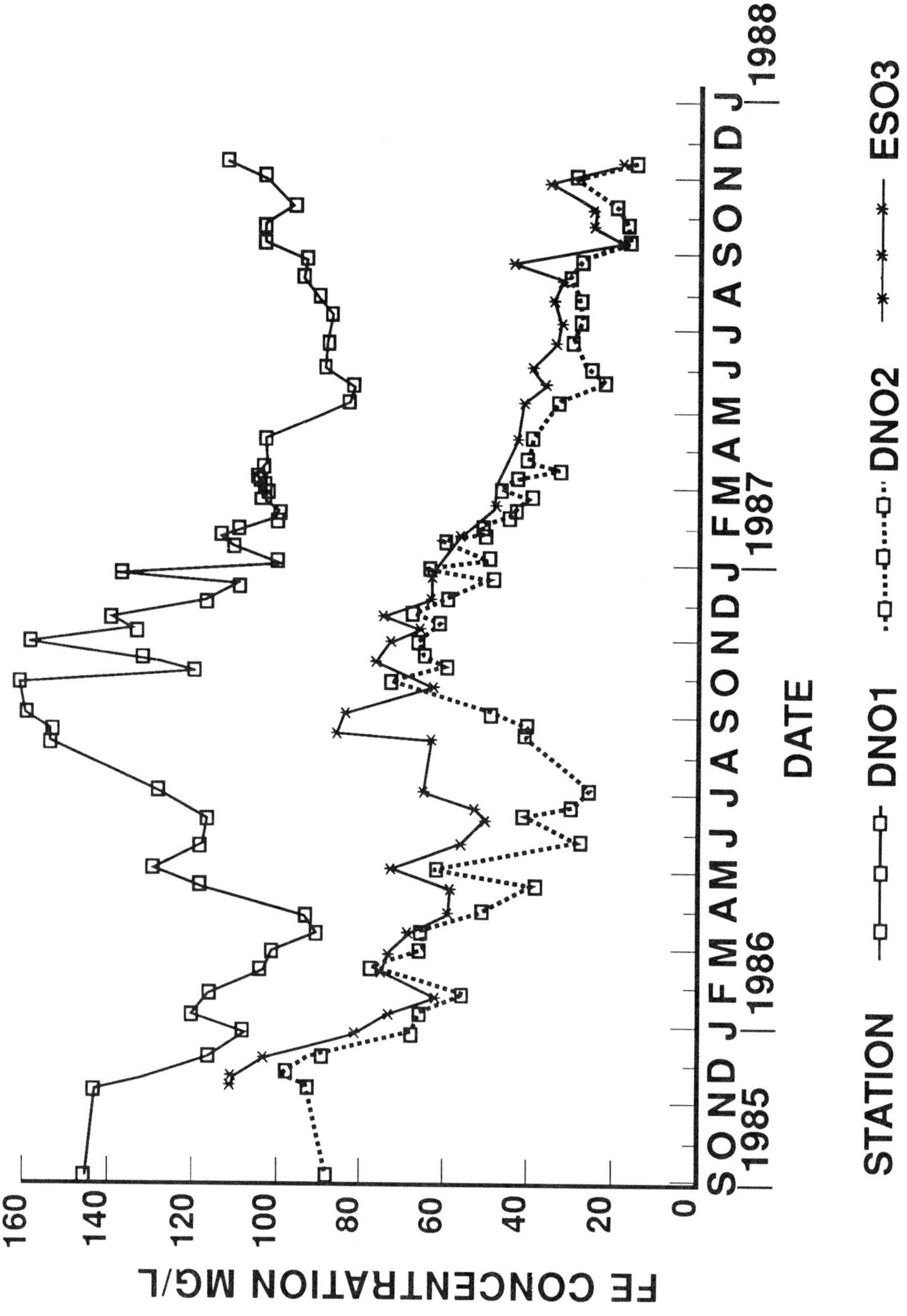

Figure 10. *Total Fe concentrations at DNO1 (seep), ES03 (wetland-proper outlet), and DNO2 (100 m (330 ft) downstream from wetland) (modified from Stark et al., 1988).*

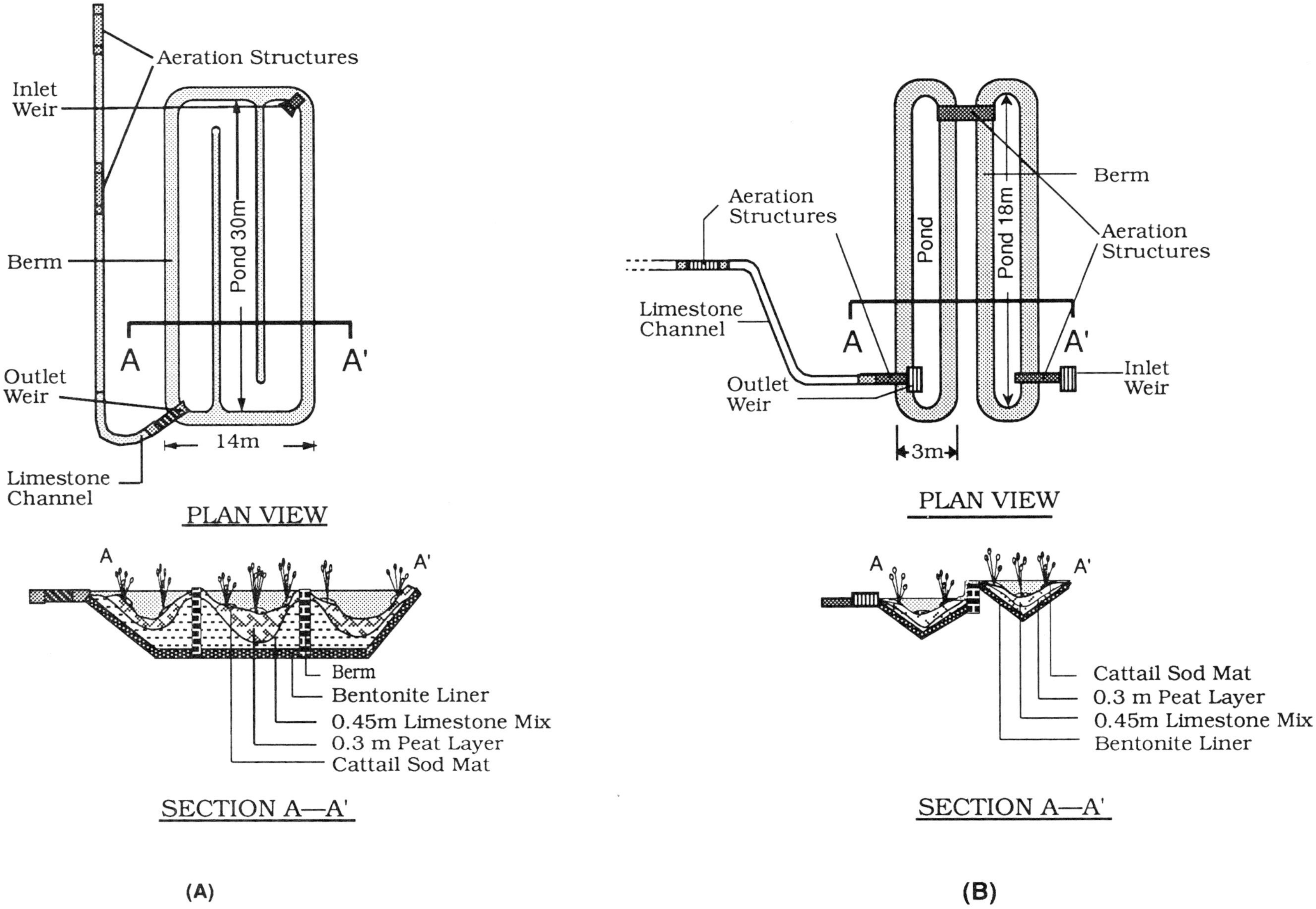

Figure 11. *Diagrams of wetlands at Tracy, Montana (modified from Hiel and kerins, 1988). A) Large wetland. B) Small wetland.*

	Large Wetland				Small Wetland			
	Inlet		Outlet		Inlet		Outlet	
	Mean	St.Dev.	Mean	St.Dev.	Mean	St.Dev.	Mean	St.Dev.
Total Al	178	17.5	180	117.91	46.7	2.1	45.7	2.1
Total Fe	284	134.9	271'	110.6	148.5	37.2	94.1	52.7
Total Mn	1.51	0.14	1.67	0.23	1.2	0.1	1.3	0.3
Sulfate	2618	285	2683	201	1560	223	1551	218
pH	2.7	0.27	2.58	0.19	3.1	0.3	2.8	0.4
Specific Conductance	3349	307	3440	194	2414	246	2559	278

Table 8. *Water quality summary (modified from Hiel and Kerins, 1988). All values in mg/L except for pH (standard units) and specific conductance (μmho/cm). All samples taken between July 10, 1986 and September 30, 1987. Number of samples taken was 21.*

Both wetlands were constructed using a limestone base, overlain by peat, and planted with cattail (*Typha latifolia*). The physical configuration and flow pattern differed in the two wetlands (Figure 11a and b). A limestone-soil mixture was used in the baffles between the channels of the two wetlands.

The retention time of water in the wetlands was much smaller than planned because the water flowed over the surface of the wetlands, rather than through the substrate. The wet peat of the wetlands was almost impermeable to water. This characteristic was noted by other researchers, including Guertin (1985). Dye tests indicated a retention time of seven hours for the large wetland and three hours for the small wetland. The limestone became coated or armored with ferric oxyhydroxide within two weeks of initiating flow through the systems.

Table 8 summarizes the water quality for the inlet and outlets of the two constructed wetlands. It is apparent that there is no detectable improvement in water quality. The case study emphasizes two parameters critical to the engineering of constructed wetlands. The retention time of the acid mine drainage must be adequate for the microbially induced processes to reach a reasonable degree of completion. An increase in permeability of the system must be achieved so that the full volume of the wetland is effective in treatment.

SUMMARY

The natural biogeochemical processes which operate in natural and constructed wetlands are reviewed, followed by summaries of case histories for constructed wetlands which have had varying degrees of success. The processes are presented in a sequence which we hypothesize will be the sequence in a multistage wetland which will exhibit the best overall performance. The first generation of wetlands have relied primarily on precipitation of iron oxyhydroxides as the main mechanism for metal removal. The next generation of wetlands likely will be staged, relying on chemical reduction followed by oxidation as a polishing step. Future wetlands are likely to remain relatively ineffective in the removal of Mn from acid mine drainage. Chemical modeling of these processes using an equilibrium computer code is supportive of these conclusions.

ACKNOWLEDGEMENTS

This work was supported in part by the U.S. Environmental Protection Agency Contract CR-815325-02-0 and the Department of Chemistry and Geochemistry of the Colorado School of Mines.

REFERENCES

Aiken, G.R., D.M. McKnight, R.L. Wershaw, and P. MacCarthy, 1985, Humic substances in soil, sediment, and water: New York, Wiley-Interscience, 692 p.

Altschuler, Z.S., M.M. Schnepfe, C.C. Silber, and F.O. Simon, 1983, Sulfur diagenesis in Everglades peat and origin of pyrite in coal: Science, v. 221, p. 221-227.

Berner, R.A., 1971, Principles of chemical sedimentology: New York, McGraw-Hill, 240 p.

Brodie, G.A., D.A. Hammer, and D.A. Tomljanovich, 1988, Constructed wetlands for acid drainage control in the Tennessee Valley, in Mine drainage and surface mine reclamation, Vol. 1, Mine water and mine waste: U.S. Bureau of Mines Information Circular IC 9183, p. 325-331.

Decker, A.D., R.W. Klusman, and D.M. Horner, 1987, Geochemical techniques applied to the identification and disposal of connate coal water, in Proceedings of the 1987 Coalbed Methane Symposium, Tuscaloosa, Alabama: Tuscaloosa, University of Alabama, p. 229-242.

Drever, J.I., 1982, The geochemistry of natural waters: Englewood Cliffs, New Jersey, Prentice-Hall, 388 p.

Ehrlich, H.L., 1987, Manganese oxide reduction as a form of anaerobic respiration: Geomicrobiology Journal, v. 5, p. 423-431.

Ehrlich, H.L., 1990, Geomicrobiology (2nd Ed.): New York, Marcel Dekker, 646 p.

Emerick, J.C., W.W. Huskie, and D.J. Cooper, 1988, Treatment of discharge from a high altitude metal mine in the Colorado Rockies using an existing wetland, in Mine drainage and surface mine reclamation, Vol. 1, Mine water and mine waste: U.S. Bureau of Mines Information Circular IC 9183, p. 345-351.

Girts, M.A., and R.L.P. Kleinmann, 1986, Constructed wetlands for treatment of acid mine drainage: A preliminary review, in D.H. Graves, ed., Proceedings of the 1986 National Symposium on Surface Mining, Hydrology, Sedimentology, and Reclamation, Lexington, Kentucky, December 8-11, 1986: Lexington, University of Kentucky Press.

Goldhaber, M.B., and I.R. Kaplan, 1975, Controls and consequences of SO_4^{2-} reduction rates in recent marine sediments: Soil Science, v. 119, p. 42-55.

Guertin, deForest, 1985, Evaluation of peat for the treatment of acid mine drainage: Golden, Colorado School of Mines, Unpublished M.Sc. Thesis T-3113, 91 p.

Gwynfryn-Jones, J., 1986, Iron transformations by freshwater bacteria, in K.C. Marshall, ed., Advances in microbial ecology: New York, Plenum Press, p. 149-181.

Hedin, R.S., D.M. Hyman, and R.W. Hammack, 1988, Implications of sulfate-reduction and pyrite formation processes for water quality in a constructed wetland: Preliminary observation, in Mine drainage and surface mine reclamation, Vol. 1, Mine water and mine waste: U.S. Bureau of Mines Information Circular IC 9183, p. 383-388.

Hedin, R.S., and R.W. Nairn, 1990, Sizing and performance of constructed wetlands: Case studies, in J. Skousen, J. Sencindiver, and D. Samuel, eds., Proceedings of the 1990 Mining and Reclamation Conference and Exhibition, Morgantown, West Virginia, April 23-26, 1990: Morgantown, West Virginia University Press, p. 385-392.

Hem, J.D., 1970, Study and interpretation of the chemical characteristics of natural water (2nd Ed.): U.S. Geological Survey Water Supply Paper 1473, 363 p.

Herlihy, A.T., G.M. Hornberger, and A.E. Bruckner, 1987, The importance of sediment sulfate reduction to the sulfate budget of an impoundment receiving acid mine drainage: Water Resources Research, v. 23, p. 287-282.

Herlihy, A.T., and A.L. Mills, 1985, Sulfate reduction in freshwater sediments receiving acid mine drainage: Applied and Environmental Microbiology, v. 49, p. 179-186.

Heil, M.T., and F.J. Kerins, Jr., 1988, The Tracy wetlands: A case study of two passive mine drainage treatment systems in Montana, in Mine drainage and surface mine reclamation, Vol. 1, Mine water and mine waste: U.S. Bureau of Mines Information Circular IC 9183, p. 352-358.

Jorgensen, B.B., 1977, The sulfur cycle of a coastal marine sediment (Limfjorden, Denmark): Limnology and Oceanography, v. 22, p. 814-832.

Jorgensen, B.B., 1982, The ecology of bacteria of the sulfur cycle with special reference to anoxic-oxic interface environments: Philosophical Transactions of the Royal Society of London, Series B, v. 298, p. 543-561.

Jorgensen, B.B., 1983, The microbial sulfur cycle, in W.E. Krumbein, ed., Microbial geochemistry: Oxford, Blackwell Scientific Publications, p. 91-124.

Klusman, R.W., 1990, Development of a computer code to model biogeochemical processes in the Big Five constructed wetland: U.S. Environmental Protection Agency, Contract Report No. CR815325, 29 p.

Laudon, L., 1988, Forms of sulfur in the Big Five constructed wetland: Golden, Colorado School of Mines, Unpublished M.Sc. Thesis T-3660, 58 p.

Lundgren, D.G. and Dean, W., 1979, Biogeochemistry of iron, in P.A. Trudinger and D.J. Swaine, eds., Biogeochemical cycling of mineral-forming elements: Amsterdam, Elsevier, p. 211-251.

Marshall, K.C., 1979, Biogeochemistry of manganese mineral, in P.A. Trudinger and D.J. Swaine, eds., Biogeochemical cycling of mineral-forming elements: Amsterdam, Elsevier, p. 253-292.

McHerron, L.E., 1986, Removal of iron and manganese from mine drainage by a wetland: Seasonal effects: University Park, Pennsylvania State University, Unpublished M.Sc. thesis, 129 p.

McIntire, P.E., and H.M. Edenborn, 1990, The use of bacterial sulfate reduction in the treatment of drainage from coal mines, in J. Skousen, J. Sencindiver, and D. Samuel, eds., Proceedings of the 1990 Mining and Reclamation Conference and Exhibition, Morgantown, West Virginia, April 23-26, 1990: Morgantown, West Virginia University Press, p. 409-415.

Morel, F.M.M., 1983, Principles of aquatic chemistry: New York, Wiley, 446 p.

Nealson, K.H., 1983, The microbial iron cycle, in W.E. Krumbein, ed., Microbial geochemistry: Oxford, Blackwell Scientific Publications, p. 159-190.

Sencindiver, J.C., and K.D. Bhumbia, 1988, Effects of cattails (Typha) on metal removal from mine drainage, in Mine drainage and surface mine reclamation, Vol. 1, Mine water and mine waste: U.S. Bureau of Mines Information Circular IC 9183, p. 359-368.

Singer, P.C., and W. Stumm, 1970, Acid mine drainage: The rate limiting step: Science, v. 167, p. 1121-1123.

Smith, R.L., and M.J. Klus, 1981, Reduction of sulfur compounds in the sediments of eutrophic lake basin: Applied and Environmental Microbiology, v. 41, p. 1230-1237.

Sparling, J.H., and B.M. Hennick, 1974, The production of hydrogen sulphide in peats: Folia Microbiology, v. 19, p. 59-66.

Stark, L.R., R.L. Kolbash, H.J. Webster, S.E. Stevens, Jr., K.A. Dionis, and E.R. Murphy, 1988, in Mine drainage and surface mine reclamation, Vol. 1, Mine water and mine waste: U.S. Bureau of Mines Information Circular IC 9183, p. 332-344.

Stark, L.R., S.E. Stevens, Jr., H.J. Webster, and W.R. Wenerick, 1990, Iron loading, efficiency, and sizing in a constructed wetland receiving mine drainage, in J. Skousen, J. Sencindiver and D. Samuel, eds., Proceedings of the 1990 Mining and Reclamation Conference and Exhibition, Morgantown, West Virginia, April 23-26, 1990: Morgantown, West Virginia University Press, p. 393-401.

Stumm, W., and J.J. Morgan, 1981, Aquatic chemistry (2nd Ed.): New York, Wiley-Interscience, 780 p.

Taylor, G.J., A.A. Crowder, and R. Rodden, 1984, Formation and morphology of an iron plaque on the roots of Typha latifolia L. grown in solution culture: American Journal of Botany, v. 71, p. 666-675.

Thurman, E.M., 1985, Organic geochemistry of natural waters: Boston, Dordrecht, 488 p.

Tuttle, J.H., P.R. Dugan, C.B. MacMillan, and C.I. Randles, 1969, Microbial dissimilatory sulfur cycle in acid mine water: Journal of Bacteriology, v. 97, p. 594-602.

Tuttle, M.L., 1988, Geochemical evaluation and depositional history of sediment in modern and ancient saline lakes: Evidence from sulfur geochemistry: Golden, Colorado School of Mines, Unpublished Ph.D. Thesis T-3240, 312 p.

Watzlaf, G.R., 1985, Comparative tests to remove manganese from acid mine drainage, in Control of acid mine drainage: Proceedings of a technology transfer seminar: U.S. Bureau of Mines Information Circular IC 9027, p. 41-52.

Wieder, R.K., 1985, Peat and water chemistry at Big Run Bog, a peatland in the Appalachian Mountains of West Virginia, USA: Biogeochemistry, v. 1, p. 277-302.

Wieder, R.K., 1988, Determining the capacity for metal retention in man-made wetlands constructed for treatment of coal mine drainage, in Mine drainage and surface mine reclamation, Vol. 1, Mine water and mine waste: U.S. Bureau of Mines Information Circular IC 9183, p. 375-381.

Wieder, R.K., 1989, A survey of constructed wetlands for acid coal mine drainage treatment in the Eastern United States: Wetlands, v. 9, p. 299-315.

Wieder, R.K., and G.E. Lang, 1986, Fe, Al, Mn, and S Chemistry of Sphagnum peat in four peatlands with different metal and sulfur input: Water, Air and Soil Pollution, v. 2, p. 309-320.

Wieder, R.K., G.E. Lang, and V.A. Granus, 1987a, Sulphur transformations in Sphagnum-derived peat during incubation: Soil Biology and Biochemistry, v. 19, p. 101-106.

Wieder, R.K., G.E. Lang, and A.E. Whitehouse, 1987b, Treatment of acid coal mine drainage using Sphagnum-dominated wetlands, in T. Paces and M. Bedrich, eds., Proceedings of the GEOMON-International Workshop on Geochemistry and Monitoring in Representative Basins, Prague, Czechoslovakia, April-May 2, 1987: Prague, Czechoslovakia Geological Survey, p. 218-220.

Wildeman, T.R., and L.S. Laudon, 1989, Use of wetlands for treatment of environmental problems in mining: Non-coal-mining applications, in D.A. Hammer, ed., Constructed wetlands for wastewater treatment: Municipal, industrial and agricultural: Chelsea, Michigan, Lewis Publishers, p. 221-231.

Wildeman, T.R., S.D. Machemer, R.W. Klusman, R.R.H. Cohen, and P. Lemke, 1990, Metal removal efficiencies from acid mine drainage in the Big Five constructed wetland, in J. Skousen, J. Sencindiver, and D. Samuel, eds., Proceedings of the 1990 Mining and Reclamation Conference and Exhibition, Morgantown, West Virginia, April 23-26, 1990: Morgantown, West Virginia University Press, p. 417-424.

THE ROLE OF GEOLOGY IN MANAGING COAL CONVERSION BYPRODUCTS IN THE NORTHERN GREAT PLAINS

Frank W. Beaver, and Gerald H. Groenewold
Energy and Environmental Research Center
ND Mining and Mineral Resources Research Institute
University of North Dakota
Box 8213, University Station
Grand Forks, ND 58202

Robert D. Koob
California Polytechnic State University
San Luis Obispo, CA 93407

ABSTRACT

Coal-conversion byproduct management relies heavily on applying geological understanding to materials that originate with coal, are transformed during energy conversion, that may be utilized as partially processed natural resources, and that ultimately may be disposed of in a geologic environment. The physical, chemical, mineralogical, and leaching characteristics of byproducts all must be understood for effective utilization and disposal. For disposal, site characterization, including topography, structure, stratigraphy, geochemistry, mineralogy, and hydrogeology, also must be embraced. Studies in North Dakota have demonstrated that understanding the waste/setting interactions is the key to ensuring environmentally safe disposal over the long term.

INTRODUCTION

Coal-to-energy conversion technologies, such as steam electric-power generation and coal gasification, result in a variety of solid, liquid, and gaseous byproduct materials. Some of these byproducts, if improperly managed, have the potential to adversely affect the environment. Management of solid residual materials originating from mineral matter in the coal is an increasingly significant problem for operators utilizing low-rank coal. In the North Dakota portion of the Northern Great Plains, approximately 1,800,000 tonnes (2,000,000 tons) of fly ash, bottom ash, and gasification slag are produced annually. This is approximately 10-15% of the coal mined and consumed. Disposal costs range from $11.00/tonne ($10.00/ton) in the Northern Great Plains to $110.00/tonne ($100.00/ton) in the Eastern United States. Liability associated with disposal sites poses additional concern and expense. Byproduct management includes a variety of options, including source reduction, recycling, utilization, and various forms of treatment as well as disposal. Understanding the interactions between geological factors is the key in choosing the appropriate utilization and/or disposal management options. The geological factors include the physical, chemical, mineralogical, and leaching character of byproducts as materials, as well as the character of the disposal site. Disposal should be the last resort.

The present acceptability of disposal options is measured by their success in protecting human health and the environment, typically through the isolation of the waste material from the geologic setting. The effectiveness of this action is questionable because isolation liners have limited long-term effectiveness. Over the long term, the waste disposal setting must be considered in terms of a reactor which potentially could play an active role in the disposal process. Such an approach is particularly appropriate in the case of coal-conversion residuals from mine-mouth facilities, because they typically are elements that originated from the same geologic setting in which they will be interred. Finally, the waste and the disposal environment both must be considered as elements of a dynamic, interactive setting. It is not a question of "if" interactions will occur, but "when" such interactions will occur, and then whether or not these interactions will result in a significant risk for a given environmental quality parameter.

HISTORICAL BACKGROUND

This paper draws, in large part, from two sources: 1) extensive hydrogeological research focused on understanding the pre- and post-mining landscape, and 2) a field investigation into the effects of fly ash and flue gas desulfurization (FGD) wastes on groundwater quality in a reclaimed Western North Dakota lignite strip mine setting. The majority of coal-conversion facilities in the Northern Great Plains are the mine-mouth type, and the majority of waste disposal occurs in mine settings which are eventually reclaimed. Both the settings and disposal/setting interactions have been studied for a number of years in active and reclaimed mine settings and related coal-conversion sites in Western North Dakota, including the Falkirk, Indian Head, Beulah-Hazen, Dunn Center, and Center areas (Groenewold, 1979; Groenewold et al., 1979, 1980, 1981, 1983; Groenewold and Rehm, 1982; Moran et al., 1975, 1976, 1978, 1979; Rehm et al., 1980, 1982; Williams et al., 1978; Beaver et al., 1987).

ENVIRONMENTALLY SOUND DISPOSAL

Key Considerations

The essence of environmentally sound disposal lies in understanding a series of phenomena. First, the chemical, physical, mineralogical, and leaching character of the waste must be defined. Second, the disposal setting geology, including topography, structure, geochemistry, mineralogy, and stratigraphy, must be defined because these elements of the geologic setting define the three-dimensional (3-D) environment into which the wastes will be placed. Third, the water moving through the wastes/geologic setting (hydrogeology) must be defined because water is the reaction and transport medium for any mobile contaminants. Fourth, the interaction between the waste and the setting must be addressed because it is the occurrence, transport, and fate of contaminants that encompass the ultimate question to be answered by designers of waste disposal facilities. The role of geology in disposal can be addressed best by considering each geological factor and its relationship to the other geological factors in order to come to a full understanding of how wastes dynamically interact with the hydrogeological environment. For example, it is in this context that leaching tests must be critically evaluated for their appropriateness in a given setting.

In summary, these factors are the following:

1) Waste Characterization

   a) Physical

   b) Chemical

   c) Mineralogical

2) Leaching Tests

   a) Synthetic Groundwater Leaching Procedure

   b) EP Tox

   c) TCLP

   d) ASTM

3) Geologic Characterization - The 3-D Geologic Setting

   a) Topography

   b) Geochemistry

   c) Mineralogy

   d) Stratigraphy

   e) Structure

4) Hydrogeology - The Water Moving through the 3-D Geologic Setting

   a) Groundwater Occurrence

   b) Groundwater Flow

   c) Groundwater Quality

5) Waste-Setting Interactions

   a) Contaminant Occurrence

   b) Contaminant Transport

   c) Contaminant Fate

Leaching Tests

Leaching characterization, a laboratory procedure, measures the tendency of a material to influence the chemistry of a liquid contacting it. Leaching has many uses, as well as many forms. Both column and batch leaching provide valuable data which have unique and differing applications. Batch leaching provides information on the effect of single variables on the equilibrium concentration of elements in a solid-liquid system. Column leaching, a dynamic technique, while not able to provide the same detailed information of the effect of single variables on equilibrium chemistry, does provide a method of modeling certain effects such as the intermediate processes that lead to the establishment of equilibrium conditions. Batch leaching has resulted in an understanding of the effects of individual variables on the evolution of groundwater chemistry, as influenced by the leaching of disposed materials. This procedure is used to gain an understanding of both the release and the attenuation of individual elements in complex solid-liquid systems. Several different batch-leaching procedures have been devised for application in the understanding of phenomena under varying circumstances.

In general, a given waste mass is subjected to leaching with a fluid under a controlled set of environmental conditions, and the leachate is analyzed to determine what has leached from the waste mass. In order to obtain meaningful data, the leaching solution must approximate the fluid present in the field setting. The ultimate test of appropriateness for a given leaching procedure is whether it will accurately and precisely predict what is shown to exist in a field study. Only after verification is the procedure truly valuable as a predictive tool.

The leaching characterization of materials prior to disposal is a valuable laboratory technique for the evaluation of hazardousness. There are two widely used standard tests of leachability. These are the "toxicity characteristic leaching procedure" (TCLP) test and the ASTM leaching procedure. The TCLP is the regulatory test for the determination of hazardousness on the basis of leachable toxic materials. The TCLP test was "designed to simulate the leaching a waste would undergo if disposed in a sanitary landfill." This leaching procedure is designed to simulate leaching at about pH = 5 using acetic acid. It is assumed that these conditions would simulate not only the pH of a sanitary landfill, but also the source of the acidity. Cellulosic and other municipal wastes would generate acetic acid and other carboxylic acids through microbial action.

There are several types of waste, however, which would, under almost all conceivable circumstances, never be disposed of in a municipal landfill. One of these materials is coal-conversion ash. It is the consideration of this class of materials which led to the development of a generic test of leachability referred to as the synthetic groundwater leaching procedure (SGLP) (Hassett, 1987). The SGLP uses a leaching solution which simulates groundwater from the disposal area, in this case, Central and Western North Dakota. The test used a sodium sulfate/bicarbonate-buffered water which is typical for the region where the water is slightly alkaline (bicarbonate-buffering), and the primary mineralization is from sodium sulfate. Significant differences, with far-reaching regulatory implications, have been demonstrated through the use of different leaching media (Hassett et al., 1987).

The TCLP, the SGLP, the original EPA-EP extraction procedure, and the ASTM leaching tests can be summarized as follows:

TCLP — 18-hr equilibration, acidic conditions, 20:1 liquid to solid ratio, end-over-end agitation

EPA-EP — 24-hr equilibration, acidic conditions, 20:1 liquid to solid ratio, rotary extraction (orbital rotation)

SGLP — 18-hr equilibration, variable composition extractant (e.g., sodium sulfate/bicarbonate-buffered water), 20:1 liquid to solid ratio, end-over-end agitation

ASTM — 48-hr equilibration, distilled water extractant, 4:1 liquid to solid ratio, reciprocal agitation

Gasification ash from the Dakota Gasification Company and facilities at Beulah, North Dakota, was chosen for comparative leaching tests. Notable trends among elements could be seen in a comparison between the EP and TCLP leaching procedures. These differences reflected the larger amount of acid which can be used in the EP procedure. Comparison of results from the SGLP, EP, and TCLP tests also indicated notable trends. It was shown that large variations in the concentrations of elements can exist when different leaching procedures are used on the same material. This can have particular significance if the mobility of select toxic trace elements is highly affected by the different tests. The most striking example of this was seen in comparing results from a specific test for arsenic (As) concentrations in the leachates. The EP and TCLP both suggested that little leachable As was present in gasification ash, but the SGLP results showed that, under simulated field conditions, more than 30 times as much As was leached. This indicated a need for a test complementary to the regulatory procedure based on leaching which more closely simulated field conditions.

There are certain criteria which also should apply for a complementary test such as the ability to directly compare data with the TCLP and the flexibility to allow for variations in local conditions. The acceptance of a generic leaching procedure as a complementary test to the TCLP would allow for more valid assessment of hazardousness as it applies to the field disposal of solid wastes.

Materials Characterization

Materials characterization is also a key activity. For example, a chemical analysis of carbon reveals nothing of the structure of the material which could be pure graphite, a soft electrical conductor, or diamond, a hard electrical insulator. Additionally, carbon also may occur as an organic life form. Thus, the structural arrangement of the carbon provides important information. The significance of mineralogical characterization in identifying potential uses for a given substance has been well established (McCarthy et al., 1987). Arsenic loosely attached to the surface of a fly ash grain may be readily liberated when contacted by water. Alternately, As combined in the structural matrix of a solid phase may be immobile in contact with water. Chemical analysis alone would not reveal this distinction. Leaching potential is largely determined by the mineral form in which the material exists.

Topography

Topography at disposal sites is an extremely important consideration because topography controls surface water flow and, therefore, has a tremendous impact on groundwater recharge. Surface water runs off topographic highs relatively rapidly, allowing less infiltration than in topographic lows or depressions in which surface water collects and is available for downward percolation and groundwater recharge. In glaciated areas, paleotopographic surfaces also must be considered because of possible perched water table conditions and subsurface drainage.

Design of waste disposal sites must consider topography. Waste disposal sites should not be placed in topographic lows where surface water may collect and provide for contaminant transport in the subsurface. Topographic highs are preferred for waste disposal settings because water can be diverted away from the site. The site can be designed and constructed to be a topographic high with a relatively impermeable material placed over the deposit to form a geologic roof which can divert flow to natural or constructed depression-focused recharge sites. In mine land reclamation, an opportunity exists to locate both disposal sites and groundwater recharge sites to eliminate contaminant movement and provide for groundwater recharge.

Mineralogy and Geochemistry

The mineralogy of both coal-conversion wastes and materials in the disposal site is important, controlling how the individual chemical components of materials exist as phases and governing their reactivity in a given environment. Waste disposal poses the fundamental question of determining the conditions that allow problematic inorganic and/or organic chemical species to become mobile and available for biologic assimilation. The answer lies in understanding the mineral phases and glass matrixes which contain the major and trace constituents of the wastes. Mineralogical analyses are critical in this regard. The following are examples which illustrate the importance of mineralogical information:

1) Arsenic is a commonly occurring trace constituent of many fly ashes. Arsenic molecules mineralogically bound in an inert phase behave very differently than As molecules located on the surface of a fly ash grain where they may be leached easily and available for transport. Thus it is important to know more than what a strictly chemical analysis for As may reveal. It also is important to know the As content and mineralogy of the undisturbed sediments for the purpose of comparison. It is not simply the presence of As that is important, but rather the form in which it is present, because some forms are more or less mobile than others in a given disposal environment. Similarly, the mineralogical form and location of cadmium (Cd), lead (Pb), mercury (Hg), barium (Ba), boron, molybdenum (Mo), and other potentially problematic species directly influence their mobility.

2) Ettringite is a calcium aluminosulfate hydroxide hydrate which tends to form in certain waste deposits under high pH conditions. This naturally occurring mineral may be critically important to the future of waste management because the mineral structure appears to favor inclusion of trace metals (Hassett, 1990).

Stratigraphy and Structure

Stratigraphy and structure define the three-dimensional distributional relationship of earth materials and waste deposited within the earth materials. Understanding the stratigraphy of a disposal site is one of the most critical elements in site selection. Typically in the coal-bearing region of the Northern Great Plains, the stratigraphy consists of nearly horizontal coal-bearing strata which may be overlain by complex glaciofluvial deposits. The most significant structural feature is the regional fracture system notable in the coal which may be of tectonic origin. Coal seams are commonly aquifers because of their fracture hydraulic conductivity. When the coal is removed and replaced with spoil material, a spoil aquifer typically is established within the disturbed strata adjacent to the undisturbed strata.

REGIONAL HYDROGEOCHEMICAL SETTING

At surface mines in the Northern Great Plains, coal-conversion wastes are disposed of in undisturbed and disturbed settings, including reclaimed and nonreclaimed spoils. Since 1974, studies in west-central North Dakota (Groenewold et al., 1979, 1980, 1981, and 1983; Groenewold and Rehm, 1982; Moran et al., 1975, 1976, 1978, 1979; Rehm et al., 1980, 1982; Williams et al., 1978) have detailed the geological, hydrogeological, and geochemical mechanisms controlling subsurface water character (see Figure 1). Information from the Indian Head and Center mines is particularly important because both disturbed and undisturbed settings were investigated at those sites.

The Center and Indian Head mines use walking draglines to remove overburden and dozers and pan scrapers to recontour spoils. The study areas at both mines are characterized by significant and consistent differences in the character of the overburden and groundwater quality in both undisturbed and spoils settings (Groenewold et al., 1979, 1981). However, the overburden in the area of investigation at the Indian Head Mine was predominantly silty and clayey, whereas the overburden in the study area at the Center Mine was predominantly silty sand to sandy silt. Because all other significant variables, such as mining and reclamation methods and climate, are the same at these sites, the Center and Indian Head mines allowed the comparison of overburden textural and mineralogical characteristics with groundwater quality and chemical evolution in both undisturbed and spoils settings.

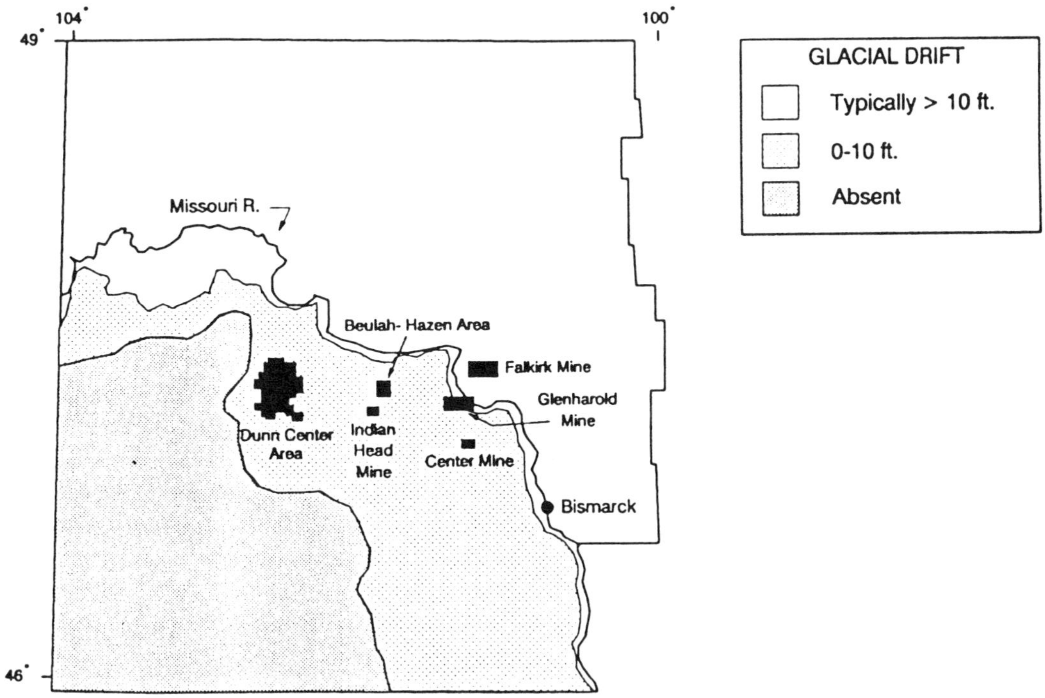

Figure 1. Location of the selected hydrogeological study sites in Western North Dakota.

Climate

The climate in the vicinity of all the North Dakota study sites is classified as semiarid-continental. The summers generally are cool with only a few hot, humid days. The winters typically are quite cold and long. The mean annual temperature for Bismarck, North Dakota, which is located approximately 50 km (46 mi.) southeast of the Center Mine, is 5.7°C, varying from -12.3°C in January to 22.1°C in July. Precipitation at Beulah, North Dakota, approximately 6 km (4 mi.) east of the Indian Head Mine, averages 43.7 cm (17.2 in.) annually.

Geology

The North Dakota study sites are representative of the geological conditions found in most surface coal mining areas in the Northern Great Plains. Glacial sediments (tills) veneer portions of some of the North Dakota sites and are absent at other sites. The potentially minable coal at all the sites occurs in the Tertiary Sentinel Butte Formation of the Fort Union Group. Although the thickness of the coal seams is less than in some other areas, the coals and associated sediments which underlie the North Dakota sites are typical of the Cretaceous and Tertiary coal deposits found throughout the plains region of Western North America.

About 60-80% of the sediment making up the coal-bearing Sentinel Butte Formation in North Dakota consists of interbedded silt and clay which occurs in beds ranging from a few millimeters to tens of meters in thickness. From 15% to 35% of the sediments making up these formations consists of silty, fine-grained to medium-grained sand in beds that range in thickness from 0.3 to 35 m (1.0 to 115 ft). Lignite is a minor constituent occurring in seams that range in thickness from 1 cm (0.4 in.) to about 12 m (39 ft) locally. The lignite generally comprises less than 5% of the total thickness of these units in North Dakota.

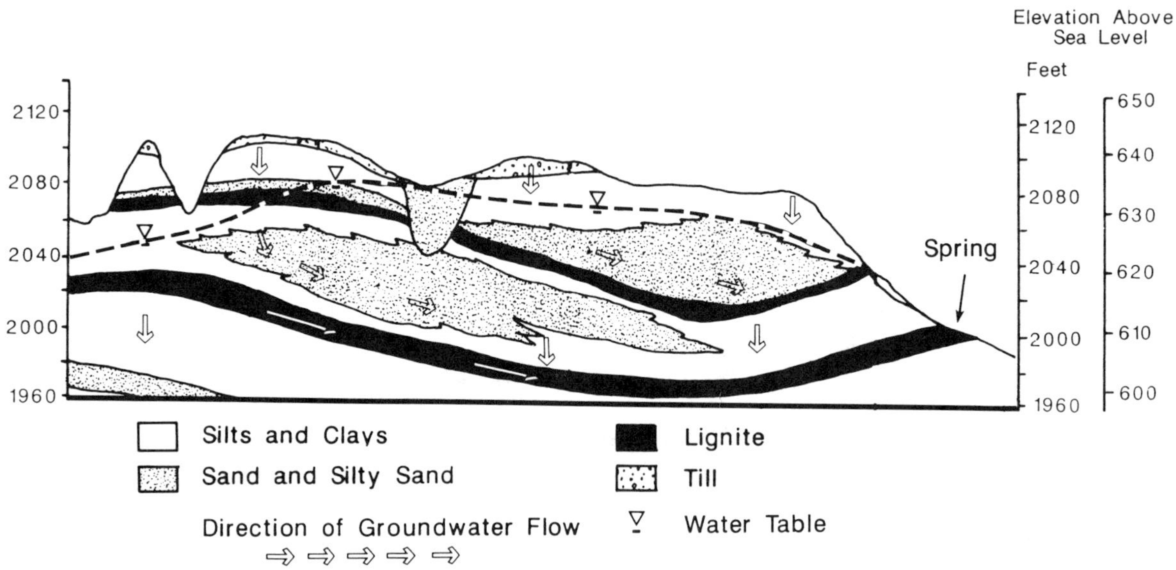

Figure 2. *Schematic diagram of subsurface water movements in much of the plains region (from Moran et al., 1978).*

The sediments of which the rocks are composed were derived from tectonically active areas to the west. The sediments were transported via low-gradient streams which terminated at the shorelines of inland seas. The coal-bearing sediments were deposited in various subenvironments within an alluvial flood-plain environment. Sand and silt were deposited in predominantly oxidizing environments on point bars and natural levees. Clayey sediments were deposited farther from the channels in largely reducing flood-basin settings (Cherven, 1978; Hemish, 1975; Jacob, 1972, 1973).

These differences in the original depositional settings of the sediments are reflected in the mineralogy of the sediments. Carbonate minerals are most abundant in the sands and silts. Clay minerals and pyrite are more abundant in the fine-textured sediments deposited in reducing environments. The predominant clay mineral at all the North Dakota sites is sodium (Na)- montmorillonite (Groenewold et al., 1983). Thus, the original depositional setting of the near-surface sediments in an area has particular significance with respect to hydrogeochemical evolution in that setting, both prior to and after disturbance by mining. Detailed discussions of the geology of the coal-bearing sediments of Western North Dakota and, in particular, the Center, Indian Head, Dunn Center, and Falkirk study sites can be found in Moran et al. (1978) and Groenewold et al. (1979, 1981), whereas Forsman (1985) details the mineralogy of the formation.

Hydrogeology

As is common throughout the plains region of North America, the hydrologic regime at any site is primarily a function of the climate and stratigraphy, as well as the configuration of the landscape. Most precipitation at the study sites is scattered both temporally and spatially and is lost through evapotranspiration.

All study sites are within groundwater recharge areas. Nearly all recharge to the groundwater system occurs during spring runoff and occasionally during periods of relatively heavy precipitation in the fall of the year (Rehm et al., 1982). The upper portion of the landscape thus is constantly subjected to alternating wet and dry cycles (Figure 2). Recharge occurs to some degree over most of the landscape, but is concentrated in restricted positions. These include areas of standing surface water (closed depressions) and ephemeral-stream channels. In most other landscape positions, the majority of the precipitation is lost through either runoff or evapotranspiration.

Subsurface water movement largely is controlled by the lithologic characteristics of the sediments. Movement of water through sand and lignite aquifers generally is lateral, whereas movement though fine-textured sediment generally is vertical and downward. The hydraulic conductivity of sediments at the North Dakota sites varies considerably, ranging from 10^{-5} to 10^{-6} m/sec in sandy sediments, from 10^{-5} to 10^{-7} m/sec in the lignite, and from 10^{-8} to 10^{-10} m/sec in the fine-textured sediments (Groenewold et al., 1979; Rehm et al., 1980). The hydrogeologic characteristics of the North Dakota study sites have been discussed in detail by Moran et al. (1978), Groenewold et al. (1979), and Rehm et al. (1980).

Subsurface Water Flow and Occurrence in Spoils

The vast majority of recharge in undisturbed settings in Western North Dakota results from sporadic depression-focused infiltration (Rehm et al., 1982). Evaluation of recharge in reclaimed spoils at the Center and Indian Head mines, based upon stable isotope data, flooding experiments, and sporadic availability of water from pressure-vacuum lysimeters in the unsaturated portion of the spoils at positions well below the rooting zone suggests recharge conditions similar to those which occur in undisturbed settings in Western North Dakota (Groenewold et al., 1983). The values for hydraulic conductivity in the spoils fall within the same range as sediments in undisturbed settings (Groenewold et al., 1983). In addition, the post-mining position of the water table, as observed in spoils at the Center and Indian Head mines, is approximately the same as in the premining settings. Other similarities between undisturbed and spoils settings include

1) hydraulic conductivities of the lignite aquifers and the corresponding rubble zone at the base of the spoils, and

2) the position of the water table as documented at the Indian Head and Center mines.

HYDROGEOCHEMICAL MODEL

A conceptual model developed to account for the observed chemical characteristics of subsurface water in Western North Dakota has been discussed in several publications (Moran et al., 1978; Groenewold et al., 1979, 1981, 1983). The following discussion is drawn largely from those publications.

Table 1 is a summary of groundwater chemical characteristics in shallow lignite aquifers at five North Dakota study sites. Groundwater of similar quality is found throughout the plains region of North America.

Important geochemical processes that are active in the soil and subsoil and their relationship to salt accumulation in the plant rooting zone, infiltration events, and groundwater chemical evolution are illustrated in Figure 3. Normal rainfall and snowmelt events (on the right side of Figure 3) result in infiltration of water below the ground surface, but not beyond the root zone. During these normal infiltration events, acidity results from the production of CO_2, the biochemical decay of organic matter, and to a lesser extent, the oxidation of pyrite which is finely but widely dispersed throughout the landscape (Groenewold et al., 1983). The pH is buffered by the dissolution of carbonate minerals. Evapotranspiration causes the concentration of salts in soil water and, consequently, the precipitation of those salts, usually within 1-3 m (3-10 ft) of the land surface. By this mechanism, soluble salts, particularly gypsum, are generated and occasionally concentrate in horizons above the water table. Areas of low permeability accentuate the salinity by retarding water movement and permitting maximum evaporation.

	Field temp., °C	Field pH	Field cond., μs/cm	TDS	Ca	Mg	Na	K	HCO_3	Cl	SO_4
Falkirk Site-Hagel Lignite Bed											
$\bar{x}$	7.9	7.3	1603.0	1169.0	122.9	66.4	174.5	8.8	575.5	16.0	447.2
s	0.7	0.5	1332.2	1210.3	104.9	53.1	228.5	5.4	184.5	18.3	680.2
n	10	7	10	10	10	10	10	10	10	10	10
high	9.0	7.75	5000	4310	400	204.8	620	20.2	932.1	46.9	2264
low	7	6.46	630	195	36.2	20.4	9.5	4.4	378	0.5	20.6
Indian Head Site-Beulah-Zap Lignite Bed											
$\bar{x}$	—	7.6	3213.4	2617.6	51.7	30.9	742.8	27.6	1099.6	20.0	937.4
s	—	0.2	1803.9	1565.4	25.9	11.7	414.4	7.3	616.0	12.0	516.0
n	—	4	5	5	5	5	5	5	5	5	5
high	—	7.83	5587	4390	75.0	44	1107	30.5	1663	37.7	1608
low	—	7.28	1497	938	19.4	12.8	278	18.5	415	8.2	409
Beulah-Hazen Site-Various Lignite Beds											
$\bar{x}$	7.5	7.48	2700.0	1480.1	39.1	24.6	415.5	10.3	818.7	14.5	427.8
s	4.2	1.01	1874.3	853.3	37.2	19.9	337.6	7.3	465.6	12.4	321.7
n	9	5	6	12	12	12	12	12	12	12	12
high	12	8.69	6250	3380	87	54.1	1039	31.4	1779	39.8	1211
low	1	6.07	1100	482	3.0	2.4	99	4.5	221	4.3	123
Center Site-Kinneman Creek Lignite Bed											
$\bar{x}$	10.4	6.7	1655.6	1229.3	84.3	51.9	249.7	18.8	651.1	8.1	372.2
s	2.8	4.6	437.1	33.3	47.0	20.3	122.7	11.3	208.7	8.0	137.5
n	9	9	9	9	9	9	9	9	9	9	9
high	16.0	7.5	2200	1631.0	165.0	94.4	441.0	48.0	1021.1	21.0	642.7
low	6.5	6.0	975	838.0	29.0	29.0	115.0	11.4	334.3	0.5	232.6
Dunn Center Site-Various Lignite Beds											
$\bar{x}$	9.3	7.06	3000	1602.5	113.0	56.1	388.0	11.1	665.3	16.5	821.1
s	2.2	—	—	831.3	108.8	39.3	226.9	7.3	375.9	18.7	591.5
n	4	1	1	4	7	7	7	7	7	6	7
high	12	—	—	2256	334	103	700	23	1405	38	1760
low	7.0	—	—	413	8.0	1.0	71	2.4	350	2.7	70

Table 1. *Summary of chemical analyses of groundwater from shallow (less than 75 m (250 ft) deep) lignite aquifiers at the five North Dakota study sites (from Groenewold et al., 1981). All concentrations are given in milligrams per liter. TDS = total dissolved solids.*

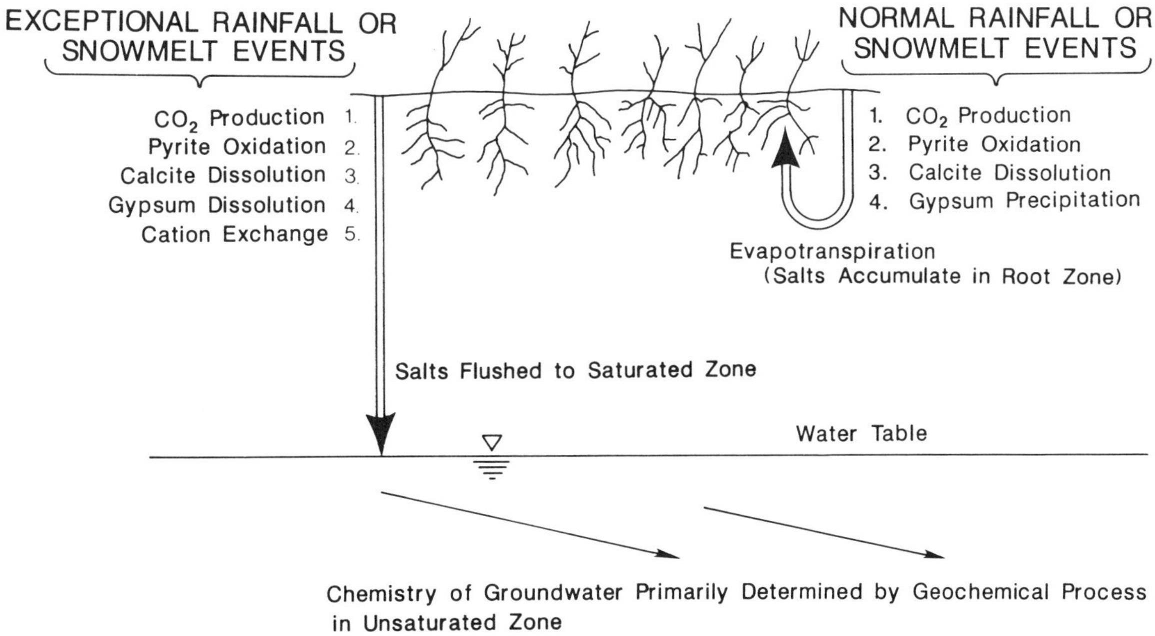

Figure 3. *Schematic diagram of chemical processes and salt movement in coal-bearing strata of the plains region (from Moran et al., 1978).*

Exceptional rainfall or snowmelt events (as demonstrated on the left of Figure 3) are those which cause infiltration below the root zone and, thus, eventually produce groundwater recharge. Carbon dioxide production, pyrite oxidation, and calcite dissolution occur as during normal events. The "flushing" action of such an event, however, results in dissolution of the gypsum, thereby generating major groundwater concentrations of SO_4^{2-}. Dissolution of calcite (and occasionally dolomite) is the main mechanism by which HCO_3^- is generated in the groundwater. Cation exchange in the form of calcium ions (Ca^{2+}) for Na^+ on Na-montmorillonitic clays results when excess Ca^{2+} from calcite and gypsum dissolution is available to the infiltrating waters. Exchange of Ca^{2+} for Na^+ on the exchange sites results in an increase in dissolved Na and further dissolution of calcite. Through the mechanism of cation exchange, high concentrations of Na^+ can be generated in the subsurface.

This framework of geochemical processes accounts for the Na-, Ca-, SO_4^{2-}-, HCO_3^--bearing water which is common in the Tertiary-age deposits of the Northern Great Plains as well as for the chemical characteristics of saturated-paste extracts. More importantly, it demonstrates the significance of geological variability as related to hydrochemical evolution in the near-surface environment. For example, in settings where Na-montmorillonite is essentially absent (e.g., areas of sandy bedrock), Ca^{2+}, HCO_3^-, and SO_4^{2-} will be the dominant ions. If pyrite or secondary gypsum are absent (e.g., some areas of glacial and sandy bedrock sediments), Na^+ and HCO_3^- will dominate.

The significant role played by gypsum is worthy of note. Elevated salinities in paste-extract samples, expressed in terms of electrical conductivity (EC), result from the dissolution of gypsum, as do the elevated levels of SO_4^{2-} in groundwater. Because gypsum does not appear to be a primary constituent of the sediment, but rather an alteration product of pyrite, the sulfate content of the saturated-paste extract is determined by the state of oxidation. Salinity and sodium adsorption ratio (SAR) values will be particularly high in settings that are characterized by a minimum of flushing. The degree of flushing is determined by landscape position and texture of the materials. Knolls and certain side-slope positions are the least likely to allow infiltration below the rooting zone. This will be compounded if the near-surface materials in these settings have very low permeabilities.

These observations have serious implications from the standpoint of reclamation design. Of particular concern is the potential for salt generation and elevated SO_4^{2-} concentrations in subsurface waters in post-mining landscapes where unoxidized overburden, characterized by relatively high concentrations of sulfides, is placed in the near-surface oxidizing environment during mining and contouring operations. In the plains region, overburden materials containing significant concentrations of sulfides also typically contain high concentrations of sodic clays (Groenewold et al., 1981). Thus, such settings are potential sites for elevated SAR values and the generation of highly sodic groundwater.

In summary, it appears that the chemical composition of groundwater and of saturated-paste extracts in much of the plains areas of Western North America is a function of three factors: a) the amount of calcite (and dolomite), Na-montmorillonite, and pyrite in the near-surface sediments; b) the degree of oxidation of pyrite and the resulting gypsum in the near-surface sediments; and c) the degree of flushing that occurs in the near-surface sediments.

HYDROGEOCHEMISTRY OF SPOILS SETTINGS

Material Redistribution and Increased Availability of Oxygen

The previous discussion summarized the hydrogeochemical model as it relates to chemical evolution of subsurface water in undisturbed settings in Western North Dakota. The processes and mechanisms described in the model account for the observed chemical characteristics of subsurface water in these settings.

The premining landscape consists of various components that are more or less in equilibrium with one another and with the forces acting upon them. Redistribution of overburden by a dragline during surface mining operations has been described in detail by Winczewski (1977). That study, which focused on the Indian Head Mine, indicated that spoils handled by dragline consist of nonhomogeneous, semistratified materials. That study also indicated that, during mining operations, sediments, from the mildly reducing environment below the position of the premining water table, often are emplaced in the near-surface oxidizing environment of the post-mining landscape. Conversely, it was observed that sediments from the oxidized portion of the premining landscape often are emplaced in the saturated basal portion of the spoils.

The redistribution of overburden sediment suggests the potential for significant changes in the chemical reactivity of the landscape:

1) Emplacement of unoxidized sediments in the oxidizing, near-surface environment increased the potential for sulfide oxidation, sulfate dissolution/precipitation, and cation exchange on clays.

2) The rapid dissolution of sulfate salts in sediments from the near-surface portion of the premining landscape emplaced at the base of the spoils or those which had been oxidized during the mining process.

3) The generation of highly mineralized water results, at least initially, from the oxidation of sulfides present in the unsaturated portion of the spoils by air (oxygen) trapped in the spoils.

Comparison of Subsurface Water Chemical Characteristics in Spoils and Undisturbed Settings

A Na-, $CaSO_4$-, HCO_3^--bearing to HCO_3^--, SO_4^{2-}-bearing groundwater was documented at both the Center and Indian Head mines prior to mining (Groenewold et al., 1983) (Table 2), with the major difference being in the degree of mineralization. The mean total dissolved solids (TDS) concentrations in groundwater from near-surface lignite aquifers at the Center and Indian Head mines are 1229.3 and 2617 mg/L, respectively (Table 1).

These differences, in turn, reflected differences in the mineralogy of the overburden at the two sites. The overburden sediments at the Indian Head Mine commonly are enriched in fine-textured, Na-montmorillonitic materials. The average clay content, based on 13 analyses of overburden from the Indian Head Mine, was 57%. In contrast, the overburden-averaging on coarser-textured material revealed sediments relatively deficient in sodic clays. At the Center Mine, the sediment averaged 30% clay on the basis of 26 overburden analyses. The hydrogeochemical model suggests that the Na^+/Ca^{2+} ratio is the major factor controlling sulfate concentrations in shallow groundwater in this region, just as the relatively sodic overburden at the Indian Head Mine produces more highly mineralized groundwater than at the Center Mine. Predominant species in solution at both mines are typically Na and sulfate.

Data generated by analyses of saturated-paste extracts of near-surface spoils materials from various locations at the Center and Indian Head mines also indicate significant differences in the chemistry of spoils at the two mines. The mean EC for spoils at the Center Mine is 2.9, with a standard deviation of 1.6. The mean EC for spoils at the Indian Head Mine is 5.2, with a standard deviation of 2.0. The mean SAR for spoils at the Center Mine is 3.6, with a standard deviation of 2.9. SAR values for spoils at the Indian Head site have a mean of 13.9, with a standard deviation of 5.6. These data indicate that the near-surface spoils at the Indian Head site have the capacity to generate much more highly mineralized pore water than spoils at the Center Mine.

Although groundwater in spoils at the Center and Indian Head mines are the same as in the undisturbed settings, concentrations in the spoil waters are typically two to three times higher than in undisturbed settings.

	Field temp., °C	Field pH	Field cond., μsec/cm	TDS	Ca	Mg	Na	K	HCO_3	CO_3	SO_4	Cl
Center												
$\overline{x}$	10.5	7.1	4969	3283	281	145	497	20	958	304	1516	7
s	2.8	0.6	2181	1326	180	103	289	11	294	2	789	15
n	98	98	98	97	98	98	98	98	98	42	98	57
Indian Head												
$\overline{x}$	12.6	7.0	8446	5479	256	164	1321	35	1087	3	2762	10
s	2.4	0.6	5339	3460	212	177	853	36	482	3	1901	9
n	40	38	40	40	39	39	39	39	40	33	40	39

Table 2. *Summary of groundwater chemical data from reclaimed spoils at the Center and Indian Head mines (from Groenewold et al., 1983). All concentrations are given in milligrams per liter. TDS = total dissolved solids.*

This suggests that the same hydrogeologic and hydrogeochemical mechanisms and processes are operating in spoils settings as in undisturbed settings, with the difference in water chemistry resulting from the increased chemical reactivity in the spoils.

INTERACTIONS BETWEEN COAL-CONVERSION WASTES AND THE DISPOSAL SETTING

Fly ash and FGD wastes were studied in an 8-yr-long field investigation in the region and were reported in Beaver et al. (1987) who dealt with waste/setting interactions in a reclaimed, Western North Dakota, lignite strip mine setting. Coal-utilization wastes were generated at a lignite-fired, steam electric-power station near Center, North Dakota, and were buried in unsaturated and saturated disturbed settings within the nearby strip mine. These settings included between spoil ridges, wet pit, and dry pit. During the course of the study (1978 to 1986), approximately 240 piezometers were used to collect groundwater hydraulic head data and groundwater quality data at specific stratigraphic horizons. In addition, pressure vacuum lysimeters were used to collect the water in the unsaturated zone. Interstitial water was extracted from Shelby tube and auger cutting samples at other critical, but uninstrumented, locations in the study area.

More than 15,000 water-level recordings allowed precise definition of groundwater flow and occurrence at the study site. Evaluation of the hydrogeological data indicates that the hydrogeological regime stabilized near the premining levels after reclamation and that the flow systems are not homogeneous. The base of the spoils commonly constituted the major aquifer in the reclaimed setting.

Leachate Character

More than 12,000 chemical analyses for 1) major cations and anions, 2) trace metals, including As, selenium (Se), Mo, Ba, Pb, chromium, and Hg, and 3) other constituents were performed on groundwater samples from selected settings and strata during this study. Analyses of field samples indicate that both the fly ash and the FGD waste initially generate highly mineralized leachates from entrained water. Leachates were characterized by relatively high Na, Mo, Se, As, and sulfate concentrations. Fly ash generates an initially high-pH leachate, whereas FGD wastes produce leachates at the upper end of the normal pH range.

Leachate pH, typically, was quickly buffered to the six to nine range within the sediments. A naturally occurring amphoteric alkaline-buffering capacity and acid-buffering capacity, provided by an abundance of naturally occurring carbonate minerals available for dissolution under acidic conditions, is proposed as the buffering mechanism. Iron oxyhydroxide coatings on the sediments, formed during previous alternate wetting and drying cycles, also may play a role in alkaline pH buffering.

Groundwater mineralization resulting from coal-conversion waste disposal is in addition to that occurring in the spoils from the strip mining process. However, it is confined to disposal sites where sufficient leachate was generated to allow actual groundwater recharge.

Selected trace element concentrations in waste leachate and groundwater were determined. Arsenic and Se concentration levels are simultaneously attenuated as pH was buffered to within the normal range. Iron oxyhydroxide coatings on the sediments may provide a mechanism for alkaline pH buffering with concurrent As and Se adsorption. Lead and Cd are maintained at low levels through direct carbonate precipitation in a system which contains abundant carbonate anions. Barium is maintained at a low, acceptable level by a sulfate precipitation reaction yielding barite. Molybdenum remains mobile in this setting. Molybdenum mobility may be related to its affinity to form stable complexes in several oxidation states.

If groundwater and/or waste leachate containing Ca encounter Na-montmorillonite clay, cation exchange may occur wherein Ca^{2+} is preferentially adsorbed on the clay particles. This would result in decreased Ca^{2+} concentration and elevated Na^{+} concentration in the water.

If groundwater and/or waste leachate containing sulfate enters a local groundwater environment containing active sulfur-reducing bacteria, the sulfate level decreases, and bicarbonate is produced in the biochemical reaction. Hydrogen sulfide is produced and is usually the first observed indicator that this process is occurring. This reaction typically occurs as the sulfate-rich water moves into the undisturbed sediments below the base of the spoils.

Disposal Site Design Criteria

The single most important factor in preventing groundwater contamination from fly ash and/or FGD waste sites is keeping the wastes as dry as possible and disposing of them in the unsaturated zone. If there is no water present, there is no leaching or transport medium and, consequently, no movement of contaminants. A disposal site well above the reestablished groundwater table, protected from recharge, is considered the optimal setting for disposal of fly ash and FGD waste in both saturated and unsaturated settings. Even if a small amount of leachate forms, natural pH buffering and trace element attenuation mechanisms provide protection against groundwater contamination. Also, a small amount of leachate typically would be eliminated through evapotranspiration and adsorption on the sediments of the unsaturated zone. Lined disposal sites, therefore, are unnecessary to control currently regulated trace elements in a dry setting, but would be necessary to contain Na, Ca, sulfate, and Mo migration in a saturated setting, if containment were deemed advisable.

In addition to the previously described field study, other closely related work has been carried out. Soil attenuation phenomena have been described by Hassett et al. (1988, 1986). Specific waste disposal site selection research has been carried out by Ronnei (1988) and Hunke (1988). Co-disposal of hazardous liquid wastes and cementitious solid wastes has been demonstrated by Hassett et al. (1989).

Ideal Waste Disposal Site Design

The understanding of wastes, geologic and hydrogeologic settings, and their interactions has established a basis for general design of coal-conversion residual-material waste disposal. This design is summarized as follows and in Figure 4:

1) Wastes from coal-conversion technologies are residuals typically returned to the same, or a very similar, setting.

2) The waste materials ultimately will interact with the setting. The question is how to meaningfully control that interaction to minimize any potentially adverse environmental impacts.

3) Synthetic disposal-site liners are short-term solutions to leachate movement and may result in a basin of concentrated leachate.

4) Naturally occurring attenuation mechanisms have a capacity to buffer and/or attenuate problem species. Disposal site design should take advantage of these phenomena.

5) Waste leachate generation can be eliminated or minimized by keeping wastes dry, which means placing the final depositional area above the saturated zone and preventing water from moving downward or laterally through the waste mass.

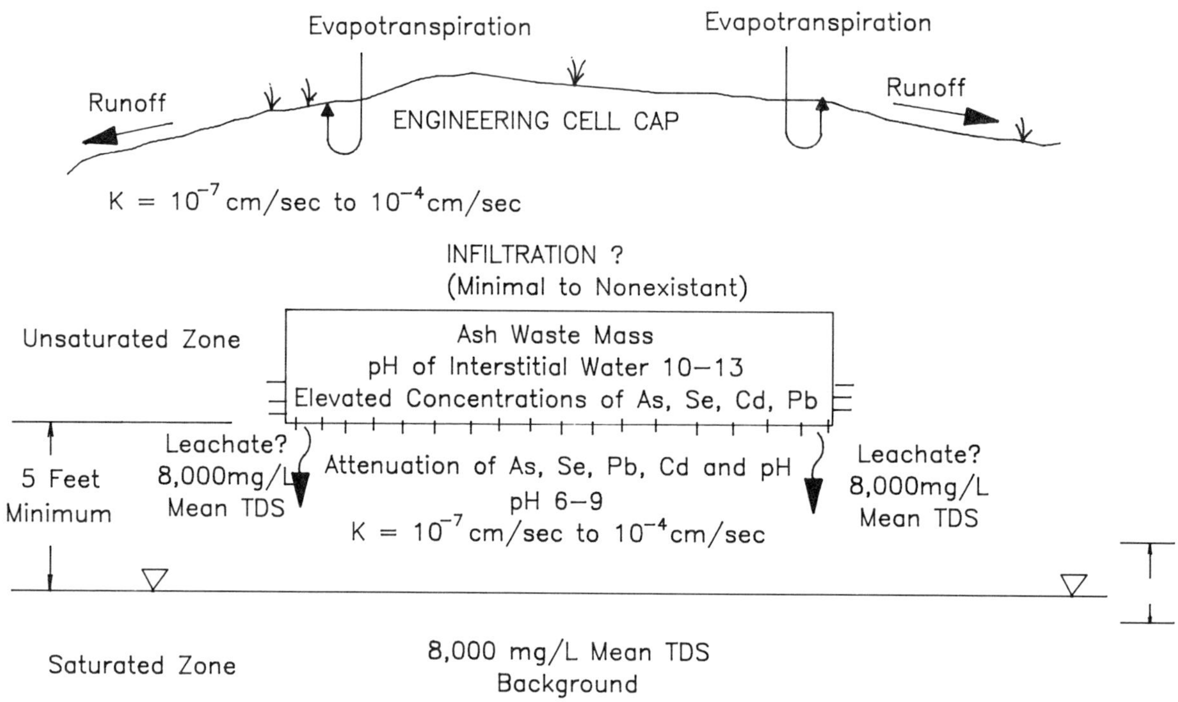

Figure 4. *Idealized waste disposal site design.*

6) The closed waste disposal site can be contoured to divert water from above or nearby toward distant uncontaminated depression-focused recharge basins.

7) Even low hydraulic conductivities in subsurface materials can allow substantial leachate movement over time.

8) Vertical fracturing can be minimized through proper construction techniques.

9) This design places a "roof" over the waste mass to divert water from percolating through the waste mass. Such a cover is easily repaired without disturbing the waste mass. Pit liners in their various forms enhance the formation of a container of concentrated leachate which may overflow. A "basement" or "bathtub" liner is not easily repairable without disturbing the waste mass. Water which does penetrate the near surface has time to be evapotranspired and does not provide downward recharge.

Research Needs

Past and ongoing research has demonstrated the need to understand waste materials as natural resources. This same understanding also is necessary for environmentally acceptable disposal. Fundamental characterization and understanding of many existing coal conversion residuals have been developed over the decades. The residuals from the new generation of conversion technologies, such as fluidized-bed combustion (FBC), require characterization to gain an understanding of their disposal requirements. The residuals from these technologies must be understood more quickly than were those from previous generations.

The greatest opportunities lie in byproduct utilization. The potential for this aspect only has been touched upon in the U.S. In the Netherlands, for example, more than 2,000 products are produced from mineral wool fibers alone, a product which research has shown can be produced from boiler slag and other residuals.

REFERENCES

Beaver, F.W., 1986, The effects of fly ash and flue gas desulfurization wastes on groundwater quality in a reclaimed lignite strip mine disposal site: Grand Forks, University of North Dakota, Unpublished Ph.D. Dissertation, 447 p.

Cherven, V.B., 1978, Fluvial and deltaic facies in the Sentinel Butte Formation, central Williston Basin: Journal of Sedimentary Petrology, v. 48, p. 159-170.

Forsman, N., 1985, Petrology of the Sentinel Butte Formation (Paleocene), North Dakota: Grand Forks, University of North Dakota, Unpublished Ph.D. Dissertation, 218 p.

Groenewold, G.H., 1979, Hydrologic and hydrogeochemical characteristics of selected strip mine spoils in Western North Dakota, in M.K. Wali, ed., Ecology and coal resource development: New York, Pergamon Press, p. 769-774.

Groenewold, G.H., L.A. Hemish, J.A. Cherry, B.W. Rehm, G.N. Meyer, L.S. Clayton, and L.M. Winczewski, 1979, Geology and geohydrology of the Knife River Basin and adjacent areas of west-central North Dakota: North Dakota Geological Survey, Report of Investigation 64, 402 p.

Groenewold, G.H., R.D. Koob, G. McCarthy, B.W. Rehm, W. Peterson, 1983, Geologic and geochemical controls on the chemical evolution of subsurface water in undisturbed and surface mined landscapes in Western North Dakota: North Dakota Geological Survey, Report of Investigation No. 79, 299 p.

Groenewold, G.H., and B.W. Rehm, 1982, Instability of contoured surface mined landscapes in the Northern Great Plains, causes and implications: Reclamation and Revegetation Research, v. 1, no. 2, p. 161-176.

Groenewold, G.G., B.W. Rehm, and J.A. Cherry, 1981, Depositional setting and groundwater quality in coal-bearing strata and spoils in Western North Dakota, in F.G. Ethridge and R.M. Flores, eds., Recent and ancient nonmarine depositional environments: Models for exploration: Society of Economic Paleontologist and Mineralogists Special Publication 31, p. 157-167.

Groenewold, G.H., B.W. Rehm, J.A. Cherry, R.D. Koob, S.R. Moran, C.D. Palmer, and W.M. Peterson, 1980, Potential hydrogeochemical impacts of surface mining in the Northern Great Plains, in Proceedings of the Symposium on Surface Mining Hydrology, Sedimentology, and Reclamation, Lexington, KY, December 1-5, 1980: Lexington, University of Kentucky, p. 465-472.

Hassett, D.J., 1987, A generic test of leachability: The synthetic groundwater leaching procedure, in Proceedings of the Waste Management for the Energy Industries Conference, Grand Forks, North Dakota, April 29-May 1, 1987: Grand Forks, University of North Dakota, p. 31-39.

Hassett, D.J., and G.H. Groenewold, 1987, Attenuation capacity of Western North Dakota overburden sediments: North Dakota Mining and Mineral Resources Research Institute, Bulletin No. 86-04-MMRRI-01, 105 p.

Hassett, D.J., G.H. Groenewold, R.J. Stevenson, C.R. Schmit, and D.J. Daly, 1988, Ion exchange and soil attenuation of Gulf Coast and Great Plains soils: Grand Forks, University of North Dakota, Energy and Environmental Research Center, 91 p.

Hassett, D.J., D.F. Hassett, P. Kumarathasan, and G.J. McCarthy, 1989, Ettringite as an agent for the fixation of hazardous oxyanions, in Proceedings of the Twelfth Annual Madison Waste Conference on Municipal and Industrial Waste, Madison, Wisconsin, September 20-21, 1989: Madison, University of Wisconsin-Madison, Department of Engineering Professional Development, p. 471-482.

Hemish, L., 1975, Stratigraphy of the upper part of the Fort Union Group in southwestern McLean County, North Dakota: Grand Forks, North Dakota, University of North Dakota, Unpublished M.S. Thesis, 160 p.

Hunke, N.T., 1988, Hydrogeologic and hydrogeochemical evaluation of a proposed flue gas cleanup dry waste disposal pit near Beulah, North Dakota: Grand Forks, University of North Dakota, Unpublished M.S. Thesis, 186 p.

Jacob, A.F., 1972, Depositional environments of parts of the Tongue River Formation, Western North Dakota, in F.T.C. Ting, ed., Depositional environments of the lignite-bearing strata in Western North Dakota: North Dakota Geological Survey, Miscellaneous Series 50, 134 p.

Jacob, A.F., 1973, Depositional environments of Paleocene Tongue River Formation, Western North Dakota: American Association of Petroleum Geologists Bulletin, v. 57, no. 6, p. 1038-1052.

McCarthy, G.J., F.W. Beaver, R.J. Stevenson, G.H. Groenewold, and D.J. Hassett, 1987, The essential role of mineralogical characterization in utilization and disposal of coal conversion ashes, in Proceedings of the Waste Management for the Energy Industries Conference, Grand Forks, North Dakota, April 29-May 1, 1987: Grand Forks, University of North Dakota, p. 41-62.

Moran, S.R., J.A. Cherry, B.W. Rehm, and G.H. Groenewold, 1979, Hydrologic impacts of surface mining of coal in Western North Dakota, in Proceedings of the Symposium on Surface Mining, University of Kentucky, Lexington, Kentucky, December 4-7, 1979: Lexington, University of Kentucky, p. 57-65.

Moran, S.R., J.A. Cherry, J.H. Ulmer, W.M. Peterson, M.H. Somerville, J.K. Schafer, D.O. Lechner, C.L. Triplett, G.R. Loken, and P. Fritz, 1976, An environmental assessment of a 250-MMSCFD dry ash Lurgi coal gasification facility in Dunn County, North Dakota: University of North Dakota Engineering Experiment Station, Bulletin 76-12-EES-01, Vol. 1 of 3, 126 p.

Moran, S.R., G.H. Groenewold, and J.A. Cherry, 1978, Geologic, hydrologic, and geochemical concepts and techniques in overburden characterization for mined land reclamation: North Dakota Geological Survey, Report of Investigation 63, 152 p.

Moran, S.R., G.H. Groenewold, L.A. Hemish, and C.A. Anderson, 1975, Development of a premining geological framework for landscape design reclamation in North Dakota, in Proceedings of the Fort Union Coal Field Symposium, Billings, Montana: v. 3, p. 308-316.

Rehm, B.W., G.H. Groenewold, and K. Morin, 1980, Hydraulic properties of coal and related materials, Northern Great Plains: Ground Water, v. 18, no. 6, p. 551-561.

Rehm B.W., G.H. Groenewold, and W. Peterson, 1982, Mechanisms, distribution, and frequency of groundwater recharge in upland areas of Western North Dakota: North Dakota Geological Survey, Report of Investigation 75, 72 p.

Ronnei, D.M., 1987, Geohydrologic evaluation of a proposed coal-ash disposal site near Mandan, North Dakota: Grand Forks, University of North Dakota, Unpublished M.S. Thesis, 206 p.

Williams, T.T., R.D. Koob, and P. Rechard, 1978, Surface and groundwater problems associated with potential strip mine sites: U.S. Environmental Protection Agency, Report R-303727.

Winczewski, L.M., 1977, Western North Dakota lignite strip mining processes and resulting subsurface characteristics: Grand Forks, University of North Dakota, Unpublished M.S. Thesis, 433 p.

GEOLOGICAL CONTROLS ON COAL SEAMS BURNING "IN PLACE"[1]

Fariborz Goodarzi
Institute of Sedimentary and Petroleum Geology
Geological Survey of Canada
3303 - 33rd Street N.W.
Calgary, Alberta T2L 2A7
Canada

Thomas Gentzis
Alberta Research Council
Coal Research Centre Devon
One Oil Patch Drive
Devon, Alberta T0C 1E0
Canada

ABSTRACT

The combustion of coal seams in Western Canada has either external causes (i.e., lightning strikes or forest fires) or internal causes (spontaneous combustion). The geology of the coal-bearing strata has great control on the susceptibility of a coal seam to combustion. The following geological factors affect the ignition of coal seams: 1) thickness of coal overburden; 2) water saturation of the coal seam; 3) lithologic composition of the coal-bearing section; 4) morphology of the coal seam; and 5) rank and composition of the coal.

The combustion of coal seams often is facilitated by thin sedimentary cover and the development of combustion vents which allow the intake of air and release of volatile matter. If a coal seam is saturated by water or if the water table is high, the combustion advance either will be reduced or prevented completely. If the coal seam is not water saturated, the combustion "front" will advance and consume the entire seam.

Sedimentary partings (sandstone and siltstone) present in coal-bearing strata may act as thermal blankets and transmit less heat to the underlying coal. Therefore, if the heat is directional from the top of the section, coal beneath sedimentary partings is less combusted than the coal above. Occasionally, buoyancy-driven convective thermal transport may be more significant than thermal conduction in moving heat through the coal and surrounding strata.

The morphology of combusted coal seams is related to the rank of the coal and the temperature in the burning zone during combustion. Heat generated by combustion often transforms the coal beneath the combustion zone into a porous, cohesive coke for coals in the coking rank and into noncohesive chars for other coals.

---

1. Geological Survey of Canada Contribution No. 37489.

Both rank and composition are important in the preservation of combusted seams. High-rank coals are transformed into coke (if their rank is in the coking coal range) or char and maintain the seam morphology, whereas low-rank coals often are transformed into ash. Finally, the higher the content of vitrinite and exinite (reactive macerals) in the coal, the higher is the propensity of the coal for spontaneous combustion.

INTRODUCTION

Combustion of a coal seam may start as a result of lightning strikes or forest fires (Pearson and Creaney, 1980; Bustin and Mathews, 1982; Goodarzi, 1987; Goodarzi et al., 1988). Combustion in coal stockpiles generally has internal causes (i.e., spontaneous combustion) (Gentzis and Goodarzi, 1989; Goodarzi and Gentzis, in press). A coal seam or coal-bearing strata may burn over a relatively long period of time. A coal seam at Aldridge Creek, British Columbia, Canada, has been burning since 1936 (Miard, 1945) and coal-bearing strata at Coalspur, Alberta, Canada, have been burning since 1922 (Fryer, 1981).

The combustion of coal and heat-affected coal seams can be observed in the field and detected by

a) the presence of rusty red-brown to burgundy-colored ash, representing the heat affected roof of a coal seam, and cracks at the top and side of burning coal seams (Figure 1),

b) appearance of water condensate (steam) emanating from the strata,

c) increase in temperature, formation of haze over the affected area, and appearance of volatile matter (Figure 2);

d) a tarry, aromatic, "scorched" scent; and

e) occurrence of sulfur crystals (Figure 3) and tar (Figure 4) on the surface.

DISCUSSION OF COMBUSTION CONTROLS

An understanding of the oxidation-combustion process is essential to the discussion of geological factors affecting the combustion of coals. Combustion involves a very complicated reaction system, but in its simplest form can be described as:

The geology of the coal-bearing strata has great control on the susceptibility of a coal seam to in-place combustion. The combustion of coal seams is geologically controlled by the following factors:

1) thickness of coal overburden

2) water saturation of the coal seam

3) lithologic composition of the coal-bearing strata

4) morphology of the coal seam

5) rank and composition of coal

Figure 1. Photograph of cracks formed on the top and side of the burning coal seam at Coalspur, Alberta.

Figure 2. Photograph of emission of volatile matter from the burning coal seam at Coalspur, Alberta.

Figure 3. Photograph of sulphur crystals (S) on the surface of loose sediments near a combustion vent (C) at Coalspur, Alberta.

Figure 4. Photograph of occurrence of tar (T) on the surface of the burning coal seam at Coalspur, Alberta.

Thickness Of Coal Overburden

Exposure of coal to oxygen (oxidation-weathering) is essential for the ignition of coal. Weathering occurs when the coal is exposed to air due to the removal of the sedimentary cover. An exposed coal seam is an easy target for ignition due to a fire hazard, such as the case of the burning coal seam at Aldridge Creek, British Columbia (Figure 5,A) which started because of a forest fire.

The thickness of sedimentary cover is an important factor in the combustion of coal seams. A coal seam with relatively thick overburden is less likely to be affected by burning due to a forest fire or ignition by a lightning impact, because the heat generated by these external sources is absorbed by the sedimentary cover first. Therefore, the sedimentary cover acts as a barrier and the coal seam often is not affected by the heat.

An example of this process is observed in Seam 15 at Eagle Mountain, British Columbia (Figure 5,C). The high-volatile bituminous coal in the upper part of this seam was transformed into high-temperature char of meta-anthracitic rank by combustion caused by a lightning strike (Pearson and Creaney, 1980). The vertical vitrinite reflectance variation of a section from this coal seam (Goodarzi and Gentzis, in press, Section E), indicates that coal at the top of the seam is converted to anthracitic char (%R_o max $\approx$ 4.0) followed by a sudden drop in reflectance in the underlying sedimentary parting. A coal layer underneath this sedimentary parting has a much lower reflectance (%R_o max $\approx$ 2.0), indicating clearly that the sedimentary parting acted as a heat barrier in this case. The thickness of the sedimentary parting between the high-temperature, anthracitic char and the low-temperature char is only 1 m (3 ft) (Figure 6).

The combustion of coal seams often is facilitated by a thin sedimentary cover and the development of combustion vents (Figure 1), which allow the intake of air into the combustion zone of the burning coal seam and the release of volatile matter into the atmosphere. The burning coal seam at Coalspur, Alberta, is an example of combustion with a thin sedimentary cover, the combustion zone being present approximately 10 cm (4 in.) beneath the ash and loose sedimentary cover (Figure 7).

Water Saturation

A sudden and sharp boundary often is observed between the burnt zone and the unaltered coal in a seam. Coal from Hat Creek, British Columbia is an example of such a sudden termination of burning of coal with no obvious cause.

The advance of the combustion "front" depends largely on the water content of the coal seam. Coal seams often are water saturated and, depending on the rank of coal, they may act as aquifers (especially subbituminous coal; Van der Flier-Keller and Fyfe, 1987). If the seam is saturated by water or if the water table is high, the combustion advance either will be reduced or prevented completely. If, on the other hand, the seam is not water saturated, combustion will advance and consume the entire seam if sufficient oxygen is available.

A good example is Seam 15 at the Fording Coal Property in British Columbia. The seam was weathered, therefore exposed, before being introduced to high temperatures during a lightning strike and subsequently carbonized (Goodarzi and Gentzis, in press). Although burning converted most of the bituminous coal to meta-anthracitic char, there are areas in the seam, only meters away from the carbonized coal, that have not been affected at all by combustion. It has been postulated (Goodarzi and Gentzis, in press) that the unaffected part of the seam may have been saturated by water, thus preventing the advance of the combustion front (Figure 8).

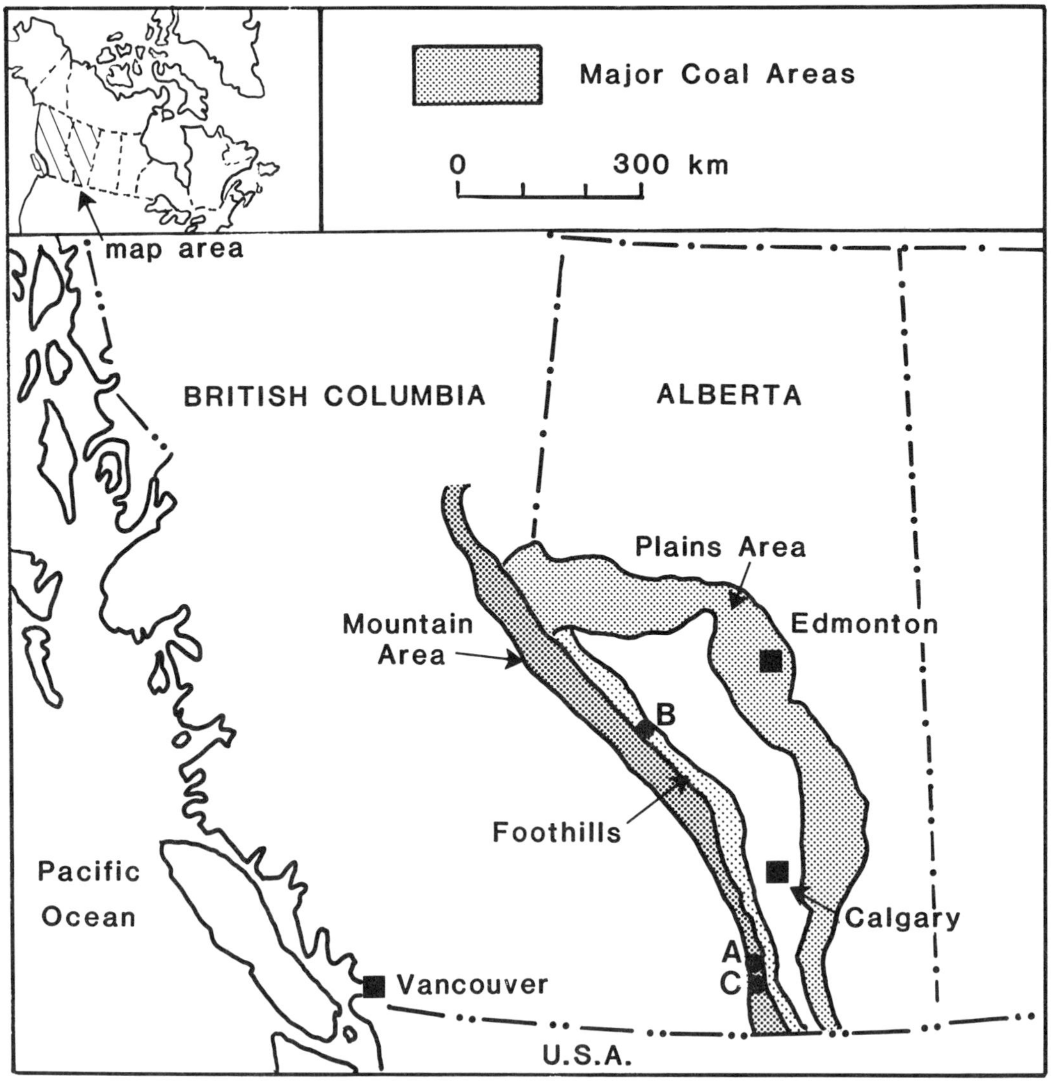

Figure 5. Locations of burning and heat-affected coal seams at Aldridge Creek (A), Coalspur (B), and Seam 15 (C).

Pyrolysis is neutral to slightly endothermic. The lightning strike resistively heated the coal to pyrolysis temperatures, but reaction stopped where the heat was used up as it would have been if a water-saturated zone were encountered by the combustion front.

Also, if there is inadequate oxygen (e.g., as in buried coal seams) no ignition takes place. This could be an alternate explanation in the case of Seam 15 at the Fording Coal Property, although the seam itself is present as surface outcrop and is not buried.

Lithology

A coal-bearing section may consist of a number of coal seams and sedimentary partings. The lithologic composition of the coal-bearing section is important in controlling the extent of combustion. During combustion, coal seams often consume the total energy produced in the combustion zone in order to convert coal into char or coke (Goodarzi, 1987; Goodarzi et al., 1988). Therefore, the coal seams act as thermal blankets.

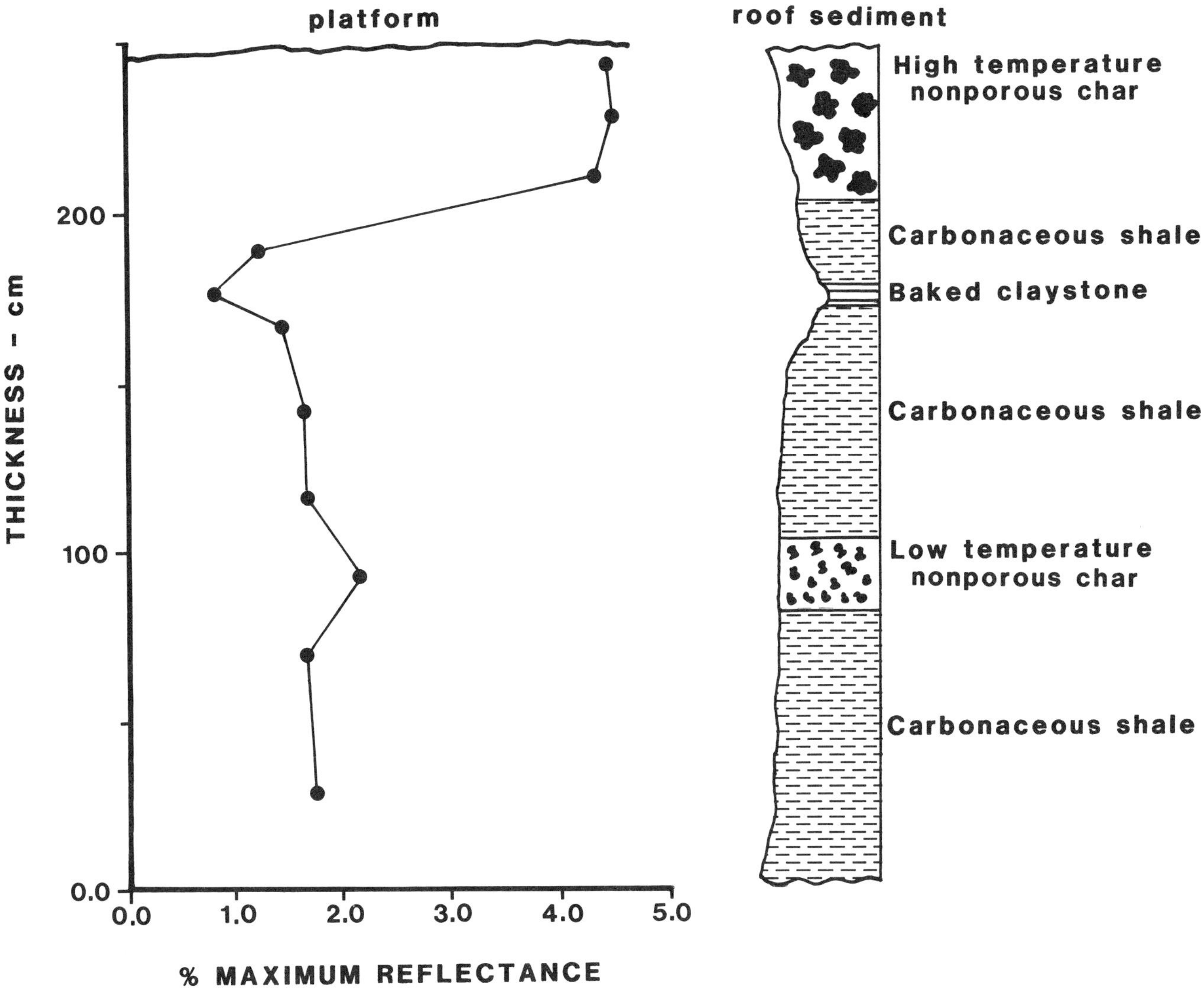

Figure 6. *Reflectance (%R_o maximum in oil) versus thickness profile for heat-affected Seam 15 at Eagle Mountain, Fording Coal Mine, British Columbia.*

The profile of a burning coal seam often shows a zonation based on the temperature and accessibility of coal to oxygen. The hottest area is in the combustion zone (Figures 7 and 9), where coal is consumed, followed by a carbonization zone, where coal is carbonized due to heat generated in the combustion zone. The carbonization zone acts as a thermal blanket, and coal beneath the carbonization zone often remains unaltered. This is similar to a cross section of coal in a coke oven before the termination of coking.

Examples of the profile of burning coal seams showing the temperature zonation of the seams and thermal blanketing of carbonization zones are given by Goodarzi (1987) and Goodarzi et al. (1988) (see also Figures 6, 7, and 9). The sedimentary partings, particularly the silicates (sandstone, shale, etc.) within coal seams or coal-bearing strata, may act as thermal blankets and transit less heat to the surrounding strata. Often coal beneath a sedimentary parting is less metamorphosed than the coal above, particularly when the heating is directional and from the top (Figures 6 and 9). However, the amount of heat transmitted by sedimentary partings is related to the thickness of the partings, the initial temperature level generated by the burning of coal, and the rate of heating.

An example of the effect of a parting on combustion is a burning bituminous coal seam (Figure 9) at Aldridge Creek, British Columbia (Figure 5,A). Heat generated from combustion of the seam was carried upwards by convecting gas, thereby coking the coal and baking the roof rocks (Goodarzi et al., 1988). The roof rocks consisted of light gray sandstone which had collapsed, formed clinker, and melted completely due to the high temperature (~1,100°C) produced by the combustion (Goodarzi et al., 1988).

However, the profile of the burning coal seam (Figure 9) shows that there is a major reflectance break at the boundary between coke and semi-coke. Some of the directional heating from the roof towards the floor has been stopped by the sedimentary parting. As a result, the coal in the upper part of the seam is in the semi-coke to coke stage (%R_o max = 1.5-10.0), whereas the coal underneath the 45-cm thick parting is unaltered (%R_o max = 1.1).

Morphology Of Heat-Affected Coal Seams

The morphology (or texture) of burnt and coked products of a burning coal seam is related to the rank of coal, temperature of combustion, and weathering of the coal.

Rank

For coals of coking rank (%R_o max = 0.9-1.6), heat generated by combustion of coal transforms the coal beneath the combustion zone into a porous, cohesive coke, whereas for non-coking coal (%R_o max = 0.7-1.0) the coal is transformed into a noncohesive char (Goodarzi, 1987, Goodarzi et al., 1988). The formation of a cohesive coke layer also can protect the surface of unburnt coal by keeping air from it, whereas the formation of noncohesive ash probably will not result in a protective layer. Therefore, the formation of protective layers depends on the forces acting to disrupt the coke, char, or ash from the unburnt surface (R.F. Chaiken, personal communication, 1989). In addition, the morphology of sediments forming the roof of the coal seam is dependent on the temperature generated by coal combustion. At high temperatures, the roof sediments melt, whereas at low temperature they remain unfused.

Temperature of Combustion

Temperature of combustion is related to the rank and availability of oxygen. In coking coals, a porous coke is formed which allows the easy access of hot gases and oxygen to the less altered coal beneath the coke. An example is the burning coal seam at Aldridge Creek, where the by-products of coking (i.e., coal tar) penetrate into the porous semi-coke underneath the coke. The depth of alteration of coal is greater for coking coals (Figure 9). The formation of coke results in the deformation of the coal seam due to the expansion and contraction of coal.

In non-coking coals (e.g., coal from Coalspur (%R_o max = 0.70)) the char is nonporous. Therefore, the depth of penetration of oxygen is limited (Figure 7). This results in the restricted alteration of coal (Goodarzi, 1987; Goodarzi et al., 1988). The samples of hot, angular char (oxidized or combusted residues of coal) from the combustion zone of the Coalspur burning coal seam, when exposed to air overnight, were converted into ash without changing their angularity. Micro-layering was evident in the char fragments (Figure 11) (Goodarzi and Jerzykiewicz, 1986).

Weathering

Weathering (oxidation) reduces the coking property of coal. Therefore, a weathered coking coal may behave as a non-coking, low-rank coal and, instead of producing a cohesive porous coke, it develops angular char. An example of the influence of weathering is Seam 15 at Eagle Mountain in British Columbia. The original rank of this seam is high-volatile bituminous A (%R_o max = 0.9). This coal would be converted to cohesive

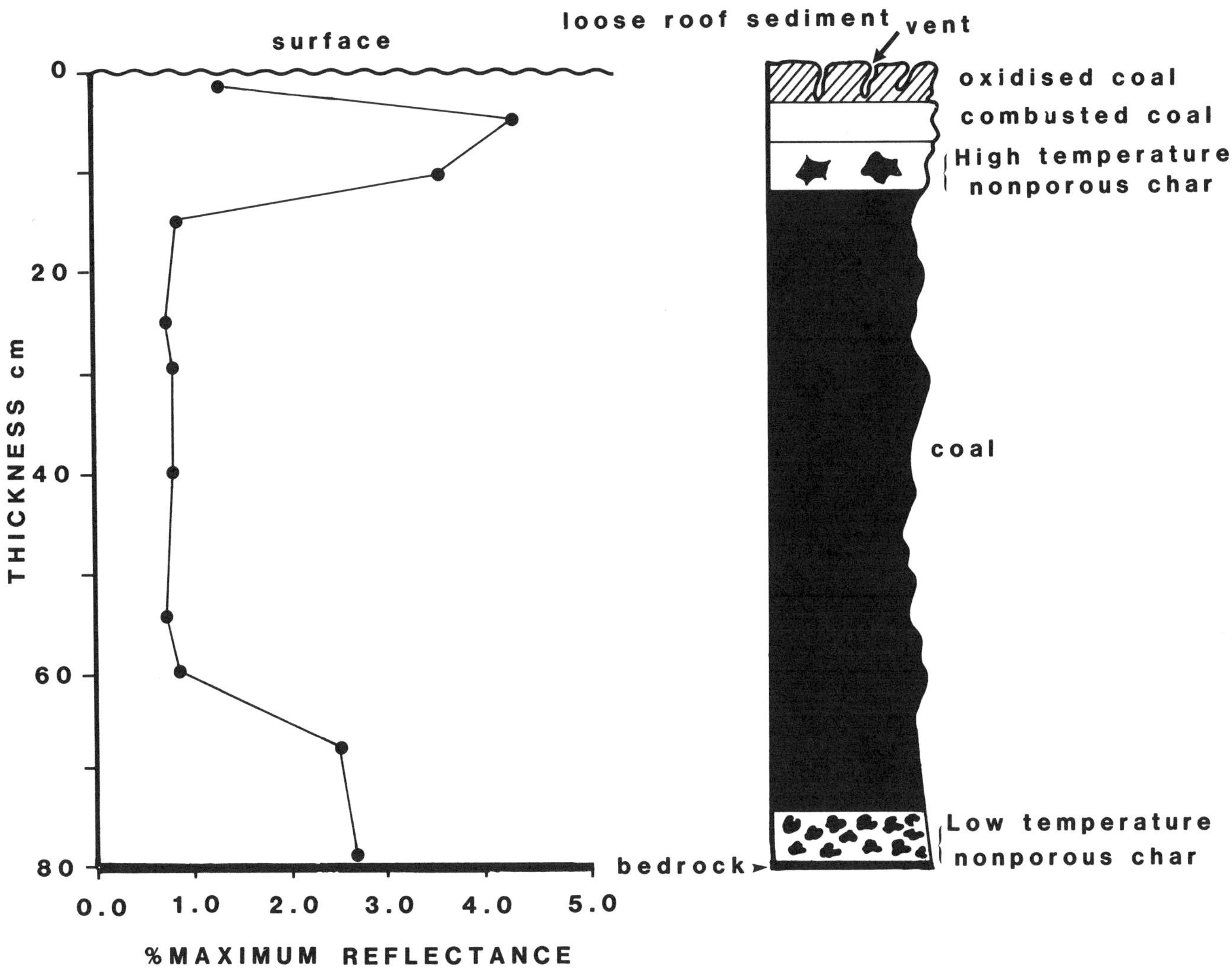

Figure 7. Reflectance (%R_0 maximum in oil) versus thickness profile for burning coal seam at Coalspur, Alberta.

coke when fresh; however, combustion of a weathered outcrop of the coal produced char of anthracite rank. The original layering of the coal seam still is evident in hand specimen, even though the coal has been transformed to a metallic char.

Rank And Composition

Rank

The effect of oxidation and the rate of consumption of oxygen is greater the lower the rank of the coal (Alpern and Maume 1969; Francis, 1961; Guney, 1971). Lignite oxidizes very rapidly and often cannot be stored without risk of spontaneous combustion (Chandra, 1958). Humphreys et al.,(1981) state that there are strong positive correlations between the increase in oxygen content of coal and its self-heating.

Oxidation of coal is related to its molecular structure. The low rank coals (e.g., lignite) are more reactive to oxygen due to the "open pore" molecular structure (Hirsh, 1954), greater surface area available for

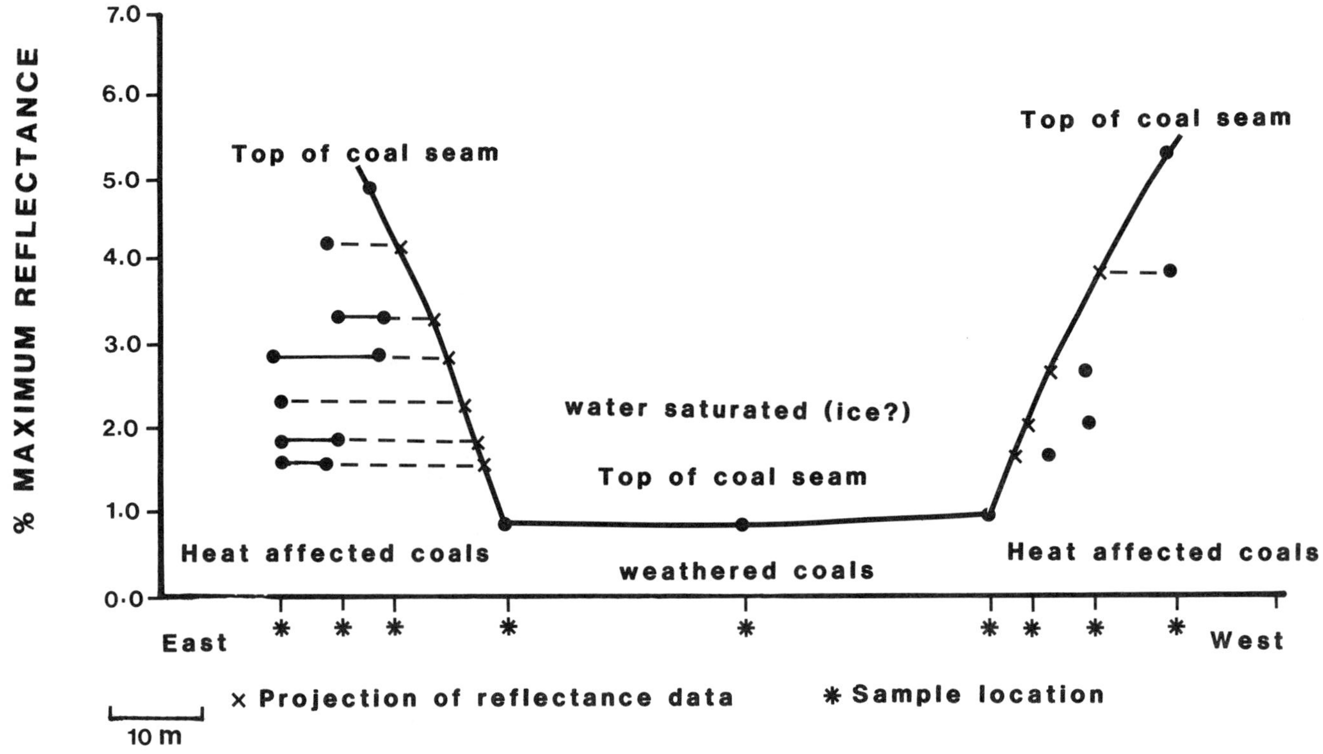

Figure 8. Reflectance (%R_0 maximum in oil) versus sample locality in section along strike of Seam 15 at Eagle Mountain, Fording Coal Mine, British Columbia.

oxygen-carbon reaction, and higher aliphatic/aromatic carbon ratio. The results of Masmadar and Lahiri (1958) indicate that the aliphatic fraction of coal is more reactive to oxygen.

The inner surface area and aliphatic/aromatic ratio of coals decrease with increasing rank of coal. Hence, the reactivity of coal with oxygen decreases with increasing rank. Anthracite is less reactive to oxygen than subbituminous coal.

The susceptibility to self-heating decreases exponentially with increasing rank (Moxon and Richardson, 1985a and b). The higher the rank, the slower the oxidation process, whereas the lower the rank, the larger the number of reactive sites available for oxygen adsorption, and the higher the risk for spontaneous ignition. In addition, moisture content and capacity to absorb moisture are highest at lower rank. Molecular porosity, internal surface area, and the number of non-aromatic hydrocarbons also are highest at lower rank (Falcon, 1986). As a result, low-rank coals have a higher propensity for spontaneous combustion.

Anthracites practically are excluded from spontaneous combustion because of their very low volatile matter content. There also are many high-volatile bituminous coals which have a low propensity to self-heat (R.F. Chaiken, personal communication, 1989). Lignites and subbituminous coals contain large amounts of volatile matter and moisture and are susceptible to spontaneous combustion.

As Morris and Atkinson (1988) state, there is no direct correlation between rank and propensity to spontaneous combustion. The same seam in different mines may react differently, or one part of a seam may be more susceptible to spontaneous combustion than another. The lateral variation of coal macerals,

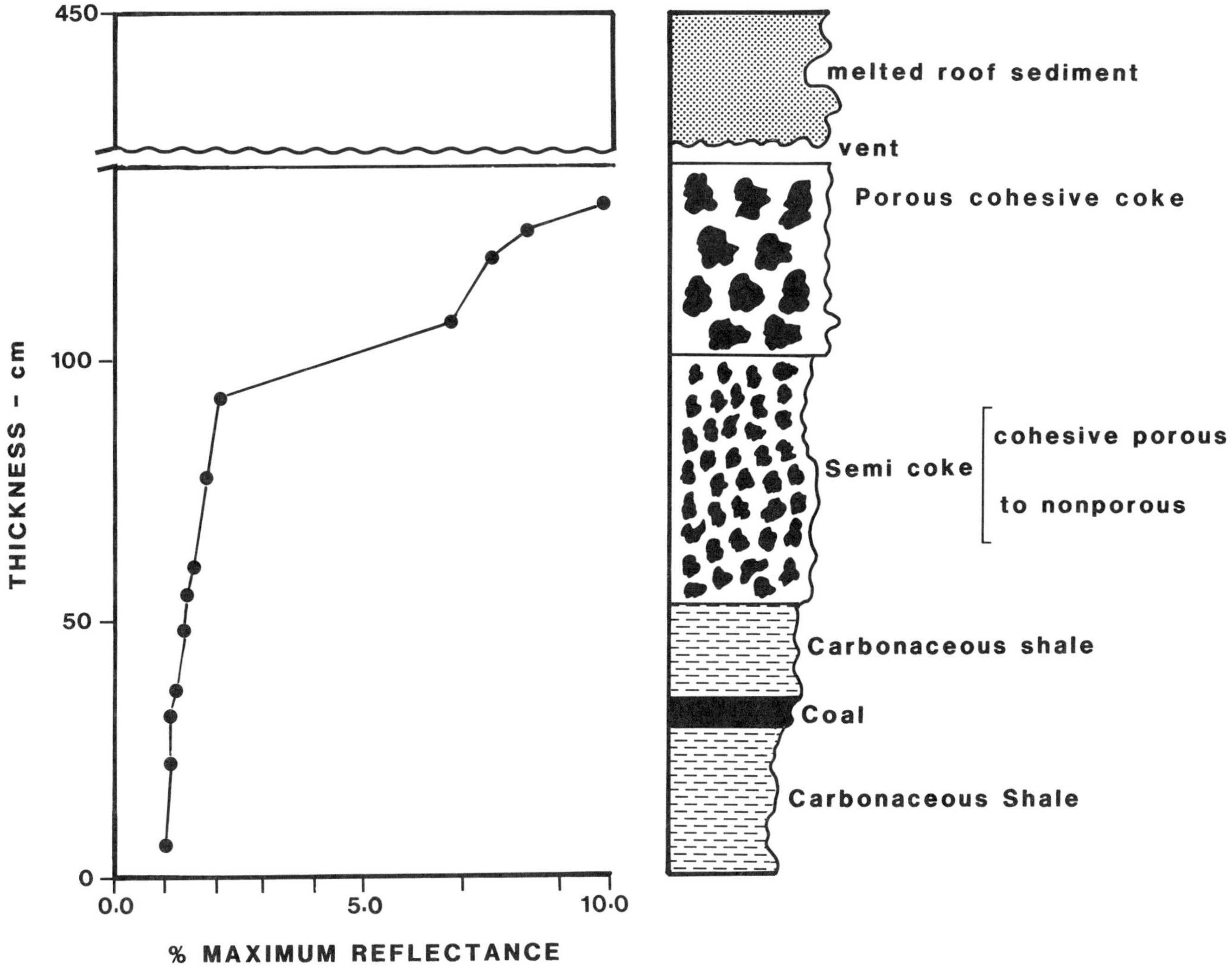

Figure 9. *Reflectance (%R_o maximum in oil) versus thickness profile for the burning coal seam at Aldridge Creek, British Columbia.*

moisture, and pyrite content may be the cause for different behavior of coal in various parts of the same seam.

Canadian burning coal seams of bituminous rank have been examined by Goodarzi (1987) and Goodarzi et al. (1988). The coal at Coalspur, Alberta, is high-volatile bituminous B (%R_o max = 0.70). Burning of this coal resulted in the collapse of the combusted coal seam because the transformation of coal to ash resulted in an almost complete loss of seam morphology. In contrast, the burning coal seam at Aldridge Creek, British Columbia, is medium-volatile bituminous (%R_o max = 1.1). This coal is transformed into hard, cohesive coke (Figure 12) which partially has retained the morphology (texture) of the coal seam. Also, the heat generated by combustion has melted the seam roof. Morphology of a coal seam often is better preserved if the coal seam is an outcrop (i.e., weathered) (Figure 13). Well preserved chars, showing micro-layering are observed

Figure 10. Photograph of natural coke formed due to burning of coal at Aldridge Creek. Photograph courtesy of Professor M. Bustin, University of British Columbia.

Figure 11. Photograph showing conversion of coal to ash in oxidation-combustion zone. Note the angular shape of individual coal lumps. Long axis of photograph is 50 cm (20 in.).

in high-volatile bituminous A coal (%R_o max = 0.9) from Eagle Mountain, British Columbia, even though it is an unweathered coal of almost the same rank as at Coalspur (%R_o max = 0.70), is transformed into ash.

Composition

Technological properties of coals have a direct relationship to the maceral composition of coal, such as with respect to the proportion of reactive to nonreactive macerals. The reactive macerals (liptinite and vitrinite) are more susceptible to oxidation-combustion than the nonreactive macerals (semifusinite-fusinite) (van Krevelen, 1961; Falcon, 1986).

Figure 12. Photograph of heat-affected Seam 15 at Eagle Mountain, Fording Coal Mine, British Columbia.

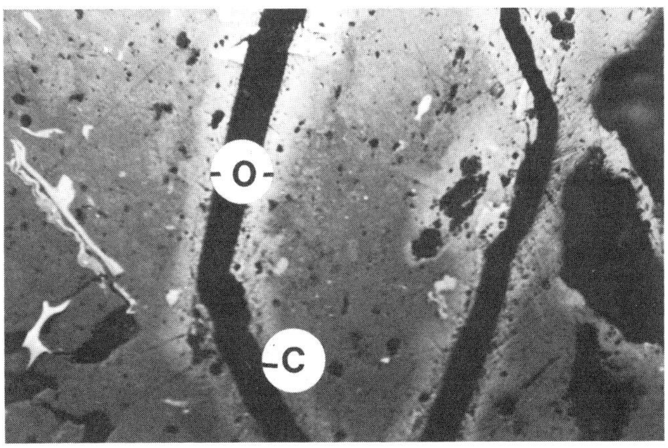

Figure 13. Photomicrograph showing oxidation rim (0) in the vicinity of cracks (C) in char from oxidation zone at Coalspur, Alberta (reflected light, oil immersion). The long axis of photomicrograph is 240 μm (0.0094 in.).

Chamberlain and Hall (1973) conducted a series of oxidation experiments on individual macerals and found exinites to be much more reactive than vitrinites, particularly at higher temperatures of about 150°C, and vitrinites were more reactive than inertinites. Falcon (1986) attributed this to larger proportions of non-aromatic components in the molecular structure of vitrinites and exinites. These non-aromatic components have lower activation energies and, thereby, are expected to oxidize easier and more rapidly. Inertinites have a densely aromatic molecular structure, and a higher activation energy is needed for their oxidation.

The results of Chamberlain et al. (1976) show that fusain is the least reactive, and in general, durain is more reactive than vitrain. Calculations were made of the reaction velocities of vitrinite, exinite, and inertinite and showed that exinites have a much greater oxidation rate than the other two constituents. More reactive macerals such as resinite and exsudatinite have been documented to oxidize earlier and at higher rates than

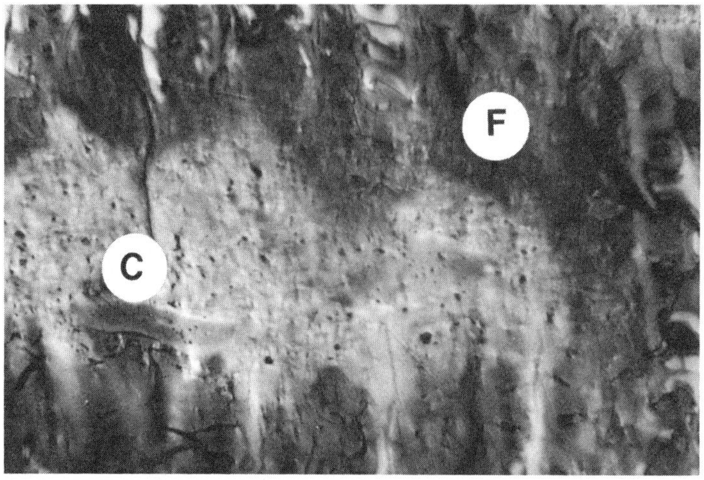

Figure 14. *Photomicrograph of combusted coal from Coalspur, Alberta, showing combustion front (F) and structureless coal matrix (C) (reflected light, oil immersion). Scale same as Figure 13.*

others (Youngberg and Rich, 1983). Unstructured collinite occurring within a botanically structured inertinite has a greater susceptibility than vitrinite alone (Wagner, 1980). A positive correlation was obtained between the vitrinite content and an increase in spontaneous combustion (Humphreys et al., 1981).

The effects of the macroporosity and microporosity of the various macerals on the susceptibility to oxidation are discussed by Falcon (1986). The highly porous (microscopic and submicroscopic) inertinite has a common botanical structure which greatly increases the effective surface area of these grains. Similarly, vitrinites retain some botanical structures and have extensive microscopic and submicroscopic porosity due to the ordered structuring of the macromolecules. Exinites, having an origin as protective outer layers, tend to have low porosity.

These descriptions are generalizations of each maceral group. Within each maceral grouping a number of structural types occur, each with their own characteristic porosities and differing reactivities (Feng, 1985; Wagner, 1980; Youngberg and Rich, 1983). The exinites consist of highly structured spore, pollen and algal remains, plant waxes and resins, finely disseminated detrital material of algal or plant origin, and dispersed liquid products such as amorphous fluorescing matrix and exsudatinite (Stach et al., 1982). Similarly, vitrinite is composed of a number of morphologically distinct groups. Not only are these maceral groups morphologically different, but they also display differences in their chemical make-up, such as the level of aromaticity. Therefore, their susceptibilities to oxidation differ. Thus, subdividing macerals into three main types is insufficient for estimating susceptibility to spontaneous combustion. A detailed maceral analysis, combined with particle size analysis and moisture content (or level of drying), may enable us to address the problems of storage, transport, and handling of various ranks of coal.

Goodarzi (1987) and Goodarzi et al. (1988) have made detailed studies of the petrology of burning coal seams. Their results indicate that the chars in the oxidation zone show oxidation rims and cracks. The reflectance of oxidation rims is higher than that of the main body of the char (Figure 13), whereas the residues in the combustion zone have a burnt appearance and combusted collinite often shows its original cellular morphology (Figure 14). The residues in the carbonization zone either are coked and form anisotropic granular structures (Figure 15) typical of coking coals, such as in the burning coal seam at Aldridge Creek, or develop non-granular anisotropy in non-coking coals (Figure 16), such as at Coalspur, Alberta.

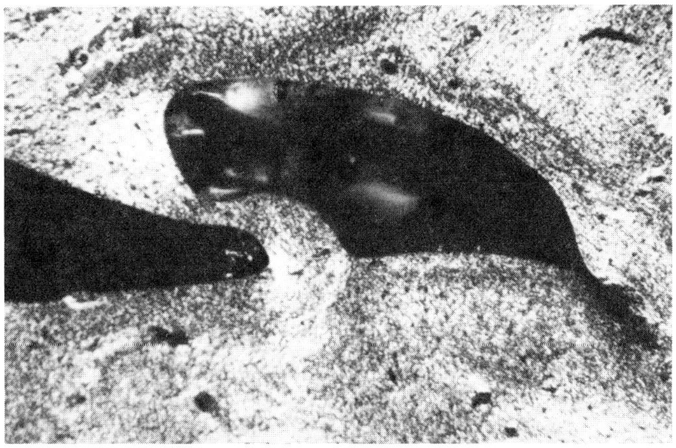

Figure 15. *Photomicrograph of coke, showing anisotropic granular texture, from Aldridge Creek, British Columbia (reflected light, oil immersion). Scale same as Figure 13.*

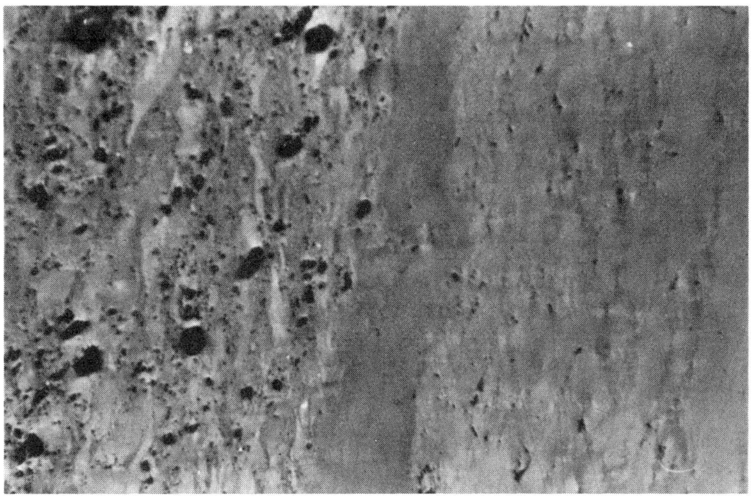

Figure 16. *Photomicrograph of char showing nongranular "basic" anisotropy, Coalspur, Alberta (reflected light, oil immersion). Scale same as Figure 13.*

SUMMARY

Combustion of coal seams starts because of the susceptibility of coal to external or self ignition. It is controlled by the following parameters: 1) thickness of coal overburden; 2) water saturation; 3) lithology; 4) morphology of coal seams; and 5) rank and composition of coal.

A thin sedimentary cover over a coal seam allows the exposure of the coal to air, which results in the easing of the ignition of the coal through the processes of chemisorption, oxidation, and combustion. Water saturation of the seam will either reduce or completely prevent the combustion.

The lithologic composition of the coal-bearing section is important in controlling the extent of combustion. The total energy produced in the combustion zone often is used to alter the coal; therefore, coal seams can act as thermal blankets. In addition, thin sedimentary partings transmit less heat to the surrounding strata.

The morphology of heat-affected sediments forming the roof of the coal seam is dependent on the temperature generated by the coal combustion. At high temperatures, the roof sediments may melt, whereas at low temperatures they do not.

Low-rank coals are more susceptible to spontaneous combustion than high-rank coals due to their "open pore" molecular structure which allows oxygen into the internal coal structure. The higher the composition of a coal in "reactive" macerals (vitrinite and exinite), the higher is the propensity of the coal for spontaneous combustion.

ACKNOWLEDGEMENTS

We would like to thank Dr. A.R. Cameron of the Institute of Sedimentary and Petroleum Geology, Geological Survey of Canada, and Dr. R.F. Chaiken of the U.S. Department of the Interior, Bureau of Mines, Pittsburgh Research Center, for critically reviewing the manuscript, and Ms. Joyce Moir for cheerfully typing several drafts of the manuscript. In addition, excellent drafting and technical assistance by Mrs. Margaret Goodarzi is very much appreciated.

REFERENCES

Adams, W.N., and G.J. Pitt, 1955, Examination of oxidized coal by absorption methods: Fuel, v. 34, p. 383-384.

Alpern, B., and F. Maume, 1969, Etude petrographique de l'a oxydation naturelle et artificielle houilles: Revue Industriale Minerale, v. 51, no. 1, p. 979-997.

Bannerjee, S.C., 1985, Mechanisms of spontaneous combustional coal, in Spontaneous combustional of coal and mine fires: Rotterdam, A.A. Balkema, p. 5-32.

Bustin, R.M., and W.H. Mathews, 1982, In situ gasification of coal, a natural example: History, petrology and mechanics of combustion: Canadian Journal of Earth Sciences, v. 19, no. 3, p. 514-523.

Chamberlain, E.A.C., G. Barress, and J.T. Thirlaway, 1976, Gases evolved and possible reaction during low temperature oxidation of coal: Fuel, v. 55, p. 217-223.

Chamberlain, E.A.C., and D.A. Hall, 1973, The liability of coals to spontaneous combustion: Colliery Guardian, v. 221, no. 2, p. 65-72.

Chandra, D., 1958, Reflectance of oxidized coals: Economic Geology, v. 53, no. 2, p. 102-108.

Falcon, R.M., 1986, Spontaneous combustion of the organic matter in discards from the Witbank Coalfield: Journal of South Africa Mining and Metallurgy, v. 86, no. 7, p. 242-250.

Feng, K.K., 1985, Spontaneous combustion of Canadian coals: Canadian Institute of Mining and Metallurgy Bulletin, v. 78, no. 877, p. 71-75.

Francis, W., 1961, Coal, its formation and composition: London, Edward Arnold Ltd., 806 p.

Fryer, H., 1981, Ghost towns of Alberta: Langley, British Columbia, Stage Coal Publishing Co., p. 149.

Gentzis, T., and F. Goodarzi, 1989, Organic petrology of a self-burning coal wastepile from Coleman, Alberta, Canada: International Journal of Coal Geology, v. 10, p. 257-273.

Goodarzi, F., 1987, Reflectance and petrology of a burning bituminous coal seam: Fuel, v. 66, p. 1073-1078.

Goodarzi, F., and T. Gentzis, 1990, Lateral and vertical reflectance and petrological variation of a heat-affected bituminous coal seam from southeastern British Columbia, Canada: International Journal of Coal Geology, v. 15, p. 317-339.

Goodarzi, F., T. Gentzis, and R.M. Bustin, 1988, Reflectance and petrology profile of a partially combusted and coked bituminous coal seam from British Columbia: Fuel, v. 67, p. 1218-1222.

Goodarzi, F., and T. Jerzykiewicz, 1986, Petrology of a burning bituminous coal seam from Coalspur, Alberta, in Current Research, Part B: Geological Survey of Canada, Paper 86-1B, p. 421-422.

Goodarzi, F., and D.G. Murchison, 1976, Petrography and anisotropy of carbonized preoxidized coals: Fuel, v. 55, p. 141-147.

Guney, M., 1971, An adiabatic study of the influence of moisture on the spontaneous heating of coal: Canadian Institute of Mining and Metallurgy Bulletin, v. 64, no. 707, p. 138-146.

Hirsh, P.B., 1954, X-ray scattering from coals: Proceedings of the Royal Society, Series A, v. 226, p. 143-169,

Humphreys, D., D. Rawlands, and J.F. Cudmore, 1981, Spontaneous combustion of some Queensland coals, in A.J. Hargraver, ed., Ignitions, Explosions and Fires in Coal Mines Symposium, Wollongong, Australia, May, 1981: Illawarra, Australian Institute of Mining and Metallurgy, p. 5-1 - 5-15.

Majundar, A. and F. W. Bachelor, 1986, A detailed study of the reaction mechanism of spontaneous combustion of coal with particular emphasis in Western Canadian low rank coals: Canada Centre for Mineral and Energy Technology, D.S.S. File No. 01SG.23440-3-9162, D.S.S. Contract Serial No. 0SG83-00109, 169 p.

Masmadar, E.K., and A. Lahiri, 1958, Dimension of side chains in coal structure: Fuel, v. 37, p. 495-501.

Miard, H., 1945, Upper Elk Valley Coalfield: British Columbia Ministry of Mines Annual Report A 170, 145 p.

Morris, R., and T. Atkinson, 1988, Seam factor and the spontaneous heating of coal: Mining Science and Technology, v. 7, p. 149-159.

Moxon, N.T. and S. B. Richardson, 1985a, Assessment and chemical inhibition of self heating in coal, in Proceedings of the International Conference on Coal Science, Sydney, Australia, October, 1985: Sydney, Pergamon Press, p. 487-490.

Moxon, N.T., and S.B. Richardson, 1985b, Development of a self heating index for coal: Coal Preparation, v. 2, p. 91-105.

Oplinski, W., 1954, Prace Glowenga Intstitute, Gornictwa: Komun, v. 139, p. 3204.

Pearson, D.E., and S. Creaney, 1980, Spontaneous carbonization of oxidized high volatile coal by a lightning strike: Canadian Journal of Earth Sciences, v. 17, no. 1, p. 36-42.

Rashid, M.A. and C. D. McIntyre, 1982, Spontaneous combustion. Part 1--Literature review: Canadian Coal Mining Research Centre, Report 81/53-T Task 81/66, 45 p.

Sevenster,, P.G., 1961, Studies on the interaction of oxygen with coal in the temperature range 0° to 90°C, part 1: Fuel, v. 40, p. 7-17.

Stach, E., M.-Th. Mackowsky, M. Teichmüller, G.H. Taylor, D. Chandra, and R. Teichmüller, 1982, Stach's Textbook of Coal Petrology (3rd Ed.): Berlin-Stuttgart, Gebrüder Borntraeger, 535 p.

Van der Flier-Keller, E., and W.S. Fyfe, 1987, Geochemistry of two Cretaceous coal sequences: James Bay Lowland, Northern Ontario and Peach River Basin, NE British Columbia: Canadian Journal of Earth Sciences, v. 24, p. 1038-1052.

van Krevelen, D.W., 1981, Coal, typology-chemistry-physics-constitution: New York, Elsevier, Coal Science and Technology, Vol. 3, 514 p.

Wagner, M., 1980, Thermal changes of hard coal in the core of mine dump fires: Krakow, Institute of Geology and Mineral Raw Materials, The Stanislaw Staszic University of Mining and Metallurgy, Wydawnictwa Geologiczne, Geologia, v. 6, no. 2, p. 5-14.

Youngberg, A.D., and F.J. Rich, 1983, Determination of underground coal gasification burn temperatures with thermal metamorphism of coal macerals as a geothermometer, in Proceedings of the Ninth Annual Underground Coal Gasification Symposium, Bloomington, Indiana, August, 1983: Morgantown, West Virginia, U.S. Department of Energy, p. 293-301.

Author Index

Bauer, Robert A.
Coping With Tectonic Stress In The Illinois Basin Coal Field 321

Beaver, F.W.
The Role Of Geology In Managing Coal Conversion Byproducts In The Northern Great Plains 541
The Role Of Hydrogeology In Underground Coal Gasification With An Example From The Rocky Mountain 1 (RM1) Test, Carbon County, Wyoming 169

Bilgesu, H.I.
Structural Properties Of Coal That Control Coalbed Methane Production 105

Brown, Jr., Robert D.
Coal As A Host And As An Indicator Of Mineral Resources 471

Campbell, Elizabeth V.M.
Restrictions To Mining: Their Effect On Available Coal Resources 81

Cavaroc, Victor V.
Lithotype Control Of Trace Element Concentrations In Some Southeastern West Virginia Coals 427

Cheel, D.B.
Environmental Impact Of Surface Coal Mining And Reclamation In The Plains Region Of Alberta, Canada 457

Collins, Bruce A.
Stress In Underground Coal Mines: A Practical Review Of Origin, Recognition, And Control 313

Daly, D.J.
The Role Of Hydrogeology In Underground Coal Gasification With An Example From The Rocky Mountain 1 (RM1) Test, Carbon County, Wyoming 169

Dana, G.F. "Pete"
Underground Coal Gasification 155

Deul, Maurice
Coal Quality And Geology 359

Dutcher, L.A.F.
Coal Utilization, The Environment, And The Role Of Geologists 453

Dutcher, R.R.
Coal Utilization, The Environment, And The Role Of Geologists 453

Ertekin, T.
Structural Properties Of Coal That Control Coalbed Methane Production 105

Fenton, Mark M.
Glacially Deformed Bedrock: Implications For Coal Exploration, Mining, And Highwall Stability ... 37
Surface Geophysics For Coal Exploration In The Alberta Plains: Recent Research ... 13

Ferm, John C.
Application Of Depositional Models To Mining Problems ... 189

Finkelman, Robert B.
Coal As A Host And As An Indicator Of Mineral Resources ... 471

Gardner, Nancy K.
Restrictions To Mining: Their Effect On Available Coal Resources ... 81

Gentzis, Thomas
Geological Controls On Coal Seams Burning "In Place" ... 559

Ghose, S.N.
Use Of Geostatistics In Planning An Optimum Drilling Program ... 69

Goodarzi, F.
Geological Controls And Constraints On The Concentration Of Elements In Western Canadian Coals ... 389
Geological Controls On Coal Seams Burning "In Place" ... 559

Goßel, Wolfgang
Analysis Of Aromatic Hydrocarbons In Overburden From Coal Mines: Assessment Of The Environmental Impact ... 483

Greb, Stephen F.
Roof Falls And Hazard Prediction In Eastern Kentucky Coal Mines ... 245

Griswold, Thomas B.
Structural Controls On The Petrology And Geochemistry Of The Pond Creek Coal Seam, Pike And Martin Counties, Eastern Kentucky ... 413

Groenewold, Gerald H.
The Role Of Geology In Managing Coal Conversion Byproducts In The Northern Great Plains ... 541
The Role Of Hydrogeology In Underground Coal Gasification With An Example From The Rocky Mountain 1 (RM1) Test, Carbon County, Wyoming ... 169

Henderson, James D.
Surface Geophysics For Coal Exploration In The Alberta Plains: Recent Research ... 13

Hoffman, Georgia L.
Surface Geophysics For Coal Exploration In The Alberta Plains: Recent Research ... 13

Hostettler, Karen K.
Restrictions To Mining: Their Effect On Available Coal Resources ... 81

Howard, A.E.
Environmental Impact Of Surface Coal Mining And Reclamation In The Plains Region Of Alberta, Canada ... 457

Hower, James C.
Structural Controls On The Petrology And Geochemistry Of The Pond Creek Coal Seam, Pike And Martin Counties, Eastern Kentucky ... 413

Hucka, Brigitte P.
Cleat And Joint System Evaluation In Coal Seams And Zones Of The Blackhawk Formation And Ferron Sandstone Member Of The Mancos Shale Formation For Mine Planning And Site Characterization In The Book Cliffs, Wasatch Plateau, And Emery Coal Fields, Utah 225

Ingram, David K.
Effects Of Structural Faults On Ground Control In Selected Coal Mines In Southwest Virginia 287

Kane, William F.
The Effects Of Geology On Coal Mine-induced Subsidence In The United States 335

Klusman, Ronald W.
Review Of Processes Operating In Constructed Wetlands To Attenuate Acid Mine Drainage 513

Koob, Robert D.
The Role Of Geology In Managing Coal Conversion Byproducts In The Northern Great Plains 541

Machemer, Steven D.
Review Of Processes Operating In Constructed Wetlands To Attenuate Acid Mine Drainage 513

Macyk, T.M.
Environmental Impact Of Surface Coal Mining And Reclamation In The Plains Region Of Alberta, Canada 457

Martin, Jeffrey W.
Lithotype Control Of Trace Element Concentrations In Some Southeastern West Virginia Coals 427

Mercier, John M.
Commentary On The Evolution Of The Exploration And Evaluation Process 3

Miller, Marshall S.
Geological Modeling Techniques For Evaluation Of Productivity-related Longwall Mining Roof Conditions: A Case Study 263

Mitchell, Terry E.
Coalbed Methane Production Potential In Complex Geologic Settings 137

Moebs, Noel N.
Geologic Diagnosis For Reducing Coal Mine Roof Failure 203

Molinda, Gregory M.
Effects Of Structural Faults On Ground Control In Selected Coal Mines In Southwest Virginia 287

Moran, S.R.
Environmental Impact Of Surface Coal Mining And Reclamation In The Plains Region Of Alberta, Canada 457

Mullennex, Ronald H.
Geological Modeling Techniques For Evaluation Of Productivity-related Longwall Mining Roof Conditions: A Case Study 263

Murray, D. Keith
Coalbed Methane: Natural Gas Resources From Coal Seams ... 97

Nelson, W. John
Coping With Tectonic Stress In The Illinois Basin Coal Field ... 321

Nelson, J. Scott
Geological Modeling Techniques For Evaluation Of Productivity-related Longwall Mining Roof Conditions: A Case Study ... 263

Oliver, Robert L.
The Role Of Hydrogeology In Underground Coal Gasification With An Example From The Rocky Mountain 1 (RM1) Test, Carbon County, Wyoming ... 169
Underground Coal Gasification ... 155

Palmer, James E.
Coal Resources and Reserve Estimates in the United States in 1991 — A Status Report ... 55

Pappajohn, Steven P.
Coalbed Methane Production Potential In Complex Geologic Settings ... 137

Pawlowicz, John G.
Glacially Deformed Bedrock: Implications For Coal Exploration, Mining, And Highwall Stability ... 37
Surface Geophysics For Coal Exploration In The Alberta Plains: Recent Research ... 13

Platt, Jeremy
Introduction: Fitting Coal Geoscience Into The Big Picture ... xiii

Pollock, James D.
Structural Controls On The Petrology And Geochemistry Of The Pond Creek Coal Seam, Pike And Martin Counties, Eastern Kentucky ... 413

Powell, Larry R.
The Effects Of Geology On Coal Mine-induced Subsidence In The United States ... 335

Püttmann, Wilhelm
Analysis Of Aromatic Hydrocarbons In Overburden From Coal Mines: Assessment Of The Environmental Impact ... 483

Sames, Gary P.
Geologic Diagnosis For Reducing Coal Mine Roof Failure ... 203

Sarnecki, Joseph C.
Formation Of Clinker And Its Effects On Locating And Limiting Coal Resources ... 29

Sartorelli, Anthony N.
Surface Geophysics For Coal Exploration In The Alberta Plains: Recent Research ... 13

Schmit, C.R.
The Role Of Hydrogeology In Underground Coal Gasification With An Example From The Rocky Mountain 1 (RM1) Test, Carbon County, Wyoming ... 169

Sites, Roy S.
Restrictions To Mining: Their Effect On Available Coal Resources ... 81

Steffens, Kai
Analysis Of Aromatic Hydrocarbons In Overburden From Coal Mines: Assessment Of The Environmental Impact 483

Soeder, Daniel J.
The Effects Of Overburden Stress On Coalbed Methane Production 125

Sung, W.
Structural Properties Of Coal That Control Coalbed Methane Production 105

Tewalt, Susan J.
Restrictions To Mining: Their Effect On Available Coal Resources 81

Tivy, Stephen V.
Commentary On The Evolution Of The Exploration And Evaluation Process 3

Treworgy, Colin G.
Techniques For Estimating The Coal Reserve Base Of A Region Using A Geographic Information System: A Case Study From Illinois 59

Trudell, M.R.
Environmental Impact Of Surface Coal Mining And Reclamation In The Plains Region Of Alberta, Canada 457

van der Flier-Keller, E.
Geological Controls And Constraints On The Concentration Of Elements In Western Canadian Coals 389

Weisenfluh, Gerald A.
Application Of Depositional Models To Mining Problems 189

Wildeman, Thomas
Drainage From Coal Mines: Chemistry And Environmental Problems 499

Winston, Richard B.
Coal Quality Of The Mary Lee Coal Group And Its Relationship To Coal Resource Development, Coalbed Methane, And Geologic History In The Warrior Coal Field, Alabama 369